DIGITAL GEOMETRY

DIGITAL GEOMETRY

Geometric Methods for Digital Picture Analysis

Reinhard Klette
The University of Auckland

Azriel Rosenfeld
University of Maryland

AMSTERDAM • BOSTON • HEIDELBERG • LONDON
NEW YORK • OXFORD • PARIS • SAN DIEGO
SAN FRANCISCO • SINGAPORE • SYDNEY • TOKYO
MORGAN KAUFMANN PUBLISHERS IS AN IMPRINT OF ELSEVIER

Acquisitions Editor	Tim Cox
Publishing Services Manager	Andre Cuello
Project Manager	Justin Palmeiro
Editorial Assistant	Richard Camp
Cover Design	Yvo Riezbos
Composition	Kolam Inc.
Copyeditor	Kolam USA
Proofreader	Kolam USA
Indexer	Kolam USA
Interior printer	Maple-Vail Book Manufacturing Group
Cover printer	Phoenix Color Corp.

Morgan Kaufmann Publishers is an imprint of Elsevier.
500 Sansome Street, Suite 400, San Francisco, CA 94111

This book is printed on acid-free paper.

© 2004 by Elsevier Inc. All rights reserved.

Designations used by companies to distinguish their products are often claimed as trademarks or registered trademarks. In all instances in which Morgan Kaufmann Publishers is aware of a claim, the product names appear in initial capital or all capital letters. Readers, however, should contact the appropriate companies for more complete information regarding trademarks and registration. No part of this publication may be reproduced, stored in a retrieval system, or transmitted in any form or by any means—electronic, mechanical, photocopying, scanning, or otherwise—without prior written permission of the publisher.

Permissions may be sought directly from Elsevier's Science & Technology Rights Department in Oxford, UK: phone: (+44) 1865 843830, fax: (+44) 1865 853333, e-mail: *Permissions@elsevier.com.uk*. You may also complete your request on-line via the Elsevier homepage (*http://elsevier.com*) by selecting "Customer Support" and then "Obtaining Permissions."

Library of Congress Cataloging-in-Publication Data
Application submitted.

ISBN: 1-55860-861-3

For information on all Morgan Kaufmann publications, visit our website at www.mkp.com.

Printed in the United States of America:
04 05 06 07 08   5 4 3 2 1

Preface

Digital geometry deals with the geometric properties of subsets of digital pictures and with the approximation of geometric properties of objects by making use of the properties of the digital picture subsets that represent the objects. It emerged in the second half of the 20th century with the initiation of research in the fields of computer graphics and digital image analysis. It has its mathematical roots in graph theory and discrete topology; it deals with sets of grid points which are also studied in number theory (since C.F. Gauss) and the geometry of numbers, or with cell complexes (which have been studied in topology since the middle of the 19th century). Studies of gridding techniques, such as those by Gauss, Dirichlet, or Jordan (for measuring the content of a set), also provide historic context for digital geometry. Digitizations on regular grids are also frequently used in numeric computation in science and engineering.

This book uses the term "picture" rather than "image," because pictures can be the result of drawing, painting, stitching, or other technologies that do not involve imaging processes. The book deals with digital geometry in the context of picture analysis. The medium on which digital pictures reside is called a *grid* which is a finite set of grid points, grid cells, or other types of discrete elements; the book discusses the geometric and topologic properties of subsets of grids.

Digital geometry can be viewed as a special branch of discrete geometry that deals with graph-theoretical or combinatorial concepts. It can also be viewed as approximate Euclidean geometry on the basis of the fact that picture analysis generally makes use of ideas about Euclidean space. However, digital geometry differs from approximation theory in its use of digitized input data (grid points that are not necessarily on the original curve) rather than sampled input data (sample points that are on the curve but that are not necessarily grid points) and in its focus on understanding the data in digital terms rather than approximating the data with the use of polynomials. Digital geometry also differs from computational geometry, which deals with finite sets of geometric objects in Euclidean space.

The book is intended to be a text that can be used in advanced undergraduate or graduate courses about image analysis in fields such as computer science or engineering. Selections from the material in this book should be sufficient to fill a one-semester course; see the course proposals in the section called "Structure of this Book" for suggested selections. Prerequisites to the use of this book are a basic knowledge of set theory and graph theory and programming experience for the suggested experimental exercises (course assignments). It should be pointed out that some of the exercises are quite difficult; see the references provided in the Commented Bibliography sections at the ends of the chapters for additional information.

The book is also designed to be a comprehensive review of research in digital geometry. The authors have chosen a mathematic viewpoint rather than a practitioner's viewpoint. However, the fundamentals of digital geometry are also of value to those who work on applications of image analysis or computer graphics, especially if they are concerned with theoretical foundations. Each chapter concludes with exercises and has references to related or more advanced work. When proofs are not given, references to the relevant literature are provided.

This book provides discussions and citations of important mathematic ideas and methodologies that are important to digital geometry and date back, in some cases, to previous centuries or even to ancient times. This information should give students and researchers a better understanding of where the subject fits into a long-term historic process of knowledge acquisition, which began long before their own work or that of their supervisors.

The authors acknowledge comments by (in alphabetical order) Valentin Brimkov, David Coeurjolly, Isabelle Debled-Rennesson, David Eberly, Atsushi Imiya, Gisela Klette, T. Yung Kong, Longin Jan Latecki, Majed Marji, Lyle Noakes, Theo Pavlidis, Christian Ronse, Garry Tee, Klaus Voss, and Jovisa Zunić. The help of Janice Perrone and Cecilia Lourdes in preparing the manuscript and providing library contacts was very important, and it is appreciated by the authors.

Reinhard Klette and *Azriel Rosenfeld*
Auckland, New Zealand and Baltimore,
Maryland, USA
October, 2003

I greatly regret that Professor Rosenfeld did not live to see our book published in final form. I have lost not just a friend, but an outstanding teacher and scientist colleague. I shall miss him.

Reinhard Klette
May, 2004

To Gisela, Kristian, and Alexander Klette, and to Abraham Rosenfeld and his family

Professor Azriel Rosenfeld
19 February 1931 - 22 February 2004

Reinhard Klette is professor of information technology in the Department of Computer Science at the University of Auckland (New Zealand). His research interests are directed toward theoretical and applied subjects in image data computing, robot vision, visualization, pattern recognition, image analysis, and image understanding. He has published more than 200 journal and conference papers on topics in computer science, and books about parallel processing, image processing, and shape recovery based on visual information. He has been a plenary speaker at conferences in Europe, America and Australasia. He is an Associate Editor of IEEE Transactions on Pattern Analysis and Machine Intelligence.

Azriel Rosenfeld was a tenured Research Professor, a Distinguished University Professor, and the Director of the Center for Automation Research at the University of Maryland in College Park, where he also held affiliate professorships in the Departments of Computer Science, Electrical Engineering, and Psychology.

Dr. Rosenfeld was widely regarded as the leading researcher in the world in the field of computer image analysis. Over a period of nearly 40 years, he made fundamental and pioneering contributions to nearly every area of that field. He wrote the first textbook in the field (1969), was founding editor of its first journal (1972), and was co-chairman of its first international conference (1987). He published over 30 books and over 600 book chapters and journal articles, and directed nearly 60 Ph.D. dissertations.

Dr. Rosenfeld's research on digital image analysis (specifically on digital geometry and topology and the accurate measurement of statistical features of digital images) in the 1960s and 1970s formed the foundation for a generation of industrial vision inspection systems that have found widespread applications from the automotive to the electronics industry.

He was a Fellow of the Institute of Electrical and Electronics Engineers (1971), won its Emanuel Piore Award in 1985, and received its Third Millennium Medal in 2000; he was a founding Fellow of both the American Association for Artificial Intelligence (1990) and the Association for Computing Machinery (1993). He was a Fellow of the Washington Academy of Sciences (1988) and won its Mathematics and Computer Science Award in 1988. He was a founding Director of the Machine Vision Association of the Society of Manufacturing Engineers (1985–1988), won its President's Award in 1987, and was a certified Manufacturing Engineer (1988). He was a founding member of the IEEE Computer Society's Technical Committee on Pattern Analysis and Machine Intelligence (1965), served as its Chairman (1985–7), and received the society's Meritorious Service Award in 1986, its Harry Goode Memorial Award in 1985, became a Golden Core member of the Society in 1996, and received its Distinguished Service Award for Lifetime Achievement in Computer Vision and Pattern Recognition in 2001. Dr. Rosenfeld received the IEEE Systems, Man, and Cybernetics Society's Norbert Wiener Award in 1995, and he received an IEEE Standards Medallion in 1990 and the Electronic Imaging International Imager of the Year Award in 1991. He was a founding member of the Governing Board of the International Association for Pattern Recognition (1978–1985), served as its President (1980–1982), won its first K.S. Fu Award in 1988, and became one of its founding Fellows in 1994. In 1998, he received the Information Science Award from the Association for Intelligent Machinery. He was a Foreign Member of the Academy of Sciences of the German Democratic Republic (1988–92) and was a Corresponding Member of the National Academy of Engineering of Mexico (1982).

Structure of this Book

Chapters 2 through 8 provide foundations for digital geometry; they discuss grids, metrics, graphs, topology, and geometry and introduce concepts and methods used in digital geometry that are related to these subjects.

This book is organized as shown below.

Basics

Chapter 1: Introduction
Chapters 2–8: Grids, Metrics, Graphs, Topology, Geometry

Selected topics

Chapters 9–12: Straight Lines, Curves, Planes, Surfaces
Chapters 13–16: Hulls and Diagrams, Transformations
                (Geometrical, Morphological, Deformations)
Chapter 17: Other Properties and Relations

Chapters 9 through 13 discuss topics in digital geometry: digital "straightness" in Chapter 9; length and curvature of arcs and curves in Chapter 10; 3D straightness and planarity in Chapter 11; area and curvature of surfaces in Chapter 12; and hulls and diagrams in Chapter 13.

Chapter 14 discusses geometric operations on pictures; Chapter 15 discusses the application of operations of mathematic morphology to pictures; Chapter 16 discusses deformations of pictures; and Chapter 17 discusses picture properties and spatial relations.

Chapter 1 provides a general introduction and should be read first. Depending on the background of the reader, the different chapters may allow more or less independent reading. However, there are some obvious "clusters", such as Chapters 4 and Chapter 5 (graph-theoretical models of pictures), Chapters 6 and 7 (basics of topology in the context of picture analysis), Chapters 8, 9, 11, 13, 14, and 17 (basics of geometry in the context of picture analysis), and Chapter 10 and 12 (performance evaluation of algorithms in digital geometry).

A third year undergraduate course about algorithms for digital pictures (in a program in electrical engineering, computer science or mathematics involving picture analysis or computer graphics) could focus on selected algorithms (see the List of Algorithms at the end of the book) and on the fundamentals that underlie these algorithms. The students would have the benefit of related mathematical topics and material for additional reading being provided in the same textbook. For example, the course could be structured as follows:

1. (1-2 lectures) Start with Section 1.1

2. (3-5 lectures) Follow this with Chapter 2, possibly shortening Section 2.3 and adding the example from Section 1.2.7 to the presentation of Section 2.4.

3. (3 lectures) Follow this with metrics and distance transforms (Chapter 3).

4. (2-3 lectures) Continue with the border tracing algorithm of Chapter 4 (with related property calculations; see, e.g., Section 8.1.6).

5. (2 lectures) Cover the frontier tracing algorithm of Chapter 5.

6. (2-3 lectures) Follow this with one or two DSS approximation algorithms (**K1990** in Chapter 9, related to frontier tracing, or **DR1995**, related to border tracing of planar regions).

7. (3-8 lectures) Facilitate an extensive discussion about methods, algorithms, and performance evaluation for different arc length and curvature estimators (see Chapter 10).

8. (3-8 lectures) If time allows, algorithms for 3D region analysis could be added. This would include surface scanning from Section 8.4 (with related property calculations; see, e.g., Section 8.3.7), DPS approximation from Chapter 11, and surface area and curvature estimation with comparative performance evaluation from Chapter 12.

9. (3-6 lectures) Algorithms from Chapter 13 (hulls and diagrams; see also Section 1.2.9) or from Chapter 15 (morphologic operations) could also be added to the course.

Note that the exercises in this book are of varying complexity and should be selected carefully for such a course; however, all of the experimental exercises can be recommended for course work (assignments). The course could also cover other algorithms from the List of Algorithms (e.g., geometric transforms, which are not difficult to implement, or simple deformations, which require a good understanding of the "more challenging" concepts discussed in Chapter 16).

Graduate courses could focus on more specific topics clustered around selected sections in the book, such as the following:

(i) *Picture Analysis and Topology* (Chapter 2 as an introduction, then Chapters 4 through 7 and Chapter 16).

(ii) *Multigrid Analysis of Property Estimators in Picture Analysis* (basics from Chapter 2 and Chapter 3, including the example from Section 1.2.7, followed by multigrid subjects in Chapters 10, 12, and 17).

(iii) *Combinatorial Picture Analysis* (combinatorial subjects in Chapters 1 and 2 as an introduction, then Chapters 4 and 5, combinatorial subjects in Chapters 9 and 11, digital tomography in Chapter 14, and digital moments in Chapter 17).

(iv) The axiomatic approaches to different subdisciplines, especially the axiomatic theory of digital geometry in Chapter 14, could provide material for a graduate research seminar about *Mathematical Fundamentals of Picture Analysis* (see also the List of Axioms at the end of this book).

The extensive bibliography, with commented bibliography sections at the ends of the chapter, should also provide support for designing graduate student research seminars based on selected readings.

Contents

1 Introduction 1
   1.1 Pictures 1
   1.2 Digital Geometry and Related Disciplines 11
   1.3 Exercises 30
   1.4 Commented Bibliography 33

2 Grids and Digitization 35
   2.1 The Grid Point and Grid Cell Models 35
   2.2 Connected Components 46
   2.3 Digitization Models 55
   2.4 Property Estimation 66
   2.5 Exercises 70
   2.6 Commented Bibliography 73

3 Metrics 77
   3.1 Basics About Metrics 77
   3.2 Grid Point Metrics 89
   3.3 Grid Cell Metrics 100
   3.4 Metrics on Pictures 105
   3.5 Exercises 112
   3.6 Commented Bibliography 115

4 Adjacency Graphs 117
   4.1 Graphs, Adjacency Structures, and Adjacency Graphs .. 117
   4.2 Some Basics of Graph Theory 125
   4.3 Oriented Adjacency Graphs 135
   4.4 Combinatorial Maps 150
   4.5 Exercises 153
   4.6 Commented Bibliography 156

5 Incidence Pseudographs 159
   5.1 Incidence Structures 159
   5.2 Boundaries, Frontiers, and the Euler Characteristic . 168
   5.3 The Regular Case 175
   5.4 Pictures on Incidence Grids 181
   5.5 Exercises 189
   5.6 Commented Bibliography 190

6 Topology ... 193
- 6.1 Topologic Spaces ... 193
- 6.2 Digital Topologies ... 197
- 6.3 Topologic Concepts ... 209
- 6.4 Combinatorial Topology ... 216
- 6.5 Exercises ... 226
- 6.6 Commented Bibliography ... 229

7 Curves and Surfaces: Topology ... 231
- 7.1 Curves in the Euclidean Topology ... 231
- 7.2 Curves in Incidence Grids ... 237
- 7.3 Curves in Adjacency Grids ... 241
- 7.4 Surfaces in the Euclidean Topology ... 251
- 7.5 Surfaces and Separations in 3D Grids ... 258
- 7.6 Exercises ... 264
- 7.7 Commented Bibliography ... 266

8 Curves and Surfaces: Geometry ... 269
- 8.1 Planar Curves and Arcs ... 269
- 8.2 Space Curves and Arcs ... 281
- 8.3 Surfaces and Solids ... 282
- 8.4 Surface Tracing and Approximation ... 300
- 8.5 Exercises ... 304
- 8.6 Commented Bibliography ... 305

9 2D Straightness ... 309
- 9.1 Basics ... 309
- 9.2 Supporting Lines ... 312
- 9.3 Self-Similarity ... 316
- 9.4 Periodicity ... 321
- 9.5 Number-Theoretic Properties ... 325
- 9.6 Algorithms ... 328
- 9.7 Exercises ... 336
- 9.8 Commented Bibliography ... 337

10 2D Arc Length; Curvature and Corners ... 341
- 10.1 The Length of a Digital Curve ... 341
- 10.2 Definitions of 2D Arc Length Estimators ... 346
- 10.3 Evaluation of 2D Arc Length Estimators ... 353
- 10.4 The Curvature of a Planar Digital Curve ... 362
  - 10.4.1 Corner detectors ... 362
  - 10.4.2 Curvature estimators ... 364
- 10.5 Exercises ... 372
- 10.6 Commented Bibliography ... 373

Contents

11 3D Straightness and Planarity 375
   11.1 3D Straightness 375
   11.2 Digital Planes in 3D Adjacency Grids 390
   11.3 Digital Planes in the 3D Incidence Grid 399
   11.4 DPS Recognition and Generation 402
   11.5 Exercises 405
   11.6 Commented Bibliography 407

12 3D Arc Length, Surface Area, and Curvature 409
   12.1 3D Arcs 409
   12.2 Surface Area Estimation 414
   12.3 Surface Curvature Estimation 422
   12.4 Exercises 424
   12.5 Commented Bibliography 426

13 Hulls and Diagrams 427
   13.1 Hulls 427
   13.2 2D Digital Convexity 436
   13.3 Diagrams 439
   13.4 Exercises 450
   13.5 Commented Bibliography 453

14 Transformations 455
   14.1 Geometries 455
   14.2 Axiomatic Digital Geometry 456
   14.3 Transformation Groups and Symmetries 459
   14.4 Neighborhood-Preserving Transformations 462
   14.5 Applying Transformations to Pictures 464
   14.6 Magnification and Demagnification 470
   14.7 Digital Tomography 475
   14.8 Exercises 477
   14.9 Commented Bibliography 479

15 Morphologic Operations 481
   15.1 Dilation 481
   15.2 Erosion 483
   15.3 Combining Dilations and Erosions 485
   15.4 Simplification 486
   15.5 Segmentation 490
   15.6 Decomposition 492
   15.7 Exercises 496
   15.8 Commented Bibliography 496

16 Deformations 499
   16.1 Topology-Preserving Deformations and Simple Pixels 499
   16.2 Shrinking 506
   16.3 Thinning 509

16.4 Deformations of Curves 513
16.5 Interchangeable Pairs of Pixels 520
16.6 Deformations of 3D Pictures 530
16.7 Deformations of Multivalued Pictures 532
16.8 Exercises 534
16.9 Commented Bibliography 535

17 Picture Properties and Spatial Relations 537
17.1 Properties 537
17.2 Moments 541
17.3 Experimental Evaluation of Moment Estimates 546
17.4 Operations on Pictures and Invariant Properties 554
17.5 Spatial Relations 555
17.6 Exercises 558
17.7 Commented Bibliography 559

List of Algorithms   **561**

List of Symbols   **565**

List of Axioms   **569**

Bibliography   **571**

Index   **645**

CHAPTER **1**

Introduction

This book deals with the concepts, methods, and algorithms of digital geometry. This introductory chapter provides a few basic definitions and a brief introduction to the subject; it discusses how digital geometry relates to other mathematic disciplines and indicates which related topics will and will not be covered in the book.

1.1 Pictures

Scientists often deal with functions defined in a space (e.g., in three-dimensional [3D] Euclidean space or in a lower-dimensional space such as a plane or a surface). Such functions are often obtained by collecting sensory data; for example, two-dimensional (2D) scanners collect data from a 2D surface; 3D sensors collect data from a volume of 3D space; and optical sensors collect 2D images by projecting a 3D scene onto a plane or surface. Because the functions are not always obtained using imaging processes, we will call them *pictures* rather than images.

When computers are used to process or analyze a 2D or 3D picture, they deal with a discrete form of the picture, obtained by a process of *digitization*, which involves *sampling* the picture and *quantizing* the sampled values. The resulting set of 2D or 3D digital data is called a *digital picture*. *Picture analysis* (more commonly called *image analysis*) derives multidimensional information about objects or scenes from sensory data. Conversely, *computer graphics* synthesizes and generates digital pictures from models for objects or scenes.

A 2D digital picture consists of a finite number of pixels, each of which is defined by a location and a value at that location. The term *pixel* is short for "picture element"; this acronym was introduced in the late 1960s by a group at Jet Propulsion Laboratory in Pasadena, California, that was processing pictures taken by space vehicles [640]. The analogous 3D term is "voxel," which is short for "volume element." In this book, we will use the terms *pixel* and *voxel* in a different sense; they will refer to the elements of the medium on which pictures reside, which is defined by a regular orthogonal *grid*. A *picture* is then a function defined on the grid that assigns values to the pixels or voxels. Thus, pixels and voxels are locations defined by grid coordinates, and they have values defined by a picture.

In this section, we introduce the standard methods of representing 2D digital pictures, in which a pixel is a grid point or a grid square. These representations, in both 2D and 3D, will be discussed in greater detail in Section 2.1.

1.1.1 Pixels, voxels, and their values

A 2D digital picture captured or constructed on a surface (usually planar) is typically defined using a finite data structure that models a regularly spaced planar orthogonal grid.

> **Definition 1.1** A (2D) *picture* P is a function defined on a (finite) rectangular subset $\mathbb{G}$ of a regular planar orthogonal array. $\mathbb{G}$ is called a (2D) *grid*, and an element of $\mathbb{G}$ is called a *pixel*. P assigns a value $P(p)$ to each pixel $p \in \mathbb{G}$.

The values of pixels can be integers, floating point numbers, vectors, or even (finite) sets. For example, the values of pixels in color pictures are usually represented by triples of scalar values, such as red, green, and blue or hue, saturation, and intensity. For the purpose of this book, it will usually be sufficient to restrict pixel values to nonnegative integers.

> **Definition 1.2** A (3D) *picture* P is a function defined on a (finite) rectangular parallelepiped (cuboid) in a regular spatial orthogonal array. $\mathbb{G}$ is called a (3D) *grid*, and an element of $\mathbb{G}$ is called a *voxel*. P assigns a value $P(p)$ to each voxel $p \in \mathbb{G}$.

A 3D picture is defined on a finite data structure that models a regular orthogonal grid in 3D space. We will sometimes also use *one-dimensional (1D) pictures* as simple examples. A 1D picture is defined on a (finite) set of regularly spaced points of a line. 2D or 3D pictures can also be defined on other grids; see Exercise 1 in Section 1.3 for the 2D case.

Pixels have grid-based coordinates; we assume integer coordinates as a default so that the regular planar orthogonal array can be identified with $\mathbb{Z}^2 = \mathbb{Z} \times \mathbb{Z} = \{(i,j) : i,j \in \mathbb{Z}\}$, where $\mathbb{Z}$ is the set of all integers. Every *grid point* in $\mathbb{Z}^2$ is the center point of a *grid square* with sides (*grid edges*) of length 1, oriented parallel to the Cartesian coordinate axes (see Figure 1.1). The corners of grid squares are called *grid vertices*. The corresponding terminology in 3D will be introduced in Section 2.1. Note that the assumption of a uniform orthogonal grid is a simplification. In practical applications (e.g., medical imaging), the spacing between pixels or voxels may vary. For example, a 3D magnetic resonance picture usually has larger spacing between slices (cross-sections) than within a slice; the units of measurement cannot be ignored in such

1.1 Pictures

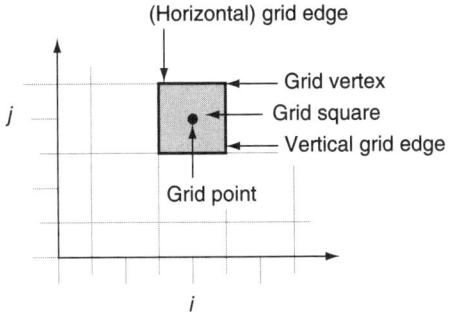

FIGURE 1.1 Grid point and grid square notations in the plane.

situations. However, in digital geometry, the spacing between pixels is often not relevant.

A 2D grid of size $m \times n$ is a rectangular array of grid points

$$\mathbb{G}_{m,n} = \{(i,j) \in \mathbb{Z}^2 : 1 \leq i \leq m \wedge 1 \leq j \leq n\} \tag{1.1}$$

or a rectangular set of grid squares

$$\mathbb{G}_{m,n} = \{\text{grid square } c : (i,j) = \text{center of } c \wedge 1 \leq i \leq m \wedge 1 \leq j \leq n\} \tag{1.2}$$

where $m, n \gg 1$ (read: "is much larger than 1").

Figure 1.2 (left) shows a magnified small rectangular portion (sometimes called a "window") of a picture. This illustrates the normal way pixels appear on a screen; after zooming in on a picture, the pixels are visible as colored squares, where the colors (or gray levels) represent the values of the pixels. In the common array data structure used to represent a picture, each pixel is represented by a pair of integers (i.e., by a grid point). Thus picture displays correspond to a grid composed of squares and array data structures to a grid composed of points (both types of grids coexist in

 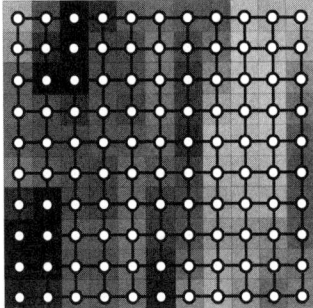

FIGURE 1.2 Magnified picture: grid of squares (left) and of points (right).

picture analysis and computer graphics). We use both representations in this book as alternative ways of representing pictures.

Pictures are quantized as well as sampled; a pixel or voxel can have only a finite number of possible values. The range of values in a (scalar) picture P will usually be of the form $\{0,\ldots,G_{\max}\}$, when $0 \leq P(p) \leq G_{\max}$ where $G_{\max} \geq 0$. $G_{\max} = 0$ is the trivial case of a constant ("blank") picture. (As mentioned earlier, the values in a color picture are triples $[u_1, u_2, u_3]$, such as red, green, and blue color components. These triples could also be mapped onto $\{0,\ldots,G_{\max}\}$, but color is of no relevance in this book, so we can think of G as short for "gray level." Color is becoming increasingly important in modern treatments of picture analysis and computer graphics, and it may also become more important in digital geometry. However, for the present, this book deals only with scalar (integer valued) pictures.

The range of values of the pixels or voxels in a *binary picture*[1] is $\{0,1\}$ (i.e., $G_{\max} = 1$). The pixels whose values are 1 (called 1s for short) define a subset $\langle P \rangle$ of the grid. These pixels are often referred to as "object" or "black" pixels. (In multivalued pictures, higher pixel values usually correspond to lighter gray levels, with gray level 0 being black; the opposite convention is often used for binary pictures: black = 1, white = 0.) Nonobject pixels in $\langle \overline{P} \rangle = \mathbb{G}_{m,n} \setminus \langle P \rangle$ are called 0s for short.

1.1.2 Picture resolution and picture size

Picture resolution is a display parameter. It is defined in dots per inch (dpi) or equivalent measures of spatial pixel density, and its standard value for recent screen technologies is 72 dpi. Recent printers use resolutions such as 300 dpi or 600 dpi, and such values can also be used for picture presentation on a screen.

Picture resolution is also a digitization parameter that is measured in samples per inch or equivalent measures of spatial sampling density. The human eye itself makes use of sampled pictures. The retina of the eye is an array of about 125 million photoreceptor cells called rods and cones. A rod is about 0·002 mm in diameter, and a cone is about 0·006 mm in diameter.

Pictures whose acquisition satisfies the traditional pinhole camera model are captured on planar surfaces. The light-sensitive array of a typical digital camera, which is a charge-coupled device (CCD) matrix, can be regarded as a rectangular set of square cells in a plane. The elements of the matrix capture a discrete set of pixel values. Other types of cameras may capture pictures on nonplanar surfaces. For example, Figure 1.3 shows a super–high-resolution panoramic picture captured with a rotating line camera. A geometric model of the panoramic picture acquisition process assumes that the picture is captured on a cylindric surface.

Discrete methods of picture generation are frequently used in both art and technology. The dots in a pointillistic painting can be as small as 1/16 of an inch in diameter. Figure 1.4 illustrates two other technologies for discrete picture generation:

---

1. These are the most common types of *two-valued pictures* that are used in picture analysis, but using the values $+1$ and -1 may be convenient in some algorithms. If there are more than two values, we call the picture *multivalued*.

1.1 Pictures

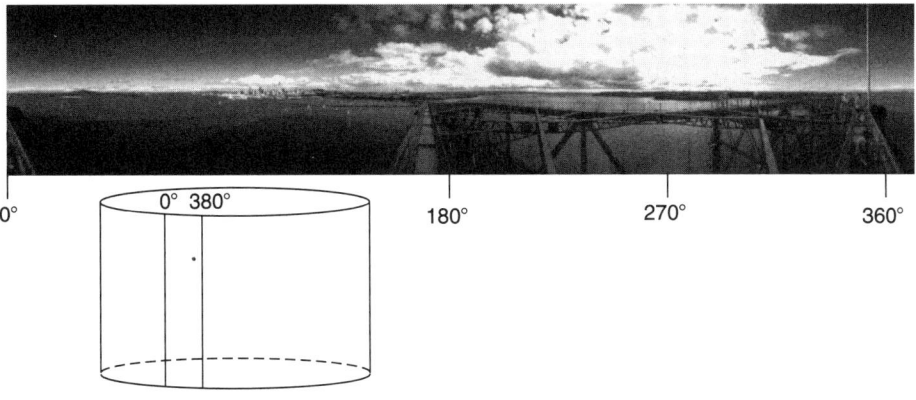

FIGURE 1.3 A $380°$ panoramic picture of Auckland, New Zealand, captured from the top of the harbor bridge. The full-resolution picture consists of about $10^4 \times 5 \cdot 10^4$ pixels captured on a cylindric surface with a rotating line camera.

pebble mosaics or tiled floors are composed of pixels, and patterns on fabrics or rugs can also be regarded as digital pictures. In 1725, B. Bouchon invented the idea of controlling a loom by perforated tape. The pattern generated by the loom was broken up into discrete areas with discrete color values. This idea was further developed by Falcon, a master silk weaver in Lyons, France. In 1738, Falcon applied for an English patent on his automatic card-controlled loom and provided a small working model as was required by English patent law. That working model continues to operate in the Science Museum in London, driven by an electric motor and weaving threads of several colors into patterned ribbon with the pattern controlled by punched cards laced together in a loop. When Falcon died in 1765, about 40 of his looms were operating. J.-M. Jacquard greatly improved the design of card-controlled looms in the early 19th century; thousands of Jacquard looms were soon in operation [835]. Thus pictures were generated by a programmed machine even before the first programmed machine (invented by C. Babbage) performed calculations on numbers! A surviving example of a pattern woven by a Jacquard loom is a black-and-white silk portrait of Jacquard himself, woven under the control of a "program" consisting of about 24,000 cards.

Today we think of pixels as tiny cells on a screen or as atomic units of a digital picture stored on a CD or DVD or in a computer. New media for representing large quantities of pictorial information will become available in the future, and contemporary screen technology may seem like pebbles only a few years from now.

Picture size is another important picture property. Pictures cannot be arbitrarily large; picture capturing, display, and printing technologies will always impose finite limits. The number of pixels in a typical picture has increased greatly since the early days of picture analysis and computer graphics (the 1950s and 1960s). In those days,

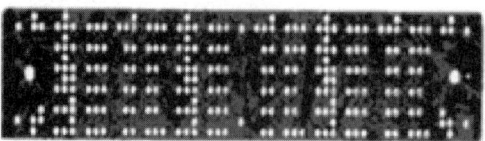

FIGURE 1.4 Lower left: a Greek pebble mosaic, detail from "The Lion Hunt" in Pella, Macedonia, circa 300 BC. Right and top: a picture of J.-M. Jacquard woven on silk on one of his "programmable" looms by digitizing a portrait of him, and one of the 24,000 punched cards that controlled the loom [669].

a picture might contain only a few thousand pixels; today, a color picture may require gigabytes of memory, as we saw in Figure 1.3.

1.1.3 Scan orders

Algorithms in picture analysis are often applied to the pixels of a picture in sequence, where the sequence is obtained by scanning the grid. A *scan* of a grid $\mathbb{G}_{m,n}$ is a one-to-one mapping ϕ of the $m \times n$ pixels of the grid into a linear sequence $\phi(1), \cdots, \phi(mn)$. Scans are used in picture processing programs to control the order of pixel accesses. A scan can also be viewed as an enumeration of the pixels; $\phi(k)$ is the k-th pixel ($1 \leq k \leq mn$).

G. Cantor (1845–1918) showed, using his famous enumeration principle, that the set of all rational numbers has the same cardinality $\aleph_0$ as the set of all natural numbers (i.e., this set is infinite but still enumerable). The enumeration scheme shown in Figure 1.5 defines a scan of the infinite set $\mathbb{Z}^2$ of grid points.

The standard method of scanning a grid $\mathbb{G}_{m,n}$ is row by row, with the pixels in each row accessed from left to right and the rows accessed from top to bottom;

1.1 Pictures

```
(-2,2)——(-1,2)——(0,2)——(1,2)——(2,2)
  |                                |
(-2,1)   (-1,1)——(0,1)——(1,1)   (2,1)
  |        |              |        |
(-2,0)   (-1,0)  (0,0)——(1,0)   (2,0)
  |        |                       |
(-2,-1)  (-1,-1)—(0,-1)—(1,-1)—(2,-1)
  |
(-2,-2)—(-1,-2)—(0,-2)—(1,-2)—(2,-2)→
```

FIGURE 1.5 Enumeration principle defining a scan of the infinite grid, starting at $(0,0)$ and proceeding in outward spiral order.

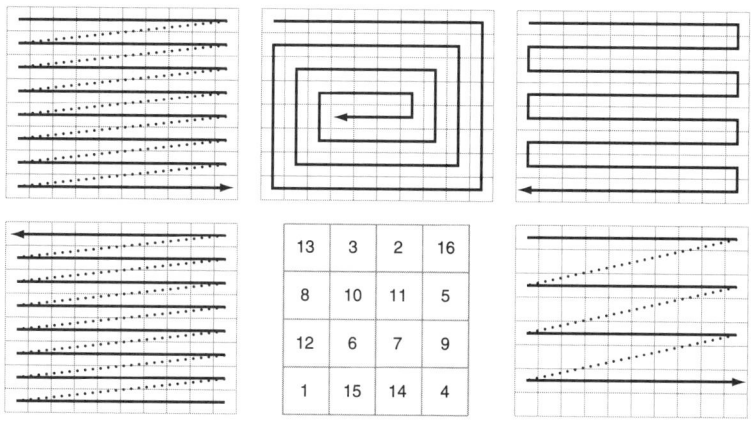

FIGURE 1.6 Scan orders: standard (upper left), inward spiral (upper middle), meander (upper right), reverse standard (lower left), magic square (lower middle), and selective, as used in interlaced scanning: standard, every second row (lower right).

this scan order is often called *row-major order*. It is a lexicographic order derived from the grid coordinates: $(1,1),(1,2),\ldots,(1,n),(2,1),(2,2),\ldots$ in the 2D case, where $(1,1)$ is in the upper left-hand corner; $(i_1,j_1) < (i_1,j_2)$ iff[2] $i_1 < i_2$, or $i_1 = i_2$ and $j_1 < j_2$. 3D scan order is defined analogously. This and five other common scan orders are illustrated in Figure 1.6. Standard and reverse standard scans are used, for example, for defining distance transforms in Section 3.4. Selective scans can be applied, for example, to speed up the search for picture objects when an estimate of the minimum size of the objects is available. In general, a scan can be split into scans

---

2. Read "if and only if"; acronym proposed by P.R. Halmos.

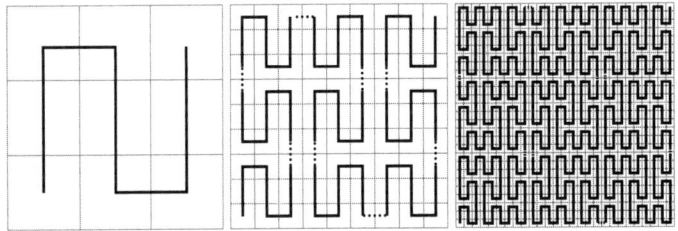

FIGURE 1.7 Peano scan: The scan pattern on the left is repeatedly used (nine times) to obtain the refined scan in the middle (with rotations at some places), and the same pattern of repetition of the scan in the middle is used to obtain the scan on the right.

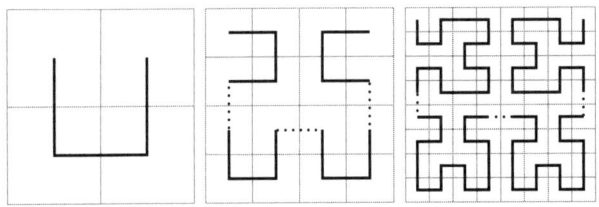

FIGURE 1.8 Hilbert scan: The scan pattern on the left is repeated only four times (with rotations) to obtain the refined scan.

of even rows followed by scans of odd rows, and this splitting process can be repeated recursively.

Scan orders play only minor roles in digital geometry, but they can be of interest in discrete mathematics. For example, the magic-square scan is an enumeration of pixels such that the sums of the pixel numbers in each row and column are equal. This scan is easily constructible for small pictures, and its scan order may appear to be "pseudorandom." Random scans can be defined using a random number generator to address one pixel at a time. (Efficiently keeping track of the set of remaining pixels is an interesting algorithmic problem.)

Scans related to the mathematic history of defining curves have also found their way into the picture-processing literature. This historic context may justify our briefly discussing two examples of "space-filling" curves: the *Peano curve* and the *Hilbert curve*.

The Peano curve was originally defined by G. Peano (1858–1932) in 1890, following a proposal by C. Jordan about a method of defining curves in parametric form. Peano showed that a curve defined in that form can completely fill a square, and Jordan therefore revised his original definition. Figure 1.7 illustrates a recursive (nonparametric) way of constructing the Peano curve by infinite repetition of the construction. Of course, in practice, we use only finitely many repetitions of the construction, in a grid of size $3^n \times 3^n$; the resulting curves are called *Peano scans*.

In 1891, D. Hilbert defined a similar curve. A finite number of repetitions of this construction, as illustrated in Figure 1.8, leads to a *Hilbert scan* in a grid of size $2^n \times 2^n$.

Any scan of a binary picture defines *runs* of 0s and 1s: maximum-length sequences of 0s or 1s that are visited by the scan. Evidently, runs of 0s must alternate with runs of 1s, and each run has a length of at least 1.

1.1.4 Adjacency and connectedness

Grid points are isolated points in the (real) plane, but, in the grid, adjacency relations between grid points can be defined. For $p = (x,y) \in \mathbb{Z}^2$, we define the *neighborhoods*

$$N_4(p) = \{(x,y), (x+1,y), (x-1,y), (x,y+1), (x,y-1)\}$$

and

$$N_8(p) = N_4(p) \cup \{(x+1,y+1), (x+1,y-1), (x-1,y+1), (x-1,y-1)\}.$$

Two grid points $p, q \in \mathbb{Z}^2$ are called *4-adjacent* or *proper 4-neighbors* (*8-adjacent* or *proper 8-neighbors*) iff $p \neq q$ and $p \in N_4(q)$ ($p \in N_8(q)$). We often use geographic language to identify the proper neighbors of a pixel: $(x, y+1)$ is called the north neighbor, $(x+1, y+1)$ is called the northeast neighbor, and so on.

Neighborhoods can also be defined for grid squares by applying 4- or 8-adjacency to the center points (grid points) of the grid squares, and they can also be defined for grids in 3D space. Adjacency relations in 2D and 3D grids will be discussed in greater detail in Chapter 2.

The reflexive, transitive closure of an adjacency relation on a set M (e.g., of grid points), which is the smallest reflexive, transitive relation on M that contains the given adjacency relation, defines a *connectedness relation*. M is called *connected* if for all $p, q \in M$ there exists a sequence $p_0, \ldots, p_n$ of elements of M such that $p_0 = p$, $p_n = q$, and p_i is adjacent to p_{i-1} ($1 \leq i \leq n$); such a sequence is called a *path* and is said to *join* p and q in M. Maximal connected subsets of M are called (connected) *components* of M. Evidently, components are nonempty and distinct components are disjoint. The concepts of 4- and 8-adjacency, connectedness, and components were introduced into picture analysis in 1966 [921], although the prefixes "4-" and "8-" were not used until a few years later.

It was observed in [921] that difficulties arise when 4-adjacency or 8-adjacency (and the corresponding type of connectedness) is used for both the 1s and 0s in a binary picture. Figure 1.9.a (an object containing two pixels) is 8-connected, as is its complementary set. Figure 1.9.b is both 4- and 8-connected, and its complement

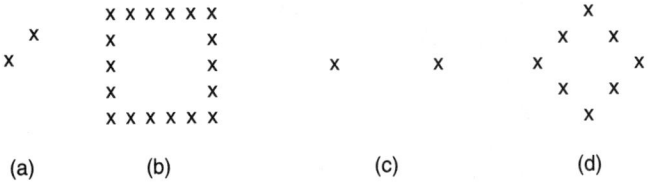

(a)   (b)   (c)   (d)

FIGURE 1.9 Examples of connected and nonconnected sets [921]. The xs stand for 1s; the 0s are not shown.

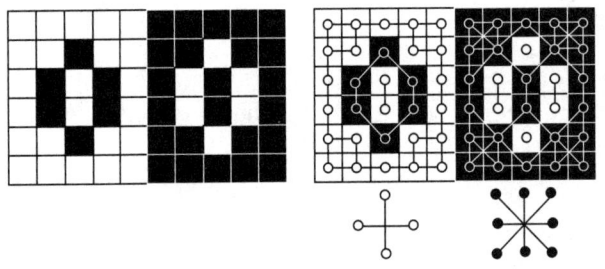

FIGURE 1.10 Left: a binary picture. Right: 4- and 8-adjacencies applied to white or black pixels.

is neither 4- nor 8-connected. Figure 1.9.c is neither 4- nor 8-connected, but its complement is both 4- and 8-connected.

> "The 'paradox' of Figure 1.9.d can be (expressed) as follows: If the 'curve' is connected ('gapless') it does not disconnect its interior from its exterior; if it is totally disconnected it *does* disconnect them. This is of course not a mathematical paradox,[3] but it is unsatisfying intuitively; nevertheless, connectivity is still a useful concept. It should be noted that if a digitized picture is defined as an array of hexagonal, rather than square, elements, the paradox disappears." [921]

The first case assumes 8-connectedness for both "curve points" and "background points," and the second case assumes 4-connectedness for both.

Commenting on [921] (in an unpublished technical report in 1967), R.O. Duda, P.E. Hart, and J.H. Munson proposed the dual use of 4- and 8-connectedness for 0s and 1s in a binary picture [286]. Figure 1.10 illustrates the use of 4-adjacency for white pixels and 8-adjacency for black pixels in a binary picture. The advantages of using opposite types of adjacency for the 0s and the 1s in binary pictures will be discussed in later chapters.

---

3. A mathematic *paradox* (antinomy) is characterized by the deduction of a contradiction within a theory. The existence of statements in digital topology that do not resemble statements in Euclidean topology is not a mathematic paradox.

1.2 Digital Geometry and Related Disciplines

Definition 1.3 *Digital geometry* is the study of geometric or topologic properties of sets of pixels or voxels. It often attempts to obtain quantitative information about objects by analyzing digitized (2D or 3D) pictures in which the objects are represented by such sets.

We usually assume that objects are represented by connected sets of pixels or voxels and that the quantitative information involves quantities studied in Euclidean or similarity geometry (see Section 1.2.2); we can then attempt to ensure that the properties computed in digital geometry adequately approximate these quantities. In this sense, we can regard digital geometry as digitized similarity geometry. As we will see in Section 1.2.2, Euclidean geometry is a special case of similarity geometry.

Digital geometry also often attempts to obtain topologic characterizations of pictures or to transform pictures into "simpler" topologically equivalent pictures. Due to the discrete nature of digital geometry, these topologic problems belong to the field of combinatorial topology (the topology of cell complexes), which is discussed in Chapter 6.

The remainder of this section briefly discusses topics and disciplines related to digital geometry and indicates the extent to which these topics will be treated in this book.

1.2.1 Coordinates and metric spaces

The concept of defining the locations of points in a plane by their distances from two straight lines ("axes") was used by Archimedes and Apollonius more than 2000 years ago. A *Cartesian coordinate system* makes use of a set of axes as introduced by R. Descartes (in Latin: Cartesius, 1596–1650) in [264] to define nonnegative coordinates in the plane. Descartes dealt with general oblique coordinates, addressing rectangular coordinates as an important special case.[4] Negative coordinates were first proposed by I. Newton, and the first use of the term "coordinates" is ascribed to G. Leibniz. When we use rectangular Cartesian coordinates, we define the (real) plane as a "Cartesian product" of two (real) lines: $\mathbb{R}^2 = \mathbb{R} \times \mathbb{R}$, where $\mathbb{R}$ is the set of reals. A Cartesian coordinate system in the plane, together with the Euclidean metric (see p. 12), defines the *Euclidean plane* $\mathbb{E}^2$. The n-dimensional real and Euclidean spaces are defined analogously using n-fold Cartesian products. The point o where the axes intersect is called the *origin* of the coordinate system; because o is at distance 0 from all the axes, its coordinates are all 0s.

A *right-handed coordinate system* is a rectangular Cartesian coordinate system in which the positive x-axis is identified with the thumb (pointing outward in the plane of the palm), the positive y-axis with the forefinger (pointing outward in the

---
4. See pages 26–27 in the English translation of [264].

plane of the palm), and, in 3D, the positive z-axis with the middle finger of the right hand (pointing downward from the plane of the palm).

Coordinate systems can also be defined using distances to points. For example, *barycentric coordinates* (homogeneous or triangular coordinates) in the plane were introduced by Möbius in 1827 [807] as a way of representing points in the plane relative to a given triple of noncollinear points. The prefix *bary-* refers to weight or mass. For any point p inside the triangle, there exist masses a, b, and c such that, if they are placed on the vertices of the triangle, their center of gravity (balancing point) will be at p. The masses a, b, and c are uniquely determined if we require that $a+b+c=1$. The triple (a,b,c) defines the barycentric coordinates of p in the given triangle.

The measurements studied in picture analysis are always related to a regular grid that defines the locations of pixels or voxels in a Cartesian coordinate system. If we fix all but one of the coordinates, the locations become regularly spaced points on a line. The distance between consecutive locations on such a line is called the *grid constant*; it is the unit or scale of measurement in the grid coordinate system.

Let S be an arbitrary nonempty set. A function $d : S \times S \mapsto \mathbb{R}$ is a *distance function* or *metric* on S iff it has the following properties:

M1: For all $p,q \in S$, we have $d(p,q) \geq 0$, and $d(p,q) = 0$ iff $p = q$ (positive definiteness).

M2: For all $p,q \in S$, we have $d(p,q) = d(q,p)$ (symmetry).

M3: For all $p,q,r \in S$, we have $d(p,r) \leq d(p,q) + d(q,r)$ (triangularity: the triangle inequality).

If d is a metric on S, the pair $[S,d]$ defines a *metric space*. M. Fréchet (1878–1973) introduced the concept of a *metric space*; the elements of S can be any mathematic objects. The name "metric space" is due to F. Hausdorff (1868–1942). Quantitative geometric measurements are often based on metrics.

It should be pointed out that property **M1** can be simplified to the following:

M1: For all $p,q \in S$, we have $d(p,q) = 0$ iff $p = q$.

Indeed, from the simplified **M1** together with **M2** and **M3**, we have as follows:

$$0 = d(p,p) \leq d(p,q) + d(q,p) = 2d(p,q) \text{ for all } p,q \in S$$

Let $[S,d]$ be a metric space, $p \in S$, and $\epsilon > 0$. The set of points $q \in S$ such that $d(p,q) \leq \varepsilon$ is called a *ball* of radius ε with *center* p. A subset M of S is called *bounded* iff it is contained in a ball of some finite radius.

Euclidean space $\mathbb{E}^n$ is defined with an orthogonal coordinate system; it is used to define a metric d_e called the *Euclidean metric*. If two points p,q have coordinates $(x_1,\ldots,x_n)$ and $(y_1,\ldots,y_n)$, we define the following:

$$d_e(p,q) = \sqrt{(x_1 - y_1)^2 + \ldots + (x_n - y_n)^2}$$

1.2 Digital Geometry and Related Disciplines

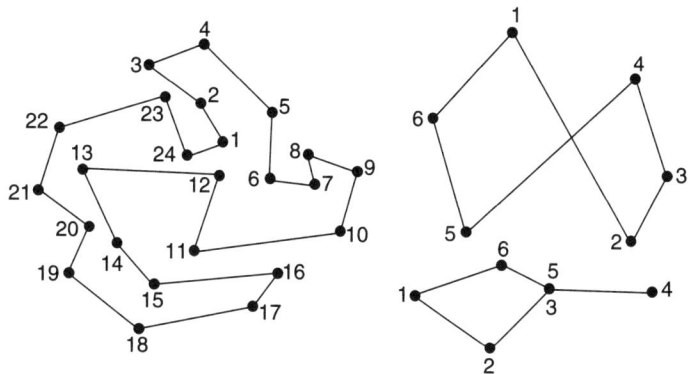

FIGURE 1.11 A simple planar polygon (left) and two nonsimple planar polygons (right). The numbers specify a cyclic order of the vertices.

It will be proved in Chapter 3 that d_e is a metric. Two-dimensional Euclidean space is called the *Euclidean plane* $\mathbb{E}^2$. *Digital space* $\mathbb{Z}^n$ is a subspace of $\mathbb{E}^n$, which is defined by the subset of all points with integer coordinates and the Euclidean metric; $\mathbb{Z}^2$ is the *digital plane*.

Metrics defined on grids or on Euclidean spaces play central roles in digital geometry. They will be frequently used in this book and will be discussed in detail in Chapter 3.

1.2.2 Euclidean, similarity, and affine geometry

The history of geometry dates back about 4000 years to societies in Egypt, Mesopotamia, and China. The word *geometry*, which means earth measurement, has been in use for more than 2500 years. The measurement of distances and the calculation of areas and volumes are among the earliest developments in mathematics. Of course, only simple 2D or 3D objects such as polygons, prisms, cuboids, and cylinders were studied in those days.

A (finite, connected) *polygonal arc* in Euclidean space $\mathbb{E}^n$ is a finite sequence of points $(p_1, p_2, \ldots, p_n)$ where $n \geq 3$, which defines $n-1$ straight line segments $p_i p_{i+1}$ ($i = 1, 2, \ldots, n-1$). The p_is are called the *vertices* of the arc, and the line segments are called its *edges*. The polygonal arc forms a *circuit* $\langle p_1, p_2, \ldots, p_n \rangle$ if we add the nth edge $p_n p_1$.

A *simple polygon* $\Pi = \langle p_1, \ldots, p_n \rangle$ (an "n-gon") is a polygonal circuit such that no point belongs to more than two edges and the only points that belong to two edges are the vertices; see Figure 1.11. If $n = 3$, the polygon is called a triangle.

TABLE 1.1 Transformations allowed in different geometries.

Transformations	Euclidean geometry	Similarity geometry	Affine geometry	Projective geometry
Rotations	Yes	Yes	Yes	Yes
Translations	Yes	Yes	Yes	Yes
Uniform scalings (all axes)	No	Yes	Yes	Yes
Nonuniform scalings	No	No	Yes	Yes
Shears	No	No	Yes	Yes
Central projections	No	No	No	Yes

TABLE 1.2 Examples of invariant quantities in these geometries.

Invariants	Euclidean geometry	Similarity geometry	Affine geometry	Projective geometry
Lengths	Yes	No	No	No
Angles	Yes	Yes	No	No
Ratios of lengths	Yes	Yes	No	No
Parallelism	Yes	Yes	Yes	No
Incidence	Yes	Yes	Yes	Yes
Cross-ratios of lengths	Yes	Yes	Yes	Yes

The concept of an angle, the decomposition of a simple planar polygon into triangles, and the calculation of simple areas and volumes (such as the volume of a frustum of a pyramid) were known in ancient Egypt. The law of Pythagoras (about right triangles) was known in ancient Mesopotamia, and the laws of similar triangles were also widely used.

It is not certain whether Euclid of Alexandria (about 325–265 BC) was an individual, the leader of a team of mathematicians, or the pseudonym of a group of mathematicians. It is certain, however, that the book *The Elements* established *Euclidean geometry* on the basis of just a few postulates (or "axioms") about straight lines, straight line segments, circles, and angles.

Similarity and *affine geometry* are "intermediate" between Euclidean and projective geometry (see Section 1.2.3) with respect to the transformations that are allowed (see Table 1.1) and the quantities or relations that remain invariant under these transformations (see Table 1.2).[5] The use of invariants under groups of transformations to characterize geometries will be discussed in Chapter 14.

In general, the transformations listed in Table 1.1 do not take grids into themselves. For example, a rotation takes an orthogonal grid into itself only if it is a

---

5. Two sets are called *incident* if one of them contains the other. In all of these geometries, incidences between sets (e.g., between points and lines) are invariant.

1.2 Digital Geometry and Related Disciplines

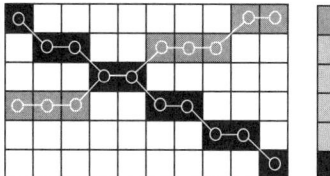 

FIGURE 1.12 Left: two lines that intersect in a segment. Right: two "arcs" (black and dark gray) and an "ellipse" (white) that have no pixel in common.

rotation by a multiple of $90°$. The application of general transformations to a grid requires approximation or interpolation, as discussed in Chapter 14. As a result, the "invariants" listed in Table 1.2 are only approximately invariant when the transformations are applied to digital pictures.

Objects in digital geometry often do not behave like Euclidean objects. For example, we can define adjacency relations between pixels (see Section 1.1.4), whereas distinct points cannot be "adjacent" in Euclidean geometry. Digital lines are sequences of pixels and can intersect in segments (see Figure 1.12, left). Nonparallel lines may have no pixel in common; if two lines are defined by sequences of 8-adjacent pixels, they can cross without intersecting. The two digital "arcs" and the digital "ellipse" on the right in Figure 1.12 have no pixel in common!

1.2.3 Projective geometry

In the fifteenth century, some Italian painters developed a system of perspective that allowed them to paint the world as it is seen. This inspired new geometric ideas and led G. Desargues (1596–1662) to invent *projective geometry*.

A picture acquisition process that maps (the visible portion of) a scene onto a surface has an ideal description in terms of projective geometry (see Figure 1.13). In actual digital pictures, the pixels vary from this ideal (e.g., in the case of a planar picture because of the finite physical dimensions and small irregularities of the cells of the CCD array; in the case of a cylindric picture [see Figure 1.3] because of rotation errors of the line camera; and in both cases because of optical aberrations).

If we use the ideal description, the *geometric resolution* of a picture is specified by the geometric laws of projection that model the picture acquisition process. Geometric resolution allows us to relate measurements in pictures (in terms of pixel coordinates) to measurements in the real world. Picture resolution, which was discussed in Section 1.1.2, is a display parameter that need not correspond to geometric resolution.

A CCD cell in a camera collects light from a scene or object surface patch (see Figure 1.13). The size of this patch depends on the size of the cell, the camera optics (e.g., focal length, aberrations), the distance and direction between the camera

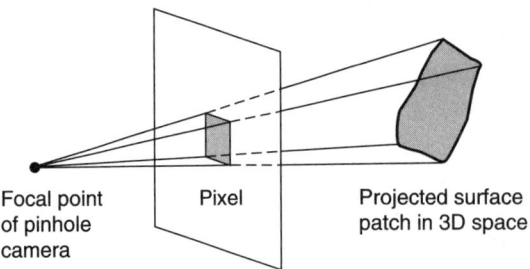

FIGURE 1.13   Central projection onto a pixel in a picture.

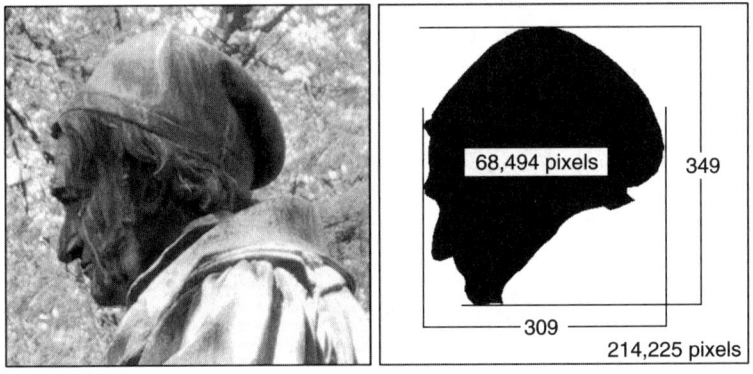

FIGURE 1.14   Left: detail of the Gauss statue in Göttingen, Germany. Right: number of pixels in the area of the head and in the entire picture, and height and width measured in numbers of pixel units (grid edges).

and the surface of the object, and the 3D shape of the object. Thus geometric resolution depends on many picture acquisition parameters. Camera calibration, which is a subject in the computer vision literature, deals with the calculation of these parameters.

If a picture shows only a flat surface (e.g., a document, a microscope slide), specification of the picture size and the pixel size or spacing in terms of coordinates on the surface is relatively simple. Similar remarks apply to aerial pictures obtained by a camera flying at constant altitude over flat terrain. In such situations, picture resolution has a simple relationship to geometric resolution, so measurements made on the picture can be easily related to geometric measurements made on the surface.

In a 3D scene such as the head of the Gauss statue in Figure 1.14, the problem of calculating geometric resolution is much more complicated. Different pixels

1.2 Digital Geometry and Related Disciplines

in the picture may correspond to surface patches of different sizes in the scene. Determining these sizes may require 3D shape recovery and the application of projective geometry to map the pixels onto the recovered surface. This book will not discuss camera calibration, 3D shape recovery, or projective geometry, and it will not deal with modeling picture acquisition processes in a 3D environment. The reader is referred to books about computer vision that deal with these subjects.

Both geometric and picture resolution have increased over the years because of progress in picture acquisition and display technologies. In this book, we will normally not be concerned with geometric picture size or resolution but only with the sizes of pictures as measured in numbers $m \times n$ of pixels.

1.2.4 Vector and geometric algebra

A grid point p can be identified with a vector $p = \vec{op}$ that starts at the origin o and ends at p; we can also consider vectors $\vec{pq}$ from one grid point to another. In this book, we deal only with 2D and 3D pictures. (We do not consider time sequences of 2D or 3D pictures as being 3D or 4D pictures in which the last coordinate is time; the time coordinate is qualitatively different from the spatial coordinates.) However, we will sometimes use n-dimensional spaces in this book (e.g., as a generalization of 2D and 3D spaces or when we deal with n-tuples of property values).

The 2D vector space $[\mathbb{R}^2, +, \cdot, \mathbb{R}]$ over the real numbers is defined by an addition operation $(x_1, x_2) + (y_1, y_2) = (x_1 + y_1, x_2 + y_2)$ for all $(x_1, y_1), (x_2, y_2)$ in $\mathbb{R}^2$ and a scalar multiplication operation $a \cdot (x, y) = (ax, ay)$ for all $a \in \mathbb{R}$ and all $(x, y) \in \mathbb{R}^2$. Higher-dimensional vector spaces are defined analogously using n-tuples rather than pairs.

A *vector space* $[S, +, \cdot, \mathbb{R}]$ over the real numbers $\mathbb{R}$ is defined by a nonempty set S and two operations, $+$ and $\cdot$, that have the following properties for all $p, q, r \in S$ and all $a, b \in \mathbb{R}$:

V0: $p + q \in S$ and $a \cdot p \in S$ (closure under vector addition and under scalar multiplication)

V1: $p + q = q + p$ (commutativity of vector addition)

V2: $(p + q) + r = p + (q + r)$ (associativity of vector addition)

V3: $a \cdot (b \cdot p) = (ab) \cdot p$ (associativity of scalar multiplication)

V4: $(a + b) \cdot p = a \cdot p + b \cdot p$ (distributivity of scalar multiplication over vector addition)

V5: $a \cdot (p + q) = a \cdot p + a \cdot q$ (distributivity of vector addition over scalar multiplication)

V6: There exists an $o \in S$ such that $o + p = p$ (identity for vector addition).

V7: $1 \cdot p = p$ (identity for scalar multiplication)

V8: For any $p \in S$, there exists $-p \in S$ such that $p + (-p) = o$ (inverse for vector addition).

A vector space is called *finite-dimensional* if there exists a finite subset B of S such that every $p \in S$ is a sum of scalar multiples of vectors in B. S is called *n-dimensional* if the smallest such B has cardinality n. The elements of S are called *vectors*, and the elements of $\mathbb{R}$ are called *scalars*.

The *norm* $\|p\|$ of a vector $p = (x_1, \ldots, x_n) \in \mathbb{R}^n$ is as follows:

$$\|p\| = \left(x_1^2 + \ldots + x_n^2\right)^{1/2}$$

Note that $\|p\| = 0$ iff $x_1 = \cdots = x_n = 0$; evidently this implies that $p = o$. If $p \neq o$, the *direction* of p is defined by the *unit vector* $p^\circ = p/\|p\|$.

The *dot product* of $p = (x_1, \ldots, x_n)$ and $q = (y_1, \ldots, y_n)$ is as follows:

$$p \cdot q = x_1 y_1 + \ldots + x_n y_n \tag{1.3}$$

It can be shown that

$$p \cdot q = \|p\| \cdot \|q\| \cdot \cos \eta$$

where η is the (smaller) angle between the unit vectors p° and q°, $0 \leq \eta < \pi$.

p and q are called *orthogonal* iff $p \cdot q = 0$, and they are called *orthonormal* iff they are orthogonal and $|p| = |q| = 1$.

The *cross product* will be defined here only for $n = 3$:

$$p \times q = (x_1, x_2, x_3) \times (y_1, y_2, y_3) = (x_2 y_3 - x_3 y_2, x_3 y_1 - x_1 y_3, x_1 y_2 - x_2 y_1) \tag{1.4}$$

It can be shown that if p and q are linearly independent (i.e., there do not exist $a, b \in \mathbb{R}, a \neq 0$, and $b \neq 0$ such that $ap + bq = 0$); then $p \times q$ is orthogonal to the plane defined by p and q. We also have

$$\|p \times q\| = \|p\| \cdot \|q\| \cdot \sin \eta \tag{1.5}$$

with η as defined above.

Vector algebra can be generalized to multidimensional oriented geometric entities such as are studied in *Clifford* or *geometric algebra*; see [772] for a review of the work by W.K. Clifford (1845–1879). This theory is based on definitions of "inner" and "outer" products of "multivectors," which are classified by their "grades." (The inner product generalizes the dot product. Multivectors of grade 0 are scalars in $\mathbb{R}$; of grade 1 are vectors in $\mathbb{R}^n$ [$n \geq 2$]; of grade 2 are bivectors that are oriented trapezoids defined by the outer product of two vectors; of grade 3 are oriented volume elements; and so forth.) Geometric algebra allows compact descriptions of distances and "angles" between geometric entities, including "degrees of parallelism." Related studies in digital geometry might be of future interest, but the subject is not forth discussed in the present edition. As a possible initial step, see Chapter 3 for digital versions of the definitions of angles and seminorms.

1.2 Digital Geometry and Related Disciplines

1.2.5 Graph theory

Euler's analysis of the bridge situation in Königsberg, Germany (i.e., is it possible to cross all of the bridges just once during a walk through the city?) initiated *graph theory*. The Königsberg bridge situation is shown in Figure 1.15; for more about the solution to the problem (i.e., the nonexistence of such a walk), see Section 4.1.1.

An adjacency relation, for example, on grid points (see Section 1.1.4), defines an undirected *graph* whose *nodes* are the grid points and where there is an *edge* between two nodes iff the corresponding grid points are adjacent. Graphs will be discussed in Chapter 4.

A *path* of length n in a graph is a finite sequence of nodes $p_0, \ldots, p_n$ such that there is an edge between p_i and p_{i-1} ($1 \leq i \leq n$). If such a path exists, we say that p_0 and p_n are *connected*. (Evidently connectedness is symmetric, because the reversal of a path from p to q is a path from q to p; it is also transitive, because, if there are paths from p to q and from q to r, we obtain a path from p to r by concatenating the two paths.) We also say that each node is connected to itself by a path of length zero; thus connectedness is reflexive as well as symmetric and transitive. A set of nodes of a graph is *connected* iff every pair of its nodes is connected. For example, the graph on the right in Figure 1.10 is not connected.

A maximal connected set of nodes of a graph is called a (connected) *component*. For example, the graph on the right in Figure 1.10 has eight components. A connected graph consists of a single component. Finite components (with respect to a given

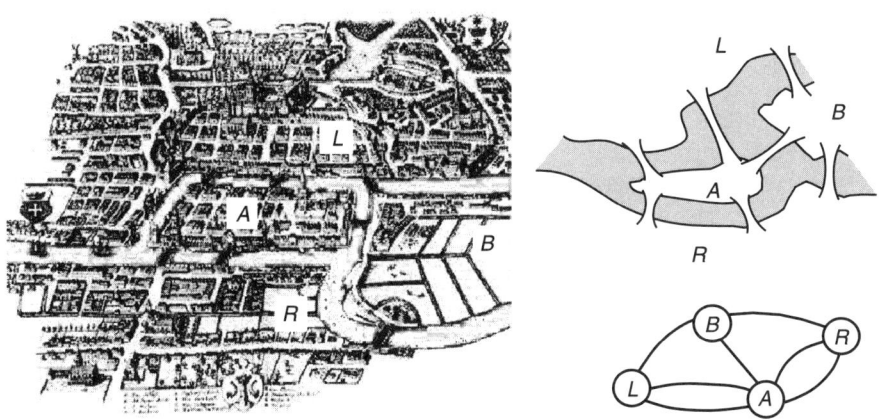

FIGURE 1.15 Three different representations of the Königsberg bridge situation. Left: The city at the time of Euler. Right, top: Simplified map (not to scale) of the islands, the left and right banks of the river, and the bridges. Right, bottom: Schematic representation; the labeled circles represent the islands and banks, and the arcs represent the bridges.

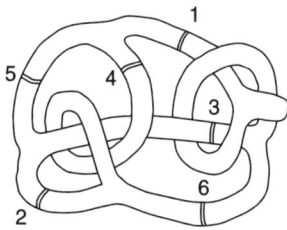

FIGURE 1.16 Positions of possible cuts in a 3D solid [660].

adjacency relation) of the pixels of voxels of a picture are called *regions*. Digital geometry is often concerned with geometric properties of regions.

Graph theory deals with many topics that are not directly relevant to digital geometry. However, many topics in digital geometry can be discussed on a graph-theoretic level, treating pixels or voxels as nodes of a graph rather than as elements of a grid. In Chapter 4 we will emphasize graph-theoretic concepts that are applicable to digital geometry.

1.2.6 Topology

The origin of *topology* is often identified with the *Descartes-Euler Theorem* $\alpha_0 - \alpha_1 + \alpha_2 = 2$, which was originally stated regarding the numbers α_2, α_1, and α_0 of faces, edges, and vertices of a convex polyhedron. (A convex *polyhedron* is a nonempty bounded set that is an intersection of finitely many half-spaces.)

J.B. Listing (1802–1882) was the first to use the word "topology" in his correspondence, beginning in 1837.[6] The term, which replaced Leibniz's "geometria situs" or "analysis situs," was introduced to distinguish "qualitative geometry" from geometric topics that emphasized quantitative measurements and relations. Topology can be informally viewed as "rubber-sheet geometry": the study of properties of objects that remain the same when the objects are (continuously) deformed. For example, incidence relations between sets are topologically invariant.

As a more specialized example of a topologic property, the *genus* of a connected set is the minimum number of "cuts" needed to transform the set into a *simply connected set*.[7] The 3D object shown in Figure 1.16 has genus 3 (the complexity of

---

6. In an 1861 publication, Listing writes that "*topologische* Eigenschaften (solche sind), die sich nicht auf die Quantität und das Maass der Ausdehnung, sondern auf den Modus der Anordnung und Lage beziehen." (*Translation*: Topologic properties are those that are related not to quantity or content but to spatial order and position.) Because Listing was a student and then a close friend of C.F. Gauss, Listing's interest in the subject may have followed the advice or example of Gauss himself.

7. This will be defined later (see Chapter 6) as a connected set that has a trivial fundamental group. Informally, a simply connected set in 3D space is connected and has no cavities or tunnels. In general, a "cut" creates a (doubly oriented) simple arc, simple curve, or simply connected face. Removing a single point from a hollow sphere transforms it into a simply connected object; this is the limiting case of a circular cut that has radius 0.

FIGURE 1.17 Sets of genus 0, 1, and 2.

this example illustrates the advanced concepts that were being studied in topology even at its birth in the mid-19th century); the three 2D objects shown in Figure 1.17 have genus 0, 1, and 2, respectively. Genus is a qualitative property that characterizes the *degree of connectedness* of a set. Aspects of topology that are relevant to digital geometry will be reviewed in Chapters 6 and 7.

Digital topology (see Chapters 6 and 7) provides topologic foundations for digital geometry. It extends the graph-theoretic concepts mentioned in Section 1.2.5 and develops a topologic theory of subsets of pictures that provides a basis for designing many algorithms for 2D and 3D picture analysis.

1.2.7 Approximation and estimation

The fundamental theorem of *approximation theory* is due to K.W.T. Weierstrass (1815–1897). It states that, for any function f that has continuous derivatives on a finite interval $[a,b]$ and for any $\varepsilon > 0$, there exists a polynomial g such that $|f(x) - g(x)| < \varepsilon$ for all $x \in [a,b]$. This illustrates the orientation of approximation theory toward continuous functions, polynomials, and arbitrarily small errors. Digital geometry differs from approximation theory in all of these respects: digital data need not be obtained from continuous functions; approximations usually involve linear functions such as straight line segments or planar surface patches; and the approximations may not be arbitrarily close.

Approximation is commonly used to estimate the values of geometric quantities. In ancient mathematics, Archimedes estimated π on the basis of inner and outer regular n-gon approximations of a circle with $n = 3, 6, 12, 24, 48, 96$; see the left side of Figure 1.18. (An n-gon is called *regular* if its edges all have the same length.) For $n = 6$, this approximation gives $3 < \pi < 3.46$; for $n = 96$, it gives the following:

$$\tfrac{223}{71} < \pi < \tfrac{220}{70} \quad \left(\text{i.e.,} \quad \pi \approx 3.14\right) \tag{1.6}$$

In this method of approximation, the perimeters $\mathcal{P}$ of the inner and outer regular n-gons converge toward the circle's perimeter as $n \to \infty$. For example, let the radius of the circle be r so that its perimeter is $2\pi r$, and let the inner n_m-gon P_{n_m} have $n_m = 3 \times 2^m$ edges of length e_m so that $\mathcal{P}(P_{n_m}) = n_m \cdot e_m$. From elementary geometry

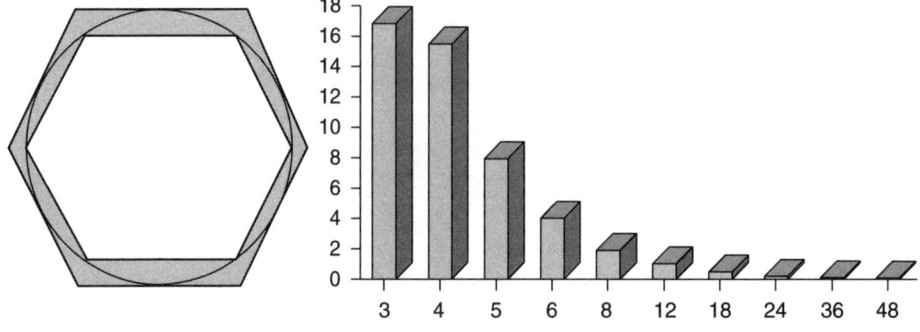

FIGURE 1.18 Inner and outer 6-gons approximating a circle, and percentage errors between the perimeters of the inner n-gons and the perimeter of the circle.

we know that

$$e_{m+1} = \sqrt{2r^2 - r\sqrt{4r^2 - e_m^2}} \qquad (1.7)$$

where $e_0 = r\sqrt{3}$ for a triangle, $e_1 = r$ for a hexagon, and so on. It follows that the *estimation error*

$$\kappa(n_m) = |\mathcal{P}(P_{n_m}) - 2\pi r| \approx \frac{2\pi r}{n_m} \qquad (1.8)$$

converges to zero as $n_m \to \infty$ (see the right side of Figure 1.18; e.g., $\kappa(3)/\mathcal{P}(P_3) = 17.3028\ldots$ and $\kappa(24)/\mathcal{P}(P_{24}) = 0.2853\ldots$). The *speed of convergence* $1/\kappa(n)$ is thus (asymptotically) a linear function of n.

An improved method of calculating π was given by the Chinese mathematician Liu Hui in 263. He calculated the areas S_n of regular n-gons inscribed in a circle and proved that, for all $n > 2$, the area S of the circle is bounded by $S_{2n} < S < S_{2n} + (S_{2n} - S_n)$. Starting with $n = 6$, he doubled n five times to 192 and got close bounds for the area S. From that area, he computed the circumference of the circle and got an approximation to π of 3.1410 [1148].

These historic examples of length estimation by approximation do not fit into the methodologic framework of digital geometry. The inner regular n_m-gons are defined by sample points on the circle, and the outer regular n_m-gons are defined by tangent lines to the circle. In digital geometry, we cannot use such samples, which can be arbitrary points and lines in the Euclidean plane and have exact relationships with the circle; we have to deal with sets of grid points, which are not arbitrary points and need not exactly lie on the circle ("digitization error").

A common task in 2D image analysis is to find a curve (e.g., an ellipse) that best fits (with respect to some error criterion) a given set of pixels. Figure 1.19 shows a best-fitting ellipse constructed using a numeric method described in [329]. Note that this ellipse does not necessarily circumscribe the set of pixels. Numeric methods

1.2 Digital Geometry and Related Disciplines

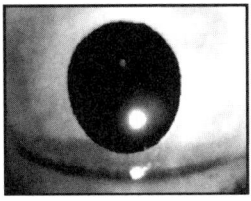 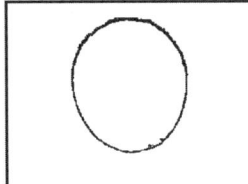 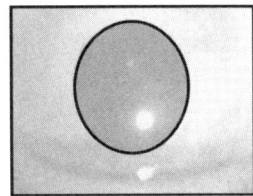

FIGURE 1.19   Left: a picture of a pupil. Middle: the set of pixels detected as the edge of the pupil. Right: the ellipse fitted to these pixels.

are not discussed in this book, except for an iterative (3D) curve approximation algorithm in Chapter 10.

The estimates of geometric quantities studied in this book will be based only on digital approximations. A methodology for studying the convergence of digital approximations to geometric quantities as the grid constant goes to zero will be described in Section 2.4.

1.2.8 Combinatorial geometry

Generalizations of Archimedes' and Liu Hui's methods of estimating perimeter are studied in combinatorial geometry, which is more than 100 years old; it started with the geometry of numbers established by H. Minkowski (1864–1909). The geometry of numbers does not have close links with digital geometry in spite of the use of grid points in both fields. However, some results in combinatorial geometry are potentially applicable to digital geometry, especially if they deal with sets of grid points. Examples that we will discuss in this book are corner counts for isothetic polygons or polyhedra[8] and asymptotic bounds on the number of convex grid polygons in an $m \times n$ grid. We also apply (in Chapter 9) the *Transversal Theorem* [957], which is a direct consequence of *Helly's First Theorem* [422]:

> Let $\mathcal{F}$ be a finite family of parallel straight line segments in $\mathbb{R}^2$. If every three segments in $\mathcal{F}$ have a common transversal, then there is a transversal common to all of the segments in $\mathcal{F}$.

A *transversal* of a straight line segment σ in $\mathbb{R}^2$ is a straight line that intersects σ but does not contain it.

We give two recent examples [789] of results in combinatorial geometry. Let $\mathcal{A}(S)$ denote the area of a bounded subset S of the plane. Let S be convex and have nonzero area. (A set S is called *convex* if for any two points p, q of S the straight line

---

8. In a Cartesian coordinate system, a line is called *isothetic* iff it is parallel to a coordinate axis; a plane is called isothetic iff it is parallel to a coordinate plane. An *isothetic polygon* has only isothetic edges, and an *isothetic polyhedron* has only isothetic faces. A polygon is called a *grid polygon* and a polyhedron is called a *grid polyhedron* if their vertices are grid points. In this book, we will often deal with isothetic grid polygons and polyhedra.

segment pq is contained in S.) Let Π_n (π_n), where $n \geq 3$, be an n-gon of minimum area circumscribed around S (of maximum area inscribed in S). The sequence of positive reals $\mathcal{A}(\Pi_n)$ ($\mathcal{A}(\pi_n)$) decreases (increases) as $n \to \infty$. Furthermore, we have

$$\mathcal{A}(\Pi_n) \leq \frac{\mathcal{A}(\Pi_{n-1}) + \mathcal{A}(\Pi_{n+1})}{2}$$

$$\text{and} \quad \mathcal{A}(\pi_n) \geq \frac{\mathcal{A}(\pi_{n-1}) + \mathcal{A}(\pi_{n+1})}{2}$$

for all $n \geq 4$; hence, if $\mathcal{A}(\Pi_m) = \mathcal{A}(\Pi_{m+1})$ ($\mathcal{A}(\pi_m) = \mathcal{A}(\pi_{m+1})$), then $\mathcal{A}(\Pi_n) = \mathcal{A}(\Pi_m)$ ($\mathcal{A}(\pi_n) = \mathcal{A}(\pi_m)$) for all $n \geq m$.

Similarly, let $\mathcal{P}(S)$ denote the perimeter of a bounded subset S of the plane. (We assume that the frontier of S is rectifiable so that its perimeter is well defined. Measurability and rectifiability will not be defined in this book; the frontier of a set will be defined in later chapters.) Let S be convex and have nonzero area. Let Λ_n (λ_n), where $n \geq 3$, be an n-gon of minimum perimeter circumscribed around S (of maximum perimeter inscribed in S). Then, for all $n \geq 4$, the following are given:

$$\mathcal{P}(\Lambda_n) \leq \frac{\mathcal{P}(\Lambda_{n-1}) + \mathcal{P}(\Lambda_{n+1})}{2}$$

$$\text{and} \quad \mathcal{P}(\lambda_n) \geq \frac{\mathcal{P}(\lambda_{n-1}) + \mathcal{P}(\lambda_{n+1})}{2}$$

1.2.9 Computational geometry

Computational geometry deals with finite collections of simple geometric objects (e.g., points, lines, circles) in Euclidean space. It studies algorithms for solving problems about such collections and the complexity of applying the algorithms as the number of objects increases. The phrase "computational geometry" was first used in the title of a 1969 book [733] about property computation, then in the early 1970s for geometric modeling by means of spline curves and surfaces [378], and finally in the mid-1970s [974] with the meaning that it has today.

Let f and g be functions from the set $\mathbb{N}$ of natural numbers (nonnegative integers) into the set $\|R^+$ of positive real numbers. Thus f is said to have the following characteristics:

- it is *upper-bounded* by g [$f \in \mathcal{O}(g(n))$] iff there exist an $n_0 > 0$ and an *asymptotic constant* $c > 0$ such that $f(n) \leq c \cdot g(n)$ for all $n \geq n_0$;

- it is *lower-bounded* by g [$f \in \Omega(g(n))$] iff there exist an $n_0 > 0$ and an *asymptotic constant* $d > 0$ such that $d \cdot g(n) \leq f(n)$ for all $n \geq n_0$;

- it is *asymptotically equivalent* to g [$f \in \Theta(g(n))$] iff there exist an $n_0 > 0$ and *asymptotic constants* $c, d > 0$ such that $d \cdot g(n) \leq f(n) \leq c \cdot g(n)$ for all $n \geq n_0$.

1.2 Digital Geometry and Related Disciplines

The *time complexity* $f(n)$ of an algorithm is measured by the number of elementary computational operations used by the algorithm when the input data have size n (for example, the size can be characterized by the number of points). Reducing the asymptotic time complexity of an algorithm is of practical interest if the asymptotic constant remains reasonably small.

As an example of the types of problems studied in computational geometry, we consider the problem of determining the convex hull of a simple planar n-gon. Let S be a subset of a Euclidean space $\mathbb{E}^n$. The *convex hull* $C(S)$ is the intersection of all of the halfspaces of $\mathbb{E}^n$ that contain S; it is the smallest convex set that contains S. If S is a finite set of points or a simple polygon in the plane, $C(S)$ is a simple polygon with vertices that are a subset of the original set of points or of the vertices of the original polygon. For example, the simple polygon $\Pi = \langle p_1, \ldots, p_{24} \rangle$ on the left in Figure 1.11 has the convex hull $C(\Pi) = \langle p_3, p_4, p_9, p_{10}, p_{17}, p_{18}, p_{19}, p_{21}, p_{22} \rangle$.

Determining the convex hull of a set of n points in the plane has an optimal time complexity of $\Theta(n \log n)$. Determining the convex hull of a simple planar n-gon has an optimal time complexity of $\Theta(n)$.

Graham's Scan, shown in Algorithm 1.1, calculates the convex hull of a set S of n points in the plane in optimal time $\mathcal{O}(n \log n)$ ([379] discusses incorrect versions of this algorithm in [375].) It uses one of the points $p = (x_0, y_0)$ that is known to be on $C(S)$ as a "pivot"; for example, p can be the uppermost-rightmost point of S, which can be identified in time $\mathcal{O}(n)$. Let $p_0 = (0, y_0)$. For every other $p_i \in S$, let η_i be the angle between the vectors $\vec{pp_0}$ and $\vec{pp_i}$. The sorting in Step 2 requires time $\mathcal{O}(n \log n)$. Backtracking means that we delete previously inserted points from $C(S)$ until we reach a point q_i that results in a left turn. For each q_i, the edge pq_i can only be added once. Thus Step 4 requires time $\mathcal{O}(n)$ so that the algorithm has $\mathcal{O}(n \log n)$ worstcase time complexity. Figure 1.20 illustrates the algorithm for the vertex set of the simple polygon shown in Figure 1.11.

---

1. Start at a point of S (called the *pivot p*) that is known to be on the convex hull.
2. Sort the remaining points p_i of S in order of increasing angles η_i; if the angle is the same for more than one point, keep only the point furthest from p. Let the resulting sorted sequence of points be $q_1, \ldots, q_m$.
3. Initialize $C(S)$ by the edge between p and q_1.
4. Scan through the sorted sequence. At each left turn, add a new edge to $C(S)$; skip the point if there is no turn (a collinear situation); backtrack at each right turn.

ALGORITHM 1.1 Graham's Scan for computing the convex hull of a finite set of points in the plane.

As discussed in Section 1.2.2, simple objects in digital geometry often do not behave like Euclidean objects. Also, for practical purposes, picture size in digital geometry is bounded so that the input data of geometric algorithms have bounded complexity. However, digital geometry is concerned with designing efficient

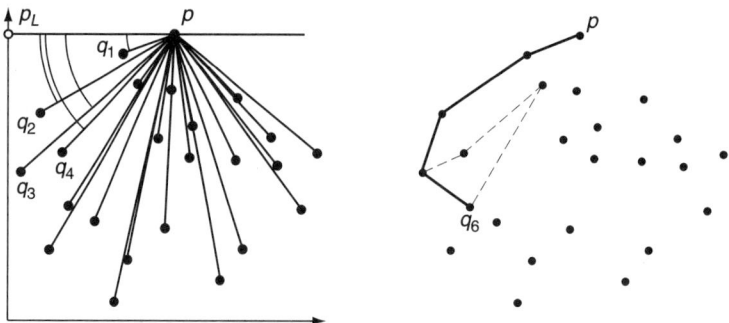

FIGURE 1.20 Left: angles η_i for the vectors defined by q_1, q_2, q_3, q_4. Right: a backtrack situation at q_6; the dashed edges are removed or not added in Step 4 of the algorithm.

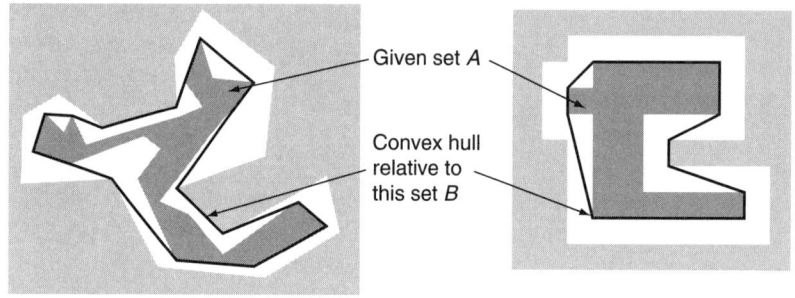

FIGURE 1.21 Relative convex hulls. Left: A, B are simple polygons. Right: A, B are isothetic simple polygons.

algorithms and has therefore benefited from developments in computational geometry, as we will see in this book. It has also occasionally contributed to computational geometry; here we give just one example.

Definition 1.4 [1001] Let $S \subseteq B \subset \mathbb{R}^2$. S is called B-*convex* iff, for all $p, q \in S$, if the straight line segment pq is in B, it is also in S. The B-convex hull of S is the intersection of all B-convex sets that contain S.

Figure 1.21 shows two examples of B-convex hulls. If A, B are simple polygons and A is contained in B, it can be shown that the frontier of the B-convex hull of A is the (uniquely determined) minimum-perimeter polygon that is contained in B and that circumscribes A. Such "relative convex hulls" are often used in robotics and computational geometry; they are also useful in digital geometry, as we will see in Chapters 10 and 11.

Errata

The following are corrections for typographical errors in the first printing of *Digital Geometry: Geometric Methods for Digital Picture Analysis* (1-5860-861-3)

All references in the book with numbers 640 through 737 should be increased by 1

Page 24, line 8 from bottom:
||R+ of positive real numbers... should read ℝ+ **of positive real numbers...**

Page 75, line 10 from bottom: **of the Association for Computing Machinery (ACM)** should read **of the Southpacific Chapter of the Association for Computing Machinery (ACM)**

Page 226, line 4 of caption for Figure 6.18: delete **hbox**

Page 232, line 10: delete **tesub**

Page 235, line 5: delete **tesub**

Page 459, line 14: delete **¡**

Page 507, first line in algorithm b): **A variant of algorithm (a) alternates between...** should read **A variant of algorithm [(a)] which keeps the test condition (2)] alternates between...**

Page 507, fourth line in algorithm b): $= 0$. **An....** should read $= 0$, **provided that the following is not true:** $(x, y+1) = (x+1, y+1) = (x+1, y) = 1$ and $(x-1, y) = (x, y-1) = (x+2, y+1) = (x+1, y+2) = 0$. **An....**

Page 508, Proposition 16.3, b8) and b4): **remains A_j-simple** should be replaced with **remains simple.**
Also note that condition c) is only required for sets of mutually adjacent 1s that are 8-components; if c) is true then condition b4) is vacuous.

Page 531, Proposition 16.11, b): **remains A_j-simple** should be replaced with **remains simple**
Also note that condition c) is only required for 26-components of 1s; if c) is true then condition b) is vacuous.

1.2.10 Fuzzy geometry

When an object in a scene is represented by a set of pixels in a picture, it may not always be obvious which pixels belong to the object. It was suggested in 1970 [825] that "a pictorial object is a fuzzy set which is specified by some membership function defined on all picture points."

> **Definition 1.5** [1157] A function from a set S into [0,1] is called a *fuzzy subset* of S. For any $x \in S$, $\mu(x)$ is called the *degree of membership* of x in μ.

If μ is into $\{0,1\}$, it defines an ordinary subset of S, namely $S_\mu = \{x \in S : \mu(x) = 1\}$; μ is then called the *characteristic function* of S_μ. Thus an ordinary subset can be regarded as a special type of fuzzy subset.

For any picture P with pixel values in the range $\{0, \ldots, G_{\max}\}$, if we divide the pixel values by $G_{\max}$, we obtain a picture P' in which the pixel values are in the range [0,1], so that they can be regarded as membership values of the pixels in a fuzzy subset μ of P. For example, $\mu(p)$ might represent the degree to which p is associated with some object in the scene. If P is a binary picture, its pixel values are in $\{0,1\}$, and the pixels whose values are 1 define an ordinary subset $\langle P \rangle$ of P.

Figure 1.22 shows a rectangle (on the left) and two "fuzzy rectangles." In the middle, μ has nonzero membership values only at the pixels of the original rectangle; on the right, some of the pixels outside of the original rectangle also have nonzero membership values.

Let μ be a fuzzy subset of S, and let $0 \leq \lambda \leq 1$. The set $\mu_\lambda = \{x \in S : \mu(x) \geq \lambda\}$ is called the λ-*level set* of μ. Evidently μ_0 is all of S, and if $0 \leq \lambda' \leq \lambda \leq 1$, we have $\mu_\lambda \subseteq \mu'_\lambda$. In later chapters, we will see how geometric properties of subsets of a picture p can be generalized to fuzzy subsets. We will sometimes define a fuzzy subset as having a geometric property (e.g., connectedness, convexity) iff all its level sets have that property.

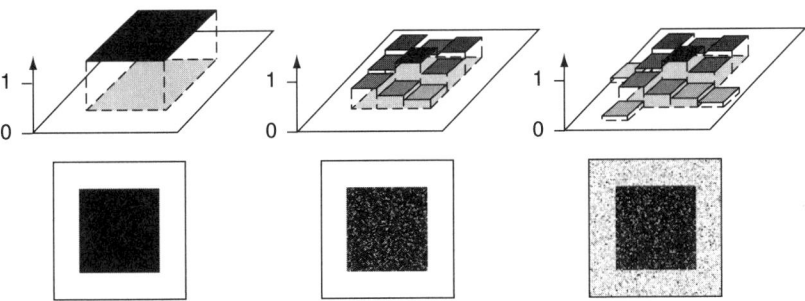

FIGURE 1.22 Lower row: a square in a binary image (left) and two fuzzy squares (middle and right). Upper row: simplified squares with levels in [0,1].

1.2.11 Integral geometry, isoperimetry, stereology, and tomography

Integral geometry measures properties of a set S in n-dimensional Euclidean space — for example, length $\mathcal{L}$ for $n=1$; area $\mathcal{A}$, perimeter $\mathcal{P}$, and curvature for $n=2$; volume $\mathcal{V}$, surface area $\mathcal{S}$, and (integral of) mean curvature $\mathcal{M}$ for $n=3$ — by computing integrals of the intersections of S with line segments or convex sets. The values of these integrals are independent of the position and orientation of S; hence they are invariant with respect to the transformations of Euclidean geometry (see Table 1.1 in Section 1.2.2).

The *isoperimetric inequality* in 2D, which has been known since ancient times, states that, for any planar set that has a well-defined area $\mathcal{A}$ and perimeter $\mathcal{P}$, we have the following:

$$\frac{\mathcal{P}^2}{4\pi\mathcal{A}} \geq 1 \tag{1.9}$$

It follows that, among all such sets that have the same perimeter, the disk has the largest area. The expression on the left-hand side of Equation 1.9 is called the *shape factor* or *isoperimetric deficit* of the set; it measures how much the set differs from a disk [102]. The first proof of the isoperimetric property of the circle is due to Zenodorus (about 200–140 BC), who is known through the fifth book of the "Mathematic Collection" by Pappus of Alexandria [565].

The study of 3D isoperimetric problems (surface minimizations under volume constraints, possibly with additional boundary or symmetry conditions) has an extensive history involving such famous mathematicians as Euler, the Bernoullis, Gauss, Steiner [1019], and Weierstrass. In 3D, we have the following [391, 746]:

$$\frac{\mathcal{S}^3}{36\pi\mathcal{V}^2} \geq 1 \tag{1.10}$$

This inequality says that, of all sets with a given surface area, the ball has the largest volume; it has led to studies of the isoperimetric deficit for 3D objects [600]. For bounded convex sets, we also have the following [391, 958]:

$$\frac{\mathcal{M}^3}{48\pi^2\mathcal{V}} \geq 1, \quad \frac{\mathcal{M}^2}{4\pi\mathcal{S}} \geq 1, \quad \text{and} \quad \frac{\mathcal{S}^2}{3\mathcal{M}\mathcal{V}} \geq 1 \tag{1.11}$$

For example, a sphere of radius r has mean curvature $\mathcal{M} = 4\pi r$. See Section 8.3.6 for further formulas involving convex sets.

Buffon's needle problem is a classic example of a problem in geometric probability that was stated by G.-L. le Comte de Buffon (1707–1788) in 1777 [146]. Draw a set of parallel lines distance d apart in a plane, and toss a needle of length l ($0 < l \leq d$) onto the plane. What is the probability that the needle will intersect one or more of the lines? It turns out that this probability is $2l/\pi d$; thus needle tossing can be used to estimate the value of π. *Stereology* uses statistical methods to estimate geometric properties of sets in n-dimensional Euclidean space by measuring intersections of m-dimensional hyperplanes ($0 \leq m < n$) with the sets. This characterizes stereology as a field that deals with *inverse problems* (i.e., determining higher-dimensional sets or functions based on lower-dimensional "observations").

1.2 Digital Geometry and Related Disciplines

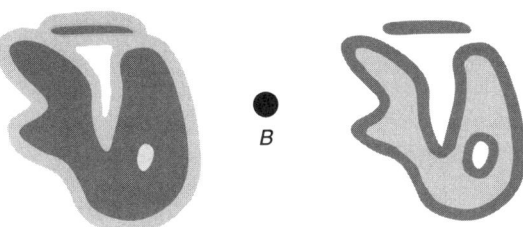

FIGURE 1.23 Examples of the Minkowski sum (left; the original dark-gray set A is contained in the light-gray sum) and the Minkowski difference (right; the original gray set A contains the light-gray difference). In these examples, the set B is the black disk shown in the middle.

This book will not usually discuss methods based on integrals, probability, or statistics, but it will be concerned with other methods of estimating properties such as area, perimeter, and so forth, for subsets of a picture. It will also briefly discuss *digital tomography* in Chapter 14; this is a subfield of *discrete tomography*,[9] which is another example of a field that deals with inverse problems. An early example of a problem studied in discrete tomography is the following: when is a planar convex set uniquely determined by its projections [406]? The case of projections along parallel lines received special attention. We define digital tomography to be the subfield in which the function to be determined or approximated has integer values only and is defined on a finite subset of $\mathbb{Z}^n$ for some $n \geq 2$.

1.2.12 Mathematic morphology

Mathematic morphology (see Chapter 15) makes use of operations based on concepts introduced by J. Steiner (1796–1863) and H. Minkowski (1864–1909) that can be defined on arbitrary vector spaces. For any set $A \subseteq \mathbb{R}^n$, $\overline{A} = \mathbb{R}^n \setminus A$ is called the *complementary set* of A, and $\underline{A} = \{-p : p \in A\}$ is called the *mirror set* of A. A set $A \subseteq \mathbb{R}^n$ is *symmetric* (with respect to the origin) iff $A = \underline{A}$.

The *Minkowski sum* of $A, B \subseteq \mathbb{R}^n$ (see Figure 1.23) is as follows:

$$A \oplus B = \{p + q : p \in A \land q \in B\}$$

The *Minkowski difference* (actually due to H. Hadwiger [390]; see also [392]) is as follows:

$$A \ominus B = \overline{\overline{A} \oplus B}$$

For example, $A \oplus \underline{A}$ defines a symmetric set with a convex hull that is equal to the Minkowski sum of the convex hulls of A and $\underline{A}$.

---

9. This was the name of the first meeting on the subject in 1994, which was organized by L. Shepp. Tomography is concerned with determining or approximating a function defined on a discrete or continuous set given a collection of weighted sums or weighted integrals of the function over subsets of its domain.

In mathematic morphology, $(A \ominus \underline{B})$ is called the *erosion* of A by B, and $(A \oplus \underline{B})$ is called the *dilation* of A by B. The *opening* of A by B is as follows:

$$A_B = (A \ominus \underline{B}) \oplus B$$

The *closing* of A by B is as follows:

$$A^B = (A \oplus \underline{B}) \ominus B$$

(For the notation used here, see [969].) These operations have many uses in digital geometry, as we will see in Chapter 15.

1.3 Exercises

1. In a *regular grid*, in the plane, the grid vertices are connected by grid edges that form simple regular polygons; all of the polygons have the same number m of edges, and every grid vertex is incident with the same number n of grid edges. There are three regular grids in the plane: the orthogonal (square) grid (used in Section 1.1), for which $(m,n) = (4,4)$; the *hexagonal grid*, for which $(m,n) = (6,3)$; and the *triangular grid*, for which $(m,n) = (3,6)$. Prove that these are the only possible regular grids in the plane.

2. A center point of one of the regular m-gons that define a regular grid (see Exercise 1) is called a *grid point*. We can define an xyz-coordinate system for the hexagonal grid; any two of the three coordinates uniquely determine a grid point. Define a test based on the signs of the three coordinates for determining to which sextant (I, II, III, IV, V, VI; see figure) a grid point belongs.

1.3 Exercises

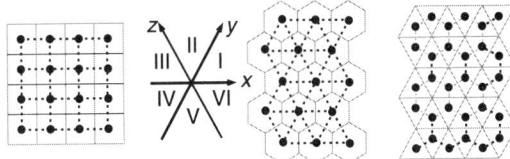

3. Implement the Hilbert scan for pictures of size $2^n \times 2^n$ $(n = 2, 3, \ldots, n_0)$.

4. A set S in the Euclidean plane $\mathbb{E}^2$ is called *polygonally connected* iff, for any $p, q \in S$, there is a polygonal arc $(p_1, p_2, \ldots, p_n)$ where $p = p_1$ and $q = p_n$ such that the edges of the arc are all contained in S.

   (i) Prove that any simple polygon is polygonally connected.

   (ii) Consider the four quadrants of the square shown on the left. Specify conditions under which the union of the gray squares or the union of the white squares is polygonally connected. Can you specify conditions under which the union of the gray squares in the upper n rows of the chessboard (on the right) and the union of the white squares in the lower m rows are polygonally connected when $4 \leq m$ and $n \leq 8$?

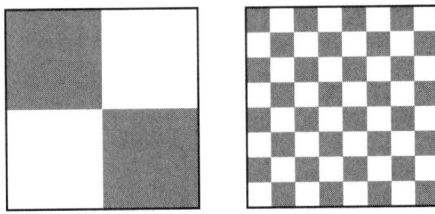

5. A set S in the Euclidean plane $\mathbb{E}^2$ is called *continuously connected* iff, for any $p, q \in S$, there is a continuous function $f : [0,1] \to \mathbb{E}^2$ such that $f(0) = p$, $f(1) = q$, and $f(x) \in S$ for any real x in the closed interval $[0,1]$. Give an example of a set $S \subseteq \mathbb{E}^2$ that is continuously connected but not polygonally connected.

6. Specify two metrics d_1, d_2 on a set S and a subset $M \subseteq S$ such that M is unbounded in $[S, d_1]$ but bounded in $[S, d_2]$.

7. Prove that property **V8** of a vector space follows from properties **V0** through **V7**.

8. Which properties of **V0** through **V8** of a vector space are satisfied by $[\mathbb{Z}^2, +, \cdot, \mathbb{R}]$? (Operations $+$ and $\cdot$ are defined as for $[\mathbb{R}^2, +, \cdot, \mathbb{R}]$ in Section 1.2.4.)

9. Let $\mathbb{G}_{n,n}$ be the $n \times n$ grid (see Equation 1.1), where $n \geq 2$, and consider vectors joining two nonidentical points of $\mathbb{G}_{n,n}$. What is the smallest possible nonzero value of the angle η between two such vectors expressed as a function of n?

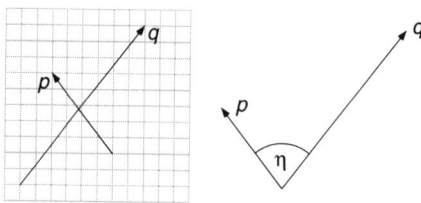

10. The two figures below appear to be congruent triangles composed of the same four grid polygons. Explain the empty grid square in the figure on the right.

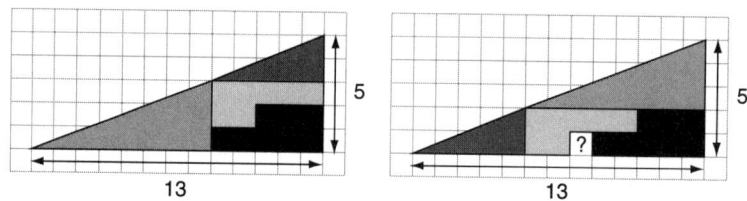

11. Implement the Graham Scan for finite sets of grid points. (Hint: Give arithmetic definitions of "left turns" and "right turns.")

12. Let $\mathcal{S}(\Pi) = \mathcal{P}^2/4\pi\mathcal{A}$ (see Equation 1.9) be the shape factor of a simple polygon Π with perimeter $\mathcal{P}$ and area $\mathcal{A}$. What is $\mathcal{S}(\Pi_n)$ if Π_n is a regular n-gon? If $\kappa(n) = |\mathcal{S}(\Pi_n) - 1|$, what is the speed of convergence $1/\kappa(n)$ of $\mathcal{S}$ to 0 as $n \to \infty$?

13. Prove that, among all simple polygons that have the same perimeter and the same number of sides, the regular polygon has the greatest area.

14. In the figure in Exercise 4, let the (closed) gray squares have size $2^{-n} \times 2^{-n}$; let the large square have size 1×1; and let A_n be the union of the gray squares. (A_n is illustrated in Exercise 4 for $n = 1$ and $n = 3$.) Let $B_m = \{(x, y) : -2^{-m} \leq x, y \leq 2^{-m}\}$ where $m \geq 1$. What are $A \oplus B$, $A \ominus B$, A_B, and A^B where A can be any A_n ($n \geq 1$) and B can be any B_m ($m \geq 1$)?

15. Prove that, if n grid points in $\mathbb{Z}^2$ form a regular n-gon ($n > 2$), then $n = 4$. The following figure shows an example on the left.

16. Prove that, if four grid points determine a nonsquare rhombus with angle η (see the figure above on the right), then the quotient η/π is irrational. Also prove that, if η is an acute angle of a right triangle with sides that have integer lengths, (called a *Pythagorean triangle*) then η/π is irrational.

17. Prove that, if all the points in an infinite set are at integer distances from each other, the points must all be collinear.

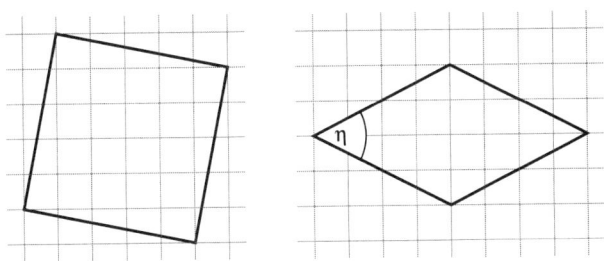

1.4 Commented Bibliography

Digital geometry is a very lively research area in which many hundreds of journal papers have been published. The study of pictures as mappings defined on rectangular arrays of grid points was initiated in the 1960s and early 1970s in, for example, [286, 342, 756, 881, 888, 921].

The book chapter [477], Chapter 3 in [992], the books [733, 802, 911], and the digital geometry chapter in [891] (all published before 1980), as well as the books [174, 430, 623, 702, 805, 860, 969, 1012, 1107], define *digital geometry* as a geometric theory of n-dimensional *digital spaces* (grid point or grid cell spaces); see also the article [1106].

Digital geometry has its mathematic roots in graph theory and discrete topology; it deals with sets of grid points, which are also studied in the geometry of numbers [732, 816], or with cell complexes [9, 10], which have also been studied in topology since its beginning [9, 10, 659, 660, 820]. Studies of gridding techniques (e.g., by Gauss, Dirichlet, or Jordan [484]) can also be cited as historic context. Digitizations on regular grids are used in many types of numeric calculations.

For the mathematic theory of space-filling curves (also in 3D), see [935]. [996] applies space-filling curves to picture printing. For the Peano curve, see [806]; for the Hilbert curve, see [435].

For more about Euclid, see [564]. Coordinates in geometry are reviewed in [335].[10] For Apollonius of Perga, see [24]. For the ancient history of length estimation

---

10. Descartes' friends pressured him into publishing some of his work. In 1637, his first scientific book was published in Leyden [264]. Descartes' method consisted of doubting all accepted knowledge (a mortal sin for orthodox thinkers), basing all thinking on clear self-evident truths, and using logical reasoning (based on mathematics) to build up a system of ideas. He explained the advantages of the top-down approach to large and difficult problems, breaking them into smaller problems in such a way that the solutions to the smaller problems could be combined to solve the large problem. The part of his book entitled *La Géomètrie* united algebra and geometry and introduced algebraic notation that was so much better than anything previous that it has remained almost unchanged since 1637. *La Géomètrie* is the earliest book that a modern mathematician can read without having to learn obsolete notation and terminology; it is generally regarded as the foundation of modern mathematics.

in mathematics (e.g., the work of Archimedes and Liu Hui on the estimation of π), see [1083]. The error measure in Equation 1.8 is discussed in [543].

For the history of graph theory, see [89]. [299, 394, 789, 1145] are monographs about combinatorial geometry, and [97, 300, 823] are textbooks about computational geometry. For Graham's Scan, see [375].

For integral geometry and stereology, see [17, 858]. "Discrete integral geometry," following [1107, 1108], will be treated in Chapters 4 and 5. For discrete tomography, see [380, 431]. For the Minkowski sum, see [731]; for the Minkowski difference, see [390].

For early work on fuzzy convexity, see [670, 1157], and for early work on fuzzy topology, see [892]. "Fuzzy shape transforms" were introduced in [644]. The review [900] discusses a variety of fuzzy geometric properties that had been studied by 1984.

A geometric model of the panoramic picture acquisition process (see Figure 1.3) is discussed in [447]. There are many papers about hexagonal grids (Exercises 1 and 2); see, for example, [132] for properties and advantages of such grids, and see also [68, 105, 1015]. For semiregular grids, see [106]. Exercise 13 is due to Zenodorus [565]. For Exercise 14, see [961]. Exercise 15 follows from a general theorem in [389]: if $0 < \eta < \pi/2$ and the number $\cos \eta$ is rational, then either $\eta = \pi/3$ or η/π is irrational. For Exercise 16, see [307].

CHAPTER 2

Grids and Digitization

This chapter begins by defining the 2D and 3D grid point and grid cell adjacency models as well as a more refined cell model called the grid (cell) incidence model, which combines cells of different dimensionalities. It then discusses connectedness (the reflexive and transitive closure of adjacency) and algorithms for identifying ("labeling") connected components. It also discusses digitization models, including the classic Gauss, Jordan, and grid intersection models, and defines a "domain" model that generalizes all of them.

Measurements made on digital pictures can only approximate the measurements that might ideally have been made on real objects or real pictures. *Digital geometry* deals with the computation of geometric measurements (or properties or relations) from digital pictures and with the study of how well these measurements approximate the corresponding ideal measurements on the real objects or pictures. This chapter introduces a methodology called *multigrid convergence* for comparing ideal and approximate measurements.

2.1 The Grid Point and Grid Cell Models

Figure 2.1 illustrates a portion of a 2D grid with grid constant θ (see Section 1.2.1). We will first assume that $\theta = 1$, and we will discuss θ as a variable later in this section.

2.1.1 Grid points and grid cells

In 2D, the *grid point* set is $\mathbb{Z}^2$, and, in 3D, the grid point set is $\mathbb{Z}^3$. A *grid vertex* is shifted by $(0.5, 0.5)$ with respect to a grid point in 2D and by $(0.5, 0.5, 0.5)$ in 3D.[1] A pair of adjacent grid vertices (i.e., vertices at Euclidean distance 1 from each other)

---

1. Half-integers can be avoided in a computer program, for example, by identifying the corners of voxel (i, j, k) with $(i+1, j, k)$, $(i, j+1, k)$, $(i+1, j+1, k)$, and so on; however, we find it convenient to use half-integer grid vertices in this book.

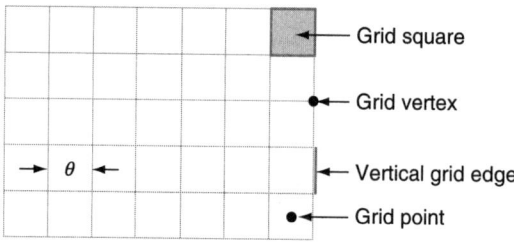

FIGURE 2.1 A regular orthogonal grid in the plane.

defines a *grid edge*. A *grid square* is defined by four grid edges that form a square, and a *grid cube* in 3D is defined by six grid squares that form a cube.

We specify grid points, vertices, edges, squares, and cubes in terms of Cartesian coordinates. In 2D, the set of positions of the grid vertices is as follows:

$$(0.5, 0.5) + \mathbb{Z}^2 = \{(i+0.5, j+0.5) : i, j \in \mathbb{Z}\}$$

A grid edge[2] connects a pair of adjacent grid vertices:

$$(i+0.5, j+0.5)(i+0.5, j+1.5) \quad \text{or} \quad (i+0.5, j+0.5)(i+1.5, j+0.5) \tag{2.1}$$

A grid square is defined by a quadruple of grid edges in which successive edges (modulo 4) share a vertex.

$$(i+0.5, j+0.5)(i+0.5, j+1.5)(i+1.5, j+1.5)(i+1.5, j+0.5) \tag{2.2}$$

In 3D, the positions of grid vertices, grid edges, grid squares, and grid cubes are specified similarly.

In both 2D and 3D, the sets of grid vertices and grid points are congruent. This supports the use in this book of *the same* planar or spatial grid for 2D or 3D pictures when we use either the grid cell or grid point model.

The dimensionalities 3, 2, 1, and 0 of grid cubes, grid squares, grid edges, and grid vertices suggest an alternative terminology:

Definition 2.1 A grid cube is also called a *3-cell*; a grid square is a *2-cell*; a grid edge is a *1-cell*; and a grid vertex is a *0-cell*.

In this terminology, a 2D grid point is the center point of a 2-cell (see Figures 2.2 and 2.3), and a 3D grid point is the center point of a 3-cell. $\mathbb{C}_2^{(i)}$ will denote the set of all i-cells in the plane ($i = 0, 1, 2$), and $\mathbb{C}_3^{(i)}$ will denote the set of all i-cells in 3D space ($i = 0, 1, 2, 3$). We also define the following:

$$\mathbb{C}_2 = \mathbb{C}_2^{(2)} \cup \mathbb{C}_2^{(1)} \cup \mathbb{C}_2^{(0)} \quad \text{and} \quad \mathbb{C}_3 = \mathbb{C}_3^{(3)} \cup \mathbb{C}_3^{(2)} \cup \mathbb{C}_3^{(1)} \cup \mathbb{C}_3^{(0)}$$

---

2. pq is the line segment with endpoints p and q, pqr is the triangle with vertices p, q, and r, and so forth.

2.1 The Grid Point and Grid Cell Models

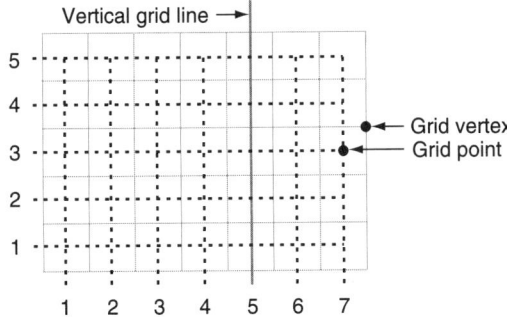

FIGURE 2.2 Grid points in the plane (grid constant $\theta = 1$).

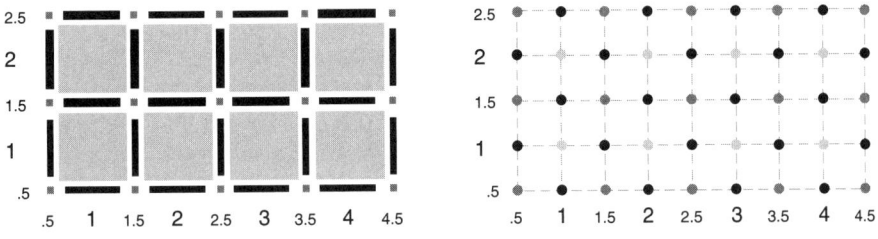

FIGURE 2.3 Left: graphic sketch of 0-, 1- and 2-cells. Right: the centers of these cells.

(The symbol $\mathbb{C}$ is often used in the mathematic literature for the set of complex numbers, but this book will make no use of complex numbers, so there is no conflict when using $\mathbb{C}$ to denote sets of cells.)

We now define the two *basic grid models* that will be used throughout this book:

- In the *grid point model*, a 2D grid $\mathbb{G}$ is either the infinite grid $\mathbb{Z}^2$ or an $m \times n$ rectangular subarray of $\mathbb{Z}^2$; see $\mathbb{G}_{m,n}$ in Equation 1.1. Similarly, a 3D grid is either $\mathbb{Z}^3$ or an $l \times m \times n$ cuboidal subarray of $\mathbb{Z}^3$.

- In the *grid cell model*, a 2D grid $\mathbb{G}$ is either $\mathbb{C}_2$ or an $m \times n$ "block" of 2-cells whose union $\bigcup \mathbb{G}$ is a rectangular region of the Euclidean plane $\mathbb{E}^2$; see $\mathbb{G}_{m,n}$ in Equation 1.2. Similarly, a 3D grid is either $\mathbb{C}_3$ or an $l \times m \times n$ set of 3-cells whose union is a cuboid in Euclidean space $\mathbb{E}^3$.

These grids can also be defined for a grid constant θ different from 1, but $\theta = 1$ is the default value throughout this book. Grid cells or grid points are the basic elements used in digital geometry. A pixel is either a 2-cell (grid square) or a grid point (the center of a 2-cell); see Figure 2.2. A voxel is either a 3-cell (grid cube) or a grid point (the center of a 3-cell). The formulation of a definition or algorithm may sometimes be more convenient when one or the other model is used.

A *grid line* in 2D is incident with two different grid points whose x- or y-coordinates are the same. The 2D grid $\mathbb{Z}^2$ can be regarded as a subset of the 3D grid $\mathbb{Z}^3$ by adding a third coordinate $z = 0$ to every 2D grid point. In 3D, a *grid plane* is incident with two orthogonal grid lines. It follows that all of the grid points of a grid plane have the same x-, y-, or z-coordinate. A *grid line* in 3D is a set of points, two with coordinates that are constant in $\mathbb{Z}$, whereas the third is a variable in $\mathbb{R}$. Grid lines intersect at grid points in either 2D or 3D.

2.1.2 Variable grid resolution

The grid constant θ is the distance between neighboring grid lines. *Grid resolution* is the inverse of the grid constant. It refers to the number of grid elements per unit of distance without specifying the physical size of the unit. We will use an integer parameter $h > 0$ to denote grid resolution. In a grid with resolution $h = 1$, we can have either one grid point in a unit or two grid points as endpoints of a unit. In general, the maximum number of grid points per unit is $h+1$.[3]

The parameters h and θ are useful when discussing the possible effects of improvements in geometric or picture resolution. They are especially relevant in theoretic studies of the convergence behavior of algorithms under refinement of grid resolution (i.e., decrease of the grid constant).

Let $\mathbb{Z}_h = \{i/h : i \in \mathbb{Z}\}$; then $\mathbb{Z}_h^2$ is the set of all 2D grid points in a grid of resolution $h > 0$, and $\mathbb{Z}_h^3$ is the set of all such 3D grid points. We similarly use the notation

$$\mathbb{C}_{2,h}^{(i)} \; (i=0,1,2),\; \mathbb{C}_{3,h}^{(i)} \; (i=0,1,2,3),\; \mathbb{C}_{2,h},\; \text{and}\; \mathbb{C}_{3,h}$$

for grid cells in a 2D or 3D grid of resolution h.

2.1.3 Adjacencies in 2D grids

Definition 2.2 Two 2-cells, c_1 and c_2, are called *1-adjacent iff* $c_1 \neq c_2$ and $c_1 \cap c_2$ is a 1-cell. Two grid points $p_1 = (x_1, y_1)$ and $p_2 = (x_2, y_2)$ are called *4-adjacent iff* $|x_1 - x_2| + |y_1 - y_2| = 1$.

In other words, two 2-cells c_1 and c_2 are 1-adjacent *iff* they are not identical but they share a grid edge. Let p_i be the center of c_i ($i = 1, 2$); then c_1 and c_2 are 1-adjacent *iff* p_1 and p_2 are 4-adjacent.

---

3. In optics, resolution is defined as the minimal distance between two point light sources that allows them to be distinguished by the optic system. In our terminology, resolution is the physical size of a pixel (i.e., the grid constant θ).

2.1 The Grid Point and Grid Cell Models

We have defined relations A_1 of *1-adjacency* and A_4 of *4-adjacency* for the grid cell and grid point models. In general, we write pRq or $(x,y) \in R$ iff p,q are in relation R. In this notation, we have $c_1 A_1 c_2$ iff $\{c_1, c_2\} \in A_1$, iff c_1 and c_2 are 1-adjacent, and $p_1 A_4 p_2$ iff $\{p_1, p_2\} \in A_4$ iff p_1, p_2 are 4-adjacent. The relation A_1 (A_4) is symmetric and irreflexive.

Definition 2.3 Two 2-cells c_1 and c_2 are called *0-adjacent iff* $c_1 \neq c_2$ and $c_1 \cap c_2$ contains a 0-cell. Two grid points $p_1 = (x_1, y_1)$ and $p_2 = (x_2, y_2)$ are called *8-adjacent iff* $\max\{|x_1 - x_2|, |y_1 - y_2|\} = 1$.

In other words, two 2-cells c_1 and c_2 are 0-adjacent *iff* they are not identical but share (at least) a grid vertex. Again, let p_i be the center of c_i ($i = 1, 2$); then c_1 and c_2 are 0-adjacent *iff* p_1 and p_2 are 8-adjacent. The relation A_0 (A_8) is symmetric and irreflexive. Figure 2.4 illustrates relations A_0, A_1, A_4, and A_8 on a square grid.

Adjacent 2-cells are transformed into adjacent grid points by mapping the 2-cells onto their center points, and adjacent grid points are transformed into adjacent 2-cells by mapping the grid points into the 2-cells that have them as center points. The existence of this one-to-one mapping gives us the following:

Proposition 2.1 The grid defined by $m \times n$ 2-cells and adjacency relation A_1 (A_0) is isomorphic to the grid defined by $m \times n$ grid points and adjacency relation A_4 (A_8). Either of these grids will be denoted by $\mathbb{G}_{m,n}$.

(In general, let R_1 be a relation on a set S_1 and R_2 a relation on a set S_2. The structures $[S_1, R_1]$ and $[S_2, R_2]$ are called *isomorphic* iff there exists a one-to-one mapping f from S_1 onto S_2 such that pR_1q iff $f(p)R_2f(q)$ for all $p,q \in S_1$. The mapping is called an *isomorphism*.)

For the relations A_1 and A_4 on $\mathbb{C}_2$ ($\mathbb{Z}^2$), the number of pixels (grid points or grid cells) adjacent to a pixel is always 4. For A_0 and A_8, it is always 8; however, for adjacency relations defined on pictures, the number can vary.

Adjacencies between pixels in multilevel pictures are defined by locations and pixel values. To start with, let us consider the values alone. Let P be a picture

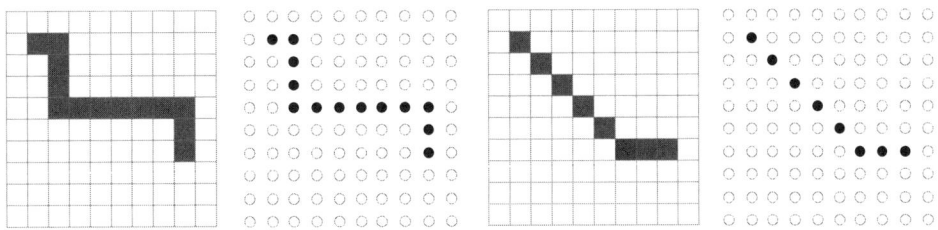

FIGURE 2.4 Consecutive sequences of adjacent pixels for adjacency relations A_1 and A_4 (on the left), A_0 and A_8 (on the right).

defined on the grid $\mathbb{G}$ in which pixel p has value $P(p)$ in $\{0, 1, \ldots, G_{max}\}$. We say that two pixels p and q are *P-equivalent* iff $P(p) = P(q)$. Let M_u be the set of all $q \in \mathbb{G}$ such that $P(q) = u$. If there exists $p \in \mathbb{G}$ such that $u = P(p)$, M_u is an equivalence class with respect to the relation of P-equivalence on $\mathbb{G}$ (in brief, a *P-equivalence class*).

These P-equivalence classes are defined only by pixel values, not by locations. Each P-equivalence class splits into components, depending on the adjacency relation used for the pixels. For example, we can define p and q to be *P-adjacent* if they are 8-adjacent (or 0-adjacent) and P-equivalent. The number of pixels P-adjacent to p can then vary between 0 and 8. Using 8-adjacency uniformly for all values u may lead to situations such as the one illustrated in Figure 1.9.d. (If there are more than two values, we cannot use 4-connectedness for one value and 8-connectedness for the other as proposed in Section 1.1.4 for binary pictures.)

More generally, we can model "uncertainties" in picture values by assuming that we are given a *similarity relation* σ (a reflexive and symmetric relation) on $V \times V$, where $V = \{0, 1, \ldots, G_{max}\}$. (For example, σ_2 is the similarity relation in which $u\sigma_2 v$ iff $|u - v| \leq 2$.) Define two pixels p, q to be (σ, α)-*adjacent* iff $pA_\alpha q$ and $P(p)\sigma P(q)$. The relation of (σ, α)-adjacency is symmetric and irreflexive. Note that p may have no (σ, α)-adjacent pixels if $P(p)$ is not σ-similar to the value of any of the pixels that are α-adjacent to p.

In the remainder of this section, we describe a method of defining adjacencies in a 2D multivalued picture P, which we call *switch adjacency* (or *s-adjacency*) and denote with A_s. Switch adjacency allows us to avoid topologic conflicts (see Section 1.1.4 about binary pictures).

Pairs of 4-adjacent pixels are always regarded as s-adjacent; in other words, $A_s \supseteq A_4$. In addition, in each 2×2 block of pixels, we call exactly one of the two diagonally adjacent pairs s-adjacent. We can think of the 2×2 block as a switch that can be in either of the two diagonal positions (see Figure 2.5); the position of the switch determines which diagonally adjacent pair of pixels in the block is regarded as s-adjacent. The states of the switches in P can be specified by a binary picture **S** (a "switch state matrix") that has pixels that correspond to the lower left-hand corners of the 2×2 blocks of pixels in P. A pixel of **S** has value 1 iff the switch in the corresponding 2×2 block of P is to the left (in the main diagonal position) and value 0 iff that switch is to the right (in the other diagonal position).

We can define specific s-adjacency relations in many ways. The switch position in a block can depend on the position of the block (i.e., on the coordinates of its lower left-hand corner) and not on the values of the pixels in the block. For example, **S** can be the binary picture in which pixel (x, y) has value 1 if y is odd and 0 if y is even. (In this case, two pixels are s-adjacent iff their hexagonal distance d_h is 1;

FIGURE 2.5 Using a "switch" to choose one pair of diagonal adjacencies.

2.1 The Grid Point and Grid Cell Models

129	61
61	129

128	61
61	129

129	61
62	129

FIGURE 2.6 Only one of these three 2×2 blocks of pixels is a flip-flop case.

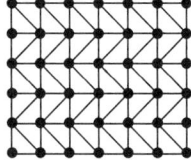
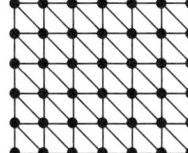
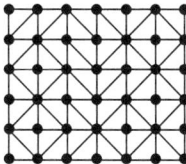

FIGURE 2.7 Two examples of (regular) s-adjacencies where switch positions depend only on block positions (left and middle), and an irregular s-adjacency relation (right).

see the leftmost image of Figure 2.7, and see Section 3.2.3 about metric d_h.) It is more "natural" to let the diagonal s-adjacencies depend on the pixel values. Let the difference between the values of one pair of diagonally adjacent pixels in a 2×2 block be d_1 and the difference between the values of the other pair be d_2; if $d_1 < d_2$, we call the first pair s-adjacent; if $d_2 < d_1$, we call the second pair s-adjacent. This determines all of the s-adjacencies, except in flip-flop cases, where $d_1 = d_2$; in such a case, the switch can be in either position. In real pictures, flip-flop cases are rare (see Figure 2.6); there are typically less than 0.5% of such cases in a grayscale picture and less than 0.2% in a color picture (see Figure 2.8 for four grayscale examples).

Evidently the s-adjacency relation is symmetric and irreflexive. The number of pixels s-adjacent to p can be anywhere between 4 (the pixels 4-adjacent to p) and 8 (up to four additional diagonal adjacencies).

In general, we can use a *Set Switches* procedure to define a switch state matrix **S**. The procedure can analyze larger neighborhoods of a pixel to determine the state of its switch. This analysis can be based on templates such as those shown in Figure 2.9 in which the state of the switch in the lower 2×2 block is also used in the upper 2×2 block. **S** can even be generated randomly (using a random number generator) or using a function of the pixel coordinates (e.g., a pseudorandom function, to avoid bias).

2.1.4 Adjacencies in 3D grids

The definitions of 1-adjacency and 0-adjacency of 2-cells (Definitions 2.2 and 2.3) can also be used in 3D, and we can also define the 0-adjacency of 1-cells. However, our

FIGURE 2.8 These pictures are of size 2014×1426 (so that they contain 2,872,964 pixels) and have $G_{max} = 255$. The picture at the upper left position has only 14,359 flip-flop cases (i.e., 0.5% of the pixels). In the pictures in the upper right, lower left, and lower right positions, the percentages of flip-flop cases are 0.38%, 0.38%, and 0.22%, respectively.

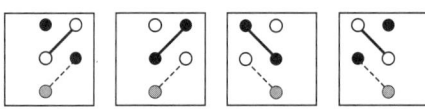

FIGURE 2.9 A set of simple templates for defining the states of flip-flop switches (rule: "copy the state from the row below").

main interest is in adjacencies between 3-cells or 3D grid points. Let d_e be the Euclidean metric (see Section 1.2.1).

Definition 2.4 Two 3-cells c_1 and c_2 are called α-*adjacent iff* $c_1 \neq c_2$ and the intersection $c_1 \cap c_2$ contains an α-cell ($\alpha \in \{0, 1, 2\}$). Two 3D grid points $p_1 = (x_1, y_1, z_1)$ and $p_2 = (x_2, y_2, z_2)$ are called 6-*adjacent iff* $0 < d_e(p_1, p_2) \leq 1$, 18-*adjacent iff* $0 < d_e(p_1, p_2) \leq \sqrt{2}$, and 26-*adjacent iff* $0 < d_e(p_1, p_2) \leq \sqrt{3}$.

2.1 The Grid Point and Grid Cell Models

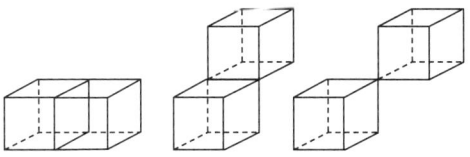

FIGURE 2.10 Left: two α-adjacent 3-cells ($\alpha = 0, 1, 2$). Middle: two α-adjacent 3-cells ($\alpha = 0, 1$). Right: two 0-adjacent 3-cells.

Let c_1 and c_2 be 3-cells and let p_i be the center of c_i ($i = 1, 2$). Then c_1 and c_2 are 0-adjacent iff p_1 and p_2 are 26-adjacent iff c_1 and c_2 are not identical but share a grid vertex; c_1 and c_2 are 1-adjacent iff p_1 and p_2 are 18-adjacent iff c_1 and c_2 are not identical but share a grid edge; and c_1 and c_2 are 2-adjacent iff p_1 and p_2 are 6-adjacent iff c_1 and c_2 are not identical but share a grid square (see Figure 2.10).

This defines symmetric and irreflexive relations A_α, where $\alpha = 0, 1, 2$ for the grid cell model and $\alpha = 6, 18, 26$ for the grid point model. The parameter α denotes the dimension of the intersection of the grid cells in the first case and the number of adjacent grid points in the second case.

Proposition 2.2 *The grid $\mathbb{G}_{l,m,n}$ defined by $l \times m \times n$ 3-cells and adjacency relation A_2, A_1, or A_0 is isomorphic to the grid defined by $l \times m \times n$ 3D grid points and adjacency relation A_6, A_{18}, or A_{26}, respectively.*

Data-dependent types of adjacencies (see Section 2.1.3) can also be generalized to 3D grids. We can now define the 2D and 3D grid point and grid cell adjacency models:

- A 2D (3D) *grid point adjacency model* combines the grid point model with an adjacency relation defined between 2D (3D) grid points.

- A 2D (3D) *grid cell adjacency model* combines the grid cell model with an adjacency relation defined between grid squares (grid cubes).

Both 2D and 3D grid point and grid cell adjacency models are called α-*adjacency grids*; the value of α determines whether we use a grid point model ($\alpha \geq 4$) or a grid cell model ($\alpha \leq 3$). In Section 1.1.4, we briefly mentioned a dual use of adjacencies for 2D binary pictures P: α_1-adjacency for $\langle P \rangle$ and α_2-adjacency for $\langle \overline{P} \rangle$; in this case we speak about (α_1, α_2)-*adjacency grids*.

Let A be a symmetric and irreflexive adjacency relation on a set S. $A(p) = \{q : q \in S \wedge qAp\}$ is called the *adjacency set* of $p \in S$. For example, for relation A_4 on grid points, we have the following,

$$A_4(x, y) = \{(x-1, y), (x+1, y), (x, y-1), (x, y+1)\} \quad (2.3)$$

provided all four of these grid points are also contained in the grid $\mathbb{G}_{m,n}$.

$N(p) = A(p) \cup \{p\}$ is called the (smallest nontrivial) *neighborhood* of $p \in S$ defined by adjacency relation A. N defines a symmetric and reflexive relation qNp

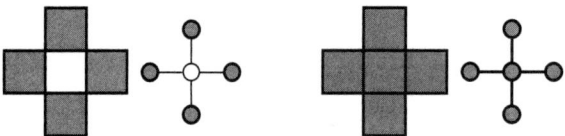

FIGURE 2.11 $A_1(c)$ and $A_4(p)$ (left) and the corresponding neighborhoods $N_1(c)$ and $N_4(p)$ (right).

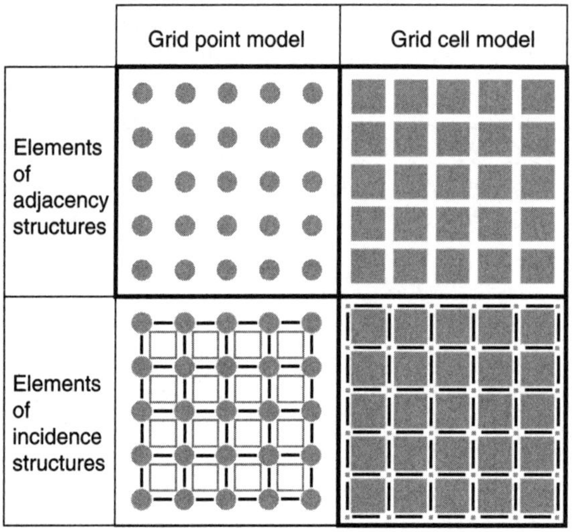

FIGURE 2.12 Four ways of representing a 2D grid. Adjacency or incidence relations can be defined on the grid points or grid cells.

on S. $p \in S$ is never in $A(p)$ but is always in $N(p)$. Figure 2.11 shows A_1 (A_4) and N_1 (N_4) for the 2D grid cell and grid point models. Analogous drawings for A_0 (A_8) and $N_0(p)$ ($N_8(p)$) would also contain the grid cells or grid points in the four corners.

2.1.5 Grid cell incidence

Two sets are called *incident* iff one of them contains the other (i.e., any set is incident with itself). For example, a 3D grid vertex (0-cell) is incident with six grid edges (1-cells); a grid square (2-cell) is incident with four grid edges; and a grid cube (3-cell) is incident with 12 grid edges.

Figure 2.12 shows four ways of representing a 2D grid. In the grid cell model, we can use only 2-cells (upper right), or we can also use 0- and 1-cells (lower right). In the first case, we can use 0- or 1-adjacency to define a grid cell adjacency model,

2.1 The Grid Point and Grid Cell Models

and, in the second case, we use the incidence relations between all of the cells to define the 2D *grid cell incidence model*.

There exists a grid point adjacency (incidence) model that is isomorphic to any of these grid cell adjacency (incidence) models. For grid points, we usually use only an adjacency model; however, for completeness, we mention that a grid point incidence model can be defined by adding isothetic edges between grid points—and "loops" consisting of quadruples of such edges—to the set of grid points. This refined model is isomorphic to the grid cell incidence model. In the correspondence between the models, a grid point p is the center of a 2-cell, an isothetic edge connecting p with another grid point intersects a 1-cell, and a loop of four such edges has a 0-cell as its center. The dimensions of the elements in the grid point incidence model are defined by the dimensions of the corresponding cells; for example, a loop is zero-dimensional.

Analogously, there are four types of models for 3D grids: (1) grid point and (2) grid cell adjacency models (see Section 2.1.4); (3) a grid cell incidence model that includes 0-, 1-, and 2-cells in addition to 3-cells (and again, for completeness, (4) the grid point model allows a structure that is isomorphic to the grid cell incidence model in which a grid point corresponds to a 3-cell [in which it is the center]); an isothetic edge between two grid points corresponds to a 2-cell (which it intersects at its center point), a loop of four isothetic edges corresponds to a 1-cell (where the loop defines a square having the 1-cell at its center), and 12 such edges forming a cube correspond to a 0-cell (at the center of the cube).

The 2D and 3D grid cell incidence models are called (2D or 3D) *incidence grids*.

Definition 2.5 A subset M of an incidence grid is called complete *iff*, for any cell c such that all cells incident with c and of larger dimension than c are in M, we also have $c \in M$.

This definition leads to a recursive test for completeness: start with the 3-cells, then check the 2-cells between them, then check the 1-cells that are incident only with 2- or 3-cells already known to be in M, and so on.

The *geometric representation* of the 2D incidence grid used in this book is illustrated in Figure 2.13: 0-cells are small squares (representing grid vertices);

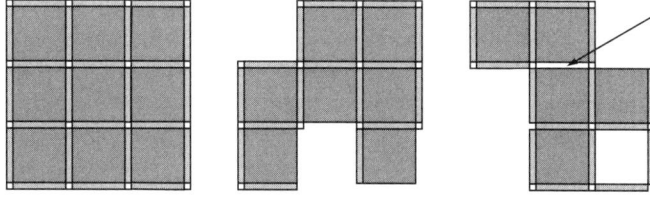

FIGURE 2.13 Representation of a 2D incidence grid by rectangles: a 3×3 grid on the left, a complete subset of it in the middle, and an incomplete subset on the right.

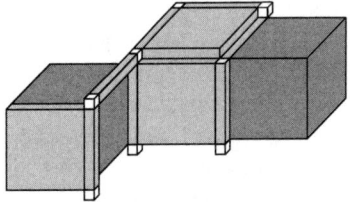

FIGURE 2.14 Representation of a 3D incidence grid by cuboids: a complete set of cells that contains three 3-cells and a few incident 0-, 1-, and 2-cells.

1-cells are thin rectangles (representing grid edges); and 2-cells are large squares.[4] In 3D, we use a similar representation, with small cubes as 0-cells, large cubes as 3-cells, and elongated (flat) cuboids as 1-cells (2-cells); see Figure 2.14. Despite these geometric representations, an i-cell is still considered to be of (abstract) dimension i.

2.2 Connected Components

The concept of connectedness defined in Section 1.1.4 applies to any of the adjacency relations defined in Section 2.1 for 2D or 3D grids. We recall that the reflexive and transitive closure of an adjacency relation is called a connectedness relation. In other words, any element of the grid is connected to itself, and two elements p and q of the grid are connected iff there exists a sequence of elements $(p_0, \ldots, p_n)$ where $n \geq 0$ such that $p_0 = p$, $p_n = q$, and p_{i+1} is adjacent to p_i ($0 \leq i < n$).

2.2.1 Connectedness and components

Let $\mathbb{G}$ be an adjacency grid. A sequence $\rho = (p_0, \ldots, p_n)$ of pixels or voxels, where $p_0 = p$, $p_n = q$, and p_{i+1} is adjacent to p_i ($0 \leq i \leq n-1$), is called a *path* of length n from p to q, and p and q are called the *endpixels* or *endvoxels* of ρ. The elements p and q are called *connected* if there is a path from p to q. In particular, for α-adjacency ($\alpha \in \{0, 1, 2, 3, 4, 6, 8, 18, 26\}$), we use the terms "$\alpha$-path" and "$\alpha$-connected."

Figure 2.15 shows a 1-path of 2-cells in the grid cell model and the corresponding 4-path in the grid point model. Figure 2.16 shows a 2-path of 3-cells in the grid cell model and the corresponding 6-path in the grid point model.

---

4. Note that the large squares and thin rectangles are topologically closed sets; however, they represent grid squares and grid edges that are open sets (i.e., the squares do not contain their edges, and the edges do not contain their endpoints). This representation will be convenient when we define topologic equivalence in Chapter 6, because the squares and rectangles used in the representation are all compact (closed and bounded) sets.

2.2 Connected Components

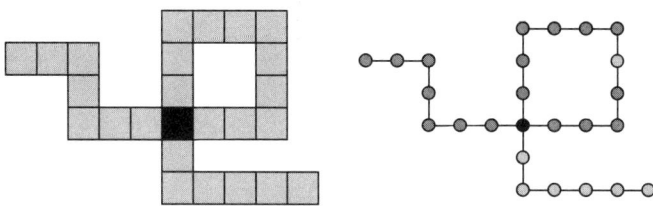

FIGURE 2.15 A 1-path in the grid cell model (left) that corresponds with a 4-path in the grid point model (right).

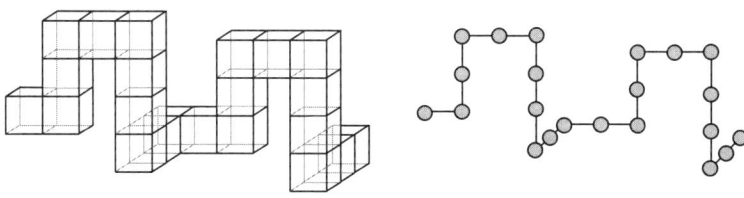

FIGURE 2.16 A 2-path in the grid cell model (left) that corresponds with a 6-path in the grid point model (right).

Let M be a finite subset of $\mathbb{G}$. A maximal connected set of pixels or voxels of M is called a *component* of M. Evidently $p, q \in M$ are in the same component of M iff there is a path completely contained in M that has p and q as endpixels (endvoxels).

Let $\mathbb{G}_{m,n}$ be an $m \times n$ grid. (Similar remarks apply to an $l \times m \times n$ grid $\mathbb{G}_{l,m,n}$.) $\mathbb{G}_{m,n}$ can be extended into the infinite discrete plane ($\mathbb{Z}^2$ in the case of the grid point model; $\mathbb{C}_2^{(2)}$ in the case of the grid cell model). Let M be a subset of $\mathbb{G}_{m,n}$; thus the complement $\overline{M}$ of M contains the complement $\overline{\mathbb{G}}_{m,n}$ of $\mathbb{G}$. The (infinite) set of pixels of $\overline{M}$ that are connected to pixels of $\overline{\mathbb{G}}_{m,n}$ is called the *background component* of M. Any other pixel of $\overline{M}$ belongs to a finite component of $\overline{M}$.

Let P be a picture defined on $\mathbb{G}$. The P-equivalence classes define subsets of $\mathbb{G}$. Components of these classes are often of interest, as we will see in later chapters. Figure 2.17 illustrates the components defined by these classes in a 5-valued picture using 1-adjacency in the grid cell model. Class 5 consists of six 1-components (also defining two complementary 1-components, one of which belongs to the background 1-component of class 5); class 4 consists of five 1-components (both complementary 1-components belong to the background 1-component of class 4); class 3 consists of four 1-components (one complementary 1-component); class 2 consists of four 1-components (two complementary 1-components, both belonging to the background 1-component of class 2); and class 1 consists of three 1-components.

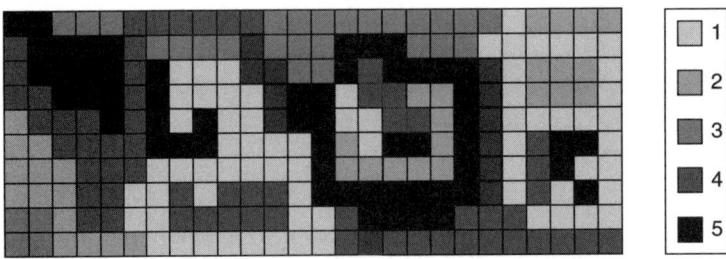

FIGURE 2.17 A picture that has five P-equivalence classes; the numbers on the right are used to refer to these classes in the text.

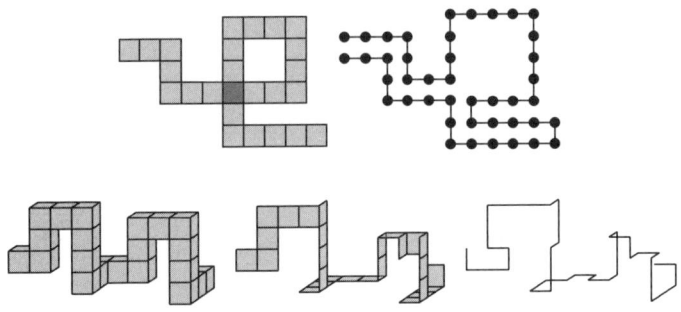

FIGURE 2.18 Top row: a 1-path of 2-cells (left) and a 0-path of 1-cells (right) in the 2D incidence grid. Bottom row: a 2-path of 3-cells (left), a 1-path of 2-cells (middle), and a 0-path of 1-cells (right) in the 3D incidence grid.

In the 2D or 3D incidence grid, two cells c_1 and c_2 are called k-*adjacent* iff they are not identical and there is a k-cell c ($c \neq c_1$ and $c \neq c_2$) that is incident with both c_1 and c_2.

Definition 2.6 Two i-cells ($i \geq 1$) are called *adjacent* iff they are k-adjacent for some $k < i$.

Figure 2.18 shows examples of paths for these adjacency relations. This definition generalizes the concepts of 1-adjacency (see Definition 2.2) and 0-adjacency (see Definition 2.3) between 2-cells or of α-adjacency ($0 \leq \alpha \leq 2$) between 3-cells (see Definition 2.4). Connectedness and components in 2D or 3D incidence grids can be defined using these adjacency relations.

The following two propositions show that for pictures defined on an incidence grid, cells can be assigned to P-equivalence classes in such a way that the components

2.2 Connected Components

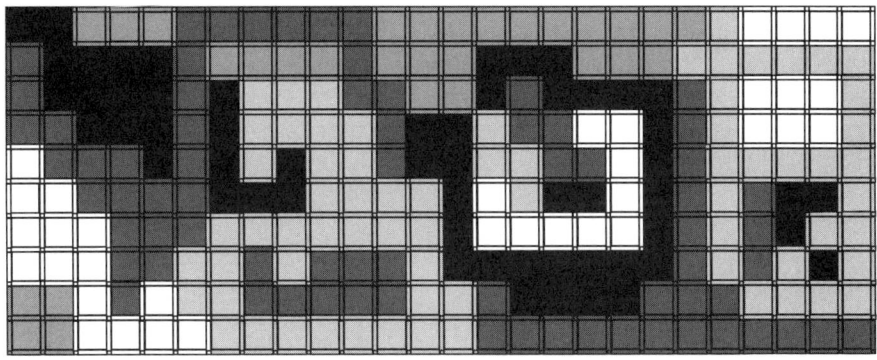

FIGURE 2.19 A picture on an incidence grid.

of the classes are complete. We give the proof for the 2D grid only; the 3D generalization is straightforward.

Proposition 2.3 Let P be a multilevel picture defined on the 2D incidence grid $\mathbb{G}$. Every 0- or 1-cell of $\mathbb{G}$ can be assigned to exactly one P-equivalence class of an incident 2-cell in such a way that all of the components of P-equivalence classes of 2-cells and their assigned 0- and 1-cells are complete subsets of $\mathbb{G}$.

Proof Assign to each 0- or 1-cell the highest value among those of the 2-cells with which it is incident. Evidently, if a 0- or 1-cell is incident only with 2-cells that all have the same value u, it will be assigned to class M_u. Hence the resulting components of P-equivalence classes of 2-cells and their assigned 0- and 1-cells are all complete. ∎

Figure 2.19 shows an example of such an assignment for the picture P shown in Figure 2.17. Note that we do not need an extended data structure to represent P on an incidence grid; all of the assignments of 0- or 1-cells are determined by the order of the picture values.

Proposition 2.4 Let P be a multilevel picture defined on the 3D incidence grid $\mathbb{G}$. Every 0-, 1-, or 2-cell can be assigned to exactly one P-equivalence class of an incident 3-cell in such a way that all of the components of P-equivalence classes of 3-cells and their assigned 0-, 1-, and 2-cells are complete subsets of $\mathbb{G}$.

FIGURE 2.20   The 12 free 5-ominoes.

2.2.2 Counting connected sets

A finite connected subset of the 1-adjacency grid $[\mathbb{C}_2^{(2)}, A_1]$ is called a *polyomino*. (Polyominoes illustrate the combinatorial complexity of finite sets of pixels; see Chapters 13 and 14 about polyominoes in the context of digital convexity and digital tomography.) If it consists of exactly n grid squares, it is called an *n-omino*.

Equivalence classes of polyominoes can be defined with respect to geometric transformations that map the grid into itself. Such transformations include (see also Section 14.4) translations by vectors (i,j), $i, j \in \mathbb{Z}$; rotations around the origin by $90°$, $180°$, or $270°$; and reflections in the x- or y-axis.

Two polyominoes are called translation-equivalent if there is a translation that maps one of them into the other. A translation-equivalence class of polyominoes is called a *fixed polyomino*. An equivalence class of polyominoes with respect to translations and rotations is called a *chiral polyomino*; an equivalence class of polyominoes under translations, rotations, and reflections (a *congruence class*) is called a *free polyomino*.[5]

Figure 2.20 shows the 12 free 5-ominoes. Figure 2.21 shows the 369 free 8-ominoes (the 8-ominoes that contain a hole are positioned in the center of the figure; the filled dots indicate 8-ominoes that contain at least one 2×2 block).

The numbers of fixed n-ominoes, chiral n-ominoes, and free n-ominoes are denoted by $t(n), r(n),$ and $s(n)$, respectively. It is not hard to show the following:

$$\frac{t(n)}{8} \leq s(n) \leq r(n) \leq t(n) \, .$$

There are no simple formulas for these functions, but the following has been demonstrated [371]:

$$\lim_{n \to \infty} t(n)^{1/n} = a \text{ where } 3{\cdot}9 < a < 4{\cdot}65 \tag{2.4}$$

Table 2.1 shows the values of $t(n), r(n)$, and $s(n)$ for $1 \leq n \leq 24$. The calculation of these functions is a topic in the theory of polyominoes. Redelmeier's algorithm [839] for calculating them has exponential time complexity; no algorithm is yet known that has subexponential time complexity. An algorithm for generating polyominoes would also be of interest; it could be used to generate data sets for testing algorithms in digital geometry.

---

5. For properties of a set that are invariant under specific types of geometric transformations, see Section 1.2.2 and Chapter 14.

2.2 Connected Components

FIGURE 2.21 The 369 free 8-ominoes [368].

2.2.3 Component labeling

The following task arises frequently in picture analysis and computer graphics; it is known as *labeling, filling,* or *region detection*. Let P be a 2D multivalued picture defined on a finite adjacency or incidence grid $\mathbb{G}$, and let the P-equivalence classes have a total of k components. We regard the (infinite) complement of $\mathbb{G}$ as consisting of pixels that all have the same value, so they belong to one of the components. The task is to assign k labels (e.g., integers) to the pixels of P in such a way that all of

TABLE 2.1 Values of $t(n)$, $r(n)$, and $s(n)$ [839].

n	t(n)	r(n)	s(n)
1	1	1	1
2	2	1	1
3	6	2	2
4	19	7	5
5	63	18	12
6	216	60	35
7	760	196	108
8	2,725	704	369
9	9,910	2,500	1,285
10	36,446	9,189	4,655
11	135,268	33,896	17,073
12	505,861	126,759	63,600
13	1,903,890	476,270	238,591
14	7,204,874	1,802,312	901,971
15	27,394,666	6,849,777	3,426,576
16	104,592,937	26,152,418	13,079,255
17	400,795,844	100,203,194	50,107,909
18	1,540,820,542	385,221,143	192,622,052
19	5,940,738,676	1,485,200,848	742,624,232
20	22,964,779,660	5,741,256,764	2,870,671,950
21	88,983,512,783	22,245,940,545	11,123,060,678
22	345,532,572,678	86,383,382,827	43,191,857,688
23	1,344,372,335,524	336,093,325,058	168,047,007,728
24	5,239,988,770,268	1,309,998,125,640	654,999,700,403

the pixels in each component have the same label and pixels in different components have different labels. To keep P unaltered, we can put the labels into an array of the same size as P.

A simple method of labeling components is as follows. Scan the picture until a pixel p is found that has not yet been labeled. Suppose $P(p) = u$ and that labels $L_1, \ldots, L_{k-1}$ have already been used. Choose a new label L_k, and call the procedure FILL(p, u, L_k) (see Algorithm 2.1). (Note that the adjacency set $A(r)$ may depend on u [e.g., if we use 1-adjacency for 1s and 0-adjacency for 0s].) After labeling the component that contains p, continue scanning the picture until all of the pixels have been labeled.[6] Figure 2.22 shows the order of the pixel visits (assuming the order

---

6. The FILL algorithm in Algorithm 2.1 uses a depth-first strategy to visit all of the pixels in a component. The books [773] and [805] discuss variants of this strategy, such as recursive and nonrecursive (i.e., "filling by connectivity"). The time complexity of this strategy can be improved by stacking horizontal runs of pixels [873]. In some versions of the FILL algorithm, the stack is replaced by a first-in-first-out queue. See Chapter 4 for a generalization and modifications.

2.2 Connected Components

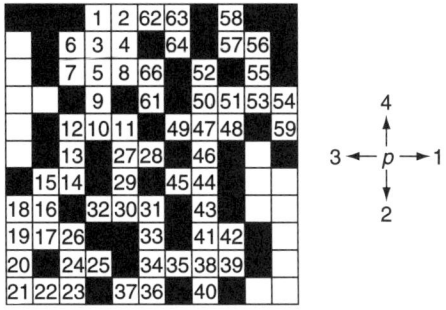

FIGURE 2.22 The numbers show the order in which the pixels are labeled, assuming a standard scan. The diagram on the right shows the order in which 1-adjacent pixels are visited.

shown on the right is used for visiting 1-adjacent pixels) when this algorithm is used to label the white pixels of the picture in Figure 2.23.

> 1. Label p with L_k.
> 2. Put p into a stack.
> 3. If the stack is empty, stop.
> 4. Pop r out of the stack.
> 5. Label with L_k all pixels $q \in A(r)$ that have value u and have not yet been labeled, and put these qs into the stack.
> 6. Go to Step 3.

ALGORITHM 2.1 Procedure FILL(p, u, L_k) for component labeling.

The *Rosenfeld-Pfaltz labeling algorithm* [921] labels all of the components of P in two scans of P (see Section 1.1.3); see Algorithm 2.2. In the first scan, we propagate smallest labels, and, whenever labels merge, we note this fact in a table of equivalent pairs of labels. In the second scan, we replace each label with a representative of its equivalence class. This algorithm replaces the use of a stack (of size mn) in procedure FILL with the use of an equivalence table; its run time can be compared with that of the simple depth-first search procedure for given classes of pictures. It also provides an illustration of the cases that can occur during component labeling.

As an example of the operation of this algorithm, consider the binary picture shown in Figure 2.23. In the label propagation step of the algorithm, we use 1-adjacency for white pixels and 0-adjacency for black pixels. At the end of the first scan (a standard scan), we have the equivalence table shown in Table 2.2. Label A

1. In the first scan, propagate the labels until the end of the picture is reached:
   1.1 If the current pixel p is adjacent to one or more previously visited pixels that all have the same label, assign that label to p, and continue the scan.
   1.2 If the current pixel p is adjacent to two or more previously visited pixels that have different labels, assign the smallest of those labels (e.g., L) to p, enter the other labels into the table as being equivalent to L, and continue the scan.
   1.3 Otherwise, assign to p a label that has not yet been used and continue the scan.
2. Determine the equivalence classes of the labels by computing the transitive closure of the equivalent pairs of labels detected in Step 1. Choose one label from each equivalence class as its pivot.
3. Scan the picture a second time, and replace every label with the pivot of its equivalence class.

ALGORITHM 2.2 The Rosenfeld-Pfaltz component labeling algorithm.

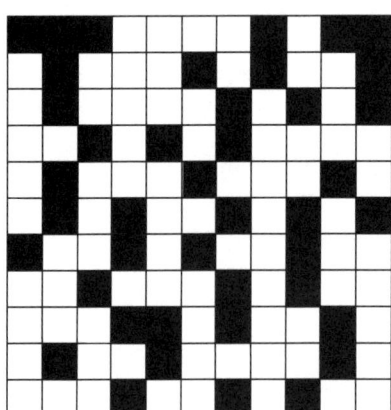

FIGURE 2.23 Let P be the binary picture shown on the left. If we use (1,0)-adjacency and the standard scan and assume that the infinite background component is black, the algorithm produces the label assignments shown on the right and the equivalence table shown in Table 2.2.

is also the label of all of the pixels of the background component. Note that some equivalences are detected more than once.

The equivalences between pairs of the 20 labels $A,\ldots,T$ are shown in Figure 2.24. We will use the smallest label in each equivalence class as the pivot of that class.

2.3 Digitization Models

TABLE 2.2 Equivalence table for the picture in Figure 2.23.

Label	A	B	C	D	E	F	G	H	I	J	K	L	M	N	O	P	Q	R	S	T
Equivalent labels					B	A	C	A	B	A	A			I	G	N	B		B	M
					B		B			A						N				
																N				

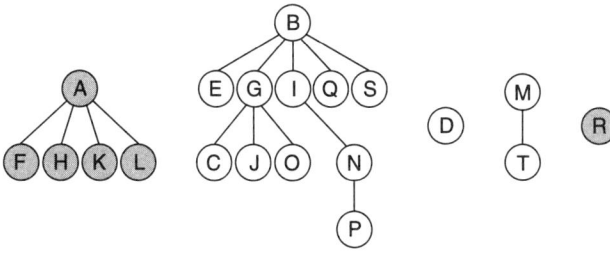

FIGURE 2.24 All equivalences between pairs of labels.

TABLE 2.3 Pivots for the example in Figure 2.23.

Label	A	B	C	D	E	F	G	H	I	J	K	L	M	N	O	P	Q	R	S	T
Pivot	A	B	B	D	B	A	B	A	B	B	A	A	M	B	B	B	B	R	B	M

To find the pivots, we scan the labels in order, smallest first. Any label that is equivalent to A has pivot A and can be marked and replaced with A. We next examine the smallest label that has not yet been marked; this label must be the pivot of its equivalence class so that all of the labels equivalent to it can be marked and replaced with it. This process is repeated until all of the labels have been marked.[7] The pivots in our example are shown in Table 2.3.

A binary picture of size $m \times n$ in which 0s and 1s alternate ("a chessboard") requires $\mathcal{O}(mn)$ labels. An a priori threshold on the number of labels can be used to limit the size of the equivalence table. However, memory limitations are no longer an issue, which they were in 1966, when the algorithm was first published.

The nontriviality of connected component labeling is illustrated in Figure 2.25. In one of these binary pictures, the black pixels are connected, and in the other one, there are two 4-components of black pixels; can you tell which is which?

2.3 Digitization Models

We use mathematically defined methods of *digitization* to create digital pictures and to compare results obtained by analyzing these pictures with corresponding

---

7. Standard graph traversal algorithms (e.g., depth-first, breadth-first) can also be used to find the labels equivalent to a given label. Graphs are the subject of Chapter 4.

FIGURE 2.25 In which of these pictures are the black pixels connected? [631].

results in Euclidean or similarity geometry (see Section 1.2.2). In this section, we describe three digitization methods: Gauss digitization and grid-intersection digitization, which were originally proposed for 2D, and Jordan digitization, which was defined more than a century ago for 3D. We generalize these methods to allow variable grid resolution, and we extend the Gauss and grid-intersection models to 3D and the Jordan model to 2D. We define these digitization methods for the grid cell model, but they can also be used with the grid point model by representing cells with their centers. We briefly discuss relationships among these digitizations, and we conclude by defining a general class of digitization methods that has all of these methods as special cases.

2.3.1 Gauss digitization

C.F. Gauss (1777–1855) studied the measurement of the area of a planar set $S \subset \mathbb{R}^2$ by counting the grid points $(i,j) \in \mathbb{Z}^2$ contained in S. This approach suggests the following:

Definition 2.7 Let S be a subset of the plane. The *Gauss digitization* $G(S)$ is the union of the grid squares with center points in S.

Figure 2.26 shows the Gauss digitizations $G(D)$ of four disks D of different diameters (measured in grid units). (The results would be the same if the disks all had unit diameter and were digitized on grids of different resolutions. The Gauss digitization of S on a grid of resolution h will be denoted by $G_h(S)$.) $G(D)$ is an isothetic polygon that has 12 vertices for diameter 5, 20 vertices for diameter 10, and 36 vertices for diameter 17. Note that the number of vertices is always a multiple of 4, because a disk that is centered at a grid point has a symmetric Gauss digitization.

Proposition 2.5 The Gauss digitization $G(S)$ of any nonempty bounded set $S \subset \mathbb{R}^2$ is the union of a finite number of simple isothetic polygons.

2.3 Digitization Models

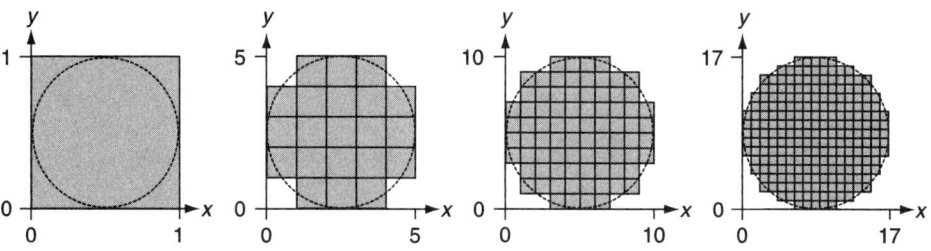

FIGURE 2.26 Four disks (dashed) and their Gauss digitizations (shaded).

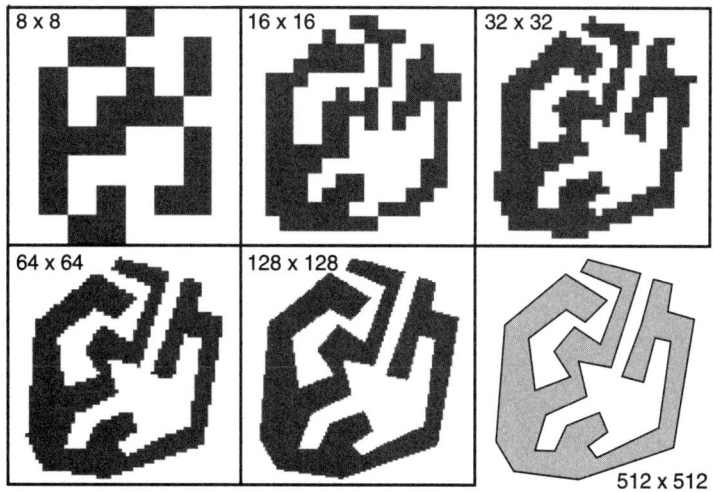

FIGURE 2.27 Gauss digitization of a simple polygon using grids of sizes from 8×8 (upper left) to 128×128 (lower middle). The original polygon was drawn on a grid of size 512×512 (lower right).

Proof A Gauss digitization $G(S)$ is a union of grid squares, all of equal size. This union contains only a finite number of grid squares, because S is bounded. Any grid square is a simple isothetic polygon.[8] ∎

Obviously, different sets can have identical Gauss digitizations. Figure 2.27 shows the Gauss digitization of a simple polygon Π with area 102,742·5 and perimeter 4,040·796,631 ... drawn on a 512×512 grid. On the upper left, each grid square

---

8. Partitions of a Gauss digitization into minimum numbers of simple isothetic polygons are considered in Exercise 4 in Section 2.5.

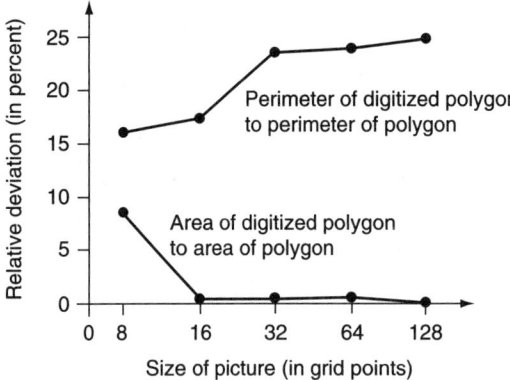

FIGURE 2.28 Relative deviations of area and perimeter for the digitized polygon in Figure 2.27.

contains 64×64 squares of the original 512×512 grid; in the upper middle, 32×32; and so on.

Figure 2.28 shows the *relative deviations* of the area and perimeter of $G_h(\Pi)$ from those of Π when Π is digitized on a $2^n \times 2^n$ grid (i.e., $h = 2^n$). The relative deviation is the absolute difference between the property values for $G_h(\Pi)$ and Π divided by the property value for Π. See Chapter 10 for an in-depth discussion about perimeter estimation; the perimeter of $G_h(\Pi)$ (i.e., the number of 1-cells on its frontier) is not a good estimate of the perimeter of Π.

Gauss digitization is defined analogously in 3D. If $S \subset \mathbb{R}^3$, the Gauss digitization $G_h(S)$ is the union of all of the 3-cells (in a grid of resolution $h > 0$) with center points in S.

2.3.2 Jordan digitization

Let $S \subset \mathbb{R}^3$ and $h > 0$. The *magnification of S by factor h* is denoted by $h \cdot S$. In terms of multiplication of vectors by a scalar (see Section 1.2.12), we have the following:

$$h \cdot S = \{(h \cdot x, h \cdot y, h \cdot z) : (x, y, z) \in S\}$$

This magnification leaves the origin $(0,0,0)$ fixed; other points of $\mathbb{R}^3$ could also be chosen as fixed points.

C. Jordan (1838–1922) [484] used grids to estimate the volumes of subsets of $\mathbb{R}^3$. Let $S \subset \mathbb{R}^3$ be contained in the union of finitely many 3-cells. Magnify S by factor h with respect to an arbitrary fixed point $p \in \mathbb{R}^3$; this transforms S into S_h^p. Let $l_h^p(S)$ be the number of 3-cells completely contained in S_h^p and $u_h^p(S)$ the number of 3-cells that have nonempty intersections with S_h^p. Then $h^{-3} \cdot l_h^p(S)$ and $h^{-3} \cdot u_h^p(S)$ converge to limits $L(S)$ and $U(S)$, respectively, as h goes to infinity; these limits are the same

2.3 Digitization Models

for any p [484]. Jordan called $L(S)$ the *inner volume* and $U(S)$ the *outer volume* of S or the *volume* $\mathcal{V}(S)$ of S if $L(S) = U(S)$.

We use Jordan's digitization model for subsets of the plane as well as of 3D space:

Definition 2.8 Let S be a nonempty subset of $\mathbb{R}^2$. Let $J_h^-(S)$ be the union of all 2-cells (for grid resolution $h > 0$) that are completely contained in S, and let $J_h^+(S)$ be the union of all such 2-cells that have nonempty intersections with S. $J_h^-(S)$ is called the *inner Jordan digitization of S* and $J_h^+(S)$ the *outer Jordan digitization* of S. For $S \subseteq \mathbb{R}^3$, we use 3-cells instead of 2-cells. For brevity, we denote J_1^- and J_1^+ with J^- and J^+, respectively.

Outer Jordan digitization is also called *super-cover digitization* [209]. (Inner Jordan digitization will be further specified in Section 3.2.5 by one additional [topologic] constraint that is still missing in the above definition.) Figure 2.29 shows a 2D example in which S is a circle of radius n (in grid units) for $n = 4$ (left), $n = 8$ (middle), and $n = 16$ (right). If the frontier of a nonempty set $S \subset \mathbb{R}^2$ does not contain any grid edge segment of nonzero length, then the frontier of $J_h^-(S)$ never intersects the frontier of $J_h^+(S)$. For example, this is the case if S has a *smooth frontier* that has continuous partial derivatives with respect to both coordinates and has positive curvature everywhere. A straight line γ has an empty $J^-(\gamma)$ and a connected $J^+(\gamma)$.

Proposition 2.6 The inner and outer Jordan digitizations $J_h^-(S)$ and $J_h^+(S)$ of any nonempty bounded set $S \subset \mathbb{R}^2$ are unions of finite numbers of simple isothetic polygons.

Propositions 2.5 and 2.6 have straightforward 3D generalizations.

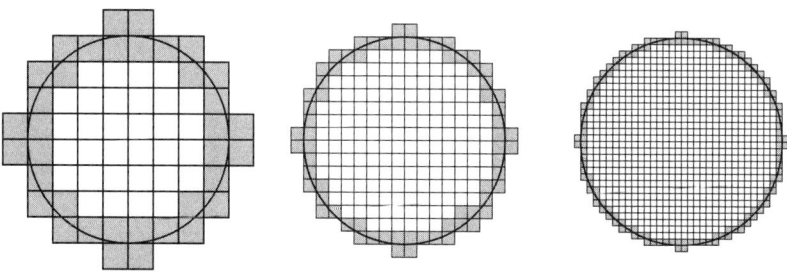

FIGURE 2.29 Inner and outer Jordan digitizations of a centered disk.

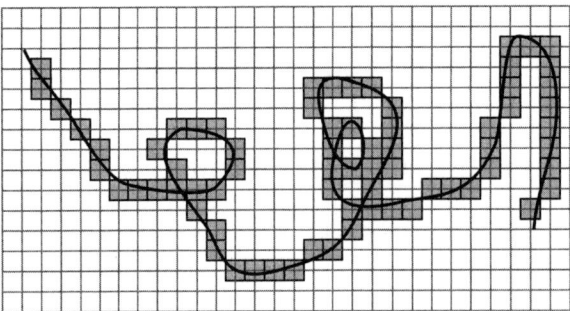

FIGURE 2.30   Grid-intersection digitization of an arc.

2.3.3 Grid-intersection digitization

Gauss digitization and inner Jordan digitization are obviously not appropriate for curves or arcs.[9] Outer Jordan digitization is appropriate, but, in this section, we will define grid-intersection digitization, which is commonly used for arcs and curves in the plane.

> **Definition 2.9** The *grid-intersection digitization* $R(\gamma)$ of a planar curve or arc γ is the set of all grid points (i, j) that are closest (in Euclidean distance) to the intersection points of γ with the grid lines.

Figure 2.30 illustrates this definition. Note that an intersection point may have the same minimum distance to two different grid points; such an intersection point contributes two grid points to $R(\gamma)$. (Alternatively, we could always choose, for example, the right point or the upper point.)

A traversal of γ defines an ordered sequence (list) of grid points in $R(\gamma)$. We assume the following for simplicity: (i) if an intersection point is at the same minimum distance from two grid points, we list only the grid point that has the larger x-coordinate, or, if their x-coordinates are equal, the one with the larger y-coordinate; and (ii) if consecutive intersection points have the same closest grid point, we list that grid point only once.

The resulting ordered sequence of grid points is called the *digitized grid-intersection sequence* $\rho(\gamma)$ of γ (see Figure 2.31). It defines a polygonal arc (or polygon) with vertices at grid points. The sequence represents $R(\gamma)$ uniquely if an intersection point is never at the same minimum distance from two grid points.

---

9. Precise definitions of curves and arcs will be given in Chapters 6, 7, and 8. For the moment, we say that a simple arc or curve intersects the lines of a 2D grid (or the planes of a 3D grid) a finite number of times.

2.3 Digitization Models

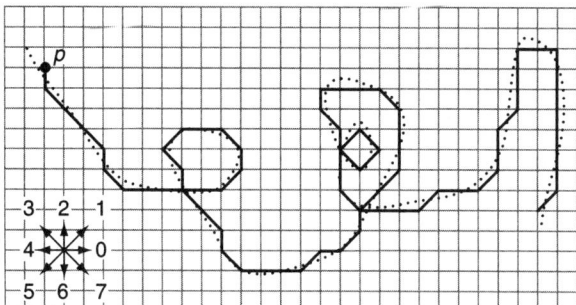

FIGURE 2.31 Directional encoding of an arc. Starting at grid point p, the arc can be represented by the sequence of codes 677767000001...65.

A similar method can be used to digitize a 3D arc or curve γ: for each intersection point of γ with a grid plane, we add the grid point(s) closest to the intersection point to the digitization.

Successive pairs of grid points in $\rho(\gamma)$ define steps of length 1 along grid lines and diagonal steps of length $\sqrt{2}$. The directions of the steps can be represented with codes $0, 1, \ldots, 7$ as shown at the lower left of Figure 2.31; code i represents a step that makes angle $(45 \cdot i)°$ with the positive x-axis. Figure 2.31 shows an example of the *directional encoding* of an arc. The directional codes are usually called *chain codes*. A *chain* is an ordered finite sequence of code numbers. In Chapter 9, such a sequence will be described as a word in an alphabet, which is the (finite) set of code numbers. The *length* of a chain is the number of code numbers in it; note that this length is not related to the geometric length of the arc or curve represented by the chain.

The (geometric) *length* of $\rho(\gamma)$ is the sum of the lengths of the steps. The question arises whether this length can be used to estimate the length of γ. In what follows, we denote the grid-intersection digitization of γ in a grid of resolution h with $R_h(\gamma)$ and the corresponding digitized grid intersection sequence by $\rho_h(\gamma)$. Note that, for any γ, we have the following:

$$J_h^-(\gamma) = \emptyset \subseteq R_h(\gamma) \subseteq J_h^+(\gamma) \tag{2.5}$$

Let γ be *rectifiable*; thus γ has a well-defined length $\mathcal{L}(\gamma)$. The length of $\rho(\gamma)$ is not a good estimate of $\mathcal{L}(\gamma)$; it does not necessarily converge to $\mathcal{L}(\gamma)$ as the grid constant goes to zero.

As a simple example, consider the straight line segment pq in Figure 2.32 that has a slope of $22.5°$ and a length of $5\sqrt{5}/2$. The length of $\rho(pq)$ is $3 + 2\sqrt{2}$ for grid constant 1 and $(5 + 5\sqrt{2})/2$ for *all* grid constants $1/2^n$ ($n \geq 1$). This shows that the length of $\rho(pq)$ does not converge to $5\sqrt{5}/4$ as the grid constant goes to zero.

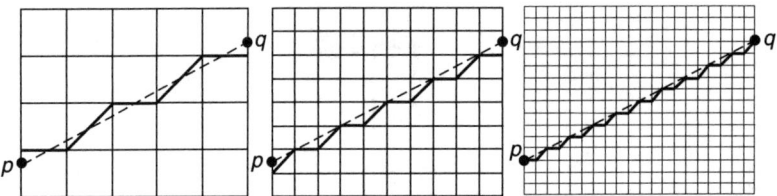

FIGURE 2.32 A straight line segment of slope 22.5° digitized using three grids of decreasing grid constant. The differences in coordinates between p and q are 10 and 5 units.

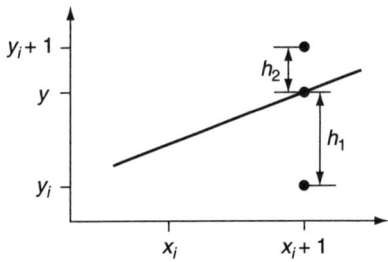

FIGURE 2.33 Differences in h_1 and h_2 from the correct y value.

As a more general example, consider a line segment γ with slope $1/(m+1)$. Its chain code representation is $(0^m 1)^k$, where k depends on the grid resolution. No matter what k is, the length of $\rho(\gamma)$ is $k(m+\sqrt{2})$ where $\mathcal{L}(\gamma)$ is $k\sqrt{1+(m+1)^2}$. The ratio $\mathcal{L}(\rho(\gamma))/\mathcal{L}(\gamma)$ is not 1 unless $m=0$ or $m \to \infty$.

We conclude this section by discussing the grid-intersection digitization of a straight line segment. *Bresenham's algorithm* [122] is a standard routine in computer graphics (Algorithm 2.3). We discuss the use of this algorithm to digitize a segment of a line $y = ax + b$ in the first octant (i.e., with slope $a \in [0,1]$). To draw the resulting *digital straight line segment*, we increase the x-coordinate stepwise by $+1$; the y-coordinate is "occasionally" increased by $+1$ and remains constant otherwise. (By interchanging the startpoints and endpoints of the segment, we can handle octants "to the left of the y-axis." In the eighth octant, we use a y-increment of -1, and in the second and seventh octants, we interchange the roles of the x- and y-coordinates.)

The digital straight line segment is a sequence of grid points $(x_i, y_i), i \geq 1$. The point (x_1, y_1) is the grid point closest to the endpoint of the real segment. If we already have point (x_i, y_i), the next point has x-coordinate x_{i+1}, and, for its y-coordinate, we must decide between y_i and $y_i + 1$. Let $y = a(x_i + 1) + b$, and define the differences

2.3 Digitization Models

---

1. Let $dx = x_q - x_p$, $dy = y_q - y_p$, $x = x_p$, $y = y_q$, $b_1 = 2 \cdot dy$, $error = b_1 - dx$, and $b_2 = error - dx$.
2. Repeat Steps 3 through 6 until $x > x_q$. Stop when $x > x_q$.
3. Change the value of (x, y) to the value of a line pixel.
4. Increment x by 1.
5. If $error < 0$, let $error = error + b_1$, or else increment y by 1 and let $error = error + b_2$.
6. Go to Step 2.

ALGORITHM 2.3 Bresenham's straight line segment algorithm (first octant only).

---

h_1 and h_2 (see Figure 2.33) with the following:

$$h_1 = y - y_i = a(x_i + 1) + b - y_i$$
$$h_2 = (y_i + 1) - y = y_i + 1 - a(x_i + 1) - b$$

To decide between y_i and $y_i + 1$, we use the following difference:

$$h_1 - h_2 = 2a(x_i + 1) - 2y_i + 2b - 1$$

We choose $(x_i + 1, y_i)$ if $h_1 < h_2$ and $(x_i + 1, y_i + 1)$ otherwise. For reasons of efficiency (integer arithmetic only), we do not use $h_1 - h_2$ for this decision. Rather, let $p = (x_1, y_1)$ and $q = (x_q, y_q)$ be the grid points closest to the endpoints of the segment, and let $dx = x_q - x_1$ and $dy = y_q - y_1$. Let $e_i = dx \cdot (h_1 - h_2)$; thus $e_i = 2(dy \cdot x_i - dx \cdot y_i) + b'$, where $b' = 2dy + 2dx \cdot b - dx$ is independent of i. Thus e_i can be updated iteratively for successive decisions at $x_i + 1$ and $x_i + 2$:

$$e_i = 2dy \cdot x_i - 2dx \cdot y_i + b'$$
$$e_{i+1} = 2dy \cdot x_{i+1} - 2dx \cdot y_{i+1} + b'$$

Thus

$$e_{i+1} - e_i = 2dy(x_{i+1} - x_i) - 2dx(y_{i+1} - y_i) = 2dy - 2dx(y_{i+1} - y_i)$$

because $x_{i+1} = x_i + 1$; this is sufficient for deciding about the y-increment. Let $x_1 = x_p = 0$ and $y_1 = y_p = 0$ give the starting value:

$$e_1 = 2dy \cdot x_1 - 2dx \cdot y_1 + 2dy + dx(2b - 1) = 2dy - dx$$

The resulting algorithm for the first octant is shown above. At Step 1, we have $error = e_1 = 2dy - dx$. The values $b_1 = 2 \cdot dy$ and $b_2 = 2 \cdot dy - 2 \cdot dx$ are used to efficiently update the variable $error$. The algorithm runs in $\mathcal{O}(x_q - x_p)$ time because, for each i, it involves only a constant number of operations: setting one pixel value, two simple logical tests, one addition, and one or two increments.

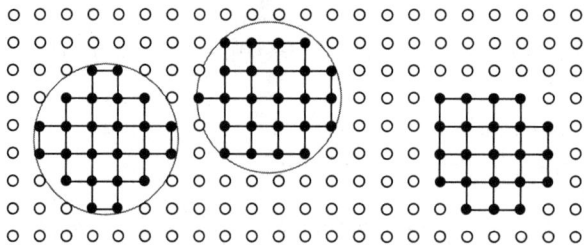

FIGURE 2.34 Gauss digitizations of the same disk at different locations. (In the example on the right, the disk is not shown: this is done to illustrate the difficulty of recognizing digital disks.)

2.3.4 Types of digital sets

If γ is, for example, a straight line, straight line segment, circle, or parabola, we call $R_h(\gamma)$ a *digital straight line*, *digital straight segment*, *digital circle*, or *digital parabola*, respectively.

If S is, for example, a disk, square, or convex set (and similarly in 3D), we call $J_h^-(S), G_h(S)$, or $J_h^+(S)$ a *digital disk*, *digital square*, or *digital convex set*, respectively, provided it is connected. We call a connected set of grid points a digital disk and so forth (with respect to a given digitization model), if there exists a disk and so forth that has that connected set as its digitization.

If Gauss or inner Jordan digitization is used, a connected set can have a digitization that consists of several connected isothetic polygons (polyhedra). On the other hand, the outer Jordan digitization of a connected set S is always a single connected isothetic polygon or polyhedron. However, J^+ does not preserve simple connectedness; it can create holes.

Figure 2.34 shows how a disk in different positions can create different digital disks by Gauss digitization. The left and center digital disks both consist of 24 grid points, but the disk on the right consists of only 22 grid points. It can be shown that the number of different digital disks (up to translation), with respect to Gauss digitization, that consist of exactly $n \geq 1$ grid points is at most the following:

$$\mathcal{O}(n^2) \qquad (2.6)$$

Gauss and Jordan digitization allow us to study methods or algorithms of digital geometry under slightly different assumptions about the relationships between sets S in the Euclidean plane and their digitizations. Evidently, $J_h^-(\emptyset) = G_h(\emptyset) = J_h^+(\emptyset) = \emptyset$ and $J_h^-(\mathbb{R}^2) = G_h(\mathbb{R}^2) = J_h^+(\mathbb{R}^2) = \mathbb{R}^2$ (and similarly for $\mathbb{R}^3$). If S is a nonempty proper subset of $\mathbb{R}^2$ or of $\mathbb{R}^3$ with a smooth frontier, we have $J_h^-(S) \subset J_h^+(S)$. Furthermore, the following is true:

$$J_h^-(S) \subseteq G_h(S) \subseteq J_h^+(S) \text{ for any } S \subseteq \mathbb{R}^2 \ (S \subseteq \mathbb{R}^3) \qquad (2.7)$$

2.3 Digitization Models

One or both relations $\subseteq$ in the left part of Equation 2.7 can be replaced by $=$, but both cannot be if S has a smooth frontier. Let S be a finite union of grid squares; then we have $J^-(S) = G(S) = J^+(S)$.

2.3.5 Domain digitizations

In this section, we define a framework for a general class of digitization models. To simplify the discussion, we formulate this framework in n dimensions ($n \geq 1$), but our main interest is of course in $n = 2$ and $n = 3$.

Let the following be the n-cell centered at the origin $o = (0, \ldots, 0)$:

$$\Pi_{\text{cube}} = \{(x_1, \ldots, x_n) : \max_{1 \leq i \leq n} |x_i| \leq \tfrac{1}{2}\}$$

Let $\emptyset \neq \Pi_\sigma \subseteq \Pi_{\text{cube}}$, and consider translates $\Pi_\sigma(q) = \{q + p : p \in \Pi_\sigma\}$ of Π_σ centered at grid points $q \in \mathbb{Z}^n$. (Note that $\Pi_\sigma(q)$ is the Minkowski sum of Π_σ and $\{q\}$.) In particular, $\Pi_{\text{cube}}(q)$ is the n-cell c_q centered at q.

We will use the translates $\Pi_\sigma(q)$ of Π_σ as the *domains of influence* for digitizations that we call dig_σ^+ and dig_σ^-. For any set $S \subseteq \mathbb{R}^n$, $dig_\sigma^+(q)$ is the union of all c_q such that $\Pi_\sigma(q)$ intersects S, and $dig_\sigma^-(S)$ is the union of all c_q such that $\Pi_\sigma(q)$ is contained in S. Thus the following is true:

$$c_q \subseteq dig_\sigma^+(S) \text{ iff } \Pi_\sigma(q) \cap S \neq \emptyset \text{ and } c_q \subseteq dig_\sigma^-(S) \text{ iff } \Pi_\sigma(q) \subseteq S \quad (2.8)$$

So $dig_\sigma^+(S)$ is called the *outer σ-digitization* of S and $dig_\sigma^-(S)$ the *inner σ-digitization* of S. Evidently, for any $S \subseteq \mathbb{R}^n$, we have $dig_\sigma^-(S) \subseteq dig_\sigma^+(S) \subseteq \mathbb{C}_n^{(n)}$.

We now show that Jordan and Gauss digitizations are all σ-digitizations and that grid-intersection digitization can also be regarded as a σ-digitization.

- If $\Pi_\sigma = \Pi_{\text{cube}}$ for $n = 2, 3$, we obtain the inner and outer Jordan digitizations such that, for $S \subseteq \mathbb{R}^2$ or $S \subseteq \mathbb{R}^3$, we have the following:

$$J^+(S) = dig_{\text{cube}}^+(S) \quad \text{and} \quad J^-(S) = dig_{\text{cube}}^-(S)$$

- If $\Pi_\sigma = \{o\}$, we have $\Pi_\sigma(q) = \{q\}$ so that $dig_\sigma^+(S) = dig_\sigma^-(S)$ for all $S \subseteq \mathbb{R}^n$. For $n = 2$ or 3, this set is just the Gauss digitization $G(S)$.

- If $\Pi_\sigma = \{(x_1, \ldots, x_n) : \exists i (1 \leq i \leq n \wedge x_i = 0) \wedge \max_{1 \leq i \leq n} |x_i| \leq \tfrac{1}{2}\}$, dig_σ^+ is essentially grid-intersection digitization. (It is a union of cross-shaped

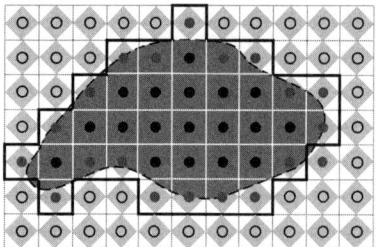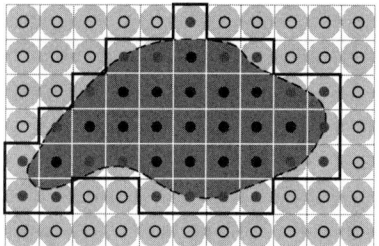

FIGURE 2.35 Inner and outer diamond and ball digitizations in the plane. The inner digitization is the union of the grid squares centered at black grid points, and the outer digitization (frontier shown as a bold black line) also contains the grid squares centered at shaded grid points. Left: diamond digitization. Right: ball digitization.

neighborhoods of grid cells rather than a union of grid points.) If γ is a planar arc or curve and $n = 2$, it is not hard to see that $R(\gamma) = dig_\sigma^+(\gamma)$, provided γ does not intersect any grid line midway between two grid points.[10]

Thus the Jordan digitizations are σ-digitizations in which Π_σ is a cube; Gauss digitization is a σ-digitization in which Π_σ is a point; and grid-intersection digitization is essentially a σ-digitization in which Π_σ is a cross. Other digitization models can be defined by using other simple sets Π_σ (e.g., "ball digitization" by using $\Pi_\sigma = \{(x_1, \ldots, x_n) : \sum_{i=1}^n x_i^2 \leq \frac{1}{4}\}$, "diamond digitization" by using $\Pi_\sigma = \{(x_1, \ldots, x_n) : \max\{|x_1|, \ldots, |x_n|, 1/(n-1)\sum_{i=1}^n |x_i|\} \leq \frac{1}{2}\}$). These digitizations are illustrated in Figure 2.35. The figure also illustrates a general property that follows directly from the definitions of dig_σ^+ and dig_σ^-:

$$\Pi_{\sigma_1} \subseteq \Pi_{\sigma_2} \text{ implies } dig_{\sigma_1}^+(S) \subseteq dig_{\sigma_2}^+(S) \text{ and } dig_{\sigma_2}^-(S) \subseteq dig_{\sigma_1}^-(S)$$

2.4 Property Estimation

In Section 2.3.3, we briefly discussed the estimation of arc length from a grid-intersection digitization. In this section, we discuss area and perimeter estimation from a Gauss digitization and introduce the general concept of multigrid convergence of property estimates.

---

10. In this situation, two grid cells must be added to $dig_\sigma^+(\gamma)$, but only one to $R(\gamma)$. However, this technical difference could be removed by a simple modification of Π_σ: we require that $-1/2 \leq x_i < 1/2$ (i.e., Π_σ is a product of $n-1$ half-open segments).

2.4 Property Estimation

2.4.1 Content estimation

Any simple grid polygon P has a well-defined area and perimeter. The *area* $\mathcal{A}(P)$ is defined by the number of grid squares contained in P multiplied by h^{-2}, which is the area of a single grid square. The *perimeter* $\mathcal{P}(P)$ is defined by the number of grid edges that form the frontier of P multiplied by $1/h$, which is the length of a single grid edge. These measurements are invariant with respect to rotations and translations.

A grid polyhedron is *simple* iff it is topologically equivalent to a closed sphere. (For topologic equivalence, see Chapter 6.) The surface area $\mathcal{S}(\Pi)$ and the volume $\mathcal{V}(\Pi)$ of a simple grid polyhedron Π are defined by the number of 2-cells that form the frontier of Π multiplied by h^{-2} and the number of 3-cells contained in Π multiplied by h^{-3}, respectively.

The question arises whether these properties of grid polygons or grid polyhedra can serve as estimates of the corresponding properties of real objects that have the polygons or polyhedra as their digitizations. Such estimates can be evaluated using criteria such as absolute error or bias (for a fixed h) or convergence (as $h \to \infty$).

Let $S \subset \mathbb{R}^n$ ($n \geq 1$) be a closed bounded set that has measurable *content* $\mathcal{C}(S)$, which is the length $\mathcal{L}(S)$ for $n=1$, the area $\mathcal{A}(S)$ for $n=2$, and the volume $\mathcal{V}(S)$ for $n=3$. We consider a fixed grid constant 1 and magnifications of S by factors $h > 1$. (Our preferred model of a fixed set S and increases in the grid constant will be used in Section 2.4.2.) Let $\mathcal{N}(S) = \mathcal{C}(G(S))$ be the number of grid points in S; this is defined by its Gauss digitization in the n-dimensional orthogonal grid.

Suppose S_h depends on only one parameter $h > 0$ (e.g., a disk or a sphere depends on its radius h). Then the following is true as $h \to \infty$:

$$\mathcal{N}(S_h) = \mathcal{C}(S_1) \cdot h^n + \mathcal{O}(h^{n-1}) \tag{2.9}$$

Magnification of S by the factor h with respect to the origin $o \in \mathbb{R}^n$ transforms S into S_h. (Magnification was defined in Section 2.3.2 only for $n=3$, but its generalization to any $n \geq 2$ is straightforward.) For $n=2$, we have [616, 1023] the upper bound

$$|\mathcal{N}(S) - \mathcal{C}(S)| \leq 4(\mathcal{P}(S) + 1) \tag{2.10}$$

where $\mathcal{P}(S) \geq 1$ is the perimeter of S, assuming that S has a rectifiable frontier.[11] The following general upper bound (for $n \geq 2$) was claimed by R. Lipschitz (1832–1903) in [658]:

$$|\mathcal{N}(S) - \mathcal{C}(S)| \leq c \cdot Q \tag{2.11}$$

---

11. H. Davenport reviews [1023] as follows in *Mathematical Reviews*, **9**:335d: "The author gives a proof of the following theorem, stated and proved by V. Jarník in a recent letter to him. Let J be a closed rectifiable Jordan curve of length l, enclosing an area a. Then, provided $l \geq 1$, the number w of points with integral coordinates inside J satisfies $|w - a| < l$. The proof is elementary and depends on the following result: if a Jordan arc S joins two points on the boundary of the square $|x| < \frac{1}{2}$, $|y| < \frac{1}{2}$, dividing the square into two regions, and if Δ is the region which does not contain the origin, then the area of Δ is less than the length of S. This is proved by simple considerations in each of four possible cases."

where Q is the greatest $(n-1)$-dimensional content of any of the projections of S into the $(n-1)$-dimensional subspaces of $\mathbb{R}^n$ defined by $x_i = 0$ $(i = 1, \ldots, n)$ and c is a positive constant. Without discussing constraints on the set S, it was pointed out in [248] that this upper bound cannot be correct; in fact, it fails for a long thin cylinder around one of the coordinate axes, where $\mathcal{N}(S)$ can be arbitrarily large and $\mathcal{C}(S)$ and Q can be arbitrarily small.

Suppose the bounded closed set $S \subset \mathbb{R}^n$ is such that the following are true: (i) any line parallel to one of the coordinate axes intersects S at a set of points that (if it is not empty) consists of at most t intervals; and (ii) the same is true (with m in place of n) for any of the m-dimensional sets $(m = 1, \ldots, n-1)$ obtained by projecting S into a subspace defined by setting any $n-m$ coordinates to zero. Then we have the following:

Theorem 2.1 (H. Davenport, 1951) If a bounded closed set S satisfies (i) and (ii), then
$$|\mathcal{N}(S) - \mathcal{C}(S)| < \sum_{m=0}^{n-1} t^{n-m} Q_m$$
where Q_m is the sum of the m-dimensional volumes of the projections of S on the subspaces obtained by setting any $n-m$ coordinates to zero; t is the maximum number of intervals in (i) for all $m = 0, \ldots, n-1$; and $Q_0 = 1$ by convention.

Note that t does not change when S is magnified by a factor h (e.g., with respect to the origin).

The n-dimensional content of $S \subset \mathbb{R}^n$ is the zero-order moment of S (see Chapter 17). Error estimates for content and for digital moments are a general topic in number theory. For information about error estimates for digital moments when $n = 2$, see Theorems 17.2 and 17.3.

2.4.2 Convergent 2D area estimates

A (real) disk of unit diameter has $\mathcal{A}(D) = \pi/4$ and $\mathcal{P}(D) = \pi$. The area of a digitized disk converges toward the area of the real disk:
$$\lim_{h \to \infty} \mathcal{A}(G_h(D)) = \mathcal{A}(D) = \pi/4 \qquad (2.12)$$

On the other hand, the perimeter of the digitized disk is always equal to 4, because the total length of the isothetic edges on its frontier remains constant as h increases; thus it trivially converges, but not toward the correct value $\mathcal{P}(D) = \pi$.

Digital geometry should provide accurate estimates of quantitative properties of digitized sets if the grid resolution is sufficiently large. The study of the accuracy of such estimates is one of the main topics in this book.

2.4 Property Estimation

Estimation of the area of a planar set by the number of grid points contained in the set has an extensive history in number theory. C.F. Gauss and his colleague P. Dirichlet (1805–1859) at Göttingen University already knew that the number of grid points $(i,j) \in \mathbb{Z}^2$ inside $h \cdot S$ (the magnification of a given planar convex set S by factor h) estimates $\mathcal{A}(h \cdot S)$ within an asymptotic order of $\mathcal{O}(\mathcal{P}(h \cdot S))$. Note that, because S is convex, $\mathcal{P}(h \cdot S)$ is $\mathcal{O}(h)$.

Let S be a convex set contained in a unit square so that $\mathcal{P}(S) \leq 4 = \mathcal{O}(1)$, and consider increases in the grid resolution h instead of magnifications by an increasing factor h. $\mathcal{A}(G_h(S))$ is estimated by the number of grid points $(i/h, j/h)$ inside S times the scale factor h^{-2}; note that h^{-2} is the size of a 2-cell in $G_h(S)$. Then we can rewrite the historic result as follows:

Theorem 2.2 (C.F. Gauss and P. Dirichlet) *For any planar convex set S and any grid resolution $h > 0$, $|\mathcal{A}(G_h(S)) - \mathcal{A}(S)| = \mathcal{O}(h^{-1})$.*

This result can be extended to nonconvex planar sets that can be partitioned into finite numbers of convex sets. Theorem 2.2 implies that counting grid points inside such an S provides a convergent estimate of $\mathcal{A}(S)$ as the grid resolution h goes to infinity.

The situation that arises when $S = D$ is a disk has received the most attention in number theory. A *centered disk* is a disk with its center at a grid point. A lower bound on the estimation error in this case is given by the following:

Theorem 2.3 (G.M. Hardy, 1915) *Let D be a centered disk. Then, for grid resolution $h > 0$, we have $|\mathcal{A}(G_h(D)) - \mathcal{A}(D)| = \Omega(h^{-1.5})$.*

A very accurate upper bound on the error, proved by M.N. Huxley in 1993, applies not only to disks but also to planar convex sets with frontiers that are *3-smooth curves*[12]:

Theorem 2.4 *Let S be a planar convex 3-smooth set. Then, for grid resolution $h > 0$, we have $|\mathcal{A}(G_h(S)) - \mathcal{A}(S)| = \mathcal{O}(h^{-\frac{100}{73}} \cdot (\log h)^{\frac{315}{146}})$.*

As we have seen, estimates of perimeter require more advanced methods than simply taking the perimeters of isothetic polygons. Such methods will be discussed in Chapter 10.

---

12. A 3-smooth curve has continuous partial derivatives with respect to both coordinates up to the third order and has positive curvature everywhere; this definition follows [559].

2.4.3 Multigrid convergence

A general scheme for comparing measurements made on a digital picture subset with the true measurements for the preimage of the subset in Euclidean space has been formalized by J. Serra [969]. (Limit properties with respect to digitizations were also studied in [740] by U. Montanari.)

Let $\mathbb{F}$ be a family of sets S in $\mathbb{R}^n$, and let $dig_h(S)$ denote a digitization of S on a grid of resolution h. Assume that a property $\mathcal{Q}$ (e.g., area, perimeter) is defined for all $S \in \mathbb{F}$. The following definition specifies also a measure of the speed of convergence of a digital quantity toward the true quantity:

Definition 2.10 An estimator $E_\mathcal{Q}$ is *multigrid convergent* for $\mathbb{F}$ and for dig_h iff, for any $S \in \mathbb{F}$, there is a grid resolution $h_S > 0$ such that the estimated value $E_\mathcal{Q}(dig_h(S))$ is defined for any grid resolution $h \geq h_S$, and

$$|E_\mathcal{Q}(dig_h(S)) - \mathcal{Q}(S)| \leq \kappa(h)$$

where κ is a function defined on the real numbers that takes on only positive real values and converges to zero as $h \to \infty$. The function κ specifies the *speed of convergence*.

In Theorems 2.2, 2.3, and 2.4, $\mathbb{F}$ is the set of planar convex 3-smooth sets, dig_h is the Gauss digitization G_h, and $\mathcal{Q}$ is the area $\mathcal{A}$. The estimate $E_\mathcal{A}$ is obtained by counting the number of grid points scaled by the area of the 2-cells. Theorems 2.2 and 2.4 demonstrate progress in obtaining upper bounds on the convergence speed, and Theorem 2.3 provides a lower bound. The actual convergence speed is between these bounds and is still unknown.

There are two ways to study convergence with respect to increased grid resolution: either consider sets $h \cdot S$ digitized on a grid with unit grid constant, or consider sets S digitized on grids with grid constant $1/h$. In both cases, $h \to \infty$ corresponds to an increase in grid resolution; we consider either a repeatedly magnified $h \cdot S$ in the grid with unit grid constant or a fixed S in a repeatedly refined grid. The repeatedly refined grid approach is commonly used in numeric mathematics and is also the preferred approach in this book. This is motivated by the assumed scenario in which the set to be analyzed remains physically the same while improvements in hardware (e.g., scanners, computing power) allow refinements in grid resolution. However, in experiments on multigrid convergence, it is often more appropriate to use repeatedly magnified sets digitized into pictures of increasing size.

2.5 Exercises

1. A spider sitting on the surface of a cubic room wants to crawl over all six faces of the cube and return to its starting point as quickly as possible. Does the starting point affect the length of the shortest path?

2.5 Exercises

2. Prove that there is no equilateral triangle that has all three of its vertices in $\mathbb{Z}^2$ (see also Exercise 15 in Section 1.3) but that there are equilateral grid triangles in $\mathbb{Z}^3$.

3. Use the 12 free 5-ominoes (see Figure 2.20) to cover a 6×10 rectangle without overlap. (Hint: All congruence transformations are allowed.)

4. The following figure shows a convex polygon S and its digitization $G_h(S)$. In this example, $G_h(S)$ consists of two rectangles. Give conditions on a bounded convex polygon S and on the grid resolution $h > 0$ that imply that $G_h(S)$ consists of exactly one simple isothetic polygon. (For example, this is true if S is an isothetic rectangle.)

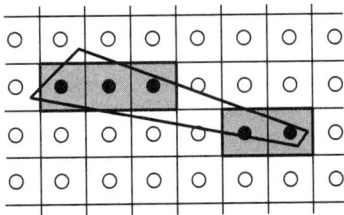

5. Construct Gauss digitizations of a square (of constant size) in different orientations using grid resolution 6. Draw a diagram showing the relative deviations of the area and perimeter of the digitized square as a function of its orientation.

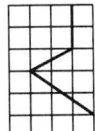 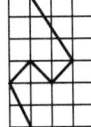 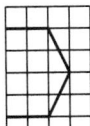 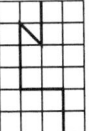 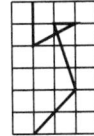 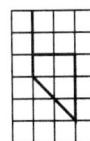

6. Consider a polygonal arc with vertices that are grid points and with endpoints that are on the border of $\mathbb{G}_{m,n}$ so that the arc cuts $\mathbb{G}_{m,n}$ into two parts. Write a program to determine whether the parts can be moved apart (as in the three cases on the left in the figure above) or not (as in the three cases on the right). (Hint: If so, they can be moved apart by sliding them in the direction of one of the arc segments.)

7. When we use the coding scheme of Figure 2.31, adding 2 modulo 8 to all of the directional codes results in a $90°$ rotation of the curve. Give an example of a digital curve for which adding 1 modulo 8 to all of the codes results in a digital arc that is not a closed digital curve. (It follows that the addition of 1 modulo 8 does not result in a $45°$ rotation.)

8. Define a code for the frontier of a free polyomino consisting of the numbers of 2-cells incident with the vertices of the polyomino when the vertices are visited by scanning the frontier of the polyomino in clockwise order. For example,

113131121312 encodes the free 5-omino shown on the right in Figure 2.21 when the scan starts at the upper left vertex. Any cyclic permutation of this code represents the same free polyomino with respect to a different starting vertex. Define the *vertex chain code* (VCC) of the free polyomino as the smallest integer that can be generated in this way. (In our example, it is 112131211313.) Prove that, for any free polyomino, the sum of the codes in a VCC of length n is $2n-4$.

9. Let n be the length of the VCC (see Exercise 2.8) of a given free m-omino M, and let c be the number of 1-adjacent pairs of 2-cells in M. (For example, $c=4$ for the free 5-omino shown on the right in Figure 2.21.) Prove that $2c+n=4m$.

10. Let γ be the straight line $y = x + \varepsilon$. How does the value of ε affect the digitizations $J^+(\gamma)$, $R(\gamma)$, and $\rho(\gamma)$?

11. Show that the 2D cross digitization of any planar convex set is either empty or a polygonally connected set (see Exercise 4 in Section 1.3).

12. How large should h be to achieve the same accuracy of $\pi = 3.14$ that Archimedes obtained if the perimeter of the convex hull of $G_h(D)$ is used for the perimeter estimation of D where D is the unit disk? (Hint: The answer can be found by implementing Graham Scan [see Section 1.2.9] using the leftmost and rightmost vertices in all the rows of the grid that intersect D.)

13. Implement Bresenham's algorithm (see Algorithm 2.3) and measure the length of the digital segment by initializing a variable $\mathcal{L}_{est}$ as 0 and adding 1 for every isothetic step and $\sqrt{2}$ for every diagonal step. Calculate the relative deviation between $\mathcal{L}_{est}$ and $d_e(p,q)$ (the length of the real segment pq) for different slopes of pq. Which slopes produce maximum deviations?

14. The circle $(x-x_c)^2 + (y-y_c)^2 = r^2$ can be digitized by using the formula $y = y_c \pm (r^2 - (x-x_c)^2)^{1/2}$, which defines (in general) two y-values for every integer x in $[x_c - r, x_c + r]$. Alternatively, each grid line that intersects the circle is perpendicular to a diameter of the circle that divides the grid line into two rays. Usually there exists a grid point on each of these rays that minimizes the residual $|(x^2 - x_c)^2 + (y - y_c)^2 - r^2|$; these minimum-residual grid points also define a digitization of the circle. Characterize positions of (x_c, y_c) and values of r for which the minimum is not unique.

15. Prove that the residual $|x^2 + y^2 - r^2|$ is never the same for adjacent grid points on a grid line if r^2 is an integer. It follows that the minimum is unique whenever $(x_c, y_c) \in \mathbb{Z}^2$ and $r^2 \in \mathbb{Z}$.

16. The digital circle produced by the minimum-residual method has a *sharp corner* iff three of its consecutive grid points form a right angle; this can happen only

at octant frontiers. The figure below illustrates a sharp corner when $r = 4$ and shows that no sharp corner exists when $r = 9$. Prove that sharp corners occur iff either there exists an integer $x \geq 0$ such that $r^2 = x^2 + x + 1$ or there exists an integer $y > 0$ such that $r^2 = 2y^2 - y + 1$.

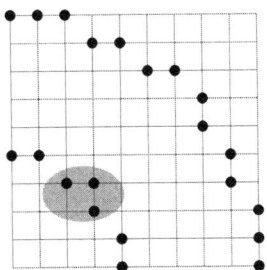

17. Let $(x_c, y_c) = (0,0)$ and $r \in \mathbb{Z}$, and suppose (x_i, y_i) ($i \geq 1$) is a grid point on the digital circle generated using the minimum-residual method, where $(x_1, y_1) = (0, r)$ and $x_i < r$. Suppose the (upper) ray at $x = x_i$ intersects the circle at (x_i, y), where $y \in \mathbb{R}$. Let $d_{1,i} = y_i^2 - y^2$, $d_{2,i} = y^2 - (y_i - 1)^2$, and $e_i = d_{1,i} - d_{2,i}$. Prove the following:

$$e_{i+1} = e_i + 4(x_i + 1) + 2 \quad \text{for } e_i < 0$$
$$e_{i+1} = e_i + 4(x_i + 1) + 2 - 4(y_i - 1) \quad \text{for } e_i \geq 0$$

Show how these error calculations can be used to generate a digital disk using the given formulas in the second octant and using reflection in the remaining octants.

18. Let S be a measurable set in $\mathbb{R}^2$ with a frontier that is a rectifiable simple closed curve. Let $\mathcal{A}(S)$ be the area of S and $\mathcal{P}(S)$ its perimeter. We recall (see Section 1.2.11) that $\mathcal{P}^2(S)/\mathcal{A}(S) \leq 4\pi$ and $= 4\pi$ iff S is a disk. Let M be a finite set of grid points; define $\mathcal{A}(M)$ as the number of grid points in M and $\mathcal{P}(M)$ as the number of border grid points in M (i.e., grid points of M that are adjacent to $\overline{M}$). Show that $\mathcal{P}^2(M)/\mathcal{A}(M)$ does not take on its minimum value when M is the set of grid points of a digitized disk.

2.6 Commented Bibliography

Both 2D and 3D regular orthogonal grids have a long history (see Figure 2.36). Adjacency grids have been basic models in digital geometry since its beginnings (see [881, 886] for the crisp case and [895] for the fuzzy case). Switch adjacency was defined in [545]. Reference [721] also defines an adjacency relation that is intermediate between 4- and 8-adjacency.

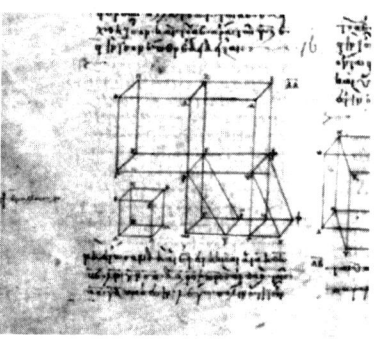

FIGURE 2.36 A figure from Euclid's "Elements," Book XI, Propositions 31 through 33 about the volumes of parallelepipeds (today we might call them "face-adjacent 3-cells").

For incidence grids (also called the *cuberille model*) see, for example, [432]. Interestingly, 2D (regular orthogonal grid) incidence was briefly introduced in an exercise in [13].

The theory of *polyominoes* (see [368, 563]) goes back about 100 years. For a brief review of this theory, see Chapter 12 (by D.A. Klarner) in [371]. See [59, 262] for reconstructions of convex polyominoes from orthogonal projections (which are further discussed in Section 14.7), and see [61] for permutations and generation of polyominoes.

[873] and [976] are reviews of component labeling and region adjacency graph construction for 2D and 3D adjacency grids. The Rosenfeld-Pfaltz algorithm was originally formulated for binary pictures only [921]; it is compared in [674] with two other techniques. Linear-time component labeling algorithms that can handle a variety of image representations are described in [269]; see also [1049]. For component labeling in 3D, see [673, 1053]. The recursive fill algorithm is standard in the picture analysis and computer graphics literature; see, for example, [802]. Component labeling for pictures of very large size can be based on the reuse of labels [506]. For other references about component labeling see [161, 260, 321, 663, 981].

Parallel algorithms for component labeling are discussed in [645, 769, 898]. Connected components can also be determined by identifying borders and computing their chain codes; see, for example, [169, 230, 602, 1034] for early work. (Border tracing is discussed in Section 4.3.4.)

The Gauss digitization model is common in number theory [598]. The Jordan digitization model [484] has been used in connection with volume estimation [732, 960]. Outer Jordan digitization is called super-cover digitization in [209].

Grid-intersection digitization is used in both computer graphics [121, 122] and picture analysis [342, 343, 804, 883]. Bresenham's algorithm [122] was published in 1965 and is one of the most frequently applied algorithms worldwide. Two versions of grid-intersection digitization are discussed in [951]. For a fast algorithm for

2.6 Commented Bibliography

grid-intersection digitization of analytically defined curves, see [1032]. The accuracy and time complexity of two line-digitization algorithms are discussed in [1084].

The directional encoding scheme illustrated in Figure 2.31 was proposed by H. Freeman in [342]. *Differential chain codes* (differences between pairs of successive codes) were proposed in [805]. [859] discusses measures for comparing chain codes.

The upper bound in Equation 2.6 is studied in [1167]. For determining whether a given finite set of grid points is a digital disk or a digital square, see [512] and [765], respectively. For characterizing digitizations of simple shapes such as disks, squares, and triangles, see also [22, 277, 763, 764, 813, 896, 1132].

Digitization schemes based on intersections with domains (e.g., a closed convex set; a square or cube, in Jordan digitization) are studied in [21, 480, 540, 739, 998, 1000]. Domain digitization is also called "digitization by dilation and by erosion" [875]. For digitization by dilation, see [416, 417, 419]; here one usually makes the covering assumption, namely that the union of all $\Pi_\sigma(p)$, for all $p \in \mathbb{Z}^2$, covers the Euclidean space. For 2D digitizations based on percentages (ratios of areas within a 2-cell), see [398]. For digitizations of line drawings, see [345, 583]; the special case of convex curves is studied in [1139]. [131] studies digitization schemes, for example, for straight lines and planes from topologic points of view. For Hausdorff digitization, which uses the Hausdorff distance (see Chapter 3), see [875, 876, 1043, 1044]. Digitization is also discussed in [361] and [624] in the context of topology and shape preservation.

Theorem 2.3 is from [412], and Theorem 2.4 from [459]. Definition 2.10 follows [498, 560]. Limit properties of digitization schemes are discussed in [740, 969]. The estimation of radii of digital disks is studied in [1132], of areas of digital disks and rectangles in [878], and of parameters of digital conics in [176]. [556] gives a set of axioms for *area measures*. For areas and perimeters of digitized objects, see [607].

Exercise 6 is from the 2002 international programming contest of the Association for Computing Machinery (ACM). Exercise 7 is from [1024]. For the vertex chain code and Exercises 8 and 9, see [126]. A chain code for 6-curves (in 3D space) based on a notation for turns is discussed in [127]. Methods of generating digital circles (see Exercises 14 and 16) are discussed in [712]. The result of Exercise 15 is from [606]; see also [178, 276, 813] for digitizations of disks. Exercise 17 is a brief sketch of the basic steps of *Bresenham's digital circle algorithm* [123]. The minimum-residual method coincides with grid-intersection digitization for circles that have integer radii and are centered at grid points; see [712]. For the digital isoperimetric inequality (Exercise 18), see [884].

CHAPTER 3

Metrics

This chapter discusses metric spaces $[S, d]$, in which S is a subset of $\mathbb{R}^n$ or a subset of a grid; it also addresses metric spaces, in which S is a family of such subsets. It also discusses metrics on pictures in which distances depend on the pixel values.

3.1 Basics About Metrics

Measurement requires a metric space. In this section, we summarize facts about metric spaces that are relevant to digital geometry. The definition of a metric (distance function) based on properties **M1** through **M3** was given in Section 1.2.1.

3.1.1 The Euclidean metric

We first consider the metric that is used in Euclidean geometry: the *distance* between two points is equal to the length of the straight line segment defined by the two points. This metric will allow us to define arc lengths, angles, and areas. Digital geometry is often concerned with estimates of such quantities.

We assume a Cartesian coordinate system on $\mathbb{R}^n$. (We treat the n-dimensional case, because this allows us to discuss the 2D and 3D cases at the same time.) Let $p = (x_1, x_2, \ldots, x_n)$ and $q = (y_1, y_2, \ldots, y_n) \in \mathbb{R}^n$ $(n \geq 1)$; then the following is true:

$$d_e(p, q) = \sqrt{(x_1 - y_1)^2 + \ldots + (x_n - y_n)^2}$$

The function d_e is the *Euclidean metric*, and $\mathbb{E}^n = [\mathbb{R}^n, d_e]$ is the *n-dimensional Euclidean space*.

It is easily seen that d_e satisfies **M1** and **M2**; we now prove that it satisfies **M3**. Let $r = (z_1, z_2, \ldots, z_n)$ be a third point. For $i = 1, \ldots, n$, let $a_i = z_i - x_i$ and $b_i = y_i - z_i$ so that $a_i + b_i = y_i - x_i$. From the *Minkowski inequality*

$$\sqrt{\sum_{i=1}^{n} a_i^2} + \sqrt{\sum_{i=1}^{n} b_i^2} \geq \sqrt{\sum_{i=1}^{n} (a_i + b_i)^2} \qquad (3.1)$$

it follows that $d_e(p,q) \leq d_e(p,r) + d_e(r,q)$. The Minkowski inequality follows from the *Schwarz inequality*

$$\sqrt{\sum_{i=1}^{n} a_i^2} \sqrt{\sum_{i=1}^{n} b_i^2} \geq \left| \sum_{i=1}^{n} a_i b_i \right| \qquad (3.2)$$

where the a_is and b_is are reals and $n \geq 1$. A proof of Equation 3.2 is as follows:

$$0 \leq f(t) = \sum_{i=1}^{n} (a_i + t \cdot y_i)^2 = t^2 \cdot \left(\sum_{i=1}^{n} b_i^2 \right) + 2 \cdot t \cdot \left(\sum_{i=1}^{n} a_i \cdot b_i \right) + \sum_{i=1}^{n} a_i^2$$

Because this inequality holds for any t, it follows that the discriminant of $f(t)$ is not strictly positive. This means that the following is true and is equivalent to Equation 3.2:

$$0 \geq \left(2 \cdot \left(\sum_{i=1}^{n} a_i \cdot b_i \right) \right)^2 - 4 \cdot \left(\sum_{i=1}^{n} a_i^2 \right) \cdot \left(\sum_{i=1}^{n} b_i^2 \right)$$

3.1.2 Norms and Minkowski metrics

Euclidean spaces are often introduced as normed spaces rather than metric spaces. A norm always defines a metric, and a metric defines (at least) a seminorm. Norms can also be related to the metrics studied in digital geometry.

Let $[S, +, \cdot, \mathbb{R}]$ be an n-dimensional vector space over $\mathbb{R}$ (see Section 1.2.4). A *norm* $\|\cdot\|$ assigns to any $p \in S$ a nonnegative real number $\|p\|$ that satisfies the following properties for all $p, q \in S$ and all $a \in \mathbb{R}$:

N1: $\|p\| = 0$ iff $p = (0, \ldots, 0)$ (identity).

N2: $\|a \cdot p\| = |a| \cdot \|p\|$ (homogeneity).

N3: $\|p + q\| \leq \|p\| + \|q\|$ (the triangle inequality: triangularity).

For example, let $S = \mathbb{R}^n$, $p = (x_1, \ldots, x_n) \in \mathbb{R}^n$, and let the following be true[1]:

$$\|p\|_m = \sqrt[m]{|x_1|^m + \ldots + |x_n|^m} \quad \text{for } m = 1, 2, \ldots$$
$$\|p\|_\infty = \max\{|x_1|, \ldots, |x_n|\}$$

These functions have properties **N1** through **N3** on the vector space $[\mathbb{R}^n, +, \cdot, \mathbb{R}]$.

Let $\|\cdot\|$ be a norm on $[S, +, \cdot, \mathbb{R}]$. It can be easily verified that

$$d(p, q) = \|p - q\| \quad (p, q \in S) \qquad (3.3)$$

defines a metric on S. Evidently, the norm $\|p\|_2$ defines the Euclidean metric d_e on $[\mathbb{R}^n, +, \cdot, \mathbb{R}]$.

---

1. The norm $\|\cdot\|_m$ is defined not only for integer m but for any real number $m \geq 1$.

3.1 Basics About Metrics

A metric defined by a norm using Equation 3.3 also has the following properties:

M4: $d(p+r, q+r) = d(p,q)$ for $p, q, r \in S$ (translation-invariance).

M5: $d(a \cdot p, a \cdot q) = |a| \cdot d(p,q)$ for $p, q \in S$ and $a \in \mathbb{R}$ (homogeneity).

The norms $\|p\|_m$ ($m \geq 1$ or $m = \infty$) define the *Minkowski metrics* L_m on $\mathbb{R}^n$; $\|\cdot\|_m$ is therefore called a *Minkowski norm*. Note that $L_2 = d_e$. Evidently we have

$$L_m(p,q) = \sqrt[m]{|x_1 - y_1|^m + \ldots + |x_n - y_n|^m} \quad (m = 1, 2, \ldots)$$
$$L_\infty(p,q) = \max\{|x_1 - y_1|, \ldots, |x_n - y_n|\}$$

where $p = (x_1, x_2, \ldots, x_n)$ and $q = (y_1, y_2, \ldots, y_n)$. All the Minkowski metrics L_m have properties **M1** through **M5**. The following can also be shown:

$$L_{m_1}(p,q) \leq L_{m_2}(p,q) \quad \text{for all } 1 \leq m_2 \leq m_1 \leq \infty \text{ and all } p, q \in \mathbb{R}^n \quad (3.4)$$

It is easily verified that two 2D grid points p_1 and p_2 are 4-adjacent iff $L_1(p_1, p_2) = 1$ and 8-adjacent iff $L_\infty(p_1, p_2) = 1$. Similarly, two 3D grid points p_1 and p_2 are 6-adjacent iff $L_1(p_1, p_2) = 1$ and 26-adjacent iff $L_\infty(p_1, p_2) = 1$.

Let $[S, d]$ be a metric space, and assume that $[S, +, \cdot, \mathbb{R}]$ is an n-dimensional vector space with additive identity o. Let the following be true:

$$\|p\| = d(p, o) \quad \text{for all } p \in S \quad (3.5)$$

If d also satisfies **M4** and **M5** on $[S, +, \cdot, \mathbb{R}]$, Equation 3.5 defines a norm on S. If d does not satisfy **M5**, the function $\|\cdot\|$ derived from d by Equation 3.5 need not be a norm, but it is a *seminorm*, which has properties **N1**, **N3**, and

N2*: $\|a \cdot p\| \leq |a| \cdot \|p\|$ (upper boundedness).

3.1.3 Scalar products and angles

A norm often allows us to define a scalar product, which in turn allows us to define angular values. A metric allows us to define a seminorm, a weak scalar product, and angular values.

A *scalar product* $\langle p, q \rangle$ is a symmetric, positive definite, such that

$$\langle p, p \rangle > 0 \quad \text{for all } o \neq p \in S$$

and linear, such that

$$\langle ap + bq, r \rangle = a \langle p, r \rangle + b \langle q, r \rangle$$

mapping of S^2 into $\mathbb{R}$. Let $\|\cdot\|$ be a seminorm or norm on an n-dimensional vector space $[S, +, \cdot, \mathbb{R}]$, and let the following be true:

$$\langle p, q \rangle = \tfrac{1}{4}\left(\|p+q\|^2 - \|p-q\|^2\right) \quad \text{for all } p, q \in S \quad (3.6)$$

A seminorm or norm always defines a *weak scalar product* in this way. It is positive definite and symmetric, and it is a scalar product iff it is linear.

For example, the norm $\|\cdot\|_1$ does not define a scalar product $\langle \cdot, \cdot \rangle_1$; this weak scalar product is not linear. There are grid points p_1 and q_1 such that $\langle 2p_1, q_1 \rangle_1 < 2 \langle p_1, q_1 \rangle_1$, and there are grid points p_2 and q_2 such that $\langle 2p_2, q_2 \rangle_1 > 2 \langle p_2, q_2 \rangle_1$. In general, the linearity of a scalar product implies the following:

$$\|p+q\|^2 = \|p\|^2 + \|p\|^2 + 2\langle p, q \rangle$$

and

$$\|p+q\|^2 + \|p-q\|^2 = 2\|p\|^2 + 2\|q\|^2$$

These are not true for all norms.

Vectors $p, q \in S$ are called *orthogonal* (notation: $p \perp q$) with respect to a weak scalar product $\langle \cdot, \cdot \rangle$ iff $\langle p, q \rangle = 0$. For example, for the Euclidean space $[\mathbb{R}^2, d_e]$, the norm $\|p\|_e$, and the scalar product $\langle p, q \rangle_e$, we have

$$\langle p, q \rangle_e = x_1 x_2 + y_1 y_2 = \cos \eta \cdot \|p\|_e \cdot \|q\|_e$$

where $p = (x_1, y_1)$, $q = (x_2, y_2)$, $p \neq o, q \neq o$, and η is the (smaller) angle between the vectors $p = \vec{op}$ and $q = \vec{oq}$. It follows that $p \perp q$ iff $\cos \eta = 0$ iff $\eta = 90°$.

We say that a weak scalar product *satisfies the generalized Schwarz inequality on S* iff the following is true:

$$|\langle p, q \rangle| \leq \|p\| \cdot \|q\| \text{ for all } p, q \in S$$

For example, the weak scalar products defined by the metrics d_4 and d_8 on $\mathbb{R}^2$ satisfy the generalized Schwarz inequality on $\mathbb{R}^2$.

Suppose the weak scalar product $\langle \cdot, \cdot \rangle$ defined by a metric d satisfies the generalized Schwarz inequality on S. Following [540], we can define an *angular value*

$$H(p, q, r) = \frac{\langle p - q, r - q \rangle}{d(p, q) \cdot d(r, q)} \text{ for all } p, q, r \in S \qquad (3.7)$$

where $p \neq q, q \neq r$; see Figure 3.1. With the generalized Schwarz inequality, we have the following:

Proposition 3.1 $H(p, q, r)$ is always in the interval $[-1, 1]$.

In the Euclidean space $[\mathbb{R}^2, d_e]$, we have $H(p, q, r) = \cos \eta$, where η is the (smaller) angle between the vectors $p - q = \vec{qp}$ and $r - q = \vec{qr}$.

3.1 Basics About Metrics

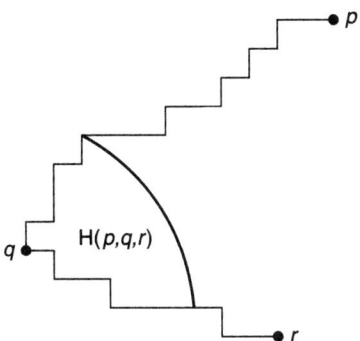

FIGURE 3.1 Illustration of angular value $\mathrm{H}(p,q,r)$. $\vec{qp}$ and $\vec{qr}$ illustrate ways of measuring the (shortest) distances $d(p,q)$ and $d(r,q)$; the sketch in the figure resembles a path in metric d_4.

3.1.4 Integer-Valued metrics

The Minkowski metrics can obviously have noninteger values, even for pairs of grid points; for example, in Figure 3.2, we have $d_e(p,r) = \sqrt{5}$. The measurements used in digital geometry are often based on integer-valued metrics.

We recall the definitions of the floor, ceiling, and nearest integer functions for all real a:

$\lfloor a \rfloor$, the largest integer less than or equal to a
$\lceil a \rceil$, the smallest integer greater than or equal to a
$\lceil a \rfloor$, the nearest integer to a if it is unique,
and $\lfloor a \rfloor$ otherwise

For any function $d : S \times S \mapsto \mathbb{R}$, we can define $\lfloor d \rfloor$ by $\lfloor d \rfloor(p,q) = \lfloor d(p,q) \rfloor$ and similarly for $\lceil d \rceil$ and $\lceil d \rfloor$. However, even if d is a metric, these integer-valued functions may not be metrics. For example, we have the following:

Proposition 3.2 $\lfloor d_e \rfloor$ and $\lceil d_e \rfloor$ are not metrics on $\mathbb{Z}^2$.

Proof Let $p = (2,3)$, $q = (-1,-1)$, and $r = (0,0)$. Then $\lfloor d_e \rfloor(p,q) = 5$, but $\lfloor d_e \rfloor(p,r) = 3$ and $\lfloor d_e \rfloor(r,q) = 1$, so property **M3** is not satisfied. For $\lceil d_e \rfloor$, use, for example, $p = (1,1)$ and q and r as before. ∎

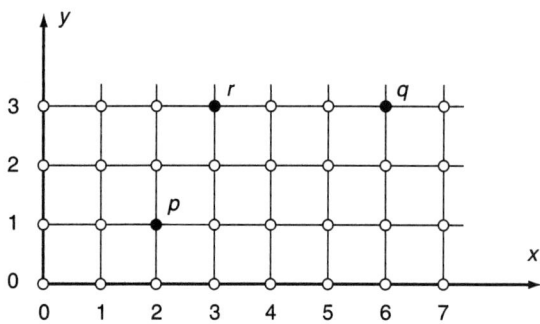

FIGURE 3.2 $p = (2,1)$, $q = (6,3)$, and $r = (3,3)$.

It follows that $\lfloor d_e \rfloor$ and $[d_e]$ are not metrics on $\mathbb{R}^n$ or $\mathbb{Z}^n$ for $n \geq 2$. Interestingly, $\lceil d_e \rceil$ is a metric. In fact, we have the following:

Theorem 3.1 If d is a metric, $\lceil d \rceil$ is also a metric.

Proof Let $p, q, r \in S$, $a = d(p,q)$, $b = d(q,r)$, and $c = d(p,r)$. We show that $\lceil d \rceil$ has properties **M1** through **M3**.

M1: For $a \geq 0$, we have $\lceil a \rceil = 0$ iff $a = 0$, such that $\lceil d(p,q) \rceil = 0$ iff $d(p,q) = 0$ iff $p = q$. Furthermore, $a \geq 0$ implies $\lceil a \rceil \geq 0$.

M2: Because $d(p,q) = d(q,p)$, we also have $\lceil d(p,q) \rceil = \lceil d(q,p) \rceil$.

M3: We have $a + b \geq c$, because d is a metric on S. First assume that a or b is an integer, for example, $a = n$; then $\lceil a+b \rceil = \lceil n+b \rceil = n + \lceil b \rceil = \lceil a \rceil + \lceil b \rceil$. Now assume that both a and b are not integers, so that $\lceil a \rceil = \lfloor a \rfloor + 1$ and $\lceil b \rceil = \lfloor b \rfloor + 1$. Because $\lceil a+b \rceil \leq \lfloor a+b \rfloor + 1$, it follows that $\lceil a \rceil + \lceil b \rceil = \lfloor a \rfloor + \lfloor b \rfloor + 2 \geq \lfloor a+b \rfloor + 1 \geq \lceil a+b \rceil$. ∎

In the example in Figure 3.2, we have $\lceil d_e \rceil(p,r) = 3$, $\lceil d_e \rceil(p,q) = 5$, and $\lceil d_e \rceil(r,q) = 3$.

Definition 3.1 An integer-valued metric d on a set S is called *regular* iff, for all $p, q \in S$ such that $d(p,q) \geq 2$, there always exists an $r \in S$ ($r \neq p$ and $r \neq q$) such that $d(p,q) = d(p,r) + d(r,q)$.

It is not hard to show that d is regular iff, for all distinct $p, q \in S$, there exists an $r \in S$ such that $d(p,r) = 1$ and $d(p,q) = 1 + d(r,q)$. $\lceil d_e \rceil$ is a regular integer-valued metric on $\mathbb{R}^2$ but not on $\mathbb{Z}^2$. For example, let p and q be grid points that differ by 4 in one coordinate and by 3 in another coordinate. The distance (both d_e and $\lceil d_e \rceil$) between

3.1 Basics About Metrics

p and q is 5, but there is no $r \in \mathbb{Z}^2$ at $\lceil d_e \rceil$ distance 1 from p and 4 from q; such a real point would have to lie on the segment pq, but it cannot then have integer coordinates.

Integer-valued metrics are of special interest in picture analysis. It can be shown that a finite metric space is isomorphic to the distance space on a graph (see Chapter 4) iff the metric is regular; this implies that d_4 and d_8 are regular. Integer-valued metrics will be discussed further in Sections 3.2 and 3.4.

3.1.5 Restricting and combining metrics

From Exercise 8 in Section 1.3, we know that $[\mathbb{Z}^n, +, \cdot, \mathbb{R}]$ is not a vector space, because it is not closed under scalar multiplication $a \cdot p$ (property **V0**). The algebraic structure $[\mathbb{Z}^n, +, \cdot, \mathbb{Z}]$ is an example of a *unitary module*: it satisfies properties **V0** trough **V8** with respect to a ring of scalars that has an additive identity.

> **Proposition 3.3** If $[S, d]$ is a metric space and A is a nonempty subset of S, then $[A, d]$ is also a metric space. If d is not a metric on A, d is also not a metric on any set S containing A.
>
> *Proof* If **M1** through **M3** hold for d on S, they also hold for d on any subset of S. The definition of a metric space requires that A be nonempty. ∎

In particular, metrics on $\mathbb{R}^n$ define metrics on $\mathbb{Z}^n$, because $\mathbb{Z}^n$ is a subset of $\mathbb{R}^n$. For example, the Minkowski metrics define metric spaces on the sets $\mathbb{Z}^2$ and $\mathbb{Z}^3$ of all 2D or 3D grid points, and they define metric spaces on rectangular grids $\mathbb{G}_{m,n}$ for all m, n (see Equation 1.1) or cuboidal grids $\mathbb{G}_{l,m,n}$ for all l, m, n, because these grids (using the grid point model) are finite subsets of $\mathbb{Z}^2$ or $\mathbb{Z}^3$.

There are ways of combining two metrics d', d'' on a set S so that the resulting function d is a metric on S. For example:

(i) A linear combination of two metrics, $d(p,q) = a \cdot d'(p,q) + b \cdot d''(p,q)$, where $0 < a, b \in \mathbb{R}$ is a metric.
(ii) The maximum of two metrics, $d(p,q) = \max\{d'(p,q), d''(p,q)\}$, is a metric.

On the other hand, the product or minimum of two metrics is not necessarily a metric.

3.1.6 Boundedness

The Minkowski metrics on $\mathbb{R}^n$ are examples of *unbounded metric spaces* $[S, d]$, where $S = \mathbb{R}^n$ is an infinite set, and the distances between points in S can exceed any finite bound. Any metric space $[S,d]$ on a unitary module S satisfying the homogeneity property **M5** is necessarily unbounded. We now give some examples of bounded metric spaces.

We first give a degenerate example. Let S be a nonempty set, and we define that

$$d_b(p,q) = \begin{cases} 0 & \text{if } p = q \\ 1 & \text{otherwise} \end{cases}$$

It can easily be verified that $[S,d_b]$ is a metric space, and it is evidently bounded. The function d_b is called the *binary metric*. The norm $\|p\|_b = d_b(p,o)$, defined as in Equation 3.5, satisfies **N2*** but not **N2**.

If $[S,d]$ is an unbounded metric space,

$$d'(p,q) = \frac{d(p,q)}{1+d(p,q)}$$

defines a metric d' on S, and $[S,d']$ is a bounded metric space.

We now give a more detailed example using the mapping that takes $p = (x,y)$, $\in \mathbb{R}^2$ into $p^\star = (x^\star, y^\star)$, where the following are true:

$$x^\star = \frac{2x}{1+\sqrt{x^2+y^2+1}} \quad \text{and} \quad y^\star = \frac{2y}{1+\sqrt{x^2+y^2+1}} \tag{3.8}$$

For any p, $p^\star$ is contained in a disk of radius 2. Indeed, for $o = (0,0)$ we have the following:

$$d_e(p^\star, o) = 2\sqrt{\frac{x^2+y^2}{x^2+y^2+a}} = 2c < 2$$

(Note: c is defined in this equation.) This is true because $a = 2 + 2\sqrt{x^2+y^2+1} > 0$ and thus $c < 1$. Thus the mapping defined by Equation 3.8 is one-to-one from $\mathbb{R}^2$ onto the open disk of radius 2. Any (x,y) on the circle with center o and radius $4/3$ is mapped into $(3x/4, 3y/4)$, which is on the circle with center o and radius 1. Hence any point p farther than $4/3$ from o is mapped into a point $p^\star$ in the open annulus defined by the circles of radii 1 and 2. (The function $(\arctan(x), \arctan(y))$ is another example of a continuous one-to-one mapping from $\mathbb{R}^2$ into a bounded set—the open square $(-\pi/2, \pi/2)^2$ in this case.)

Bounded distances between points $p, q \in \mathbb{R}^2$ can now be defined using the distances, for any metric d on $\mathbb{R}^2$, between $p^\star$ and $q^\star$ in the disk of radius 2. In other words, for any metric space $[\mathbb{R}^2, d]$ we define the following:

$$d^\star(p,q) = d(p^\star, q^\star) \text{ for all } p, q \in \mathbb{R}^2$$

If d is a metric on $\mathbb{R}^2$, so is $d^\star$, and $d^\star(p,q) < 4$ for all $p, q \in \mathbb{R}^2$. For example, for the integer-valued metric $\lceil d_e^\star \rceil$ (see Theorem 3.1), all distances between points $p, q \in \mathbb{R}^2$ are integers in the set $\{0,1,2,3,4\}$. We sometimes have $\lceil d_e^\star \rceil(p,q) = \lceil d_e^\star \rceil(q,r) = \lceil d_e^\star \rceil(p,r)$; note that this does not contradict the triangularity property. We can change the cardinality of the range of d^* by increasing or decreasing the parameter 2 in Equation 3.8. Such metrics might be of interest for classifying pixels (pairs of integers) using finite numbers of distance values.

3.1.7 The topology induced by a metric

A metric induces a *topology* defined by a family of open or closed sets; this section briefly addresses such issues. For a more extensive discussion of topologic subjects, see Chapter 6.

For any metric space $[S,d]$, any $p \in S$, and any $\varepsilon > 0$, let the following be true:

$$U_\varepsilon(p) = \{q : d(p,q) < \varepsilon\}$$

$U_\varepsilon(p)$ is called the (open) *ε-neighborhood* of p in S; evidently $p \in U_\varepsilon(p)$. The family of all ε-neighborhoods defines a *basis of a topology* and allows us to generate open sets by taking (finite or infinite) unions of ε-neighborhoods. The Euclidean metric d_e on $\mathbb{R}^n$ defines the *Euclidean topology*. For the binary metric d_b, we have $U_\varepsilon(p) = \{p\}$ for $0 < \varepsilon \leq 1$ and $U_\varepsilon(p) = S$ for $\varepsilon > 1$.

Definition 3.2 *$p \in S$ is a frontier point of $A \subseteq S$ iff, for any $\varepsilon > 0$, $U_\varepsilon(p)$ contains points of A as well as points of $\bar{A} = S \setminus A$. The frontier ϑA of A consists of all frontier points of A.*

For example, the frontier of a disk is a circle. The *interior* A° of A is $A \setminus \vartheta A$, and the *closure* $A^\bullet$ of A is $A \cup \vartheta A$. A is *closed* iff $A = A^\bullet$ and *open* iff $A = A^\circ$. The empty set $\emptyset$ and the set S are both closed and open.

The interior of A is the largest open set contained in A, and the closure of A is the smallest closed set that contains A. A set is closed iff its complement is open.

$A \subseteq S$ is *bounded* iff $A \subseteq U_\varepsilon(p)$ for some $p \in S$ and some $\varepsilon > 0$. (A bounded set need not be of finite cardinality.) A is called *compact* iff, whenever it is contained in the union of a set of open sets, it is contained in a finite union of these sets. The Heine-Borel-Lebesgue theorem [112] says that a subset of $\mathbb{R}^n$ is compact in the topology defined by any Minkowski metric on $\mathbb{R}$ iff it is bounded and closed. A continuous real-valued function defined on a compact set always has a minimum and a maximum on that compact set.

Two metrics d and d' on S are called *topologically equivalent* iff a subset of S is open with respect to d iff it is open with respect to d'. For example, the Minkowski metrics on $\mathbb{R}^n$ are all topologically equivalent. A bounded metric on an infinite subset of $\mathbb{R}^n$ (see the examples in Section 3.1.6) can be topologically equivalent to an unbounded metric. For example, the bounded metric $d_e(p,q)/[1+d_e(p,q)]$ is topologically equivalent to the unbounded metric d_e; the ball of radius r in the unbounded metric corresponds to the ball of radius $r/(1+r)$ in the bounded metric, and the ball of radius $s < 1$ in the bounded metric corresponds to the ball of radius $s/(1-s)$ in the unbounded metric.

3.1.8 Distances between sets

Any metric d on a set S can be extended to a *Hausdorff metric* on the family of all nonempty compact subsets A, B of S by defining

$$d(A,B) = \max\left\{\sup_{p \in A} \inf_{q \in B} d(p,q), \sup_{p \in B} \inf_{q \in A} d(p,q)\right\} \quad (3.9)$$

where sup and inf denote the least upper bound and greatest lower bound, respectively. (For compact sets A and B we can replace inf and sup with min and max, respectively.) Figure 3.3 shows an example of this metric in which d is the Euclidean metric d_e.

The definition of the Hausdorff metric can be broken up into steps. We first define the *closest distance* from $p \in S$ to $T \subseteq S$ using the following:

$$d(p,T) = \inf_{q \in T} d(p,q) \quad (3.10)$$

Let $A, B \subset S$; let $h_p(B) = d(p,B)$ for all $p \in A$; let $h_p(A) = d(p,A)$ for all $p \in B$; and define the following:

$$h_A(B) = \sup_{p \in A} h_p(B); \quad h_B(A) = \sup_{p \in B} h_p(A)$$

Then here we have the Hausdorff metric of Equation 3.9:

$$d(A,B) = \max\{h_A(B), h_B(A)\}$$

In Figure 3.3, we have $h_A(B) = h_p(B) = \sqrt{34}$ and $h_B(A) = h_q(A) = \sqrt{26}$ so that $d(A,B) = h_A(B)$.

An alternative way of defining the Hausdorff metric makes use of the definition of an (open) ε-*neighborhood* of a set. We define that

$$U_\varepsilon(A) = \{q : q \in S \wedge h_q(A) < \varepsilon\} = A \oplus U_\varepsilon(o)$$

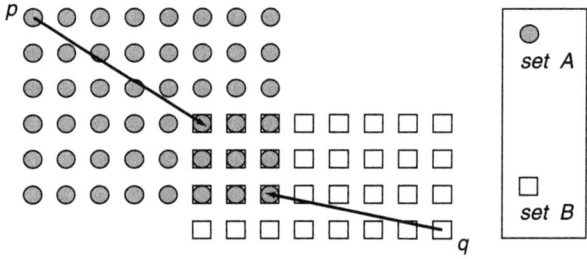

FIGURE 3.3 The Hausdorff distance between A and B.

3.1 Basics About Metrics

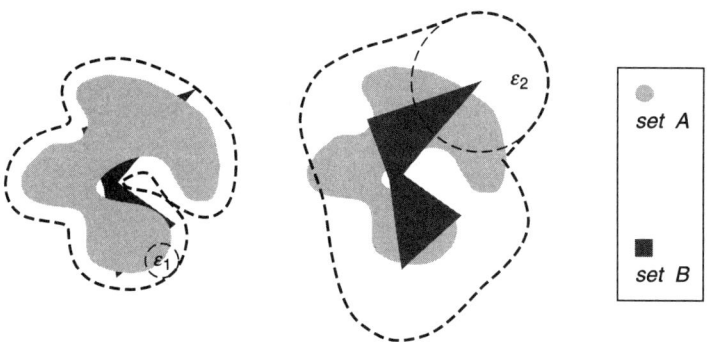

FIGURE 3.4 Left: B (a simple polygon) is completely contained in $U_{\varepsilon_1}(A)$. Right: A is not completely contained in $U_{\varepsilon_2}(B)$, showing that $d(A, B) > \varepsilon_2$.

Let h_q be defined by a metric d on S; $\varepsilon > 0$; $A \subset S$; and $\oplus$ be the Minkowski sum (see Section 1.2.12). Then, if A and B are nonempty compact subsets of S, we have the following:

$$h_A(B) = \inf\{\varepsilon : A \subseteq U_\varepsilon(B)\}$$

Figure 3.4 illustrates this method of defining a Hausdorff distance. $U_\varepsilon(B)$ is a *dilation* of B; dilations will be studied in Chapter 15. Note that, for $D = U_\varepsilon(o)$, we have $\underline{D} = D$ where $\underline{D}$ is the mirror set used in the definition of dilation; see Section 1.2.12. (If A and B are compact and we dilate by a closed ball of radius ε instead of an open ball, $h_A(B)$ is defined by min instead of inf.)

The Hausdorff distance is not a metric in the family of all nonempty bounded subsets of S. For example, consider the closed unit square $[0,1]^2 = [0,1] \times [0,1]$ and the open unit square $(0,1)^2 = (0,1) \times (0,1)$ in the Euclidean topology of the plane.[2] Then $d_e([0,1]^2, (0,1)^2) = 0$, but the sets are not identical, so property **M1** of a metric is violated.

A *generalized metric* satisfies the axioms of an ordinary metric but can also have value ∞. The Hausdorff distance is a generalized metric in the family of closed sets (bounded or not). We can also include the empty set; in the definition of Hausdorff distance, we replace an empty supremum with 0 and an empty infimum with ∞ so that the empty set is at distance 0 from itself but at distance ∞ from any nonempty set.

The Hausdorff metrics are based on maximum distances between sets; a single point (an "outlier") in a set can strongly influence these distances. Distances between sets defined by set-theoretic differences are less sensitive to single points. The *symmetric difference* between two subsets A, B of a set S is as follows:

$$A \Delta B = (A \setminus B) \cup (B \setminus A)$$

---

2. In the mathematics literature, a *unit circle* or *unit sphere* is traditionally defined as having radius 1 (i.e., diameter 2), whereas a *unit square* or *unit cube* is (inconsistently!) defined as having side 1.

Let the following be true:

$$d_{\text{sym}}(A,B) = \text{card}(A \Delta B) \text{ and } d'_{\text{sym}}(A,B) = \frac{\text{card}(A \Delta B)}{\text{card}(A \cup B) + 1}$$

Let S be any nonempty set, and let $\wp_{\text{fin}}(S)$ be the family of all finite subsets of S.

Proposition 3.4 d_{sym} and d'_{sym} are metrics on $\wp_{\text{fin}}(S)$.

Proof Let $A, B, C \in \wp_{\text{fin}}(S)$, $a = d_{\text{sym}}(A,B)$, $b = d_{\text{sym}}(B,C)$, and $c = d_{\text{sym}}(A,C)$. We show that d_{sym} has properties **M1** through **M3**.

M1: $a = 0$ iff $A \Delta B = \emptyset$ iff $A = B$.

M2: Because $A \Delta B = B \Delta A$, we have symmetry.

M3: Let $A = A_1 \cup D \cup E \cup F$, $B = B_1 \cup D \cup E \cup G$, and $C = C_1 \cup E \cup F \cup G$ (see Figure 3.5). Let $a_1 = \text{card}(A_1)$, $d = \text{card}(D)$, and so forth. It follows that $a = a_1 + b_1 + f + g$, $b = b_1 + c_1 + d + f$, and $c = a_1 + c_1 + d + g$. This shows that $a + b = a_1 + 2b_1 + c_1 + d + 2f + g \geq c$.

For d'_{sym}, **M1** and **M2** follow by arguments similar to those for d_{sym}. Regarding **M3**, let $a = d'_{\text{sym}}(A,B)$, $b = d'_{\text{sym}}(B,C)$, and $c = d'_{\text{sym}}(A,C)$, and consider the intersections of A, B, and C as before. Let $h = b_1 + f + g$, $k = d + e + 1$, and $i = a_1 + c_1 - b_1$. Then the following are true:

$$a = \frac{a_1 + h}{a_1 + h + k}, \quad b = \frac{c_1 + h}{c_1 + h + k}, \quad \text{and } c = \frac{i + h}{i + h + k}$$

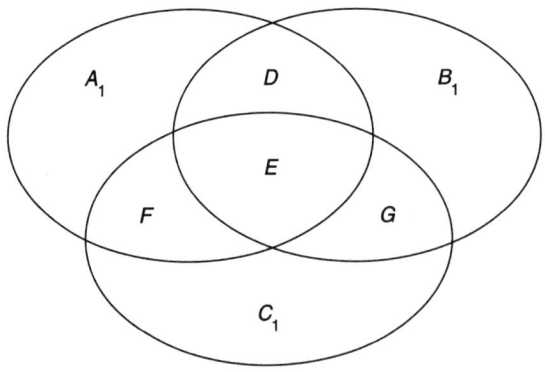

FIGURE 3.5 Intersections between three sets.

Let $a_2 = a_1 + h$, $b_2 = c_1 + h$, and $c_2 = i + h$. Because $h \geq 0$ and $0 \leq i \leq a_1 + c_1$, we have $0 \leq c_2 \leq a_2 + b_2$. Together with $k \geq 1$, this allows us to show that $a + b - c \geq 0$ so that $a + b \geq c$. ∎

d_{sym} is the L_∞ distance of the characteristic functions of sets; this provides a direct proof that it is a metric. For d'_{sym}, we could also derive Proposition 3.4 from the more general fact that, for any metric d, the function d' defined by $d'(p,q) = d(p,q)/[1+d(p,q)]$ is a topologically equivalent bounded metric; note that our proof covers the general case.

3.2 Grid Point Metrics

In this section, we discuss integer-valued metrics that are related to the grid point adjacency models defined in Sections 2.1.3 and 2.1.4. We also discuss methods of defining neighborhoods and closeness; grid point metrics that approximate the Euclidean metric; paths, geodesics, and intrinsic distances; and distances between sets of grid points.

3.2.1 Basic grid point metrics

Let $p, q \in \mathbb{R}^2$, $p = (x_1, y_1)$, $q = (x_2, y_2)$, and define the following:

$$d_4(p,q) = |x_1 - x_2| + |y_1 - y_2|$$

Then $[\mathbb{R}^2, d_4]$ is a metric space; in fact, d_4 is the Minkowski metric L_1. We call d_4 the *city-block metric* or *Manhattan metric* because, when we restrict it to $\mathbb{Z}^2$, $d_4(p,q)$ is the minimal number of isothetic unit-length steps from p to q; it resembles a shortest walk in a city with streets that are laid out in an orthogonal grid pattern. In the example in Figure 3.2, we have $d_4(p,r) = 3$, $d_4(p,q) = 6$, and $d_4(r,q) = 3$.

Let $p, q \in \mathbb{R}^2$, $p = (x_1, y_1)$, and $q = (x_2, y_2)$, and define the following:

$$d_8(p,q) = \max\{|x_1 - x_2|, |y_1 - y_2|\}$$

Then $[\mathbb{R}^2, d_8]$ is a metric space; in fact, d_8 is the Minkowski metric L_∞. Thus d_8 is called the *chessboard metric* because, when we restrict it to $\mathbb{Z}^2$, $d_8(p,q)$ is the minimal number of moves from p to q by a king on a chessboard. In the example in Figure 3.2, we have $d_8(p,r) = 2$, $d_8(p,q) = 4$, and $d_8(r,q) = 3$.

Let $S \subseteq \mathbb{R}^2$, let $o = (0,0) \in S$, and let d be a metric on S. The set $\{p : p \in S \wedge d(p,o) \leq 1\}$ is called a *unit disk* in $[S, d]$. Figure 3.6 shows the unit disks in $\mathbb{R}^2$ for the metrics d_4, d_e, and d_8.

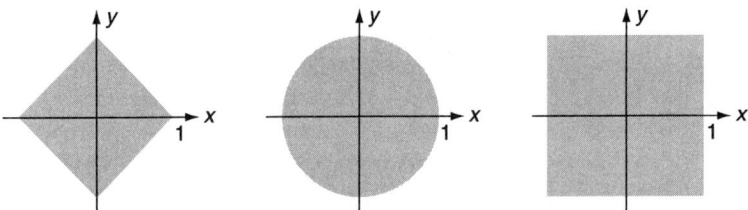

FIGURE 3.6 The city block, Euclidean, and chessboard unit disks in the real plane.

For any grid point p, the smallest neighborhood of p in $[\mathbb{Z}^2, d_\alpha]$ ($\alpha \in \{4,8\}$) is defined by the following:

$$N_\alpha(p) = \{q \in \mathbb{Z}^2 : d_\alpha(p,q) \leq 1\}$$

The notations d_4 and d_8 are suggested by the fact that $N_4(p) - \{p\}$ has cardinality 4 and $N_8(p) - \{p\}$ has cardinality 8. Using Equation 3.4, we have the following:

Theorem 3.2 $d_8(p,q) \leq d_e(p,q) \leq d_4(p,q) \leq 2 \cdot d_8(p,q)$ for all $p, q \in \mathbb{R}^2$.

The last inequality is an easy consequence of the definitions of d_8 and d_4: without loss of generality, let $d_8(p,q) = \max\{|x_1 - x_2|, |y_1 - y_2|\} = |x_1 - x_2|$; then $d_4(p,q) \leq 2 \cdot |x_1 - x_2| = 2 \cdot d_8(p,q)$.

Let $p, q \in \mathbb{R}^3, p = (x_1, y_1, z_1)$, and $q = (x_2, y_2, z_2)$, and define the following:

$$d_6(p,q) = L_1(p,q) = |x_1 - x_2| + |y_1 - y_2| + |z_1 - z_2|$$

$$d_{26}(p,q) = L_\infty(p,q) = \max\{|x_1 - x_2|, |y_1 - y_2|, |z_1 - z_2|\}$$

We also define the following:

$$d_{18}(p,q) = \max\{d_{26}(p,q), \lceil d_6(p,q)/2 \rceil\}$$

(This definition is equivalent to the one based on 18-paths; see Theorem 3.6.) For any grid point p, the smallest neighborhood of p in $[\mathbb{Z}^3, d_\alpha]$ ($\alpha \in \{6, 18, 26\}$) is defined by the following:

$$N_\alpha(p) = \{q \in \mathbb{Z}^3 : d_\alpha(p,q) \leq 1\}$$

Note that $N_\alpha(p) - \{p\}$ has cardinality α for $\alpha = 6, 18$, and 26. Analogous to Theorem 3.2 and using Equation 3.4, we have the following (see also Exercise 3 in Section 3.5):

Theorem 3.3 $d_{26}(p,q) \leq d_e(p,q) \leq d_6(p,q) \leq 3 \cdot d_{26}(p,q)$ for all $p, q \in \mathbb{R}^3$, and $d_{26}(p,q) \leq d_{18}(p,q) \leq d_e(p,q)$ for all $p, q \in \mathbb{Z}^3$ such that $d_e(p,q) \neq \sqrt{3}$.

3.2 Grid Point Metrics

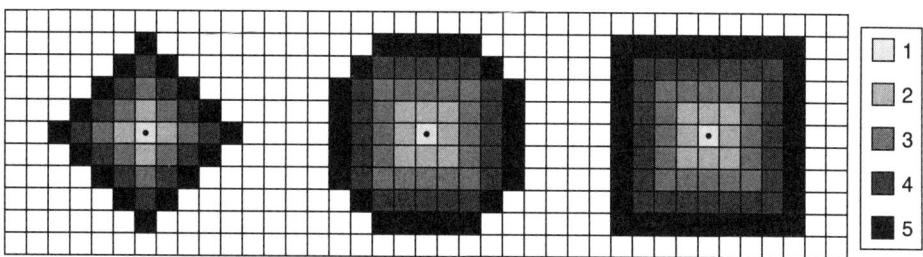

FIGURE 3.7 The e-neighborhoods for $e = 1, 2, 3, 4,$ and 5 in the 2D grid cell model defined by the city block (left), Euclidean (middle), and chessboard (right) metrics.

Proof $d_{26}(p,q) \leq d_{18}(p,q)$ follows from the definition of d_{18}. For $d_{18}(p,q) \leq d_e(p,q)$, we need only to show that $\lceil d_6(p,q)/2 \rceil \leq d_e(p,q)$, because $L_\infty(p,q) = d_{26}(p,q) \leq d_e(p,q) = L_2(p,q)$.

If $0 < d_6(p,q) < 1$, then $\lceil d_6(p,q)/2 \rceil = 1$ and $\lceil d_6(p,q)/2 \rceil > d_6(p,q) \geq d_e(p,q)$. If $p, q \in \mathbb{Z}^3$ and $d_6(p,q) = 0$ or $d_6(p,q) = 1$, we have $\lceil d_6(p,q)/2 \rceil = d_6(p,q) = d_e(p,q)$.

If $1 < d_6(p,q)$, then $\lceil d_6(p,q)/2 \rceil < d_6(p,q)$. If $p, q \in \mathbb{Z}^3$ and $d_6(p,q) = 2$, we have $d_e(p,q) = 2$ or $d_e(p,q) = \sqrt{2}$, such that $\lceil d_6(p,q)/2 \rceil = 1 < d_e(p,q)$. If $p, q \in \mathbb{Z}^3$ and $d_6(p,q) = 3$, we have $d_e(p,q) = 3$, $d_e(p,q) = \sqrt{5}$, or $d_e(p,q) = \sqrt{3}$, such that $\lceil d_6(p,q)/2 \rceil = 2 > d_e(p,q)$ in the latter case. However, if $p, q \in \mathbb{Z}^3$ and $d_6(p,q) \geq 4$, then $\lceil d_6(p,q)/2 \rceil \leq d_e(p,q)$. ∎

3.2.2 Neighborhoods and degrees of closeness

The ε-neighborhood $U_\varepsilon(p)$ (see Section 3.1.7) is defined for any metric space $[S, d]$, any $p \in S$, and any $\varepsilon > 0$. In some cases we have $U_\varepsilon(p) = \{p\}$; for example, this is true for $\varepsilon \leq 1$ and any of the metrics $\lceil d_e \rceil, d_4$, and d_8 on $\mathbb{Z}^2$ or for the binary metric.

If the range of d is countable so that U_ε is of interest only for discrete values of ε (e.g., for metrics on a grid $\mathbb{G}$ that is a subset of $\mathbb{Z}^2, \mathbb{Z}^3, \mathbb{C}_2,$ or $\mathbb{C}_3$), we use the notation *e-neighborhood* instead of ε-neighborhood. The e-neighborhoods for three metrics on $\mathbb{C}_2^{(2)}$ are illustrated in Figure 3.7; see Figure 3.6 for the corresponding disks in the real plane. The metrics d_4, d_e, and d_8 are translation-invariant; hence the sets $U_\varepsilon(p)$ have identical "shapes" for all p.

For any metric d on a grid $\mathbb{G}$, there is an interval of values $e > 0$ such that $U_e(p)$ contains as few grid points as possible in addition to p itself. This minimal set of grid points is called the *smallest (nontrivial) neighborhood* $N(p)$ of p with respect to d. (In the grid cell model, we use the notation $\eta(c)$, where c is a cell.) For example, for

d_α ($\alpha \in \{4,6,8,18,26\}$), the smallest neighborhoods $N_\alpha(p)$ defined in Section 3.2.1 are obtained for $1 < e \leq 2$. For $e = 1$, we have $U_1(p) = \{q \in \mathbb{Z}^2 : d_\alpha(q,p) < 1\} = \{p\}$ for any of these d_αs.

With fuzzy geometry (see Section 1.2.10), we can define the *degree of closeness* of two points p and q of a metric space $[S,d]$ as a monotonically nonincreasing function of the distance between p and q. For example, we can define $c(p,q) = 1/[1+d(p,q)]$. It follows that $0 < c(p,q) \leq 1$ for all p,q; hence, for any p, $c(p,q)$ defines a fuzzy subset μ_p of $S \setminus \{p\}$ that we can think of as a fuzzy neighborhood of p.

Degrees of closeness between pixels or voxels p and q in a picture P can be defined using monotonically nonincreasing functions of the absolute difference between $P(p)$ and $P(q)$. For example, we can define $c'(p,q) = 1/(|P(p) - P(q)| + 1)$. Note that, for any p and q, we have $1/(G_{max}+1) \leq c'(p,q) \leq 1$ so that $c'(p,q)$ defines a fuzzy subset μ'_p of the picture. We can also define degrees of closeness between pixels or voxels that depend on both the distance between them and the absolute difference between their values. In Section 3.4, we will define a metric on a picture in which the distance from p to q depends on the sums of the pixel or voxel values along paths from p to q.

3.2.3 Approximations to the Euclidean metric

We saw in Figure 3.6 that the set of points within a given d_4 or d_8 distance from a given point is a square. These distances depend on direction; their "disks" are not good approximations to Euclidean disks. If we restrict d_4 and d_8 to $\mathbb{Z}^2$ the set of grid points q such that $d_4(p,q) \leq k$ is a diagonally oriented square (a "diamond") of odd diagonal length $2k+1$ centered at p, and the set of grid points q such that $d_8(p,q) \leq k$ is an upright square of odd side length $2k+1$ centered at p; see Figure 3.7.

Better approximations to Euclidean disks can be obtained by combining d_4 and d_8, for example, by taking the following:

$$d(p,q) = \max\{d_8(p,q), \tfrac{2}{3} \cdot d_4(p,q)\} \tag{3.11}$$

It is not hard to see that the set of grid points such that $d(p,q) \leq k$ is the intersection of an upright square of side length k with a diamond of diagonal length $3k/2$; this intersection is an upright octagon. Varying the coefficient of d_4 in Equation 3.11 causes the shape of the octagon to vary between a diamond and a square. The octagon can be made arbitrarily close to regular by choosing the coefficient appropriately.

Metrics with disks that are "hexagons" can be defined by using a modification of the standard orthogonal grid in which, for example, the odd-numbered rows are shifted half a unit to the right; see Figure 3.8. This is equivalent to working with an unshifted grid but treating a grid point (i,j) on an odd-numbered row as having the six neighbors $(i \pm 1, j), (i, j \pm 1)$, and $(i+1, j \pm 1)$ and a grid point on an even-numbered row as having the six neighbors $(i \pm 1, j), (i, j \pm 1)$, and $(i-1, j \pm 1)$.

In the hexagonal grid shown in Figure 3.8, we can introduce a coordinate system by using any two of the three axes shown on the right, for example, x and y. We can

3.2 Grid Point Metrics

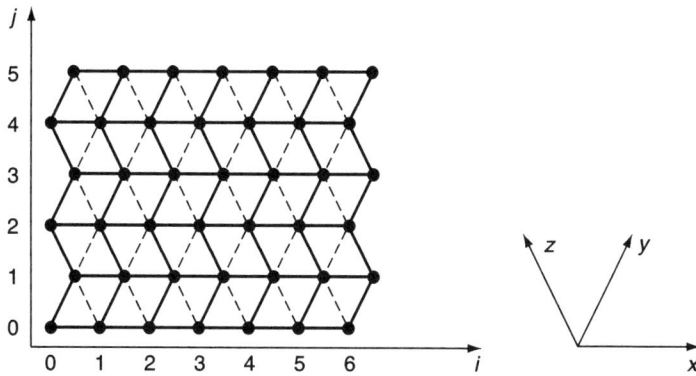

FIGURE 3.8 Modification of the standard grid in which the odd-numbered rows are shifted half a unit to the right.

TABLE 3.1 Signs of the coordinates in the six sextants of the hexagonal grid.

Sextant	u	v	$u + v$
I	≥ 0	≥ 0	≥ 0
II	≤ 0	≥ 0	≥ 0
III	≤ 0	≥ 0	≤ 0
IV	≤ 0	≤ 0	≤ 0
V	≥ 0	≤ 0	≤ 0
VI	≥ 0	≤ 0	≥ 0

reach any grid point p from the origin by making an (positive, negative, or zero) integer number u of moves in the $+x$ direction and then an integer number v of moves in the $+y$ direction; the resulting (u,v) are the coordinates of p. We will use these coordinates in the remainder of this discussion.

The $x, y,$ and z axes divide the plane into six sextants that we number counterclockwise beginning at the $+x$ axis; see the figure in Exercise 2 in Section 1.3. It is easily verified that the signs of the (u,v) coordinates of the points lying in these sextants can be characterized as shown in Table 3.1. Note that the z-axis is the locus of points such that $u+v=0$.

The *hexagonal distance* between two points p and q of the hexagonal grid is the minimum number of unit moves in the x and y directions needed to go from p to q. If $p = (i,j)$ and $q = (h,k)$, it can be shown that this number is given as follows:

$$d_h((i,j),(h,k)) = \begin{cases} |i-h|+|j-k| & \text{if } \text{sgn}(i-h) = \text{sgn}(j-k) \\ \max\{|i-h|,|j-k|\} & \text{otherwise} \end{cases}$$

(The *signum function* $\text{sgn}(a)$ is 1 if $a \geq 0$ and 0 otherwise.)

Proposition 3.5 d_h is a metric on $\mathbb{Z}^2$.

Proof Positive definiteness and symmetry are easily verified. To prove triangularity, assume without loss of generality that the three grid points are $(0,0), (i,j)$, and (h,k), and consider all possible values of the signs of i, j, h, k, $(i-h)$, and $(j-k)$. ∎

It can be shown that $d_h((i,j),(h,k))$ is also equal to the following:

$$\max\left\{|j-k|, \frac{1}{2}(|j-k|+j-k) + \left\lceil\frac{j+1}{2}\right\rceil - \left\lceil\frac{k+1}{2}\right\rceil + h - i,\right.$$

$$\left.\frac{1}{2}(|j-k|-j+k) - \left\lceil\frac{j+1}{2}\right\rceil + \left\lceil\frac{k+1}{2}\right\rceil - h + i\right\}$$

It can also be shown that hexagonal coordinates (u,v) are related to Cartesian coordinates (i,j) by the following for j even:

$$u = i - \left\lceil\frac{j}{2}\right\rceil, \quad v = j$$

and by the following for j odd:

$$u = i - \left\lceil\frac{j+1}{2}\right\rceil, \quad v = j$$

Obviously, $\lceil d_e \rceil$ is the integer-valued metric that best approximates d_e. However, "incremental" algorithms for distance computation on a grid (see Section 3.4) normally use local neighborhoods; this makes it easy to compute metrics such as d_4, d_h, or d_8 or octagonal metrics, but not $\lceil d_e \rceil$. For a method of computing a good approximation to d_e, see Section 3.4.3.

We conclude this section by describing a general method of defining approximations to Euclidean distance by counting moves in different directions (e.g., isothetic moves, diagonal moves). Let $p, q \in \mathbb{Z}^2$, and let ρ be a sequence of king's moves from p to q. Let $l_{a,b}(\rho) = ma + nb$ where m is the number of isothetic moves and n the number of diagonal moves, and let the following be true:

$$d_{a,b}(p,q) = \min_\rho l_{a,b}(\rho) \tag{3.12}$$

Thus $d_{a,b}$ is called the (a,b) *chamfer distance* (or *weighted distance*) from p to q. Chamfer distances that closely approximate Euclidean distance can be defined by appropriately choosing a and b. If the following is true,

$$0 < a \leq b \leq 2a \tag{3.13}$$

3.2 Grid Point Metrics

the (a,b) chamfer distance $d_{a,b}$ is a metric [738], which also defines a norm $\|p\|_{a,b} = d_{a,b}(p,o)$ (the distance of p from the origin $o = (0,0)$). This metric is a nonnegative linear combination of d_4 and d_8. Convex linear combinations of d_4 and d_8 also give useful chamfer distances; for example, $(d_4 + 2d_8)/3$ is the $(3,4)$ chamfer distance (see Exercise 12 in Section 3.5).

[761] formulated necessary and sufficient conditions for a 2D chamfer distance d to define a norm $\|p\| = d(p,o)$ on $\mathbb{Z}^2$.

We can similarly define 3D chamfer distances $d_{a,b,c}$ where a, b, and c correspond to moves in which only one coordinate changes (isothetic moves), two coordinates change, and all three coordinates change, and we can obtain good approximations to Euclidean distance by appropriately choosing a, b, and c.

Generalized chamfer distances can be defined using additional types of moves that are not necessarily moves between 8-neighbors or 26-neighbors.

3.2.4 Paths, geodesics, and intrinsic distances

A sequence ρ of grid points $(p_0, p_1, \ldots, p_n)$ such that $p_0 = p$, $p_n = q$, and p_{i+1} is α-adjacent to p_i ($0 \leq i \leq n-1$) is called an α-*path* of length n from p to q; p and q are called the *endpoints* of ρ.

> **Proposition 3.6** If ρ is a shortest α-path from p to q, the p_is must all be distinct, and nonconsecutive p_is cannot be α-adjacent.
>
> **Proof** If we had $p_h = p_k$ with $h < k$, $(p_0, \ldots, p_h, p_{k+1}, \ldots, p_n)$ would be a shorter α-path with the same endpoints. Similarly, if p_h were α-adjacent to p_k where $h < k$ and $k - h > 1$, $(p_0, \ldots, p_h, p_k, \ldots, p_n)$ would be a shorter α-path. ∎

An α-path is called an α-*geodesic* if no shorter α-path with the same endpoints exists.

> **Proposition 3.7** If $(p_0, \ldots, p_n)$ is an α-geodesic, $(p_h, \ldots, p_k)$ is an α-geodesic for all $0 \leq h \leq k \leq n$.
>
> **Proof** If $(q_0, \ldots, q_m)$ were a shorter α-path from $q_0 = p_h$ to $q_m = p_k$, $(p_0, \ldots, p_{h-1}, q_0, \ldots, q_m, p_{k+1}, \ldots, p_n)$ would be a shorter α-path from p_0 to p_n. ∎

> **Theorem 3.4** The length of a shortest α-path from p to q is $d_\alpha(p,q)$.

Proof We give the proof in 2D for $\alpha = 4$; the proofs for other cases are similar. If the length of the path is 1 (e.g., the path is (p,q)), p and q are 4-adjacent, so $d_4(p,q) = 1$. We proceed by induction on the shortest length. Let $(p_0, p_1, \ldots, p_n)$ be a shortest 4-path; then, using Proposition 3.7, $(p_0, p_1, \ldots, p_{n-1})$ is a shortest path from p to p_{n-1}, so $d_4(p, p_{n-1}) = n - 1$ by the induction hypothesis. Because q is 4-adjacent to p_{n-1}, we have $d_4(p_{n-1}, q) = 1$ so that, by the triangle inequality, $d_4(p,q) \leq (n-1) + 1 = n$. If $d_4(p,q) = m < n$, we can easily construct a 4-path of length m from p to q. For example, suppose $p = (i,j)$ and $q = (h,k)$ where $i \leq h$ and $j \leq k$; the argument is similar if $i \geq h$ and/or $j \geq k$. Because $d_4(p,q) = m$, we have $(h-1) + (h-j) = m$, and we can construct a 4-path from (i,j) to (h,k) by first increasing i by 1 until it reaches h and then increasing j by 1 until it reaches k; this 4-path has length $|i - h| + |j - k| = m < n$, which is contrary to our assumption that a shortest 4-path from (i,h) to (j,k) has length n. ∎

It follows that an α-path ρ of length n is an α-geodesic iff the d_α-distance between the endpoints of ρ is n.

In Euclidean space, there is a unique shortest arc between any two points p and q, which is namely the straight line segment pq. In a grid, there can be many shortest α-paths between two grid points, and these paths need not be digital straight line segments (see Section 2.3.4 and Chapter 9). In what follows, we consider only the 2D cases $\alpha = 4$ and 8, and we assume that p_i ($0 \leq i \leq n$) has coordinates (x_i, y_i).

Proposition 3.8 The following properties of a 4-path ρ are equivalent:

(a) ρ is a 4-geodesic.

(b) ρ cannot turn right (or left) twice in succession; left and right turns must alternate.

(c) $x_0 \leq x_1 \leq \cdots \leq x_n$ (or all $\geq$), and $y_0 \leq y_1 \leq \cdots \leq y_n$ (or all $\geq$).

(d) $|x_0 - x_n| + |y_0 - y_n| = n$.

Proof To see that (a) implies (b), suppose that ρ made two successive turns in the same direction:

$$\begin{array}{ccc} & p_{r+1} \quad p_{r+2} \cdots p_{s-1} & \\ p_r & & p_s \\ \vdots & & \vdots \end{array}$$

(The argument in other cases is analogous.) Then the subpath of ρ from p_r to p_s has length $s - r$, but there is a horizontal path from p_r to p_s of length $s - r - 2$, so Proposition 3.7 is violated, and ρ cannot be a 4-geodesic.

3.2 Grid Point Metrics

We next show that (b) implies (c). Suppose the initial direction of ρ is horizontal toward the right and its first turn (if any) is a left turn; the proofs in other cases are analogous. Let the turns be at $p_{n_1}, p_{n_2}, \ldots$ where $0 < n_1 < n_2 < \cdots < n$; then $x_0 < x_1 < \cdots < x_{n_1}$, and $y_0 = y_1 = \cdots = y_{n_1}$. After the first turn, ρ is headed vertically upward; thus $x_{n_1} = x_{n_1+1} = \cdots = x_{n_2}$, and $y_{n_1} < y_{n_1+1} < \cdots < y_{n_2}$. By (b), the second turn must be a right turn, after which ρ is again horizontal and headed rightward, so $x_{n_2+1} < \cdots < x_{n_3}$, $y_{n_2+1} = \cdots = y_{n_3}$, and so on, proving (c). Note that, by (c), if we take any p_m on ρ as origin, the subpaths $(p_0, \ldots, p_m)$ and $(p_m, \ldots, p_n)$ must lie in a pair of opposite quadrants.

Next we prove that (c) implies (d). At each step along a 4-path, either x or y (but not both) changes by 1. Hence, if (c) holds (e.g., $x_0 \leq x_1 \leq \cdots \leq x_n$ and $y_0 \leq y_1 \leq \cdots \leq y_n$ and similarly in the other cases), the number of steps at which the xs increase and the number at which the ys increase must add up to n, which implies (d).

Finally we show that (d) implies (a). Any 4-path from (x_0, y_0) to (x_n, y_n) must have length at least $|x_0 - x_n| + |y_0 - y_n|$, because a coordinate can change by only 1 at each step, and only one coordinate can change at a time. If (d) holds, this length is n, and because ρ has length n, its length is the shortest possible, thus proving (a). ∎

Proposition 3.9 The following properties of an 8-path ρ are equivalent:
(a) ρ is an 8-geodesic.
(b) $x_0 < x_1 < \cdots < x_n$ (or all $>$), or $y_0 < y_1 < \cdots < y_n$ (or all $>$).
(c) $|x_0 - x_n| = n$ or $|y_0 - y_n| = n$; because each of them must be $\leq n$, this is equivalent to their max being n.

Proof The x and y coordinates can each change by at most 1 at each step along an 8-path. Hence, to achieve $|x_0 - x_n| = n$, successive x_is must all differ by 1 in the same direction (i.e., $x_0 < x_1 < \cdots < x_n$ [or all $>$]), which proves that (c) implies (b). Conversely, $x_0 < x_1 < \cdots < x_n$ means that the successive x_is must differ by 1 in the same direction so that $|x_0 - x_n| = n$; thus (b) implies (c).

On the other hand, because ρ has length n, we must have $|x_0 - x_n| \leq n$ and $|y_0 - y_n| \leq n$; an 8-path of length n cannot involve coordinate changes of more than n, because each coordinate changes by at most 1 at each step. Any 8-path from p_0 to p_n must have length of at least $\max\{|x_0 - x_n|, |y_0 - y_n|\}$. If (c) holds (e.g., $|x_0 - x_n| = n$), the max is n; thus ρ (which has length n) is a shortest 8-path, proving (a).

Conversely, if (c) fails, and $|y_0 - y_n| < n$. Suppose $x_0 \leq x_n, y_0 \leq y_n$, and $x_n - x_0 \leq y_n - y_0$; the argument is analogous if any of these $\leq$s is $\geq$. Then $((x_0, y_0), (x_0 +$

$1, y_0 + 1), \ldots, (x_n, y_0 + (x_n - x_0)), (x_n, y_0 + (x_n - x_0) + 1), \ldots, (x_n, y_n))$ is an 8-path from (x_0, y_0) to (x_n, y_n) of length $y_n - y_0 < n$, so ρ is not a shortest 8-path, thus completing the proof. ∎

These results imply that digital straight line segments (see Chapter 9) are geodesics. It is not hard to show that the only possible turns in an 8-geodesic are 45° right and left turns in alternation.

If S is an α-connected set of grid points (see Chapter 4), for any $p, q \in S$, there exists an α-path $\rho = (p_0, p_1, \ldots, p_n)$ from $p_0 = p$ to $p_n = q$ such that the p_is are all in S. The length $d_\alpha^S(p,q)$ of a shortest such path is called the *intrinsic α-distance* in S from p to q. The ordinary α-distance from p to q will sometimes be called *extrinsic* to contrast it with intrinsic α-distance.[3]

Proposition 3.10 $d_\alpha(p,q) \leq d_\alpha^S(p,q)$ for all $p, q \in S$.

Proof The length of a shortest α-path that lies in S cannot be less than the length of an unrestricted shortest α-path. ∎

We will see in Chapter 13 that a set S of grid points is digitally convex iff any two points of S are the endpoints of a digital straight line segment that is contained in S. Because a digital straight line segment is a geodesic, it follows that, if S is digitally convex, $d_\alpha^S(p,q)$ is equal to $d_\alpha(p,q)$ for all $p, q \in S$.

3.2.5 Distances between sets

Integer-valued metrics d on a grid, such as $\lceil d_e \rceil, d_4$, and d_8 in 2D, define Hausdorff metrics in the family of all finite subsets of the grid; see Section 3.1.8. For any such d, any grid point p, and any finite set of grid points S, the distance from p to S is $h_p(S) = \min d(p,q)$ where the min is taken over all $q \in S$. Evidently, $h_p(S) = 0$ iff $p \in S$. For example, for the p, q, and r in Figure 3.2, let $A = \{p,q\}$ and $B = \{q,r\}$; then $d_4(A,B) = d_4(p,r) = d_4(r,q) = 3, d_8(A,B) = d_8(p,r) = 2$, and $\lceil d_e \rceil(A,B) = \lceil d_e \rceil(p,r) = \lceil d_e \rceil(r,q) = 3$. Similarly, for the sets A and B in Figure 3.3, we have $d_4(A,B) = h_p(B) = 8 > h_q(A) = 6$ (h_p and h_q with respect to d_4) and $d_8(A,B) = h_p(B) = h_q(A) = 5$ (h_p and h_q with respect to d_8).

A Hausdorff metric can be used to measure the distance between the frontiers of the inner and outer Jordan digitizations of a set; see Figure 2.29 for an example. Section 3.1.7 allows us to complete Definition 2.8: the inner Jordan digitization $J_h^-(S)$

---

[3]. Intrinsic distance is sometimes called "geodesic distance," but, to avoid confusion with the noun "geodesic," we will not use that term.

3.2 Grid Point Metrics

---

1. Calculate a distance field $F(A)$ in an array of size $m \times n$.
2. Calculate a distance field $F(B)$ in an array of size $m \times n$.
3. Let a be the maximum value in $F(A)$ at all positions belonging to B.
4. Let b be the maximum value in $F(B)$ at all positions belonging to A.
5. $H(A,B) = \max\{a,b\}$.

ALGORITHM 3.1 Algorithm for calculating the Hausdorff distance between two subsets A and B of an $m \times n$ grid.

of a set $S \subseteq \mathbb{R}^n$ is actually the union of all n-cells (for grid resolution $h > 0$ and $n = 2$ or $n = 3$) that are contained in the *interior* of set S.

Theorem 3.5 For any compact set $S \subset \mathbb{R}^2$ such that $J_h^-(S) \neq \emptyset$, the d_4 or d_8 Hausdorff distance between the (polygonal) frontiers $\partial J_h^-(S)$ and $\partial J_h^+(S)$ is at least $1/h$.

Proof Let p be an arbitrary grid vertex on $\partial J_h^-(S)$. Thus p cannot be on $\partial J_h^+(S)$, because the frontier of $J_h^-(S)$ never intersects the frontier of $J_h^+(S)$ if S is a nonempty compact subset of $\mathbb{R}^2$. The d_4 (or d_8) distance from p to any point q on $\partial J_h^+(S)$ is at most $1/h$. It follows that

$$\max_{p \in \partial J_h^-(S)} \min_{q \in \partial J_h^+(S)} d_4(p,q) \geq 1/h$$

and thus $d_4(\partial J_h^-(S), \partial J_h^+(S)) \geq 1/h$; this is analogous for d_8. ∎

Finally, we discuss algorithms for calculating the Hausdorff distance between two finite sets $A, B \subset \mathbb{G}_{m,n}$ of grid points (see Algorithm 3.1). We assume that m and n are the dimensions of the smallest isothetic rectangle that contains $A \cup B$.

We first describe a brute-force approach. For every point in A, calculate the minimum distance to a point in B, and, for every point in B, calculate the minimum distance to a point in A. Take the maxima of these two sets of distances; then the maximum of the two maxima is the desired Hausdorff distance. The points of A and B can be located by scans of $\mathbb{G}_{m,n}$ (see Section 1.1.3); on one scan, we find all of the points p in A, and, for each p, we scan $\mathbb{G}_{m,n}$ again to find all of the points in B and calculate the distances from p to these points. If $\text{card}(A)$ and $\text{card}(B)$ are $\mathcal{O}(mn)$, this brute-force algorithm takes $\mathcal{O}(m^2 n^2)$ computation steps.

A much more efficient algorithm is shown above. For any $S \subset \mathbb{G}_{m,n}$, the *distance field* $F(S)$ is an array of size $m \times n$ such that $F(S)(p) = h_p(S)$; in particular, $F(S)(p) = 0$ iff $p \in S$. It can be shown (see Section 3.4.2 for grid metrics and Section 3.4.3 for the Euclidean metric) that $F(S)$ can be calculated in $\mathcal{O}(mn)$ computation steps for any Minkowski metric on $\mathbb{G}_{m,n}$. This allows us to calculate the

Hausdorff distance in $\mathcal{O}(mn)$ computation steps by computing distance fields for A and B and scanning each of these fields.

3.3 Grid Cell Metrics

In Sections 3.1.2 and 3.1.3, we defined metric-related concepts such as norms, scalar products, and angular values in n-dimensional vector spaces. In this section, we apply these concepts to the n-dimensional unitary modules defined by the grid cell model. Results involving grid cell adjacency models easily translate into results involving the isomorphic grid point adjacency models.

3.3.1 Basic grid cell metrics

We first consider the 2D grid cell model. Let d be a metric defined on the set $\mathbb{Z}^2$ of grid points. For any two 2-cells $c_1, c_2 \in \mathbb{C}_2^{(2)}$, we define $\partial(c_1, c_2)$ by the value of d for the center points of c_1 and c_2; ∂ is thus a metric on $\mathbb{C}_2^{(2)}$. When $d = d_4$, we call this metric ∂_1, and, when $d = d_8$, we call it ∂_0. Evidently, $\partial_1(c_1, c_2) \leq 1$ iff $c_1 \cap c_2$ contains (at least) one 1-cell, and $\partial_0(c_1, c_2) \leq 1$ iff $c_1 \cap c_2$ contains (at least) one 0-cell; the subscript indicates the dimension of the cells that have to be contained in the intersection. Smallest neighborhoods in the grid cell model are denoted by the following:

$$\eta_\alpha(c) = \{c' \in \mathbb{C}_2^{(2)} : \partial_\alpha(c, c') \leq 1\} \qquad (\alpha \in \{0, 1\})$$

The 2D (grid cell) incidence model also includes 1-cells and 0-cells; it was illustrated in Figure 2.3, and in a more abstract way in Figure 2.13. The set of centers of the 2-, 1-, and 0-cells in $\mathbb{C}_2$ is the grid with grid constant $\theta = 0.5$. For any metric d on this grid and any $b, c \in \mathbb{C}_2$, $\partial(b, c)$ is defined by the value of d for the center points of b and c; thus ∂ is a metric on $\mathbb{C}_2$. For example (see Figure 2.3), the 2-cell with its center at $(1, 2)$ and the 0-cell at $(1.5, 4.5)$ are at Euclidean distance $\sqrt{26}/2$, city block distance 3, and chessboard distance $5/2$.

Metrics on $\mathbb{C}_3^{(3)}$ or $\mathbb{C}_3$ can be defined as they were in the 2D case by identifying cells with their centers. For any two 3-cells $c_1, c_2 \in \mathbb{C}_3^{(3)}$, $\partial(c_1, c_2)$ is defined by the value of a grid point metric d for the center points of c_1 and c_2; thus ∂ is a metric on $\mathbb{C}_3^{(3)}$. The metrics defined in this way by d_6, d_{18}, and d_{26} will be denoted by ∂_2, ∂_1, and ∂_0, respectively. Evidently, $\partial_2(c_1, c_2) \leq 1$ iff $c_1 \cap c_2$ contains (at least) one

3.3 Grid Cell Metrics

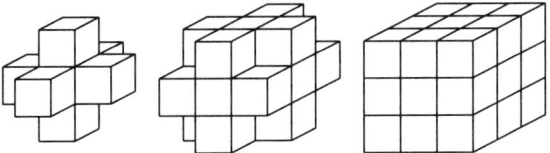

FIGURE 3.9 Balls of 3-cells of radius 1 with respect to d_6 (left), d_{18} (middle), and d_{26} (right).

2-cell, $\partial_1(c_1, c_2) \leq 1$ iff $c_1 \cap c_2$ contains (at least) one 1-cell, and $\partial_0(c_1, c_2) \leq 1$ iff $c_1 \cap c_2$ contains (at least) one 0-cell. The smallest neighborhoods are as follows:

$$\eta_\alpha(c) = \{c' \in \mathbb{C}_3^{(3)} : \partial_\alpha(c, c') \leq 1\} \qquad (\alpha \in \{0, 1, 2\})$$

Figure 3.9 shows the (smallest) neighborhoods of 3-cells at d_6-, d_{18}-, and d_{26}-distance ≤ 1 from a given 3-cell; they are "balls" of 3-cells of radius 1. A general definition of such metrics will be given in Section 3.3.2 for the n-dimensional grid cell model $\mathbb{C}_n^{(n)}$.

3.3.2 Seminorms

An *n-cell* ($n \geq 2$) is an n-dimensional grid (hyper)cube with edges of length 1 with its center at a grid point $p \in \mathbb{Z}^n$. $[\mathbb{C}_n^{(n)}, +, \cdot, \mathbb{Z}]$ is a unitary module. We identify cells by their centers; hence, if p is the center of n-cell $c \in \mathbb{C}_n^{(n)}$, $k \cdot c$ is the n-cell with its center at $k \cdot p$, and $-c$ is the n-cell with its center at $-p$.

Let the following be true ($0 \leq i < n$):

$$B_i = \{(x_1, \ldots, x_n) : \forall k (1 \leq k \leq n \rightarrow x_k \in \{-1, 0, +1\}) \wedge \mathrm{card}\{k : x_k = 0\} = i\}$$

B_i is a subset of the frontier of the n-dimensional cube $[-1, +1]^n$. For example, $B_0 = \{-1, +1\}^n$ is the set of all vertices of this cube. We always have $o = (0, \ldots, 0) \notin B_i$. Because $i < n$, we have $x_k \neq 0$ for at least one coordinate k.

Let $c \in \mathbb{C}_n^{(n)}$ be an n-cell, let $0 \leq \alpha < n$, and let

$$\eta_\alpha(c) = \{c\} \cup \{c\} \oplus \bigcup_{i=\alpha}^{n-1} B_i \tag{3.14}$$

where $\oplus$ is the Minkowski sum defined in Section 1.2.12. Then the following is true:

$$\mathrm{card}(\eta_\alpha(c)) = 1 + \sum_{i=\alpha}^{n-1} 2^{n-i} \binom{n}{i}$$

For example, for $n = 3$, we have $\mathrm{card}(\eta_0(c)) = 27$. Two n-cells c_1 and c_2 are called α-*neighbors* iff $c_1 \in \eta_\alpha(c_2)$. This defines the n-dimensional *grid cell adjacency models*.

Let the following be true for $0 \leq \alpha < n$ and $t \geq 0$:

$$A_\alpha^{(0)} = \{(0,\ldots,0)\} \quad \text{and} \quad A_\alpha^{(t+1)} = A_\alpha^{(t)} \oplus \bigcup_{i=\alpha}^{n-1} B_i$$

Let $\|c\|_\alpha$ be the smallest t such that $c \in A_\alpha^{(t)}$, where $c \in \mathbb{C}_n^{(n)}$. Here, $\|c\|_\alpha$ is called the *α-value* of c. For example, for $n=2$, we have $\|c\|_0 = \max\{|x|,|y|\}$ and $\|c\|_1 = |x|+|y|$ where x and y are the coordinates of the center of c.

Theorem 3.6 Let $c \in \mathbb{C}_n^{(n)}$, $c = (x_1,\ldots,x_n)$, and $0 \leq \alpha < n$. Then the following is true:

$$\|c\|_\alpha = \max\left\{|x_1|,\ldots,|x_n|, \left\lceil \frac{\sum_{i=1}^n |x_i|}{n-\alpha} \right\rceil \right\}$$

Proof By induction on t, it can be shown that the following is true:

$$\{\|c\|_\alpha \leq t\} \text{ iff } \left\{ \sum_{i=1}^n |x_i| \leq t(n-\alpha) \wedge \max_{1 \leq i \leq n} |x_i| \leq t \right\}$$

Consequently, we have $\|c\|_\alpha = t$ iff the following is true:

$$\sum_{i=1}^n |x_i| \leq t(n-\alpha) \wedge \max_{1 \leq i \leq n} |x_i| \leq t$$

$$\wedge \left\{ (t-1)(n-\alpha) < \sum_{i=1}^n |x_i| \vee \max_{1 \leq i \leq n} |x_i| = t \right\} \quad \blacksquare$$

For $\alpha = 0$ and $\alpha = n-1$, we obtain the following norms:

$$\|c\|_0 = \max_{1 \leq i \leq n} |x_i|$$

This coincides with the Minkowski norm $\|\cdot\|_\infty$; see Section 3.1.2. We also obtain the following:

$$\|c\|_{n-1} = \sum_{i=1}^n |x_i|$$

This coincides with the Minkowski norm $\|\cdot\|_1$. For $1 \leq \alpha \leq n-2$, $k = n-\alpha$, and $c = (k-1, k, \ldots, k)$, we have $\|c\|_\alpha = n$ (recall that $n \geq 2$), and thus we have the following:

$$\|k \cdot c\|_\alpha = nk - 1 < nk = k \cdot \|c\|_\alpha$$

3.3 Grid Cell Metrics

Thus, for $1 \leq \alpha \leq n-2$, the α-value $\|\cdot\|_\alpha$ is not a norm but rather a seminorm, which also satisfies $\|-c\|_\alpha = \|c\|_\alpha$ for all $c \in \mathbb{C}_n^{(n)}$ and all $0 \leq \alpha < n$. As in Equation 3.3, we define the following for all $c_1, c_2 \in \mathbb{C}_n^{(n)}$:

$$\partial_\alpha(c_1, c_2) = \|c_1 - c_2\|_\alpha \tag{3.15}$$

For example, for $n = 3$ and $\alpha = 1$, ∂_1 is identical to the metric d_{18} defined in Section 3.2.1.

Theorem 3.7 ∂_α $(0 \leq \alpha < n)$ is a regular metric on $\mathbb{C}_n^{(n)}$.

Proof An α-path of length $m - 1$ from c_1 to $c_m \neq c_1$ is a sequence of n-cells $(c_1, c_2, \ldots, c_{m-1}, c_m)$ such that c_i is an α-neighbor of c_{i+1} and $c_i \neq c_{i+1}$ ($1 \leq i \leq m-1$). First, we show that $\|c\|_\alpha$ is equal to the length of an α-geodesic $(c_1, c_2, \ldots, c_m)$ from the origin o to $c = c_m$. Let $b_{i+1} = c_{i+1} - c_i$ ($1 \leq i \leq m-1$). Then we have that

$$b_{i+1} \in \bigcup_{i=\alpha}^{n-1} B_i$$

and let $c = b_2 + b_3 + \ldots + b_m$ so that $c \in A_\alpha^{(m)}$ and $\|c\|_\alpha \leq m - 1$. On the other hand, for any $c \in A_\alpha^{(m)}$, there exists an α-path of length $m - 1$ from o to c so that the length of an α-geodesic from o to c is at most $\|c\|_\alpha$.

We have shown that, for any $c_1 \neq c_2$, $\partial_\alpha(c_1, c_2) = \|c_2 - c_1\|_\alpha = m - 1 \geq 1$ is the length of an α-geodesic from c_1 to c_2. Let $(b_1, b_2, \ldots, b_m)$ be an α-geodesic from $b_1 = o$ to $b_m = c_2 - c_1$; then $(b_1 + c_1, b_2 + c_1, \ldots, b_m + c_1)$ is an α-geodesic from c_1 to c_2. Let $c_3 = b_2 + c_1$; then $\partial_\alpha(c_1, c_3) = 1$ and $\partial_\alpha(c_1, c_2) = 1 + \partial_\alpha(c_3, c_2)$, because $(b_2 + c_1, \ldots, b_m + c_1)$ is an α-geodesic from c_3 to c_2. ∎

Theorem 3.6 shows that the metrics ∂_α satisfy the following:

$$\partial_{\alpha_1}(b, c) \leq \partial_{\alpha_2}(b, c) \text{ for any } n \geq 2, \ \alpha_1 \leq \alpha_2, \text{ and } b, c \in \mathbb{C}_n^{(n)} \tag{3.16}$$

This complements Theorems 3.2 and 3.3.

3.3.3 Scalar products and angles

The seminorms $\|\cdot\|_\alpha$ define weak scalar products $\langle \cdot, \cdot \rangle_\alpha$ (see Section 3.1.3), which satisfy the generalized Schwarz inequality on $\mathbb{Z}^n$. It follows that they define angular values $H_\alpha(c_1, c_2, c_3)$ (see Equation 3.7). Following Proposition 3.1, $H_\alpha(c_1, c_2, c_3)$ is always in the range of the arccos function.

TABLE 3.2 Rounded angular values $H_0(o,b,c)$ at cell b. Positions with nonpositive values are shaded.

.33	.25	.17	.13	.14	.15	.16	.18	.20	.23	.26	.30	.36	.45	.45	.45	.45	.41	.39	.50	.75	1
.36	.27	.18	.09	.05	.05	.06	.06	.07	.08	.09	.10	.12	.15	.20	.20	.19	.17	.33	.69	1	1
.40	.30	.20	.10	0	-.05	-.06	-.06	-.07	-.08	-.09	-.10	-.12	-.15	-.20	-.28	-.26	0	.50	1	1	1
.44	.33	.22	.11	0	-.11	-.17	-.19	-.21	-.24	-.27	-.32	-.38	-.48	-.58	-.78	-1	0	1	1	1	1
.50	.38	.25	.13	0	-.13	-.25	-.33	-.36	-.41	-.47	-.55	-.66	-.73	-.88	-1	-1	c	1	1	1	1
.57	.43	.29	.14	0	-.14	-.29	-.43	-.53	-.59	-.68	-.79	-.83	-.92	-1	-1	-1	0	1	1	1	1
.67	.50	.33	.17	0	-.17	-.33	-.50	-.67	-.80	-.91	-.91	-.96	-1	-1	-1	-.50	0	.50	1	1	1
.80	.60	.40	.20	0	-.20	-.40	-.60	-.80	-1	-.98	-.98	-1	-1	-1	-.67	-.33	0	.33	.67	1	1
1	.75	.50	.25	0	-.25	-.50	-.75	-1	-1	-1	-1	-1	-1	-.75	-.50	-.25	0	.25	.50	.75	1
1	1	.67	.33	0	-.33	-.67	-1	-1	-.98	-.98	-1	-.80	-.60	-.40	-.20	0	.20	.40	.60	-.80
1	1	1	.50	0	-.50	-1	-1	-1	-.96	-.91	-.91	-.80	-.67	-.50	-.33	-.17	0	.17	.33	.50	.67
1	1	1	1	0	-1	-1	-1	-.92	-.83	-.79	-.68	-.59	-.53	-.43	-.29	-.14	0	.14	.29	.43	.57
1	1	1	1	o	-1	-1	-.88	-.73	-.66	-.55	-.47	-.41	-.36	-.33	-.25	-.13	0	.13	.25	.38	.50
1	1	1	1	0	-1	-.78	-.58	-.48	-.38	-.32	-.27	-.24	-.21	-.19	-.17	-.11	0	.11	.22	.33	.44
1	1	1	.50	0	-.26	-.28	-.20	-.15	-.12	-.10	-.09	-.08	-.07	-.06	-.06	-.05	0	.10	.20	.30	.40
1	1	.67	.33	.17	.19	.20	.20	.15	.12	.10	.09	.08	.07	.06	.06	.05	.05	.09	.18	.27	.36
1	.75	.50	.39	.41	.45	.45	.45	.45	.36	.30	.26	.23	.20	.18	.16	.15	.14	.13	.17	.25	.33

Angular values can be used to characterize 3-cell configurations. We say that c_2 is *between* c_1 and c_3 according to the α-metric (notation: $(c_1, c_2, c_3)_\alpha$) iff $\|c_1 - c_3\|_\alpha = \|c_1 - c_2\|_\alpha + \|c_2 - c_3\|_\alpha$. We call c_1, c_2, and c_3 α-*cogeodetic* iff they are contained in an α-geodesic; this is equivalent to $(c_1, c_2, c_3)_\alpha$, $(c_1, c_3, c_2)_\alpha$, or $(c_2, c_1, c_3)_\alpha$.

Conjecture 3.1 If $H_\alpha(c_1, c_2, c_3) = -1$, then $(c_1, c_2, c_3)_\alpha$; if $H_\alpha(c_1, c_2, c_3) = 1$, then $(c_2, c_1, c_3)_\alpha$.

Some values of H_0 and H_1 for $n = 2$ are given in Tables 3.2 and 3.3. In both tables, $o = (0,0)$ and $c = (13, 8)$ are fixed 2-cells, and the values $H_0(o, b, c)$ or $H_1(o, b, c)$ are given by the positions of the variable 2-cell b. From these examples, it is clear that $(c_1, c_2, c_3)_\alpha$ does not imply $H_\alpha(c_1, c_2, c_3) = -1$. $H_1(o, b, c) = 0$ indicates a position for which $\vec{o - b} = \vec{bo}$ and $\vec{c - b} = \vec{bc}$ are orthogonal with respect to metric ∂_α.

We saw in Section 3.3.2 that, if $\alpha = 0$ or $\alpha = n - 1$, $\|\cdot\|_\alpha$ is a norm. Hence we always have $\langle c, c \rangle_0 = \|c\|_0^2$ and $\langle c, c \rangle_{n-1} = \|c\|_{n-1}^2$, but, for $1 \leq \alpha \leq n-2$, there exist n-cells c such that $\langle c, c \rangle_\alpha < \|c\|_\alpha^2$. Because $\langle b, c \rangle_\alpha = -\langle b, -c \rangle_\alpha$ and $\|-c\|_\alpha = \|c\|_\alpha$ for

3.4 Metrics on Pictures

TABLE 3.3 Rounded angular values $H_1(o,b,c)$ at cell b.

.69	.65	.61	.56	.49	.35	.22	.10	0	-.09	-.19	-.20	-.11	0	.14	.33	.60	1	1	1	1	1
.65	.61	.56	.49	.41	.26	.12	0	-.11	-.22	-.32	-.33	-.25	-.14	0	.20	.50	1	1	1	1	1
.61	.56	.49	.41	.31	.14	0	-.13	-.25	-.36	-.47	-.50	-.43	-.33	-.20	0	.33	1	1	1	1	1
.56	.49	.41	.31	.17	0	-.15	-.29	-.42	-.54	-.67	-.71	-.67	-.60	-.50	-.33	0	1	1	1	1	1
.49	.41	.31	.17	0	-.19	-.35	-.49	-.63	-.77	-.92	-1	-1	-1	-1	-1	-1	c	1	1	1	1
.36	.27	.15	0	-.20	-.37	-.50	-.62	-.73	-.83	-.94	-1	-1	-1	-1	-1	-1	0	.33	.50	.60	
.24	.14	0	-.18	-.42	-.55	-.65	-.74	-.82	-.89	-.96	-1	-1	-1	-1	-1	-1	-.33	0	.20	.33	
.12	0	-.16	-.37	-.68	-.76	-.82	-.87	-.91	-.95	-.98	-1	-1	-1	-1	-1	-1	-.50	-.20	0	.14	
0	-.14	-.33	-.60	-1	-1	-1	-1	-1	-1	-1	-1	-1	-1	-1	-1	-1	-.60	-.33	-.14	0	
.14	0	-.20	-.50	-1	-1	-1	-1	-1	-1	-1	-.98	-.95	-.91	-.87	-.82	-.76	-.68	-.37	-.16	0	.12
.33	.20	0	-.33	-1	-1	-1	-1	-1	-1	-1	-.96	-.89	-.82	-.74	-.65	-.55	-.42	-.18	0	.14	.24
.60	.50	.33	0	-1	-1	-1	-1	-1	-1	-1	-.94	-.83	-.73	-.62	-.50	-.37	-.20	0	.15	.27	.36
1	1	1	1	o	-1	-1	-1	-1	-1	-1	-.92	-.77	-.63	-.49	-.35	-.19	0	.17	.31	.41	.49
1	1	1	1	1	0	-.33	-.50	-.60	-.67	-.71	-.67	-.54	-.42	-.29	-.15	0	.17	.31	.41	.49	.56
1	1	1	1	1	.33	0	-.20	-.33	-.43	-.50	-.47	-.36	-.25	-.13	0	.14	.31	.41	.49	.56	.61
1	1	1	1	1	.50	.20	0	-.14	-.25	-.33	-.32	-.22	-.11	0	.12	.26	.41	.49	.56	.61	.65
1	1	1	1	1	.60	.33	.14	0	-.11	-.20	-.19	-.09	0	.10	.22	.35	.49	.56	.61	.65	.69

all $b, c \in \mathbb{C}_n^{(n)}$ ($0 \leq \alpha < n$), the angular values are symmetric, as we see in Tables 3.2 and 3.3.

The weak scalar products $\langle \cdot, \cdot \rangle_\alpha$ are not homogeneous; for any α ($0 \leq \alpha < n$), there exist pairs of cells b_1 and b_2 and c_1 and c_2 such that $2 \cdot \langle b_1, b_2 \rangle_\alpha < \langle 2b_1, b_2 \rangle_\alpha$ and $2 \cdot \langle c_1, c_2 \rangle_\alpha > \langle 2c_1, c_2 \rangle_\alpha$. It follows that these products are not linear.

3.4 Metrics on Pictures

3.4.1 Value-weighted distance

Let P be a picture with pixel or voxel values that have been divided by G_{max} so that they are in the range [0,1]. Let p and q be pixels or voxels of P, and let $\rho = (p_0, \ldots, p_n)$ be an α-path from $p = p_0$ to $q = p_n$. We define the *value-weighted length* of ρ as follows:

$$l_P(\rho) = \begin{cases} 0 & \text{if } n = 0 \\ \frac{1}{2} \sum_{i=1}^{n} (P(p_i) + P(p_{i-1})) & \text{if } n > 0 \end{cases}$$

We define the *value-weighted distance* $d_P(p,q)$ as $\min_\rho l_P(\rho)$ where the min is taken over all α-paths ρ from p to q.

Because the reversal of a path from p to q is a path from q to p and the concatenation of a path from p to q with a path from q to r is a path from p to r, d_P is symmetric and satisfies the triangle inequality, and evidently it is nonnegative. However, d_P is not positive definite (metric property **M1**); for example, if p and q are α-adjacent and $P(p) = P(q) = 0$, we have $d_P(p,q) = 0$ even though $p \neq q$. On the other hand, if $P(p) \neq 0$, we have $d_p(p,q) \neq 0$ for any $q \neq p$, because any α-path from p to q must go from p to some $p' \neq p$, and the pair (p,p') contributes a nonzero quantity $\frac{1}{2}(P(p)+P(p'))$ to the sum. Thus d_P is a metric if we restrict it to the set of pixels or voxels with values that are nonzero or even if we restrict it to pairs (p,q) with values that are not both zero. (The latter is not a restriction of d_P to a set of pixels or voxels, but it justifies our studying the [value-weighted] distance from non-0s to 0s in the next section.)

When we restrict d_P to $\langle P \rangle$ (the set of pixels or voxels with values that are 1), we evidently have $l_P(\rho) = n$ and $d_P(p,q) = d_\alpha(p,q)$ for all $p,q \in \langle P \rangle$. In the next sections, we will study 4- and 8-distances from the pixels of $\langle P \rangle$ to the subset $\langle \overline{P} \rangle$ in a 2D binary picture.

3.4.2 Distance transforms

Let P be a binary picture in which $\langle P \rangle$ and $\langle \overline{P} \rangle$ are proper subsets of the grid. For any grid metric d_α, the d_α *distance transform* of P associates with every pixel p of $\langle P \rangle$, the d_α distance from p to $\langle \overline{P} \rangle$.[4] The d_α distance transform of the set of gray pixels in the picture shown on the left in Figure 3.10 is shown for $\alpha = 4$ in the middle and for $\alpha = 8$ on the right.

We will assume in the rest of this section that the pixels outside of a rectangular region $\mathbb{G}$ all have value 0. We will now show that the d_4 or d_8 distance transform of P can be computed by performing a series of local operations while scanning $\mathbb{G}$ twice. (A *local operation* gives each pixel p a new value that depends only on the old values of the neighbors of p.)

For any $p \in \mathbb{G}$, let $B(p)$ ("before") be the set of pixels (4- or 8-) adjacent to p that precede p when $\mathbb{G}$ is scanned row by row from top to bottom when each row is scanned from left to right (see Section 1.1.3); thus, if p has coordinates (x,y), B contains $(x,y+1)$ and $(x-1,y)$, and if we use 8-adjacency, it also contains $(x-1,y+1)$ and $(x+1,y+1)$. Let $A(p)$ ("after") be the remaining (4- or 8-) neighbors of p.

Let the following be true:

$$f_1(p) = \begin{cases} 0 & \text{if } p \in \langle \overline{P} \rangle \\ \min\{f_1(q)+1 : q \in B(p)\} & \text{if } p \in \langle P \rangle \end{cases}$$

---

4. The distance transform is essentially the same as the distance field $F(\langle \overline{P} \rangle)$; see Section 3.2.5.

3.4 Metrics on Pictures

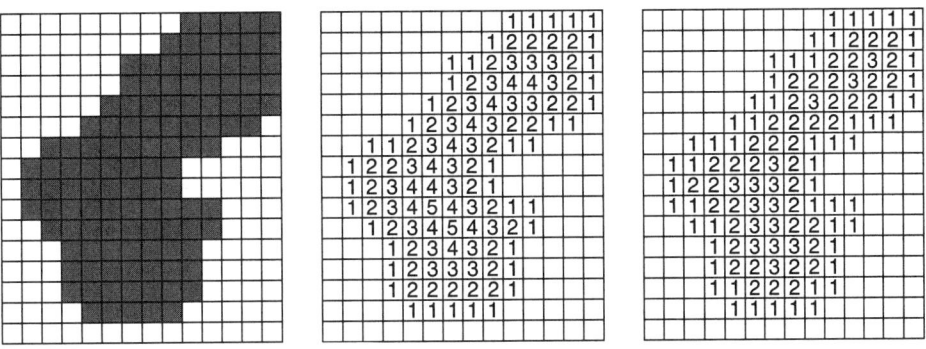

FIGURE 3.10 Distance transforms: Left: Picture. Center: d_4 transform. Right: d_8 transform.

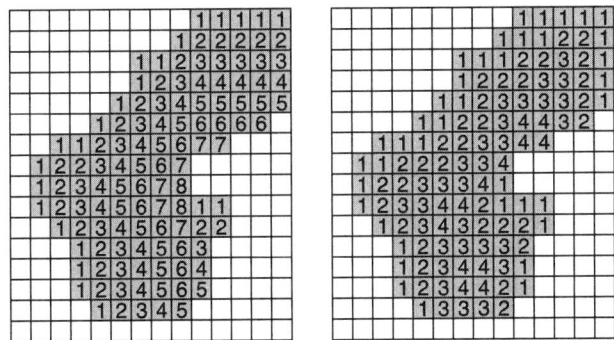

FIGURE 3.11 The first stage in computing the distance transform of the binary picture on the left in Figure 3.10. Left: d_4 transform. Right: d_8 transform.

We can compute $f_1(p)$ for every pixel in $\mathbb{G}$ in a single left-to-right, top-to-bottom scan of $\mathbb{G}$, because for each p, f_1 has already been computed for all of the qs in $B(p)$. (If p is on the top row or in the left column of $\mathbb{G}$, some of these qs are outside $\mathbb{G}$, but we know that $f_1 = 0$ for these qs because they are in $\langle \overline{P} \rangle$.)

Now let the following be true:

$$f_2(p) = \min\{f_1(p), f_2(q) + 1 : q \in A(p)\}$$

We can compute $f_2(p)$ for every pixel in $\mathbb{G}$ in a single right-to-left, bottom-to-top scan of $\mathbb{G}$, because for each p, f_2 has already been computed for all of the qs in $A(p)$ or is known because they are outside of $\mathbb{G}$.

The f_1s that use 4- and 8-adjacency are shown in Figure 3.11 for the picture on the left in Figure 3.10. The f_2s are not shown in Figure 3.11, because they are the same as the d_4s (d_8s) shown in Figure 3.10, as we see from the following:

Theorem 3.8 $f_2(p) = d(p, \langle \overline{P} \rangle)$ for all $p \in \mathbb{G}$ where $d = d_4$ for the 4-adjacency version of the algorithm and $d = d_8$ for the 8-adjacency version.

Proof Evidently, if $f_2(p) = 0$, p must be in $\langle \overline{P} \rangle$. Suppose $f_2(p) = d(p, \langle P \rangle)$ for all p such that $f_2(p) < n$, and let $f_2(p) = n > 0$. Then either some $q \in A(p)$ has $f_2(q) = n - 1$ or else $f_1(p) = n$, which implies that some $q \in B(p)$ has $f_1(q) = n - 1$. In either case, using the induction hypothesis, $d(q, \langle \overline{P} \rangle) = n - 1$ so that $d(p, \langle \overline{P} \rangle) \leq n$. If $d(p, \langle \overline{P} \rangle) < n$, we must have $d(q, \langle \overline{P} \rangle) < n - 1$ for some neighbor of p. For this neighbor, we must have either $f_1(q)$ or $f_2(q) < n - 1$, which implies that $f_2(p) < n$, thereby contradicting our assumption that $f_2(p) = n$. ∎

Note that a two-pass distance transform algorithm is valid for any chamfer distance that satisfies Montanari's inequalities 3.13, not just for d_4 and d_8.

Let $P^{(1)} = P$, and, for $k = 1, 2, \ldots$, let $P^{(k+1)}$ be the integer-valued picture in which $P^{(k+1)}(p) = 0$ if $P(p) = 0$, and otherwise let $P^{(k+1)}(p) = \min P^{(k)}(q) + 1$, where the min is taken over the pixels q that are α-adjacent to p. It is not hard to see that, for all $k \leq d_\alpha(p, \langle \overline{P} \rangle)$, we have $P^{(k)}(p) = k$ and, for all $k \geq d_\alpha(p, \langle \overline{P} \rangle)$, we have $P^{(k)}(p) = d_\alpha(p, \langle \overline{P} \rangle)$. Let $D_\alpha = \max_{p \in \langle P \rangle} d_\alpha(p, \langle \overline{P} \rangle)$; we call D_α the α-*radius* of $\langle P \rangle$. Then, for any $k \geq D_\alpha$, we have $P^{(k)}(p) = d_\alpha(p, \langle \overline{P} \rangle)$ for all $p \in \langle P \rangle$ so that $P^{(k)}$ is the d_α distance transform of $\langle P \rangle$. Note that computing the d_α distance transform in this way requires performing local operations at every pixel during $D_\alpha - 1$ scans of P to successively compute $P^{(2)}, P^{(3)}, \ldots, P^{(D_\alpha)}$, whereas the method used in Theorem 3.8 requires performing local operations during only two scans of P.

3.4.3 The Euclidean distance transform

The d_4 and d_8 distance transforms of Section 3.4.2 are easy to compute, but, as we saw in Section 3.2.3, d_4 and d_8 are not good approximations to Euclidean distance. We will now describe *Danielsson's algorithm* [229] for computing a distance transform in which the distances differ from Euclidean distance by at most a fraction of the grid constant.

To each pixel $p = (x, y)$ of P, we assign a pair of integers $(f(x), f(y))$ that is initially $(0,0)$ if $p \in \langle \overline{P} \rangle$ and (D, D) if $p \in \langle P \rangle$, where D is greater than the diameter of P (the greatest distance between any two pixels of P). We then scan P and update the $(f(x), f(y))$ values as described in Algorithm 3.2. In the min computations, we pick the pair (u, v) for which $u^2 + v^2$ is smaller; if they are equal, we pick the one for which u is smaller. Note that, in both sets of scans, the values of $(f(x), f(y))$ are first modified by a single comparison with a vertical neighbor and then by a set of comparisons with left and right horizontal neighbors.

3.4 Metrics on Pictures

---

1. For each row of P (from top to bottom), replace each $(f(x), f(y))$ (from left to right) with

   $$\min((f(x), f(y)), ((f(x), f(y-1))+(0,1)));$$

   then replace each $(f(x), f(y))$ (from left to right) with

   $$\min((f(x), f(y)), ((f(x-1), f(y))+(1,0)));$$

   then replace each $(f(x), f(y))$—except the rightmost one (from right to left)—with

   $$\min((f(x), f(y)), ((f(x+1), f(y))+(1,0))).$$

2. For each row of P—except the bottom row (from bottom to top)—replace each $(f(x), f(y))$ (from left to right) with

   $$\min((f(x), f(y)), ((f(x), f(y+1))+(0,1)));$$

   then replace each $(f(x), f(y))$ (from left to right) with

   $$\min((f(x), f(y)), ((f(x-1), f(y))+(1,0)));$$

   then replace each $(f(x), f(y))$—except the rightmost one (from right to left)—with

   $$\min((f(x), f(y)), ((f(x+1), f(y))+(1,0))).$$

ALGORITHM 3.2 Danielsson's algorithm for calculating the Euclidean distance transform.

When the scans are complete, the value of $f(x)$ at p should be the difference between the x coordinates of p and the nearest pixel q of $\langle \overline{P} \rangle$, and the value of $f(y)$ should be the difference between their y coordinates. The Euclidean distance between p and q would then be $\sqrt{f^2(x)+f^2(y)}$. In fact, we will see in the next paragraph that the $(f(x), f(y))$ values are not always exactly equal to the nearest-pixel coordinate differences. Figure 3.12 shows the $(f(x), f(y))$ values for the pixels in the gray area in Figure 3.10; in this simple example, the values are all correct.

To see how the distances computed by Danielsson's algorithm can be incorrect, consider circles of radius a centered at $(x-1, y)$ and $(x, y-1)$ (see Figure 3.13). Let $q = (x-a-1, y)$ and $s = (x, y-a-1)$, and let r be the point where the circles intersect. If q, r, and s are in $\langle \overline{P} \rangle$, the algorithm gives value $a+1$ to the pixel p at (x, y), because its neighbors $(x-1, y)$ and $(x, y-1)$ are at distance a from q and s, respectively; however, p is actually at distance $b < a+1$ from r. Indeed, as we see from Figure 3.13, the distance a from $(x, y-1)$ to r is the hypotenuse of a right triangle with legs that are $b/\sqrt{2}$ and $(b/\sqrt{2})-1$; thus $a^2 = b^2/2+(b^2/2)+1-b\sqrt{2} = b^2-b\sqrt{2}+1 = (b-\frac{1}{\sqrt{2}})^2+\frac{1}{2}$, so that the following is true:

									0,1	0,1	0,1	0,1	0,1
								0,1	1,1	0,2	0,2	0,2	1,0
						0,1	0,1	1,1	1,2	2,2	0,3	0,2	1,0
						0,1	0,2	1,2	2,2	1,3	2,2	0,2	1,0
					0,1	1,1	1,2	1,3	2,2	1,2	0,2	1,1	0,1
				0,1	1,1	1,2	2,2	1,2	0,2	1,1	0,1	0,1	
		0,1	0,1	1,1	1,2	2,2	1,2	1,1	0,1	0,1			
	0,1	1,1	0,2	1,2	2,2	3,0	2,0	1,0					
	1,0	0,2	2,2	1,3	2,3	3,0	2,0	1,0					
	0,1	1,1	1,2	2,2	3,2	3,1	2,1	1,1	0,1	0,1			
		0,1	1,1	2,1	3,1	3,2	2,2	1,2	1,1	0,1			
			1,0	2,0	3,0	0,4	3,0	2,0	1,0				
			1,0	2,0	2,2	0,3	2,2	2,0	1,0				
			0,1	1,1	0,2	0,2	0,2	1,1	1,0				
				0,1	0,1	0,1	0,1	0,1					

									1	1	1	1	1
								1	1.41	2	2	2	1
						1	1	1.41	2.24	2.83	3	2	1
						1	2	2.24	2.83	3.16	2.83	2	1
					1	1.41	2.24	3.16	2.83	2.24	2	1.41	1
				1	1.41	2.24	2.83	2.24	2	1.41	1	1	
		1	1	1.41	2.24	2.83	2.24	1.41	1	1			
	1	1.41	2	2.24	2.83	3	2	1					
	1	2	2.83	3.16	3.61	3	2	1					
	1	1.41	2.24	2.83	3.61	3.16	2.24	1.41	1	1			
		1	1.41	2.24	3.16	3.61	2.83	2.24	1.41	1			
			1	2	3	4	3	2	1				
			1	2	2.83	3	2.83	2	1				
			1	1.41	2	2	2	1.41	1				
				1	1	1	1	1					

FIGURE 3.12 Computation of the Euclidean distance transform. Left: the pair of (final) values at a pixel of $\langle P\rangle$ are its x and y coordinate differences from the nearest pixel of $\langle\overline{P}\rangle$. Right: corresponding values of the Euclidean distance, rounded to two decimal places.

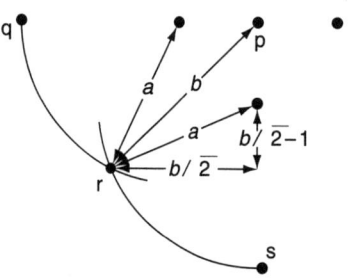

FIGURE 3.13 A worst case for the Euclidean distance transform.

$$b = \frac{1}{\sqrt{2}} + \sqrt{a^2 - \frac{1}{2}} \approx a + \frac{1}{\sqrt{2}} < a+1$$

It can be shown that this is a worst case; note that even in this case the error is only a fraction of the grid constant.

3.4.4 Medial axes

For any grid metric d_α, any pixel p, and any $k \geq 0$, let $P^{(k)}(p) = \{q : d_\alpha(p,q) \leq k\}$; thus $P^{(0)}(p) = \{p\} \subset P^{(1)}(p) \subset P^{(2)}(p) \subset \cdots$. We recall that, for $\alpha = 4$, $P^{(k)}(p)$ is a diagonally oriented square centered at p, and, for $\alpha = 8$, $P^{(k)}(p)$ is an upright square centered at p. If $p \in \langle P \rangle$ and $k < d_\alpha(p, \langle \overline{P} \rangle)$, evidently $P^{(k)}(p) \subseteq \langle P \rangle$, and all of the $P^{(k)}(p)$s contain p, so $\langle P \rangle$ is the union of all balls $P^{(k)}(p)$ with $p \in \langle P \rangle$ and $k < d_\alpha(p, \langle \overline{P} \rangle)$.

In the d_α distance transform of P, each pixel p of $\langle P \rangle$ has value $d_\alpha(p, \langle \overline{P} \rangle)$. Evidently, $P^{(d_\alpha(p, \langle \overline{P} \rangle))}(p)$ is not contained in $P^{(d_\alpha(q, \langle \overline{P} \rangle))}(q)$ for any neighbor q of p iff $p \in M_\alpha(\langle P \rangle)$. Hence $\langle P \rangle$ is the union of the balls $P^{(d_\alpha(p, \langle \overline{P} \rangle))}(p)$ for all $p \in M_\alpha(\langle P \rangle)$. The picture in which the value of p is $d_\alpha(p, \langle \overline{P} \rangle)$ if $p \in M_\alpha(\langle P \rangle)$ and 0 otherwise is called the d_α *medial axis transform* (MAT) of $\langle P \rangle$.

Definition 3.3 We say that p belongs to the *medial axis* $M_\alpha(\langle P \rangle)$ of $\langle P \rangle$ if $d_\alpha(p, \langle \overline{P} \rangle)$ is a local maximum of the d_α distance transform of P (i.e., $d_\alpha(q, \langle \overline{P} \rangle) \leq d_\alpha(p, \langle \overline{P} \rangle)$ for all α-neighbors q of p).

The medial axis transforms for the distance transforms of Figures 3.10 and 3.12 are shown in Figure 3.14. Note that the pixels of $M_\alpha(\langle P \rangle)$ are centrally located in $\langle P \rangle$, so they constitute a kind of "skeleton" of $\langle P \rangle$; however, this skeleton may not be connected even if $\langle P \rangle$ is simply connected, and it may be two pixels thick if $\langle P \rangle$ has even width. Methods of constructing thin connected skeletons will be discussed in Section 16.3.

For any $p \in \langle P \rangle$ and any grid metric d_α, let $D_\alpha(p)$ be the largest ball $P^{(k)}(p)$ centered at p that is contained in $\langle P \rangle$. If $p \in M_\alpha(\langle P \rangle)$, $D_\alpha(p)$ can be α-adjacent to $\langle \overline{P} \rangle$ at only one pixel. For example, let $\langle P \rangle$ be a "vertical strip" of even width; then M_α is of width 2, and the largest balls "touch" the strip's border on only one side, at a

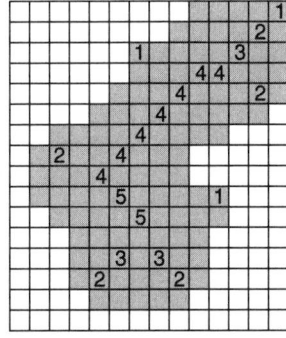
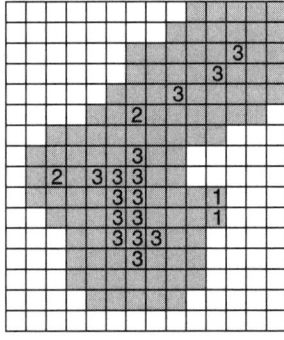
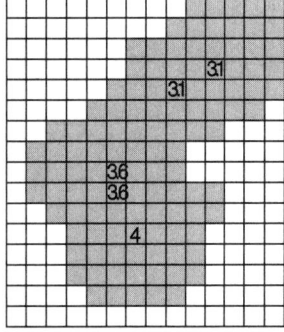

FIGURE 3.14 Medial axis transforms for the distance transforms of Figures 3.10 and 3.12. Left: d_4. Center: d_8. Right: approximate d_e (values shown to one decimal place).

single border pixel. Thus $p \in M_\alpha(\langle P \rangle)$ implies only that there is at least one shortest α-path from p to $\overline{\langle P \rangle}$. The original definition of $M(\langle P \rangle)$ described the process of constructing $M(\langle P \rangle)$ in terms of a "grass fire" ignited along the border of $\langle P \rangle$ and defined $M(\langle P \rangle)$ as the locus of points at which the grass fire meets itself. However, as we have seen, a pixel on the medial axis is not necessarily characterized by two shortest α-paths from p to $\overline{\langle P \rangle}$. (This is true also in the continuous case, for example, for a parabolic set, where the endpoint of the medial axis is at distance 0 only from itself.) On the other hand, if there are at least two shortest α-paths from p to $\overline{\langle P \rangle}$, then p is on the medial axis.

Our definitions of the distance transform and the medial axis transform assumed that P is a binary picture. We conclude this section by mentioning several generalizations of the medial axis transform to multivalued pictures.

In the SPAN [7], P is approximated by a set of maximal "disks" (e.g., squares) in which the pixel values are "homogeneous," and the generalized medial axis is the set of centers of the disks. If P is binary, the disks have constant value 1, the approximation is exact, and the set of centers of the disks is $M(P)$. (A medial axis based on fuzzy disks [fuzzy sets with membership, that depend only on distance from an origin] is described in [792].)

In the GRAYMAT [644], the gray-weighted length of a path is defined as proportional to the sum of the values of the pixels on the path (see Section 3.4.1), and the generalized medial axis is the set of pixels p that do not lie on any minimal-length path from any other pixel to the set of 0s.

The GRADMAT [1118] assigns a score to a pixel p by summing the gradient magnitudes (the maximal rates of change) of the pixel values at pairs of pixels that have p as their midpoint; the generalized medial axis consists of pixels that have high scores. Note that, in a binary picture, the gradient of the pixel values is nonzero only at the frontier of $\langle P \rangle$.

A definition of the medial axis based on morphologic operations will be given in Section 15.6.3; this definition too applies to multivalued pictures.

3.5 Exercises

1. Let $p = (x_1, y_1, z_1)$ and $q = (x_2, y_2, z_2)$ be points in $\mathbb{R}^3$, and let the following be true:

$$d_t(p,q) = \begin{cases} d_e((x_1,y_1),(x_2,y_2)) + z_1 + z_2 & \text{if } d_e((x_1,y_1),(x_2,y_2)) > 0 \\ |z_1 - z_2| & \text{otherwise} \end{cases}$$

Prove that d_t is not a metric on $\mathbb{R}^3$ but rather that it is a metric on the subset $\{p : p = (x,y,z) \in \mathbb{R}^3 \wedge z \geq 0\}$. We call it the *forest metric*, because it corresponds to the distance in which moves are of the form "climb down the first tree, walk to the second tree, and climb up the second tree."

3.5 Exercises

2. Prove that d has properties **M1** through **M3** on S iff it has the following two properties:

    (i) For all $p, q \in S$, $d(p,q) = 0$ iff $p = q$.

    (ii) For all $p, q, r \in S$, $d(p,r) \leq d(q,p) + d(q,r)$.

3. Let $[S,d]$ be a metric space. A sequence $\{p_i\}_{i=0,1,2,\ldots}$ of points of S is called *convergent* iff there is a $p \in S$ such that, for any $\varepsilon > 0$, there is an $i_0 \in \mathbb{N}$ such that $p_i \in U_\varepsilon(p)$ for all $i \geq i_0$. It is not hard to see that p must be unique; it is called the *limit point* of the sequence. Show that a sequence $p_i = (x_i, y_i) \in \mathbb{R}^2$ ($i = 0, 1, 2, \ldots$) is convergent in $[\mathbb{R}^2, L_m]$ ($1 \leq m \leq \infty$) iff it is convergent in $[\mathbb{R}^2, d_e]$.

4. Is the sequence $1, -2, 3, -4, \ldots, (-1)^{i+1}i, \ldots$ convergent in $[\mathbb{R}, d_e]$? Are the sequences

    $$(\frac{i}{i+1}, \frac{2-i}{3+i}), i \geq 0 \text{ and } (\frac{i^2+2}{3i^2+3}, \frac{3i+2}{4i+3}), i \geq 0$$

    convergent in $[\mathbb{R}^2, d_e]$? Are they convergent if we use the binary metric d_b? If so, what are their limit points?

5. Let C be a disk of integer radius centered at a grid point, and let D be the frontier of the union of the grid cells that are contained in C. Prove that the Hausdorff distance between the frontiers of C and D is equal to the grid constant.

6. Prove that d_4 and d_8 are metrics on $\mathbb{Z}^2$.

7. Prove that, on a $(k+1) \times (k+1)$ grid, $d_e - d_8$ can be as great as $(\sqrt{2}-1)k \approx 0.41k$ and $d_4 - d_e$ can be as great as $(2-\sqrt{2})k \approx 0.59k$. Prove that, for the "octagonal" distance d (Equation 3.11), we have $|d_e - d| \leq ((\sqrt{5}/2) - 1)k \approx 0.12k$.

8. Let $\|p\|_m$ be the norms defined in Section 3.1.2. Prove that, for any $p \in \mathbb{R}^2$, we have $\|p\|_2 \leq \|p\|_1 \leq \sqrt{2}\|p\|_2$, $\|p\|_\infty \leq \|p\|_2 \leq \sqrt{2}\|p\|_\infty$, and $\|p\|_\infty \leq \|p\|_1 \leq 2\|p\|_\infty$. Express these inequalities in terms of the metrics d_4, d_e and d_8 on $\mathbb{R}^2$. For $\mathbb{R}^3$, prove that $d_{26}(p,q) \leq d_e(p,q) \leq \sqrt{2} \cdot d_{26}(p,q)$ and $d_e(p,q) \leq d_6(p,q) \leq \sqrt{2} \cdot d_e(p,q)$.

9. Define "hyperoctagonal" distances d by combining d_6 with d_{18} or d_{26}, and find bounds on $|d_e - d|$ on a $(k+1) \times (k+1)$ grid.

10. The 3D grid is composed of grid planes $z = k$, where k is an integer. Construct a modified grid in which odd-numbered grid planes are shifted half a unit in the $+x$ and $+y$ directions. Each point in this grid can be regarded as having 12 neighbors: four on its own plane and four on each of the planes above and below it. Define a "d_{12}" metric on this grid in analogy with the d_h metric on the hexagonal 2D grid defined in Section 3.2.3.

11. Prove that $d_{1,1} = d_8$ and $d_{1,\infty} = d_4$.

12. Prove that the following is true:
$$\left|d_e - d_{1,\sqrt{2}}\right| \leq (\sqrt{2\sqrt{2}-2})k \approx 0{\cdot}9k$$
Prove that the chamfer distance $d_{1,b}$ that best approximates d_e has $b = (1/\sqrt{2}) + \sqrt{\sqrt{2}-1} \approx 1{\cdot}351$ and that, for this $d_{1,b}$, we have the following:
$$|d_e - d_{1,b}| \leq ((1/\sqrt{2}) - \sqrt{\sqrt{2}-1})k \approx 0{\cdot}06k$$
This optimal b is close to 4/3; we therefore get a good approximation to $3d_e$ by using $a = 3$ and $b = 4$ ("(3,4) chamfer distance"). (Other simple combinations of basic moves can be used to give even better approximations to Euclidean distance.)

13. Find bounds for $|d_e - d_{a,b,c}|$ on a $(k+1) \times (k+1) \times (k+1)$ grid where $d_{a,b,c}$ is a 3D chamfer distance, and find values for a, b, and c that minimize these bounds.

14. A knight in chess can move two steps in an isothetic direction and one step in a perpendicular direction. For any $p,q \in \mathbb{Z}^2$, let $d_k(p,q)$ be the minimum number of knight's moves required to go from p to q. Prove that d_k is a metric on $\mathbb{Z}^2$.

15. Define a concept of intrinsic distance for fuzzy subsets. (Hint: Define the length of a path $(p_1, \ldots, p_n)$ as the sum of $f(\mu(p_i))$ where f is a monotonic function that maps 0 into ∞ and 1 into 0.)

16. An integer-valued picture P is called α-*smooth* iff $|P(p) - P(q)| \leq 1$ whenever p and q are α-neighbors. Prove that the d_α distance transform of a binary picture P is the lowest-valued α-smooth picture that has value 0 at all pixels of $\langle \overline{P} \rangle$.

17. Give a "two-scan" algorithm for computing the value-weighted distance from every non-0 to the set of 0s in a multivalued picture.

18. Give "two-scan" algorithms for computing 3D distance transforms for d_6, d_{18}, and d_{26}.

19. Give an algorithm for computing a 3D Euclidean distance transform.

20. Prove that $p \in M_\alpha(P)$ iff p does not lie on a shortest α-path from any $q \neq p \in \langle P \rangle$ to $\langle \overline{P} \rangle$.

21. Define algorithms analogous to those in Theorem 3.8 for constructing the binary picture with a set of 1s that is $\langle P \rangle$ given the medial axis transform of P.

22. An *oval* is a bounded closed convex subset of $\mathbb{R}^2$; it is said to be *proper* if it has interior points. Two sets *do not overlap* iff their interiors are disjoint. Let M be a proper oval, and let $n(M)$ denote the maximum number of nonoverlapping translates of M that can be arranged so as not to be disjoint from M. Prove that $7 \leq n(M) \leq 9$, and show that $n(M) = 7$ if M is a disk and $n(M) = 9$ if M is a square.

23. Design an efficient algorithm for calculating the intrinsic diameter of a simple polygon or polyhedron and the intrinsic distances between pairs of its vertices.

24. Prove that the intrinsic (4- or 8-)diameter of a connected set of pixels S (the greatest intrinsic [4- or 8-]distance between any two pixels of S) is at most half the total (4- or 8-)perimeter of S (the sum of the [4- or 8-]perimeters of all of the frontiers of S).

3.6 Commented Bibliography

There are many textbooks about metric, normed, and Hilbert spaces; see, for example, [144, 490, 1071]. For relationships between topologies and metrics, see [352]. For a survey of publications about digital metrics, see [720].

Metrics on $\mathbb{Z}^2$ were studied in [922], which is the source of Proposition 3.2. The integer-valued metrics d_4 and d_8, the "octagonal" metrics obtained by combining d_4 and d_8, and the "hexagonal" metric d_h were all introduced in [922]; an improved treatment of d_h was given in [672]. For characterizations of d_4 and d_8, see [411, 718, 719, 723, 850] and [284]; for rounded Euclidean distance, see [851]; for additional results about octagonal distances, see [243, 244, 752].

Metric d_{18} was defined and studied in [785]. [365] also calculates d_{18} and counts the number of all minimum-length 18-paths between the origin and a grid point $(i, j, k) \in \mathbb{Z}^3$. For other references about numbers of paths, see [235, 241, 365, 922].

The n-dimensional case (see Section 3.3) was treated in [540]. The metrics ∂_α on $\mathbb{C}_n^{(n)}$, $0 \leq \alpha < n$ (see Theorem 3.6 and Equation 3.15) define balls (all n-cells at distances $\leq r \in \mathbb{N}$ from the origin) and spheres (all n-cells at distance $= r \in \mathbb{N}$ from the origin). [242] studied the volumes of these balls and the "surface areas" of these spheres (the numbers of n-cells contained in the ball or sphere).

For metrics defined by arbitrary neighborhood sequences, see [233, 239, 1146, 1147]. Chamfer distances in arbitrary dimensions were popularized in [103]. See [104, 108, 182, 236, 237, 238, 239, 241, 245, 526, 701, 785, 852, 1037] for related work. For criteria for optimizing chamfer distances, see [66, 67, 107, 156, 1093, 1114]. Distance functions were used to define "continuous" functions on pictures in [767, 903]. For linear metrics on discrete sets, see [247]. For metric-preserving transforms, see [233].

For geodesic distances, see [619]; for geodesic distances on fuzzy subsets, see [91]. Much of the material in Section 3.2.4 is from [889]. Average distances in digital sets are studied in [904], and metric bases in the grid are studied in [724]. [525] estimates distances between borders of components of voxels by calculating geodesics that contain only border voxels.

See [63] for properties of the Hausdorff metric on compact sets. The algorithm discussed in Section 3.2.5 was proposed in [989]. For a linear-time algorithm for calculating the Hausdorff distance between convex polygons, see [52]. The maximum Euclidean distance between two finite planar sets can be calculated in $\mathcal{O}(n \log n)$ time [1063]. Algorithms for calculating Hausdorff distances (defined by arbitrary Minkowski metrics) between finite planar sets are discussed in [457]. Distances between sets (as in Proposition 3.4) were studied in [556] under the name *measures of correspondence*. Minimization of the Hausdorff distance between a bounded set $S \subset \mathbb{R}^n$ and a set $M \subset \mathbb{Z}^n$ defines the *Hausdorff digitization* of S; see [131, 1116]. See [458] for more about distances between pictures. Hausdorff distances between fuzzy sets (or multilevel pictures) are studied in [179, 181, 901].

For a generalization of the concept of a distance transform, see [897]. The two-scan algorithm for computing the d_4 and d_8 transforms was introduced in [921], which also introduced the digital medial axis (under the name "distance skeleton"). This was further explored in [933]. For fast computation of distance transforms, see [833]. Generalized distance transforms are discussed in [1119]. For computing distance and distance-related transforms in nonrectangular domains, see [818]. For constrained distances, see [1095]. For 3D distance transforms and their uses, see [1040]. For other references about distance transforms, see [520, 929, 977]. [224] contains a detailed review of distance transforms and also covers algorithmic and application aspects.

The generalization of distance transforms to multivalued pictures (using the value-weighted distance to the set of pixels that have value 0) was studied in [644]. For the computation of such transforms, see [880]; for other weighted distances, see, for example, [92, 829, 1011, 1054, 1057]. For fuzzy distance transforms, see [945].

Algorithms for computing Euclidean distance transforms are discussed in [103, 125, 225, 229, 304, 306, 647, 754, 982, 983, 987]. Figure 3.13 and the accompanying discussion follows [229]. Other references for Euclidean distance transforms are [115, 263, 608, 717, 745, 831, 832, 1010, 1097]. For the 3D case, see [834, 1056, 1068].

For the medial axis (also called the "symmetric axis") and its mathematic theory, see [93, 160, 707]. For other references about digital medial axes, see [2, 28, 34, 40, 85, 197, 199, 200, 213, 322, 349, 476, 521, 634, 686, 774, 777, 953, 972, 956, 975, 1038, 1142]. For medial axes for chamfer and Euclidean distances, see [39, 41, 42, 303, 359, 845, 846, 953, 984, 1094, 1133].

For the "knight's distance" (Exercise 10), see [240, 244]. Exercise 22 is from [394], and Exercise 24 is from [886].

CHAPTER **4**

Adjacency Graphs

This chapter treats pictures as graph-theoretic objects and introduces graph-theoretic concepts that will be used throughout the book. It defines 2D and 3D adjacency graphs based on the grid point and grid cell models and on the assumption that pixels and voxels are the smallest units ("atoms") of a 2D or 3D grid. By specifying local circular orders, these graphs become oriented adjacency graphs; such graphs are related to 2D combinatorial maps, which provide descriptions of spatial subdivisions.

4.1 Graphs, Adjacency Structures, and Adjacency Graphs

The study of geometric properties of regions in 2D or 3D pictures requires specification of conditions under which pixels or voxels are considered to be adjacent or connected so that they can be regarded as belonging to the same region. Adjacencies are important in picture analysis at different levels of abstraction. Sections 1.1.4, 2.1.3, and 2.1.4 described adjacency grids of pixels or voxels. To unify the treatment of adjacency concepts in digital geometry, we generalize from these basic examples.

4.1.1 Graphs and adjacency structures

Let S be a set and R a relation on S (i.e., a set of ordered pairs of elements of S). R is called *reflexive* iff $(p,p) \in R$ for all $p \in S$ and *irreflexive* if $(p,p) \notin R$ for all $p \in S$. R is called *symmetric* iff $(p,q) \in R$ implies $(q,p) \in R$ for all $p,q \in S$. If R is irreflexive and symmetric, the elements of R can be regarded as unordered pairs (i.e., as subsets $\{p,q\} \subseteq S$ where $p \neq q$).

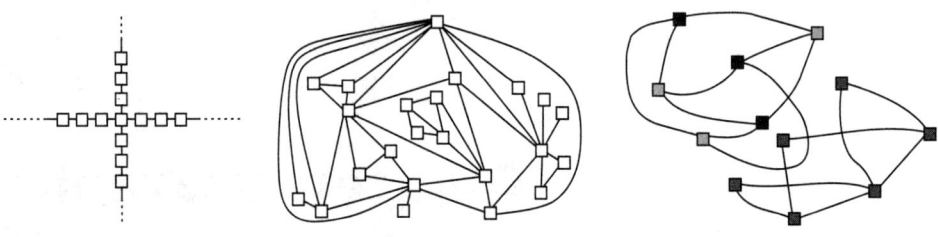

FIGURE 4.1 Three examples of adjacency structures: (left) infinite; (middle) finite, connected, and planar; and (right) finite, disconnected, and nonplanar.

We call $[S, R]$ a *graph* (more fully, a simple undirected graph) if R is irreflexive and symmetric. If R is not necessarily symmetric, we call $[S, R]$ a *directed graph* (*digraph* for short). If R is not necessarily irreflexive, we call $[S, R]$ a *pseudograph*.

If $G = [S, R]$ is a graph, S is called the set of *nodes* of G, and R is called the set of *edges* of G.[1] Edge $\{p, q\}$ is said to *join* nodes p and q or to be *between* p and q. We also say that $\{p, q\}$ is *incident* with p and q and that p and q are incident with $\{p, q\}$. Nodes p and q are called *adjacent* iff they are joined by an edge; we also say that p is adjacent to q and vice versa.

In a digraph $[S, R]$, an ordered pair $(p, q) \in R$ is called a *directed edge* from p to q; p is called its *initial node* and q its *terminal node*. Ignoring the ordering maps the digraph onto its *underlying* undirected graph. In a pseudograph, an edge that joins a node to itself is called a *loop*.

We will frequently use graph-theoretic notation in this book. Graphs provide representations of relations and tools for studying them. However, digital geometry often does not follow traditional directions of study in the theory of graphs; it poses its own problems in the context of picture analysis. For example, combinatorial problems like the historic Königsberg bridge problem illustrated in Figure 1.15 are not typically studied in digital geometry.[2]

A graph $[S, R]$ is called an *adjacency structure* if S is countable. The relation in an adjacency structure is denoted by A rather than R and is called an *adjacency relation*. (From now on, we will use $[S, A]$ rather than $[S, R]$ to denote a graph.) If p and q are adjacent, we call $\{p, q\}$ an *adjacency pair*, and we use the notation pAq. For any $p \in S$, the set $\{q : pAq\}$ is called the *adjacency set* of p and is denoted with $A(p)$. The set $\{p\} \cup A(p)$ is called the *neighborhood* of p and is denoted with $N(p)$.

Figure 4.1 shows three examples of adjacency structures. (For connectedness, see Sections 1.1.4 and 1.2.5; "planarity" will be defined later in this chapter.) Figure 4.2 shows a geographic map and indicates how the regions of the map can be represented by nodes in a graph. (The figure also illustrates the fact that four colors suffice

---

1. In graph theory, nodes are often called *vertices*, and edges are sometimes called *arcs*. To avoid confusion with *grid vertices*, we will use the term "node."

2. The abstract representation of the Königsberg bridges in Figure 1.15 is not a graph, because it has multiple "edges" between some pairs of nodes. In such a *multigraph*, R is a "bag" (a "set" in which elements can be repeated) of pairs of nodes.

4.1 Graphs, Adjacency Structures, and Adjacency Graphs

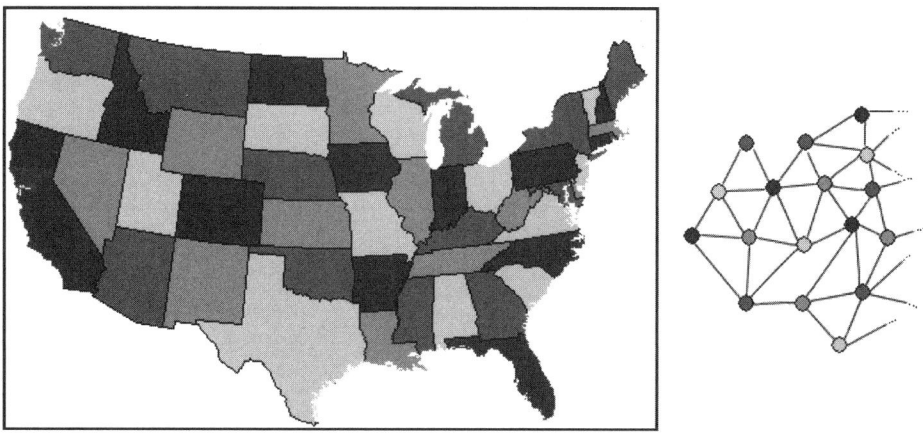

FIGURE 4.2 A map in which adjacent regions are differently colored (courtesy of Robin Thomas, Georgia Institute of Technology). On the right: a (partial) graph representation of the map.

for coloring a map in such a way that adjacent regions are colored differently; see Section 4.2.2.) The physical realization of adjacency in Figure 1.15 is via a bridge; in Figure 4.2, it is via a joint border of nonzero length.

4.1.2 Connectedness with respect to a subgraph

A *subgraph* of a graph $G = [S, A]$ is a graph with nodes that are in S and with edges that are in A. Two subgraphs are *disjoint* iff they have no nodes in common. It follows that disjoint subgraphs also have no edges in common. In this section, we generalize the definition of connectedness in Sections 1.2.5 and 2.1.1 to connectedness with respect to a subgraph.

Let $[S, A]$ be a graph and $M \subseteq S$. M defines a subgraph with node set M and edge set A_M consisting of the pairs $\{p, q\} \in A$ such that $p, q \in M$. We will sometimes use "M" as a shorthand for the subgraph $[M, A_M]$. (Such a subgraph is also called *induced by M*.)

> **Definition 4.1** Two nodes $p, q \in S$ are *connected with respect to* $M \subseteq S$ iff there is a *path* $(p_0, p_1, \ldots, p_n)$ in $[S, A]$ where $p_0 = p$, $p_n = q$, and the p_is are either all in M or all in $\overline{M} = S \setminus M$.

A single node $p \in S$ is also connected with respect to M, because a single node is a path of length 0. Adjacency with respect to a subset M of an incidence grid was defined in Definition 2.6. $[S, A]$ is called connected iff every pair of nodes $p, q \in S$ is connected with respect to S.

In Section 2.1.1, we defined adjacency relations A_α in 2D and 3D grids, where $\alpha = \{0, 1, 4, 8\}$ in 2D and $\alpha = \{2, 6, 18, 26\}$ in 3D. If A is one of these relations A_α, we use the terms α-*path* and α-*connected*. Other examples of adjacency relations introduced in Section 2.1.1 are (α, σ)-adjacency, switch adjacency, and adjacencies between cells in incidence grids.

Figure 2.4 shows a 4-path (1-path) and an 8-path (0-path) in the grid point (grid cell) model. The P-equivalence classes in Figure 2.19 consist of components of 2-cells in the incidence grid. Figure 4.12 shows 10 4-components in the grid point model.

Nodes that are connected with respect to $M \subseteq S$ are said to be in relation Γ_M. For any $M \subseteq S$, we have $\Gamma_M \subseteq S \times S$ and $\Gamma_M = \Gamma_{\overline{M}}$. Γ_M is an equivalence relation on S (i.e., it is reflexive, symmetric, and transitive). It therefore defines equivalence classes $\Gamma_M(p) = \{q : q \in S \wedge (p, q) \in \Gamma_M\}$; here p is called a *representative* of $\Gamma_M(p)$. It is easy to see that $\{p, q, \ldots\}$ is an equivalence class of Γ_M iff $\Gamma_M(p) = \Gamma_M(q)$ iff $(p, q) \in \Gamma_M$ and that $p \in \Gamma_M(p)$ for any $p \in S$. It follows that any equivalence class $\Gamma_M(p)$ is a connected subset of either M or $\overline{M}$, and these equivalence classes define a partition of S.

Definition 4.2 The equivalence classes $\Gamma_M(p)$ are called the *components* of M if they are contained in M and the complementary components of M if they are contained in $\overline{M}$. $\Gamma_M(p)$ is called *the component of p* with respect to M.

If we use one of the adjacency relations A_α to define Γ_M, the components are called α-*components* or *complementary α-components* of M.

In Section 2.2.3, we gave two algorithms for component labeling in pictures represented on adjacency or incidence grids. These algorithms generalize straightforwardly to component labeling in arbitrary adjacency graphs. In the Rosenfeld-Pfaltz labeling algorithm (see Algorithm 2.2), we need only replace the scan of the picture with a scan of all of the nodes of the adjacency graph; and in procedure FILL (see Algorithm 2.2), we need only replace "pixel" with "node" and "value u" with "node label u."

Procedure FILL can be further generalized by replacing the stack with a list data structure that may, for example, be a stack ("first-in-last-out") or a queue ("first-in-first-out"). Let $p \in S$ be a node of a (not necessarily connected) subgraph of

---

1. Label node p.
2. Put node p on a list.
3. If the list is empty, stop.
4. Take r off of the list.
5. Label all nodes $q \in A(r)$ that have not yet been labeled, and put (only!) them on the list.
6. Go to Step 3.

ALGORITHM 4.1 FILL algorithm.

4.1 Graphs, Adjacency Structures, and Adjacency Graphs

an adjacency graph $G = [S, A]$. The task is to label all of the nodes in the component $\Gamma(p)$ of this subgraph, which contains p (Algorithm 4.1). When a stack is used, the nodes in the component are visited depth first; when a queue is used, they are visited breadth first.

4.1.3 Adjacency graphs

An adjacency structure $[S, A]$ is called an *adjacency graph* iff it has the following properties:

A1: $A(p)$ is finite for any $p \in S$.
A2: S is connected with respect to A.
A3: Any finite subset $M \subseteq S$ has at most one infinite complementary component.

These properties are independent of each other:

(1) Let $S = \mathbb{Z}^2$, and let any $p \in S$ except (0,0) be adjacent to (0,0). This relation has properties **A2** and **A3** but not **A1**.

(2) Let $S = \{p, q\}$ and $A = \emptyset$. This relation has properties **A1** and **A3** but not **A2**.

(3) Let $S = \{(x, 0) : x \in \mathbb{Z}\} \cup \{(0, y) : y \in \mathbb{Z}\}$, and let A be A_4. (See the adjacency structure shown on the left in Figure 4.1.) Properties **A1** and **A2** hold, but **A3** does not hold for $M = \{(0, 0)\}$.

Statements about adjacency graphs $[S, A]$ will be based on the definition of an adjacency structure given in Section 4.1.1 and properties **A1** through **A3**; further assumptions will be made only in special cases. For example, a finite $M \subset S$ has a finite number of complementary components, because (by **A1**) the set of points of $S \setminus M$ adjacent to a point in M is finite. According to **A2**, every complementary component contains a point (in $S \setminus M$) adjacent to a point in M.

Definition 4.3 Any finite component of an adjacency graph is called a *region*. S itself is a region if it is finite. Picture analysis often involves geometric or topologic description of regions.

Adjacency grids (see Section 2.1.4) are examples of adjacency graphs. In these examples, S is either a finite $m \times n$ rectangular subset $\mathbb{G}_{m,n}$ of the discrete plane ($\mathbb{Z}^2$ or $\mathbb{C}_2^{(2)}$) or a finite $l \times m \times n$ cuboidal subset $\mathbb{G}_{l,m,n}$ of 3D discrete space ($\mathbb{Z}^3$ or $\mathbb{C}_3^{(3)}$) or the infinite discrete plane or 3D space; A can be one of the adjacency relations A_α, (α, σ)-adjacency, or switch adjacency. A set of pixels or voxels in an incidence grid is an adjacency graph iff it is nonempty, connected, and satisfies **A3**. Note that the nodes in an adjacency or incidence grid have assigned locations in a Euclidean space, but the nodes in an adjacency graph do not.

We conclude this section by giving two other examples of adjacency graphs. In the next section, we will discuss another class of examples: region adjacency graphs.

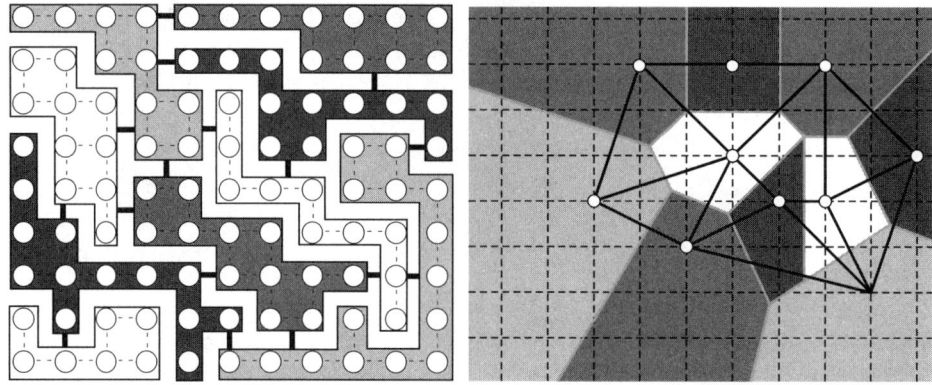

FIGURE 4.3 Left: 4-adjacencies between regions of a picture. Right: Voronoi adjacencies between selected grid points.

It can be verified that, in all of these examples, the adjacency relations are irreflexive and symmetric.

Given a set of points $\{p_1, \ldots, p_n\}$ in a plane (e.g., grid points), we associate with each p_i a *Voronoi cell*

$$V(p_i) = \{q \in \mathbb{R}^2 : d_e(p_i, q) \leq d_e(p_j, q) \text{ for } 1 \leq j \leq n\}$$

in which d_e is Euclidean distance. Points p_i and p_k are called *Voronoi-adjacent* iff $p_i \neq p_k$ and $V(p_i) \cap V(p_k)$ is a nontrivial straight line segment; the segment may be finite or infinite, but it must consist of more than one point. Figure 4.3 (right) shows Voronoi adjacencies between 10 grid points. Voronoi adjacencies in $\mathbb{R}^2$ will be discussed further in Chapter 13.

Polygonal tilings of the plane and partitions of 3D space into polyhedra also define adjacency graphs. Such a partition is called *regular* iff the intersection of two nondisjoint tiles or polyhedra is either a vertex, an edge, or a face.

Figure 4.3 (left) shows 4-adjacency between nine regions in a picture. An adjacency between two pixels p and q in a picture P is called *valid* iff p and q are P-equivalent (i.e., $P(p) = P(q)$); otherwise it is called *invalid*. The valid adjacencies do not define an adjacency graph, because P is not generally connected with respect to these adjacencies.

4.1.4 Types of nodes; region adjacencies

Nodes in a subset of an adjacency graph are classified as follows:

Definition 4.4 $p \in M \subseteq S$ is called an *inner node* iff $A(p) \subseteq M$; otherwise it is called a *border node*.[3] The set of inner nodes of M is called the *inner set* M^∇ of

---

3. In graph theory [1073], it is called a *node of attachment* of M in S.

M, and the set of border nodes of M is called the *border* δM of M. A border node of $S \setminus M$ is sometimes called a coborder node of M.

Evidently, M^∇ and δM are disjoint. These concepts resemble the topologic concepts of the interior and frontier of a subset of a topologic space (see Section 3.1.7 and further discussion in Chapter 6). In a topologic space, a closed subset contains its frontier, but a proper open subset does not. A proper subset M of an adjacency graph has a nonempty border, but it may have an empty inner set. Class 5 in Figure 2.17 has five 1-components; only the 1-component M on the left has a nonempty inner set M_1^∇, which consists of two nodes.

If the adjacency relation is one of the A_αs defined in Section 2.1, we use the terms α-*region*, α-*inner set*, or α-*border* or the symbols M_α^∇ and $\delta_\alpha M$ with $\alpha \in \{0, 1, 2, 4, 8, \ldots\}$.

The connectedness relation Γ_M partitions M^∇ and δM into components called *inner* and *border components* of M. A component M such that $M^\nabla = \emptyset$ consists of one border component. Figures 4.4 and 4.5 show the 1-border and 1-inner components of the set M defined by the union of the classes 2, 3, and 5 shown in Figure 2.17. Note that, in these figures, we assume a finite rectangular grid; for an extended (infinite) grid, we would have a few more 1-border components.

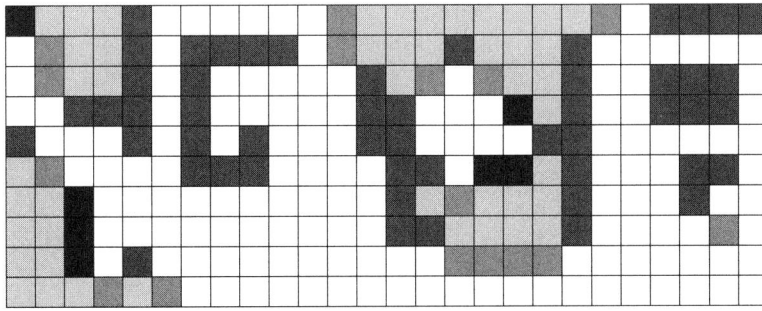

FIGURE 4.4   25 1-border components.

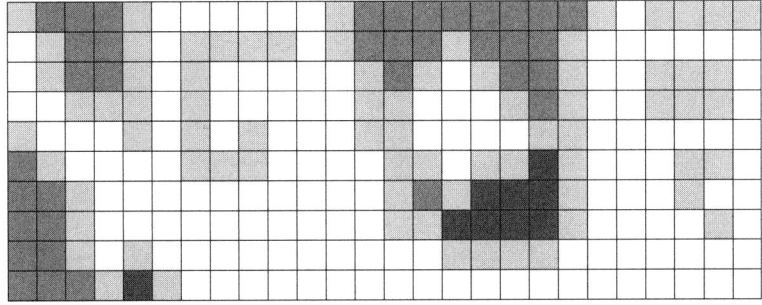

FIGURE 4.5   Six 1-inner components.

In an adjacency graph S, the operation **E** of *erosion* (see Section 1.2.12 and Chapter 15) transforms $M \subseteq S$ into M^∇, and the operation **D** of *dilation* transforms M into $M \cup A(M)$ (see below). For any $L, M \subseteq S$, we have the following:

$$\mathbf{D}M = M^\nabla \cup \mathbf{D}(\delta M) \tag{4.1}$$

In addition, $\mathbf{D}(M \cup L) = \mathbf{D}M \cup \mathbf{D}L$. For any $M \subseteq S$, we have the following:

$$\mathbf{E}M \subseteq \mathbf{DE}M \subseteq M \subseteq \mathbf{ED}M \subseteq \mathbf{D}M \tag{4.2}$$

In this case, **DE** means that we first apply **E** and then **D**. **O** = **DE** is called *opening*, and **C** = **ED** is called *closing*. Figures 4.6 and 4.7 show examples of $\mathbf{D}M$ and $\mathbf{C}M = \mathbf{ED}M$ based on 1-adjacency of 2-cells. Figure 4.8 shows an example of $\mathbf{D}M$, $\mathbf{D}^2 M$, and $\mathbf{D}^3 M$ based on 2-adjacency of 3-cells; here M contains only one 3-cell.

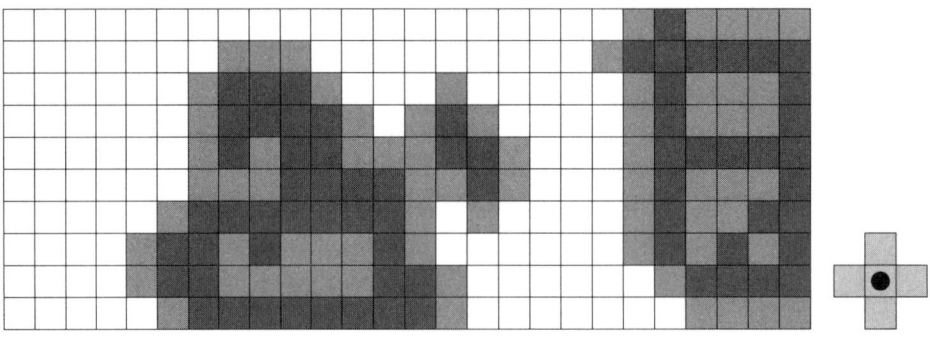

FIGURE 4.6   1-dilation of class 1 shown in Figure 2.17.

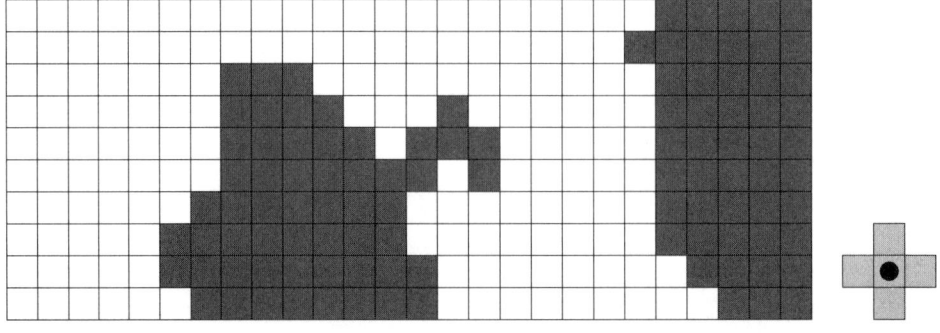

FIGURE 4.7   1-closing of class 1 shown in Figure 2.17.

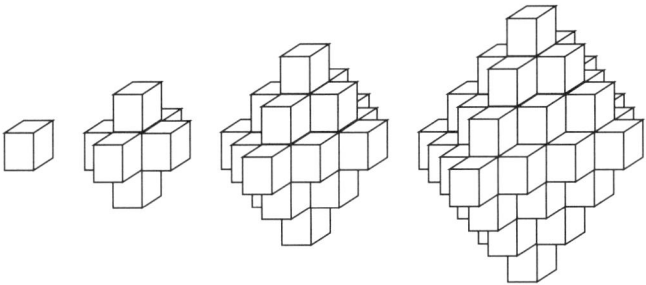

FIGURE 4.8 Repeated 2-dilation starting with one 3-cell: balls of 3-cells of radii 0, 1, 2, and 3 with respect to metric ∂_2.

The set $A(M)$ of all nodes adjacent to $M \subseteq S$ (i.e., the set of all $p \in \overline{M}$ such that $A(p) \cap M \neq \emptyset$) is called the *adjacency set* of M. Evidently, if M is finite, so is $A(M)$, and $A(S) = A(\emptyset) = \emptyset$. $A(M)$ is the set of border nodes of $\overline{M}$; this set is sometimes called the *coborder* of M. The number of complementary components of M is at most equal to the number of components of $A(M)$, because every complementary component of M intersects $A(M)$, and every component of $A(M)$ is included in a unique complementary component.

For any $p \in M^\nabla$, there is at least one $q \in \delta M$ such that p and q are connected with respect to M. If $A(M)$ consists of only one component, the same is true for $\overline{M}$. We also have $A(M) = \overline{M^\nabla} \cap A(\delta M)$.

Definition 4.5 If $[S, \mathcal{A}]$ is an adjacency graph, two disjoint subsets M_1 and M_2 of S are called *adjacent* ($M_1 \mathcal{A} M_2$ or $(M_1, M_2) \in \mathcal{A}$) iff $A(M_1) \cap M_2 \neq \emptyset$.

Because $\mathcal{A}$ is symmetric, we have $A(M_1) \cap M_2 \neq \emptyset$ iff $A(M_2) \cap M_1 \neq \emptyset$ so that $\mathcal{A}$ is symmetric. Because M_1 and M_2 are disjoint, $\mathcal{A}$ is irreflexive, so it is an adjacency relation on any partition of S.

Let $\mathcal{R}$ be a partition of S into regions and (possibly) the infinite background component. The undirected graph $[\mathcal{R}, \mathcal{A}]$ is an adjacency graph; it is called the *region adjacency graph* of $\mathcal{R}$. Figure 4.9 shows an example of a region adjacency graph defined by 1-adjacency of 2-cells. Figure 4.10 (left) shows the region adjacency graph for the picture in Figure 4.3 (left), and Figure 4.10 (right) shows the region adjacency graph for the picture in Figure 2.23, where node A represents the infinite background component. Region adjacency graphs will be discussed further in Chapter 7; see also Theorem 4.2.

4.2 Some Basics of Graph Theory

Before continuing our discussion of adjacency graphs, we review some basic graph-theoretic concepts that are (potentially) relevant to digital geometry. In this section, $G = [S, A]$ is a graph.

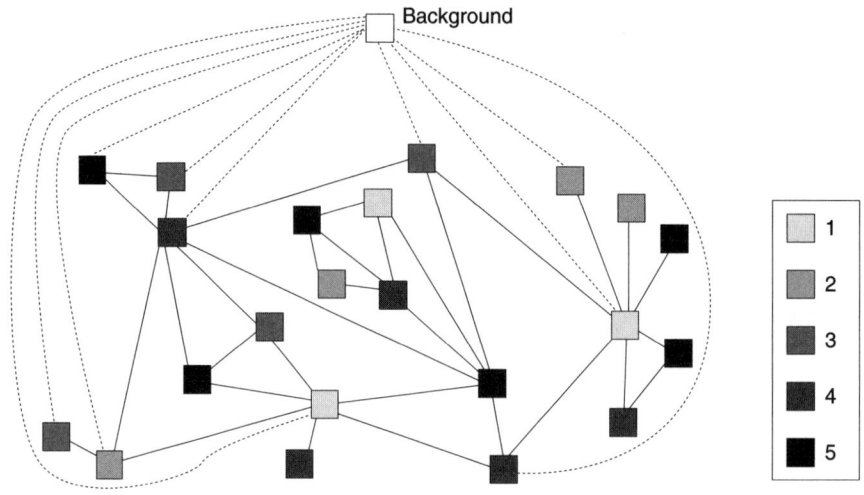

FIGURE 4.9 The region adjacency graph for the 1-regions shown in Figure 2.17.

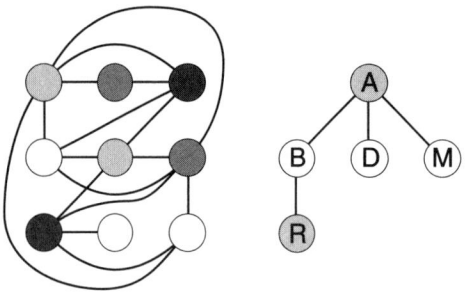

FIGURE 4.10 Two region adjacency graphs. Left: for the picture on the left in Figure 4.3. Right: for the picture in Figure 2.23.

4.2.1 Nodes, paths, and distances

Let $\alpha_0 = \mathrm{card}(S)$ and $\alpha_1 = \mathrm{card}(A)$. G is finite iff $\alpha_0 < \aleph_0 = \mathrm{card}(\mathbb{N})$. For any finite graph, we have the following:

$$0 \leq \alpha_1 \leq \sum_{n=0}^{\alpha_0 - 1} n = \frac{\alpha_0 (\alpha_0 - 1)}{2} = \binom{\alpha_0}{2} \qquad (4.3)$$

The *degree* $\nu(p)$ of node p is the number of edges that are incident with p; thus $\nu(p) = \mathrm{card} A(p)$. For example, all nodes in the infinite graph on the left in Figure 4.1 have degree 2 except for one, which has degree 4. For any finite graph, we have the

4.2 Some Basics of Graph Theory

following:

$$\sum_{p \in S} \nu(p) = 2\alpha_1 \qquad (4.4)$$

Nodes in a digraph have an *in-degree* and an *out-degree*, which are defined by the numbers of edges for which the node is initial or terminal. The maximum and sum of the in-degree and out-degree define a lower and an upper bound (respectively) for the degree of the node in the underlying undirected graph.

A *path* ρ is a sequence $(p_0, p_1, \ldots, p_n)$ of nodes of G in which consecutive nodes are adjacent. The *length* $\lambda(\rho)$ of ρ is the number n of consecutive pairs of nodes (i.e., the number of edges in the sequence).

A single node is a path $\rho = (p)$ of length zero. A *circuit* is a path that begins and ends at the same node (i.e., $p_n = p_0$) and in which no edge occurs twice. A circuit of length n will be denoted with $\langle p_1, p_2, \ldots, p_n \rangle$. A loop in a pseudograph will be regarded as a circuit of length 1. There can be a circuit of length 2 in a multigraph but not in a simple graph. A proper subpath of a circuit is called an *arc*. A single node of a circuit is an arc of length 0. Two nodes p and q of G are called *connected* if there is a path $(p_0, \ldots, p_n)$ such that $p_0 = p$ and $q_n = q$. G is called connected iff any two of its nodes are connected. Maximal connected sets of nodes of G are called *components* of G.

The *distance* $d_G(p, q)$ between two connected nodes p and q of G is the length of a shortest path connecting p and q. (If p and q are not connected, the distance between them is said to be infinite.) If G is connected, this distance is a metric, which is called the *graph metric*. Evidently, $d_G(p,q) = 0$ iff $q = p$, and $d_G(p,q) = 1$ iff pAq. Hence $A(p) = \{q \in S : d_G(p,q) = 1\}$ and $N(p) = \{q \in S : d_G(p,q) \leq 1\}$. Note that, if G is not connected, d_G is a generalized metric in the sense that the axioms of a metric are satisfied (except that a distance value can also be infinite).

A shortest path connecting two nodes of G is sometimes called a *geodesic*. It can be found by simple breadth-first search [743]. Figure 2.4 illustrates two geodesics: in $[\mathbb{Z}^2, A_4]$ on the left and in $[\mathbb{Z}^2, A_8]$ on the right.

In a *weighted graph* $G = [S, A, w]$, a positive real *weight* $w(p,q) > 0$ is assigned to every edge $\{p,q\} \in A$. Note that $w(p,q) = w(q,p)$. If p and q are not adjacent, we define $w(p,q)$ to be infinite. A graph can be regarded as a weighted graph in which $w(p,q) = 1$ for all edges $\{p,q\} \in A$.

If we identify the nodes of a graph with points in Euclidean space, we can assign to each edge $\{p,q\}$ a weight equal to the Euclidean distance between p and q. For example, in $[\mathbb{Z}^2, A_8]$, we have $w(p,q) = \sqrt{2}$ if p and q are 8-adjacent but not 4-adjacent, and $w(p,q) = 1$ if they are 4-adjacent. In a 3D picture with nonuniform spacing between voxels (e.g., different spacing between slices and within a slice), these spacings can be used to define weights.

The *total weight of a path* in a weighted graph is the sum of the weights of all the edges on the path. A path between two nodes that has minimum total weight is called a *shortest path*.

The *Shortest Path Problem* is as follows: Given a connected weighted graph $G = [S, A, w]$ and a node $p_0 \in S$, find a shortest path from p_0 to each $q \in S$. This is a special case of the distance field problem (see Section 3.1.8) in which we use minimum distances to a set of nodes instead of distance to the single node p_0.

Dijkstra's algorithm [268] (see Algorithm 4.2) solves the shortest path problem with a computational complexity of $\mathcal{O}(\alpha_0^2)$. This can be transformed into $\mathcal{O}(\alpha_1 \log \alpha_0)$ if a heap data structure is used for the set $\{q \in S \setminus V_i : D(q) < \infty\}$ of remaining nodes. Using the labels assigned in Step 2, we can construct a shortest path from any node $q \in S$ back to p_0. These labels also give the remaining shortest path lengths.

Figure 4.11 shows an application of Dijkstra's algorithm to a simply 4-connected region (shown in the grid cell model; for an informal definition of "simply connected," see Section 1.2.6). We choose a start node p on the border of the component and find all shortest 4-paths from p to nodes in the component with

1. Let $i = 0$, $V_0 = \{p_0\}$, $D(p_0) = 0$, and $D(q) = \infty$ for $q \neq p_0$. If $\alpha_0 = 1$, then stop; otherwise, go to Step 2.
2. For each $q \in S \setminus V_i$, update $D(q)$ by $\min\{D(q), D(p_i) + w(q, p_i)\}$. If $D(q)$ is replaced, put a label $[D(q), p_i]$ on q. (This allows for the tracking of shortest paths.) Overwrite the previous label, if there is one.
3. Let p_{i+1} be a node that minimizes $\{D(q) : q \in S \setminus V_i\}$.
4. Let $V_{i+1} = V_i \cup \{p_{i+1}\}$.
5. Replace i with $i + 1$. If $i = \alpha_0 - 1$, then stop; otherwise, go to Step 2.

ALGORITHM 4.2 Dijkstra's algorithm.

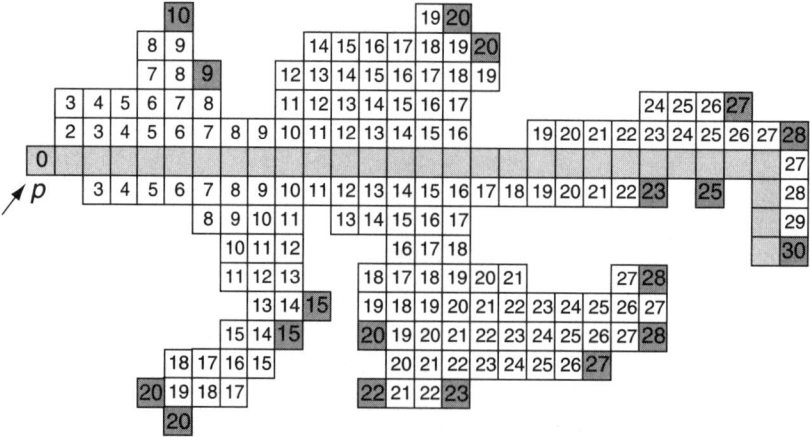

FIGURE 4.11 A start node p; nodes (shaded) with distances to p that are local maxima; and a shortest path to the shaded node farthest from p.

4.2 Some Basics of Graph Theory

lengths that are local maxima; this maps the component into a set of 4-arcs. The figure shows a shortest path to a node farthest from p. ("Thinning" algorithms that map components into sets of 4-arcs are discussed in Section 16.3.)

The *eccentricity* $e(p)$ of a node p of a finite connected graph $G = [S, A]$ is the greatest distance $\max\{d_G(p,q) : q \in S\}$ from p to any node of S. Dijkstra's algorithm can be used to calculate $e(p)$. The *radius* $r(G)$ and the *diameter* $d(G)$ are (respectively) the minimum and maximum eccentricities of all of the nodes of S. Thus p is called a *central node* of G iff $e(p) = r(G)$; the set of central nodes is called the *center* of G. Figure 4.12 shows the centers of some 4-connected sets of grid points in the graph defined by A_4.

An *Eulerian path* in G is a path that contains all of the edges of G; see the historic comments at the beginning of Section 1.2.5. A graph is called *Eulerian* iff it has an Eulerian circuit.

A connected graph G is Eulerian iff every node of G has even degree, and G has an Eulerian path iff it has at most two nodes of odd degree (see Figure 4.13); in the latter case, the path starts at one of the nodes that has odd degree and ends at the other. (Note that there cannot be exactly one node of odd degree.) It follows that the existence of Eulerian paths or circuits can be determined, and Eulerian paths can be found in computation time-linear in the number of nodes.

A path that visits each node of G exactly once is called *Hamiltonian*. (In 1857, the London toymaker J. Jaques manufactured a puzzle called the "icosian game" that involved finding such a path along the edges of a dodecahedron. The toymaker

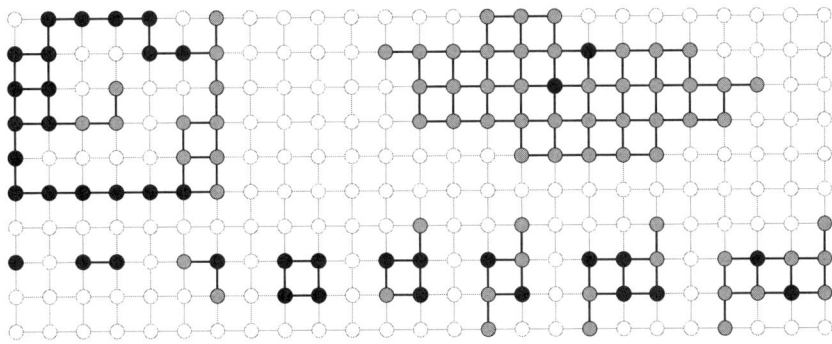

FIGURE 4.12 Centers of 4-connected sets of grid points (shown as black dots).

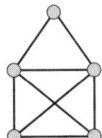

FIGURE 4.13 A graph that has an Eulerian path but is not Eulerian.

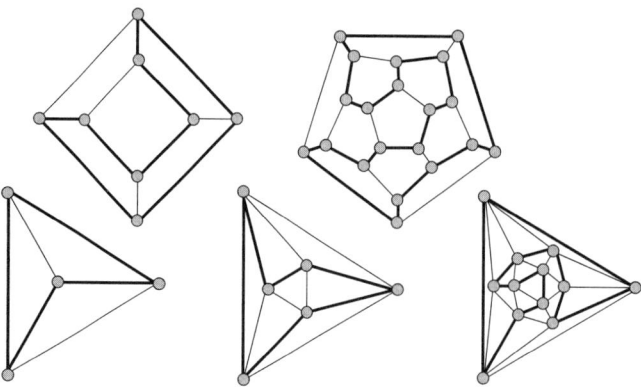

FIGURE 4.14 Five examples of Hamiltonian graphs: planar graphs representing the five Platonic solids.

paid 25 pounds to Sir W.R. Hamilton for the rights to the puzzle. Hamilton proudly told Jaques that his bankers considered him at this point to be a man of business, because he had obtained *cash* for one of his inventions [408].) A graph that has a Hamiltonian circuit is called a *Hamiltonian graph*. Figure 4.14 shows examples of Hamiltonian graphs; they are planar representations (see Figure 4.17) of the vertices, edges, and faces of the five *Platonic solids*: convex polyhedra with faces that are congruent convex regular polygons. There are exactly five such solids (*Euclid*): the tetrahedron, the octahedron, the cube, the icosahedron, and the dodecahedron, which have 4, 6, 8, 12, and 20 vertices, respectively. The 4-adjacency grid $\mathbb{G}_{m,n}$ is a Hamiltonian graph if either m or n is even; to obtain a Hamiltonian circuit, start at the lower left corner, go all the way to the right, then zig-zag up ("on the right") and down ("on the left") until the start node is reached again. Note that, for $m,n \geq 4$, such a Hamiltonian circuit is not uniquely defined.

The problem of finding a Hamiltonian circuit in a graph or deciding whether one exists is *NP-complete*.[4] A connected graph G with $n \geq 3$ nodes has a Hamiltonian circuit if $\nu(p) + \nu(q) \geq n$ for every pair of nonadjacent nodes p and q of G [272].

4.2.2 Special types of nodes, edges, and graphs

A node of degree 0 is called *isolated* and is a component of G. A node of degree 1 is called an *end node*. The edge incident with an end node is called a *pendant edge*.

---

[4]. A computational problem is called NP-complete iff it is both NP (solvable in nondeterministic polynomial time) and NP-hard (any other NP problem can be translated into it) [356]. Algorithms that solve NP-complete problems have time complexities that may be between polynomial and exponential, exponential (e.g., using exhaustive search), or even bigger than exponential (e.g., on the order of n^n). "Approximate" solutions to NP-complete problems can sometimes be found that have lower computational cost.

4.2 Some Basics of Graph Theory

A *cut node* is a node such that, if it is removed from S along with all of the edges incident with it, the number of components of G increases. For example, the graph in the middle in Figure 4.1 has two cut nodes.

If G is connected, its *node connectivity* is the smallest number of nodes for which removal from S disconnects G or completely deletes it. G is called *k-strong* iff its node connectivity is at least k. If G is connected, it is at least 1-strong; if G has no cut nodes, it is at least 2-strong.

A *bridge* is an edge of G for which removal from A disconnects G. An edge of a connected graph is a bridge iff it does not belong to any circuit.

A finite graph that has no circuits is called a *forest*, and, if it is also connected, a *tree*. Any tree that has more than one node has at least one pendant edge.

Theorem 4.1 (C. Jordan, 1869) The center of a tree is either a single node or a pair of adjacent nodes.

Theorem 4.2 Let P be a binary picture defined on a (4,8)- or (8,4)-adjacency grid (see Section 2.1.3) that is extended into $\mathbb{Z}^2$. Then the region adjacency graph of P is a tree.

A proof will be given in Section 7.3.2.

A *rooted tree* is a tree with a distinguished node called the *root*. We recall (Section 4.2.2) that a region adjacency graph has at most one node that represents an infinite component. If the graph is a tree, it is usual to choose this node as the root of the tree. Figure 4.15 shows the region adjacency trees defined by the (4,8)- and (8,4)-adjacency structures of a binary picture P in which the pixels of $\langle P \rangle$ are represented by filled dots.

The distances of the nodes from the root of a rooted tree define *layers* or *levels* in the tree. Layer -1 is the empty set, and Layer 0 contains only the root. For $n \geq 0$, Layer $n+1$ contains all nodes that are not in Layer $n-1$ but that are adjacent to a node in Layer n. Nonroot nodes of a rooted tree that have degree 1 are called *leaves*.

A *spanning tree* of a graph $G = [S, A]$ is a subgraph of G that is a tree and that has vertex set S. A *minimum-length spanning tree* is a spanning tree with the fewest possible edges. For example, the minimum-length spanning tree of a 4-adjacency grid $\mathbb{G}_{m,n}$ is defined in terms of the Manhattan distance d_4. Starting the construction of a minimum-length spanning tree at the center of a graph minimizes the diameter of the tree [997].

$G = [S, A]$ is called *bipartite* iff S can be partitioned into two disjoint subsets S_1 and S_2 such that each edge in A joins a node in S_1 to a node in S_2. For example, one component of the graph on the right in Figure 4.1 is bipartite. The 4-adjacency grid $\mathbb{G}_{m,n}$ is bipartite; the nodes can be partitioned like the squares on a chessboard.

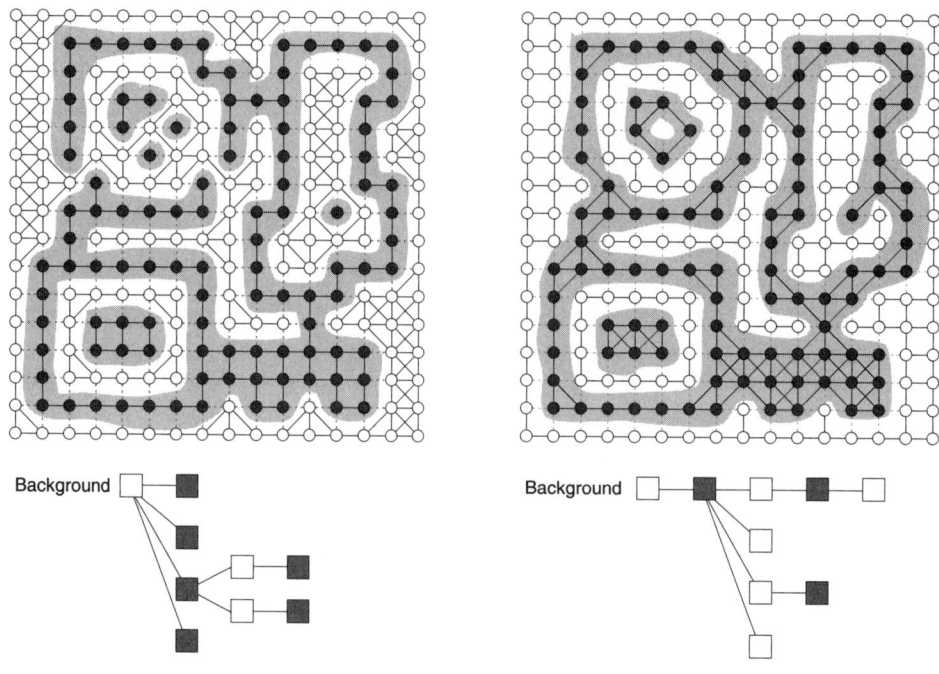

FIGURE 4.15 (4,8)- (on the left) and (8,4)- (on the right) adjacency structures defined by the set $\langle P \rangle$ of 1s of a binary picture. Below: The rooted region adjacency trees for the regions and complementary regions of $\langle P \rangle$.

$\mathbb{G}_{m,n}$ is even *complete bipartite* (i.e., there exists a partition of its node set into two subsets S_1 and S_2 such that every node in S_1 is adjacent to every node in S_2). If S_1 has n nodes and S_2 has m nodes, the complete bipartite graph is denoted by $K_{n,m}$. One component of the graph on the right in Figure 4.1 is $K_{3,3}$. This graph arises in the puzzle of "three houses, each connected to the electricity, water, and gas companies."

A graph is called *complete* if every pair of its nodes is adjacent. For a finite complete graph, we have $\alpha_1 = \frac{\alpha_0(\alpha_0-1)}{2}$. The complete graph that has $\alpha_0 = n$ nodes is denoted by K_n. Evidently, K_n is n-strong. K_1 consists of one isolated node, K_2 consists of one pair of adjacent nodes, and K_3 consists of a circuit of length 3 (a *triangle*).

Renaming the nodes in a graph leads to an *isomorphic* graph. Formally, two graphs $[S_1, A_1]$ and $[S_2, A_2]$ are isomorphic iff there is a one-to-one mapping f from S_1 onto S_2 such that each edge $\{p,q\}$ in A_1 maps onto an edge $\{f(p), f(q)\}$ in A_2 and this mapping of A_1 onto A_2 is one-to-one. We can now say more precisely that one of the components of the graph on the right in Figure 4.1 is isomorphic to $K_{3,3}$. Figure 4.16 illustrates graphs that are isomorphic to $K_{3,3}$ and K_5. The computational problem of determining whether or not two finite graphs are isomorphic is NP-complete.

4.2 Some Basics of Graph Theory

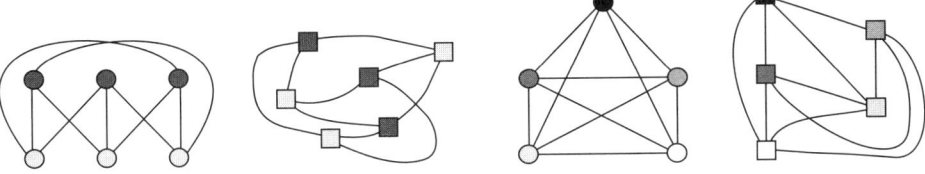

FIGURE 4.16 Two isomorphic representations of $K_{3,3}$ on the left and of K_5 on the right.

The nodes of a graph have no coordinates; there is no specific way to draw a graph, for example, in the plane. A graph is called *planar* iff it allows a *planar drawing*[5] (i.e., it can be drawn in a plane in such a way that its edges are drawn as simple arcs that intersect only at nodes). For example, K_1, K_2, K_3, and K_4 are planar, but K_5 is not; the adjacency structure $[\mathbb{Z}^2, A_4]$ is planar, but $[\mathbb{Z}^2, A_8]$ is not. There are algorithms with linear run times (in the number of nodes) for deciding whether or not a given finite graph is planar; the first such algorithm was published in [443], and it is an iterative version of a method proposed in [53] and correctly formulated in [367]. The *crossing number* of a graph is the smallest possible number of intersections of edges (other than at nodes) for any drawing of the graph in a plane. A graph with crossing number 0 is planar. Determining the crossing number of a graph is an NP-complete problem.

With any polyhedron Π, one can associate a graph G as follows: the vertices of Π are represented by the nodes of G, and two nodes of G are joined by an edge iff the corresponding vertices of Π are joined by an edge of Π. The following theorem was proved by E. Steinitz (1871–1928):

Theorem 4.3 (Steinitz's Theorem) *The graph associated with any simple polyhedron is a 3-strong planar graph. Conversely, every finite 3-strong planar graph is the graph associated with some simple polyhedron.*

Figure 4.17 shows a graph associated with a convex polyhedron.

A planar drawing of a finite planar graph partitions the plane into *faces*. The frontier of each face defines a *cycle* of consecutive arcs that join pairs of points p and q where $\{p,q\}$ is an edge in the graph; this cycle corresponds to a circuit in the graph. Let α_2 be the number of faces in such a drawing. Using *Euler's formula*,[6]

$$\alpha_2 = 2 - \alpha_0 + \alpha_1 \tag{4.5}$$

---

5. This is an informal definition; a formal definition will be given in Section 4.3.2.

6. This formula appeared in an earlier (unpublished) fragment by R. Descartes; it is sometimes called the *Descartes-Euler polyhedron theorem*. It was discovered by L. Euler (1707–1783) around 1750 for convex polyhedra and was first proved in 1794 by A.-M. Legendre (1752–1833).

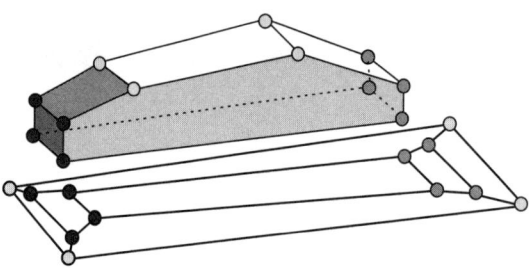

FIGURE 4.17 The planar graph (below) represents all the faces, edges and vertices of the convex polyhedron (above).

we know that α_2 is uniquely defined for any finite planar graph (i.e., it depends on the graph but not on the drawing). For example, for K_1 and K_2, we have $\alpha_2 = 1$, and, for K_3, we have $\alpha_2 = 2$.

A planar drawing of a finite planar graph has $\alpha_2 - 1$ bounded faces (its *internal faces*) and one infinite face (its *external face*). (Formula 4.5 also counts the external face; otherwise, 2 needs to be replaced with 1. The formula is also correct for drawing a planar graph on a sphere, which results in bounded faces only.) For example, any planar drawing of K_3 has one internal and one external face. For any finite planar graph, we have the following:

$$\sum_\rho \lambda(\rho) = 2\alpha_1 \qquad (4.6)$$

where the sum is taken over all of the cycles ρ defined by all of the faces of a planar representation of the graph. (We recall that $\lambda(\rho)$ is the length of ρ.)

The *merging of two adjacent nodes* p and q in a graph $[S, A]$ is defined by replacing p and q with a new node r that is adjacent to every node in $S \setminus \{p, q\}$ to which p or q was originally adjacent. A finite sequence of merging operations is called a *contraction*.

Theorem 4.4 (C. Kuratowski, 1930) A graph is planar iff it has no subgraph that can be contracted into either $K_{3,3}$ or K_5.

Efficient planarity testing algorithms are not based on this theorem.

The *chromatic number* of a (finite) graph is the smallest number of colors needed to color the nodes of the graph so that adjacent nodes have different colors. For example, any bipartite graph has chromatic number 2, and the complete graph K_n has chromatic number $n - 1$. The problem of calculating the chromatic number of a finite graph is NP-complete.

A geographic map (see Figure 4.2) can be represented by a finite planar graph in which each country is represented by a node, and there is an edge between two nodes iff the intersection of the frontiers of any two countries consists of more

4.3 Oriented Adjacency Graphs

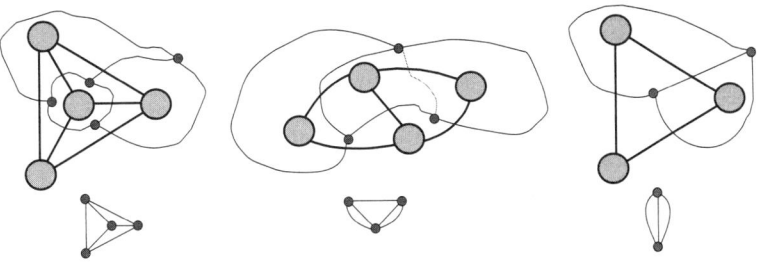

FIGURE 4.18 Planar drawings of three planar graphs and their geometric duals.

than a set of isolated points. (The intersection may consist of several disjoint arcs.) Any planar drawing of a finite planar graph is a planar drawing of such a "map graph."

Theorem 4.5 (The fourcolor theorem) *The chromatic number of any finite planar graph is at most 4.*

This theorem has a long and interesting history of published "proofs" that were later found to be incorrect [89, 410]. The problem was first stated in 1852, and many mathematicians contributed to its eventual solution [25].

Given a planar drawing of a (finite or infinite) planar graph, the *geometric dual* of the drawing is constructed by choosing a point in each face of the drawing (including the external face), and, if the frontiers of two faces have a simple arc γ in common, joining those faces' points with a simple arc that crosses γ. The result of this construction is a planar drawing of a multigraph. Figure 4.18 shows three examples. The planar drawing on the left is *self-dual*, because the drawing and its geometric dual are isomorphic. The planar drawing of $[\mathbb{Z}^2, A_4]$ is also self-dual. The other two planar drawings in Figure 4.18 have duals that are multigraphs but not graphs. (Note that multigraphs occur only if the original planar graph is not connected or because of the external face.) Thus the dual of a planar drawing of a finite planar graph is more general than a planar graph representation of a geographic map, where we excluded multiple edges.

4.3 Oriented Adjacency Graphs

There are infinitely many ways to draw a finite planar graph. We call two drawings equivalent iff the clockwise order of the edges around each node is the same in both drawings. This defines equivalence classes of planar drawings, which are called *combinatorial embeddings*.

4.3.1 Local circular orders

In this section, we introduce the concept of a "local" circular order of the edges incident with a node of a graph and generalize it so that it can be applied to any (not necessarily finite or planar) adjacency graph.

The directional code shown in Figure 2.31 specifies a local circular order on the grid points of $[\mathbb{Z}^2, A_8]$ (i.e., a cyclic ordering of the grid points that are 8-adjacent to any given grid point). Such orders provide a basis for defining border tracing routines.

Let $[S, A]$ be an adjacency graph. In a *local circular order* $\xi(p)$ at node $p \in S$, the nodes $\langle q_1, \ldots, q_n \rangle$ of $A(p)$ appear exactly once each. We can use these local orders to trace (directed) edges in $[S, A]$ as follows: if we arrive at p from $q_i \in A(p)$, we move next to q_k, where $k = i + 1$ (modulo n).

Figure 4.19 shows all possible ways of defining a local circular order on $[\mathbb{Z}^2, A_4]$. For example, in case (A), for any grid point $p = (x, y) \in \mathbb{Z}^2$, we have $\xi(p) = \langle (x - 1, y), (x, y + 1), (x + 1, y), (x, y - 1) \rangle$. Let us denote the directions from (x, y) to these four neighbors with 1, 2, 3, and 4 (modulo 4). If we arrive at p from below (direction 4), we proceed (modulo 4) in direction 1 to a new grid point. At the new grid point, we arrive from direction 3 and proceed in direction 4. At the next grid point, we arrive from direction 2 and proceed in direction 3, and, at the next grid point, we arrive from direction 1 and proceed in direction 2; this closes a circuit of length 4 in $[\mathbb{Z}^2, A_4]$. Note that, in this example, we used the same local circular (A) at every node; we will do this from now on.

Any move from a grid point p to one of its neighbors q *initiates a (directed) path* defined by the local circular order. Figure 4.20 shows all possible initiated paths using

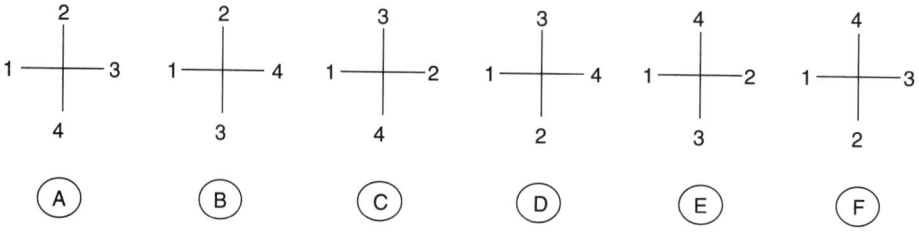

FIGURE 4.19 All possible local circular orders for 4-adjacency.

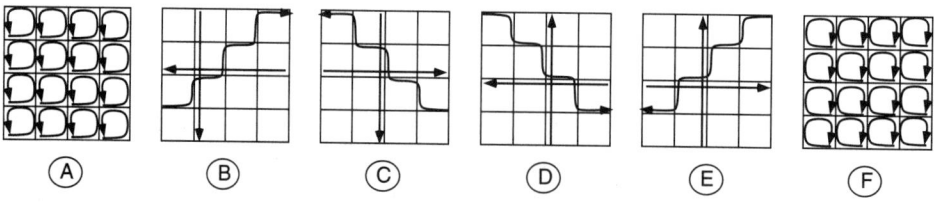

FIGURE 4.20 Initiated 4-paths in the infinite grid point plane.

4.3 Oriented Adjacency Graphs

each of the local circular orders shown in Figure 4.19. The initiated paths are infinite in cases (B) to (E); only cases (A) and (F) lead to finite paths (circuits), which are always of length 4. We call these finite paths *cycles*.

Local circular orders (A) and (F) generalize the cycles used to trace borders in planar drawings, where we use either the clockwise or counterclockwise order of the edges incident with each node.

An *oriented adjacency graph* $[S, A, \xi]$ is defined by an adjacency graph $[S, A]$ (having properties **A1** through **A3**; see Section 4.1.3) and an *orientation* ξ, defined by local circular orders of the adjacency sets, which satisfies the following:

A4: Any directed edge initiates a cycle (and not an infinite path).

Any finite adjacency graph obviously has this property (i.e., any finite $[S, A]$ and any ξ define an oriented adjacency graph).

For $[\mathbb{Z}, A_4]$, circuits of length 4 are the only possible cycles, and only the orientations (A) and (F) of Figure 4.19 lead to oriented adjacency graphs. For any $[\mathbb{G}_{m,n}, A_4]$, in cases (A) and (F), we obtain cycles of length 4 plus one additional cycle, which we say *circumscribes* $\mathbb{G}_{m,n}$ (and "defines" an external face).

4.3.2 The Euler characteristic and planarity

Let α_2 be the number of cycles of a finite oriented adjacency graph $[S, A, \xi]$. (Note that a cycle cannot be decomposed into two or more cycles.) If $[S, A]$ is planar, it has α_2 faces.

We recall that $\alpha_0 = \mathrm{card}(S)$, $\alpha_1 = \mathrm{card}(A)/2$, and $\nu(p) = \mathrm{card}(A(p))$. Equation 4.4 is true for any finite graph; hence, for any finite oriented adjacency graph, the following is given:

$$\sum_{p \in S} \nu(p) = 2\alpha_1$$

Equation 4.6 also generalizes to arbitrary finite oriented adjacency graphs,

$$\sum_{\rho} \lambda(\rho) = 2\alpha_1 \qquad (4.7)$$

where $\lambda(\rho)$ is the length of cycle ρ and the sum is taken over all cycles in $[S, A, \xi]$.

From property **A2**, it follows that there are at least $\alpha_0 - 1$ edges (i.e., $\alpha_1 \geq \alpha_0 - 1$). Two nodes can be connected by at most one undirected edge so that $\alpha_1 \leq \alpha_0(\alpha_0 - 1)/2$. A single node ($\alpha_0 = 1$ and $\alpha_1 = 0$) defines a degenerate cycle ($\alpha_2 = 1$). A given (nonplanar) finite adjacency graph $[S, A]$ can have different numbers α_2 of cycles depending on its orientation ξ.

The *Euler characteristic* χ, which is also known as the *Euler number*, will be characterized in Section 6.4.5 as a topologic invariant for finite Euclidean complexes that consist of bounded sets. From Equation 4.5 (Euler's formula), we know that $\chi = 2$ for the surface of a simple polyhedron so that $\chi = 1$ (not counting the unbounded external face) for the finite planar graph that represents the surface of the polyhedron. In terms of the concepts in Chapter 6, a graph defines a finite one-dimensional complex of nodes and arcs. A planar graph defines a finite number of bounded internal faces and one unbounded external face. (If we project the planar graph onto a sphere, the external face becomes a bounded face.) In this section, we deal with graphs that are not necessarily planar. To simplify the notation, we regard all of the cycles of an oriented adjacency graph as "faces," including unbounded external faces, and indicate this with the superscript $^+$.

The Euler characteristic of a finite oriented adjacency graph $[S, A, \xi]$ is defined to be $\chi = \alpha_0 - \alpha_1 + (\alpha_2 - 1)$. Let $\chi^+ = \chi + 1 = \alpha_0 - \alpha_1 + \alpha_2$. Then we have the following:

Theorem 4.6 $\chi^+ \leq 2$ for any finite oriented adjacency graph.

Proof The proof is by induction based on repeated addition of new edges to a finite oriented adjacency graph. If $\alpha_0 = 1$, $\alpha_1 = 0$, and $\alpha_2 = 1$, we have $\chi^+ = \alpha_0 - \alpha_1 + \alpha_2 = 2$. Suppose $\chi^+ \leq 2$ for $G = [S, A, \xi]$. We can add an edge to G, thereby producing a graph G_{new} in two ways:

1. Add a new node q to S, and connect q with a new edge to an existing node p of S.

2. Add a new edge to A by connecting two nodes p and q of S that were not connected by an edge in A.

In the first case, we increase both α_0 and α_1 by 1. The local circular order $\xi(p) = \langle \ldots, p_i, p_j, \ldots \rangle$ becomes $\xi_{new}(p) = \langle \ldots, p_i, q, p_j, \ldots \rangle$. (p_i, p) initiated a cycle $\langle \ldots p_i, p, p_j, \ldots \rangle$ in G and initiates a cycle $\langle \ldots p_i, p, q, p, p_j, \ldots \rangle$ in G_{new}. All other cycles in G remain unchanged. Thus α_2 does not change, so we have the following:

$$\chi^+_{new} = (\alpha_0 + 1) - (\alpha_1 + 1) + \alpha_2 = \chi^+ \leq 2$$

In the second case, suppose first that some cycle of G contains both p and q. This cycle $\langle \ldots, p_i, p, p_j, \ldots, q_k, q, q_l, \ldots \rangle$ splits into two cycles, $\langle \ldots, p_i, p, q, q_l, \ldots \rangle$ and $\langle p, p_j, \ldots, q_k, q \rangle$, so that the following is true:

$$\chi^+_{new} = \alpha_0 - (\alpha_1 + 1) + (\alpha_2 + 1) = \chi^+ \leq 2$$

Finally, suppose p and q are in two different cycles of G, for example, $\langle \ldots, p_i, p, p_j, \ldots \rangle$ and $\langle \ldots, q_k, q, q_l, \ldots \rangle$. The new edge links these two cycles, thereby resulting in the cycle $\langle \ldots, p_i, p, q, q_l, \ldots, q_k, q, p, p_j, \ldots \rangle$, so that the following is given:

$$\chi^+_{new} = \alpha_0 - (\alpha_1 + 1) + (\alpha_2 - 1) = \chi^+ - 2 \leq 0 \qquad \blacksquare$$

4.3 Oriented Adjacency Graphs

This proof also implies that the Euler characteristic $\chi+1$ of a finite oriented adjacency graph is always a multiple of 2 $(\ldots,2,0,-2,-4,\ldots)$.

Theorem 4.6 suggests the following generalization (and formalization) of the concept of planarity, which was defined in Section 4.2.2 in a more informal way:

Definition 4.6 An oriented adjacency graph is called *planar* iff it either is finite and has $\chi^+ = 2$ or is infinite and any of its nonempty finite connected oriented subgraphs has $\chi^+ = 2$.

This definition is consistent with graph theory and combinatorial topology: $\chi = 1$ for any planar drawing of a planar graph, and each such drawing defines an orientation for the graph.

If local circular orders are defined for all of the adjacency sets of $K_{3,3}$ or K_5, it can be verified that the resulting finite oriented graph always has $\chi^+ < 2$. It follows that any finite undirected graph that has a subgraph that can be contracted into either $K_{3,3}$ or K_5 has $\chi^+ < 2$ for any orientation defined on it. By Kuratowski's theorem, it follows that a finite connected graph is planar iff it has at least one orientation for which $\chi^+ = 2$.

The infinite 4-adjacency grid $[\mathbb{Z}^2, A_4]$, using either the clockwise (A) or counterclockwise (F) local circular order (see Figure 4.19), is planar. Any connected subset of $[\mathbb{Z}^2, A_4]$ is also planar; thus the Euler characteristic remains $\chi^+ = 2$ for any connected region of a grid under 4-adjacency. An infinite switch-adjacency grid $[\mathbb{Z}^2, A_s]$ (see Section 2.1.3) is also planar, and its connected regions too are planar.

The infinite 8-adjacency grid $[\mathbb{Z}^2, A_8]$ (e.g., with counterclockwise local circular orders) is nonplanar. In fact, let $\mathbb{G}_{m,n}(m,n \geq 2)$ be a rectangular oriented subgraph of $[\mathbb{Z}^2, A_8]$. Then (see Figure 4.21) $\chi^+ = -2(m-2)$ for $m \geq 2$ and $n = 2$ ($\alpha_0 = 2m$, $\alpha_1 = 5m - 4$, and $\alpha_2 = m$); and $\chi^+ = -2(3m-4)$ for $m \geq 2$ and $n = 4$ ($\alpha_0 = 4m$, $\alpha_1 = 13m - 10$, and $\alpha_2 = 3m - 2$). For example, for $m = 6$ and $n = 4$, we have $\chi^+ = -28$. In a picture defined on an 8-adjacency grid, the Euler characteristic of a connected region (its "degree of planarity") decreases with the size of the region.

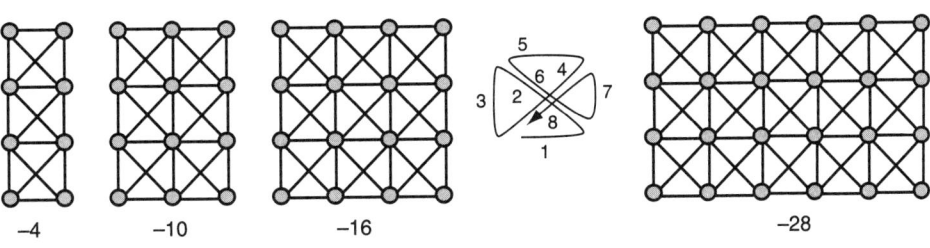

FIGURE 4.21 Unlimited decrease of the Euler characteristic χ^+ in the infinite 8-adjacency grid. The numbers below the rectangular oriented graphs are values of χ^+.

Definition 4.7 An oriented adjacency graph is called *regular* if $\lambda(\rho)$ and $\nu(p)$ are constants for all cycles ρ and all nodes p.

The planar graphs representing the Platonic solids (see Figure 4.14) are examples of finite regular planar adjacency graphs. Of course, their sets of nodes are far too small to serve as 2D grids for digital pictures. For infinite sets of nodes, there are only three regular planar adjacency graphs: the *orthogonal grid* on $\mathbb{Z}^2$ with $\lambda(g) = \nu(p) = 4$, the *triangular grid* with $\lambda(g) = 3$ and $\nu(p) = 6$, and the *hexagonal grid* with $\lambda(g) = 6$ and $\nu(p) = 3$. We also have their isomorphic structures such as the dual of the 4-adjacency grid in the grid cell model with nodes in $\mathbb{C}_2^{(2)}$.

Rectangular subsets of planar or nonplanar infinite regular adjacency graphs are "popular" 2D grids for digital pictures. However, note that nonplanarity allows a finite simply 8-connected set to have (any of) infinitely many Euler characteristics, as illustrated in Figure 4.21.

4.3.3 Atomic and border cycles

A subset $M \subseteq S$ induces a *substructure* $[M, A_M, \xi_M]$ of an oriented adjacency graph $[S, A, \xi]$ where A_M contains only those adjacency pairs $\{p,q\}$ such that $p, q \in M$ and $\{p,q\} \in A$, and where, for any $p \in M$, $\xi_M(p)$ is the *reduced local circular order* defined by deleting from $\xi(p)$ all nodes that are not in M. Such a substructure is an oriented adjacency graph iff M is connected with respect to A_M.

Let $\alpha_0^M = \text{card}(M)$ be the number of nodes, $\alpha_1^M = \text{card}(A_M)$ the number of edges or adjacency pairs, and α_2^M the number of cycles in $[M, A_M, \xi_M]$. Then the Euler characteristic $\chi_M^+ = \alpha_0^M - \alpha_1^M + \alpha_2^M$ of $[M, A_M, \chi_M^+]$ is at least equal to the Euler characteristic χ^+ of $[S, A, \xi]$. On the other hand, we have the following, where β_0^M is the number of components of M in $[S, A]$:

$$\chi_M^+ \leq 2\beta_0^M \qquad (4.8)$$

The cycles of $[M, A_M, \xi_M]$ may differ from the cycles of $[S, A, \xi]$. Let (p,q) be a directed edge in $[M, A_M, \xi_M]$, let ρ_1 be the cycle generated by (p,q) in $[M, A_M, \xi_M]$, and let ρ_2 be the cycle generated by (p,q) in $[S, A, \xi]$.

Definition 4.8 ρ_1 is an *atomic cycle* iff $\rho_1 = \rho_2$ and a *border cycle* otherwise.

For example, there are five border cycles in Figure 4.22, with $\alpha_0 = 37$, $\alpha_1 = 46$, and $\alpha_2 = 11$ for the component on the left; $\alpha_0 = 4$, $\alpha_1 = 3$, and $\alpha_2 = 1$ for the component in the middle; and $\alpha_0 = 28$, $\alpha_1 = 28$, and $\alpha_2 = 2$ for the component on the right.

For any $M \subset S$, $[M, A_M, \xi_M]$ has at least one border cycle. Each border cycle of $[M, A_M, \xi_M]$ contains at least one border node of M, and each border node is incident with at least one border cycle.

4.3 Oriented Adjacency Graphs

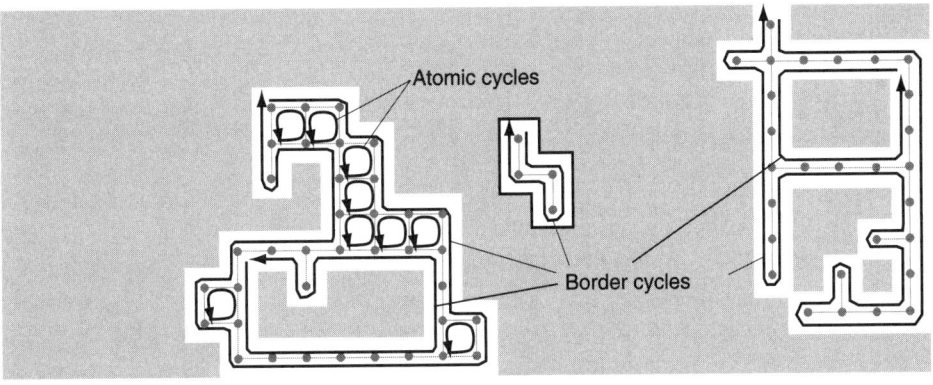

FIGURE 4.22   Nine atomic and five border cycles.

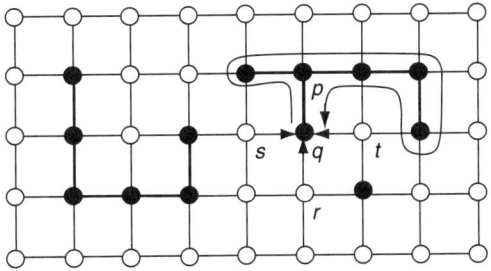

FIGURE 4.23   The directed invalid edges (r,q), (s,q), and (t,q) all point to the border cycle initiated by (q,p) in the substructure defined by the filled dots.

4.3.4 The separation theorem

Let (r,q) be a directed edge in $[S,A]$, $M \subseteq S$, $q \in \delta M$, and $r \in (S \setminus M)$. We call (r,q) a *directed invalid edge* from $\overline{M} = S \setminus M$ to M. An undirected edge between $\overline{M}$ and M is *invalid* iff one of its two directions is invalid.

A directed invalid edge (r,q) *points to a cycle* in $[M, A_M, \xi_M]$ if (q,p) is the directed edge such that p is the first node of M that follows r in the (original) local circular order $\xi(q)$. Figure 4.23 shows an example; for the directed invalid edges (r,q), (s,q), and (t,q), p is the next (and in $\xi_M(q) = \langle p \rangle$, the only) node in M, and $\xi(q) = \langle p,t,r,s \rangle$ initiates the border cycle. Every directed invalid edge points to exactly one border cycle in $[M, A_M, \xi_M]$. This defines a partition of all (directed or undirected) invalid edges into equivalence classes; each class is the set of all (directed or undirected) invalid edges that are *assigned* to a given border cycle.

Theorem 4.7 Let $[S, A, \xi]$ be a (finite or infinite) planar oriented adjacency graph and M a nonempty finite connected proper subset of S. Then $[S, A, \xi]$ splits into at least two nonconnected substructures when we delete all undirected invalid edges that are assigned to border cycles of M.

This separation theorem can be compared with the Jordan-Veblen curve theorem (see Chapter 7) in the Euclidean plane. Let M be connected, let (r, q) be a directed invalid edge from $\overline{M}$ to M, and let $\rho = \langle q_1, \ldots, q_n \rangle$ be the border cycle in $[S, A, \xi]$ to which (r, q) points. Then, for any $p \in M$ and any path $\rho = (p, \ldots, r)$, we have $G(\rho) \cap \{q_1, \ldots, q_n\} \neq \emptyset$ where $G(\rho)$ is the set of all nodes of ρ. In other words, a border cycle establishes *separations* in a planar oriented adjacency graph based on "tracing" border cycles.

Let $[S, A, \xi]$ be an oriented adjacency graph (not necessarily planar), and let $M \subset S$ be a finite proper subset of S. We consider the computational problem of finding ("tracing") all border cycles of M.

Note first that the border cycles of different components of M are disjoint, so we can consider each component of M independently. To determine the components of M we can use the recursive FILL procedure in Algorithm 4.1. We choose a node p in M and find ("fill" or "label") all of the nodes connected to p; this provides the first component of M. As long as unlabeled nodes still remain in M, we repeat this procedure to obtain all of the components of M, one for each repetition of the procedure.

Every directed invalid edge points to exactly one border cycle. For each component A of M, we generate a list L of all directed invalid edges. We choose any of these edges and trace the border cycle to which it points (see the following discussion). During tracing, we delete all of the assigned directed invalid edges from L. If there is still an undeleted directed invalid edge on L, we repeat the process; this generates all of the border cycles of A. L can be generated in a scan through A if all of the detected border nodes $p \in \delta A$ form directed invalid edges (q, p) with all nodes $q \in A(p) \cap \overline{A}$; these edges will be inserted into L.

---

1. Let $(q_0, p_0) := (q, p), i := 0$, and $k := 0$.
2. Let $\xi(p_i) = \langle \ldots, q_k, q, \ldots \rangle$ be the local circular order at p_i. If $q \in \overline{A}$, go to Step 4.
3. Node q is another node on the border cycle. Let $i := i+1$ and $p_i := q$. Let $\xi(p_i) = \langle \ldots, p_{i-1}, q, \ldots \rangle$ be the local circular order at p_i. If $q \in A$, go to Step 3; otherwise, let $k := i - 1$, and go to Step 4.
4. If $(q, p_i) = (q_0, p_0)$, go to Step 5. Otherwise, let $k := k + 1$ and $q_k := q$, and go to Step 2.
5. We are back at the original directed invalid edge (q, p). The border cycle is $\langle p_0, p_1, \ldots, p_i \rangle$.

ALGORITHM 4.3 The border tracing algorithm [1111].

4.3 Oriented Adjacency Graphs

The algorithm for tracing the border cycle pointed to by a given directed invalid edge (q,p) is given in Algorithm 4.3. The algorithm terminates because of property **A4**. The deletion of all directed invalid edges may lead to a separation if $[S, A, \xi]$ is nonplanar, but the border tracing procedure applies even to nonplanar graphs.

4.3.5 Holes

Let $G = [S, A, \xi]$ be an infinite oriented adjacency graph, and let M be a finite connected subset of S. Using property **A3** (see Section 4.1.3), M has exactly one infinite complementary component. Any finite complementary component of M is called a *hole* of M. If $S = \mathbb{Z}^n$ and the hole is α-connected, we call it an α-*hole*. For example, finite connected subsets of $[\mathbb{Z}^2, A_4]$ can have 8-holes that consist of several 4-holes.

If G is planar, M has exactly one border cycle, called its *outer border cycle*, which separates M (see Theorem 4.7) from its infinite complementary component. All other border cycles of M are called *inner border cycles*. If complementary component A of M is separated from M by border cycle ρ of M, we say that A is assigned to ρ.

Definition 4.9 Let M be a subset of an infinite planar oriented adjacency graph. A complementary component of M that is assigned to one of the inner border cycles of M is called a *proper hole* of M, and a finite complementary component that is assigned to the outer border cycle of M is called an *improper hole* of M.

Figure 4.24 illustrates this definition. Two complementary 1-components belong to the background 1-component of class 1; two complementary 1-components are proper 1-holes (separated by a 1-component from the background 1-component); and two complementary 1-components are improper 1-holes (not separated by a 1-component from the background 1-component).

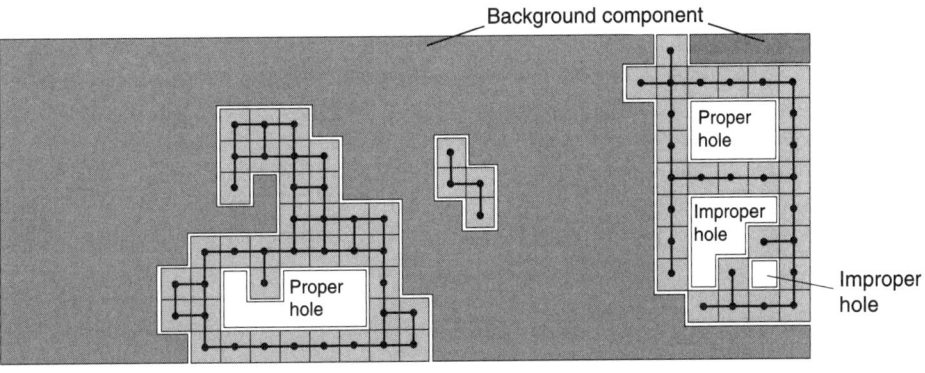

FIGURE 4.24 Three 1-components and six complementary 1-components (of class 1 in Figure 2.17). 1-holes in the grid cell model are 4-holes in the grid point model.

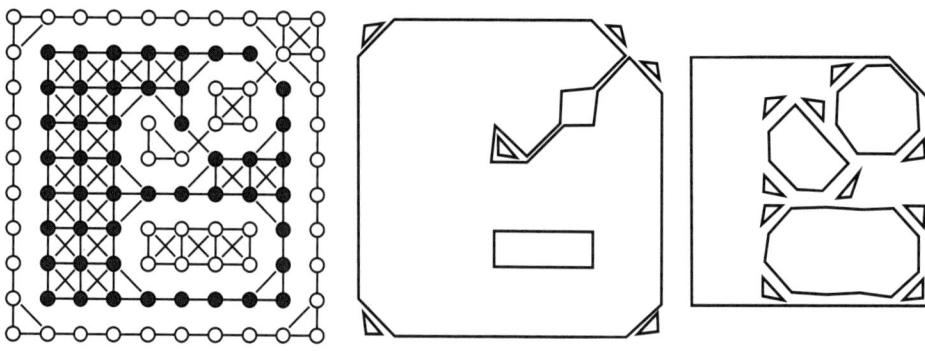

FIGURE 4.25 Left: a finite 8-component M in the infinite nonplanar oriented 8-adjacency graph. Middle: all border cycles of $\overline{M}$. Right: all border cycles of M.

The *cover* $c(M)$ of M is the union of M with all of its proper holes. $c(M)$ has no inner border cycles, but it has an outer border cycle that may separate improper holes from $c(M)$. Finite complementary components of a finite or infinite connected subset of an infinite planar oriented adjacency graph do not have proper or improper holes.

Planarity of the oriented adjacency graph is crucial for the definition of inner and outer border cycles. A border cycle *circumscribes* the proper or improper holes assigned to it. The set M in Figure 4.25 has 14 border cycles: 10 "triangles" (small artifacts of atomic cycles); three border cycles that circumscribe one proper 4-hole and two improper 4-holes; and one border cycle ρ that circumscribes the union of M with its proper and improper 4-holes. $\overline{M}$ has eight border cycles: six "triangular artifacts"; one border cycle for the proper 4-hole in M; and one border cycle that "penetrates" ρ (i.e., ρ does not separate M from its background).

4.3.6 Boundaries

The following definition applies to planar and nonplanar oriented adjacency graphs. In Figure 4.25 (left), all undirected invalid edges assigned to any border cycle of M have been removed.

> **Definition 4.10** Let $G = [S, A, \xi]$ be an oriented adjacency graph, and let M be a nonempty finite connected proper subset of G. The *boundary* of M is the set of undirected invalid edges that are assigned to border cycles of M.

This definition resembles the discussion in [202] about the "boundary between layers of white and black marbles." (We use light gray in Figure 4.26 instead of black.) Such a boundary is neither a set of white nor of black marbles and is better described as the "space between the white and black layers" (see the left of Figure 4.26) or as neighboring pairs of white and black marbles (see the right of Figure 4.26).

4.3 Oriented Adjacency Graphs

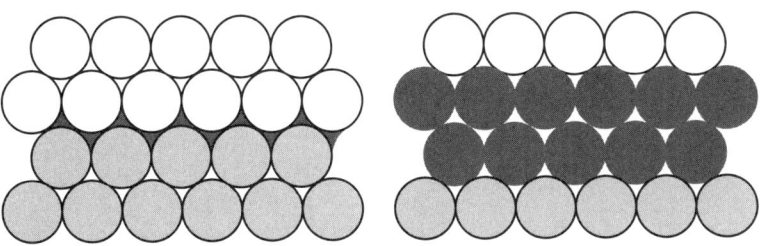

FIGURE 4.26 Example of layers of white and light-gray marbles as discussed in [202]. Left: the space between the layers is dark gray. Right: marbles that have neighboring marbles of the other color are dark gray.

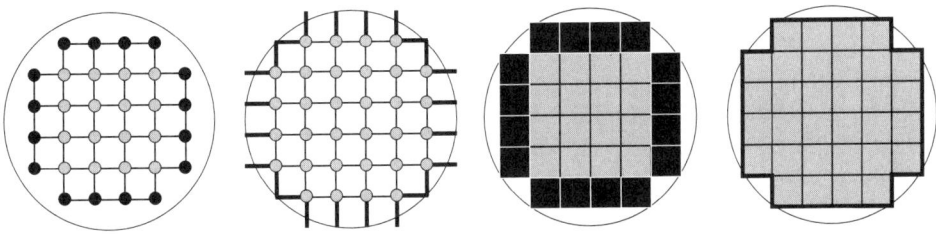

FIGURE 4.27 Left: the border (black vertices) and boundary (bold edges) of $G(D)$ in the grid point model, where D is a disk. Right: the border and boundary of $G(D)$ in the grid cell model.

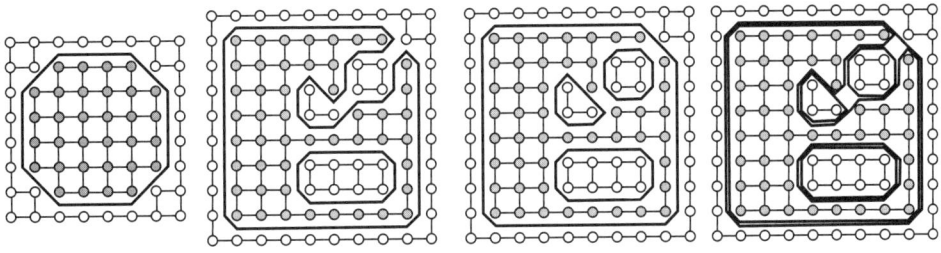

FIGURE 4.28 Left: no problem occurs in the case of the disk. The other three pictures show midpoint sequences that "prefer" connections between background points or between object points and a doubly oriented midpoint sequence that treats background and object points equally.

When we use Gauss digitization (see Section 2.3.1), the boundary of $G(S)$ (in the grid point model) consists of edges that cross the frontier of $G(S)$; see Figure 4.27. Such a boundary can be represented by an ordered sequence of midpoints of those edges. Figure 4.28 shows examples of such polygonal representations of boundaries.

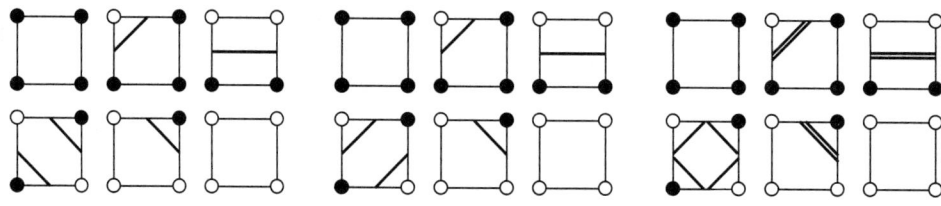

FIGURE 4.29 Left: look-up table with a preference for connections between background points. Middle: preference for connections between object points. Right: doubly oriented edges.

The edges can be found by comparing every 2×2 block of pixels with a set of templates (see Figure 4.29), and, for each match, copying (generating) the "detected" edge(s).

If we use the left or middle look-up table in Figure 4.29, the plane is segmented into polygons. The object polygons can be regarded as closed sets in the Euclidean topology, and the background polygons as open (or vice versa). This defines a partition of the plane into pairwise disjoint sets with frontiers that are the polygonal boundaries. We can assume that each doubly oriented edge shown on the right in Figure 4.29 is actually two directed edges with a small space between them. This leads to a segmentation of the plane into pairwise disjoint (even "nontouching") polygons. The space between the polygons is a representation of the boundary; this is similar to a situation where a wall is built out of stones and the boundary is the mortar between the stones.

4.3.7 Some combinatorial results

This section gives formulas for some topologic invariants (see Definition 6.9) of isothetic polygons using the more abstract language of infinite planar regular oriented adjacency graphs. We use the hexagonal and trigonal grids (see Definition 4.7) as well as the orthogonal grid.

We begin by recalling Equations 4.4 and 4.7. Let $G = [S, A, \xi]$ be a finite oriented adjacency graph with $\alpha_0 = \text{card}(S)$ and $\alpha_1 = \text{card}(A)/2$. Let the following

$$\overline{\nu} = \frac{1}{\alpha_0} \sum_{p \in S} \nu(p) \quad \text{and} \quad \overline{\lambda} = \frac{1}{\alpha_1} \sum_{\rho} \lambda(\rho)$$

be the mean outdegree of a node and the mean length of a cycle where the second sum is over all cycles of G. Then the following is true,

$$\alpha_0/\alpha_1 = 2/\overline{\nu} \quad \text{and} \quad \alpha_2/\alpha_1 = 2/\overline{\lambda}$$

which implies the following:

$$2/\overline{\nu} + 2/\overline{\lambda} = 1 + 2/\alpha_1 \tag{4.9}$$

4.3 Oriented Adjacency Graphs

This equation also applies to infinite oriented adjacency graphs if both means are well defined, because $2/\alpha_1$ goes to zero for infinite graphs.

A *regular oriented adjacency graph* $G_{\nu,\lambda}$ represents a regular tiling (of some surface) and vice versa. For examples of finite regular tilings, see Section 7.6—Exercises 7 and 8 (for the surface of a sphere), 9 and 10 (for the plane), 11 (for the surface of a torus), and 12 (for a closed surface). For infinite graphs (e.g., on a plane), Equation 4.9 has only three integer-valued solutions: $\nu = \lambda = 4$; $\nu = 3$ and $\lambda = 6$; and $\nu = 6$ and $\lambda = 3$. We will assume that $S = \mathbb{Z}^2$ in these three infinite planar $G_{\nu,\lambda}$s (see Exercise 10 in Section 7.6).

Consider a finite subgraph of one of these three $G_{\nu,\lambda}$s defined by a finite subset $M \subseteq \mathbb{Z}^2$. We first consider the case in which M has only one border cycle, for example, of length l. Following Definition 6.13, we call such an M *simply connected*. Let k be the number of invalid undirected edges between $\overline{M}$ and M; see Figure 4.30 for an example. Because $G_{\nu,\lambda}$ is planar, Equations 4.4 and 4.7 give us the following:

$$\alpha_0 - \alpha_1 + \alpha_2 = 2 \qquad (4.10)$$
$$\nu\alpha_0 - k = 2\alpha_1 \qquad (4.11)$$
$$\lambda(\alpha_2 - 1) + l = 2\alpha_1 \qquad (4.12)$$

so that the following is true:

$$\nu l - \lambda k + \nu\lambda = (2\nu + 2\lambda - \nu\lambda)\alpha_1 \qquad (4.13)$$

If we regard ν and λ as parameters of $G_{\nu,\lambda}$, the relationship between l and k is defined by the number α_1 of undirected edges in the subgraph defined by M. For the $G_{\nu,\lambda}$s, we have $2\nu + 2\lambda - \nu\lambda = 0$.

Theorem 4.8 For any simply connected set M in any of the $G_{\nu,\lambda}$s, we have $k = \nu + \nu l/\lambda$.

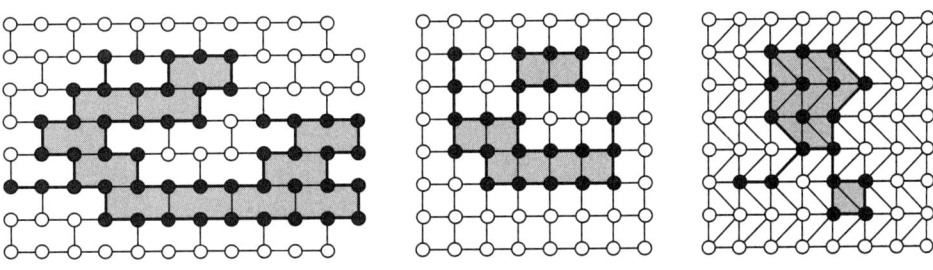

FIGURE 4.30 $\nu = 3$, $\lambda = 6$, $\alpha_0 = 49$, $\alpha_1 = 59$, $\alpha_2 = 12$, $l = 52$, $k = 29$, $f = 11$ (left); $\nu = 4$, $\lambda = 4$, $\alpha_0 = 23$, $\alpha_1 = 30$, $\alpha_2 = 9$, $l = 28$, $k = 32$, $f = 8$ (middle); and $\nu = 6$, $\lambda = 3$, $\alpha_0 = 18$, $\alpha_1 = 32$, $\alpha_2 = 16$, $l = 19$, $k = 44$, $f = 15$ (right).

In other words, the relationship between the number of invalid edges and the length of the border cycle depends only on the parameters ν and λ.

Now suppose M has $r \geq 1$ border cycles. Instead of Equations 4.10, 4.11, and 4.12, we have the following:

$$\alpha_0 - \alpha_1 + \alpha_2 = 2 \tag{4.14}$$

$$\nu\alpha_0 - \sum_{i=1}^{r} k_i = \nu\alpha_0 - K = 2\alpha_1 \tag{4.15}$$

$$\lambda(\alpha_2 - r) + \sum_{i=1}^{r} l_i = \lambda(\alpha_2 - r) + L = 2\alpha_1 \tag{4.16}$$

In these, l_i is the length of border cycle i ($1 \leq i \leq r$) and k_i is the number of invalid edges that connect $\overline{M}$ with nodes on border cycle i. For the $G_{\nu,\lambda}$s, we have $2\nu + 2\lambda - \nu\lambda = 0$, so that the following is true:

$$\nu\sum_{i=1}^{r} l_i - \lambda\sum_{i=1}^{r} k_i - (r-2)\nu\lambda = \nu L - \lambda K - (r-2)\nu\lambda = 0 \tag{4.17}$$

The outer border cycle of M is the same as the outer border cycle of its cover $c(M)$ (the union of M with all its proper holes), so Theorem 4.8 is valid for $c(M)$. It follows that we always have $\nu l_r - \lambda k_r + \nu\lambda = 0$ where r is the index of the outer border cycle of M. Subtracting this equation from Equation 4.17 gives the following:

$$\nu\sum_{i=1}^{r-1} l_i - \lambda\sum_{i=1}^{r-1} k_i - (r-1)\nu\lambda = \nu(L-l_r) - \lambda(K-k_r) - (r-1)\nu\lambda = 0 \tag{4.18}$$

The $r-1$ inner border cycles of M can be regarded as independent events, and Equation 4.18 splits into $r-1$ equations $\nu l_i - \lambda k_i + \nu\lambda = 0$, $1 \leq i \leq r-1$.

Theorem 4.9 For any connected set M in one of the $G_{\nu,\lambda}$s and any of its border cycles, we have $k = \pm\nu + \nu l/\lambda$, where the outer border cycle has the positive sign and any inner border cycle has the negative sign, k is the number of invalid edges assigned to the border cycle, and l is the length of the border cycle.

This theorem provides a simple algorithm for deciding whether a traced border cycle (see Algorithm 4.3) is inner or outer by keeping track of k and l during border cycle tracing.

From Equation 4.17, it follows that $r = 2 + L/\lambda - K/\nu$. The total length L of all border cycles and the total number K of all invalid edges allow us to calculate the number r of border cycles, which is a topologic invariant of M. Note that L and K can be accumulated by examining all 4-neighborhoods of points in M; border cycle tracing is not necessary.

4.3 Oriented Adjacency Graphs

Let f be the number of atomic cycles of M. Combining Theorem 4.8 and Equations 4.10, 4.11, and 4.12 gives the following:

$$\alpha_0 = \frac{\lambda}{\nu} f + \left(\frac{1}{\lambda} + \frac{1}{\nu}\right) l + 1$$

For any of the $G_{\nu,\lambda}$s, we have $1/\lambda + 1/\nu = 1/2$. This proves the following:

Theorem 4.10 For a connected subset M of an infinite planar $G_{\nu,\lambda}$ that has no proper holes, we have $\alpha_0 = \lambda f/\nu + l/2 + 1$, where $\alpha_0 = \text{card}(M)$, there are f atomic cycles in M, and l is the length of the outer border cycle of M.

This theorem is a graph-theoretic generalization of a result proved by G. Pick (1899)[7] for the regular orthogonal grid ($\nu = \lambda = 4$): the area f of a *simple grid polygon* (a simple polygon that has grid points as vertices) satisfies $f = \alpha_0 - l/2 - 1$, where α_0 is the number of grid points in the polygon and l is the number of grid points on the frontier of the polygon. For example, for a grid square, we have $\alpha_0 = l = 4$ and $f = 1$.

Theorem 4.10 applies to outer border cycles. For an inner border cycle ρ, let α_0 be the number of nodes of $G_{\nu,\lambda} \setminus M$ surrounded by ρ ("in the interior of ρ"), and let f be the number of atomic cycles of $G_{\nu,\lambda}$ defined by these α_0 nodes and the nodes of ρ. Then we have the following:

Theorem 4.11 For an inner border cycle ρ of a connected subset M of any of the $G_{\nu,\lambda}$s, we have $\alpha_0 = \lambda f/\nu - l/2 + 1$ where l is the length of ρ.

Note that an inner border cycle can separate several proper holes from M; Theorem 4.11 applies to the union of these holes.

Figure 4.31 shows examples; k is the number of invalid edges assigned to the inner border cycle, and we have shaded the atomic cycles of the proper holes assigned to the inner border cycle. On the left, we have three proper holes and one atomic cycle in one of them; in the middle, we have one proper hole and one atomic cycle; on the right, we have one proper hole and two atomic cycles. Let m be the total number of atomic cycles of all the proper holes assigned to the given inner border cycle. The remaining $f - m$ atomic cycles, defined by nodes in the complementary set $\overline{M}$ and on the inner border cycle, are *boundary cycles*, where the boundary specifies the "space between." The λm edges of the boundary cycles include k invalid edges, the length l of the inner border cycle, and the sum of the lengths l_i of the outer border cycles of

---

7. Georg Pick was a professor of mathematics in Prague. He was in close contact with A. Einstein when Einstein worked at Prague University in 1911 and 1912. They played music together, discussed philosophy, and established the mathematic basis of general relativity theory. Pick was murdered by the Nazis in the Theresienstadt concentration camp.

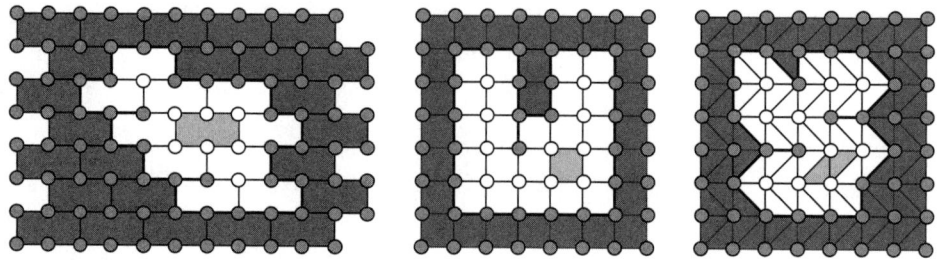

FIGURE 4.31 Inner border cycles with $\nu = 3$, $\lambda = 6$, $\alpha_0 = 8$, $l = 30$, $k = 12$, $f = 11$ (left); $\nu = 4$, $\lambda = 4$, $\alpha_0 = 11$, $l = 26$, $k = 22$, $f = 23$ (middle); and $\nu = 6$, $\lambda = 3$, $\alpha_0 = 10$, $l = 22$, $k = 38$, $f = 40$ (right).

the $n > 0$ proper holes assigned to the inner border cycle:

$$\lambda m = 2k + l + \sum_{i=1}^{n} n l_i$$

Together with Theorems 4.10 and 4.11, this implies the following:

$$n = 1 + k - m$$

This is another example of how a topologic invariant (the number n of proper holes) can be calculated by accumulating local counts (of invalid edges and boundary cycles).

For $G_{6,3}$, which has $\nu = 6$ and $\lambda = 3$, exactly one complementary component is assigned to any (inner or outer) border cycle of a finite connected set M. It follows that such an M has no improper holes; the region adjacency graph generated by M is a tree; and, for any inner border cycle of M, any directed invalid edge generates exactly one border cycle.

4.4 Combinatorial Maps

Finite oriented adjacency graphs are known in graph theory as (2D) combinatorial maps. Finite spatial subdivisions, such as a subdivision of a cuboidal subset of $\mathbb{R}^3$ into grid cubes, can be represented by 3D combinatorial maps, and such maps also generalize to n dimensions.

4.4.1 2D maps

A finite graph is planar iff it can be drawn in the plane or on a sphere so that edges intersect only at nodes. Such a drawing is called an *embedding*. In general, an *embedding of a graph* is a representation of the graph on a closed compact surface (e.g., the

4.4 Combinatorial Maps

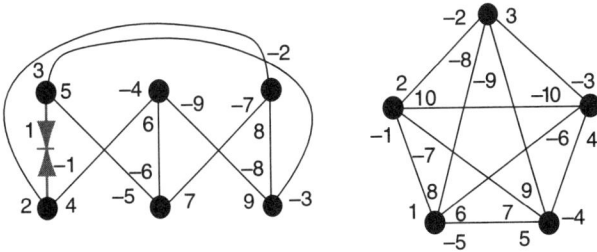

FIGURE 4.32 Combinatorial maps for the undirected graphs $K_{3,3}$ and K_5; only two darts are shown (for edge 1) in $K_{3,3}$.

surface of a sphere or torus) such that no two edges intersect except at their endpoints. An embedding is characterized by the local circular orders $\xi(p)$ of the edges around the nodes p. A 2D *combinatorial map* is a finite graph together with a set of local circular orders (i.e., it is what we have called an oriented adjacency graph $[S, A, \xi]$).

A combinatorial map can be defined by a graph and two permutations σ and θ. Let the edges of the graph be numbered $1, 2, \ldots, m$. Split each undirected edge i into two directed edges $+i$ and $-i$; these directed edges are sometimes called *darts*. Then σ is the mapping of $+i$ into $-i$ and $-i$ into $+i$ (i.e., σ is a product of cycles of length 2):

$$\sigma = \begin{pmatrix} 1 & 2 & \ldots & m & -1 & -2 & \ldots & -m \\ -1 & -2 & \ldots & -m & 1 & 2 & \ldots & m \end{pmatrix}$$
$$= (1, -1)(2, -2) \cdots (m, -m)$$

Let $S = \{p_1, p_2, \cdots, p_n\}$; then

$$\theta = \xi(p_{i_1})\xi(p_{i_2}) \cdots \xi(p_{i_n})$$

for any of the $n!$ possible permutations of the n nodes:

$$\begin{pmatrix} 1 & 2 & \ldots & n \\ i_i & i_2 & \ldots & i_n \end{pmatrix}$$

Figure 4.32 shows two examples of graphs in dart representation. For the graph $K_{3,3}$ on the left, assume anticlockwise local circular orders $\xi(p)$; then the following are true:[8]

$$\sigma = (1, -1)(2, -2)(3, -3)(4, -4)(5, -5)(6, -6)(7, -7)(8, -8)(9, -9)$$
$$\theta = (5, 3, 1)(-4, 6, -9)(-7, 8, -2)(-1, 2, 4)(-6, -5, 7)(-8, 9, -3)$$
$$\varphi = \theta \circ \sigma = (1, 2, -7, -6, -9, -3)(-1, 5, 7, 8, 9, -4)(-2, 4, 6, -5, 3, -8)$$

---

[8]. In this example, 1 goes into -1 in σ, -1 goes into 2 in θ (1 goes into 2 in $\theta \circ \sigma$), 2 goes into -2 in σ, -2 goes into -7 in θ (2 goes into -7 in $\theta \circ \sigma$), and so on.

Note that the cycles in $\varphi = \theta \circ \sigma$ are the same as the cycles in the graph. For the graph K_5 on the right, assume clockwise local circular orders $\xi(p)$; then the following are true:

$$\sigma = (1,-1)(2,-2)(3,-3)(4,-4)(5,-5)(6,-6)(7,-7)(8,-8)(9,-9)(10,-10)$$
$$\theta = (2,10,-7,1)(-2,3,-9,-8)(-3,4,-6,-10)(9,-4,5,7)(8,6,-5,1)$$
$$\varphi = \theta \circ \sigma = (-2,10,-3,-9,-4,-6,-5,7,-1,8)(1,2,3,4,5)(6,-10,-7,9,-8)$$

Permutation φ lists the three circuits of the graph. Based on the definitions of σ and θ, this is valid for any product $\varphi = \theta \circ \sigma$ for both counterclockwise and clockwise local circular orders, as long as we start with a finite connected $[S, A]$, split its undirected edges into directed edges via σ, and define local circular orders via θ. Of course, handling permutations of edges or adjacency sets is impractical when we deal with high-resolution pictures.

4.4.2 3D maps

In our 2D examples of combinatorial maps, we split each undirected edge into two darts and used the permutation σ to identify pairs of darts that represent the two orientations of the same edge (see also edge 10 in Figure 4.33). The local circular orders of the darts, defined by the cycles of the permutation θ, then allow any dart to be used to initiate a cycle in $\phi = \theta \circ \sigma$. The permutation σ is an *involution* on the set S of all darts (i.e., a one-to-one mapping such that $\sigma = \sigma^{-1}$). An involution σ is either the identity $\sigma(a) = a$ for all $a \in S$ or the product of a finite sequence of cycles (b, c) of length 2.

$[S, \sigma_1, \ldots, \sigma_n] (n \geq 2)$ is an *n-dimensional combinatorial map* iff S is a finite set; σ_1 is a permutation of S; and $\sigma_2, \ldots, \sigma_n$ are involutions on S such that $\sigma_i \circ \sigma_j$ is an involution on S for $1 \leq i \leq n-2$ and $i+2 \leq j \leq n$. For example, in 3D,

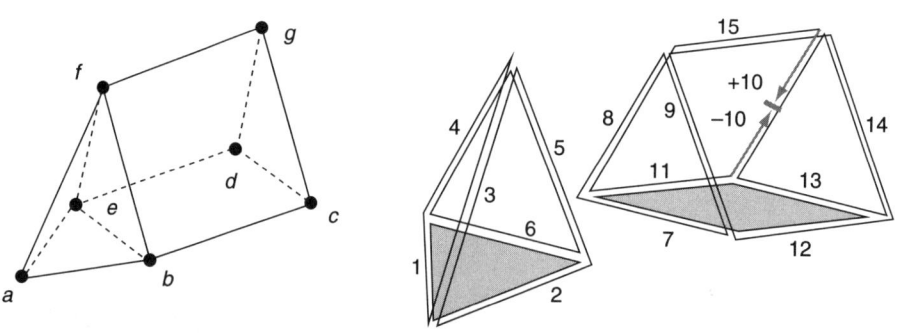

FIGURE 4.33 A spatial subdivision (left) combining a tetrahedron with a wedge, and a sketch (right) of its combinatorial map; only edge 10 is represented by two darts.

$\sigma_1 \circ \sigma_3$ is an involution on S. In 2D, we had $\sigma_1 = \theta$ and $\sigma_2 = \sigma$; the additional condition is vacuous.

Figure 4.33 shows a spatial subdivision. Let the following be given:

$$\sigma_1 = (\ -1,2,4)(-2,6,5)(1,3,-6)(-5,-3,-4)$$
$$(\ 7,8,11)(-11,-10,-13)(-7,12,9)(-8,-9,15)(14,10,-15)(-14,-12,13)$$
$$\sigma_2 = (\ 1,-1)(2,-2)(3,-3)(4,-4)(5,-5)\ldots(13,-13)(14,-14)(15,-15)$$
$$\sigma_3 = (\ 2,-1)(4,2)(-1,4)(6,-2)(5,6)\ldots(-12,-14)(13,-12)(-14,13)$$

These permutations define a 3D combinatorial map on a set S of 30 darts.

In general, let $\{\theta_1, \ldots, \theta_m\}$ be a set of permutations defined on a set S. A dart $e \in S$ and a permutation θ of S define a cycle $\theta(e)$ of darts that are "reachable" by starting at e and using only transitions contained in θ. For example, for the σ_1, σ_2, and σ_3 defined above, we have $\sigma_1 \circ \sigma_2(12) = (12, 13, -11, 7)$ and $\sigma_3 \circ \sigma_1 \circ \sigma_2(-6) = (-6, 6)$. The *orbit* of a dart e is the set of all darts contained in cycles $\theta(e)$, where θ is any finite sequence of permutations θ_i or θ_i^{-1}. For example, the orbit of a dart with respect to σ_1, σ_2, and σ_3 is either the set of all darts of the tetrahedron or the set of all darts of the wedge.

3D combinatorial maps can be used to represent spatial subdivisions. For example, merging the tetrahedron and wedge shown in Figure 4.33 means deleting the doubly represented face between them; similarly, splitting can be done by creating new separating faces. Spatial segmentation can be based on repeated merging operations that reduce the total number of darts.

4.5 Exercises

1. Show that $\mathbf{E}M \subseteq \mathbf{O}M \subseteq M \subseteq \mathbf{C}M \subseteq \mathbf{D}M$ for any $M \subseteq S$.

2. Show that $\mathbf{E}$ and $\mathbf{D}$ are *dual* operations (i.e., that $\overline{\mathbf{D}M} = \mathbf{E}\overline{M}$ and $\overline{\mathbf{E}M} = \mathbf{D}\overline{M}$ for any $M \subseteq S$) where $\overline{L} = S \setminus L$.

3. Show that $\mathbf{O}$ and $\mathbf{C}$ are *idempotent* operations (i.e., that $\mathbf{O}^2 M = \mathbf{O}M$ and $\mathbf{C}^2 M = \mathbf{C}M$ for any $M \subseteq S$).

4. Prove that the center of a simply 4-connected set of grid points, using the graph distance defined by A_4, is either a 2×2 square, a diagonal line segment, or a diagonal staircase; note that, in the latter two cases, the center can contain arbitrarily many grid points (see Figure 4.12). What happens if we use A_8?

5. Eulerian paths can be defined for an infinite graph provided it is connected and its set of edges is countably infinite. P. Erdös, T. Grünwald, and E. Vàzsonyi [308] showed in 1938 that an infinite Eulerian path can be either of the following:

   (i) one-way (i.e., with one terminal node) if at most one node of the graph has odd degree; if no node has odd degree, there must be at least one node of infinite degree, and the complement of any finite subgraph must have only one infinite component; or

(ii) two-way (i.e., with no terminal node) if no node has odd degree; the complement of any finite subgraph has at most two infinite components; and any finite subgraph that contains only nodes of even degree has only one infinite component.

Consider the three infinite graphs in Exercise 1 in Section 1.3. Do they have Eulerian paths? If so, are they one-way or two-way?

6. Do the 4-adjacency grids $\mathbb{G}_{n,n}$ ($n \geq 2$) have Hamiltonian circuits? Do they have Hamiltonian paths?

7. Find a Hamiltonian path for the infinite 4-adjacency grid $[\mathbb{Z}^2, A_4]$.

8. Show that the following two graphs are isomorphic:

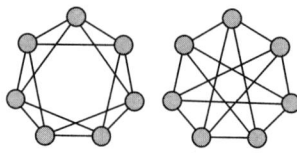

9. Is this graph planar? Verify your answer.

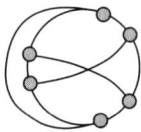

10. Let P be a picture with values in $\{0,\ldots,G_{max}\}$. Define a relation A_P on the set of pixels of P using the following:

$$p \in A_P(q) \quad \text{iff} \quad d_e((p, P(p)), (q, P(q))) = \min_{r \in A_4(q)} \{d_e((r, P(r)), (q, P(q)))\}$$

Is the graph $[S, A_P]$ connected? complete? planar? Eulerian?

11. Show that any local circular order on $K_{3,3}$ or K_5 (see Figure 4.32) results in an oriented graph that has the Euler characteristic $\chi^+ < 2$.

12. The solid dots in the following figure are the grid points in a subset M of the infinite 4-adjacency graph $[\mathbb{Z}^2, A_4]$. Draw the region adjacency graph of M. Calculate the Euler characteristics of the components of this graph assuming clockwise orientation ((A) in Figure 4.19).

4.5 Exercises

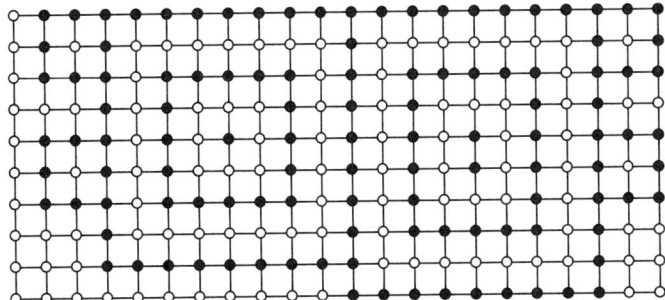

13. Can a finite oriented adjacency graph have $\alpha_0 = \alpha_1 = \alpha_2$?

14. A complete graph G with α_0 nodes has $\alpha_1 = \alpha_0(\alpha_0-1)/2$ edges. If $\alpha_0 \geq 3$, prove that, for any local circular order on G, we have $\chi^+ \leq -\alpha_0(\alpha_0-7)/6$.

15. Specify permutations of 2D combinatorial maps that represent the following 8-adjacency graphs:

 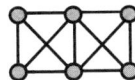 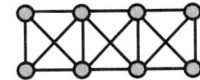 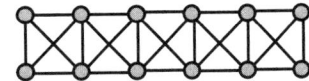

16. Let $L, M \subseteq S$ be subgraphs of the graph $[S, A]$. Show that $p \in \delta(L \cup M)$ or $p \in \delta(L \cap M)$ implies that $p \in \delta(L)$ or $p \in \delta(M)$. Furthermore, if $p \in \delta(L)$ and $M \subseteq S$, then either $p \in \delta(L \cup M)$ or $p \in \delta(L \cap M)$.

17. Let L, M be subgraphs of a finite graph $[S, A]$. Define $Q(L, M)$ as the set of all nodes $p \in \delta(L) \cap \delta(M)$ that are not in $\delta(L \cup M)$. Prove the following:

$$\operatorname{card} \delta(L \cup M) + \operatorname{card} \delta(L \cap M) = \operatorname{card} \delta(L) + \operatorname{card} \delta(M) - \operatorname{card} Q(L, M)$$

18. Let L be a subgraph of a subgraph M of a graph $[S, A]$, and let $\delta_S(M)$ denote the border of M. Prove the following:

$$\delta_M(L) \subseteq \delta_S(L)$$

Also, prove the following for all $p \in S$:

$$\text{if } p \in \delta_S(L) \text{ then } p \in \delta_M(L) \text{ or } p \in \delta_S(M)$$

19. A subset M of a graph $G = [S, A]$ is called a D-cluster $(D \geq 1)$ iff, for all $p, q \in M$, there exists a sequence $p = p_0, p_1, \ldots, p_n = q$ of nodes in S such that $d_G(p_{i-1}, p_i) \leq D$ $(i = 1, \ldots, n)$. (Here d_G is the graph metric, in which each edge has unit length; it follows that M is a 1-cluster iff it is connected.) The D-hull $H_D(M)$ of M is the set of nodes of G within distance $D/2$ of M. The geodesic D-hull $G_D(M)$ of M

is the set of nodes of G that lie on shortest paths of length $\leq D$ between pairs of nodes of M. Prove that M is a D-cluster iff $H_D(M)$ is connected iff $G_D(M)$ is connected.

4.6 Commented Bibliography

Region adjacency graphs on regular grids were introduced in [151, 881]. The notions of connectedness and components as used in graph theory had to be adapted to the special needs of picture analysis by defining connectedness with respect to a subgraph. For information about using "local properties" to compute topologic properties of pictures, see [377, 733]. Exercise 19 is from [889], which also defines "D-holes" and "D-borders" for sparse sets of 1s in binary pictures. [574, 1109] initiated studies of general adjacency models in picture analysis.

The history of graph theory is reviewed in [862]. There are many textbooks about graph theory; see, for example, [786], which is the source of Exercise 3, and [1073], which is the source of Exercises 16, 17, and 18. Periodic subgraphs of the infinite 8-adjacency graph on $\mathbb{Z}^2$ are studied in [1070]. For graph algorithms, see [997]. For Steinitz's theorem, see [1026]; for Kuratowski's theorem, see [611].

I. Lakatos's book of dialogues [614] (published after he died in 1974) discusses the very diverse opinions about polyhedra that do not obey the Descartes-Euler polyhedron theorem that were expressed by many mathematicians, including Legendre, Gergonne, Cauchy, Lhuilier, Crelle, Hessel, Becker, Listing, Möbius, Schläfli, Grunert, Becker, Poinsot, Steiner, Hoppe, de Jonquières, Matthiessen, Poincaré, Hilbert, Steinhaus, Pólya, and Forder. For example, in 1852, L. Schläfli [962] declared about Kepler's urchin[9] that "this is not a genuine polyhedron, for it does not satisfy the condition $V - E + F = 2$"!

Theorem 4.2 is from [886]. In general, let P be an n-dimensional binary picture defined on a finite hypercuboidal $(2n, 3^n - 1)$- or $(3^n - 1, 2n)$-adjacency grid that is extended into $\mathbb{Z}^n$; then the region adjacency graph of P is a tree [502].

Dijkstra's algorithm was independently discovered in [1124]. The application of shortest path algorithms to picture analysis is discussed in [452, 1095]. Exercise 7 is from [507]. For the complexity of algorithms and computational problems (e.g., NP-completeness), see [356]. For efficient planarity tests, see [443]. [687] discusses components in 2D and 3D adjacency grids that have uniquely defined Hamiltonian circuits. For centers of graphs defined by 4- or 8-regions, see [247] and [507].

The directional code introduced by H. Freeman [342] can be regarded as an example of an orientation on an adjacency graph. Oriented adjacency graphs were studied in a series of publications in the 1980s (e.g., [553, 1104, 1111]). Theorems 4.6 and 4.7 are from [1111]. Border tracing can also be defined in picture data structures

---

9. Also "echinus," which is Kepler's original name for the small stellated dodecahedron that has 12 faces, each a regular star pentagon. It is one of the Kepler-Poinsot solids with the dual polyhedron (see the end of Section 4.2.2) that is the great dodecahedron [220, 614]; for both, we have $\alpha_0 - \alpha_1 + \alpha_2 = -6$.

4.6 Commented Bibliography

such as quadtrees [290]. The book [65] provides an overview of curve tracing methods in picture analysis. The formula in Exercise 14 is proved in [1104]. Many of the combinatorial results about the $\mathbb{G}_{\nu,\lambda}$s in Section 4.3.7 are from [1103].

The theory of *combinatorial maps* was initiated by L. Heffter at the end of the 19th century [415]. He introduced maps and proved a dual characterization theorem for them. Maps were used, for example, by G. Ringel starting in the 1950s, were reinvented in [301, 1153], and finally became popular in discrete mathematics [383]. Their relevance for modeling 2D or 3D (segmented) pictures is discussed in [82, 117, 287, 326, 651, 959].

CHAPTER 5

Incidence Pseudographs

This chapter treats pictures as graph-theoretic objects; it represents spatial subdivisions with incidence pseudographs. We recall that, in 2D or 3D incidence grids (see Section 2.1.5), pixels or voxels are further refined (e.g., a 2-cell [grid square] has grid edges and grid vertices as additional structural components). We define open and closed subsets of incidence pseudographs and their frontiers in preparation for discussing topologies in the following chapters. We give combinatorial formulas for such subsets in regular incidence grids and give a graph-theoretic treatment of frontier tracing.

5.1 Incidence Structures

In this chapter, we discuss a graph-theoretic generalization of the grid cell incidence model (see Section 2.1.5). Because self-incidence is allowed, the graphs can have loops (i.e., they are pseudographs).

Let I be a reflexive and symmetric relation on a set S. We say that c and c' are *incident* (notation: cIc') iff $\{c,c'\} \in I$.

> **Definition 5.1** An *incidence structure* $[S, I, \dim]$ is defined by a countable set S of nodes, an *incidence relation* I on S that is reflexive and symmetric, and a function $\dim$ defined on S and into a finite set $\{0, 1, ..., m\}$ of natural numbers.

The function $\dim$ partitions S into pairwise disjoint *classes*. Its definition depends on the context. For example, the elements of S can be *cells* of a discrete spatial subdivision (e.g., convex bounded polyhedra or polygons, line segments, vertices); Figure 5.1 shows a 2D example. A cell c that has $\dim(c) = i$ is called an *i-cell*. Such spatial subdivisions are studied in combinatorial topology. In 1813, A. Cauchy generalized the Descartes-Euler polyhedron theorem by studying intercellular faces in simple polyhedra. Finite or infinite 2D or 3D incidence grids (see Section 2.1.5)—or subsets of such grids that form 2D or 3D regions—are other examples of discrete spatial subdivisions.

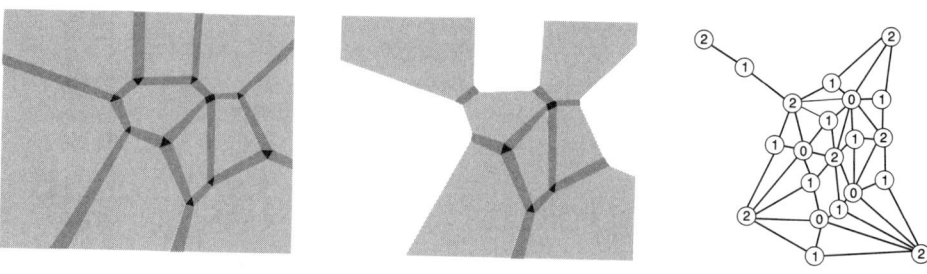

FIGURE 5.1 Left: geometric representation of the incidence structure of the Voronoi diagram shown in Figure 4.3. Middle: a substructure. Right: its graph; loops are omitted, and labels give the dimensions of the Voronoi cells, edges, and vertices.

Incidence pseudographs provide a method of describing discrete spatial subdivisions by local relations. In this section, we formulate the general n-dimensional case to illustrate potential generalizations, but our main focus continues to be on 2D and 3D grids.

5.1.1 Adjacency and completeness; incidence pseudographs

Let $G = [S, I, \dim]$ be an incidence structure. $c \in S$ is called an *i-node* if $\dim(c) = i$. The maximum value of $\dim(c)$ for any $c \in S$ is called the *index dimension* $\text{ind}(G)$ of G.[1] If $\text{ind}(G) = n$, G is called an *n-incidence structure*. Any node of G that has dimension $\text{ind}(G)$ is called a *principal node*, and any node that has smaller dimension is called a *marginal node*.

For example, 2D and 3D incidence grids define incidence structures $[\mathbb{C}_2, I, \dim]$ and $[\mathbb{C}_3, I, \dim]$ of index dimensions 2 and 3, respectively, in which pixels and voxels are the principal nodes. We generalize the adjacency definitions given for incidence grids in Section 2.2.1. Let $I(c) = \{c' : c' \in S \wedge c'Ic\}$.

Definition 5.2 Two nodes c_1 and c_2 of an incidence structure $[S, I, \dim]$ are called *i-adjacent* (notation: $c_1 A_i c_2$ or $\{c_1, c_2\} \in A_i$ or $c_1 \in A_i(c_2)$), where $0 \leq i \leq \text{ind}(S)$, iff $c_1 \neq c_2$ and there is an i-node $c \in S$ ($c \neq c_1$ and $c \neq c_2$) such that $c_1 \in I(c)$ and $c \in I(c_2)$.

A relation A_i of i-adjacency between nodes of an incidence structure defines an adjacency structure $G_i = [S, A_i]$, which is an adjacency graph if properties **A1** through **A3** are satisfied. All of the definitions and results regarding adjacency graphs in Chapter 4 apply to these G_is. For example (see Definition 4.1), two nodes $c, c' \in S$ are *i-connected* with respect to $M \subseteq S$ iff there is an i-path $(c_1, c_2, \ldots, c_k)$ in $[S, I, \dim]$

---

1. This value is often called "the dimension" of G. Later we will use nodes of G to form one-dimensional, 2D, or multidimensional subsets; the dimensions of these subsets will be defined in Definition 6.7.

5.1 Incidence Structures

where $c = c_1$, $c' = c_k$, and the nodes of the path are either all in M or all in $\overline{M} = S \setminus M$. This allows us to define i-components of M.

Figure 5.1 (middle or right) shows (on the upper left) that, unlike the situation for adjacency grids, i-adjacency of two nodes does not necessarily imply h-adjacency for $h < i$.

Definition 5.3 Two j-nodes ($j > 0$) are called *adjacent* iff there exists an $i < j$ such that the nodes are i-adjacent.

This (symmetric and irreflexive) adjacency relation defines paths of nodes of equal dimension $j > 0$, and the reflexive and transitive closure of this relation defines connectedness and components for such nodes. Components of an incidence structure will be defined in the next section; these components need not contain only nodes of a single dimension.

Our main interest is in principal nodes. Let $G = [S, I, \dim]$ be an n-incidence structure. The set of n-nodes of G is called the *core* of G. G is called an *incidence pseudograph* iff it has the following properties:

I1: For any node c of G, $I(c)$ is finite.

I2: The core of G is connected.

I3: Any finite set of principal nodes of G has at most one infinite complementary component of principal nodes.

I4: If $c' \in I(c)$ and $c' \neq c$, then $\dim(c') \neq \dim(c)$.

I5: Any marginal node of G is incident with at least one principal node of G.

Properties **I1** through **I5** are independent of one another. This is easy to verify; see Section 4.1.3. Figure 5.2 shows a finite incidence pseudograph with core $\{a, e, f\}$.

An incidence pseudograph is called *monotonic* (short for "in transitive correspondence with a monotonic chain of dimensions") iff the following is true:

I6: If $c' \in I(c)$ and $c'' \in I(c')$, where $\dim(c) \leq \dim(c') \leq \dim(c'')$, then $c'' \in I(c)$.

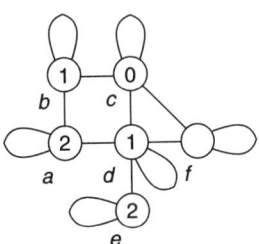

FIGURE 5.2 An incidence pseudograph with nodes $a, \ldots, f$. The dimensions of the nodes are indicated by the labels.

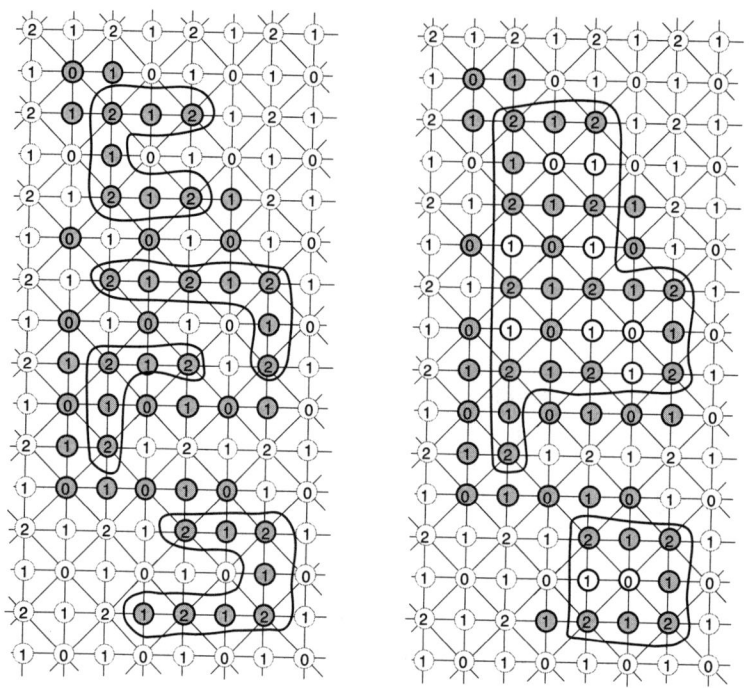

FIGURE 5.3 Left: a set M that has four 1-components of 2-nodes. Right: its completion M^+, which has two 1-components of 2-nodes (loops omitted).

For example, the pseudograph in Figure 5.2 is not monotonic because, for example, $c \in I(b)$ and $b \in I(a)$ but $c \notin I(a)$. Properties **I4** and **I6** indicate that the incidence corresponds to a partial order relation on S: $c < c'$ iff c and c' are incident and $\dim(c) < \dim(c')$. This allows us to define a poset topology; see Section 6.1.2.

Finally, we generalize Definition 2.5 (completeness) from incidence grids to arbitrary incidence pseudographs. Let $G = [S, I, \dim]$ be an incidence pseudograph, and let $M \subseteq S$. We define the *completion* M^+ of M with respect to G as the smallest subset of S that has the following properties:

(i) $M \subseteq M^+$.

(ii) If $c' \in M^+$ for all $c' \in I(c)$ such that $\dim(c') > \dim(c)$, then $c \in M^+$.

Property (ii) leads to a recursive procedure for adding nodes: first $(\text{ind}(G) - 1)$-nodes, then $(\text{ind}(G) - 2)$-nodes, and so forth. Figure 5.3 shows a rectangular subset of the 2D incidence grid[2] that is completed by adding six 1-nodes and then adding three 0-nodes.

---

2. By contrast with Figure 2.3, there are now edges between 2-nodes and 0-nodes.

5.1 Incidence Structures

Definition 5.4 A subset M of an incidence pseudograph G is called *complete* (with respect to G) iff $M = M^+$.

Figure 5.1 (middle) is an example of a complete subset.

The core of the subset M on the left in Figure 5.3 is connected; it consists of four 1-components of 2-nodes. M is a proper subset of M^+, so M is not complete; the core of M^+ contains two 1-components of 2-nodes.

5.1.2 Incidence grids

$[S, I, \dim]$ is called a *2D or 3D incidence grid* iff one of the following is true: (i) it is $[\mathbb{C}_2, I, \dim]$ or $[\mathbb{C}_3, I, \dim]$; (ii) it is a finite complete subpseudograph of $[\mathbb{C}_2, I, \dim]$ defined by a core that is an $m \times n$ rectangular subset of $\mathbb{C}_2^{(2)}$; or (iii) it is a finite complete subpseudograph of $[\mathbb{C}_3, I, \dim]$ defined by a core that is an $l \times m \times n$ cuboidal subset of $\mathbb{C}_3^{(3)}$.

Any incidence grid has properties **I1** through **I6**, as well as two other properties:

I7: If $c \in S$ and $\dim(c) > 0$, there is a $c' \in S$ such that $\dim(c') < \dim(c)$ and $\{c, c'\} \in I$.

I8: If cIc' and $\dim(c) - \dim(c') > 1$, there is a $c'' \in S$ such that $\dim(c') < \dim(c'') < \dim(c)$, cIc'', and $c'Ic''$.

In addition, the nodes of an incidence grid have locations in Euclidean space. Incidence grids are the "default examples" of incidence pseudographs, just as α- or (α_1, α_2)-adjacency grids are the "default examples" of adjacency graphs.

The cores of incidence grids are sets of pixels (2-cells) or voxels (3-cells). Definition 5.2 allows two 1-cells of $[\mathbb{C}_3, I, \dim]$ to be 0-, 2-, or 3-adjacent, but, in accordance with property **I4**, they cannot be 1-adjacent. A 1-path in $[\mathbb{C}_3, I, \dim]$ can contain 0-, 2-, or 3-cells but not 1-cells. An i-path or i-component can be restricted to contain only cells of a given dimension; see Figure 2.18. A path of cells of dimension i is defined by 0-, 1-, ..., or $(i-1)$-adjacency.

Example 5.1 Any adjacency graph $G = [S, A]$ can be represented by a one-dimensional incidence pseudograph G_I with a set of nodes that is $S \cup A$. Elements of S have dimension 0, and elements of A have dimension 1. Every $c \in A$ is incident with its two endnodes in S, and every $c \in S$ is incident with all of the edges $\{c, c'\}$ in A. Every node of G_I is also self-incident. ∎

Figure 5.4 shows the one-dimensional incidence pseudograph of $[\mathbb{Z}^2, A_4]$. Repeated merging operations (contraction; see Section 4.2.2), as shown in the middle of Figure 5.4 (including the deletion of all loops), can be used to transform this graph into a graph isomorphic to $[\mathbb{Z}^2, A_4]$.

Let $G = [S, I, \dim]$ be an n-incidence grid ($n \geq 1$), and let $0 \leq n_0 < n$. By deleting all i-nodes from S ($n_0 < i \leq n$) and all edges that have these i-nodes as endnodes, we obtain an n_0-incidence pseudograph that is a subpseudograph of $[S, I, \dim]$. Such a subpseudograph is called a *downward restriction* of $\mathbb{G}$. For example, let $\mathbb{G}$ be the

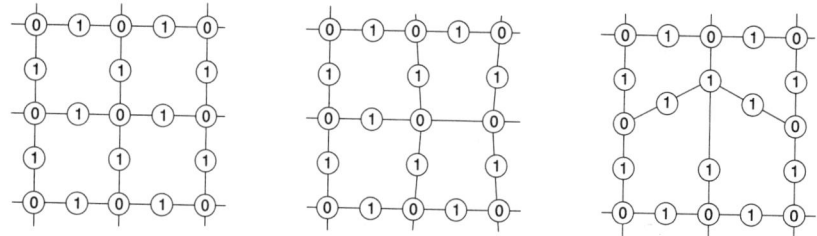

FIGURE 5.4 The one-dimensional incidence pseudograph of the 1-adjacency grid (left; all loops are omitted). Two merging operations are illustrated: a 1-node merges with a 0-node (middle); a 0-node merges with a 1-node (right).

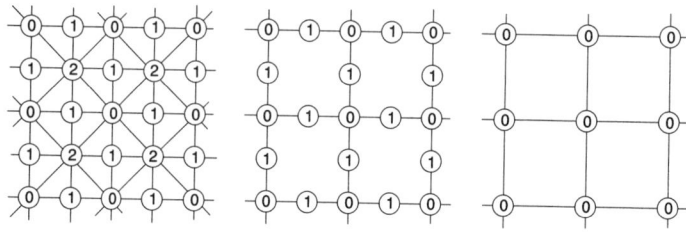

FIGURE 5.5 A 2D incidence grid, its downward restriction (both without loops), and a contraction of the downward restriction.

3D incidence grid $[\mathbb{C}_3, I, \dim]$, and let $n_0 = 2$. We can delete all 3-cubes from $\mathbb{C}_3$ to obtain a subpseudograph that consists of only grid squares, grid edges, and grid vertices. Figure 5.5 shows a 2D example: a one-dimensional downward restriction of a rectangular subgraph of $[\mathbb{C}_2, I, \dim]$, which can be transformed into $[\mathbb{Z}^2, A_4]$ by repeated merging operations.

A finite nonempty complete subset $M \subseteq S$ of an incidence grid $[S, I, \dim]$ defines an incidence structure $[M, I', \dim']$, where I' and $\dim'$ are the restrictions of I and $\dim$ to M. Such a structure satisfies properties **I1** through **I3** but need not satisfy properties **I4** through **I8**.

5.1.3 Components and regions; borders

In this section, M is a complete subset of an n-incidence structure $G = [S, I, \dim]$.

> **Definition 5.5** Let $C \subseteq M \subseteq S$. C is called a *component* of M iff the core of C is a nonempty maximal connected subset of the principal nodes of M, C also contains all nodes of M that are incident with principal nodes of M, and C is complete (with respect to G).

5.1 Incidence Structures

For example, M^+ on the right in Figure 5.3 is a component. A nonempty M may have no components, and the union of all of the components of M may not be all of M. Even a complete M is not always a union of components; there may be i-nodes in M ($i < n$) that are not incident with any n-node of M. Our main interest is in finite Ms that are unions of components.

Proposition 5.1 A node cannot be in more than one component of M.

Proof The cores of components of M are maximal connected sets of n-nodes; hence they are pairwise disjoint. If an i-node ($i < n$) were in two components, it would be incident with at least one n-node in each component; hence the n-nodes of the two components would be connected, thereby contradicting Definition 5.5. ∎

The following definitions are analogous to the definitions for adjacency graphs in Section 4.1.3:

Definition 5.6 A *region* is a finite component.

For example, the incidence pseudograph shown in Figure 5.2 is a region. The completion of its core $\{a, e, f\}$ is $\{a, d, e, f\}$. $\{a, b, d, e, f\}$ and $\{a, c, d, e, f\}$ are regions. M^+ on the right in Figure 5.3 is a region, but M on the left is not a region, because it is not complete.

Definition 5.7 A node c (of any dimension) of a region M is called an *inner node* iff $I(c) \subseteq M$; otherwise, it is called a *border node*. The set of inner nodes of M is called the *inner set* M^∇ of M, and the set of border nodes of M is called the *border* δM of M.

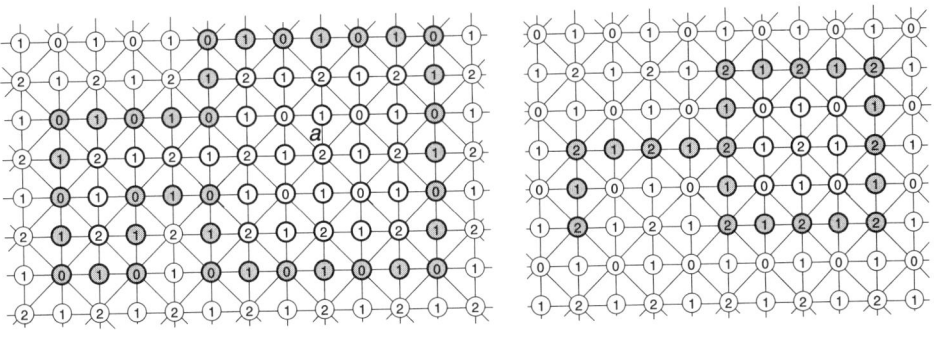

FIGURE 5.6 Inner sets (bold unfilled circles) and borders (bold filled circles) of a closed (left) and an open (right) region. The left region remains closed after deleting 2-node a; this deletion creates an open hole.

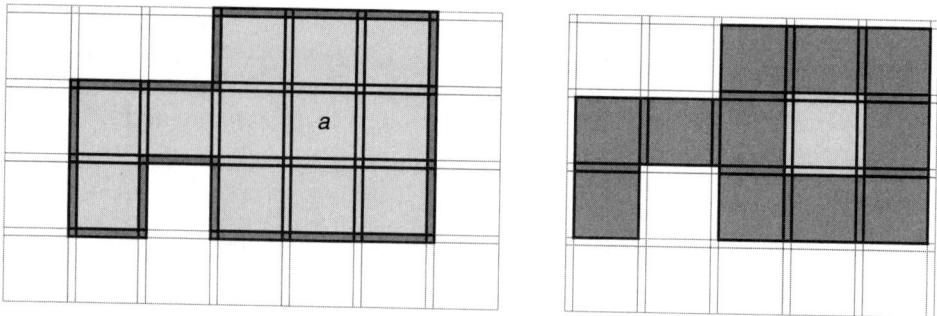

FIGURE 5.7 Geometric representations of the regions shown in Figure 5.6. Label a identifies the same 2-node as in Figure 5.6.

If $M = S$, it has only inner nodes. Figure 5.6 shows two examples of inner sets and borders. (Closed and open regions will be defined in the next section.) The same sets are shown in Figure 5.7 using the geometric representation of 2D incidence grids that was defined in Section 2.1.5.

5.1.4 Closed and open regions

The following definitions do not have analogs for adjacency graphs:

Definition 5.8 A subset $M \subseteq S$ of an incidence structure $G = [S, I, \dim]$ is called *closed in G* iff, for any $c \in M$ and any $c' \in I(c)$ such that $\dim(c') < \dim(c)$, we have $c' \in M$. M is called *open in G* iff $\overline{M} = S \setminus M$ is closed in G.

Figure 5.6 shows examples of a closed and an open region. In Figure 5.2, $\{a,d,e,f\}$ is an open region, because $\{b,c\}$ is a closed set (but not a region); $\{a,b,d,e,f\}$ is open; and $\{a,c,d,e,f\}$ is neither closed nor open. In Figure 5.3, M^+ on the right is neither closed nor open. Evidently, S and $\emptyset$ are both closed and open. Any closed or open M is complete. Closed and open sets are not necessarily components; they may contain only nodes of dimensions $< n$, or they may be disconnected.

The proof of Proposition 2.3 gave a procedure for assigning every 0- or 1-cell of the 2D incidence grid to exactly one incident 2-cell in a 2D picture P, and Proposition 2.4 generalized this procedure to the 3D incidence grid. The procedure was based on a decomposition of the grid into P-equivalence classes, which define incidence substructures that are unions of components (and hence are complete). These substructures, except for the one that contains the infinite background component, are all finite. Figure 5.8 shows an example in which the 2D procedure produces either "black closed components" and "white open components" or vice versa.

5.1 Incidence Structures

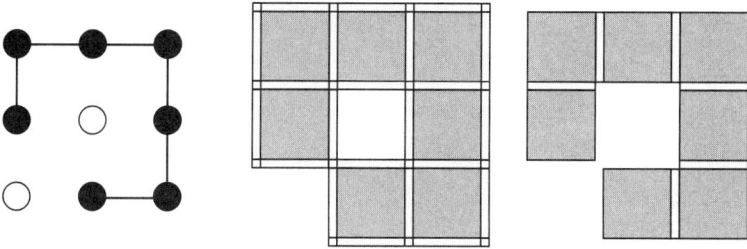

FIGURE 5.8 The black pixels (left), shown as a closed (middle) and an open (right) region of the 2D incidence grid.

As a direct consequence of the definition of closedness, we have the following:

Theorem 5.2 A finite subset $M \subseteq S$ of an n-incidence pseudograph $G = [S, I, \dim]$ is a closed region in G iff the core of M is nonempty and connected; any marginal node in M is incident with a principal node in M; and (A) for any $c \in M$ and any $c' \in I(c)$ such that $\dim(c') < \dim(c)$, we have $c' \in M$.

Any cell that is "enclosed" by higher-dimensional cells that belong to an open region is also in the region:

Theorem 5.3 A finite subset $M \subseteq S$ of an n-incidence pseudograph $G = [S, I, \dim]$ is an open region in G iff the core of M is nonempty and connected, and (B) an i-cell $c \in \mathbb{C}_n$ $(i < n)$ is in M iff every j-cell c' such that cIc' and $i < j \leq n$ is in M. If G is monotonic, M is an open region in G iff its core is nonempty and connected, and (C) an i-cell $c \in \mathbb{C}_n$ is in M iff all of the n-cells in $I(c)$ are also in M.

Proof The complement of an open set is closed; (B) is the complementary formulation of (A) in Theorem 5.2. (C) uses only the n-cells in $I(c)$, but (B) uses all of the j-cells in $I(c)$ such that $i < j \leq n$ so that (B) implies (C). Let M satisfy (C), let the i-cell c be in M, and suppose there exists a j-cell $c' \in I(c), i < j \leq n$ such that $c' \notin M$. Then at least one n-cell in $I(c')$ is not in M and hence is not in $I(c)$, which is impossible if G is monotonic. ■

The incidence pseudographs in Figures 5.3 and 5.6 are monotonic.

The *closure* $M^{\bullet}$ of a finite subset M of an n-incidence pseudograph is the smallest closed region that contains M. We can construct the closure of M by adding all cells c' such that $c' \in I(c)$ and $\dim(c') < \dim(c)$ for some $c \in M$.

5.2 Boundaries, Frontiers, and the Euler Characteristic

5.2.1 Boundaries, chains, and frontiers

Let $G = [S, I, \dim]$ be an n-incidence pseudograph. A node $c \in S$ is *invalid* with respect to $M \subseteq S$ iff $c \notin M$, but there is an n-node $c' \in M$ such that $c' \in I(c)$. This definition generalizes the concepts of invalid edges and boundaries in adjacency grids. We recall that any adjacency graph can be transformed into a one-dimensional incidence pseudograph; see Example 5.1.

Definition 5.9 The set of all nodes that are invalid with respect to $M \subseteq S$ is called the *boundary* of M.

Figure 5.9 shows two examples of boundaries for $n = 2$. The boundary of S is empty, and the boundary of a region contains no principal nodes.

Let $M \subset S$ be finite. A cell c of M will be denoted by $c^{(i)}$ if $\dim(c) = i$. An *i-chain* is an expression of the form $C^{(i)} = \sum b_k c_k^{(i)}$ where each $b_k = 0$ or 1 and the sum is taken over all of the i-dimensional cells of M. We say that $c_k^{(i)}$ is in $C^{(i)}$ iff $b_k = 1$; we can identify $C^{(i)}$ with the set of i-nodes that are in $C^{(i)}$. We define the addition of i-chains modulo 2: $\sum b_k c_k^{(i)} + \sum b_k' c_k^{(i)} = \sum (b_k + b_k') c_k^{(i)}$ where the bs are added modulo 2. Thus $c_k^{(i)}$ is in the sum of two i-chains iff it is in exactly one of the chains. We write $C^{(i)} = 0$ if all of its bs are 0 so that it corresponds to an empty set of i-nodes.

Let $C^{(i)}$ be an i-chain ($i > 0$). The *chain frontier* $\vartheta C^{(i)}$ is the $(i-1)$-chain $\sum a_k c_k^{(i-1)}$ where a_k is the number (modulo 2) of i-cells in $C^{(i)}$ that are incident with

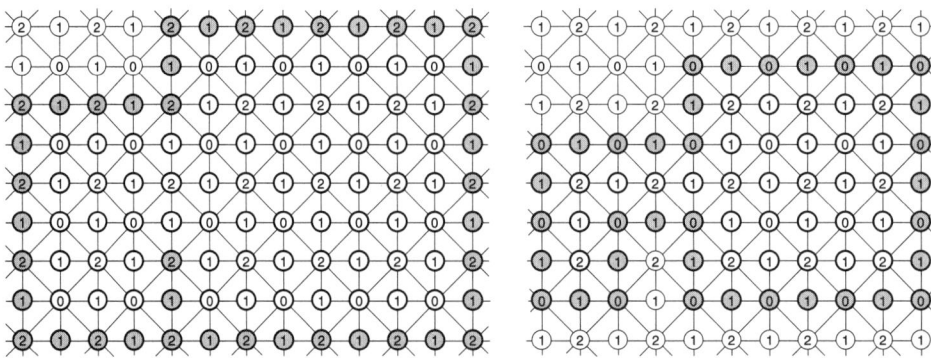

FIGURE 5.9 Two regions (bold, unfilled circles) and their boundaries (bold, shaded circles). Left: closed region. Right: open region.

5.2 Boundaries, Frontiers, and the Euler Characteristic

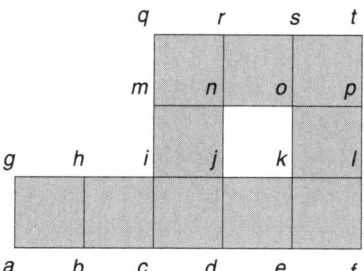

FIGURE 5.10 Example of a 2D Euclidean complex.

$C_k^{(i-1)}$. For $i = 0$, we define $\vartheta C^{(0)} = 0$. It is not hard to show that ϑ is a linear operator (i.e., $\vartheta(C^{(i)} + C\prime^{(i)}) = \vartheta C^{(i)} + \vartheta C\prime^{(i)}$ [modulo 2]).

$C^{(i)}$ is called an *i-cycle* if $\vartheta C^{(i)} = 0$. Evidently, a 0-chain is a cycle; a sum of i-cycles is an i-cycle; and the frontier of any chain is a cycle (i.e., for any i-chain, we have $\vartheta\vartheta C^{(i)} = 0$).

Figure 5.10 shows a subset M of the 2D incidence grid, with 2-cells $abhg$, $bcih$, and so on; 1-cells ag, ab, and so on; and 0-cells a, b, and so on. Using the above definitions, we have $\vartheta a = 0$; $\vartheta ab = a + b$; $\vartheta abgh = ab + bh + gh + ag$; $\vartheta(ij + jn + no + ko + jk) = i + 3j + 2n + 2o + 2k = i + j$; and $jk + ko + no + jn$ is a 1-cycle.

Definition 5.10 An open (closed) region $A \subseteq S$ is *circumscribed* by its boundary (border) M iff the set of all $(n-1)$-nodes in M is an $(n-1)$-cycle.

For example, the open region A consisting of the single 2-node $jkno$ in Figure 5.10 has boundary $M = \{j, k, n, o, jk, ko, no, jn\}$. The set of 1-nodes in M has an empty frontier, so it is a 1-cycle.

Let M be a closed region in an infinite incidence pseudograph. $\overline{M} = S \setminus M$ is the union of a finite number of pairwise disjoint open regions and one infinite open subset of S. A (finite) open region is either an *open hole* (if it is circumscribed by its boundary) or an *open finite background region*; the infinite open subset is the *open background*. The border nodes of the closed region in Figure 5.11 are shown in Figure 5.12. If we assume that the incidence pseudograph in Figure 5.11 is a subset of the 2D incidence grid and use nodes of the incidence grid not shown in the figures, the border also contains the two 1-nodes in the top row and three more 1-nodes and five more 0-nodes that "close" the border on the right and at the bottom.

Similarly, if M is an open region, we obtain *closed holes* (circumscribed by their borders), *closed finite background regions*, and a *closed background*. Figure 5.6 (or Figure 5.7) shows an open hole (after removing node a), and Figure 5.13 shows an open region for which the infinite background is a closed component (but not a region because it is infinite). The boundary nodes in Figure 5.14 coincide with the boundary nodes of the open region in Figure 5.13.

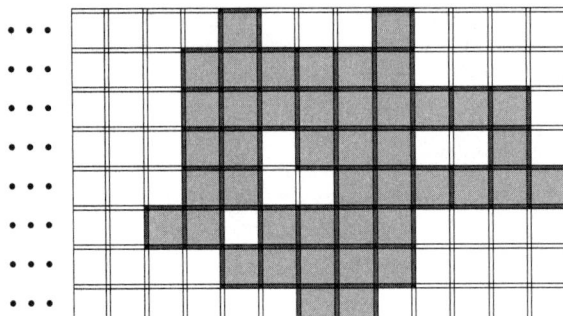

FIGURE 5.11 A closed region in an infinite incidence pseudograph (a "stripe" bounded on the right and unbounded on the left) that has three open holes (in the 1-adjacency graph of 2-nodes, two of these would be improper holes and one a proper hole), three open finite background regions, and an (infinite) open background.

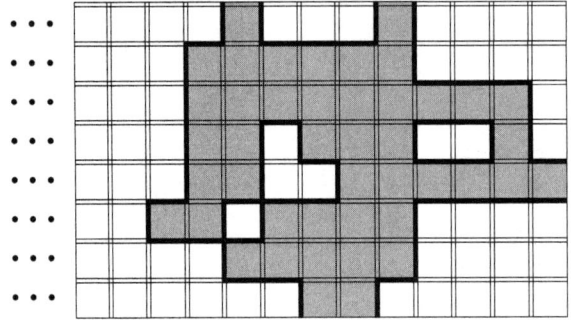

FIGURE 5.12 Boundary nodes of the closed region shown in Figure 5.11.

Theorem 5.2 If the closed component M_1 is the closure of the open component M_2, the border of M_1 coincides with the boundary of M_2.

Proof The border δM_1 of a closed component M_1 consists of all nodes c for which $c \in M_1$ and $I(c) \not\subseteq M_1$. c cannot be a principal node, because M_1 is closed so that $I(c') \subset M_1$ for any principal node c'; however, c is incident with at least one principal node in M_1 because of Definition 5.5. It follows that the nodes of δM_1 constitute the boundary of the open set $M_1 \setminus \delta M_1$.

Node c is in the boundary of an open component M_2 iff $c \notin M_2$, but there exists a principal node $c' \in M_2$ such that $c' \in I(c)$. It follows that all boundary nodes of M_2 are in the closure $M_2^{\bullet}$ of M_2, hence in the border of this closed component.

5.2 Boundaries, Frontiers, and the Euler Characteristic

FIGURE 5.13 An open region that has one closed hole (in the 1-adjacency graph of 2-nodes this would be a proper hole), three closed finite background regions, and an (infinite) closed background component with a core that consists of three 1-components.

FIGURE 5.14 Boundary nodes of the complementary open regions of the closed region shown in Figure 5.11.

Conversely, let c be in the border of $M_2^\bullet$; because M_2 is a component, c is incident with a principal node of M_2, so c is in the boundary of M_2. ∎

We recall Definition 3.2 for a subset A of a Euclidean space: the frontier ϑA is the set-theoretic difference between the closure $A^\bullet$ and the interior A°. The frontier of A°, of $A^\bullet$, or of any set containing A° and contained in $A^\bullet$ is also ϑA.

Definition 5.11 Let $M \subseteq S$ be a subset of an incidence structure $G = [S, I, \dim]$. The *frontier* ϑM of M is the border of $M^\bullet$.

In analogy with the Euclidean topology, we call $M^\circ = M^\bullet \setminus \vartheta M$ the *interior* of M. Note that, if M is a component, M° is an open component.

Section 4.3.6 discussed borders and boundaries in adjacency grids at the abstract level of adjacency graphs. There we were able to state alternatives, but were unable to unify "border" and "boundary" as in Theorem 5.2, which led to Definition 5.11.

5.2.2 The matching theorem

The *matching theorem* in this section is a basic combinatorial formula for finite incidence pseudographs of index dimension $n \geq 0$.

Definition 5.12 Let $G = [S, I, \dim]$ be an incidence pseudograph, and let $c \in S$. The number shown here is called the *incidence count* of c:

$$a_{ij}(c) = \begin{cases} \text{card}\{c' \in S : \dim(c') = j \wedge \{c,c'\} \in I\} & \text{if } i = \dim(c) \\ 0 & \text{otherwise} \end{cases}$$

Because of self-incidence and property **I4** (see Section 5.1.1), if $\dim(c) = i$, we have $a_{ii}(c) = 1$. Table 5.1 gives examples of incidence counts.

Theorem 5.2 (the matching theorem): $\sum_{c \in S} a_{ij}(c) = \sum_{c \in S} a_{ji}(c)$ for $0 \leq i, j \leq n$.

Proof If $i \neq j$, all of the edges between i-nodes and j-nodes (and only those edges) are counted in the sum; see Figure 5.15. All edges are undirected, and the number of endpoints in both sums is the same. If $i = j$, the sum is equal to the number of i-nodes. ∎

Equation 4.4 follows from the matching theorem, because $\nu(p) = a_{01}(p)$ for any node p of an adjacency graph $[S, A]$. Equation 4.4 can also be derived using the incidence counts $a_{10}(e)$ for the edges e:

$$\sum_{p \in S} \nu(p) = \sum_{p \in S} a_{01}(p) = \sum_{e \in S} a_{10}(e) = 2\alpha_1$$

TABLE 5.1 Incidence counts for the pseudograph in Figure 5.2.

node	a	b	c	d	e	f
i	2	1	0	1	2	2
a_{i2}	1	1	1	3	1	1
a_{i1}	2	1	2	1	1	1
a_{i0}	0	1	1	1	0	1

5.2 Boundaries, Frontiers, and the Euler Characteristic

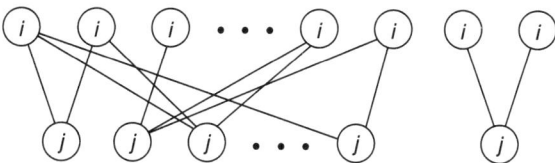

FIGURE 5.15 The undirected edges that connect i-nodes to j-nodes in an incidence pseudograph.

Definition 5.13 If G is regular, $a_{ij}(c) = a_{ij}$ is constant for all $c \in S$ where $\dim(c) = i$ and $j \geq 0$.

This definition generalizes Definition 4.7. The complete graphs K_n are examples of finite regular one-dimensional incidence pseudographs. The nodes of K_n have dimension 0, the edges of K_n have dimension 1, and every node is self-incident. The infinite incidence grids $[\mathbb{C}_2, I, \dim]$ and $[\mathbb{C}_3, I, \dim]$ are also regular.

If G has index dimension $n \geq 1$ and M is a finite subset of G, the numbers shown here are called the *class cardinalities* of M:

$$\alpha_i^M = \operatorname{card}\{c: c \in M \wedge \dim(c) = i\}, \ 0 \leq i \leq n \tag{5.1}$$

We usually omit the superscript M.

From the Matching Theorem, we know that, for finite regular incidence pseudographs, we have $\alpha_i a_{ij} = \alpha_j a_{ji}$ for $0 \leq i, j \leq n$. It follows that this equation holds true

$$\alpha_i a_{ik} - \alpha_k a_{ki} = 0 \text{ for } 0 \leq i \leq n \tag{5.2}$$

for any (e.g., fixed) index k ($0 \leq k \leq n$). Hence the possible integer values of the class cardinalities α_k and α_i define constraints (Diophantine equations) on the incidence counts a_{ki} and a_{ik} (and vice versa).

5.2.3 The Euler characteristic

In this section, we generalize the definition given in Section 4.3.2 for finite oriented adjacency graphs in 2D. Note that an edge is incident with a cycle iff it is listed in the cycle, and a node is incident with an edge iff it is one of the edge's endnodes.

Definition 5.14 Let $G = [S, I, \dim]$ be a finite n-incidence pseudograph ($n \geq 1$). The *Euler characteristic* of G is as follows, where the α_is are the class cardinalities of S:

$$\chi(G) = \sum_{i=0}^{n} (-1)^i \alpha_i$$

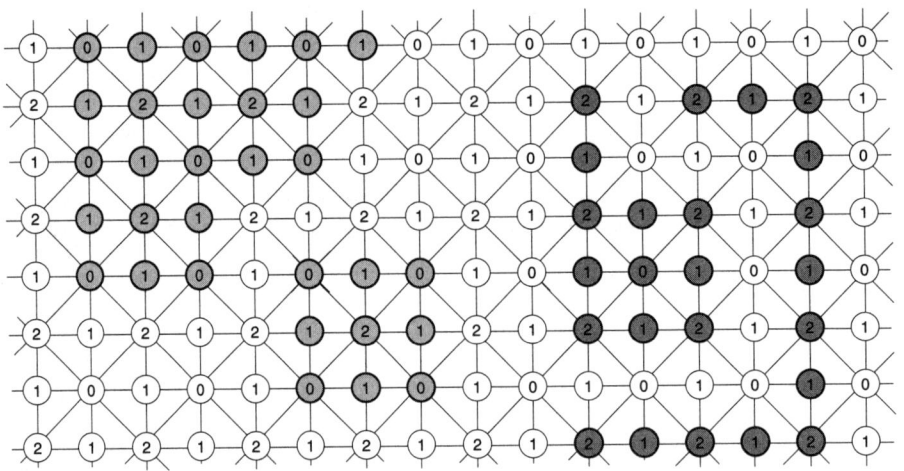

FIGURE 5.16 Four components that define subpseudographs: two closed regions on the left and two open regions on the right. The Euler characteristic is 1 in all four cases.

For example, for the G shown in Figure 5.2, we have $\chi(G) = 1 - 2 + 3 = 2$. Adding an edge (e.g., between nodes b and e) does not change the Euler characteristic. However, deleting a node (e.g., 2-node e) results in an incidence pseudograph G' for which $\chi(G') = 1 - 2 + 2 = 1$.

If G is regular, the Matching Theorem gives us the following:

$$\frac{\chi(G)}{\alpha_k} = \sum_{i=0}^{n} (-1)^i \frac{a_{ki}}{a_{ik}}$$

These $n+1$ equations ($k = 0, 1, \ldots, n$) are rational multiples of one another.

Figure 5.16 shows regions in the infinite regular incidence grid $[\mathbb{C}^2, I, \dim]$ that define subpseudographs (loops omitted). For the upper left region, we have $\chi = 3 - 10 + 8 = 1$; for the lower left region, we have $\chi = 1 - 4 + 4 = 1$; for the region on their right, we have $\chi = 5 - 5 + 1 = 1$; and, for the remaining 1-path, we have $\chi = 7 - 6 + 0 = 1$. For the region M^+ shown on the right in Figure 5.3, we have $\chi = 15 - 28 + 14 = 1$; note that removing marginal border nodes changes this value. In Figure 5.6, for the closed region on the left, we have $\chi = 12 - 33 + 22 = 1$, and, for the open region on the right, we have $\chi = 12 - 15 + 4 = 1$. Removing the "central" 2-node a from the closed region creates an "open hole" and gives $\chi = 11 - 33 + 22 = 0$. These examples illustrate the topologic invariance of the Euler characteristic; see Section 6.4.5.

5.3 The Regular Case

An incidence pseudograph provides a more "refined" representation of a grid than does an adjacency graph representation. We begin by discussing regular infinite incidence pseudographs defined in $\mathbb{R}^n$, but as usual, the cases of interest involve incidence relations on finite or countable sets of 0-, 1-, 2-, or 3-cells.

5.3.1 Regular infinite incidence pseudographs

The incidence grids $[\mathbb{C}_2, I, \dim]$ and $[\mathbb{C}_3, I, \dim]$, when regarded as pseudographs $[S, I, \dim]$, are monotonic (property **I6**) and also have properties **I7** and **I8**.

Definition 5.15 The set of all *0-cells* is the set $(0.5, \ldots, 0.5) + \mathbb{Z}^n$. Let c be a k-cell in a k-dimensional subspace of $\mathbb{R}^n$ ($0 \leq k < n$). Let e_j be the straight line segment with endpoints that are the origin o and the point $(0, \ldots, 0, 1, 0, \ldots, 0)$ where the 1 is in position j ($1 \leq j \leq n$). Let e_j be in an $n-k$ dimensional subspace of $\mathbb{R}^n$. The Minkowski sum $c \oplus e_j$ in $\mathbb{R}^n$ defines a $(k+1)$-cell. The set of all *i-cells* is denoted by $\mathbb{C}_n^{(i)}$ ($0 \leq i \leq n$), and the union of the $\mathbb{C}_n^{(i)}$ is denoted by $\mathbb{C}_n$.

This generalizes Definition 2.1. An informal description is as follows: A 1-cell is "created" by translating a 0-cell along any of the segments e_j. A 2-cell is the area occupied by a 1-cell while it shifts along an orthogonal segment e_j, and a 3-cell is the volume occupied by a 2-cell while it shifts along an orthogonal segment e_j. Figure 5.17 illustrates these processes up to the creation of a 4-cell.

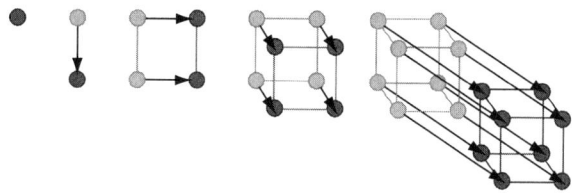

FIGURE 5.17 Examples of how a Minkowski sum ("union during translation along an orthogonal line segment e_j") of an i-cell creates an $(i+1)$-cell.

$[\mathbb{C}_n, I, \dim]$ ($n \geq 1$) is called the *regular n-incidence grid*. It can be verified that it has properties **I6** and **I8**. Its incidence counts are as follows:

$$a_{ij} \begin{cases} 2^{j-i} \binom{n-i}{n-j} & \text{if } i < j \\ 1 & \text{if } i = j \\ 2^{i-j} \binom{i}{j} & \text{if } i > j \end{cases} \quad (5.3)$$

For example, an i-cell $c \in \mathbb{C}_n$ is incident with 2^{n-1} n-cells. Equation 5.3 implies the following for $0 \leq i, j \leq n$:

$$\frac{a_{ij}}{a_{ji}} = \frac{\binom{n}{j}}{\binom{n}{i}} \tag{5.4}$$

Equation 5.4 in turn implies the following for $0 \leq i \leq n$:

$$\sum_{j=0}^{n} (-1)^j \frac{a_{ij}}{a_{ij}} = \frac{1}{\binom{n}{i}} \sum_{j=0}^{n} (-1)^j \binom{n}{j} = \frac{1}{\binom{n}{i}} \cdot 0 = 0 \tag{5.5}$$

Both of these equations will later prove to be very useful.

5.3.2 The region matching theorem

In this section, α_i^M are class cardinalities, and a_{ij} are incidence counts in the regular incidence grid $[\mathbb{C}_n, I, \dim]$. Note that the a_{ij} are constants. For $n = 2$, they are as follows:

$$\begin{aligned}
a_{00} &= 1, & a_{01} &= 4, & a_{02} &= 4, \\
a_{10} &= 2, & a_{11} &= 1, & a_{12} &= 2, \\
a_{20} &= 4, & a_{21} &= 4, & a_{22} &= 1
\end{aligned}$$

For $n = 3$, they are as follows:

$$\begin{aligned}
a_{00} &= 1, & a_{01} &= 6, & a_{02} &= 12, & a_{03} &= 8, \\
a_{10} &= 2, & a_{11} &= 1, & a_{12} &= 4, & a_{13} &= 4, \\
a_{20} &= 4, & a_{21} &= 4, & a_{22} &= 1, & a_{23} &= 2, \\
a_{30} &= 8, & a_{31} &= 12, & a_{32} &= 6, & a_{33} &= 1
\end{aligned}$$

The numbers shown here are called *boundary counts* for the cells in M:

$$b_{ij}^M(c) = \begin{cases} \text{card}\{c' \in I(c) : \dim(c') = j \land c' \text{is invalid}\} \\ \quad \text{if } c \in M \text{ and } i = \dim(c) \\ 0 \quad \text{otherwise} \end{cases}$$

The numbers shown here are called total boundary counts for M:

$$b_{ij}^M = \sum_{c \in S} b_{ij}^M(c) \quad \text{for } 0 \leq i, j \leq \text{ind}(G)$$

From now on, we omit the superscript M.

5.3 The Regular Case

Theorem 5.2 (the Region Matching Theorem): Let M be an open or closed region in the regular incidence grid $[\mathbb{C}_n, I, \dim]$. For $0 \leq i, j \leq n$, we have the following:

$$\alpha_i a_{ij} - b_{ij} = \alpha_j a_{ji} \text{ for } i < j \text{ if } M \text{ is closed or for } i > j \text{ if } M \text{ is open}$$

$$\alpha_i a_{ij} = \alpha_j a_{ji} \text{ for } i = j$$

$$\alpha_i a_{ij} + b_{ji} = \alpha_j a_{ji} \text{ for } i > j \text{ if } M \text{ is closed or for } i < j \text{ if } M \text{ is open.}$$

Proof Let M be closed. In the equations for $i < j$, the right sides are the number of j-cells in M times the number of i-cells that are incident with any of these j-cells (i.e., the number T of incidences between j-cells in M and i-cells). All of these i-cells are also elements of M; see (A) in Theorem 5.2. The total number of incidences between i-cells in M and j-cells in M is evidently T. However, the i-cells in M are also incident with b_{ij} j-cells, which are not in M; hence the count $\alpha_i a_{ij}$ on the left side of the equation must be reduced by b_{ij}.

The second equation is trivial and is given only for completeness.

The equations for $j < i$ are proved for a closed region M by simply swapping i and j in the discussion of the $i < j$ case. For an open region M, we use (B) from Theorem 5.3. ∎

Note that the proof of this theorem makes no use of the connectedness of a region, but only of its being either closed or open. Figure 5.18 shows an open and a closed region. For the open region, we have $\alpha_0 = 0$, $\alpha_1 = 1$, $\alpha_2 = 2$, $b_{10} = 2$, $b_{20} = 8$, and $b_{21} = 6$; for example, $1 \cdot 2 + 6 = 2 \cdot 4$ if $i = 1$ and $j = 2$. For the closed region, we have $\alpha_0 = 7$, $\alpha_1 = 8$, $\alpha_2 = 2$, $b_{01} = 12$, $b_{02} = 20$, and $b_{12} = 8$; for example, $7 \cdot 4 - 20 = 2 \cdot 4$ if $i = 0$ and $j = 2$.

The formulas of the Region Matching Theorem hold for any finite union of pairwise disjoint closed (or open) regions.

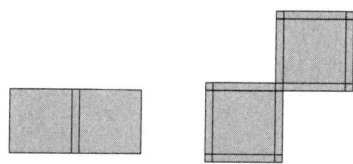

FIGURE 5.18 Left: open region. Right: closed region.

For a closed region M or a finite union of pairwise disjoint closed regions, the Region Matching Theorem implies the following:

$$\alpha_i = \alpha_j \frac{a_{ji}}{a_{ij}} + \frac{b_{ij}}{a_{ij}} \quad \text{if } i < j$$

$$\alpha_i = \alpha_j \frac{a_{ji}}{a_{ij}} \quad \text{if } i = j$$

$$\alpha_i = \alpha_j \frac{a_{ji}}{a_{ij}} - \frac{b_{ji}}{a_{ij}} \quad \text{if } i > j$$

For example, α_n specifies the contents of an isothetic polyhedron in $\mathbb{C}_n$ ($n \geq 2$). These formulas allow us to calculate α_n based on counts of, for example, vertices; see Section 8.1.6 for $n = 2$ and Section 8.3.7 for $n = 3$. Using Equation 5.5, it follows that the Euler characteristic, as follows,

$$\chi(M) = \sum_{i=1}^{n} (-1)^i \alpha_i$$

$$= \alpha_j \left[\sum_{i=0}^{j-1} (-1)^i \frac{a_{ji}}{a_{ij}} + (-1)^j + \sum_{i=j+1}^{n} (-1)^i \frac{a_{ji}}{a_{ij}} \right] + \sum_{i=1}^{j-1} (-1)^i \frac{b_{ij}}{a_{ij}} - \sum_{i=j+1}^{n} (-1)^i \frac{b_{ji}}{a_{ij}}$$

$$= \sum_{i=1}^{j-1} (-1)^i \frac{b_{ij}}{a_{ij}} - \sum_{i=j+1}^{n} (-1)^i \frac{b_{ji}}{a_{ij}}$$

can be calculated for any j ($0 \leq j \leq n$) by counting only invalid cells (cells on the boundary of the region). Similarly, for an open region M or a finite union of pairwise disjoint open regions, we have the following for any $0 \leq j \leq n$:

$$\chi(M) = -\sum_{i=1}^{j-1} (-1)^i \frac{b_{ji}}{a_{ij}} + \sum_{i=j+1}^{n} (-1)^i \frac{b_{ij}}{a_{ij}}$$

Indices $j = 0$ and $j = n$ give the simplest expressions. In the next section, we will show that these expressions can be further simplified.

5.3.3 Euler characteristics

We apply the Region Matching Theorem and Conclusion 5.3.2:

Lemma 5.1 Let M be a finite union of pairwise disjoint closed regions in the n-incidence grid. For $0 \leq i, j \leq n$, we have the following:

$$\alpha_i = \alpha_n \frac{a_{ni}}{a_{in}} + \sum_{j=i}^{n-1} \frac{b_{j,j+1}}{a_{j+1,j}} \cdot \frac{a_{j+1,i}}{a_{i,j+1}}$$

5.3 The Regular Case

Proof The proof is by downward induction starting at $i = n$:

$$\alpha_n = \alpha_n \frac{a_{nn}}{a_{nn}} + 0$$

Assuming that the equation is correct for $i \geq 1$, we show that it is also correct for $i - 1$. From the Region Matching Theorem, for a closed region we have the following, where Equation 5.4 can be used to simplify products of b-values:

$$\begin{aligned}
\alpha_{i-1} &= \alpha_i \frac{a_{i,i-1}}{a_{i-1,i}} + \frac{b_{i-1,i}}{a_{i-1,i}} \\
&= \alpha_n \frac{a_{ni} a_{i,i-1}}{a_{in} a_{i-1,i}} + \sum_{j=i}^{n-1} \frac{b_{j,j+1} a_{j+1,i} a_{i,i-1}}{a_{j+1,j} a_{i,j+1} a_{i-1,i}} + \frac{b_{i-1,i}}{a_{i-1,i}} \\
&= \alpha_n \frac{a_{n,i-1}}{a_{i-1,n}} + \sum_{j=i-1}^{n-1} \frac{b_{j,j+1} a_{j+1,i-1}}{a_{j+1,j} a_{i-1,j+1}}
\end{aligned}$$

∎

Analogously, for open regions, we have the following:

Lemma 5.2 Let M be a finite union of pairwise disjoint open regions in the n-incidence grid. For $0 \leq i, j \leq n$, we have the following:

$$\alpha_i = \alpha_0 \frac{a_{0i}}{a_{i0}} - \sum_{j=1}^{i} \frac{b_{j,j-1}}{a_{j-1,j}} \cdot \frac{a_{j-1,i}}{a_{i,j-1}}$$

The following theorem was proved by K. Voss in 1993 for open regions:

Theorem 5.2 Let M be a finite union of pairwise disjoint closed (or open) regions in $[\mathbb{C}_n, I, \dim]$. Then the Euler characteristic of M is as follows:

$$\chi(M) = \frac{1}{2n} \sum_{i=1}^{n} (-1)^{i+1} b_{i,i-1} \text{ for open regions}$$

and

$$\chi(M) = \frac{1}{2n} \sum_{i=0}^{n-1} (-1)^{i+1} b_{i,i+1} \text{ for closed regions}$$

Proof We prove the theorem for open regions; closed regions can be treated analogously. Lemma 5.2 and the equation $\sum(-1)^i a_{0i}/a_{i0} = 0$ show that the following is true:

$$\chi(M) = \sum_{i=0}^{n}(-1)^i \left(-\sum_{j=1}^{i} \frac{b_{j,j-1}a_{j-1,i}}{a_{j-1,j}a_{i,j-1}} \right)$$

The double sum can be rearranged: first take the sum for all j-values and then for all i-values. It follows that what is shown here is true:

$$\chi(M) = \sum_{j=1}^{n} \frac{b_{j,j-1}}{a_{j-1,j}} \sum_{i=j}^{n}(-1)^{i+1} \frac{a_{j-1,i}}{a_{i,j-1}}$$

The formula for closed regions then follows from Equation 5.4 and for $0 \leq m < n$:

$$-\sum_{i=m+1}^{n}(-1)^i \binom{n}{i} = \sum_{i=0}^{m}(-1)^i \binom{n}{i} = (-1)^m \binom{n-1}{m} \tag{5.6}$$

∎

The $b_{i-1,i}$s and $b_{i+1,i}$s in Theorem 5.2 can be replaced with class cardinalities and (globally known) incidence counts because the following are given:

$$b_{i,i-1} = \alpha_{i-1}a_{i-1,i} - \alpha_i a_{i,i-1} \text{ for open regions}$$
$$b_{i,i+1} = \alpha_i a_{i,i+1} - \alpha_{i+1}a_{i+1,i} \text{ for closed regions}$$

Let $n = 3$. For a closed region, b_{01} is the number of invalid grid edges incident with grid vertices in the region, b_{12} is the number of invalid grid squares incident with grid edges in the region, and b_{23} is the number of invalid grid cubes incident with grid squares in the region. For open regions, we use b_{10}, which is the number of invalid grid vertices incident with grid edges in the region; b_{21}, the number of invalid grid edges incident with grid squares in the region; and b_{32}, the number of invalid grid squares incident with grid cubes in the region. Note that the total boundary counts are sums over all cells in M so that invalid cells may be counted repeatedly if they are incident with several cells in M.

Figure 5.19 shows a 2D example. For the left closed region, we have $b_{12} = 16$ and $b_{01} = 16$, and, for the right closed region, we have $b_{12} = 4$ and $b_{01} = 8$. For the left open region, we have $b_{21} = 16$ and $b_{10} = 16$, and, for the right open region (a single node), we have $b_{21} = 4$ and $b_{10} = 0$.

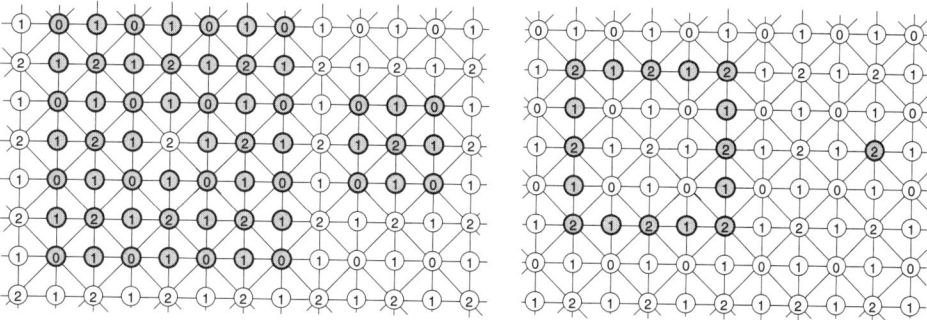

FIGURE 5.19 Nodes in regions are represented by bold filled circles. Left: two closed regions with Euler characteristics 0 and 1. Right: two open regions, also with Euler characteristics 0 and 1.

5.4 Pictures on Incidence Grids

5.4.1 Ordered labeling

In a *labeled incidence pseudograph* $G = [S, I, \dim]$, each node in G has a label, and the labels belong to a finite set $\{L_0, \ldots, L_{\max}\}$.

A picture assigns labels (values of pixels or voxels) to all of the principal nodes of S. We assume that the set of these labels is totally ordered by "significance," so nodes labeled $L_{\max}$ are the "most significant" (e.g., "object pixels") and nodes labeled L_0 are the "least significant" (e.g., "background pixels"). For example, a 2D picture assigns labels (pixel values) to all 2-cells in $\mathbb{G}_{m,n}$ and label L_0 to all 2-cells in the infinite background component.

An *ordered labeling* of the nodes of an incidence pseudograph is defined by assigning labels to the principal nodes and extending this labeling to the other nodes by applying the following *maximum-label rule*:

Give every marginal node the largest label of any of its incident principal nodes.

According to **I5**, any marginal node has candidate labels, and, according to **I1**, the set of candidates is finite, which allows us to choose the largest of them. The rule ensures that regions labeled with $L_{\max}$ are closed and regions labeled with L_0 are open.

In picture processing or analysis, we do not need to perform actual labeling but only to detect adjacencies between labeled principal nodes, assuming that all marginal nodes have been labeled. During a scan through the set of principal nodes, we have to decide whether two nodes c_1 and c_2 such that $I(c_1) \cap I(c_2) \neq \emptyset$ are still adjacent after ordered labeling. An algorithm for this is given in Algorithm 5.1. An order by decreasing node dimension (Step 2) has been assumed for efficiency reasons; the scan through the marginal nodes in $I(c_1) \cap I(c_2)$ should visit nodes with $I(c)$s that have increasing cardinalities, and the ordering by node dimension may ensure this.

If G is a binary picture and we assume that black (or white) is the most significant label, the resulting adjacency is just the (4,8)- or (8,4)-adjacency defined in Section 2.1.4. These two adjacencies are illustrated in Figure 5.20, together with the corresponding representations of the components in the incidence grid.

The situation is more complicated if G is a multilevel picture, because there are many possible total orderings of the labels. Figure 5.21 illustrates three such orderings for a three-valued picture; each ordering yields a different adjacency structure.

In the next section, we will describe a general method of defining adjacencies for a given ordered labeling. In a 2D picture, this method yields adjacencies in which diagonal adjacencies never "cross," as illustrated in Figure 5.22. These adjacencies can be defined using the 2×2 masks shown in Figure 5.23. Note that these masks accept all 4-adjacencies between pixels that have the same label and that they also accept diagonal adjacencies between two pixels that have the same label as a third pixel that is adjacent to both of them. Only in case (g) is it necessary to choose be-

---

1. If c_1 and c_2 have different labels, stop (FALSE). Otherwise, let L be the label of c_1 and c_2.
2. Scan through all of the marginal nodes in $I(c_1) \cap I(c_2)$; assume that this set is ordered by decreasing dimension, and start with a node c of maximal dimension.
    (a) If $I(c)$ does not contain a principal node with a larger label than L, stop (TRUE).
    (b) Let c be the next marginal node in $I(c_1) \cap I(c_2)$. If it does not exist, stop (FALSE); otherwise, go to Step 2.a.

ALGORITHM 5.1   Local adjacency decision procedure, assuming an ordered labeling.

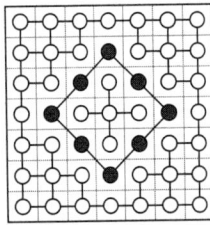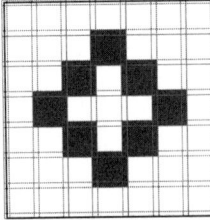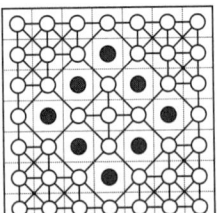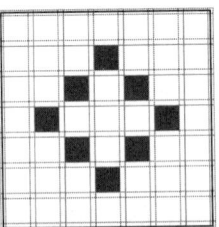

FIGURE 5.20   (4,8)- and (8,4)-adjacencies and the corresponding representations of the components in the incidence grid.

5.4 Pictures on Incidence Grids

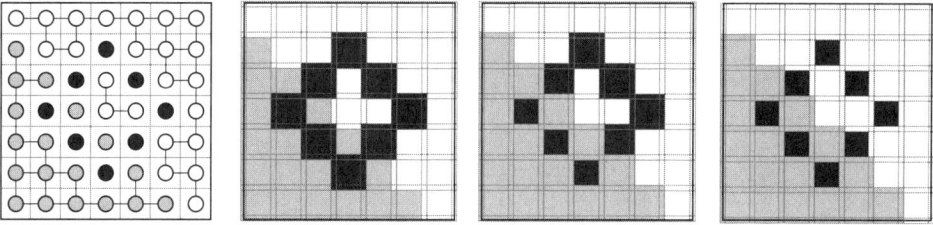

FIGURE 5.21 A three-valued input picture (from left to right): all of the 4-components in the grid point model; closed (open) regions are black (gray); closed (open) regions are gray (white); closed (open) regions are gray (black).

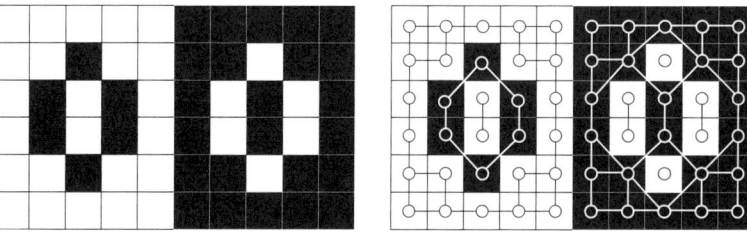

FIGURE 5.22 Left: the binary picture from Figure 1.10. Right: adjacencies resulting from ordered labeling, assuming "white is less significant than black" and dropping crossing pairs of diagonals.

tween the (black, black) and (white, white) diagonal adjacencies. When this method of defining adjacencies is used, the adjacencies define a planar graph, and, as we will see in Chapter 7, the resulting region adjacency graph is a tree.

5.4.2 The ordered adjacency procedure

We now describe a general procedure for the ordered labeling of incidence grids that applies in particular to 2D or 3D multilevel pictures. Let $G = [S, I, \dim]$ be a labeled n-incidence grid, and assume a total ordering of the set of labels. Our *ordered adjacency procedure* defines adjacencies between principal nodes of G that depend on this total order.

Let S be finite, let c_0 be a 0-cell, and let $I^{(n)}(c_0)$ be the set of all principal nodes in $I(c_0)$. We begin with any 0-cell c_0 in S and any pair c_1 and c_2 of nodes in $I^{(n)}(c_0)$, and we proceed as described in Algorithm 5.2. We then continue for all of the remaining 0-cells in S.

This procedure guarantees that components are defined by $(n-1)$-adjacency between principal nodes if they have the same label or by i-adjacency ($i < n$) if their label "wins" over at least one "competing" label. It simplifies the switch approach

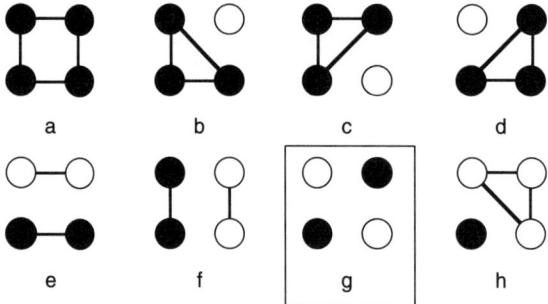

FIGURE 5.23 In case (g), decide whether "white is less significant than black" or vice versa.

described in Section 2.1.3 by eliminating redundant adjacencies and generalizes it to incidence grids of arbitrary dimension.

The result of applying the procedure to the multilevel picture shown in Figure 2.19 is shown in Figure 5.24. The scan of $I^{(n)}(c_0)$ to list all pairs (c_1, c_2) in it can be based on a uniform local cyclic order, but the result is independent of the order in which the pairs are chosen.

The adjacency defined by applying the procedure to a picture depends only on the total order of the labels $0, \ldots, G_{\max}$. As a default, we assume the order $0 < 1 < \ldots < G_{\max}$, for which we have the following:

Proposition 5.2 For a binary picture, ordered adjacency and (8,4)-adjacency generate the same sets of object and nonobject components.

1. If c_1 and c_2 are already adjacent, go to the next pair. Otherwise:
2. If they have the same label L, and they are
    (a) $(n-1)$-adjacent; or
    (b) there is a marginal node $c' \in I(c_1) \cap I(c_2)$ with $\dim(c') \leq n-2$ such that $I(c')$ is not contained in the union of all the incidence sets $I(c)$, $c \in I^{(n)}(c_0)$, and L is the maximal label of all the principal nodes in $I(c')$,
   then accept adjacency between c_1 and c_2.

ALGORITHM 5.2 Ordered Adjacency Procedure: local test for a pair of 0-adjacent principal nodes.

5.4 Pictures on Incidence Grids

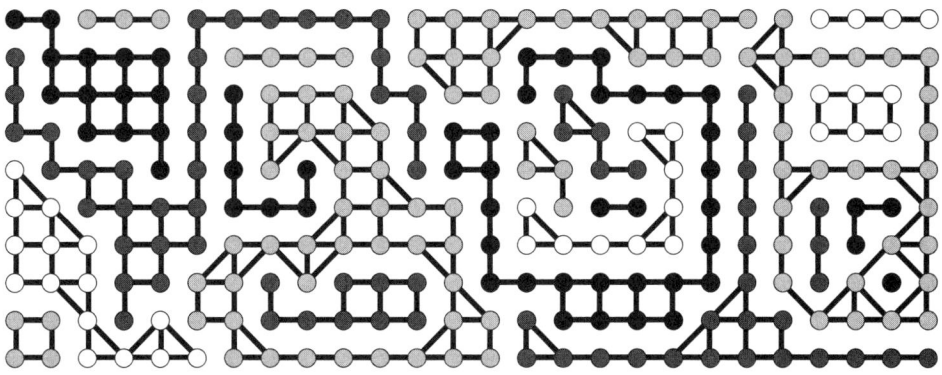

FIGURE 5.24 Ordered adjacencies for the picture shown in Figure 2.19, assuming the same order of the picture values as in Figure 2.19.

5.4.3 Frontiers in 2D incidence grids

In a 2D incidence grid, the boundary of a closed region consists of one or more 1-components of 2-cells (grid squares), and the boundary of an open region consists of one or more 0-components of 1-cells (grid edges). The union of the 1-cells in the boundary of an open region is the same as the boundary of the region in the 1-adjacency grid. Figure 4.27 (right) shows an example.

The set of 1-cells that are incident with exactly one 2-cell of a region defines the frontier of the region. Every 0-component of these 1-cells has a Hamiltonian circuit that visits each 1-cell in the 0-component exactly once.[3] The border tracing algorithm (see Chapter 4) can be used to trace the border cycles in the 0-adjacency graph of these 1-cells.

We now describe a method of representing these sets of 0- and 1-cells. Let P be a picture on an $m \times n$ grid $\mathbb{G}$, which we call the *picture grid*. We extend this grid to an $(m+1) \times (n+1)$ *frontier grid* $\mathbb{F}$, where each grid point in $\mathbb{F}$ represents a grid vertex (0-cell) in $\mathbb{G}$; specifically, grid vertex $(x-0.5, y-0.5)$ in $\mathbb{G}$ represents grid point (x,y) in $\mathbb{F}$.[4] Figure 5.25 shows the picture grid and the frontier grid for the picture shown in Figures 1.10 and 5.22.

The geometric representation of the frontiers in $\mathbb{G}$ is the union of the small squares and thin rectangles that represent 0-cells and 1-cells, respectively. Figure 5.25 shows counterclockwise frontier traversals for the interior regions and clockwise traversals for the exterior regions and the (infinite) background component. This

---

3. We assume an infinite incidence grid or a finite grid "expanded" into the infinite background component so that "missing" border cells (see Figure 5.12) can be excluded (i.e., a region's frontier always circumscribes it).

4. Interpixel frontiers (in the grid cell model) were first used by R. Brice and C.L. Fennema [128] for picture segmentation. The frontier grid is also known as a "half-integer grid" and has been popularized in digital topology; see, for example, [326, 505, 591].

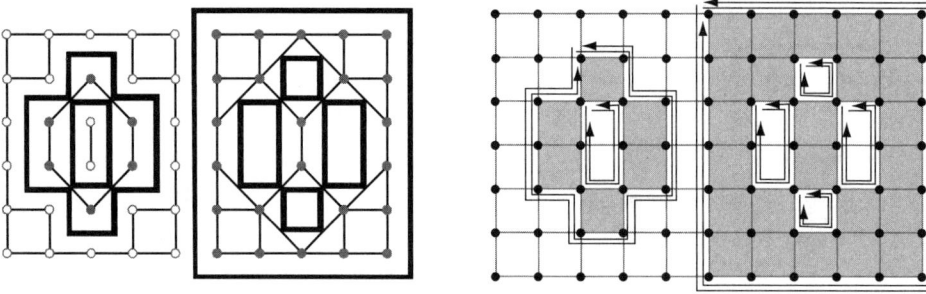

FIGURE 5.25 Left: picture grid showing the ordered adjacencies and frontiers for all components. Right: frontier grid showing the Hamiltonian circuits for all 0-components of 1-cells of all frontiers.

reflects the use of local circular order (A) in Figure 4.19. We use the following notation when q_2 is the direct successor of q_1 in the local circular order at p_i:

$$q_1 \underset{i}{\rightarrow} q_2$$

To traverse all of the frontiers, we scan the picture (e.g., using a standard scan [see Section 1.1.3]) until we arrive at a pixel (x_0, y_0) in $\mathbb{G}$ that has an 0-cell on a "new" frontier component of a region or of the background. We then generate a 4-path in $\mathbb{G}$ that traverses this frontier. To do this, we start at the upper left 0-cell of (x_0, y_0) (i.e., at grid point $(x_0, y_0 + 1)$ in $\mathbb{F}$). Each step from a grid point to a 4-adjacent grid point on the frontier goes around a pixel, keeping the pixel on the right as shown in Figure 5.26. A step $\sigma \in \{\text{UP}, \text{RIGHT}, \text{DOWN}, \text{LEFT}\}$ specifies how the coordinates of this pixel τ_σ are chosen.

We assume the default order for the pixel values u and v. In the flip-flop case (g) (see Figure 5.23), adjacency between pixels that have value v is preferred over adjacency between pixels that have value u if $u < v$. The frontier tracing algorithm is given in Algorithm 5.3. It is a special case of the general border tracing algorithm of Figure 4.26. The important difference is that here we test whether the step sequences σ, σ' are possible at a grid point. If a step sequence is not a left turn, it is possible iff the pixel $\tau_{\sigma'}(q)$ has value $u = P(x_0, y_0)$; if it is a left turn, it is impossible iff the two diagonal pixels other than $\tau_\sigma(p_i)$ and $\tau_{\sigma'}(q)$ have the same value $w > u$. The algorithm could be further optimized by removing redundant tests, but its time complexity is linear in the number of 1-cells on the 0-component of the frontier that is being traced.

5.4.4 Frontiers in 3D incidence grids

In a 3D incidence grid, the boundary of a closed region consists of one or more 2-components of 3-nodes (grid cubes), and the boundary of an open region consists

5.4 Pictures on Incidence Grids

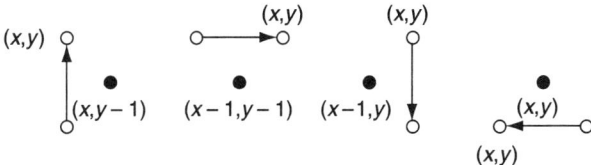

FIGURE 5.26 A step in the frontier grid (hollow dots) is shown by an arrow pointing to grid point (x,y); the corresponding pixel in the picture grid (filled dot) is on the right of this arrow.

---

1. Let $q_0 := (x_0 - 1, y_0 + 1)$, $p_0 := (x_0, y_0 + 1)$, $\sigma_0 := \text{RIGHT}$, $i := 0$, and $k := 0$.
2. Let $q_k \underset{i}{\to} q$ describe a step σ. If the step sequence σ_i, σ is not possible at p_i, go to Step 4.
3. q is the next grid point on the frontier circuit. Let $i := i+1$, $p_i := q$, and $\sigma_i := \sigma$. Let $p_{i-1} \underset{i}{\to} q$ describe a step σ. If the step sequence σ_i, σ is possible at p_i, go to Step 3; otherwise, let $k := i - 1$, and go to Step 4.
4. If $(q, p_i) = (q_0, p_0)$, go to Step 5. Otherwise, let $k =: k+1$, $q_k := q$, and go to Step 2.
5. We are back at the original directed invalid edge (q_0, p_0). The frontier circuit is $\langle p_0, p_1, \ldots, p_i \rangle$.

ALGORITHM 5.3 Frontier tracing algorithm in the frontier grid.

---

of one or more 1-components of 2-nodes (grid faces). In the latter case, the boundary defines a 1-adjacency graph of grid faces, all incident with grid cubes in the open region and with grid cubes in the complementary set. Figure 5.27 shows the simplest example; the open region consists of a single grid cube, and the 1-adjacency graph of its frontier has a Hamiltonian circuit.[5] If we add one cube at a time that is incident with exactly one face of the union of the existing cubes (note that we also have to add that face itself to ensure that the region remains complete), we obtain a 2-connected set of grid cubes that forms a *simple 2-tree* that contains no circuits (and therefore is a tree) and no 2×2 cube configurations (and therefore is simple).

Proposition 5.3 *The 1-adjacency graph of the frontier faces of a simple 2-tree of grid cubes is a Hamiltonian graph (i.e., it has a Hamiltonian circuit).*

---

5. The vertices and edges of a k-cell ($k \geq 0$) form a graph that defines a k-*dimensional hypercube* [997]. All hypercubes are Hamiltonian bipartite graphs.

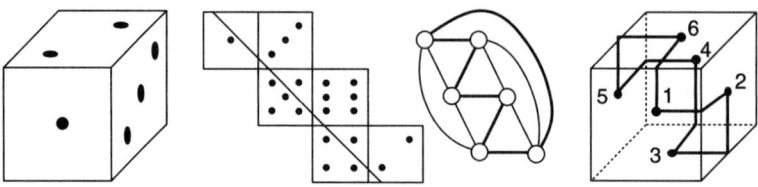

FIGURE 5.27 Left: cube. Middle: map of the cube's faces (grid squares) in the plane and 1-adjacency graph of these faces, with a Hamiltonian circuit corresponding to the straight path. Right: isothetic polygonal circuit (on the cube's frontier) that represents the Hamiltonian circuit.

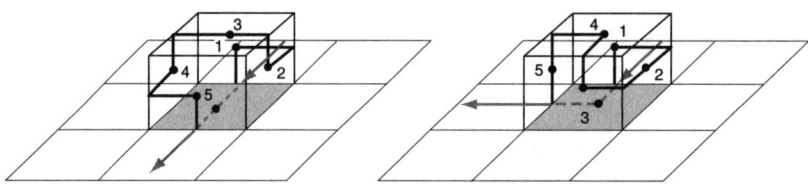

FIGURE 5.28 Left: entry and exit points on opposite edges. Right: entry and exit points on 0-adjacent edges.

Proof A Hamiltonian circuit for the simplest case of a single grid cube is shown in Figure 5.27. Suppose we have a tree of n cubes and a Hamiltonian circuit (an isothetic polygonal circuit) on the frontier of the tree, and we attach an $(n+1)$st cube to one of the existing cube faces, for example, face c_0. There are only two possibilities for the entry and exit points of the isothetic polygonal circuit on c_0 (see Figure 5.28): they are either on opposite edges or on 0-adjacent edges. In both cases, we can replace the isothetic segment in c_0 with an isothetic polygonal path (see Figure 5.28) to obtain a Hamiltonian circuit for the tree of $n+1$ cubes. ∎

Arbitrary regions in a 3D incidence grid may not have Hamiltonian circuits or paths through the 1-components of 2-nodes on their frontiers. A complete traversal of the faces on the frontier requires in general that some faces be visited repeatedly.

Let $F = [S, I, \dim]$ be a downward restriction of the 3D incidence grid that contains all of the cells in the frontier of a region. The FILL procedure (see Algorithm 4.1) provides a way of visiting all of the faces in F.

Let L be a 3D array of the same size as the array that represents the voxels. We assign labels to L that correspond to voxel positions. We use six bit-positions in each label, which correspond to the six faces of a voxel. A 1 in the ith position indicates that the face has already been visited. The tracing algorithm starts at some face in S and applies FILL based on the adjacency defined for the faces in F by available 0- and 1-cells. We can use a stack or queue in Figure 4.3 to implement a depth-first or breadth-first traversal of all of the faces on the frontier.

This *frontier tracing algorithm* is easy to implement, but the faces are visited in an "unordered" sequence. The purpose of the tracing (e.g., coloring all of the frontier faces, approximating the frontier by segments of digital planes) may determine whether the algorithm is of interest. Note that the algorithm is based on graph-theoretic concepts only; it makes no use of an embedding of the 3D incidence grid into a Euclidean or metric space.

5.5 Exercises

1. Which of the following relational properties are true in general for adjacency, (smallest nontrivial) neighborhood, and incidence: reflexivity, irreflexivity, symmetry, and transitivity?

2. A simple polygon has vertices (cells of dimension 0), edges (cells of dimension 1), and one face (a cell of dimension 2). Suppose we tile the plane with simple polygons in an arbitrary, irregular way, adding one simple polygon at a time that is disjoint from all previous polygons except for sharing an edge with one of them. This defines an incidence pseudograph with respect to set-theoretic incidence. What is the Euler characteristic of this pseudograph, assuming that the union of all of the polygons is a simple polygon (i.e., it has no holes)?

3. A *Gray code* [376] for a sequence of consecutive integers has the property that the codes for successive integers differ by only one bit. For example, $0 \to 00, 1 \to 01, 2 \to 11$, and $3 \to 10$ is a Gray code for the integers 0, 1, 2, and 3 (in that order). In analogy to Definition 5.15, we can construct hypercubes as follows: We label the vertices of a 1-cell 0 and 1. Given a labeled k-dimensional hypercube, we make two copies of it, append 0 and 1 to the node labels of the first and second copy, respectively, and connect corresponding nodes in the two copies with new edges. Prove that the sequence of node labels on any Hamiltonian circuit of the resulting $(k+1)$-dimensional hypercube defines a Gray code for the integers $0, 1, \ldots, 2^{k+1} - 1$.

4. Prove that the family of open (closed) regions in $[\mathbb{C}_2, I, \dim]$ is closed under finite unions and intersections.

5. Let the unions of the following sets of cells be either open or closed regions in $[\mathbb{C}_3, I, \dim]$. Calculate the Euler characteristics of these regions using the equation in Definition 5.14 or using Theorem 5.2.

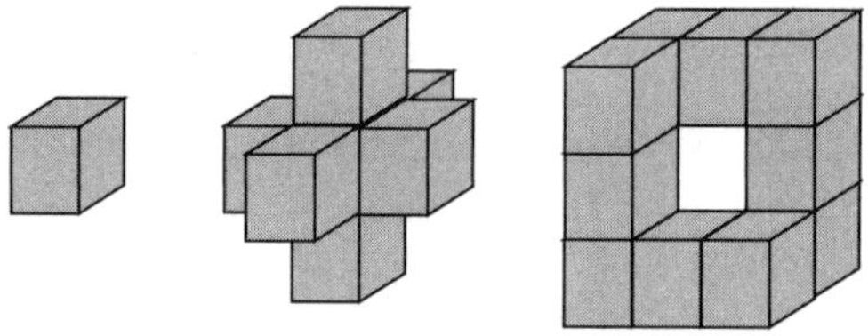

6. Prove Equation 5.6 by induction on m.

7. Suppose the ordered adjacency procedure is applied to the two multilevel pictures shown below. Identify the components of the P-equivalence classes, trace their frontiers, and calculate their Euler characteristics.

7	6	7	6	7	7	7	7	7	7
6	7	6	7	4	4	4	7	8	7
7	6	4	4	4	2	4	4	7	8
6	4	3	4	2	2	4	5	4	7
7	6	4	4	2	2	4	4	2	2
6	7	4	2	4	4	4	2	2	2
7	6	2	2	6	6	4	2	6	2
6	2	6	6	2	2	2	2	2	2

1	2	1	2	1	1	1	1	1	1
2	1	2	1	4	4	4	1	0	1
1	2	4	4	4	6	4	4	1	0
2	4	5	4	6	6	4	3	4	1
1	2	4	4	6	6	4	4	6	6
2	1	4	6	4	4	4	6	6	6
1	2	6	6	2	2	4	6	2	6
2	6	2	2	6	6	6	6	6	6

8. Implement the frontier tracing algorithm for 2D multilevel pictures, assuming the default total order of the picture values.

9. Extend the frontier tracing program of Exercise 8 by counting local properties during frontier tracing and combining the counts to determine the Euler characteristic. The program should have runtime complexity linear in the number of 1-cells visited during frontier tracing.

5.6 Commented Bibliography

Incidence pseudographs have been studied from the viewpoints of graph theory, geometry, and combinatorial topology; see [1107] by K. Voss, parts of which are summarized in Sections 5.1 and 5.3. For independent publications of Equation 5.3, see [220, 532, 927], all three of which were published between 1971 and 1973. Interestingly, the book [1107] discussed incidence pseudographs in the situation shown in the lower left of Figure 2.12 (e.g., pixels or voxels as grid points in 2D or 3D incidence structures)—which are not studied in this book—without referring to the regions as being "open." The discussion of combinatorial formulas for closed regions in this

5.6 Commented Bibliography

chapter is new. Open and closed regions will also be studied in the following chapters about topology.

For an early discussion of components of incidence grids see, for example, [432, 536]. In [591] V.A. Kovalevsky proposed the "maximum-label rule" for ordered labeling and a local adjacency decision rule for 2D pictures; see the local adjacency decision procedure (Algorithm 5.1) for the n-dimensional case. The frontier tracing algorithm (Algorithm 5.3) has its roots in [424, 591].

Edge adjacencies between frontier faces of a 2-component define a surface graph in which all nodes have constant degree 4. This allows frontier tracing using the simple FILL approach; see [1107], Section 3.1. [550] uses a breadth-first search strategy in the FILL procedure to grow "disk-like" faces on the frontier of a region in the 3D incidence grid. The "classic" algorithm [48] for traversing all frontier faces of a 6-connected region is explained in detail in [430] and will be described in Section 8.4.1. The geometric locations of the faces (2-cells) are used in this algorithm to identify two "in-faces" and two "out-faces." In comparison with the simple FILL procedure, this makes it possible to reduce the numbers of visits to faces to a maximum of 2. Unlike the rest of the material in this chapter, this is not a purely graph-theoretic approach, because it makes use of coordinates. The FILL procedure can also be applied to 18- or 26-connected regions.

Gray codes (see Exercise 3) are discussed in [997].

CHAPTER **6**

Topology

Digital geometry is often concerned with analyzing topologic properties of sets of pixels or voxels. The topology of Euclidean space was briefly discussed in Section 3.1.7. The definitions of open and closed regions in Section 5.1.4 indicated that these topologic concepts are applicable to discrete structures. This chapter summarizes basic topologic concepts and properties that are relevant to adjacency and incidence grids, defines digital topologies, and provides a brief introduction to combinatorial topology.

6.1 Topologic Spaces

In the last third of the 19th century, H. Poincaré and others established topology as a branch of modern mathematics. *Point-set topology* studies topologic spaces. In early publications about topology, the underlying set S of a topologic space was a Euclidean space, but, in modern topology, it can be an abstract set.

6.1.1 General definitions

$[S, \mathcal{G}]$ is called a *topologic space* iff $\mathcal{G}$ is a family of subsets of S that has the following three properties:

T1: $\{\emptyset, S\} \subseteq \mathcal{G}$

T2: Let $M_1, M_2, \ldots$ be a finite or infinite family of sets in $\mathcal{G}$; then the union of these sets is also in $\mathcal{G}$.

T3: Let $M_1, M_2, \ldots, M_n$ be a finite family of sets in $\mathcal{G}$; then the intersection of these sets is also in $\mathcal{G}$.

$\mathcal{G}$ is called a *topology* on S, and its elements are called *open sets*. $M \subseteq S$ is called *closed* iff its complement $\overline{M} = S \setminus M$ is open. It follows from **T2** and **T3** that the family of closed subsets of S is closed under finite unions and arbitrary intersections.

The *interior* $M°$ of $M \subseteq S$ is the union of all open subsets of M. The *closure* $M^{\bullet}$ of M is the intersection of all closed subsets of S that contain M. Set M is open iff it coincides with its interior and closed iff it coincides with its closure. $\partial M = M^{\bullet} \cap (\overline{M})^{\bullet}$ is called the *frontier* of M.

The *degenerate topology* $[S, \wp(S)]$, in which every subset of S is open as well as closed, is a topologic space. In this topology, we have $M^{\bullet} = M° = M$ and $\partial M = \emptyset$ for every $M \subseteq S$. (The degenerate topology is called the "discrete topology" in most books about topology, but this would not be appropriate in a book about digital geometry.)

Any metric space $[S, d]$ *induces* a topology on S; see Section 3.1.7. In this topology, the closure M^* is the set of all $p \in S$ such that $d(p, M) = 0$.

The Euclidean metric d_e induces the *Euclidean topology* $\mathbb{E}^n$ on $\mathbb{R}^n$ ($n \geq 1$). For example, for $n = 1$, an *open interval* $(x, y) = \{z \in \mathbb{R} : x < z < y\}$ ($x, y \in \mathbb{R}$; $x < y$) is an open set, and a *closed interval* $[x, y] = \{z \in \mathbb{R} : x \leq z \leq y\}$ is a closed set.

The binary metric d_b induces the degenerate topology on any set S. The metrics d_α, $\alpha \in \{4, 6, 8, 18, 26\}$ (see Section 3.2.2) and ∂_α, $\alpha \in \{0, 1, 2\}$ (see Section 3.3.1) also induce the degenerate topology on $\mathbb{Z}^2$ or $\mathbb{Z}^3$ (the grid point model) and on $\mathbb{C}_2^{(2)}$ or $\mathbb{C}_3^{(3)}$ (the grid cell model).

Nondegenerate topologies can be defined on discrete sets. In particular, the open sets of Section 5.1.4 define a topology on any incidence pseudograph. We will discuss these topologies in detail later in this chapter.

Let $[S, \mathcal{G}]$ be a topologic space and $M \subseteq S$; then $[M, \mathcal{G}_M]$ is also a topologic space, where $\mathcal{G}_M = \{A \cap M : A \in \mathcal{Z}\}$. $\mathcal{G}_M$ is called the *inherited topology* on M, and M is called a *topologic subspace* of S. Note that M itself is both open and closed in $\mathcal{G}_M$. (The inherited topology is also called "relative topology" in topology textbooks.)

Definition 6.1 M is called *(topologically) connected* iff it is not the union of two disjoint nonempty open subsets (or, equivalently, closed subsets) of M.

Maximum connected subsets of M are called *components* of M.

$[S, \mathcal{G}]$ is called an *Aleksandrov*[1] *space* or *Aleksandrov topology* iff any intersection of open sets is open. If S is finite, any $[S, \mathcal{G}]$ is Aleksandrov.

Definition 6.2 Let $[S, \mathcal{G}]$ be a topologic space and $p \in S$; then any subset of S that contains an open superset of p is called a *(topologic) neighborhood* of p.

For example, in the Euclidean topology $\mathbb{E}^1$ on $\mathbb{R}$, an open interval (x, y) is a neighborhood of any $z \in (x, y)$, and a closed interval $[x, y]$ is a neighborhood of $z \in [x, y]$ iff $z \neq x$ and $z \neq y$.

In an Aleksandrov space, the intersection $U(p)$ of all neighborhoods of $p \in S$ is the smallest neighborhood of p, which is also called the *star* of p. Evidently, $U(p)$ must be an open set. U may not define a symmetric relation on S; we can have $q \in U(p)$ but $p \notin U(q)$.

---

1. Also spelled *Alexandroff*.

6.1 Topologic Spaces

$[S, \mathcal{G}]$ is called a *Kolmogorov space* or T_0-*space* iff, for any two distinct points of S, at least one of them has a neighborhood that does not contain the other.[2] The degenerate topology is a T_0-space. An Aleksandrov topology is T_0 iff, for every two distinct points p and q, $U(p)$ and $U(q)$ are distinct.

Definition 6.3 $\mathcal{G}' \subseteq \mathcal{G}$ is called a *basis* of $[S, \mathcal{G}]$ iff any nonempty set in $\mathcal{G}$ is a union of (possibly infinitely many) sets in $\mathcal{G}'$.

For example, in the topology induced by a metric space, the set of all ε-neighborhoods is a basis. The set of all open intervals of $\mathbb{R}$ is a basis of the Euclidean topology $\mathbb{E}^1$. In an Aleksandrov topology, the set of all smallest neighborhoods is a basis.

Example 6.1 (N. Bourbaki, 1961) The family $\{[x, +\infty) : x \in \mathbb{R}\}$ is a basis of a topology on $\mathbb{R}$ called the *right topology* on $\mathbb{R}$. It follows, for example, that any $(-\infty, x)$ is closed. Analogously, the family $\{(-\infty, x] : x \in \mathbb{R}\}$ is a basis of a topology on $\mathbb{R}$ called the *left topology* on $\mathbb{R}$. Note that the sets $[x, +\infty)$ and $(-\infty, x]$ are not open in the Euclidean topology on $\mathbb{R}$.

A topologic space has a *countable basis* iff it has a basis of cardinality of at most $\aleph_0$, where $\aleph_0$ is the cardinality of the set $\mathbb{N}$ of natural numbers. For example, the set of all open intervals with rational endpoints is a countable basis of $\mathbb{E}^1$.

6.1.2 Poset topologies

A reflexive, antisymmetric (for all pairs p and q such that $p \neq q$), transitive binary relation on a set S is called a *partial order* and is denoted by $\triangleleft$. $[S, \triangleleft]$ is called a *partially ordered set* (*poset*, for short).

Definition 6.4 In the *poset topology* on $[S, \triangleleft]$, $M \subseteq S$ is open iff $p \in M$ and $p \triangleleft q$ implies $q \in M$ for all $p, q \in S$.

Any poset topology is Aleksandrov. Following [9], we know that there is a one-to-one correspondence between Aleksandrov topologies on a set S and quasiorders (reflexive and transitive relations) on S, and a quasiorder is a partial order iff the corresponding Aleksandrov topology is T_0. An open set defines an "upper set" of the quasiorder (in the sense of Definition 6.4), and the order "less than or equal to" corresponds to "is in the closure of."

The incidence grid topology (see Section 6.2.3) is an example of a poset topology. Another example is the following:

$$\mathbb{Z}_a = [\{\{i\} : i \in \mathbb{Z}\} \cup \{\{i, i+1\} : i \in \mathbb{Z}\}, \subseteq] \tag{6.1}$$

---

2. For completeness, we mention that S is called a T_1-*space* iff, for any two distinct points of S, each of the points has a neighborhood that does not contain the other. S is called a T_2-*space* or *Hausdorff space* iff any two distinct points of S have disjoint neighborhoods. For example, Euclidean space is a Hausdorff space.

In this example, $\{i\} \subseteq \{i\}$, $\{i,i+1\} \subseteq \{i,i+1\}$, $\{i\} \subseteq \{i,i+1\}$, and $\{i+1\} \subseteq \{i,i+1\}$ are the only instances of the partial order ⊲. This $\mathbb{Z}_a$ is a T_0-space. In this topology, $\{\{i,i+1\}\}$ and $\{\{i\},\{i,i+1\},\{i,i-1\}\}$ are open. The union of the open sets $\{\{2,3\},\{3\},\{3,4\}\}$ and $\{\{4,5\},\{5\},\{5,6\}\}$ is the open set $\{\{2,3\},\{3\},\{3,4\},\{4,5\},\{5\},\{5,6\}\}$. The union of the open sets $\{\{i\},\{i,i+1\},\{i,i-1\}\}$ such that $i \leq 2$ or $i \geq 6$ is the open set $\mathbb{Z}_a \setminus \{\{4\}\}$. The complement of this open set is the closed set $\{\{4\}\}$. Let us consider the following sets ($i \in \mathbb{Z}$, $k \geq 1$):

$$C_{ik} = \{\{i\}, \{i,i+1\}, \{i+1\}, \ldots, \{i+k-1\}, \{i+k-1, i+k\}, \{i+k\}\}$$

It is easy to see that $C_{ik}^\circ = C_{ik} \setminus \{\{i\}, \{i+k\}\}$, $C_{ik}^\bullet = C_{ik}$, and $\partial C_{ik} = \{\{i\}, \{i+k\}\}$.

The sets $\{\{i\}, \{i,i+1\}, \{i,i-1\}\}$, and $\{\{i,i+1\}\}$ ($i \in \mathbb{Z}$) are a countable basis of $\mathbb{Z}_a$. $M = \{\{i\}, \{i+1\}, \{i,i+1\}\}$, which is neither open nor closed in $\mathbb{Z}_a$, has the following inherited topology,

$$\mathcal{G}_M = \{\emptyset, \{\{i,i+1\}\}, \{\{i\},\{i,i+1\}\}, \{\{i+1\},\{i,i+1\}\}, M\}$$

in which the closed sets are as follows:

$$\{\emptyset, \{\{i\}\}, \{\{i+1\}\}, \{\{i\},\{i+1\}\}, M\}$$

Thus M is connected, because it is not a union of two disjoint nonempty closed sets.

6.1.3 Topologies on incidence pseudographs

Let $[S, I, \dim]$ be an n-incidence pseudograph. The set of all open subsets of S (consisting of nodes of any dimension) defines a topology on S. Our particular interest is in components that consist of principal nodes. A complete subset $M \subseteq S$ is not always a union of components, because there may be i-nodes in M that are not incident with any n-node in M. M is called *purely n-dimensional* iff it is a union of components.

Theorem 6.1 The family of purely n-dimensional complete open subsets of S defines a topology $\mathcal{G}$ on S. The family of open regions is a basis of this topology.

Proof S and $\emptyset$ are complete, purely n-dimensional, and open (and closed). By deMorgan's rules we have the following, where the index set is finite or countably infinite:

$$L = \bigcup_i M_i \Leftrightarrow \overline{L} = \bigcap_i \overline{M_i} \quad \text{and} \quad L = \bigcap_i M_i \Leftrightarrow \overline{L} = \bigcup_i \overline{M_i} \tag{6.2}$$

Let M_i be closed, purely n-dimensional, and complete. Then $c \in M_i, c' \in I(c)$, and, let $\dim(c') < \dim(c)$ imply $c' \in M_i$. If c is in all of the M_is, c' is also in all of the M_is; hence the family of purely n-dimensional complete closed subsets of S is closed under arbitrary intersections. The family of purely n-dimensional complete open sets, which are the complements of purely n-dimensional complete closed sets, is therefore closed under arbitrary unions. Let $M_1, \ldots, M_m$ be purely n-dimensional complete closed sets. For any c in any M_i, any $c' \in I(c)$ with $\dim(c') < \dim(c)$ is also in M_i, and hence is in their union; hence the family of purely n-dimensional complete closed sets is closed under finite unions.

Because S is an incidence structure, it is countable; hence any purely n-dimensional complete open set $M \subset S$ is the union of countably many pairwise disjoint components, each of which can be finite or countably infinite. Let C be one of the components, let $c \in C$ be an i-node ($0 \leq i < n$), and let the following be true:

$$U(c) = \{c' : c' \in I(c) \wedge \dim(c') \geq \dim(c)\}$$

$U(c)$ is a subset of C by the definition of a closed set. By property **I5**, there is at least one n-node in each $U(c)$. It follows that every $U(c)$ is an open region and that C is the (finite or countable) union of these regions. ∎

Let M be a purely n-dimensional complete subset of S. Then the following is true where the frontier ∂M and interior $M°$ are defined by the topology $\mathcal{G}$ of Theorem 6.1:

$$\delta M = \partial M \quad \text{and} \quad M^\nabla = M° \quad \text{if } M \text{ is closed}$$
$$M = \delta M \cup M^\nabla = M° \quad \text{if } M \text{ is open}$$

If M is closed, we also have $M = M^\bullet$.

6.2 Digital Topologies

In this section, we define digital topologies on sets of grid points or grid cells. We will see in Section 6.2.4 that there exist very few digital topologies on grid point or grid cell spaces of dimension ≤ 3.

6.2.1 General definition

$[S, \mathcal{G}]$ is called a *digital topology* in the grid cell model iff $S = \mathbb{C}_n^{(n)} (n \geq 1)$ and $\mathcal{G}$ is a family of open sets that satisfies **T1** through **T3**, as well as the following:

D1: All connected sets are 0-connected.
D2: All disconnected sets are $(n-1)$-disconnected.
D3: The closure of any singleton is $(n-1)$-connected.

(A singleton is a set of cardinality 1.) These properties exclude, for example, the degenerate topology $[\mathbb{C}_n^{(n)}, \wp(\mathbb{Z}^n)]$, in which every set that contains at least two pixels is disconnected.

In the grid point model, let d be a metric on $\mathbb{Z}^n$ ($n \geq 1$), and let $A_d(p) = \{q : q \in \mathbb{Z}^n \wedge d(p,q) = 1\}$. A subset of $\mathbb{Z}^n$ is called *connected with respect to d* iff it is connected with respect to the adjacency relation A_d. $[\mathbb{Z}^n, \mathcal{G}]$ is called a *digital topology* in the grid point model if $S = \mathbb{Z}^n$ ($n \geq 1$), and $\mathcal{G}$ is a family of open sets that satisfies **T1** through **T3**, as well as the following:

D1: All connected sets are connected with respect to L_∞.

D2: All disconnected sets are disconnected with respect to L_1.

D3: The closure of any singleton is connected with respect to L_1.

It can be shown that, in the 2D grid point space, **D1** and **D2** imply **D3**.

6.2.2 The grid point topology

Let $p = (x_1, \ldots, x_n) \in \mathbb{Z}^n$ ($n \geq 1$), and let $\eta_1(p)$ be the smallest nontrivial neighborhood of p with respect to the Minkowski metric L_1 on $\mathbb{Z}^n$ (see Section 3.2.2). Define the following:

$$U_{\text{GP}}(p) = \begin{cases} \{p\} & \text{if } x_1 + \ldots + x_n \text{ is odd} \\ \eta_1(p) & \text{if } x_1 + \ldots + x_n \text{ is even} \end{cases}$$

We call p an odd (even) grid point if $x_1 + \cdots + x_n$ is odd (even). Note that the relation U_{GP} is asymmetric on $\mathbb{Z}^n$.

$\{U_{\text{GP}}(p) : p \in \mathbb{Z}^n\}$ is a countable basis of a topology that we call the *grid point topology* $[\mathbb{Z}^n, U_{\text{GP}}]$. For example, for $n = 1$, this basis consists of the sets $\{2i+1\}$ and $\{2i-1, 2i, 2i+1\}$ ($i \geq 1$). This topology is called the *alternating topology* on $\mathbb{Z}$. It is sketched in Figure 6.1.

U_{GP} defines an adjacency relation A_{GP} on $\mathbb{Z}^n : p, q \in A_{\text{GP}}$ iff $p \neq q$ and $p \in U_{\text{GP}}(q)$ or $q \in U_{\text{GP}}(p)$. It is not hard to see that $A_{\text{GP}} = A_4$ if $n = 2$ (see Figure 6.2) and $A_{\text{GP}} = A_6$ if $n = 3$. This topology is an Aleksandrov space and a T_0-space.

Theorem 6.2 *The grid point topology on $\mathbb{Z}^n (n \geq 1)$ is a digital topology.*

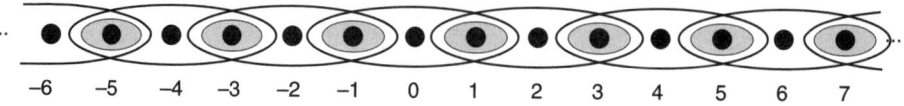

FIGURE 6.1 A sketch of the grid point topology for $n = 1$.

6.2 Digital Topologies

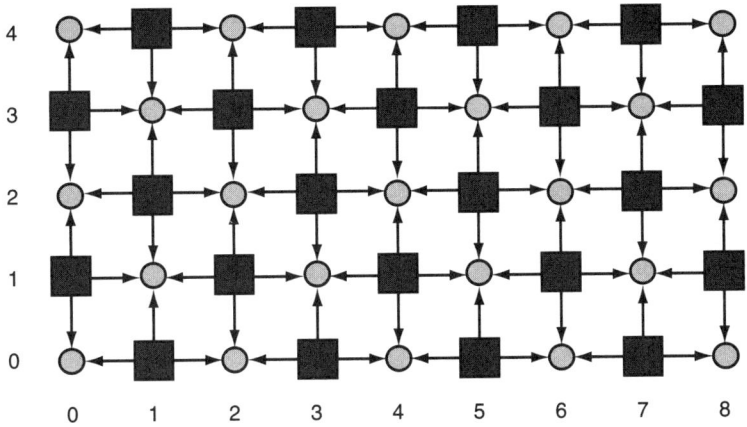

FIGURE 6.2 A directed graph showing the asymmetric neighborhood relation U_{GP} for $n=2$. Loops (corresponding to the reflexivity of U_{GP}) are omitted.

Proof $M \subset \mathbb{Z}^n$ is connected iff it is not the union of two disjoint nonempty closed sets in the induced topology on M. It is not hard to see that M is (topologically) connected iff it is connected with respect to L_1. It follows that the grid point topology has properties **D1** through **D3**. ∎

Open and closed sets, the interior and closure of a set, and the frontier of a set (the difference between its closure and its interior) can all be defined in the grid point topology. For example, any set that contains an even grid point and at most three of its 4-neighbors (which are odd grid points) is not open, and any set that contains only even grid points is closed.

Figure 6.3 shows two embeddings of the 2D grid point topology into the plane for which the odd grid points map onto the pixel positions in a regular orthogonal grid. Figure 6.4 is based on the embedding shown on the left in Figure 6.3. It shows on the left the two possible configurations of the smallest neighborhood of an even grid point and on the right an example of a closed set. In accordance with the definition of complete subsets of incidence pseudographs (see Section 5.1.1), we can define complete subsets of $\mathbb{Z}^n$ in the grid point topology; an even grid point must be in such a subset if its smallest neighborhood is in the subset.

Let d_s be the graph metric defined by a switch state matrix **S** on $\mathbb{Z}^2$ (see Section 2.1.3). The relation of s-adjacency defines sets $A_s(p) = \{q \in \mathbb{Z}^2 : d_s(p,q) = 1\}$. Let the following be true:

$$U_s(p) = \begin{cases} \{p\} & \text{if } p \text{ is odd} \\ \{p\} \cup A_s(p) & \text{if } p \text{ is even} \end{cases}$$

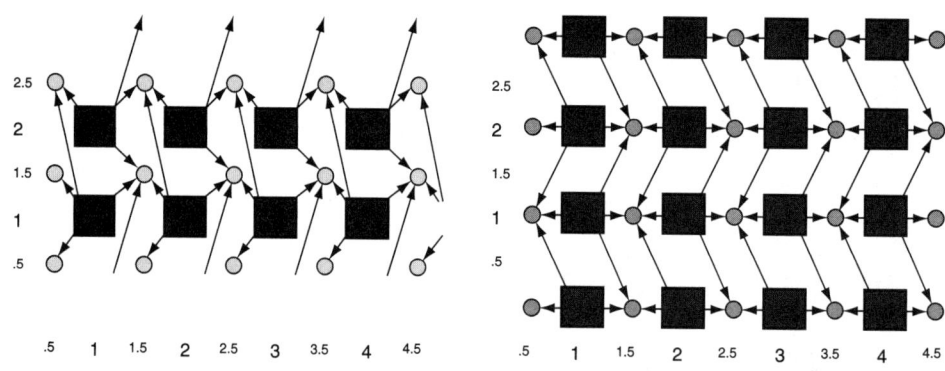

FIGURE 6.3 Two embeddings of the grid point topology into the plane such that the odd grid points (shown as squares) are in pixel positions.

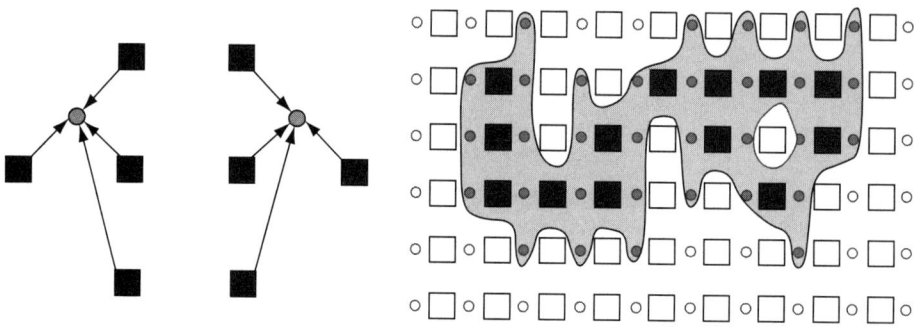

FIGURE 6.4 Left: the smallest neighborhoods of even grid points. Right: an example of a closed set. Both are for the embedding shown in Figure 6.3 on the left.

The family of sets $U_s(p)$, $p \in \mathbb{Z}^2$, is not the basis of a topology on $\mathbb{Z}^2$. To see this, consider h-adjacency, which is an example of an s-adjacency, and consider two "diagonally adjacent" even grid points p and q. $U_s(p) \cap U_s(q)$ contains exactly four grid points and so is not one of the sets $U_s(p)$. If r is another even grid point "diagonally adjacent" to p, we have $U_s(p) \cap U_s(q) \cap U_s(r) = \{p\}$. It follows that all subsets of $\mathbb{Z}^2$ are open, so the family of sets $U_s(p)$ is a basis for the degenerate topology in which only singletons are connected.

6.2.3 The grid cell topology

The incidence grid $[\mathbb{C}^n, I, \dim](n \geq 1)$ defines a poset $[\mathbb{C}^n, \leq]$ such that the following is true for all $c_1, c_2 \in \mathbb{C}^n$:

$$c_1 \leq c_2 \text{ iff } c_1 \in I(c_2) \wedge \dim(c_1) \leq \dim(c_2)$$

6.2 Digital Topologies

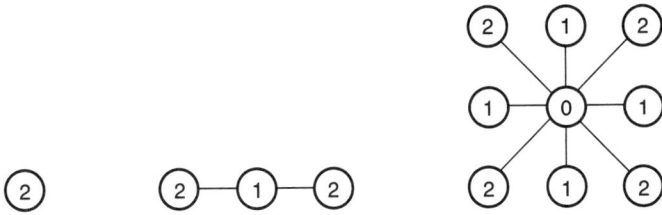

FIGURE 6.5   Stars in the 2D grid cell topology.

The poset topology on $[\mathbb{C}^n, \leq]$ is called the *grid cell topology*. This topology is a T_0-space, because it is defined by a partial order. A complete purely n-dimensional subset of $\mathbb{C}^n$ is topologically connected in this topology iff it is connected for the adjacency relation in Definition 5.3. We will see at the end of this section that $[\mathbb{C}^n, \leq]$ is a digital topology.

The *star* of a cell c is the set of all cells $c' \geq c$; it is the smallest neighborhood of c in the grid cell topology. Figure 6.5 shows examples of stars. The stars are a countable basis of the grid cell topology.

The 2D grid cell topology $[\mathbb{C}_2, \leq]$ defines a partition of the plane into grid cells, grid edges, and grid vertices. Figure 2.3 (left) shows a graphic sketch and Figure 2.13 the geometric realization of the 2D incidence grid used throughout this book. P.S. Aleksandrov and H. Hopf used this 2D example in 1935 to illustrate the poset topology.

The following is another way of defining a topology on $\mathbb{Z}$. We begin by partitioning $\mathbb{R}$ into a countable number of pairwise disjoint intervals. This partition defines a one-to-one mapping $f : \mathbb{R} \to \mathbb{Z}$ such that, for each $i \in \mathbb{Z}$, $f^{-1}(i)$ is a connected subset of $\mathbb{R}$ with respect to the Euclidean topology. We then use the Euclidean topology on $\mathbb{R}$ to induce a topology on $\mathbb{Z}$ based on these intervals. A subset M of $\mathbb{Z}$ is open in this topology iff $f^{-1}(M)$ is open in the Euclidean topology on $\mathbb{R}$.

The topology induced on $\mathbb{Z}$ in this way may not be of interest. For example, let $f(x)$ be the integer nearest to x; if x is a half-integer $i + \frac{1}{2}$, let $f(x) = i$. Then $f^{-1}(i) = (i - \frac{1}{2}, i + \frac{1}{2})$ for all $i \in \mathbb{Z}$ (i.e., $f^{-1}(i)$ is neither open nor closed in $\mathbb{R}$), and the same is true if we take $f(i + \frac{1}{2}) = i + 1$. As a result, no proper subset of $\mathbb{Z}$ can be open or closed (i.e., the induced topology on $\mathbb{Z}$ is the trivial topology that has only the empty set $\emptyset$ and $\mathbb{Z}$ itself as open and closed sets). However, we have the following:

Example 6.2 Modify f by defining $f(i + \frac{1}{2})$ as the nearest even integer [503]. This f induces the alternating topology on $\mathbb{Z}$ (see Figure 6.1). $f^{-1}(2i)$ is a closed subset of $\mathbb{R}$ in the Euclidean topology and $f^{-1}(2i + 1)$ is an open subset, so $\{2i\}$ is a closed subset of $\mathbb{Z}$ and $\{2i + 1\}$ is an open subset.

Definition 6.5 The *product* $S_1 \times S_2$ of two topologic spaces S_1 and S_2 is the set of ordered pairs (p_1, p_2) where $p_1 \in S_1$ and $p_2 \in S_2$, endowed with the *product topology*: $M \subseteq S_1 \times S_2$ is open iff, for each $(p_1, p_2) \in M$, there are open sets M_1 in S_1 and M_2 in S_2 such that $(p_1, p_2) \in M_1 \times M_2 \subseteq M$.

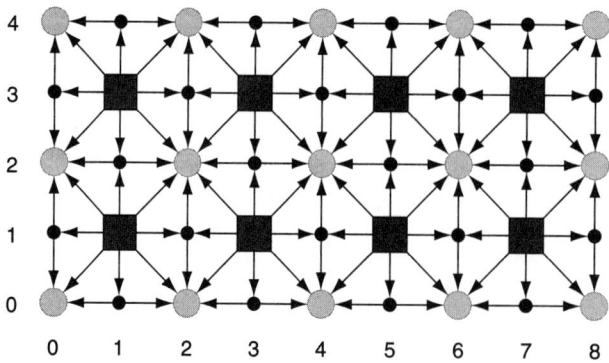

FIGURE 6.6 The asymmetric neighborhood relation U_{GC}. The hollow dots represent closed sets $\{(2i, 2j)\}$, the squares represent open sets $\{(2i+1, 2j+1)\}$, and the solid dots represent sets $\{(2i, 2j+1)\}$ or $\{(2i+1, 2j)\}$, which are neither open nor closed. Loops (corresponding to the reflexivity of U_{GC}) are not shown.

For example, the 2D Euclidean topology is the product of two one-dimensional Euclidean topologies. The topologies on $\mathbb{Z}$ defined previously can be used to define product topologies on $\mathbb{Z}^n$ ($n \geq 2$). A product of two Aleksandrov topologies is Aleksandrov, and the corresponding partial order is the product of the two partial orders.

The alternating topology is an Aleksandrov space, and so are finite products of alternating topologies. Let $U_{GC}(p)$ be the smallest neighborhood of $p \in \mathbb{Z}^n$ in such a product topology. (We use the subscript "GC," because this discussion will lead us back to the grid cell topology.)

Figure 6.6 shows the asymmetric neighborhood relation U_{GC} in the product of two alternating topologies on $\mathbb{Z}$.

Theorem 6.3 *Finite products of $n \geq 2$ alternating topologies are digital topologies on $\mathbb{Z}^n$.*

Proof We will see in Definition 6.6 how a symmetric adjacency relation A_{GC} can be defined on $\mathbb{Z}^n$ by the neighborhood relation U_{GC}. $A_{GC}(p)$ is always contained in $A_\infty(p) = \{q : L_\infty(p, q) = 1\}$ and always contains $A_1(p) = \{q : L_1(p, q) = 1\}$. This implies that **D1** through **D3** are valid for these product topologies. ∎

$M \subseteq \mathbb{Z}^2$ is open in the product of two alternating topologies iff the following is open in $\mathbb{R}$, where f is as it is in Example 6.2:

$$S_M = \bigcup_{(i,j) \in M} f^{-1}(i) \times f^{-1}(j) \quad (6.3)$$

6.2 Digital Topologies

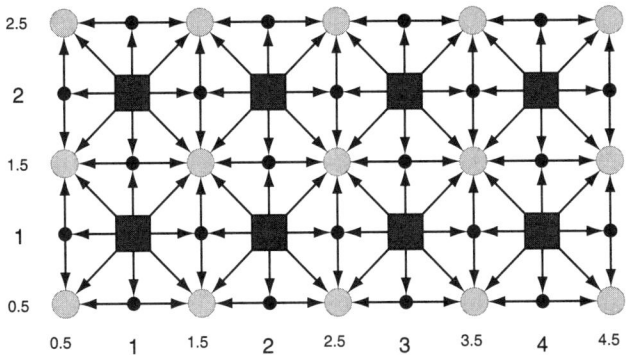

FIGURE 6.7 An embedding of the product of two alternating topologies into the plane such that the open singletons are in pixel positions.

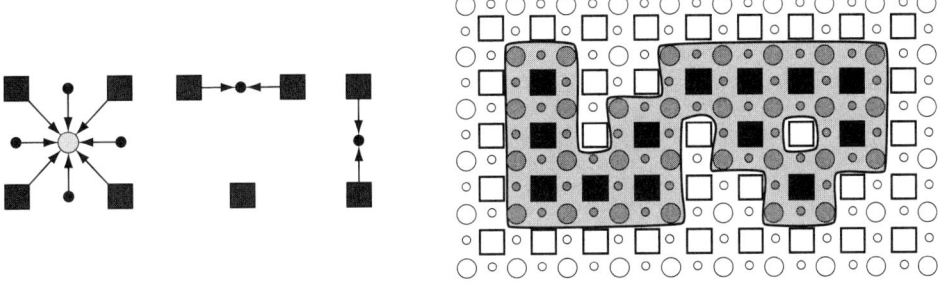

FIGURE 6.8 Left: smallest (topologic) neighborhoods of grid points. Right: an example of a closed set. These figures use the embedding shown in Figure 6.7.

Figure 6.7 shows an embedding of this 2D product topology into the plane such that all grid points $\{(2i+1, 2j+1)\}$ (shown as squares) are in pixel positions. Figure 6.8 shows on the left the smallest neighborhoods of grid points (compare Figure 6.5) and on the right an example of a closed set. The smallest neighborhood of a grid point $(2i+1, 2j+1)$ contains only that point; the smallest neighborhood of a grid point $(2i, 2j)$ is its 8-neighborhood; and all other grid points have smallest neighborhoods of cardinality 3, arranged either horizontally or vertically.

Products of alternating topologies are T_0-spaces. For example, the neighborhood of a grid point $p = (2i+1, 2j+1)$ does not contain any of the grid points that are 8-adjacent to p.

Proposition 6.1 Any product of $n \geq 2$ alternating topologies is homeomorphic (see Definition 6.8) to the n-dimensional grid cell topology $[\mathbb{C}^n, \leq]$.

Proof The base set $\mathbb{Z}$ of the alternating topology consists of an alternating sequence of *open integers* $2i+1$ and *closed integers* $2i$. The coordinates of a grid point p in the base set $\mathbb{Z}^n$ of the product topology consist of a_p open and b_p closed integers such that $a_p + b_p = n$. Consider the one-to-one mapping Φ, which maps $p \in \mathbb{Z}^n$ into an a_p-cell in $\mathbb{C}^n$ such that $p \in U_C(q)$ iff $[\Phi(p), \Phi(q)] \in I$ and $\dim(\Phi(p)) \leq \dim(\Phi(q))$. (It is not hard to show that U_C and I allow such a one-to-one mapping.) Φ maps open subsets of the product topology into open subsets of the incidence grid, and Φ^{-1} maps open subsets of the incidence grid into open subsets of the product topology. (In 2D, this follows, for example, from Equation 6.3.) Thus Φ is a homeomorphism between the two topologic spaces. ∎

From this and Theorem 6.3, it follows that $[\mathbb{C}^n, \leq]$ $(n \geq 1)$ is a digital topology.

6.2.4 The number of digital topologies

There is only one digital topology for $n = 1$: the alternating topology on $\mathbb{Z}$.

Theorem 6.4 Let S be a subset of $\mathbb{Z}^2$ that contains a translate of the set G_0 shown in Figure 6.9. Then there is no topology on S in which connectivity is the same as 8-connectivity.

Let $[\mathbb{Z}^2, \mathcal{G}]$ be a digital topology, and let $U(p)$ be the intersection of all of the open sets of $\mathcal{G}$ that contain p. Then, for all $p \in \mathbb{Z}^2$, we must have $U(p) \subseteq N_8(p)$. If not, the 8-disconnected set $\mathbb{Z}^2 \setminus A_8(p)$ would be connected in $[\mathbb{Z}^2, \mathcal{G}]$, because any open set containing p would also contain a point in $\mathbb{Z}^2 \setminus N_8(p)$.

This result limits the possible topologic neighborhoods $U(p)$ that can be used to define a basis for a digital topology:

Theorem 6.5 Up to homeomorphism (see Definition 6.8), there are only two digital topologies on $\mathbb{Z}^2$.

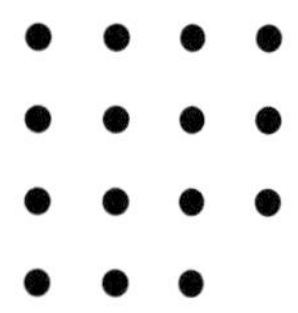

FIGURE 6.9 The set G_0 used in Theorem 6.4.

6.2 Digital Topologies

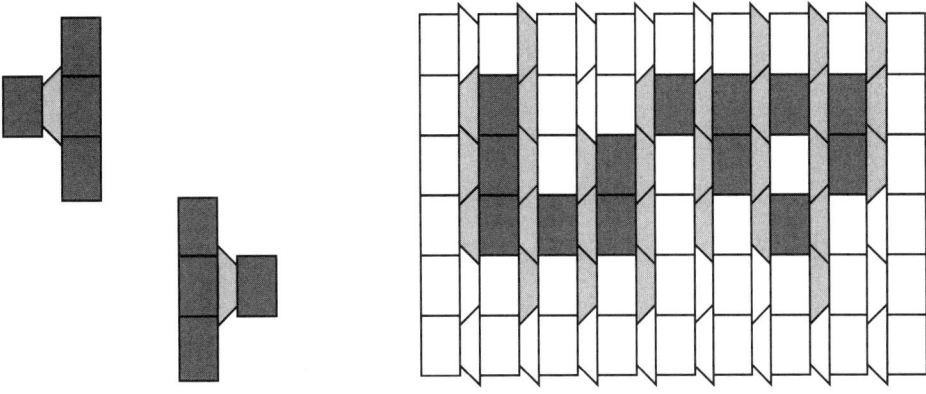

FIGURE 6.10 Left: geometric representation of the smallest neighborhoods of even grid points. Right: the closed set of Figure 6.4, which is now shown as a union of convex polygons.

It follows that the 2D grid point and grid cell topologies are the only two possible 2D digital topologies.

Theorem 6.6 Up to homeomorphism, there are only five digital topologies on $\mathbb{Z}^3$.

In addition to the 3D grid point and grid cell topologies, we also have the product of two one-dimensional grid cell topologies and one one-dimensional grid point topology; a topology $[\mathbb{Z}^2, \mathcal{G}]$ in which alternate grid planes have the grid point and grid cell topologies; and a topology with open sets that are the closed sets of $[\mathbb{Z}^2, \mathcal{G}]$.

Theorem 6.7 Up to homeomorphism, there are only 24 digital topologies on $\mathbb{Z}^4$.

The grid point and grid cell topologies exist on $\mathbb{Z}^n$ for all n; they are the "most regular" digital topologies.

Geometric realizations such as those shown in Figures 2.13 and 2.14 are models for the 2D and 3D grid cell topologies; these models can be generalized to n dimensions.[3] The grid point topology can also be geometrically represented by a tessellation of the Euclidean plane or space into convex polygons or polyhedra. For example, in Figure 6.10, the squares represent odd grid points (2-cells), and the trapezoids represent even grid points (1-cells); there are no 0-cells in this digital topology.

---

3. Following [577], these geometric realizations could also be called *continuous analogs*. We prefer to avoid the term "continuous" for geometric realizations that are defined by discrete sets.

The algorithms described in Section 5.4.1 can be used to assign 1-cells to their adjacent 2-cells that have the largest pixel values.

The values of topologic properties of a set M of pixels or voxels depend in general on which digital topology is being used. These values remain the same if the topologic spaces are homeomorphic (see Definition 6.8) and M is interpreted in the same way (e.g., as an open set).

6.2.5 Topologic adjacency and dimension

The concept of a smallest neighborhood $U(p)$ in an Aleksandrov space allows us to define a symmetric adjacency relation on the space (see Section 2.1.4):

Definition 6.6 Two distinct points p and q of an Aleksandrov space S are called *topologically adjacent* (notation: $\{p,q\} \in A$) iff $p \in U(q)$ or $q \in U(p)$.

This relation defines an adjacency structure $[S, A]$ that does not necessarily satisfy properties **A1** through **A3** of an adjacency graph. For example, if S is not topologically connected, then $[S, A]$ does not satisfy **A2**. The degenerate topology is an Aleksandrov space with $U(p) = \{p\}$ for all $p \in S$; it generates the degenerate adjacency relation $A = \emptyset$. Obviously, the adjacency relations A_α defined by the metrics d_α ($\alpha \in \{4,6,8,18,26\}$) are not the same as the adjacency relation $A = \emptyset$ in the degenerate topology induced by the metric spaces $[\mathbb{Z}^n, d_\alpha]$ ($n = 2$ or 3). Definition 6.6 relates to Definition 5.2 as follows: let two i-adjacent nodes c_1 and c_2 both be incident with the i-node c, which differs from both c_1 and c_2. In the Aleksandrov space $[S, \leq]$, we have two adjacency pairs $\{c, c_1\}$ and $\{c, c_2\}$, and c_1 and c_2 are connected via c.

The smallest neighborhoods U_{GP} and U_{GC} of grid points $p \in \mathbb{Z}^n$ ($n \geq 2$) define adjacency relations A_{GP} and A_{GC} as in Definition 6.6. $A_{\text{GP}} = A_4$ in 2D and $= A_6$ in 3D. $A_{\text{GC}}(p) = A_4(p)$ in 2D iff p has one odd and one even coordinate, and it equals $A_8(p)$ otherwise.

Any adjacency relation on a set S defines paths, connectedness, components, and so forth, as discussed in Sections 1.1.4 and 1.2.5. Let $A^\star(p)$ be the union of $A(p)$ with all points $r \in S$ for which there exist $q_1, q_2 \in A(p)$ such that a shortest A-path from q_1 to q_2 not passing through p passes through r. For example, $A_4^\star(p) = A_8^\star(p) = A_8(p)$ for $p \in \mathbb{Z}^2$, and $A_6^\star(q) = A_{18}^\star(q) = A_{18}(q)$ and $A_{26}^\star(q) = A_{26}(q)$ for $q \in \mathbb{Z}^3$. We have $A(c) = A^\star(c)$ for any cell c in the grid cell topology.

A set M is called *totally disconnected* with respect to adjacency A iff there is no pair of distinct points $p, q \in M$ such that $\{p, q\}$ is A-connected.

Definition 6.7 Let M be a subset of an adjacency structure $[S, A]$ that satisfies property **A1**. The *dimension* $\dim_A(M)$ of M is defined as follows:

(i) $\dim_A(M) = -1$ if $M = \emptyset$;

(ii) $\dim_A(M) = 0$ if M is a totally disconnected nonempty set;

6.2 Digital Topologies

(iii) $\dim_A(M) = 1$ if $\mathrm{card}(A(p) \cap M) \leq 2$ for all $p \in M$ and $\mathrm{card}(A(p) \cap M) > 0$ for at least one $p \in M$; and

(iv) $\dim_A(M) = \max\limits_{p \in M} \dim_A(A^\star(p) \cap M) + 1$ otherwise.

Figure 6.11 shows examples of one- and two-dimensional sets in the 4- and 8-adjacency grids and in the 2D incidence grid.

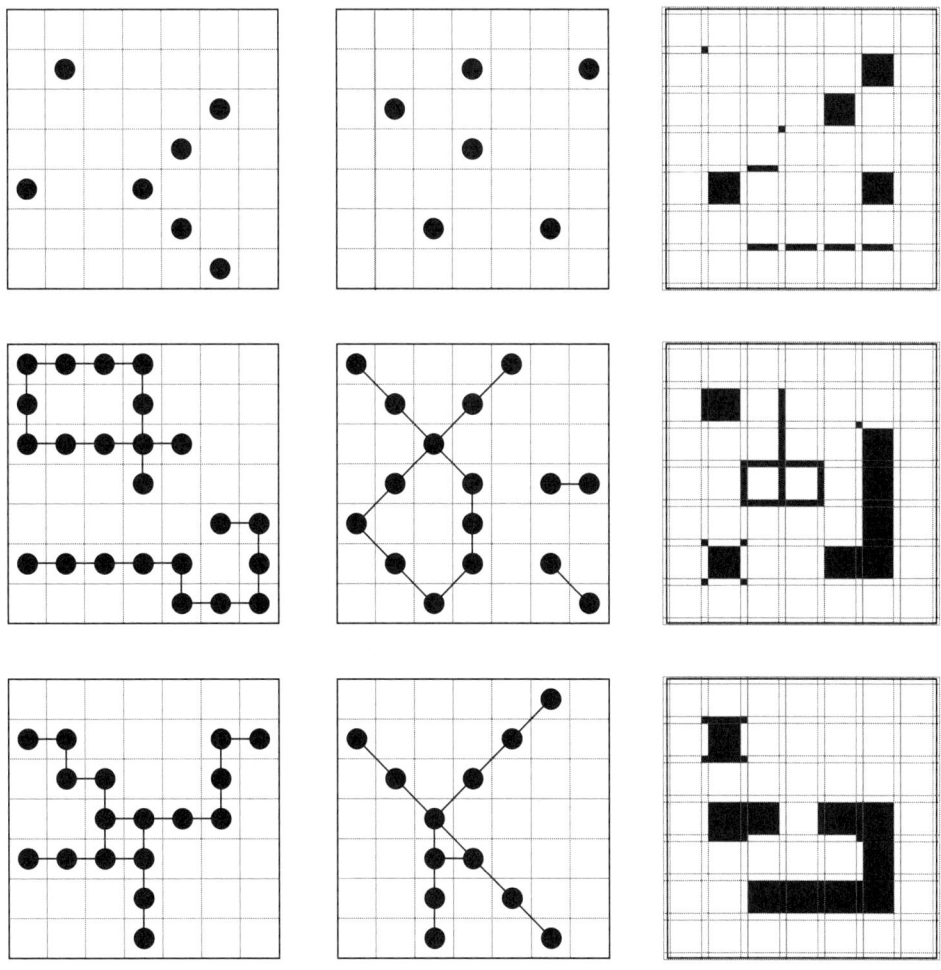

FIGURE 6.11 Upper row: zero-dimensional subgraphs in the (left) 4-adjacency grid, (middle) 8-adjacency grid, and (right) 2D incidence grid (for topologic adjacency). Middle row: one-dimensional subgraphs. Bottom row: two-dimensional subgraphs.

Proposition 6.2 A 4-connected set $M \subseteq \mathbb{Z}^2$ is 2D iff it has a 2×2 square of grid points as a proper subset.

Proof Let M be a set of four grid points that form a "T". There exists a grid point p_0 such that $\text{card}(A(p_0) \cap M) = 3$ so that case (iii) does not apply. However, from case (iv), we have $\dim_A(A_8(p_0) \cap M) = 0$, because case (ii) applies; thus $\dim_A(M) = 1$. Similarly we can show that $\dim_A(M) \leq 1$ for any set M that does not contain a 2×2 square of grid points by considering all possible such Ms in a 3×3 square of grid points.

If M is a 2×2 square of grid points, for any of its four points, we have $\text{card}(A_4(p) \cap M) = 2$; thus case (iii) applies, so $\dim_A(M) = 1$.

Finally, suppose M properly contains a 2×2 square of grid points. If, for at least one of these points p_0 we have $\text{card}(A_4(p_0) \cap M) = 3$, we can apply case (iv) to obtain $\dim_A(A_8(p_0) \cap M) = 1$. Hence, in accordance with case (iv), we have $\dim_A(M) = 2$. ∎

Figure 2.20 shows all 4-connected sets of grid points of cardinality 8; the two-dimensional sets (with respect to 4-adjacency) are shown with filled dots.

An *elementary grid triangle* is a set $T = \{(i,j), (i+1,j), (i,j+1)\}$ or a $90°$, $180°$, or $270°$ rotation of it. Such a T is one-dimensional for A_4 in accordance with case (iii).

Proposition 6.3 An 8-connected set $M \subseteq \mathbb{Z}^2$ is 2D iff it has an elementary grid triangle as a proper subset.

This proposition can be proved using a case discussion similar to that used in the proof of Proposition 6.2.

In 3D, consider adjacency relation A_6, and let $M = N_{26}(o) = A_{26}(o) \cup \{o\}$ where $o = (0,0,0)$. Without using the extended set $A^\star$ in case (iv), we obtain the following:

$$\dim_A(M) = \max_{p \in M} \dim_A(A_6(p) \cap M) + 1$$

For any $p \in M$ and $L = A_6(p) \cap M$, L is a totally disconnected nonempty set with respect to A_6 so that $\dim_A(L) = 0$ and $\dim_A(M) = 1$.

Using the extended set $A^\star$ in Definition 6.7 makes M 3D for A_6, A_{18}, and A_{26}. For example, for $o \in M$ and $L = A_{18}(o) \cap M$, we have $\text{card}(A_{18}(q) \cap L) \geq 4$ for any $q \in L$; for example, $\text{card}(A_{18}(q) \cap L) = 4$ for $q = (0,0,1)$, and $\text{card}(A_{18}(q) \cap L) = 6$ for $q = (1,0,1)$. By repeated application of case (iv), we obtain the following:

$$\dim_A(A_{18}(q) \cap L) = \max_{p \in A_{18}(q) \cap L} \dim_A(A_{18}(p) \cap (A_{18}(q) \cap L)) + 1$$

For any $p \in A_{18}(q) \cap L$, we have $\dim_A(A_{18}(p) \cap (A_{18}(q) \cap L)) = 0$ using case (ii); hence $\dim_A(A_{18}(q) \cap L) = 1$, $\dim_A(L) = 2$, and $\dim_A(M) = 3$.

6.3 Topologic Concepts

Note that the dimensions of cells in incidence grids $[\mathbb{C}_n, I, \dim]$ are the dimensions of subsets of $\mathbb{R}^n$ and that the Euclidean topology is not an Aleksandrov space. For example, the countably infinite intersection of the open intervals $(-\varepsilon, +\varepsilon)$, for all rational ε such that $1 \geq \varepsilon > 0$, is the closed singleton $\{0\}$.

6.3 Topologic Concepts

6.3.1 Homeomorphy

Let Φ be a mapping of a topologic space S_1 into a topologic space S_2. Φ is called *continuous* iff, for any open subset M of S_2, the set $\Phi^{-1}(M) = \{p \in S_1 : \Phi(p) \in M\}$ is open in S_1.

Definition 6.8 (H. Poincaré, 1895) A mapping Φ of a topologic space S_1 into a topologic space S_2 is called a *homeomorphism* iff it is one-to-one, onto S_2, continuous, and Φ^{-1} is also continuous.

Two topologic spaces are called *homeomorphic* or *topologically equivalent* iff each of them can be mapped by a homeomorphism onto the other.

Two spaces can be homeomorphic only if they have the same cardinality, because a homeomorphism is one-to-one.

The Euclidean plane $\mathbb{R}^2$ is homeomorphic to an open halfsphere. In fact, the gnomonic azimuthal projection (perspective projection from the center of the halfsphere onto a plane tangent to the halfsphere) defines a homeomorphism between the halfsphere and the plane; see the left of Figure 6.12. A triangle is homeomorphic to a circle; see the right of Figure 6.12. The surfaces of a sphere, a cube, and a cylinder are pairwise homeomorphic, but they are not homeomorphic to the surface of a torus.

A circle with one point removed is homeomorphic to $\mathbb{R}^1$; the surface of a sphere with one point removed is homeomorphic to $\mathbb{R}^2$; and the (hyper)surface of a 4D

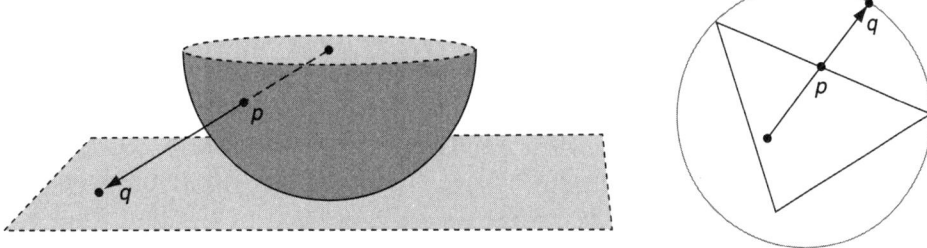

FIGURE 6.12 Left: gnomonic azimuthal projection of an open halfsphere (point p) onto the Euclidean plane (point q). Right: projection of a triangle onto a circle.

hypersphere $x_1^2 + x_2^2 + x_3^2 + x_4^2 = 1$ with one point removed is homeomorphic to $\mathbb{R}^3$. A disk with one point on its frontier removed is homeomorphic to a closed halfplane. A sphere (or cube) with one point on its surface removed is homeomorphic to a closed halfspace of $\mathbb{R}^3$. A disk with two points on its frontier removed is homeomorphic to a closed strip between two parallel straight lines in $\mathbb{R}^2$. A compact subset of $\mathbb{R}^n$ can be homeomorphic only to a compact subset of $\mathbb{R}^n$; this suggests that we should use compact subsets as geometric representations of discrete spaces.

Figure 6.13 (left) illustrates in gray two open squares in the Euclidean plane (each of these squares is homeomorphic to the infinite plane) and, in black, the union of two closed squares that have exactly one point in common (this union is homeomorphic to the union of two disks that "touch" each other at exactly one point, but it is not homeomorphic to the unit disk). Figure 6.13 (right) shows only closed subsets of the Euclidean plane: in gray, two large squares (each homeomorphic to the unit disk) and, in black, the union of two large squares, eight elongated rectangles (representing edges of grid squares), and seven small squares (representing vertices of grid squares); this union is also homeomorphic to the unit disk. Evidently, on the right, we have a geometric representation of two open regions and one closed region of the 2D incidence grid.

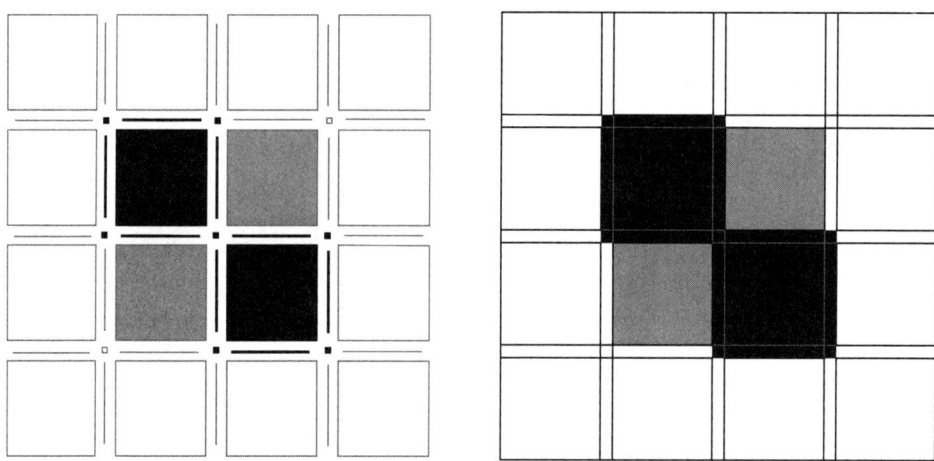

FIGURE 6.13 Left: a disjoint partition of the real plane into open squares (represented by large squares not containing their frontiers), open line segments (not containing their endpoints), and the remaining points (small squares). Right: the geometric representation of subsets of the 2D incidence grid as introduced in Section 2.1.5: a regular tessellation of the plane into (nondisjoint) large closed squares, elongated closed rectangles, and small closed squares, where nondisjoint polygons share an edge or a vertex.

6.3 Topologic Concepts

Definition 6.9 (H. Poincaré, 1895) A property of a subset M of a topologic space S is called a *topologic invariant* iff, for any homeomorphism Φ, the property is also valid for $\Phi(M)$.

For example, the property of being the empty set is a topologic invariant, and so is the property of being a nonempty set. Dimension (see Definition 6.7) is a nontrivial example of a topologic invariant. Finding topologic invariants is a central problem in topology. Similarly, calculating topologic invariants is a central problem in digital topology. For example, the number of components is a topologic invariant. Poincaré has shown that the Euler characteristic of a regular tessellation of an n-dimensional space is a topologic invariant; we will discuss this result in Section 6.4.5.

Nonhomeomorphy can be proved by comparing topologic invariants; if two sets have different values of a topologic invariant (e.g., different dimensions), they cannot be homeomorphic.

In Figure 6.13 (left), the two gray squares, which are open in $\mathbb{E}^2$, are nonhomeomorphic to the union of the two (closed) black squares. In Figure 6.13 (right), each of the two (closed) gray squares is homeomorphic to the union of the black (closed) squares and rectangles.

Definition 6.10 Two regions in a 2D (3D) picture are *topologically equivalent* (*homeomorphic*) iff their geometric representations in the incidence grid are homeomorphic in $\mathbb{E}^2$ ($\mathbb{E}^3$).

Note that topologic equivalence between picture regions depends on how marginal cells are assigned to principal cells. For example, the two regions of solid dots shown on the left in Figure 6.14 are homeomorphic in the (4,8)-adjacency grid, and their geometric representations in $\mathbb{E}^2$ are homeomorphic to the connected set shown on the right. (As we saw in Section 5.4.1, assuming (4,8)-adjacency is equivalent to assuming ordered adjacency with black $>$ white.) For all of these regions, we have $\chi = 1$. Even if we considered the two regions on the left (incorrectly) as graphs (with $v = 48$, $e = 78$, and $f = 31$ on the left and $v = 35$, $e = 37$, and $f = 3$ on the right,

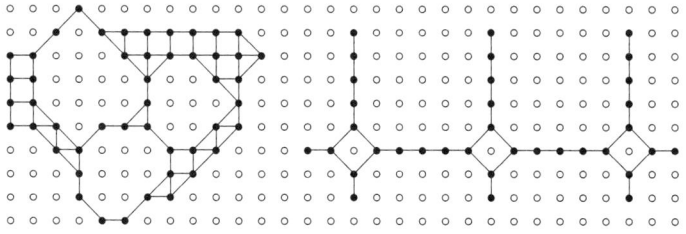 

FIGURE 6.14 Left: two homeomorphic regions of 0s in the (4,8)-adjacency grid. Right: a set in the plane homeomorphic to the geometric realization of the regions shown on the left.

counting all faces equally whether they are holes or atomic cycles), we would obtain $\chi = +1$ in both cases, because these graphs are planar.

6.3.2 Isotopy

Two subsets L and M of a topologic space S are called *isotopic* iff there exists a homeomorphism Φ from S onto itself such that $\Phi(L) = M$.

Isotopy is a stronger concept than homeomorphy. For example, suppose S contains a circle C and a rectangle R (see Figure 6.15). In the picture on the left, (A) R is surrounded by C, and, in the picture on the right, (B) R is outside of C. The two pictures are not isotopic in S; there exists no homeomorphism of S onto itself that maps (A) into (B).[4]

It can be shown that (A) and (B) are isotopic in $\mathbb{E}^3$. The two bands on the left in Figure 6.16 are homeomorphic subsets of $\mathbb{E}^3$, but they are not isotopic in $\mathbb{E}^3$. The two curves γ_1 (a meridian) and γ_2 (a parallel of latitude) on the surface of the torus shown on the right in Figure 6.16 are isotopic. The curves γ_1 and γ_3 are isotopic in $\mathbb{E}^3$ but not on the surface of the torus.

We can think of regions as being characterized by homeomorphy and pictures as being characterized by isotopy. The following definition makes use of the geometric representations defined in Section 2.1.5.

Definition 6.11 Two 2D (3D) binary pictures are *topologically equivalent* (*isotopic*) iff their geometric representations in the incidence grid are isotopic in $\mathbb{E}^2$ ($\mathbb{E}^3$).

As seen in Section 6.3.1, topologic equivalence between pictures depends on how marginal cells are assigned to principal cells. The two binary pictures in the

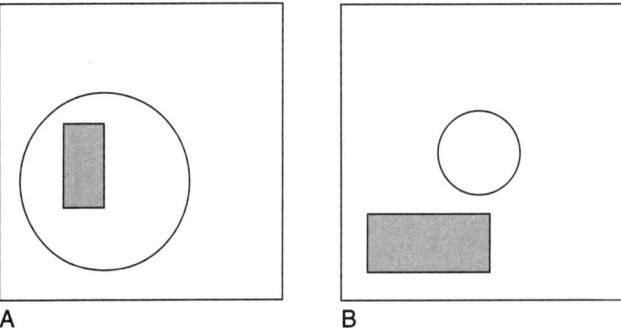

FIGURE 6.15 A rectangle inside of (A) and outside of (B) a circle.

---

4. A formal proof of this can be given on the basis of the local dimensionalities of points in S; for the necessary concepts, see Section 7.1.2.

6.3 Topologic Concepts

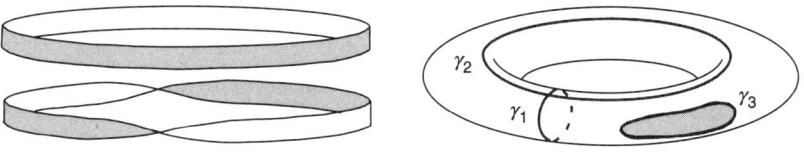

FIGURE 6.16 Left: two homeomorphic bands; the one below is twisted twice. Right: three curves on a torus.

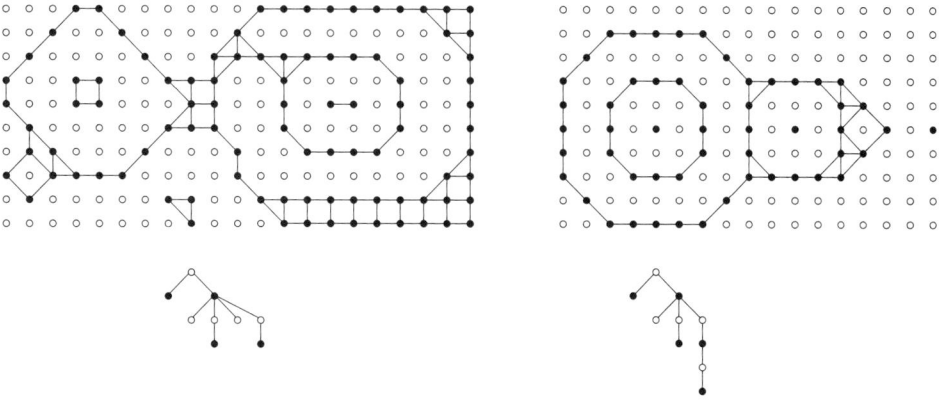

FIGURE 6.17 Two nonisotopic binary pictures in the (4,8)-adjacency grid; their rooted region adjacency trees are shown below.

(4,8)-adjacency grid shown in Figure 6.17 can be made isotopic by changing a few pixel values.

We saw in Theorem 4.2 that a 2D binary picture in the (4,8)- or (8,4)-adjacency grid has a rooted region adjacency tree in which the root represents the infinite background component.

Proposition 6.4 Two 2D binary pictures in the (4,8)- or (8,4)-adjacency grid are isotopic iff they have isomorphic rooted region adjacency trees.

This result generalizes to multivalued 2D pictures. A picture defines a partition of the plane into sets M_u ($0 \le u \le G_{\max}$) in which M_u is the P-equivalence class defined by value u. Two pictures P and P' (having equivalence classes M_u and M'_u) are isotopic iff there exists a homeomorphism Φ from $\mathbb{E}^2$ into $\mathbb{E}^2$ such that $M_u = \Phi(M'_u)$ for each u.

6.3.3 Homotopy

Homotopy allows us to give a precise definition of the topologic structure of a region. In particular, it allows us to define "simply connected."

To define the notion of homotopy, we must first define the *fundamental group* of a subset M of a topologic space $[S, \mathcal{G}]$. A continuous function $\phi : [0,1] \to M$ with $\phi(0) = p$ and $\phi(1) = q$ defines a *parameterized path* γ from p to q in M. Two parameterized paths γ_1 and γ_2 in M that have the same endpoints are called *homotopic* iff γ_1 can be continuously transformed into γ_2 in M. More precisely, let the paths γ_1 and γ_2 be defined by the functions $\phi_1 : [0,1] \to M$ and $\phi_2 : [0,1] \to M$. A continuous transformation of γ_1 into γ_2 is a continuous function $\psi : [0,1] \times [0,1] \to M$ such that $\psi(x,0) = \phi_1(x)$ and $\psi(x,1) = \phi_2(x)$ for any real x in $[0,1]$. If γ_2 is a single point, ψ defines a *contraction of* γ_1 *in M into a single point*.

Homotopy defines an equivalence relation on the class of all parameterized paths in M. For example, Figure 6.18 shows three paths from p to q in M; γ_1 and γ_2 are homotopic, but γ_3 is not homotopic to γ_1 or γ_2.

Let $p_0 \in M$ be the endpoint of γ_1 and the starting point of γ_2. The *product* $\gamma_1 \otimes \gamma_2$ is defined by concatenation: $\varphi_1 : [0,1] \to M$ and $\varphi_2 : [0,1] \to M$ are combined into a single function

$$\varphi(x) = \begin{cases} \varphi_1(2x) & \text{if } 0 \leq x < 0.5 \\ \varphi_2(2x-1) & \text{if } 0.5 \leq x \leq 1 \end{cases}$$

This product is compatible with homotopy; if γ_1 is homotopic with γ_3 and γ_2 is homotopic with γ_4, then $\gamma_1 \otimes \gamma_2$ with $\gamma_3 \otimes \gamma_4$.

Let $[\gamma]$ be the class of all paths homotopic in M to γ with respect to a given point p_0 of γ. The set $\pi(M)$ of all of these classes is a group under $\otimes$ called the *fundamental group* of M with respect to p_0. A path that is contractible in M into the single point p_0 is called *zero-homotopic*. The set of zero-homotopic paths is the identity ϵ of the group $\pi(M)$ (i.e., for any $\xi \in \pi(M)$, we have $\xi \otimes \epsilon = \epsilon \otimes \xi = \xi$). If the path γ is defined by ϕ and the path $[\gamma^{-1}]$ is defined by $\psi(x) = \phi(1-x)$, then $[\gamma^{-1}]$ is the inverse of $[\gamma]$ (i.e., $[\gamma] \otimes [\gamma^{-1}] = [\gamma^{-1}] \otimes [\gamma] = \epsilon$). Although $\otimes$ is associative, in general it is not commutative (Abelian).

If p_0 and p_1 can be connected by a path in M, the fundamental groups of M with respect to p_0 and p_1 are isomorphic. It follows that, if M is connected, its

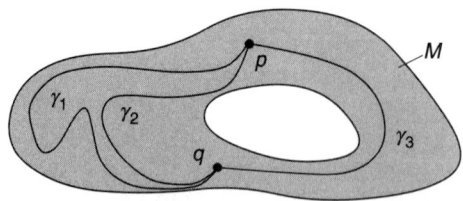

FIGURE 6.18 Three paths in a planar set M.

6.3 Topologic Concepts

fundamental group does not depend on p_0. The fundamental group is a topologic invariant in homotopy theory.

Definition 6.12 (M. Dehn and P. Heegard, 1907) Two topologically connected closed sets in a Euclidean space are called *homotopic* iff they have isomorphic fundamental groups.

This definition is based on work by C. Jordan (1866, on the topologic equivalence of curves) and H. Poincaré (1892, on the fundamental group).

Definition 6.13 A subset M of a topologic space is called *simply connected* iff $\pi(M) = \{\epsilon\}$.

In other words, M is simply connected iff any parameterized path in M that starts and ends at the same point is contractible in M into a single point. For example, a disk and a ball are both simply connected.

Example 6.3 The fundamental group of a circle is the *free cyclic group*. A path that goes around the circle clockwise $n \geq 0$ times defines a homotopy class α^n; a path that goes around the circle counterclockwise $m \geq 0$ times defines a homotopy class α^{-m}; the class α^0 is the identity ϵ. The free cyclic group is isomorphic to the additive group $[\mathbb{Z}, +, 0]$ of the integers and is commutative. A solid torus, an annulus, and the set M shown in Figure 6.18 also have fundamental groups isomorphic to $[\mathbb{Z}, +, 0]$. Thus a circle and an annulus are homotopic, but they are not homeomorphic.

The surface of a torus (i.e., a hollow torus) has a fundamental group that contains the identity, the classes of (repeated) "meridians" and (repeated) "parallels" (see Figure 6.16), and the classes defined by products of these classes. This group is commutative, because the product of a meridian cycle followed by a parallel cycle can be homotopically deformed into the product of a parallel cycle followed by a meridian cycle.

As another example, consider two circles that touch at a point p_0 and "form figure eight." The fundamental group of this set is not commutative. The product of a cycle in the upper part followed by a cycle in the lower part cannot be homotopically deformed into the product of a cycle in the lower part followed by a cycle in the upper part.

The *linear skeleton* of a set $M \subseteq \mathbb{E}^n$ (see Figure 6.19) is defined by continuous contractions; hence a set is homotopic to its linear skeleton.[5] For example, the linear skeleton of a simply connected set is a point and that of a torus is a simple closed curve.

---

5. In 1861, J.B. Listing introduced the linear skeleton under the name *cyclomatic diagram*. Because the term "skeleton" has become popular in picture analysis in the context of distance transforms, thinning operations, and so forth (see Sections 3.4.2 and 16.3), it may be preferable to use Listing's original term "cyclomatic diagram" instead of "linear skeleton."

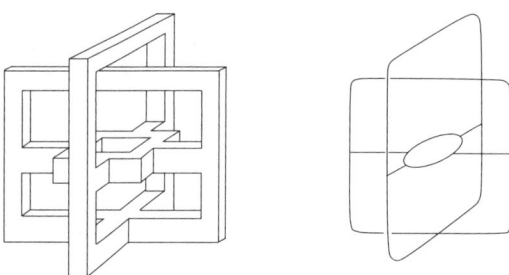

FIGURE 6.19 (J.B. Listing, 1861) A 3D object (left) and its linear skeleton. These two sets are homotopic but not homeomorphic.

6.4 Combinatorial Topology

Combinatorial topology studies partitions of objects into "complexes." A *polyhedron* is a finite union of *simplexes* (in $\mathbb{E}^3$: points, edges, triangles, or tetrahedra). A digitized subset of $\mathbb{R}^2$ or $\mathbb{R}^3$ can be regarded as a partition into convex subsets. A partition into simplexes or convex sets defines a *geometric complex*. The study of such partitions is of central interest in combinatorial topology.

6.4.1 Geometric complexes and the Euler characteristic

An elementary curve γ (to be defined in Chapter 7) can be partitioned into *one-dimensional geometric complexes* S that consist of arcs (also called *1-cells*) and their endpoints (also called isolated points or *0-cells*). Let α_0 be the number of 0-cells in S and α_1 the number of 1-cells in S that contain at least one 0-cell. The difference $\chi = \alpha_0 - \alpha_1$ is called the *Euler characteristic* of S. We will see in Section 6.4.5 that any partition of the same elementary curve has the same value of χ.

For example, a circle is partitioned by $n \geq 1$ vertices into n arcs; hence $\chi = 0$. For a simple arc with two endpoints, we have $\chi = 1$. Figure 6.20 shows the linear skeleton of a tetrahedron, which has $\alpha_0 = 4$ vertices, $\alpha_1 = 6$ arcs, and $\chi = -2$. Adding a vertex on any of the arcs increases α_0 and α_1 by 1 and thus leaves χ unchanged.

The *connectivity* β_1 of a one-dimensional geometric complex S is as follows, where β_0 is the number of components of S:

$$\beta_1 = \beta_0 - \alpha_0 + \alpha_1$$

β_1 is equal to the number of *atomic cycles* (sets of components with a union that is a simple curve) of S. β_0 and β_1 are called the first two *Betti numbers*[6] of S. (A general

---

6. The Italian mathematician E. Betti (1823–1892) published, in 1871, a memoir that defined Betti numbers. Betti's work inspired H. Poincaré to study topology; Poincaré introduced the term "Betti numbers."

6.4 Combinatorial Topology

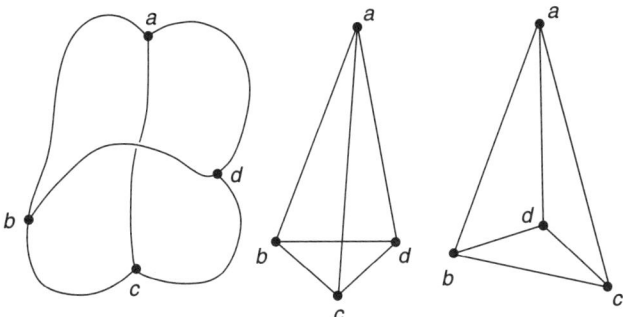

FIGURE 6.20 (J.B. Listing, 1861) The elementary curve in 3D space (left) is topologically equivalent to both graph representations of a tetrahedron (middle and right).

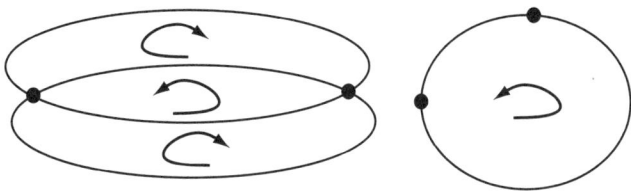

FIGURE 6.21 Example of a one-dimensional geometric complex with four atomic cycles.

definition of Betti numbers will be given in Section 6.4.5.) Figure 6.21 shows an example with four 0-cells, six 1-cells, two components, and four atomic cycles.

A planar drawing of a one-dimensional geometric complex (see Figure 6.20 [right] and Figure 6.21) can be regarded as a planar-oriented adjacency graph in which we assume clockwise local circular orders at the nodes. This graph has Euler characteristic $\chi = \beta_0 = \alpha_0 - \alpha_1 + \alpha_2$. It follows that $\beta_1 = \alpha_2$ (the number of faces of the graph) is equal to the total number of cycles minus the number of outer cycles, each of which defines the border of a component.

A *2D geometric complex*, as introduced by J.B. Listing in 1861, contains a finite number of bounded closed subsets of $\mathbb{E}^2$, which may be faces, curves, arcs, or isolated points (e.g., endpoints of arcs). The complement of the union of all of these compact sets is an open set called the *unbounded exterior*.

For example, the frontier (surface) of a simple polyhedron in $\mathbb{E}^3$ can be partitioned into elements of a 2D geometric complex called a *surface complex* of the polyhedron. A single face can be represented by a simple polygon, its edges, and its vertices (the endpoints of the edges). The surface of a simple polyhedron is topologically equivalent to the surface of a sphere. The unbounded exterior of this 2D complex splits into two disjoint open sets (see the separation theorems in Section 7.4), one bounded (the interior of the polyhedron), and the other unbounded. Let α_0, α_1, and α_2 be the number of vertices, edges, and faces of a simple polyhedron. The Descartes-Euler polyhedron theorem (see Equation 4.5) $\alpha_0 - \alpha_1 + \alpha_2 = 2$ was originally established (by Descartes and Euler) only for convex polyhedra in 3D space, but it is correct for any simple polyhedron.

Example 6.4 (M.H.A. Newman, 1939) A rectangular grating (see Figure 6.22) defines a 2D geometric complex. In the grating shown in Figure 6.22, we have $\alpha_0 = 49$ vertices, $\alpha_1 = 84$ edges, and $\alpha_2 = 36$ bounded faces, resulting in $\chi = 1$. The grating is a tessellation of a rectangle into vertices, edges, and rectangles. The sizes and shapes of the cells are unimportant for topologic purposes. Such complexes are of interest for modeling partitions of 2D pictures or of surfaces of 3D objects (see Chapter 11).

In 1813, A. Cauchy generalized the Descartes-Euler polyhedron theorem (see Equation 4.5) by introducing intercellular faces into the polyhedron. This results in the following,

$$\alpha_0 - \alpha_1 + \alpha_2 = \alpha_3 + 1 \qquad (6.4)$$

where α_3 is the number of polyhedral cells; see Figure 6.23 (left). Cauchy considered only convex polyhedra.

In 1812, A.-J. Lhuilier suggested a generalization that also allowed "tunnels" and "bubbles." He claimed that the following was true,

$$\alpha_0 - \alpha_1 + \alpha_2 = 2(b - t + 1) + p \qquad (6.5)$$

where b is the number of bubbles, t the number of tunnels, and p the number of polygons ("exits of tunnels") on faces of the polyhedron. His discussion of the number of tunnels did not cover the full range of possibilities. For example, Figure 6.19

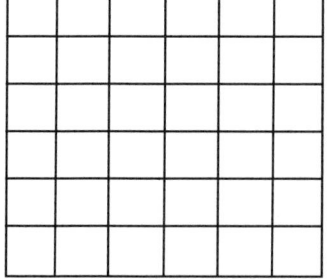
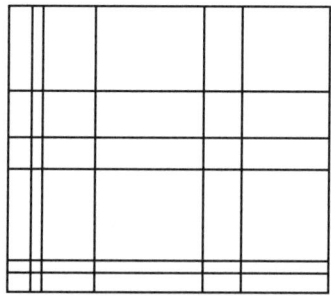

FIGURE 6.22 Two topologically equivalent rectangular gratings formed by subdividing a rectangle using finite numbers of line segments parallel to its sides.

6.4 Combinatorial Topology

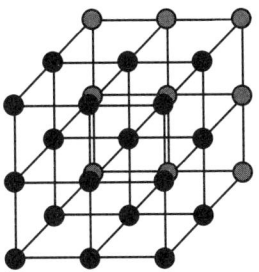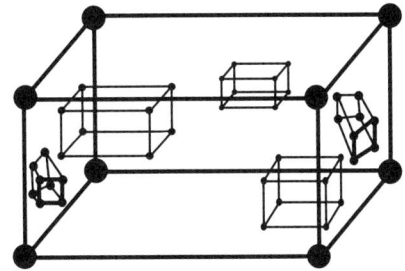

FIGURE 6.23 Left: a cube partitioned into eight subcubes has $\alpha_0 = 27$ vertices, $\alpha_1 = 54$ edges, $\alpha_2 = 36$ faces, and $\alpha_3 = 8$ subpolyhedra. Right: a parallelepiped with $b = 3$ cube-shaped bubbles, $t = 2$ tunnels, and $p = 4$ polygons on its faces; here $\alpha_0 = 48$, $\alpha_1 = 72$, and $\alpha_2 = 32$.

illustrates the difficulty of defining a tunnel; in fact, an object such as a sponge can have a "network" of tunnels. Numbers of cavities (bubbles) and tunnels will be discussed further in Section 6.4.5.

3D geometric complexes are used to model partitions of 3D sets into convex polyhedra or simplexes. A partition must be "complete" (e.g., if a polyhedron is in the complex, its surface complex must also be in the complex). The object in Figure 6.19 is a 3D geometric complex that has $\alpha_0 = 88$ vertices, $\alpha_1 = 132$ edges, $\alpha_2 = 36$ faces (assuming a straightforward partition of the surface into a surface complex), and one solid cell ($\alpha_3 = 1$). The Euler characteristic of the 2D surface complex is $\chi = \alpha_0 - \alpha_1 + \alpha_2 = -8$ (regarding the object as a hollow surface), and that of the 3D object is $\chi = \alpha_0 - \alpha_1 + \alpha_2 - \alpha_3 = -9$.

6.4.2 Euclidean complexes

The geometric complexes discussed so far are Euclidean complexes, which are partitions into compact convex sets. As a default, we assume that these sets are convex polyhedra. Surface complexes will be defined in Chapter 7.

The notion of dimension plays an important role in the definition of complexes. Dimension allows us to distinguish isolated points from line (or arc) segments and edges from faces. (In a vector space, the dimension of a set is the greatest number of linearly independent vectors in the set, but this is not a topologic characterization.)

Let $C \subset \mathbb{E}^n$ be a convex polyhedron and let P be an m-dimensional subspace ($m < n$) of $\mathbb{E}^n$. $P \cap C$ is called an $(n-1)$-*side* of C if $\dim(P \cap C) = n-1$. A nonempty intersection of finitely many $(n-1)$-sides is called a *proper side*; if it has dimension k, it is called a k-*side*. Every $(n-2)$-side is a side of exactly two $(n-1)$-sides. The 0-sides of C are its vertices. C is an *improper side* of itself. If C is bounded, it is called a *convex cell*.

Let $M \subseteq \mathbb{E}^n$ be the union of a finite number of convex cells. A *Euclidean complex* is a partition S of M into a nonempty finite set of convex cells that has the following properties:

E1: If p is a cell of S and q is a side of p, then q is a cell of S.

E2: The intersection of two cells of S is either empty or a side of both cells.

The union of the cells of S need not be connected, and a subset of the cells of S need not define a Euclidean complex.

Figure 6.24 shows four partitions of a (nonsimply connected) polygon. For the partition into squares on the left, we have $\alpha_0 = 106$, $\alpha_1 = 165$, and $\alpha_2 = 59$ (not counting the hole); for the partition into triangles on the right, we have $\alpha_0 = 106$, $\alpha_1 = 224$, and $\alpha_2 = 118$. Both counts result in $\chi = 0$. Grid cell complexes formed by cells in $\mathbb{C}^n$ are examples of Euclidean complexes. The partition on the left in Figure 6.24 is a 2D grid cell complex, and the partition on the right is a triangulation (see Section 6.4.3). For the partitions in the middle (where we also count the hole as a face), we have $\alpha_0 = 59$, $\alpha_1 = 76$, and $\alpha_2 = 18$ and $\alpha_0 = 61$, $\alpha_1 = 79$, and $\alpha_2 = 19$ so that $\chi = 1$.

Figure 6.25 shows the 1-component illustrated in Figure 6.24 (in the grid point model), as well as another 1-component. The 1-component on the left has 12 atomic

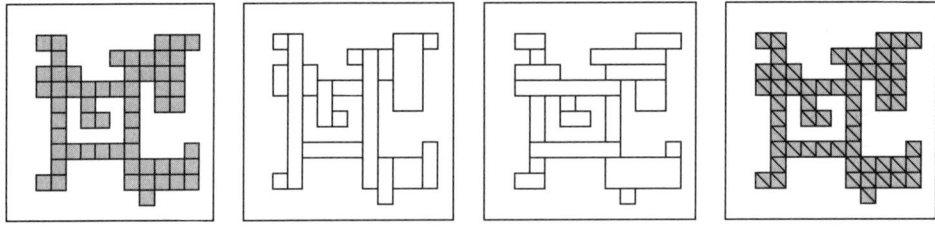

FIGURE 6.24 Four partitions of a polygon. The two in the middle require vertices at the branching points to become Euclidean complexes that satisfy **E2**.

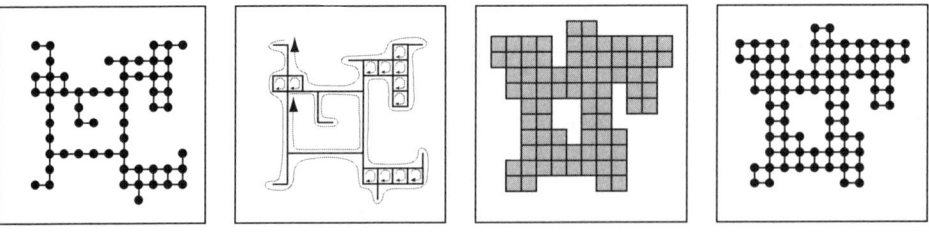

FIGURE 6.25 From left to right: the 1-component (shown in Figure 6.24) in the grid point model; a sketch of the contributing grid points, with edges representing adjacencies, and atomic cycles; another 1-component; its grid point representation.

6.4 Combinatorial Topology

cycles, one inner border cycle, and one outer border cycle (i.e., $\alpha_2 = 14$, $\alpha_1 = 71$, and $\alpha_0 = 59$) so that $\chi = \alpha_0 - \alpha_1 + (\alpha_2 - 1) = 1$. The single inner border cycle defines a hole; it is not counted in the Euler characteristic of the Euclidean complex, so $\chi = 0$ for the complex. The second 1-component has the same values of χ (0 for the Euclidean complex and 1 for the oriented adjacency graph); the two 1-components are homeomorphic (see Definition 6.10). Note that the adjacency graphs of the components are different; the graph of the component on the left is not 2-strong, but the graph of the other component is 2-strong.

6.4.3 Simplicial complexes; triangulations

A Euclidean complex is called *simplicial* iff all of its cells are simplexes. An n-dimensional *simplex* (n-simplex) is the convex hull of $n+1$ vertices $p_0, p_1, \ldots, p_n$, where the vectors $\vec{p_0 p_1}, \ldots, \vec{p_0 p_n}$ are linearly independent. For example, a 3-simplex is a solid tetrahedron. Complexes with cells that are tetrahedra are of special interest for modeling the boundaries of 3D isothetic grid polyhedra [496].

Definition 6.14 A *polyhedron* is the union of the cells of a finite simplicial complex.

A finite Euclidean complex that contains only triangles, line segments, and points is called a *triangulation*; evidently, such a complex is simplicial. The frontier of any polyhedron can be triangulated.

Figure 6.26 shows three examples of triangulations. We will now illustrate how polyhedral surfaces can be constructed by identifying vertices in the triangulation on the right. If we identify P_5, P_8, P_{11}, and P_{14} (the four corners of the large square), the (directed) line segment $P_5 P_{14}$ becomes identified with ("glued to") the directed

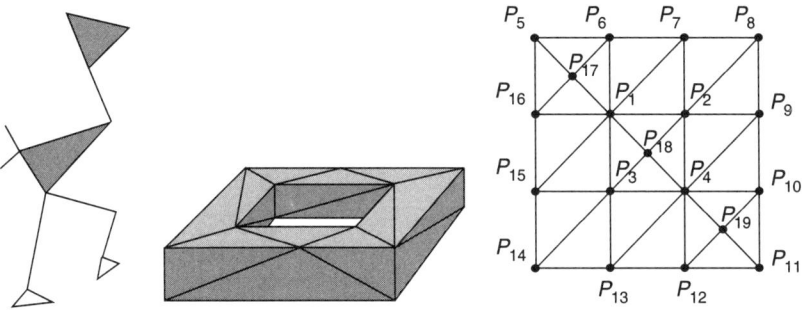

FIGURE 6.26 (P.S. Aleksandrov, 1956) A planar triangulation (left), a polyhedral surface homeomorphic to the surface of a torus (middle), and a triangulated rectangle (right; see text for details).

line segment P_8P_{11}, as does P_5P_8 with $P_{14}P_{11}$. If we then identify P_6 with P_{13}, P_7 with P_{12}, P_9 with P_{16}, and P_{10} with P_{15}, it can be shown that the resulting triangulation is homeomorphic to the surface of a torus. If instead we identify P_6 with P_{12}, P_7 with P_{13}, P_9 with P_{16}, and P_{10} with P_{15}, the resulting triangulation is homeomorphic to a *Klein bottle*.

If we identify vertices $P_1, P_2, P_3, P_4, P_{17}, P_{18}$, and P_{19}; P_5 and P_{11}; P_6 and P_{12}; P_7 and P_{13}; P_8 and P_{14}; P_9 and P_{15}; and P_{10} and P_{16}, we have a triangulation of the *projective plane* (P.S. Aleksandrov, 1956). These examples can serve as a warning not to underestimate the complexity of triangulations, even if they have small cardinalities. A triangulation used in a surface analysis algorithm (in 3D picture analysis) should certainly not contain a Klein bottle or a projective plane!

Any polyhedral surface is compact. The frontier of a simple polygon (as defined in Section 1.2.2) is an example of the union of a one-dimensional triangulation; it too can be regarded as a polyhedral surface.

The *dimension* of a triangulation is the maximum dimension of its elements. A 2D triangulation T is called *pure* iff every point or line segment in T precedes some triangle in T with respect to the side-of relation. The surface of a simple polyhedron is the union of a pure 2D triangulation.

A *path* of sets $(M_1, M_2, \ldots, M_n)$ is a finite sequence of sets such that $M_i \cap M_{i+1} \neq \emptyset$ ($i = 1, \ldots, n-1$). Such a path *connects* M_1 with M_n. A *polygonal path* is a path of line segments. A pure 2D triangulation T is called *strongly connected* iff every two triangles T_1 and T_2 in T are connected by a path of triangles, all of which are in T. The surface of a simple polyhedron is a strongly connected triangulation. The analysis of simple polygons and simple polyhedra is often done with the aid of triangulations.

6.4.4 Abstract complexes

The theory of finite *abstract complexes* is based on a partial ordering that generalizes a reflexive, transitive, antisymmetric *"bounded-by" relation* rather than the reflexive, symmetric, and in general nontransitive incidence relation.

Let S be a finite or countably infinite set of points. Assign a nonnegative integer $\dim(c)$, which is called the dimension of c, to each $c \in S$. The following definition goes back to the axiomatic definition of geometric complexes by E. Steinitz in 1908 and to the topologic study of abstract complexes by A.W. Tucker in 1933. An *abstract complex* $[S, \leq, \dim]$ has two properties:

C1: $\leq$ is a partial order on S.

C2: If $c_1 \leq c_2$ and $c_1 \neq c_2$, then $\dim(c_1) < \dim(c_2)$.

An abstract complex is an example of a poset topology. The elements of S are called the *cells* of the complex. The study of abstract complexes does not depend on interpreting cells as vertices, edges, faces, and so forth. Any subset of an abstract complex is also an abstract complex and defines a topologic subspace of the complex.

6.4 Combinatorial Topology

If $\dim(c) = i$, we call c an i-cell. The *index dimension* of an abstract complex is n iff $\dim(c) \leq n$ for all $c \in S$ and $\dim(c) = n$ for at least one $c \in S$.

The "side-of" relation in a Euclidean complex is an example of the relation $\leq$ and has property **C2**. If $c_1 \leq c_2$ and $c_1 \neq c_2$, we say that c_1 is a *proper side* of c_2. Two cells are *incident* iff $c_1 \leq c_2$ or $c_2 \leq c_1$.

Let $[S, I, \dim]$ be an incidence pseudograph. We say that $c < c'$ if $c' \in I(c)$, $c \neq c'$, and $\dim(c) < \dim(c')$. If we define $c \leq c'$ iff $c < c'$ or $c = c'$, $[S, \leq, \dim]$ is an abstract complex. Thus our discussion about incidence pseudographs in Chapter 5 can be translated into a language of abstract complexes.

6.4.5 The Poincaré formula

We conclude this chapter with a brief treatment of homology theory. Polyhedra are characterized in combinatorial topology by homology (which is easier to define than homotopy), homology groups, and Betti numbers. The Poincaré formula[7] describes the Euler characteristic of a partition of a polyhedron (a complex) by its Betti numbers. This provides a general proof that Euler characteristics are topologic invariants. A precise statement of the Poincaré formula requires a brief introduction to homology theory.

Let $[S, \leq, \dim]$ be a Euclidean complex. We write $c \in S$ as $c^{(i)}$ if $\dim(c) = i$. In Section 5.2.1, we defined i-chains in S as expressions of the form $C^{(i)} = \sum b_k c_k^{(i)}$ where $b_k = 0$ or 1, and the sum is over all i-dimensional cells in S. We recall that i-chains can be added modulo 2. We say that $c_k^{(i)}$ is in $C^{(i)}$ iff $b_k = 1$, and we write $C^{(i)} = 0$ if every b_k is 0. Let $\bigcup C^{(i)}$ be the union of the cells in $C^{(i)}$. Evidently, $C^{(i)} = 0$ iff $\bigcup C^{(0)} = \phi$.

In Section 5.2.1, we defined the frontier $\vartheta C^{(i)}$ of an i-chain $C^{(i)}$: $\vartheta C^{(0)} = 0$, and, if $i > 0$, $\vartheta C^{(i)}$ is the $(i-1)$-chain $\sum a_k c_k^{(i-1)}$ where a_k is the number of i-cells in $C^{(i)}$ that are incident with $c_k^{(i-1)}$. $C^{(i)}$ is called an i-cycle if $\vartheta C^{(i)} = 0$. Any 0-chain is a 0-cycle, and the frontier of any chain is a cycle, because $\vartheta \vartheta C^{(i)} = 0$.

We write the following if there is an i-chain $C^{(i)}$ such that $C^{(i-1)} = \vartheta C^{(i)}$:

$$C^{(i-1)} \sim 0$$

We write $C^{(i)} \sim C'^{(i)}$ iff $C^{(i)} + C'^{(i)} \sim 0$. Evidently, $\sim$ is an equivalence relation; it is called a *homology*.

Let $K = [S, \leq, \dim]$ be a Euclidean complex with index dimension n. A set B of i-cycles of k is called a *basis* for the i-cycles of K with respect to homology if every i-cycle $C^{(i)}$ of K is homologous to a sum of i-cycles of B; in essence, there are coefficients a_k such that the following imply $b_k = 0$ (modulo 2) for all k:

$$C^{(i)} \sim \sum a_k C_k^{(i)} \text{ (modulo 2)}$$

and

$$\sum a_k C_k^{(i)} \sim 0 \text{ (modulo 2)}$$

---

7. This is often called the "Euler-Poincaré formula" but in fact is entirely due to H. Poincaré; it is inspired by the Euler-Descartes theorem about polyhedra.

The cardinality β_i of a basis for the cycles of K is called the ith *Betti number* of K.

Theorem 6.8 (J.W. Alexander, 1915) Let K_1 and K_2 be two Euclidean complexes defined by partitions of the polyhedra $\bigcup K_1$ and $\bigcup K_2$. If $\bigcup K_1$ and $\bigcup K_2$ are homeomorphic, K_1 and K_2 have the same Betti numbers.

Hence the Betti numbers are topologic invariants; see Definition 6.9. We also have the following:

Theorem 6.9 (the Poincaré formula):

$$\chi(K) = \sum_{i=0}^{n}(-1)^i \cdot \beta_i \qquad (6.6)$$

This theorem was originally proved by H. Poincaré only for $n=2$ (i.e., for 2D manifolds [surfaces] in 3D Euclidean space).

The Euler characteristic is defined by the following where α_i is the number of i-cells in K:

$$\chi(K) = \sum_{i=0}^{n}(-1)^i \cdot \alpha_i$$

Theorems 6.8 through 6.9 imply the following:

Proposition 6.5 The Euler characteristic of a Euclidean complex is a topologic invariant.

Example 6.5 If K is the surface of a sphere, we have $\beta_0 = 1$, $\beta_1 = 0$, and $\beta_2 = 1$ so that $\chi = 2$. For the surface of a torus, we have $\beta_0 = 1$, $\beta_1 = 2$, and $\beta_2 = 1$ so that $\chi = 0$. If K is a sphere with $n \geq 1$ handles (informally, tori) attached to it, we have $\beta_0 = 1$, $\beta_1 = 2n$, and $\beta_2 = 1$ so that $\chi = 2 - 2n$. These three examples are illustrated in Figure 6.27 on the right. For any closed surface, we have $\beta_0 = 1$ and $\beta_2 = 1$; hence closed surfaces differ only in β_1, which is even for orientable surfaces and odd for nonorientable surfaces. For a solid sphere, we have $\beta_0 = 1$, $\beta_1 = 0$, and $\beta_2 = 0$ so that $\chi = 1$. If the solid sphere has $n \geq 1$ (pairwise nonconnected) cavities so that it has $n+1$ closed orientable surfaces, we have $\beta_0 = 1$, $\beta_1 = 0$, and $\beta_2 = n$ so that $\chi = hbox1 - n$. Three solids are illustrated in Figure 6.27 on the left.

Let K be a bounded 3D set that consists of several connected components and with a surface that consists of several orientable closed surfaces. For this K, β_0 is the

6.4 Combinatorial Topology

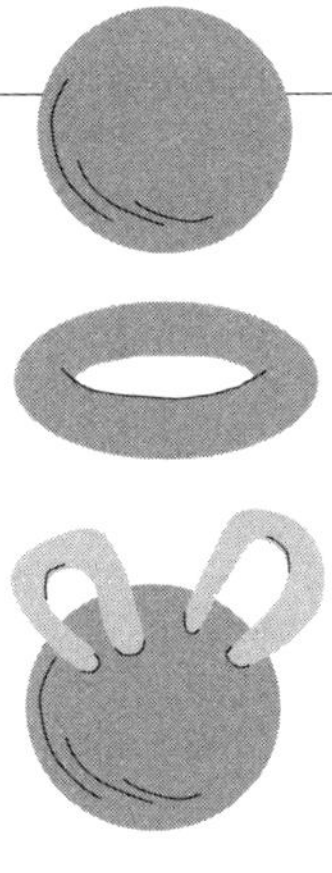

b_2	b_2	b_2	χ		b_0	b_1	b_2	χ
1	0	0	1		1	0	1	2
1	2	0	−1		1	2	1	0
1	4	0	−3		1	4	1	−2

FIGURE 6.27 A sphere, a torus, and a sphere with two handles. On the left, we assume solids; on the right, we assume surfaces only ("hollow sets").

number of connected components, $\beta_1/2$ is the *number of tunnels*, and $\beta_0 + \beta_2$ is the number of closed surfaces (so that there are β_2 cavities).

Let K be the edge skeleton (a one-dimensional complex) of a tetrahedron; it has $v = 4$ vertices, $e = 6$ edges, and $f = 0s$ faces. Here we have $\beta_0 = 1$, $\beta_1 = 3$ (note that the 1-cycle around each triangle is the sum of the three 1-cycles around the other three triangles), and $\beta_2 = 0$ so that $\chi = -2$. Note that we do not have "1·5 tunnels" in this case; the edge skeleton is not a surface. For one-dimensional complexes in 3D space, β_0 is the number of components, and β_1 is the *number of tunnels*. A one-dimensional complex in the plane has β_1 holes (the 2D analog of tunnels). Note that a Euclidean one-dimensional complex is always a planar graph (i.e., faces are not counted in its Euler characteristic, because it is only one-dimensional), and β_1 is its number of internal faces, excluding the unbounded external face; see Figure 6.28. Oriented adjacency graphs allow us to generalize the Euler characteristic and not be limited to planar graphs. See Section 17.5 about tunnel-free subsets of $\mathbb{G}_{m,n,l}$.

6.4.6 Homology groups

Betti numbers can also be defined using homology groups. Let $[G, +, 0]$ be an additive abelian group with identity 0. $[H, +, 0]$ is called a subgroup of $[G, +, 0]$ iff $H \subseteq G$ and $f, g \in H$ imply $f - g \in H$. A *homomorphism* ϕ from $[G, +, 0]$ onto $[H, +, 0]$ is a function from G onto H such that $\phi(f) - \phi(g) = \phi(f - g)$ for all $f, g \in G$. If ϕ is one-to-one, it is called an *isomorphism*. If ϕ is a homomorphism, $H = \phi^{-1}(0)$ is a subgroup of G called the *kernel* of ϕ. The kernel of an isomorphism is the singleton $\{0\}$.

Let H be a subgroup of G and $f \in G$. $f_H = \{f + g : g \in H\}$ is called a *coset* of G relative to H. Note that two cosets are either disjoint or identical. The set of cosets

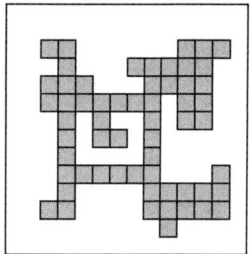 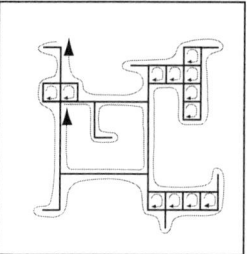 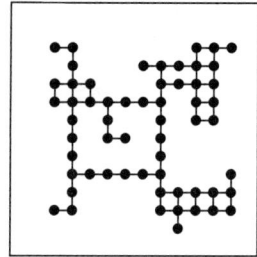

FIGURE 6.28 Left: a 2D incidence grid representation of a closed region in a 2D picture with $\alpha_0 = 106$, $\alpha_1 = 165$, $\alpha_2 = 59$, $\beta_0 = 1$, $\beta_1 = 1$, and $\beta_2 = 0$ so that $\chi = 0$. Middle: the same region in the oriented 4-adjacency graph with $\alpha_0 = 59$, $\alpha_1 = 71$, $\alpha_2 = 14$, $\beta_0 = 1$, $\beta_1 = hbox1$, and $\beta_2 = 0$ so that $\chi^+ = \chi + 1 = 1$. Right: a planar one-dimensional complex (note: no faces) with $\alpha_0 = 59$, $\alpha_1 = 71$, $\beta_0 = 1$, and $\beta_1 = 13$ so that $\chi = -12$.

of G relative to H is an additive group called the *factor group* G/H of G modulo H; we define $f_H + g_H = (f+g)_H$. $\phi : G \to G/H$ is a homomorphism and H is its kernel.

The set $\mathcal{C}^{(i)}$ of i-chains of a Euclidean complex K is an abelian group under addition modulo 2. Let $\mathcal{G}^{(i)}$ be the subgroup of $\mathcal{C}^{(i)}$ that consists of all i-cycles, and let $\mathcal{F}^{(i)}$ be the subgroup of $\mathcal{G}^{(i)}$ that consists of all frontiers $\vartheta \mathcal{C}^{(i+1)}$ of $(i+1)$-chains. The factor group $\mathcal{H}^{(i)} = \mathcal{G}^{(i)}/\mathcal{F}^{(i)}$ is called the i-th *homology group (Betti group)* of K. The homology $\sim$ partitions the cycles $\mathcal{G}^{(i)}$ into equivalence classes called *homology classes*. It can be shown that the dimension of $\mathcal{H}^{(i)}$ is the i-th Betti number.

Example 6.3 showed that the surface of a torus and a planar drawing of an 8 have different fundamental groups (i.e., they are not identical with respect to homotopy). On the other hand, the homology groups of both of these sets are isomorphic to the direct sum $[\mathbb{Z}, +, 0] \oplus [\mathbb{Z}, +, 0]$. Thus isomorphism of homology groups does not imply homotopy (isomorphism of fundamental groups).

6.5 Exercises

1. Does the adjacency grid $[\mathbb{C}_2^{(2)}, \eta_1]$ have a topology in which 1-connectedness is the same as topologic connectedness?

2. Describe the Aleksandrov topology of the poset $[\mathbb{R}, \leq]$.

3. Explain why the product of one one-dimensional grid cell topology and two one-dimensional grid point topologies is already contained in the set of 3D digital topologies listed after Theorem 6.6.

6.5 Exercises

4. Show that the Euler characteristic $\chi(\Pi)$ of a polyhedron $\Pi \subset \mathbb{R}^3$ can also be defined by the following axioms:

   1. $\chi(\emptyset) = 0$.

   2. $\chi(\Pi) = 1$ if Π is nonempty and convex.

   3. If Π_1 and Π_2 are polyhedra, $\chi(\Pi_1 \cup \Pi_2) = \chi(\Pi_1) + \chi(\Pi_2) - \chi(\Pi_1 \cap \Pi_2)$.

   (Hint: Prove that $\chi(\Pi)$ is equal to the alternating sum $\alpha_0 - \alpha_1 + \alpha_2$ where α_0 is the number of vertices, α_1 the number of edges, and α_2 the number of triangles in any triangulation of Π's frontier.)

5. Prove that, for any (planar) polyhedron $\Pi \subset \mathbb{R}^2$, $\chi(\Pi)$ is equal to the number of components of Π (in the Euclidean topology) minus the number of holes in Π.

6. Let n_i be the number of 2×2 patterns of pixels in a binary picture P that contain exactly i 1s ($i = 0,1,2,3,4$), and let n_D be the number of such patterns that contain two diagonally adjacent 1s. Let the following be true:

   $$E_{8,4}(P) = (n_1 - n_3 - 2n_D)/4 \quad \text{and} \quad E_{4,8}(P) = (n_1 - n_3 + 2n_D)/4$$

   The geometric realization of the 2D grid cell topology based on the label order "$0 < 1$" or "$1 < 0$" maps the union of the 1s in P into a (planar) polyhedron Π. Prove that $\chi(\Pi) = E_{8,4}(P)$ for label order "$0 < 1$" and that $\chi(\Pi) = E_{4,8}(P)$ for label order "$1 < 0$" (see Figure 17.2 for an example).

7. Prove that any convex set in $\mathbb{E}^n$ (e.g., a single point, a straight line segment, a sphere, a convex polyhedron) is simply connected (i.e., its fundamental group is $\{\epsilon\}$).

8. Prove that the free cyclic group is the fundamental group of the annulus and also of $\mathbb{R}^2 \setminus \{p\}$, where p is any point in $\mathbb{R}^2$.

9. Prove that $\alpha_0 - \alpha_1 = 1$ for any tree (i.e., that the Euler characteristic of any tree is 1).

10. We have given finite and countable examples of abstract complexes. The following is a noncountable example: Let $S = \{s_a : a \in [0,1]\}$ be a noncountable set of abstract cells defined on the real index interval $[0,1]$, where $s_a \neq s_b$ iff $a \neq b$. For $a \in [0,1]$, let $a = 0$ and $a_1 a_2 a_3 \ldots$ where $a_i \in \{0,1,\ldots,9\}$. Let $s_a \leq s_b$ iff $a = b$ or $a_1 < b_1$ and $a_i = b_i$ for all $i \geq 11$. Let $\dim(s_a) = a_1$. Show that $[S, \leq, \dim]$ is an abstract complex.

11. Calculate the Betti numbers of the following three tilings (2D complexes):

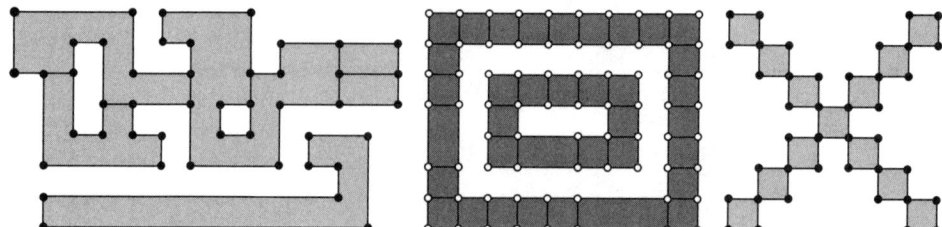

12. Suppose deletion of all of the invalid edges in a picture leads to the nonconnected undirected graph shown below. Regard each pixel as the centroid of an isothetic unit square, and take the union of all of the squares that correspond to pixels in the same 4-component. The frontiers of the resulting sets are isothetic simple curves. Find the Betti numbers β_i of these sets ($i = 0, 1, 2$).

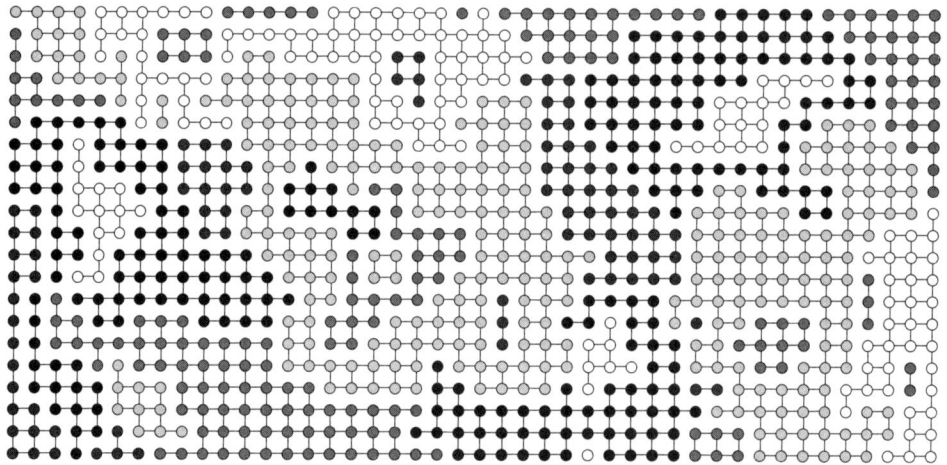

13. Calculate the number of tunnels in the one-dimensional complex that consists of the vertices and edges of a tetrahedron.

14. Consider three "systems of pathways" in a solid sphere: (1) $n \geq 3$ straight pathways that start at the surface of the sphere and meet at its center; (2) four pathways that form a hollow tetrahedron inside the sphere, with an exit to the surface of the sphere, at each vertex of the tetrahedron; and (3) a pathway that forms a hollow torus inside the sphere, with $n \geq 1$ exits to the surface of the sphere. What are the numbers of tunnels in these three cases?

15. The *Arens-Fort space* is defined on the (countable) set $S = \mathbb{N}^2$ by a set $\mathcal{G}$ that is the union of $\mathbb{N}^2 \setminus \{(0,0)\}$ and the family of all sets $U \subset \mathbb{N}^2$ with *projections*

$\{n : (m,n) \in U\}$ that are finite for all but finitely many $m \in \mathbb{N}$. Show that $[S, \mathcal{G}]$ is a topologic space and does not have a countable basis.

16. Let $M \subseteq \mathbb{N}$ be closed iff either $1 \in M$ or $\sum_{n \in M} 1/n$ is finite (note that this excludes $0 \in M$). Show that this defines a topologic space on $\mathbb{N}$ that does not have a countable basis for the neighborhoods of 1.

6.6 Commented Bibliography

There are many textbooks about topology and, in particular, combinatorial topology; see, for example, [9, 10, 13, 100, 112, 639, 771, 844, 857]. [580] is an edited book about digital topology.

Research regarding topologic approaches to picture analysis began before 1970. In 1966 and 1970, A. Rosenfeld initiated the study of connected subsets of the grid [881, 921]. The relevance of Definition 6.7 to picture analysis was discussed in 1971 by J. Mylopoulos and T. Pavlidis [756]. Figure 7.4 is from [100]. The grid point topology (Section 6.2.2) was described in [1140].

Axiomatic digital topology (axioms for connectedness) was studied in [970]; [872] extended this work. Properties **D1** and **D2** are due to U. Eckhardt and L. Latecki [297]. Property **D3** was added by Y. Kong (private communication, March 2003); see also his discussion of digital topologies in [570]. Digital topology on graphs is studied in [124, 158, 780]; see also [6] and [925]. Theorem 6.4 is proved in [172]; see also [620]. Introductory and review papers about digital topology include [571, 572, 576, 577, 579, 581, 891, 893]; see also [136]. For the relationship between continuous and digital topology, see [95]. For "well-composed" pictures for which 4- and 8-connectedness are equivalent, see [621, 622, 625].

The two-volume book by P.S. Aleksandrov [9, 10] provides a broad coverage of Euclidean and abstract complexes. For more about abstract complexes, see also [1025] and [1069]. For topologies on such complexes, see for example, [9, 844, 1069]. Geometric complexes were introduced into the mathematical literature by J.B. Listing in 1861; see also Section 1.2.6. For Theorem 6.8, see [12]. [999] introduced cellular complexes into the picture analysis literature. See [116] about oriented simplicial complexes and about the encoding of finite subsets of such complexes.

For the topology of polyhedra, see the monographs [709, 844, 1026]. Exercises 4 and 5 are discussed in [577]. See [631] about counting tunnels by counting nonseparating cuts, which gives β_1. For the difficulty of defining tunnels for polyhedra in $\mathbb{R}^3$, see [575]. A polygonal loop in $\mathbb{R}^3$ that is not contractible to a point in the complement of a polyhedron Π obviously defines one tunnel in Π. However, as indicated in Section 6.4.5, there exist dependencies between such loops, and we can only define *numbers of tunnels* (see also Exercise 14).

Exercise 6 follows [377]. Calculations of Euler characteristics based on local pattern counts in 2D or 3D pictures are reviewed in [577]. For other references about the calculation of the Euler characteristic, see [88, 186, 266, 646, 938]. Local patterns in binary pictures can also be used to provide more detailed topologic classifications of components [14].

Fundamental groups of digital pictures are discussed in [567, 692]. Effective methods of computing presentations of fundamental groups of arbitrary polyhedra are described in [709]. The problem of determining whether two polyhedra have isomorphic fundamental groups is undecidable. (For any finite group G, one can construct a polyhedron with a fundamental group that is isomorphic to G [see Theorem 3.3.20 in [709]], and isomorphism of finite groups is undecidable.) However, the restriction of polyhedra to subsets of $\mathbb{R}^3$ may define a "simpler" subproblem, and a decision algorithm for this subproblem could be used to determine whether a polygonal loop in $\mathbb{R}^3$ is knotted by determining whether the complement of the loop has the trivial fundamental group. An algorithm for the homotopy classification of binary pictures is given in [158]. Homotopy in digital spaces is studied in [54]. Topologic equivalence, which is called "component equivalence" and defined as a counterpart of homeomorphism for quantized spaces, was discussed in [757]. Topologic equivalence between preimages and "continuous analogs" of digitized sets—and its dependency on the grid constant—is studied in [624].

For Exercise 15, see [1016].

CHAPTER 7

Curves and Surfaces: Topology

On the basis of the concepts introduced in Chapter 6, this chapter discusses curves and surfaces in topologic spaces, with emphasis on the digital case.

7.1 Curves in the Euclidean Topology

The definition of a curve has an interesting history in mathematics. Jordan curves are defined by parameterization; see the "parameterized paths" in Section 6.3.3. Urysohn-Menger curves are defined using a topologic approach. The two definitions are equivalent, and they both define separations of the plane.

7.1.1 Jordan curves

Let ϕ be a parameterized continuous path $\phi : [a,b] \to \mathbb{R}^2$ such that $a \neq b$, $\phi(a) = \phi(b)$, and let $\phi(s) \neq \phi(t)$ for all s,t ($a \leq s < t < b$). The following set was defined by C. Jordan in 1893 to be a *Jordan curve* in the plane[1]:

$$\gamma = \{(x,y) \,:\, \phi(t) = (x,y) \land a \leq t \leq b\} \tag{7.1}$$

Similarly, a *Jordan arc* γ in the plane is defined by a subinterval $[c,d]$ where $a \leq c < d \leq b$. A Jordan curve is topologically equivalent (homeomorphic) to a unit circle; it does not have "crossings" or "touchings." A *rectifiable Jordan arc* γ has a bounded *arc length* as follows, where d_e is the Euclidean metric:

$$\mathcal{L}(\gamma) = \sup_{n \geq 1 \land c = t_0 < \cdots < t_n = d} \sum_{i=1}^{n} d_e(\phi(t_i), \phi(t_{i-1})) < \infty \tag{7.2}$$

---

[1]. For the first appearance of the term "Jordan curve," see [787]: "By a *Jordan curve* is meant a curve of the general class of continuous curves without multiple points, considered by Jordan, *Cours d'Analyse*, vol. I, 2d edition, 1893, p. 90 ..."

C. Jordan proposed, in 1883, a definition of a parameterized curve in the following form:

$$\gamma = \{(x,y) : x = \alpha(t) \land y = \beta(t) \land a \leq t \leq b\} \tag{7.3}$$

Using a parameterization that satisfied Equation 7.3, G. Peano defined, in 1890, a curve known as the *Peano curve* that fills the whole unit square. (One iteration of the construction of the Peano curve is shown in Figure 1.7.) Note that Equation 7.1 excludes the Peano curve. However, Equation 7.3 is still in common use for arc length calculation. (Evidently, Equation 7.3 follows from Equation 7.1; α and β can be defined by projecting ϕ on the x- and y-axis, respectively. However, Equation 7.3 allows for $(x,y) \in \gamma$ such that $x = \alpha(t_1)$ and $y = \beta(t_{tesub2})$, with $t_1 \neq t_2$.) If α and β are differentiable functions, the arc is called *differentiable*, and its arc length is as follows:

$$\mathcal{L}(\gamma) = \int_a^b \sqrt{\left(\frac{d\alpha(t)}{dt}\right)^2 + \left(\frac{d\beta(t)}{dt}\right)^2}\, dt \tag{7.4}$$

The Jordan definition applies to curves that have parametric forms. Not all curves have such forms, and, even if they do, the forms may be difficult to find. (See Chapter 8 about the geometry of Jordan curves.) In picture analysis, we have to deal with curves that are given in digitized pictorial form and for which a parametric description is often not of interest. Topologic methods of defining curves are therefore more relevant for our purposes.

7.1.2 Urysohn-Menger curves

The local (adjacency-based) approaches used for curve tracing in picture analysis are related to nonparametric curve characterizations based on topologic connectedness.

Let $[S, d]$ be a metric space. $p \in S$ is called a *point of accumulation* of $M \subseteq S$ iff, for any $\epsilon > 0$, the ε-neighborhood $U_\varepsilon(p)$ of p contains a point $q \neq p$ of M. It can be shown that M is topologically connected in S iff, for any partition of M into two disjoint subsets, some point in one of the subsets is a point of accumulation of the other subset.

Definition 7.1 A *continuum* is a nonempty subset of a topologic space S that is compact (closed and bounded) and topologically connected.

G. Cantor was the first to suggest a topologic definition of a curve, but his definition had to be revised. P. Urysohn provided a correct definition in 1923, and K. Menger did so independently in 1932. Let $M \subseteq L \subseteq \mathbb{R}^n$. $p \in \mathbb{R}^n$ is called an *L-frontier point* of M iff, for any $\epsilon > 0$, the ε-neighborhood $U_\varepsilon(p)$ of p contains points of M as well as points of $L \setminus M$. The set of L-frontier points of M is called the *L-frontier* of M. A continuum $M \subseteq \mathbb{R}^n$ is called *one-dimensional at* $p \in M$ iff, for some

7.1 Curves in the Euclidean Topology

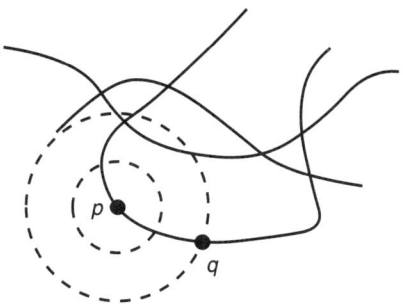

FIGURE 7.1 A sufficiently small $\varepsilon > 0$ allows us to analyze the situation at point p.

$\varepsilon > 0$, any continuum C contained in the M-frontier of $U_\varepsilon(p) \cap M$ is a singleton $\{q\}$; see Figure 7.1. M is called *one-dimensional* iff it is one-dimensional at every $p \in M$. Note that this allows "one-dimensional" to be defined in a non-Aleksandrov space. (In an Aleksandrov space, we can use Definition 6.7.)

Definition 7.2 P. Urysohn, 1923; K. Menger, 1932) A *curve* $\gamma \subseteq \mathbb{R}^n$ is a one-dimensional continuum.

This definition was originally given only for the Euclidean topology. Note that an isolated point in $\mathbb{R}^n$ satisfies this definition.

A Urysohn-Menger curve is more general than a Jordan curve; it may be a simple curve (see Definition 7.3) that forms a loop without any self-intersections, but it may also be a union of finitely many bounded arcs.

Dimension theory, as established by P. Urysohn and K. Menger, is based on a generalization of the definition of one-dimensionality at p given above. A metric space (or manifold) S has dimension n at $p \in S$ if S can be disconnected by removing an arbitrarily small set of dimension $n-1$ that contains p but not by removing a set of smaller dimension. We will return to this approach when we discuss surfaces (2D manifolds).

Let $\gamma \subset \mathbb{E}^2$ be a Jordan curve, and let $\varepsilon > 0$. The ε-*tube* of γ is the set of all points p such that $d_e(\{p\}, \gamma) \leq \varepsilon$, where d_e is Hausdorff distance. The frontier of any ε-tube of a Jordan curve that is topologically equivalent to an annulus (see Figure 7.2) consists of two disjoint Jordan curves, and the d_e Hausdorff distance between these curves is 2ε.

P.S. Aleksandrov proved that Urysohn-Menger curves can be approximated by polygonal chains:

Theorem 7.1 (P.S. Aleksandrov) A compact set $\gamma \subseteq \mathbb{R}^n$ is a Urysohn-Menger curve iff, for arbitrarily small $\varepsilon > 0$, there is a mapping Φ of γ onto a polygonal chain such that $d_e(p, \Phi(p)) < \varepsilon$ for all $p \in \gamma$.

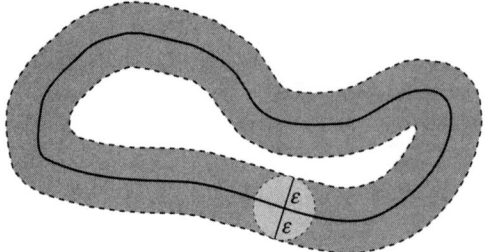

FIGURE 7.2 An ε-tube of a Jordan curve that is homeomorphic to an annulus.

Thus a Urysohn-Menger curve γ is in the ε-tube of a polygonal chain for arbitrarily small $\varepsilon > 0$.

7.1.3 Simple curves and arcs

A curve γ has *branching index* $m \geq 0$ at $p \in \gamma$ iff, for any $r > 0$, there is a positive real $\varepsilon < r$ such that the cardinality of the γ-frontier of $U_\varepsilon(p) \cap \gamma$ is at most m. For a sufficiently small real $r > 0$, it follows that, for any positive real $\varepsilon < r$, the cardinality of the γ-frontier of $U_\varepsilon(p) \cap \gamma$ is at least m. Note that the definition of a curve allows m to be countably infinite.

> **Definition 7.3** (P. Urysohn, 1923; K. Menger, 1932) A *simple curve* is a curve in which every point p has branching index 2. A *simple arc* is either a curve in which every point p has branching index 2 except for two *endpoints*, which have branching index 1, or a simple curve with one of its points labeled as an endpoint.

A basic theorem in the mathematic theory of curves is as follows:

> **Theorem 7.2** In $\mathbb{R}^2$, Jordan curves and simple Urysohn-Menger curves are the same.

This theorem shows that Theorem 7.1 also applies to Jordan curves.
A *regular point* of a curve has branching index 2 and is not an endpoint. A *branch point* has branching index 3 or greater. A *singular point* is either an endpoint or a branch point. The topologic concept of branching index is relevant to the study of skeletons (see Chapter 16).

7.1.4 Elementary curves and the Euler characteristic

An *elementary curve* is the union of a finite number of simple arcs, each pair of which have at most a finite number of points in common. It consists of a finite number

7.1 Curves in the Euclidean Topology   235

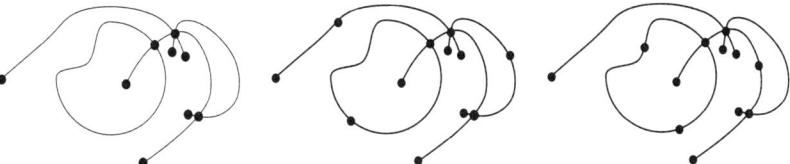

FIGURE 7.3 Left: an elementary curve with Euler characteristic $9 - 10 = -1$; it has nine singular points and 10 regular components. Right: two partitions of the curve into one-dimensional geometric complexes, both with Euler characteristic $12 - 13 = -1$.

of singular points and a finite number of *regular components*; the latter are either simple curves or simple arcs. Every regular point $p \in \gamma$ is on a uniquely determined subcurve $\gamma_p \subseteq \gamma$ that is either a simple curve (the component of p in γ), a simple arc that has only one endpoint, or a simple arc that has two endpoints.

If we partition a simple curve γ (e.g., by choosing two points of γ [$\alpha_{tesub0} = 2$] that divide γ into two simple arcs [$\alpha_1 = 2$]), we find that the Euler characteristic $\chi(\gamma)$ is $\alpha_0 - \alpha_1 = 0$. An elementary curve with at least one singular point on each of its simple curves is a one-dimensional Euclidean complex. The Euler characteristic of such a complex is equal to the number of singular points minus the number of regular components. For example, a simple arc that has only one endpoint, which is also a complex of a simple curve, has Euler characteristic 0.

Figure 7.3 (left) shows an elementary curve that is a union of simple arcs. Figure 7.4 shows the capital letters of the German alphabet; the letters are assumed to be elementary curves. (Note that a rectangle of nonzero width [e.g., a thick line segment] is not homeomorphic to a line segment.) The three (pairwise nonhomeomorphic) letters Ä, Ö, and Ü have three components each as compared with only one component for each of the remaining letters; hence these three letters cannot be homeomorphic to any of the other letters.

The homeomorphy of the letters in Figure 7.4 depends on whether we consider a letter to be a set that is not one-dimensional at any of its points or to be an elementary curve. Figure 7.5 shows both alternatives for two letters. The elementary curve

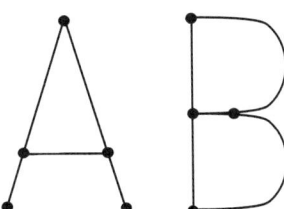

FIGURE 7.4 Capital letters of the German alphabet. We assume that the letters are elementary curves that may have endpoints or branch points, as indicated on the right for the letters A and B.

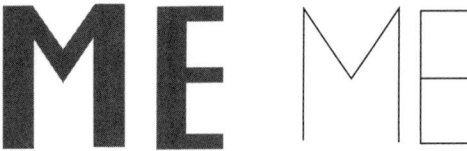

FIGURE 7.5 The two sets on the left are homeomorphic, but the two elementary curves on the right (which are not linear skeletons of the sets on the left, although they may be regarded as "skeletons" in a picture-analysis context) are not.

version of the letter E has a *decomposition vertex* (i.e., deletion of this point partitions the set into more than two connected parts), but the elementary curve version of the letter M has no decomposition vertex. The number of decomposition vertices is a topologic invariant (i.e., it is preserved by homeomorphism).

As an illustration of isotopy, consider the sets of curves 05 and 20. These sets are isotopic to each other but not to the set 97, because 9 is not homeomorphic to 0.

7.1.5 Separation theorems

The *Jordan-Veblen curve theorem* of Euclidean topology says that any set that is topologically equivalent to a unit circle decomposes the Euclidean plane into two disjoint sets.

> **Theorem 7.3** (C. Jordan, 1887; O. Veblen, 1905) Let γ be a Jordan curve in the Euclidean plane $\mathbb{E}^2$. The open set $\mathbb{R}^2 \setminus \gamma$ consists of two disjoint topologically connected open sets with the common frontier γ.

This theorem was first stated by C. Jordan in 1887. His proof (which was incorrect) attempted to use a sequence of polygons that converged to the curve.[2] The first correct proof was given by O. Veblen in 1905 using the parametric characterization of the curve. This proof left open the question of whether the inside and outside of the curve are always topologically equivalent to the inside and outside of a circle. The stronger *Schönflies-Brouwer curve theorem* is as follows:

> **Theorem 7.4** (A. Schönflies, 1906; L.E.J. Brouwer, 1910) For any planar Jordan curve γ, there is a one-to-one mapping Φ of the Euclidean plane into itself such that Φ and Φ^{-1} are continuous and $\Phi(\gamma)$ is the unit circle.

---

2. For a first use of the term "Jordan curve theorem," see [1129]; D.W. Woodard was the second African American who received a PhD in mathematics.

The proof by A. Schönflies in 1906 contained some errors, which were fixed by L.E.J. Brouwer in 1910. Both theorems also apply to simple Urysohn-Menger curves (see Theorem 7.2). By the Schönflies-Brouwer theorem, any two simple curves in $\mathbb{E}^2$ are isotopic; this is a stronger result than the Jordan-Veblen theorem.

Finally, we have a general result for continua in $\mathbb{R}^2$:

Theorem 7.5 (S. Straszewicz, 1923) Let $G_1, G_2 \subset \mathbb{R}^2$ be two continua, each of which does not divide $\mathbb{R}^2$ into two connected regions, and let $G_1 \cap G_2$ consist of two connected components. Then $G_1 \cup G_2$ divides $\mathbb{R}^2$ into two connected regions.

7.2 Curves in Incidence Grids

The Urysohn-Menger method of defining curves (see Definition 7.2) can be adapted to the grid cell topologies $[\mathbb{C}_n, \leq]$. We make use of the dimension $\dim_A$ defined by adjacency relation A; see Definition 6.7. This definition applies to arbitrary subsets of incidence grids (e.g., subsets that define frontiers and do not contain principal nodes).

7.2.1 Frontier grids; curves of marginal nodes

Section 5.4.3 defined a frontier grid for a 2D picture. Such grids can also be defined for 3D (or even nD) pictures; they were proposed and used (with different terminology and notation) for 2D and 3D picture analysis by V. Kovalevsky [594].

Definition 7.4 An $m_1 \times m_2 \times \ldots \times m_n$ incidence grid (picture grid) uniquely defines an isomorphic $(m_1+1) \times (m_2+1) \times \ldots \times (m_n+1)$ *frontier grid*. i-cells of the picture grid are mapped isomorphically (with respect to cell incidence) into $(n-i)$ cells of the frontier grid ($i = 0, \ldots, n$).

Equation 5.3 guarantees the existence of such an isomorphism.

Figure 7.6 illustrates frontiers in a 2D incidence grid (left) and their representations in the 2D frontier grid using a 2D incidence grid model representation (middle) and a (simpler) 4-adjacency grid model representation (right).

Proposition 7.1 Every frontier of a nonempty region in a 2D incidence grid is one-dimensional, closed, and bounded.

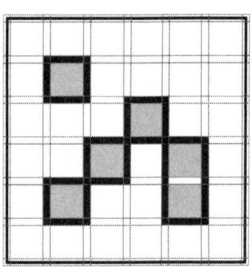 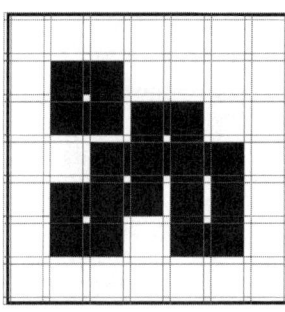 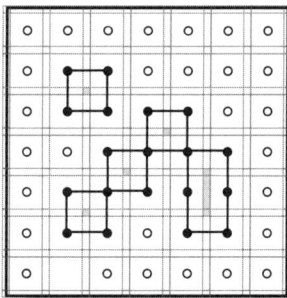

FIGURE 7.6 Left: two frontiers in the 2D incidence grid. Middle: representations of these frontiers in an incidence grid model of the frontier grid. Right: their representations in a 4-adjacency grid model of the frontier grid.

Proof The frontier contains only 0- and 1-cells. A 0-cell is incident with up to four 1-cells, and each 1-cell is incident with two 0-cells. These are one-dimensional configurations; see Definition 6.7. The frontier is closed, because it always contains the two 0-cells that are incident with each of its 1-cells, and it is bounded because the region is finite. ∎

This proposition shows that the frontiers of regions in the 2D incidence grid are curves in the sense of Definition 7.2. The definition of branching points also applies, so that simple curves, simple arcs, and elementary curves can be defined. The curves are 4-curves in the 4-adjacency representation of the frontier grid. The 4-curves that represent the frontiers of two regions in the picture can be 4-adjacent in the frontier grid; see Figure 7.6.

7.2.2 Curves of principal nodes

We are particularly interested in one-dimensional connected sets of pixels or voxels in the picture grid; see the middle of Figure 6.11. In this section, we describe a way of defining curves in an incidence grid without making use of the frontier grid.

We recall that a region M is a complete finite component of $\mathbb{C}_n$ and that its core is a nonempty maximal connected set of principal nodes; see Section 5.1.3. A closed region in $\mathbb{C}_2$ is 2D when we use the definition of dimension based on adjacency (Definition 5.3); hence we cannot use closed regions to model curves. Rather, we define a *curve* ρ of principal nodes as a one-dimensional region in $[\mathbb{C}_n, \leq]$ ($n \geq 2$). It follows that any curve has finite cardinality, so the union $\bigcup \rho$ of the cells of ρ is bounded. $\bigcup \rho$ is an isothetic polygon if ρ is a curve in $\mathbb{C}_2$ and an isothetic polyhedron if ρ is a curve in $\mathbb{C}_3$.

7.2 Curves in Incidence Grids

Figure 7.7 illustrates this definition. The upper row shows curves in which the connectedness between principal cells is based on 1-connectedness; they are therefore called *1-curves*. Note that the example in the upper left corner also contains 0-cells, but it is still one-dimensional, and these 0-cells add no further connectivities to the curve. From now on, we exclude 0-cells from 1-curves; see the definition of minimal curves below. Pixels (2-cells) in a 1-curve can "touch" (i.e., they can be 0-adjacent); see the examples in the middle and on the right in the upper row of Figure 7.7.

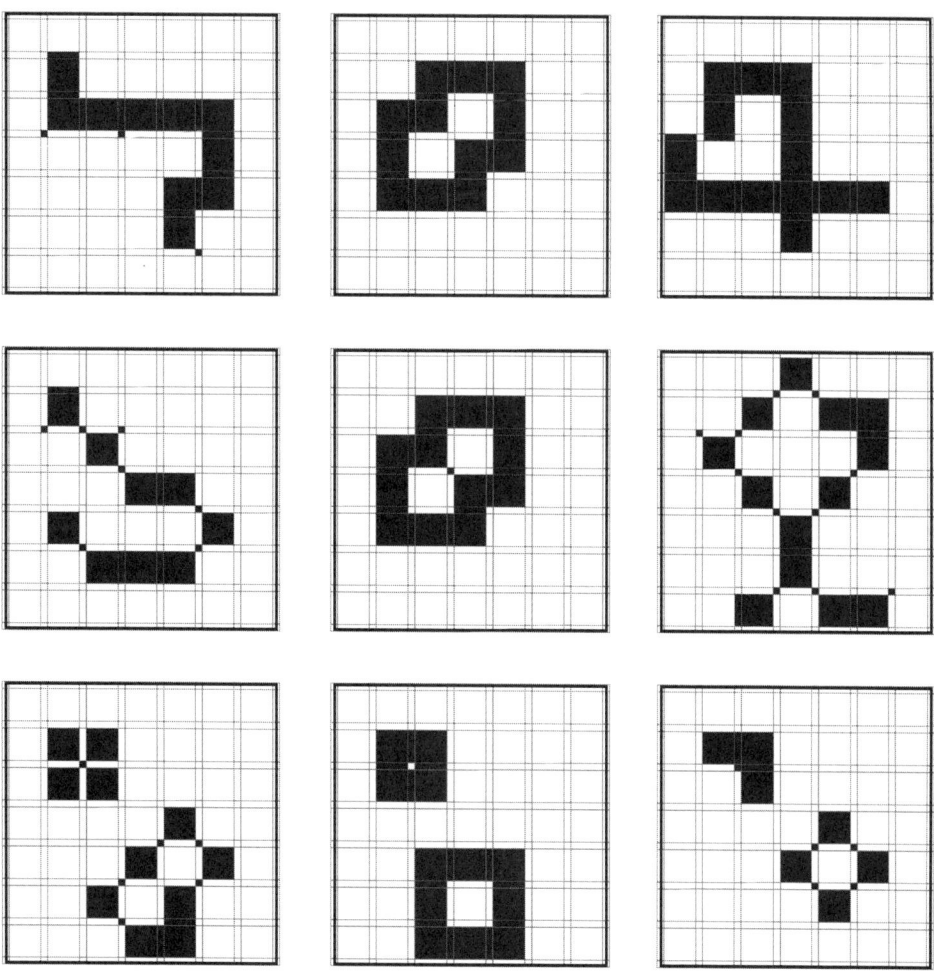

FIGURE 7.7 Upper row: 1-curves; the curve on the left is a 1-curve after removing three 0-cells. Middle row: 0-curves. Bottom row, upper part: sets that are not curves; lower part: see text.

The middle row of Figure 7.7 shows curves in which connectedness between principal nodes is based on either 1-adjacency or 0-adjacency (in the sense of Definition 5.2). Such curves are called *0-curves*. The example in the middle forms an "8".

Definition 7.5 A region in $\mathbb{C}_n$ is called an *α-curve* iff it is one-dimensional and contains only i-cells ($\alpha \leq i \leq n$).

The upper examples in the bottom row of Figure 7.7 are not curves. The examples on the left and in the middle are not regions, because they are not complete, and the example on the right is not one-dimensional. (The three examples in the lower positions will be explained later in this section.)

An α-curve ρ is called *minimal* iff it contains only marginal cells that are required for its connectivity. For example, in the upper left of Figure 7.7, we obtain a minimal 1-curve after removing three 0-cells and one "terminal" 1-cell. (Note that removal of a 1-cell instead of a 0-cell may preserve connectivity between principal cells but will result in a noncomplete set.) It follows that an α-curve ρ is minimal iff it contain only i-cells ($\alpha \leq i \leq n-1$) that are i-adjacent to two principal nodes in the core of ρ. For example, in the upper right of Figure 7.7, we obtain a minimal 1-curve after removing two "terminal" 1-cells. It follows that a minimal $(n-1)$-curve in $\mathbb{C}_n$ is always an open region. (Note that a 0-curve cannot be closed, because it is one-dimensional.)

A vertex of a 0-cell in a 0-curve ρ is called a *decomposition vertex* if its deletion partitions $\bigcup \rho$ into more than two connected components in the Euclidean topology. Minimal 1-curves do not have decomposition vertices.

An α-curve ρ is called *confirmative* iff, for any two principal nodes in ρ that are adjacent via some i-cell ($\alpha \leq i \leq n$), there is also a marginal cell in ρ that ensures this adjacency, and adding this cell does not contradict one-dimensionality. For example, the 0-curve on the lower left in Figure 7.7 is not confirmative, but adding one 0-cell makes it confirmative, and it remains one-dimensional. Note that completeness of curves does not imply confirmativeness. The 1-curves in the middle and right of the upper row in Figure 7.7 are confirmative, because $0 < \alpha = 1$. The 0-curve at the center of Figure 7.7 is confirmative, but adding the related 0-cell would destroy its one-dimensionality. From now on, we will assume that α-curves in the incidence grid are minimal and confirmative.

A curve ρ has branching index $m \geq 0$ at a principal node p iff exactly m principal nodes $q \in \rho$ are adjacent to p. For $n = 2$ (0- or 1-curves), possible branching indices are 1 (an endnode), 2 (a regular node), and 3 or 4 (a branch node). For $n = 3$ (0-, 1- or 2-curves), a branch node can have branching indices 3, 4, 5, or 6. A node is called singular if it is either an endnode or a branch node.

Definition 7.6 A curve ρ in $[\mathbb{C}_n, \leq]$ ($n \geq 3$) is called *simple* iff every n-cell in ρ is regular (i.e., has branching index 2). ρ is a *simple arc* iff either (1) every n-cell in ρ has branching index 2 except for two endnodes that have branching index 1 or (2) ρ is a simple curve and one of its n-cells is labeled as an endnode. A one-dimensional union of a finite number of simple arcs is called an *elementary curve*.

7.3 Curves in Adjacency Grids

Elementary curves differ from curves, because some of their principal nodes can be explicitly labeled as being endnodes of arcs. A simple curve (arc) has a circuit (sequence) of principal nodes. The number of principal nodes in the circuit (sequence) defines the length of the curve (arc). The bottom row of Figure 7.7 shows that the shortest simple 1-curve in $\mathbb{C}_2$ has length 8 and that the shortest simple 0-curve in $\mathbb{C}_2$ has length 4.

Proposition 7.2 A curve ρ in $[\mathbb{C}_2, \leq]$ is a simple 1-curve iff $\bigcup \rho$ is homeomorphic to an annulus; it is a simple 0-curve iff $\bigcup \rho$ is homotopic to a circle.

The union of the cells in an elementary plane curve ρ is a (not necessarily connected) isothetic polygon $\bigcup \rho$ that has a frontier $\gamma = \vartheta(\bigcup \rho)$ that is an elementary isothetic curve. If ρ is a simple 1-curve, this frontier splits into two nonempty simple curves γ_1 and γ_2. We recall that $\circ$ denotes the topologic interior (see Section 3.1.7). Let θ_0 be the side length of the large squares in the geometric representation of $\mathbb{C}_2$ (θ_0 is "slightly smaller" than the grid constant θ in our topologic representation and is equal to θ in the original grid cell representation).

Proposition 7.3 Let ρ be a simple 1-curve in $[\mathbb{C}_2, \leq]$. The frontier $\vartheta(\bigcup \rho)$ is the union of two nonempty simple curves γ_1 and γ_2 that are the frontiers of simple isothetic polygons P_1 and P_2 such that $P_1 \subset P_2^\circ$. Moreover, the Hausdorff distance $L_\infty(\gamma_1, \gamma_2)$ is θ_0.

7.3 Curves in Adjacency Grids

A curve can be regarded as a sequence of pixels or voxels in an adjacency grid (in either the grid point or grid cell model). Figure 7.8 shows examples of simple $(n-1)$-curves and $(n-hbox1)$-arcs in 2D and 3D. Note that this simple representation does not uniquely characterize the curve. For example, the 2D object on the left in Figure 7.9 may be either a simple 8-curve or a nonsimple 0-curve, and this is the same for the 3D object on the right. Note that, in the incidence grid, the ambiguities in these figures can be removed by deleting a marginal cell (the 0-cell at which the object "touches itself").

7.3.1 Euler characteristics of curves

The Euler characteristic of a simple curve in the Euclidean plane is 0; see Section 7.1.4. The Euler characteristic of a simple 0- or 1-curve can be calculated by considering the curve in the incidence grid (see Definition 5.14) as a 2D Euclidean complex. In

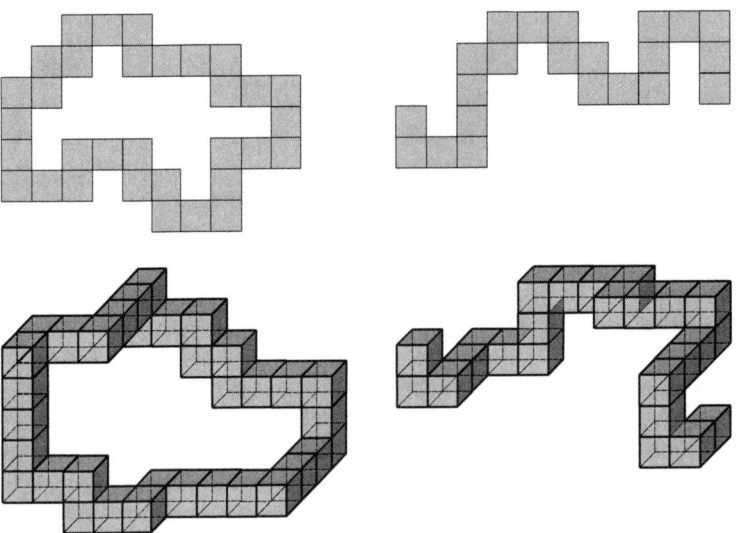

FIGURE 7.8 Simple $(n-1)$-curves (left) and simple $(n-1)$-arcs (right) for $n=2$ in the upper row and $n=3$ in the lower row.

general, the Euler characteristic of a region M is calculated by decomposing $\bigcup M$ into a Euclidean complex and applying Definition 5.14. For example, the simple 1-curve ρ shown on the upper left in Figure 7.8 has $\alpha_0 = 64$ grid vertices, $\alpha_1 = 96$ grid edges, and $\alpha_2 = 32$ grid squares so that its Euler characteristic is $\chi = \alpha_0 - \alpha_1 + \alpha_2 = 0$.

The dual interpretation of a set of 2-cells in the grid point model leads in general to a 2D Euclidean complex. In our 1-curve example, however, we obtain only a one-dimensional Euclidean complex that has $\alpha_0 = 32$ 0-cells (centers of grid squares) connected by $\alpha_1 = 32$ 1-cells, so we get the following:

$$\chi(\rho) = \alpha_0 - \alpha_1 = 32 - 32 = 0$$

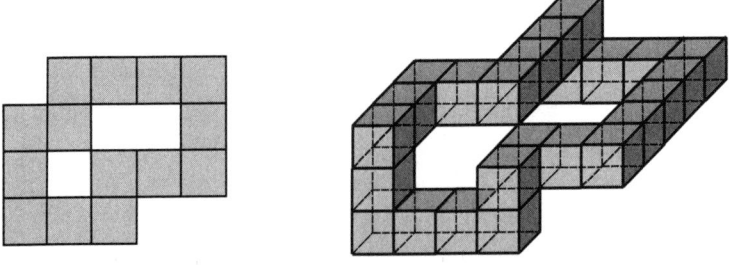

FIGURE 7.9 Left: the 2D object is either a simple 1-curve or a nonsimple 0-curve. Right: a similar case in 3D.

7.3 Curves in Adjacency Grids

The interpretation of ρ in the grid point model can also be regarded as a planar-oriented adjacency subgraph G of $\mathbb{Z}^2$. G has $\alpha_0 = 32$, $\alpha_1 = 32$, and $\alpha_2 = 2$; there are no atomic cycles in this example, but we must count one inner and one outer border cycle. However, the Euler characteristic of G is *not* the quantity we are interested in for characterizing ρ. For our 1-curve example, we have the following:

$$\chi^+(G) = \alpha_0 - \alpha_1 + \alpha_2 = 32 - 32 + 2 = 2$$

The Euler characteristic χ^+ of a planar-oriented adjacency graph is always 2 (see Theorem 4.6). The curve ρ can also be interpreted as a 2-strong planar-oriented adjacency subgraph in the frontier grid (see Section 5.4.3) of grid vertices, with 32 atomic cycles and two border cycles (i.e., $\alpha_2 = 34$) and with $\alpha_0 = 64$ and $\alpha_1 = 96$, which also leads to $\chi^+ = 2$. This is a graph-theoretic characterization only; it does not make distinctions between faces.

7.3.2 Simple 2D curves

Definition 7.7 An α-path is a *simple α-curve* iff **(C1)** its length n is greater than a threshold t_α; **(C2)** it consists of $n+1$ distinct pixels $p_0, p_1, \ldots, p_n$; and **(C3)** p_i is α-adjacent to p_k iff $i = k \pm 1$ (modulo $n+1$). A *simple α-arc* is an α-connected proper subset of a simple α-curve.

In 2D, we have $t_4 = 4$ and $t_8 = 3$. In fact, a simple 4-curve must have a length of at least 8.

A pixel $p = (i,j)$ is *inside* a simple 4-curve ρ iff p is not on ρ and the grid lines i and j cross ρ an odd number of times on each side of p. For an illustration of this property in the grid point model, see Figure 7.10 (left). Evidently, the grid line crosses the curve to the left of p. On the right of p, the grid line only "touches" the curve, because it "comes in from" and "goes out to" the same halfplane; however, further to the right, it crosses the curve, because it "comes in from" and "goes out to" different halfplanes. (We will not give a formal definition.) p is *outside* ρ if it is neither on nor inside ρ. The inside and outside of a simple 8-curve are defined analogously. Note that the set of pixels inside a simple 4-curve may not be 4-connected; see Figure 7.10 (left).

Let $G(\rho)$ be the set of all pixels on a path ρ.[3] Two 4- or 8-paths ρ_1 and ρ_2 *intersect* iff $G(\rho_1) \cap G(\rho_2) \neq \emptyset$. Figure 7.11 shows that two 8-paths can "cross" without intersecting. Let p and q be the pixels that are not in the set S. We say that S *8-separates* (*4-separates*) p and q iff any 8-path (4-path) from p to q intersects S. In particular, this definition can be applied to the set $S = G(\rho)$ of pixels on a path ρ.

Theorem 7.6 A simple 4-curve (8-curve) ρ 8-separates (4-separates) all pixels inside ρ from all pixels outside ρ.

---

3. This notation is consistent with that for Gauss digitization.

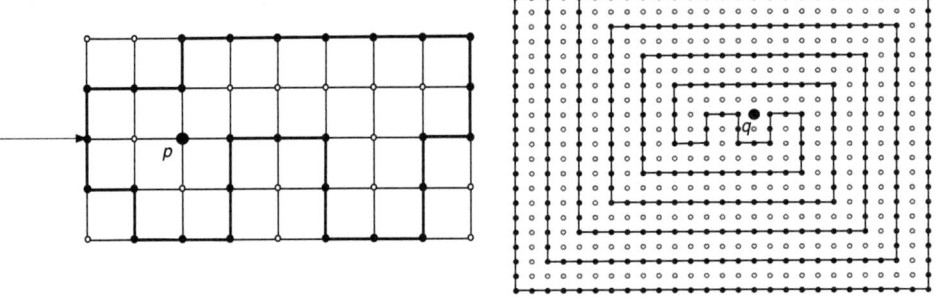

FIGURE 7.10 Left: a pixel p inside a simple 4-curve. Right: a simple 4-curve for which decisions about inside and outside are more difficult (e.g., is q inside or outside?).

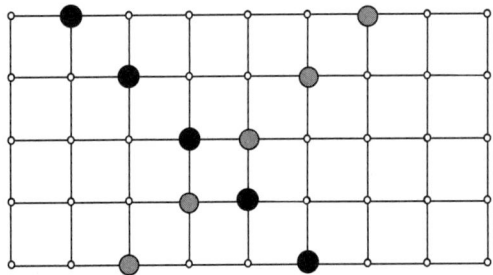

FIGURE 7.11 Two nonintersecting 8-paths.

Proof We prove the theorem for simple 4-curves and 8-separation; the case of simple 8-curves and 4-separation can be treated analogously. For simplicity, in this proof we identify an (ordered) path ρ with its (unordered) pixel set $G(\rho)$, and we denote the complement of $G(\rho)$ with $\overline{\rho}$.

We first prove that, for any simple 4-curve ρ, there is at least one pixel inside and at least one pixel outside ρ. Because ρ is finite, there are infinitely many pixels outside ρ. To see that the inside of 8 is nonempty, let $p_m = (x,y)$ be the uppermost of the rightmost pixels on ρ; then $(x-1,y)$ and $(x,y-1)$ must both be on ρ, because they are the only possibilities for p_{m-1} and p_{m+1}. If $(x-1,y-1)$ were on ρ, according to **(C3)** in Definition 7.7, it would have to be both p_{m-2} and p_{m+2} so that by **(C2)** we would have $m-2 = m+2$ (modulo $n+1$), which is impossible, because $n \geq 4$; hence $(x-1,y-1)$ is in $\overline{\rho}$. It follows that $(x,y-2)$ is on ρ. Indeed, if $(x-1,y)$ is p_{m-1}, it is the only possibility for p_{m-2}, whereas, if $(x-1,y)$ is p_{m+1}, it must be p_{m+2}. Let $H_p = \{(x+i,y) : i \in \mathbb{N}\}$ be the horizontal digital ray that emanates from pixel $p = (x,y)$ to the right. We have just proved that $H_{(x-1,y-1)}$ crosses ρ exactly once, namely at $(x,y-1)$; hence $(x-1,y-1)$ is inside ρ.

7.3 Curves in Adjacency Grids

Let p_0 be inside ρ and p_m outside ρ. Let $\varphi = (p_0, p_1, \ldots, p_m)$ be an 8-path, and suppose that every p_j is in $\overline{\rho}$. Because p_0 is inside ρ and p_m is outside ρ, there must exist some i ($0 < i \leq m$) such that p_{i-1} is inside and p_i is outside. If p_{i-1} and p_i are horizontally adjacent, this is impossible by definition of "inside" and "outside". On the other hand, if they are vertically or diagonally adjacent, H_{i-1} and H_i (where we have omitted the ps for brevity) are vertically adjacent rays. Let $R = \{q_{k+1}, \ldots, q_{k+r}\}$ be a run (maximal consecutive sequence) of pixels in which ρ intersects $H_{i-1} \cup H_i$; thus each of q_k and q_{k+r+1} is either above or below both Hs. (Because p_{i-1} and p_i are 8-adjacent, a run of 4-neighbors cannot leave $H_{i-1} \cup H_i$ but rather remain on the union of the rows that contain them; the run can leave only by passing to the row above or below.) Suppose they are both above; if they are both below, we can use exactly analogous arguments. We can assume, without loss of generality, that H_{i-1} is the upper of the two Hs. If all of $q_{k+1}, \ldots, q_{k+r}$ are on H_{i-1}, then ρ touches H_{i-1} in R and does not even touch H_i in R, so it crosses neither of the Hs in any subset of R. On the other hand, if some of these qs are on H_i, let q_s be the first one and q_t the last one. Then H_{i-1} crosses ρ in $\{q_{k+1}, \ldots, q_{s-1}\}$ and H_i crosses it in $\{q_{v+1}, \ldots, q_{k+1}\}$, but neither of them can cross it anywhere between q_u and q_v by the same argument as given in the preceding paragraph. Thus, in this case, both H_{i-1} and H_i cross ρ just once in subsets of R.

We have thus shown that, in any case, the difference between the number of times that H_{i-1} and H_i cross ρ in subsets of R is even. Because this is true for every R, it follows that the difference between the total number of times that H_{i-1} and H_i cross ρ is even; in other words, p_{i-1} and p_i are either both inside or both outside of ρ, which is a contradiction. ∎

Theorem 7.6 is a separation theorem for (4,8)- or (8,4)-adjacency grids that resembles the Jordan-Veblen curve theorem of Euclidean topology. It can also be applied to simple 0- or 1-curves in the 2D incidence grid; see Figure 7.12. (The 1-curve is open and separates two closed regions; the 0-curve is neither open nor closed, and its closure separates two open regions.)

Let $M \subseteq \mathbb{G}_{m,n}$. The background component of M is unbounded; it consists of $\mathbb{Z}^2 \setminus \mathbb{G}_{m,n}$ and all complementary 4-components of M (i.e., 4-components of $\overline{M}$) in $\mathbb{G}_{m,n}$ that are 4-adjacent to $\mathbb{Z}^2 \setminus \mathbb{G}_{m,n}$. Any remaining finite 4-components of $\overline{M}$ are (proper or improper) 4-holes in M and are separated from the background component by the outer or (one or more) inner border cycles.

Definition 7.8 An α-region M in the 2D (3D) incidence grid is *simply α-connected* iff the geometric representation of M in the incidence grid is simply connected in $\mathbb{E}^2$ ($\mathbb{E}^3$).

A more informal definition was given in Section 1.2.6. See Section 6.3 for examples of geometric representations in incidence grids. We also apply this definition in the grid point model. For example, a finite 4-connected set M without proper

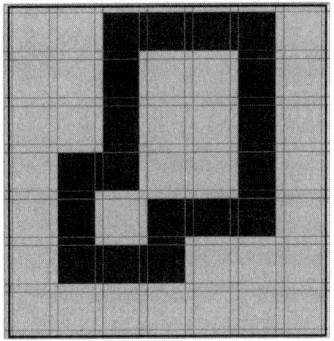 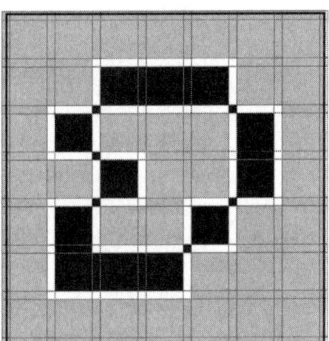

FIGURE 7.12 Left: a simple 1-curve that separates two closed regions. Right: a simple 0-curve with a closure that separates two open regions.

4-holes is simply 4-connected. An 8-hole of M is a finite 8-component of $\overline{M}$ and may consist of several proper 4-holes of M. An 8-hole of M cannot contain an improper 4-hole of M. (This discussion can be generalized to α-components and α-holes for any adjacency relation α.)

> **Proposition 7.4** A simple 4-curve has exactly one 8-hole and a simple 8-curve has exactly one 4-hole. A simple 4-curve 8-separates its 8-hole from the background, and a simple 8-curve 4-separates its 4-hole from the background.

7.3.3 Good pairs for 2D binary pictures

Theorem 7.6 can be generalized from (4,8)- and (8,4)-adjacency to "good pairs" of adjacencies:

> **Definition 7.9** (β_1, β_2) is called a *good pair*[4] in the 2D grid iff (for (i, k) in $\{(1, 2), (2, 1)\}$) any simple β_i-curve β_k-separates its (at least one) β_k-holes from the background and any totally β_i-disconnected set cannot β_k-separate any β_k-hole from the background.

As Figure 1.9 shows, (4,4) and (8,8) are not good pairs. It is not hard to see that (6,6) is a good pair. (See the discussion of the metric d_h in Section 3.2.3; two pixels are called 6-adjacent if they are at d_h-distance 1 from each other.)

For any adjacency relation A on $\mathbb{Z}^2$, a subset $M \subseteq \mathbb{G}_{m,n}$ defines a partition $\mathcal{R}$ of $\mathbb{Z}^2$ into one infinite background component, finite components, and complementary components (holes), which are all regions. Let these components be $M_1, M_2, \ldots$, and define $M_i \mathcal{A} M_j$ iff $A(M_i) \cap M_j \neq \emptyset$; this adjacency relation on $\mathcal{R}$ defines the region adjacency graph $[\mathcal{R}, \mathcal{A}]$ of M. (Note that, if $M_i \mathcal{A} M_j$ and M_i is a component, M_j

---

4. T.Y. Kong proposed this term in a talk in 2001 [569]. It is used in another sense in topology.

7.3 Curves in Adjacency Grids

must be a complementary component.) This graph need not be a tree; see Figure 4.9. However, we have the following:

Theorem 7.7 If A is generated by a good pair (α, β), the region adjacency graph $[\mathcal{R}, \mathcal{A}]$ of any M is a tree.

Proof We must show that $[\mathcal{R}, \mathcal{A}]$ does not contain a cycle. Let W be a component of M and U and V be two components of $\overline{M}$ that are $\mathcal{A}$-adjacent to W. We must show that any β-path from U to V must meet W; otherwise the regions encountered by the path, together with W, would constitute a cycle. (Similarly, if W is a component of $\overline{M}$ and U and V are components of M that are $\mathcal{A}$-adjacent to W, we must show that any α-path from U to V must meet W; the proof in this case is analogous.)

Suppose there is a β-path in $\mathbb{Z}^2$ from U to V that does not meet W; then U and V are in the same β-component D of $\overline{W}$, and an inner border cycle of D separates W from $\overline{W}$. Because U and V are $\mathcal{A}$-adjacent to W, there are undirected edges assigned to this border cycle that are incident with pixels in U and in V. The set of pixels in D, incident with these undirected edges, is β-connected. It follows that U and V must be in two different β-components of $\overline{M}$. ∎

A good pair (α, β) and a subset M of $\mathbb{G}_{m,n}$ define (via deletion of all invalid grid edges) an adjacency structure $G^M_{\alpha,\beta} = [S, A]$ such that the following are true:

1. $S = \mathbb{Z}^2$ is the set of all nodes in $G^M_{\alpha,\beta}$.

2. For all nodes p in $\mathbb{Z}^2 \setminus \mathbb{G}_{m,n}$ and all β-components of $\overline{M} = \mathbb{G}_{m,n} \setminus M$, we have $q \in A(p)$ iff $q \in A_\beta(p)$ and $q \notin M$ or $q \in A_\beta(p), p \in A_\alpha(q)$, and $q \in M$.

3. For all nodes $p \in M$, we have $q \in A(p)$ iff $q \in A_\alpha(p)$ and $q \in M$ or $q \in A_\alpha(p)$, $p \in A_\beta(q)$, and $q \notin M$.

Figure 4.15 shows examples of a good pair (4,8) on the left and a good pair (8,4) on the right. The figure shows on the left all border cycles of M in $G^M_{4,8}$ and on the right all border cycles of M in $G^M_{8,4}$. The undirected edges assigned to any border cycle of M are the same for (4,8) and (8,4); they are the invalid edges defined by M.

Using the good pair (4,8) or (8,4) is equivalent to regarding 4-components of white pixels as open regions and 8-components of black pixels as closed regions in the incidence grid (or vice versa); see Figure 5.20.

Let P be a binary picture defined on a grid $\mathbb{G}_{m,n}$. Because (4,8) and (8,4) are good pairs, this suggests using 4-connectedness for $\langle P \rangle = P^{-1}(1)$ and 8-connectedness for $\overline{\langle P \rangle} = P^{-1}(0)$, or vice versa. If C is a 4-component of $\langle P \rangle$ and D is an 8-component of $\overline{\langle P \rangle}$ (or vice versa), it can be shown that C and D are 4-adjacent iff they are 8-adjacent (see Exercise 8 in Section 7.6). The set of pixels of C that are adjacent to pixels of D is called the *D-border* of C; the *C-border* of D is defined analogously. The analysis of binary pictures can be based on a good-pairs approach if the

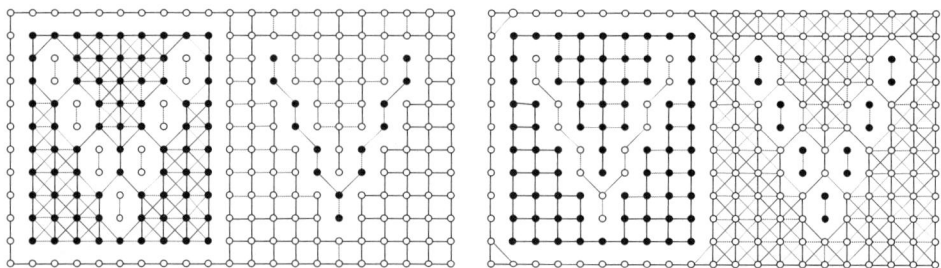

FIGURE 7.13 Left: good pair (8,4). Right: good pair (4,8). This example is discussed in [591]. In both cases, the 8-adjacencies result in "cuts" in the "V" shape.

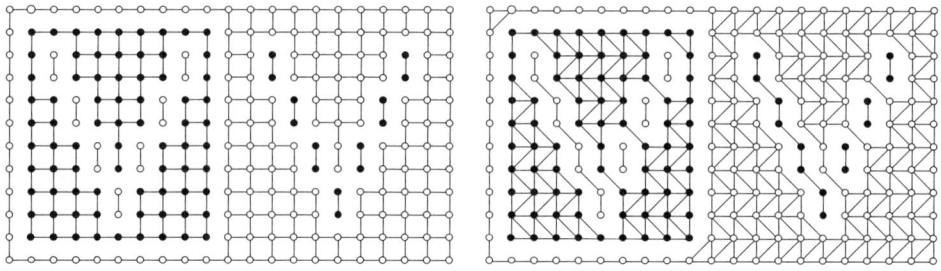

FIGURE 7.14 Pair (4,4) and good pair (6,6); there are "cuts" (as in Figure 7.13), even for (6,6).

assumption of different adjacencies for white and black pixels is acceptable. For example, M in Figure 4.15 has either six or three components, depending on whether we use good pair (4,8) or (8,4). However, there may be situations in which it is questionable to make the commitment that "black is object" and "white is background or hole" (or vice versa). For example, consider Figure 7.13, which does not show invalid adjacencies between pixels in different P-equivalence classes. The left half of Figure 7.13 shows a "hole V," and the right half shows an "object V." Because of the symmetry of this example, we can only guess which is the "object" and which is the "background or hole." There is only one connected "V" whether we use good pair (8,4) or good pair (4,8). Note that the second "V" is "cut" by 8-adjacent pixels. As Figure 7.14 shows, disconnection can occur even if we use 6-adjacency. (6-adjacency also introduces a systematic directional bias.)

7.3.4 s-Adjacencies in 2D multivalued pictures

The concepts of good pairs or of open and closed sets cannot be extended to multivalued pictures P; there is no "consistent" way to define adjacencies for $P^{-1}(u)$

7.3 Curves in Adjacency Grids

$(0 \leq u \leq G_{max})$ if $G_{max} > 1$. Figure 5.21 shows how different total orders of the pixel values influence the topology of a picture. The picture analysis context may have an impact on which order to use. However, in any case, we need "topologically sound" adjacencies in which connectedness and separation are incompatible. This is especially important at higher levels of picture analysis, where we often deal with inhomogeneous grids (e.g., picture subsets, object surfaces, objects); see Figure 4.3 for examples.

Switch-adjacency A_s (s-adjacency for short; see Section 2.1.3) is a general method of defining planar adjacency graphs for multilevel 2D pictures. It was generalized in Section 5.4.2 to ordered adjacency, which also applies to 3D pictures. The graph $[\mathbb{Z}^2, A_s]$ is in general an irregular planar graph. If we assume clockwise local circular orders at all grid points, it is an oriented adjacency graph. The s-adjacency graph is always planar because, in each 2×2 block of pixels, there is only one diagonal adjacency. The same is true for the adjacencies defined by the good pair (6,6) and for the adjacency in the product of two alternating topologies (see Figure 6.6).

A flip-flop case in s-adjacency is analogous to a situation in which two planar curves intersect at a point and the assignment of the intersection point to one of the curves determines how the curves subdivide the plane. Fortunately, such cases are unlikely; see Figures 2.8 and 7.15.

Figure 7.15 shows that it is possible to define s-adjacencies so that both "V" shapes remain connected. Figure 7.16 shows the graph of s-adjacencies for the switch state matrix shown on the right in Figure 7.15.

A switch state matrix **S** must be available when an operation such as border tracing (see Algorithm 4.3) or thinning (see Section 16.3) is performed to ensure that the adjacency graph of the picture is planar. It is important that once a switch state has been set it not be changed during a topologic operation on the picture. On the other hand, any picture processing operation that changes the pixel values may create a need to change **S**.

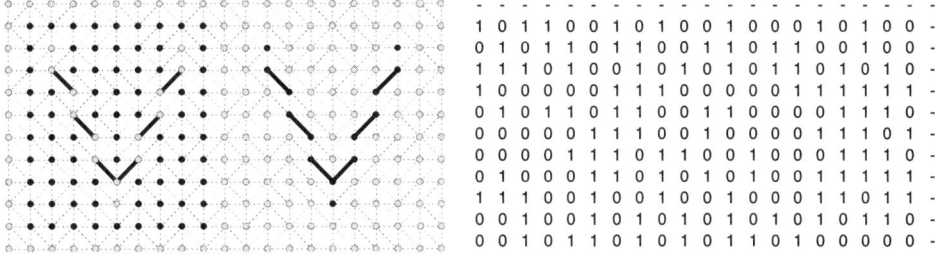

FIGURE 7.15 The states of switches are uniquely defined or can be chosen arbitrarily in most cases. Only in a few flip-flop cases (12 in this example) are the switches set using the templates shown in Figure 7.16. The binary matrix **S** on the right represents the states of all of the switches.

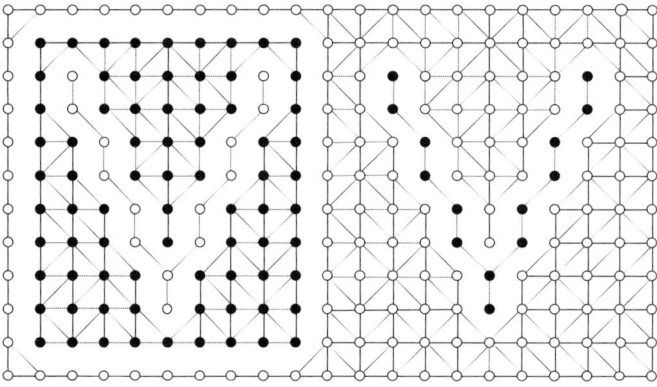

FIGURE 7.16 s-Adjacencies for the switch state matrix shown on the right in Figure 7.15.

In terms of s-adjacency, we can define s-curves, s-separation (using s-paths), and region (s)-adjacency graphs. Because an s-adjacency graph is planar, Theorems 7.6 and 7.7 are valid when we use s-adjacency:

Theorem 7.8 Any simple s-curve ρ has exactly one s-hole and s-separates this hole (the set of pixels inside ρ) from the background (the set of pixels outside ρ).

Theorem 7.9 The region adjacency graph $[\mathcal{R}, \mathcal{A}]$ defined by a nonempty subset M of an oriented s-adjacency graph is always a tree.

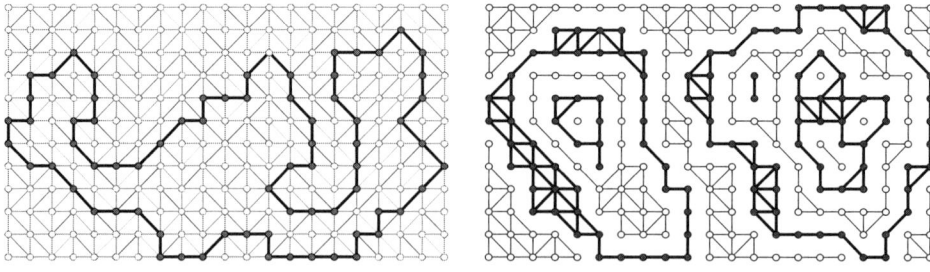

FIGURE 7.17 A simple s-curve (left). An s-adjacency graph (right) is partitioned into regions by the deletion of all undirected edges assigned to border cycles (filled nodes and bold adjacency edges); the resulting region adjacency graph is a tree.

The proofs of these theorems are the same as the proofs of the corresponding theorems for the good pair (6,6), which defines a special case of s-adjacency. Figure 7.17 shows (on the left) a simple s-curve and (on the right) a partition into regions defined by a subset M.

7.4 Surfaces in the Euclidean Topology

7.4.1 Manifolds

A topologic space S is called *locally compact* iff every $p \in S$ has a topologic neighborhood with a closure that is compact.

Definition 7.10 A topologic space S is called an *n-manifold* iff it is locally compact, has a countable basis, and every $p \in S$ has a topologic neighborhood (in S) that is homeomorphic to an *open n-sphere*.

Evidently, $\mathbb{E}^n$ is an n-manifold and so is an open n-sphere in $\mathbb{E}^n$; thus an n-manifold can be bounded or unbounded. A bounded 1-manifold is either a simple (Urysohn-Menger) curve or a simple arc without its endpoints and is homeomorphic to an open line segment. In general, a bounded manifold has an "open frontier." (We recall that the empty set is also an open set.) If $M \subseteq \mathbb{E}^3$ and $U_\varepsilon(p) \cap M$ is homeomorphic to three open semicircular areas (see Figure 7.18), p is called a *bifurcation point* of M. A 2- or 3-manifold cannot have a bifurcation point.

An n-manifold is called *hole-free* iff it is compact (i.e., bounded and closed).[5] For example, the surface of a sphere is a hole-free 2-manifold, and so is the surface of a torus.

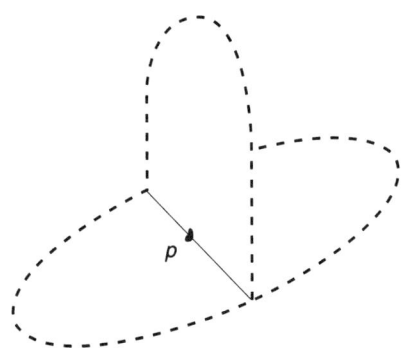

FIGURE 7.18 A bifurcation point.

---

5. We use the term "hole-free manifold" instead of "closed manifold" to avoid confusion when we discuss closed complexes defined by tilings of manifolds.

Theorem 7.10 (I. Gawehn, 1927) Any hole-free 2-manifold is homeomorphic to a polyhedral surface.

This allows us to study triangulations of hole-free 2-manifolds. Let Φ be a homeomorphism of such a 2-manifold M onto a polyhedral surface. Let Z be a triangulation of $\Phi(M)$ so that $\Phi(M)$ is the union of all of the triangles $T \in Z$. The sets $\Phi^{-1}(T)$ define a *triangulation of the 2-manifold* M that consists of curvilinear triangles, their sides (which are simple arcs), and their vertices (points). This allows us to discuss geometric (particularly simplicial) complexes on curved surfaces.

7.4.2 Surfaces

A *Jordan surface* (C. Jordan, 1887) is defined by a parameterization that establishes a homeomorphism with the surface of a unit sphere. In picture analysis, we are usually not interested in parameterizations of surfaces. (See Chapter 8 about the geometry of Jordan surfaces.) Instead, we use a topologic definition of a surface that can be compared with the topologic definition of a Urysohn-Menger curve (see Section 7.1.2):

Definition 7.11 A hole-free 2-manifold is called a *hole-free surface*. Let S be homeomorphic to a polyhedral surface, and let S be partitioned into two nonempty subsets S° and ϑS in which every $P \in S^\circ$ has a topologic neighborhood in S that is homeomorphic to an open disc and every p in ϑS has a topologic neighborhood in S that is homeomorphic to the union of the interior of a triangle and one of its sides (without endpoints) such that p is mapped onto the side. Then S is called a *surface with frontiers*. The points of S° are called *interior points* of S, and the points of ϑS are called *frontier points* of S.

Evidently, $S = S^\bullet = S^\circ \cup \vartheta S$, where $S^\bullet$ denotes the closure of S. If $\vartheta S = \emptyset$, S is called a hole-free surface. A *surface* is either a hole-free surface or a surface with frontiers. A *simple hole-free surface* is a hole-free surface that is homeomorphic to the surface of a sphere (i.e., it is a Jordan surface).

The frontier ϑS of a surface S with frontiers is an elementary curve that is a union of pairwise disjoint simple curves. Figure 7.19 shows two examples: the surface of a sphere with a few circular areas removed and the surface of a torus with one circular area removed (a *handle*). A *simple surface with r contours* is a surface with frontiers and is homeomorphic to the surface of a sphere with r holes that has frontiers that are pairwise disjoint.

In what follows, we assume that any surface with frontiers can be defined by a triangulation (i.e., it is a union of finitely many points, simple arcs, and curvilinear triangles). More generally, we say that a finite or infinite connected graph drawn on a surface S with frontiers defines a *tiling* of S iff every edge of the graph is on a circuit (encircling a face), there is a vertex at any point at which edges intersect, and ϑS is contained in the union of all of the edges and vertices. (A planar graph was

FIGURE 7.19 The surface of a sphere without a few circular areas (left). The surface of a torus without one circular area (right); this surface is called a *handle*.

defined based on a drawing on a planar surface. Similarly, a *toroidal graph* could be defined based on a drawing on a toroidal surface, and so on.) We identify a tiling with its set of faces (2-cells), edges (1-cells), and vertices (0-cells). The side-of relation defines a partial ordering on the set of 0-, 1-, and 2-cells of a tiling. Thus, it follows:

Corollary 7.1 A tiling of a surface and the side-of relation between its cells define a Euclidean complex and its poset topology.

Triangulations are examples of tilings. A tiling is called *regular* iff every face of the tiling is an *n-gon* (a simple polygon that has $n \geq 1$ vertices on its frontier) and every vertex is incident with exactly $k \geq 1$ faces.

Example 7.1 In accordance with Theorem 7.10, a Jordan surface is homeomorphic to a polyhedron. Suppose the faces and vertices of this polyhedron define a finite regular tiling. By the Descartes-Euler polyhedron theorem, we have $\alpha_0 - \alpha_1 + \alpha_2 = 2$. For a regular tiling, we have $2\alpha_1 = k\alpha_0$ (because each vertex is incident with k edges and an edge is defined by two vertices) and $2\alpha_1 = n\alpha_2$ (because each edge is incident with two faces and each face has n edges). Hence we have the following:

$$\frac{1}{k} + \frac{1}{n} = \frac{2+\alpha_1}{2\alpha_1} = \frac{1}{2} + \frac{1}{\alpha_1}$$

The only solutions to this Diophantine equation are given by the five regular polyhedra. In particular, there are no solutions for $\alpha_1 > 30$; thus no regular finite tiling on a Jordan surface can be of interest for defining a picture grid. There are triangulations ($n = 3$) of a Jordan surface that have $\alpha_1 > 30$, but the degrees of the vertices of these triangulations cannot have a constant value. Interestingly, there are finite regular tilings on the surface of a torus (which is a hole-free

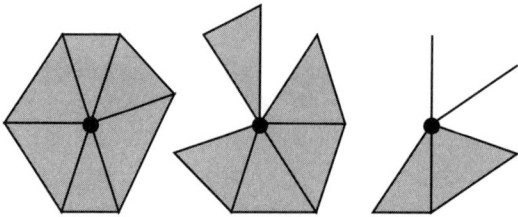

FIGURE 7.20   Left: a cyclic vertex. Middle and right: acyclic vertices.

2-manifold) in which the size of α_1 is not limited. Thus nontrivial picture grids can be defined on the surface of a torus but not on the surface of a sphere.

Let Z be a triangulation of a hole-free surface or of a surface with frontiers. The relation $\leq$ defined by the side-of relationship defines a poset that has an Aleksandrov topology. Let $U_Z(t)$ denote the smallest neighborhood of t in Z. For example, the smallest neighborhood $U_Z(p)$ of a vertex p in Z contains p and may also contain curvilinear triangles pr_1r_2 and simple arcs pq. The set of vertices q of these arcs and the set of arcs r_1r_2 of these triangles define a subcomplex $\vartheta_Z(p)$ called the *frontier* of $U_Z(p)$. $U_Z(p)$ is called *cyclic* iff the union of the simple arcs on its frontier $\vartheta_Z(p)$ is a simple curve; otherwise it is called *acyclic*. Figure 7.20 shows examples. If $U_Z(p)$ is cyclic, it is homeomorphic to a closed disk.

P. S. Aleksandrov (1956) proved several theorems about surface triangulations. We cite three of them:

(i) Z is a triangulation of a hole-free surface iff Z is connected and $U_Z(p)$ is cyclic for all vertices p of Z.

(ii) Every simple arc in a triangulation Z of a hole-free surface is a side of exactly two triangles in Z.

(iii) Any triangulation of a surface is strongly connected.

(For more about the term "strongly connected," see the end of Section 6.4.3.) Note that (i) gives a local criterion for testing a global property of a triangulation. It can be used, for example, in a marching cubes algorithm to test whether the isosurface generated by the algorithm is without frontiers (hole-free). According to (ii), "modulo 2" homology theory (see the definition of a chain frontier in Section 6.4.5) can be applied to triangulations of surfaces.

7.4.3   Orientable surfaces

A simple arc with endpoints p and q will be denoted by $[p, q]$ and the corresponding open arc by (p, q). $[p, q]$ is homeomorphic to the closed interval $[0,1]$. The *direction* of

7.4 Surfaces in the Euclidean Topology

$[p,q]$ is defined by a homeomorphism $\Phi : [0,1] \to [p,q]$ such that $\Phi(0) = p$ and $\Phi(1) = q$. Let the following be true where $r_1, r_2 \in [p,q]$:

$$r_1 \triangleleft r_2 \quad \text{iff} \quad r_1 = \Phi(x_1) \wedge r_2 = \Phi(x_2) \wedge x_1 < x_2$$

The relation $\triangleleft$ defines an order on $[p,q]$; it can be shown that this order is independent of the homeomorphism Φ. Analogously, we can define a direction on a simple curve such as the frontier of a triangle. A direction on a simple curve induces a direction on the simple arcs contained in the curve.

An *oriented triangle* is a triangle with a direction on its frontier (e.g., clockwise, counterclockwise), that is called the *orientation* of the triangle. The orientation of a triangle induces orientations of its sides. Two triangles in a triangulation that have a common side are *coherently oriented* if they induce opposite orientations on their common side. These standard definitions in combinatorial topology are consistent with our definitions of oriented adjacency graphs in Section 4.3.

Definition 7.12 (P.S. Aleksandrov, 1956) A triangulation of a surface is *orientable* iff it is possible to orient all of the triangles in such a way that every two triangles that have a common side are coherently oriented; otherwise it is called *nonorientable*.

If Z is a strongly connected orientable triangulation, the orientation of any triangle of Z determines the orientation of all of Z; hence any such triangulation has exactly two orientations. (We had an analogous result for orientations on the regular 2D grid [see Figure 4.19]: Only options (A) and (F) could be used.)

Theorem 7.11 (P.S. Aleksandrov, 1956) If Z_1 and Z_2 are triangulations of the same surface, Z_1 is orientable iff Z_2 is orientable.

The fact that orientability does not depend on the triangulation allows us to define a surface as *orientable* iff it has an orientable triangulation. Theorem 7.11 implies that orientability of a surface is a topologic invariant.

Example 7.2 (J.B. Listing, 1861; A.F. Möbius, 1865) A famous example of a nonorientable surface is the *Listing band* originally described by J.B. Listing in 1861; see Figure 7.21. (This band is usually called the "Möbius band" after A.F. Möbius, who discovered it independently in 1865. An example of an isothetic Listing band is shown in Figure 7.22.) The frontier of the Listing band is a simple curve that is homeomorphic to a circle. We can obtain a hole-free surface that is not a Jordan surface by starting with the surface of a sphere from which some circular areas have been removed (Figure 7.19) and identifying the frontier of each circular area with the frontier of a Listing band.

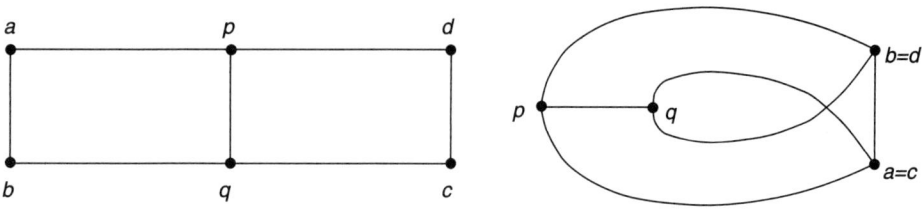

FIGURE 7.21 A Listing band as drawn by O. Veblen in 1922. The rectangle (left) is transformed into the Listing band (right) if a corresponds to c and b to d.

FIGURE 7.22 An isothetic Listing band [632].

This example illustrates the possible topologic complexity of surfaces.[6] We can also cover any of the holes with copies of a handle (a torus with a hole). This process of *gluing* a frontier of one surface to a frontier of another requires the two frontiers to be homeomorphic, and the result may also depend on the orientations of the frontiers. For example, we can glue the frontiers of two handles together and obtain a hole-free surface that is homeomorphic to the result of gluing two handles to the frontiers of a sphere with two holes.

7.4.4 The connectivity and genus of a surface

Let α_0, α_1, and α_2 be the numbers of points, simple arcs, and curvilinear triangles in a triangulation Z of a surface. The Euler characteristic of Z is $\chi(Z) = \alpha_0 - \alpha_1 + \alpha_2$.

---

6. Surfaces will be further discussed in Chapters 8 (surface geometry), 11 (planarity), and 12 (estimation of surface area and curvature).

7.4 Surfaces in the Euclidean Topology

Theorem 7.12 (P.S. Aleksandrov, 1956) Two triangulations of the same surface or of two homeomorphic surfaces have the same Euler characteristic.

This theorem allows us to speak about "the" Euler characteristic of a surface, which is a topologic invariant. It also follows that the Euler characteristic of any finite tiling of a surface (e.g., a regular tiling), where α_0, α_1, and α_2 are the numbers of vertices, edges and faces of the tiling, is equal to $\chi(S)$.

For example, each face of a cube can be triangulated into two triangles. This results in 8 vertices, 18 edges, and 12 triangles so that the Euler characteristic of the triangulation is 2. The surface of a sphere can be, for example, subdivided into four curvilinear triangles; this results in 4 vertices and 6 simple arcs so that the Euler characteristic is again 2. This also follows from Theorem 7.12, because the surfaces of the cube and the sphere are homeomorphic.

We say that a one-dimensional closed subcomplex M of a triangulation Z of a surface *does not separate* Z iff the open subcomplex $Z \setminus M$ is strongly connected. The *connectivity* $q(Z)$ of Z is the n for which Z has a closed one-dimensional subcomplex of connectivity n that does not separate Z.

We recall that the Poincaré formula (Equation 6.6) for S is as follows:

$$\chi(S) = \beta_0(S) - \beta_1(S) + \beta_2(S)$$

For a hole-free surface S, we have $\beta_0(S) = 1$ (because S is connected) and $\beta_2(S) = 1$ (because there exists a triangulation of S that forms a cycle; see (i) at the end of Section 7.4.2), and S is not contractible into a single point. Hence we have the following, where Z is any triangulation of S:

$$\chi(S) = 2 - \beta_1(S) = 2 - \beta_1(Z)$$

Z defines a graph G (a one-dimensional Euclidean complex), and we have $\chi(Z) = 1 + \chi(G) = 2 - \beta_1(G)$. $\beta_1(S)$ is called the *connectivity* of S. Finite triangulations or square tilings of homeomorphic surfaces have the same connectivities. We can therefore speak about "the" connectivity of a surface; it is a topologic invariant.

For example, the Euler characteristic of a sphere with r holes is $2 - r$. Indeed, we saw above that the sphere has Euler characteristic 2. Every deletion of a triangle from the triangulation such that its vertices and edges remain in the triangulation decreases the Euler characteristic by 1.

From the topology of complexes (see [639]), we know that the Betti number $\beta_1(S)$ of a hole-free orientable surface is even and is equal to twice the genus of the surface (see Section 1.2.6), and that, for a hole-free nonorientable surface, it is odd and equal to the genus plus 1. For example, if we start with a sphere with $r = 0$ holes and identify the frontier of each hole with the frontier of a Listing band, we obtain a hole-free nonorientable surface L_r of genus r; if we identify the frontier of each hole with the frontier of a handle, we obtain a hole-free orientable surface S_r of genus r. Any hole-free surface S has the same Euler characteristic (or genus) as one of the surfaces S_r (if S is orientable) or L_r (if it is nonorientable).

7.4.5 Separation theorems

The separation theorems for planar simple curves discussed in Section 7.1.5 were generalized around 1900, first to 3D space and then to n dimensions.

A Jordan curve is the homeomorphic image of a circle, and a Jordan surface is the homeomorphic image of the surface of a sphere. This was generalized by L.E.J. Brouwer in 1912 to the n-dimensional case. He defined a *Jordan manifold* in $\mathbb{R}^n$ ($n \geq 2$) as the homeomorphic image of a closed $(n-1)$-dimensional sphere and proved the following:

Theorem 7.13 (L.E.J. Brouwer, 1911) *A Jordan manifold separates $\mathbb{R}^n$ into two connected subsets and coincides with the frontier of each of these subsets.*[7]

In a paper published that same year, O. Veblen showed (without using a parameterization) that the surface of a simple bounded n-dimensional polyhedron decomposes the n-dimensional space (defined by axioms) into two connected subsets, that it is the frontier of each of the subsets, and that any polygonal arc that joins a point of one subset to a point of the other contains a point of the polyhedron.

7.5 Surfaces and Separations in 3D Grids

Surfaces in picture analysis are defined by sets of voxels. Surfaces in the grid can be characterized as 2D sets of voxels in a 3D adjacency grid or as 2D Euclidean complexes of marginal cells (2-, 1-, and 0-cells given by the frontier of a region). In the first case, we assume the grid point model.

7.5.1 Surfaces in the grid point model

p is called an (α, α') *surface voxel* of $S \subseteq \mathbb{Z}^3$, where $(\alpha, \alpha') = (26,6)$ (the (6,26) case can be treated analogously), iff the following are true:

a) $A_{26}(p) \cap S$ has exactly one α-component that is α-adjacent to p.

b) $A_{26}(p) \cap \overline{S}$ has exactly two α'-components C_p, D_p that are α'-adjacent to p.

c) For all $q \in N_\alpha(p) \cap S$, $N'_\alpha(q)$ intersects both C_p and D_p.

An α-connected set $S \subseteq \mathbb{Z}^3$ is called an α-*surface* iff every $p \in S$ is an (α, α') surface voxel. Let $N(p)$ be a $5 \times 5 \times 5$ block of voxels centered at p. In [747], a surface voxel p is called *orientable* if $N(p) \cap \overline{S}$ has exactly two α'-components that are α'-adjacent

---

7. Interestingly, he proved the case $n = 3$ in a footnote and the general case in the remaining four pages of the article.

7.5 Surfaces and Separations in 3D Grids

to p, and an α-connected set S is called an α-surface if it consists entirely of orientable (α, α') surface voxels; however, it was shown in [841] and [840] that the assumption of orientability was unnecessary.

For $(\alpha, \alpha') = (26,6)$, we have the following:

Proposition 7.5 A 26-connected set $S \subseteq \mathbb{Z}^3$ is a *digital surface* iff each $p = (i, j, k) \in S$ has at most two 8-adjacent grid points in at least two of the sets $\{(y, z) : (x, y, z) \in S \land x = i\}$, $\{(x, z) : (x, y, z) \in S \land y = j\}$, or $\{(x, y) : (x, y, z) \in S \land z = k\}$; if it has two, they are not mutually 8-adjacent; and if, in one of the sets (e.g., $\{(x, y) : (x, y, z) \in S \land z = k\}$), p has more than two 8-adjacent grid points or two 8-adjacent grid points that are mutually 8-adjacent, then $(i, j, k-1)$ and $(i, j, k+1)$ are not in S.

$p = (i, j, k) \in S$ is called a *border point* of S iff it has only one 26-neighbor in $\{(x, y, z) \in S : x = i\}$, $\{(x, y, z) \in S : y = j\}$, or $\{(x, y, z) \in S : z = k\}$; it is called an *inner point* of S iff it is not a border point. A 26-surface is called *simple* iff it has no border points. A simple 26-surface can be unbounded, or it can be bounded and *hole-free*. A bounded digital surface with border points that are 26-connected is called a *digital surface patch*.

Theorem 7.14 If S is an α-surface, $\overline{S}$ has exactly two α'-components. If S is bounded, exactly one of these components is bounded.

If S is an α-surface, every voxel of S is adjacent to both components of $\overline{S}$. For α-curves in $\mathbb{Z}^2((\alpha, \alpha') = (4,8)$ or $(8,4))$, the converse is also true: if S is α-connected, $\overline{S}$ has two α'-components, and every pixel of S is adjacent to both of these components, then S is an α-curve (see Exercise 5 in Section 7.6); however, the analogous statement about α-surfaces is not true, because a surface can touch itself without affecting the connectedness of its complement.

In the example in Table 7.1, the 1s are 6-connected, the (nonbackground) 0s are 26-connected, and every 1 is adjacent to both 26-components of 0s (the central 1s in the third and fifth planes are adjacent to background 0s in the fourth plane), but the central 1 in the fourth plane is adjacent to four components of 0s in its 27-neighborhood.

7.5.2 Surfaces in the grid cell topology

We recall that a connected set in $[\mathbb{C}_n, \leq]$ is 2D if it contains at least one 2×2 block of cells (e.g., one 2-cell with two incident 1-cells and one 0-cell; see Proposition 6.2). In a discrete grid, we cannot ask for a one-dimensional separation of each point of a set from the rest of the set. A hole-free surface (e.g., the frontier of a simply connected 6-region) can be transformed into a region of a 2D incidence pseudograph

TABLE 7.1 Example of a set of 1s that is not a 6-surface (see text); the blanks are background 0s.

1st Plane	2nd Plane	3rd Plane	4th Plane	5th Plane	6th Plane	7th Plane
1 1 1	1 1 1	1 1 1	1 1 1	1 1 1	1 1 1	1 1 1
1 1 1 1	1 0 1 1	1 0 1 1	1 0 1	1 0 1 1	1 0 1 1	1 1 1 1
1 1 1 1 1	1 1 0 1 1	1 1 1 1 1	1 1 1 1 1	1 1 1 1 1	1 1 0 1 1	1 1 1 1 1
1 1 1 1	1 1 0 1	1 1 0 1	1 0 1	1 1 0 1	1 1 0 1	1 1 1 1
1 1 1	1 1 1	1 1 1	1 1 1	1 1 1	1 1 1	1 1 1

by removing a few of its cells. Note that we are using "6-simply connected region" and "simply connected 6-region" as equivalent. Cells in the frontier of this 2D region cannot be separated from the rest of the region by one-dimensional circuits of cells. An important feature of Definition 6.7 is that it allows a definition of "2D at cell c" without requiring the existence of such one-dimensional separations.

Any 6-region of voxels in the grid cell model is a finite union of simply connected 6-regions. Due to 18- or 26-adjacencies, there may be "touchings" between voxels of a 6-region. A 6-connected region corresponds with an open 2-region in the grid cell topology $[\mathbb{C}^3, I, \dim]$. A simply 6-connected region M is defined by a simply connected geometric representation M_g in the incidence grid; see Definition 7.8. We characterize the surface $\vartheta(M)$ of a simply connected 6-region by the frontier ∂M_g of the geometric representation M_g of the corresponding open 2-region.

Proposition 7.6 *The frontier ∂M_g of the geometric representation of a simply connected 6-region is homeomorphic to the surface of the unit sphere (and is therefore orientable).*

Proof The geometric representation of a single 3-cell is a cube; the surface of a cube is homeomorphic to the unit sphere. Any simply connected 6-region can be obtained by adding one 3-cell at a time in such a way that each added cell is 2-adjacent to (i) exactly one 3-cell, (ii) exactly two 1-adjacent 3-cells, (iii) exactly three pairwise 0-adjacent 3-cells, (iv) exactly four 3-cells, or (v) exactly five 3-cells of the previous (already 2-connected) 3-cells. Homeomorphy of the frontier ∂M_g of the updated geometric representation M_g of the open 2-region to the unit sphere is invariant with respect to these five operations. Note that in case (iii), all three 2-adjacent 3-cells are incident with exactly one 0-cell. Further options for 2-adjacencies with exactly two or three 3-cells would allow us to construct nonsimply connected sets and, for that reason, can be excluded from the list of possible merging operations. (Here we only need one method of merging; of course, [random] removals of simple voxels can produce final sets that consist of more than one nonsimple voxel. Operations (i) through (iii) may actually be sufficient to construct any 6-connected set.) ∎

7.5 Surfaces and Separations in 3D Grids

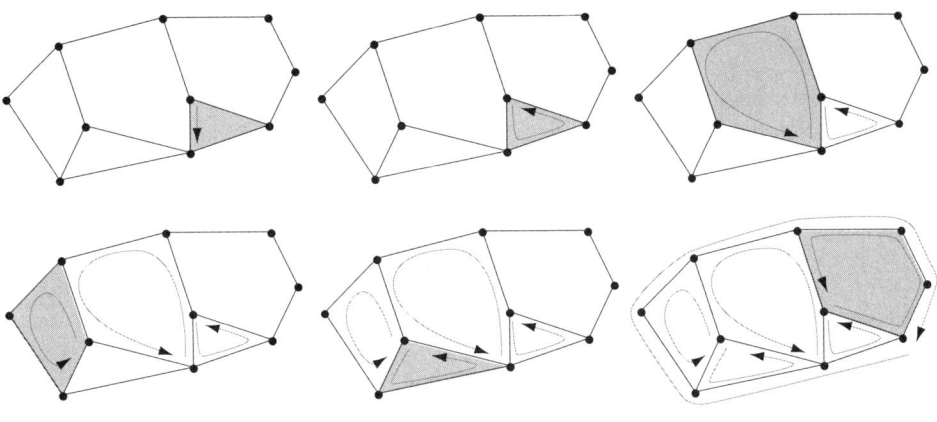

FIGURE 7.23 Propagation of orientations in an oriented planar 2-strong graph.

This proposition guarantees the existence of an orientation on ∂M_g that can be "copied" (from each face of a cube in ∂M_g to the corresponding face of the cube in ϑM) onto the surface $\vartheta(M)$. Moreover, an orientation on such a surface ∂M_g is easy to construct. Assume that the simple surface is mapped into a finite 3-strong planar graph G (see Steinitz's Theorem in Section 4.2.2). An orientation of all faces of G (including the unbounded face) can be defined by choosing an orientation for any edge in any face; this defines a cyclic orientation in the face. Any edge in any face that is edge-adjacent to an already oriented face is then oriented in the opposite direction, thereby defining an orientation in this new face (see Figure 7.23). There are only two possible orientations for the first edge; hence there are only two possible orientations for G. (For the construction of the orientation, it is actually sufficient that G be a planar 2-strong graph.)

Now consider the preimage of G on the surface ∂M_g, which defines a tiling of ∂M_g. We map the constructed orientation of G onto this tiling and obtain an orientable Euclidean complex; its faces are polygons, and their proper sides are their edges and vertices. There are only two possible orientations for this surface complex.

Now consider a purely n-dimensional simplicial complex that is $(n-1)$-connected (i.e., between any two n-simplexes c_1 and c_2 and there is a path of n-simplexes from c_1 to c_2 such that any two consecutive n-simplexes on the path share an $(n-1)$-simplex). We can define one of the two possible orientations for this simplicial complex by propagating orientations as in the planar case: choose an initial n-simplex c_0 and an initial edge $p_0 p_1$ in one of the 2-simplexes (triangles) that bound c_0, and choose one of the two possible orientations of this edge. This defines oriented triangles $p_0 p_1 p_2$, $p_1 p_3 p_2$, $p_2 p_3 p_0$, and $p_3 p_1 p_0$, which are the four 2-sides of a (now oriented) tetrahedron $p_0 p_1 p_2 p_3$. For $n > 3$, we would continue by orienting another 3-side of c_0 that is 2-adjacent to $p_0 p_1 p_2 p_3$ and that inherits its orientation by inverting the orientation on the joint 2-side. For $n = 3$, we have $c_0 = p_0 p_1 p_2 p_3$, and we propagate orientations to all of the 2-adjacent tetrahedra. For example, if

$p_0p_1p_2$ is shared with the 2-adjacent tetrahedron c_1, we have the orientation $p_2p_1p_0$ for this face on c_1, so $c_1 = p_2p_1p_0p_4$ where p_4 is the fourth vertex of c_1.

If we have a purely n-dimensional $(n-1)$-connected grid cell complex, we can propagate orientations in the same way. We start with an edge $e_0 = p_0p_1$; this defines an orientation for a face $f_0 = e_0e_1e_2e_3$. Let f_1 be an edge-adjacent face, for example, $f_0 \cap f_1 = e_0$. Then $f_1 = e_0^{-1}e_4e_5e_6$ where $e_0^{-1} = p_1p_0$. We continue orienting all of the 2-faces of a grid cube c_0, for example, in the order $f_0f_1f_2f_3f_4f_5$, which defines a Hamilton cycle through the faces (see Figure 5.27). (Note that we cannot represent grid cells solely by sequences of vertices as we did in case of simplicial cells, because the order of the vertices is now relevant.) We continue with a cube c_1 that is face-adjacent to c_0, for example $c_1 = f_0^{-1}f_6f_7f_8f_9f_{10}$ with $f_0^{-1} = e_3e_2e_1e_0$. The process continues until all of the edges, faces, cubes, and so forth of the complex have been oriented. In fact, we can define an orientation on all of $\mathbb{C}_n$ in this way, and any oriented n-dimensional grid cell complex is then just a subcomplex of $\mathbb{C}_n$ equipped with the chosen default orientation.

For both simplicial and grid cell complexes, there are only two possible orientations for any $(n-1)$-connected purely n-dimensional complex. In the case of the simplicial complex K, we choose an orientation for the triangulated surface $\vartheta(K)$. In the case of the grid cell complex G, it defines an orientation for the square tiling of the surface $\vartheta(G)$.

From Proposition 7.6, we know that the frontier ∂G_g of the geometric representation of an open 2-region $G \subseteq \mathbb{C}_3^{(3)}$ (a union of finitely many simply connected 6-regions) is always hole-free and orientable. This frontier is bounded, because a region is finite, and it is closed because, for every 2-cell c in the frontier ϑG, all of the 1-cells and 0-cells incident with c are in the frontier, and, for every 1-cell c' in the frontier, all of the 0-cells incident with c' are in the frontier.

The frontier $\vartheta(G)$ of a region $G \subseteq \mathbb{C}_3^{(3)}$ can be represented in the frontier grid; see Definition 7.4 for $n = 3$. This defines an isothetic polyhedron (see Chapter 8). The separation theorems of Euclidean topology apply to these polyhedra.

7.5.3 Separations in adjacency grids

$M \subseteq \mathbb{Z}^3$ is called an (α,β)-*separator* iff M is α-connected, M divides $\mathbb{Z}^3 \setminus M$ into (exactly) two β-components, and there exists a $p \in M$ such that $\mathbb{Z}^3 \setminus (M \setminus \{p\}) = (\mathbb{Z}^3 \setminus M) \cup \{p\}$ is β-connected.

In Figure 7.24, the set M_1 on the left is a (2,2)-, (2,1)-, and (2,0)-separator. If any voxel is removed from M_1, its complementary set becomes 0-connected. There also exist voxels in M_1 such that removal of one of them makes the complementary set 1- or 2-connected. The set M_2 in the middle is a (1,2)- and (1,1)-separator, and the set M_2 on the right is a (0,2)-separator.

(α,β)- and (β,α)-separators exist for $(\alpha,\beta) = (0,2), (2,0), (1,2), (2,1)$, and $(1,1)$. However, there are some difficulties with $(\alpha,\beta) = (1,1)$. For a 3D connected set M, we expect that $\chi(M) = 1 - t + c$, where t is the number of tunnels and c the number of cavities. For the set M in Figure 7.25, there is no cavity and one tunnel, so we

7.5 Surfaces and Separations in 3D Grids

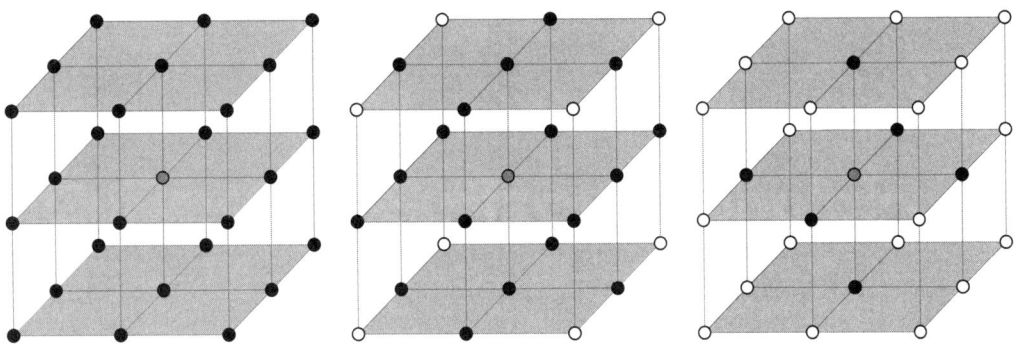

FIGURE 7.24 Left: M_1. Middle: M_2. Right: M_3.

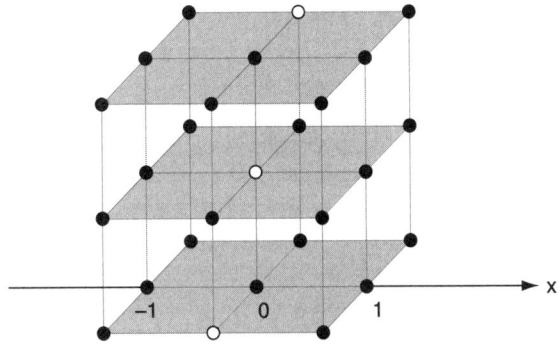

FIGURE 7.25 A set M of voxels with one "diagonal tunnel."

expect that $\chi(M) = 1 - 1 + 0 = 0$. However, the set M_0 of voxels of M that have x-coordinate 0 is simply 1-connected, so $\chi(M_0) = 1$. Let M^+ (M^-) be the set of voxels of M that have nonnegative (nonpositive) x-coordinates. By symmetry, we have $\chi(M^+) = \chi(M^-) = a$. However, because $M_0 = M^+ \cap M^-$ and $M = M^+ \cup M^-$, we must have $\chi(M) = \chi(M^+) + \chi(M^-) - \chi(M_0) = 2a - 1 \neq 0$. (Here we have used a sum formula for the Euler characteristics of unions of Euclidean complexes; it is true for arbitrary unions of finite complexes.) Note that we would encounter the same difficulty for $(\alpha, \beta) = (0,0)$, but this case is not on our list of separators. (0,2), (2,0), (1,2), and (2,1) are considered to be "good pairs of separators" in the grid cell model; they correspond with (6,26), (26,6), (6,18), and (18,6) in the grid point model.

Definition 7.13 Let $M \subseteq S \subseteq \mathbb{Z}^n$ ($n = 2,3$). M is called α-*separating* in S iff $S \setminus M$ is not α-connected ($\alpha = 4,6,8,18,26$). Let M be α-separating in some superset of S but not β-separating in S, where $\beta = 4,6,8,18,26$ and $\alpha < \beta$. Then M is said

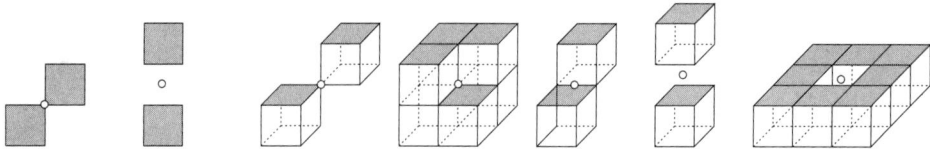

FIGURE 7.26 From left to right: for $n = 2$, a 0-gap and a 1-gap; for $n = 3$, two 0-gaps, a 1-gap, and two 2-gaps [130].

to have β-*gaps*. A set M that has no β-gaps is called β-*gapfree*, and a set that has no β-gaps for any β is called *gapfree*.

The empty set is α-connected; it follows that M is not α-separating in itself. If $S = M \cup \{p\}$, then M can also not be α-separating in S. Let p and q be any two points in $\mathbb{Z}^n \setminus M$ that are not α-connected; then M is α-separating in $M \cup \{p, q\}$.

In the grid cell model, we have $\alpha, \beta \in \{0,1,2\}$, and M has β-gaps iff it is α-separating in some superset of S but not β-separating in S for $\alpha > \beta$.

Figure 7.26 illustrates gaps. An (α, β)-separator M does not have β-gaps. See Exercise 12 for more about gaps in digital lines and Section 11.2.4 for more about gaps in digital planes.

7.6 Exercises

1. Prove that the property of being a (Urysohn-Menger) curve is a topologic invariant in the Euclidean plane.

2. Let γ be a simple (Urysohn-Menger) curve on the surface S of a sphere. Prove that the complementary set $S \setminus \gamma$ consists of two open subsets of S with the common frontier γ.

3. Classify all of the elementary curves ("letters") in Figure 7.4 with respect to topologic equivalence.

4. Suppose $n > 0$ finite connected polygonal chains in $\mathbb{E}^2$ start at p, end at $q \neq p$, and intersect only at p and q. Prove that the chains partition $\mathbb{E}^2$ into n disjoint sets.

5. Prove that a set of grid points C is a simple 4- (8-)curve iff C is 4- (8-)connected and every pixel of C is 4- (8-)adjacent to exactly two other pixels of C.

6. Prove that a subset of $\mathbb{Z}^2$ cannot be both a 4-curve and an 8-curve and that a subset of $\mathbb{Z}^3$ cannot be both a 6-surface and a 26-surface.

7.6 Exercises

7. If the paths shown in Figure 7.11 are regarded as simple curves in the poset topology of $[\mathbb{C}_2, \leq]$, is it possible that they do not intersect?

8. Prove that, if a 4-component C of 1s and an 8-component D of 0s (or vice versa) in a binary picture are 8-adjacent, they are also 4-adjacent, and either D 4-surrounds C (i.e., D 4-separates C from the infinite background component B) or C 8-surrounds D (i.e., C 8-separates D from B).

9. Suppose the chessboard pattern in Exercise 4 (ii) in Section 1.3 is digitized into an 8×8 binary picture. What are its components if we use good pairs (4,8), (8,4), and (6,6)? How do these results compare with your discussion of Exercise 4 (ii) in Section 1.3? Can you specify a switch state matrix **S** such that all of the "black pixels" in the upper four rows and all of the "white pixels" in the lower four rows are s-connected?

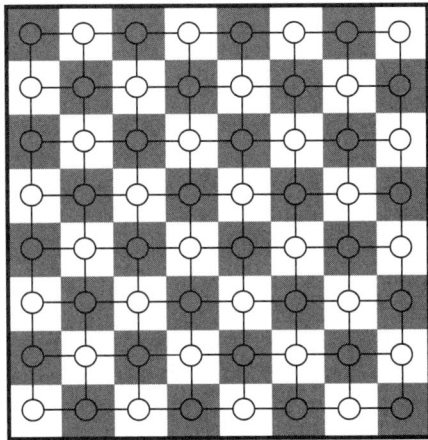

10. Prove that (6,6) is a good pair for $\mathbb{Z}^2$.

11. Define a set of 4×4 templates for choosing the values of flip-flop switches so as to give preference to line-like patterns, whether the lines are black or white.

12. Let S be a set of voxels in the grid cell model. A 2-cell that is a common face of a voxel of S and a voxel of $\overline{S}$ is called a *surface pixel* of S. Two surface pixels are called *edge-adjacent* if they share a 1-cell, and they are called *vertex-adjacent* if they share a 0-cell but not a 1-cell. Prove that any surface pixel of S must be edge-adjacent to an even number of surface pixels of S, that this number must be between 4 and 12, and that a surface pixel of S can be vertex-adjacent to at most 20 surface pixels of S.

13. Two surface pixels (see Exercise 12 above) are called *strongly edge-adjacent* if they are edge-adjacent and are surface pixels of voxels of S with intrinsic

6-distances apart that are at most 3. Prove that a surface pixel of S is strongly edge-adjacent to exactly four surface pixels of S. Prove that, if T is a maximal strongly edge-connected set of surface pixels of S, then there exist a 6-component U of S and an 18-component V of $\overline{S}$ such that T is the set of surface pixels that are common faces of voxels in U with voxels in V.

14. What are the Euler characteristic and genus of the surface of a sphere, a torus, a handle, and a Listing band?

15. Let S_1 and S_2 be two surfaces with frontiers. Glue S_1 and S_2 together by identifying a simple curve in ϑS_1 with a simple curve in ϑS_2. Prove that the resulting surface has Euler characteristic $\chi(S_1) + \chi(S_2)$.

16. Show that a handle has a tiling defined by one vertex, three edges, and one face.

17. Prove that any regular tiling of the Euclidean plane is topologically equivalent to a regular tiling of the plane in which the set of vertices coincides with $\mathbb{Z}^2$.

18. Prove that the Euclidean plane allows regular tilings (with $\alpha_1 = \aleph_0$) only for $n = k = 4$, $n = 3$, and $k = 6$, or $n = 6$ and $k = 3$.

19. Prove that the graph of a regular tiling of the surface of a sphere is homeomorphic to one of the five *Platonic graphs* (tetrahedron, octahedron, icosahedron, hexahedron, or dodecahedron) or to a graph defined either by $n = 2$ and $\alpha_1 = k \geq 2$ or by $n = \alpha_1 \geq 2$ and $k = 2$.

20. Prove that the surface of a torus has regular tilings (with $\alpha_1 < \aleph_0$) for only $n = k = 4$, $n = 3$, and $k = 6$, and $n = 6$, and $k = 3$.

21. Suppose there exists a regular tiling on a hole-free surface with $n = 5$ and $k = 4$. Prove that, if the number of faces of the tiling is not a multiple of 8, the surface is not orientable.

22. Prove that a surface with frontiers is not simply connected if its frontiers consist of more than one simple curve.

7.7 Commented Bibliography

For the original work on Urysohn-Menger curves, see [725, 1081, 1082]. Curves are studied in [798]; for a review of their history, see also [1006]. Curves and complexes are discussed in [9, 10]; see also Theorem 7.1. Theorem 7.3 was proved in [1085] (see also [483]), Theorem 7.4 in [962] and [138], and Theorem 7.5 in [1030]. For Brouwer's work on separation theorems, see [138, 139]. For Definition 7.12 and Theorems 7.11 and 7.12, see [9]. For Theorem 7.10, see [358].

7.7 Commented Bibliography

For the topology of surfaces, see [9, 10, 639]. Theorem 7.13 was published in [139], and the cited work of O. Veblen was published in [1086]. Listing's band was published in [660]; its independent publication by Möbius in [734] came four years later. Figure 7.21 is from [1087]. For a digital formulation of Theorem 7.13 for the 3D grid cell topology, see [587].

Characterizations of simple arcs and curves in 2D adjacency grids are given in [882]; see also [1029]. The term "good pair" was introduced in [569]. See [576] for a discussion of other pairs of adjacencies, and see [336] for an axiomatic treatment of good pairs. Theorem 7.7 is proved in [884]. Definitions of digital surfaces based on the (6,26)-adjacency grid are studied in [211, 689, 747, 841], related region adjacency graphs in [894], and calculations of 3D Euler characteristics (with subsequent studies of homotopy) in [694]. For "digital Urysohn curves" in 2D adjacency grids, see [1112]. Curves and surfaces in the 1- and 2-adjacency grid are discussed in [1005]. Adjacencies between convex cells (generalizing the regular orthogonal grid cell model) are studied in [942, 943, 944]. For other work on digitized curves and boundaries, see [915, 1121].

[632] shows that "simply connected" and "contractible" are locally computable properties in 2D good pair grids but not in 3D good pair grids. (The Euler characteristic is locally computable; see Exercise 6 in Section 6.5 and [577] for a related review.) Local property detection in "grid topologies" that are not necessarily defined on regular grids was studied in [320]. Local property calculation based on incidence matrices of graphs was studied in [474]. [81] characterizes surfaces in 3D adjacency grids by "simple" global properties, and [690] studies equivalent local characterizations. See [1059] for connectivities of $3 \times 3 \times 3$ voxel sets and [578] for Euler characteristics in (4,4)- or (8,8)-adjacency grids.

The approximation of n-dimensional manifolds by graphs is studied in [1060, 1061], with a special focus on topologic properties of such graphs defined by homotopy and on homology or cohomology groups. Definition 7.5 is from [511]. The approximation of boundaries of finite sets of grid points (in n dimensions) based on "continuous analogs" was proposed and studied in [613]. [496] discusses local topologic configurations (stars) for surfaces in incidence grids. Digital surfaces in the context of arithmetic geometry [848] were studied in [131].

The results of [747] are generalized in [573] to all α and α' in $\{6, 18, 26\}$ except $\alpha = \alpha' = 6$. A Jordan surface theorem for the Khalimsky topology is proved in [587]. [685] introduces a general classification of the voxels of a set S using the numbers of components of S and $\overline{S}$ in the neighborhood of the voxel. For discrete combinatorial surfaces, see [339]. For obtaining α-surfaces by the digitization of surfaces in $\mathbb{E}^3$, see [209] and [211]. It is proved in [688] that there is no local characterization of 26-connected subsets S of $\mathbb{Z}^3$ such that $\overline{S}$ consists of two 6-components and every voxel of S is adjacent to both of these components. [689] defines a class of 18-connected surfaces in $\mathbb{Z}^3$, proves a Jordan surface theorem for these surfaces, and studies their relationship to the surfaces defined in [747]. [80] introduces a class of *strong surfaces* and proves that both the 26-connected surfaces of [747] and the 18-connected surfaces of [689] are strong. [688] and [690] give local characterizations of strong surfaces. For 6-surfaces, see [194].

Frontiers in cell complexes (and related topologic concepts such as components and fundamental group) were studied in [11]. For characterizations of and algorithms for curves and surfaces in frontier grids, see [425, 426, 427, 428, 594, 596, 917, 1075, 1076, 1078]. G.T. Herman and J.K. Udupa used frontiers in the grid cell model, and V. Kovalevsky generalized these studies using the model of topologic abstract complexes that was briefly introduced in Section 6.4.4. Our definitions of curves in incidence grids resemble work in [217].

Gaps were introduced and studied in [23, 848] using the term "tunnel" instead of "gap." This was later generalized in [130, 133] to a definition of "j-tunnels" in arbitrary sets of grid points. This notion of a tunnel is not the same as the tunnels defined in Section 6.4.5. [130, 133] also introduced and studied k-separability.

For a review of curves and surfaces in digital topology, see [577], and, more recently, [569]. Exercise 12 and Example 7.1 are from [100].

CHAPTER 8

Curves and Surfaces: Geometry

This chapter discusses geometric properties of curves and surfaces such as length, area, volume, and curvature. It reviews analytic representations, the construction of digital representations, and the estimation of geometric properties from digitized data, with emphasis on area estimation in 2D and volume estimation in 3D. Digital estimation of other curve and surface properties will be treated in later chapters.

8.1 Planar Curves and Arcs

Section 7.1.2 gave topologic definitions of curves and arcs. This section reviews analytic representations and properties of planar curves, including arc length, slope, curvature, and area, as well as geometric properties of isothetic grid polygons.

8.1.1 Analytic representations

A planar curve can be defined by an equation based on a Cartesian coordinate system or by a parameterization such as Equation 7.3. (This was Jordan's original definition, which turned out to allow space-filling curves, such as the Peano curve.) For example, a circle can be defined by the equation $x^2 + y^2 = r^2$ or by the parameterization $x = x(t) = r\cos t$, $y = y(t) = r\sin t$, where $t \in [0, 2\pi)$. In vector notation, we can write $\gamma(t) = (x(t), y(t)) = x(t)\mathbf{e}_1 + y(t)\mathbf{e}_2$, where $\mathbf{e}_1 = (1, 0)$ and $\mathbf{e}_2 = (0, 1)$ are the basis vectors of the Cartesian coordinate system.

An equation of a curve determines the geometric locations of all of the points on the curve (i.e., the shape of the curve). A parameterization provides additional information: an orientation of the curve (a direction along the curve) and a *speed* (or rate of evolution) $v(t)$ as t increases.

Section 3.1.2 introduced the (Euclidean) norm $\|p\|_2$ as the distance between the origin $o = (0,0)$ and the point $p = (x,y)$. This distance is equal to the length of the straight line segment op. In norm notation, we can formally define the speed of a parameterization as follows,

$$v(t) = \|\dot{\gamma}(t)\|_2 = \|(\dot{x}, \dot{y})\|_2 = \|(\dot{x}(t), \dot{y}(t))\|_2$$

where the following is true:

$$\dot{x}(t) = \frac{dx(t)}{dt} \text{ and } \dot{y}(t) = \frac{dy(t)}{dt}$$

For example, the speed of the circle parameterization given above is as follows:

$$v(t) = \|(-r\sin t, r\cos t)\|_2 = r\sqrt{\sin^2 t + \cos^2 t} = r$$

A curve γ can have more than one parameterization. A parameterization of γ is called *regular* iff $v(t) \neq 0$ for $a \leq t \leq b$. We are often not interested in the speed of a parameterization; the parameterization that has unit speed $v(t) = 1$ can be used as the default. In a nonregular parameterization, points of the curve at which $v(t) = 0$ are called *singular*.

Section 7.1.1 defined a parametrizable Jordan curve or arc γ and defined its length $\mathcal{L} = \|\gamma\|_2$ as follows, where $a \leq t \leq b$:

$$\mathcal{L}(t) = \int_a^t \sqrt{\dot{x}^2 + \dot{y}^2}\, ds = \int_a^t v(s)\, ds \tag{8.1}$$

For example, for the circle, we have the following:

$$\mathcal{L} = \int_0^{2\pi} \left(r^2 \sin^2 t + r^2 \cos^2 t\right)^{1/2} dt = 2\pi r$$

This also provides a method of estimating the length of a digital curve by summing estimates of $\|(\dot{x}, \dot{y})\|_2$.

Equation 8.1 also says that the speed of the parameterization is the rate of change of the length of the arc from $t = a$ to a general point on the curve:

$$v(t) = \frac{d\mathcal{L}(t)}{dt} \tag{8.2}$$

The arc length between $\gamma(a)$ and $p = \gamma(t)$ is fixed, but p can have different t values in different parameterizations. We obtain unit speed if the curve is parameterized by arc length l ($0 \leq l \leq \mathcal{L} = \|\gamma\|$):

$$v(l) = \left\| \frac{dx(l)}{dl}\mathbf{e}_1 + \frac{dy(l)}{dl}\mathbf{e}_2 \right\|_2 = 1$$

8.1.2 Arc length

In digital geometry, curves are not given by analytic representations but rather in digitized form. The pixels (grid points or centers of 2-cells) of a digital curve cannot be assumed to be exactly on the (unknown) "real" curve. Section 1.2.11 briefly discussed the problem of estimating geometric quantities such as the length of a curve from digital data.

H. Steinhaus (1930) suggested an integral geometry-based method of estimating the length of a curve, which is still popular in stereology. Let γ be a plane Jordan curve that is rectifiable so that it has finite length $\mathcal{L} = \|\gamma\|_2$. Let $g(\psi)$ be the straight line through the origin that makes angle ψ with, for example, the positive x-axis. The orthogonal projection of γ onto this line can be regarded as a superposition of intervals, each of which is a projection of an arc of γ; see Figure 8.1 for an example. Let $\mathcal{L}(\psi)$ be the sum of the lengths of these intervals, and let $\overline{\mathcal{L}}$ be the mean of $\mathcal{L}(\psi)$ taken over all directions. From integral geometry, the following is known:

$$\mathcal{L} = \frac{1}{2}\pi\overline{\mathcal{L}} = \frac{1}{4}\int_0^{2\pi} \mathcal{L}(\psi)\,d\psi \tag{8.3}$$

Here $\gamma(t) = (x(t), y(t))$ is assumed to be $C^{(2)}$-*regular*: $x(t)$ and $y(t)$ have bounded second derivatives in a finite number of open intervals, so $\gamma(t)$ is the union of a finite number of arcs with continuously turning tangents. On the basis of Equation 8.3, H. Steinhaus suggested measuring $\mathcal{L}(\psi)$ for $n > 0$ values of ψ equally spaced in the interval $(0, 2\pi)$ and taking the mean $\hat{\mathcal{L}}$ of these values. He showed that the following is true if n is even and that these bounds are the best possible:

$$\frac{\pi}{n}\cos(\frac{\pi}{n})[\sin(\frac{\pi}{n})]^{-1} \leq \hat{\mathcal{L}}\overline{\mathcal{L}}^{-1} \leq \frac{\pi}{n}[\sin(\frac{\pi}{n})]^{-1} \tag{8.4}$$

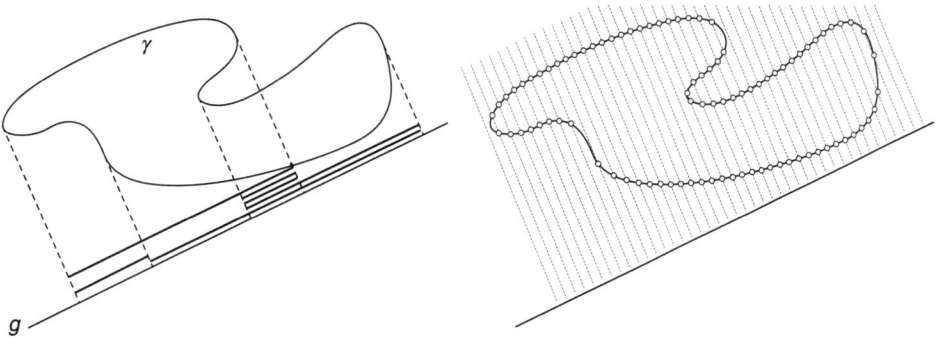

FIGURE 8.1 Left: projection of a simple curve onto a line. Right: estimation of lengths of arcs by numbers of intersection points.

This equation gives a comparison of the mean $\hat{\mathcal{L}}$ on sampled angles and the mean $\overline{\mathcal{L}}$ on continuous angles.

In numeric analysis, we usually estimate geometric properties of a curve in $\mathbb{R}^n$ from finitely many samples of the curve by fitting curves of known types to the samples and calculating the lengths of these approximating curves. Let $\gamma : [0,1] \to \mathbb{R}^n$ be a $C^{(k)}$-regular parametric curve; our task is to estimate $\mathcal{L}(\gamma)$ from $m+1$ points $q_i = \gamma(t_i)$ on γ. This estimation is easiest when the t_is are uniformly spaced (i.e., $t_i = \frac{i}{m}$ [see Figure 8.2]). We will approximate γ with a curve $\tilde{\gamma}$ that is piecewise polynomial of some degree $a \geq 1$.

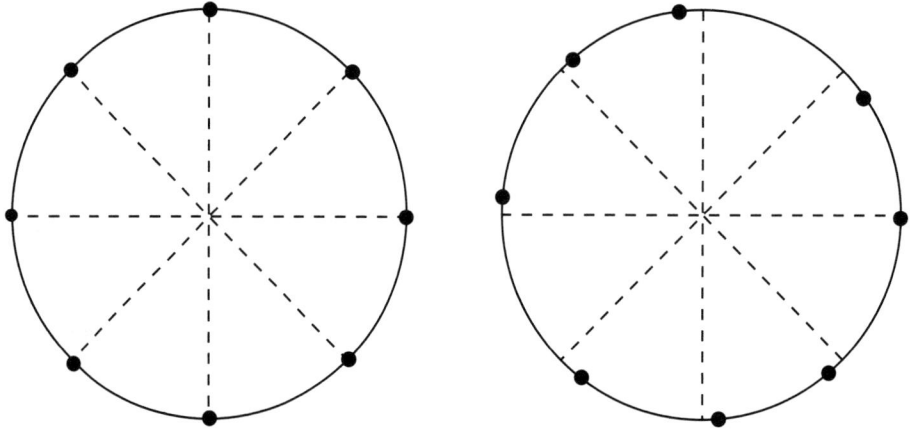

FIGURE 8.2 Left: uniform samples on a circle. Right: nonuniform samples.

Theorem 8.1 Let γ be $C^{(a+2)}$ and $t_i = i/m$. Then the t_is determine a function $\tilde{\gamma}$, which is a piecewise degree-a polynomial such that the following is true, where $a_0 = 1$ or 2 according to whether a is odd or even:

$$\mathcal{L}(\tilde{\gamma}) = \mathcal{L}(\gamma) + O(\frac{1}{m^{a+a_0}})$$

$\mathcal{L}(\tilde{\gamma})$ is called a *Lagrange estimate* of $\mathcal{L}(\gamma)$. We say that the t_is are (ε, k)-*uniformly sampled* ($\varepsilon \geq 0$ and $k \geq 1$) iff there is a $C^{(k)}$ reparameterization $\phi : [0,1] \to [0,1]$ such that the following is true:

$$t_i = \phi(\frac{i}{m}) + O(\frac{1}{m^{1+\varepsilon}})$$

$\mathcal{L}(\tilde{\gamma})$ can behave badly for $(0,k)$-uniform samplings, but for $0 < \varepsilon \leq 1$, we have the following:

8.1 Planar Curves and Arcs

Theorem 8.2 Let the t_is be (ε, k)-uniformly sampled, where $0 < \varepsilon \leq 1$ and $k \geq 4$. Then, for $a = 2$, we have the following:

$$\mathcal{L}(\tilde{\gamma}) = \mathcal{L}(\gamma) + O(\frac{1}{m^{4\varepsilon}})$$

In digital geometry, we must deal with digitization rather than sampling; pixel coordinates are known only to the accuracy imposed by the grid resolution. Piecewise linear curve approximation ($a = 1$) is usually preferred because of the simplicity of adding the lengths of line segments (see Chapter 10). However, Theorem 8.1 indicates that higher-order approximation may (!) have the potential for more accurate estimation.

8.1.3 Curvature

A Jordan curve is called *smooth* if it is continuously differentiable. Evidently, a polygon is not smooth, and a nonregular parameterization of a curve is not smooth at singular points. Curvature can be defined only at nonsingular points of a curve; in this section, we therefore assume a regular parameterization.

To define curvature, it is convenient to use the *Frenet frame*,[1] which is a pair of orthogonal coordinate axes (see Figure 8.3) with the origin at a point $p = \gamma(t)$ on the curve. One axis is defined by the *unit tangent vector*,

$$\mathbf{t}(t) = \dot{\gamma}(t)/|\dot{\gamma}(t)| = (\cos \psi(t), \sin \psi(t)) = \cos \psi(t)\mathbf{e}_1 + \sin \psi(t)\mathbf{e}_2$$

where ψ is the *slope angle* between the tangent and the positive x-axis. The other axis is defined by the *unit normal vector*:

$$\mathbf{n}(t) = (-\sin \psi(t), \cos \psi(t)) = -\sin \psi(t)\mathbf{e}_1 + \cos \psi(t)\mathbf{e}_2$$

Evidently, $\langle \mathbf{t}(t), \mathbf{n}(t) \rangle_e = 0$. It is assumed that $\mathbf{t}(t)$ and $\mathbf{n}(t)$ define a right-handed coordinate system.

Figure 8.3 also illustrates the fact that $l = \mathcal{L}(t)$ is the arc length between the starting point $\gamma(a)$ and the general point $p = \gamma(t)$. As t increases, the changes in the tangent and normal are as follows:

$$\dot{\mathbf{t}} = (-\sin \psi \frac{d\psi}{dt}, \cos \psi \frac{d\psi}{dt}) = \frac{d\psi}{d\mathcal{L}} \frac{d\mathcal{L}}{dt} (-\sin \psi, \cos \psi) \quad (8.5)$$

$$\dot{\mathbf{n}} = (-\cos \psi \frac{d\psi}{dt}, -\sin \psi \frac{d\psi}{dt}) = \frac{d\psi}{d\mathcal{L}} \frac{d\mathcal{L}}{dt} (-\cos \psi, -\sin \psi) \quad (8.6)$$

The curvature of γ is its "rate of turn" (i.e., the rate of change of ψ as p moves along γ).

---

[1] Named after the French mathematician J.F. Frenet (1816–1900).

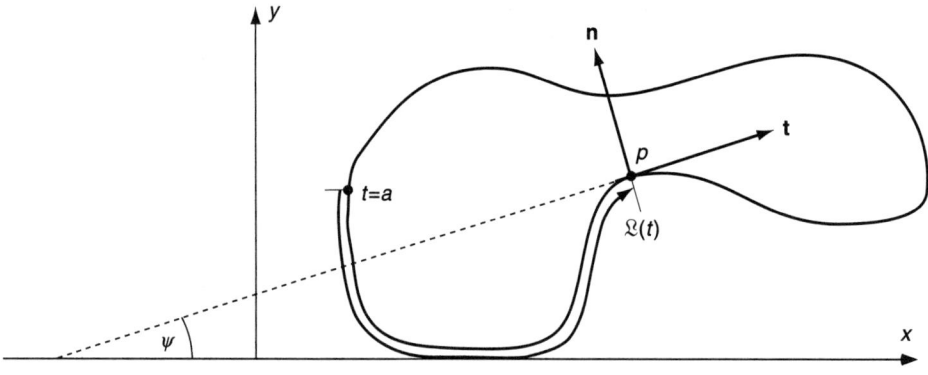

FIGURE 8.3   The Frenet frame at $p = \gamma(t)$.

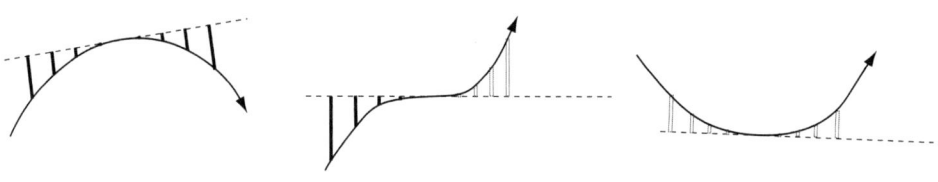

FIGURE 8.4   Three tangents to a curve. The curvature is positive on the left, has a zero crossing in the middle, and is negative on the right.

p is called a *convex point* of γ if the curvature of γ at p is positive and a *concave point* if it is negative. Figure 8.4 shows on the left a positive point, on the right a negative point, and in the middle a *point of inflection* where the curvature changes sign. As Figure 8.4 shows, the situation at p can be approximated by measuring the distances between γ and the tangent to γ at p along equidistant lines that are perpendicular to the tangent. In Figure 8.4, positive distances are represented by bold line segments and negative distances by "hollow" line segments. The area between the curve and the tangent line can be approximated by summing these distances; it is positive on the left, negative on the right, and zero in the middle, where the positive and negative distances cancel.

Definition 8.1   The curvature of a smooth Jordan curve γ at $\gamma(t) = (x(t), y(t))$ is as follows:

$$\kappa(t) = \frac{d\psi(t)}{d\mathcal{L}(t)} = \frac{d\psi(t)}{dl}$$

8.1 Planar Curves and Arcs

If an arc of γ is given explicitly as $y = f(x)$, this definition evaluates to the following:

$$\kappa(x) = \frac{d^2 f(x)}{dx^2} \cdot \left(1 + \left(\frac{df(x)}{dx}\right)^2\right)^{-1.5} \tag{8.7}$$

If γ is given explicitly in polar coordinates as $r = f(\eta)$, we have the following:

$$\kappa(\eta) = \left(r^2 + 2\left(\frac{df(\eta)}{d\eta}\right)^2 - r\frac{d^2 f(\eta)}{d\eta^2}\right) \cdot \left(r^2 + \left(\frac{df(\eta)}{d\eta}\right)^2\right)^{-1.5} \tag{8.8}$$

Finally, in the general case of a parametric representation $\gamma(t) = (x(t), y(t))$, we have the following,

$$\kappa(t) = (\dot{x}(t)\ddot{y}(t) - \dot{y}(t)\ddot{x}(t)) \cdot (\dot{x}(t)^2 + \dot{y}(t)^2)^{-1.5} \tag{8.9}$$

where the following are true:

$$\ddot{x}(t) = \frac{d^2 x(t)}{dt^2} \text{ and } \ddot{y}(t) = \frac{d^2 y(t)}{dt^2}$$

If we use the arc length parameterization, we have $\kappa(t) = \dot{x}(t)\ddot{y}(t) - \dot{y}(t)\ddot{x}(t)$.
From Definition 8.1 and Equations 8.2 and 8.6, we have the following:

$$\dot{\mathbf{t}}(t) = \kappa(t) \cdot v(t) \cdot \mathbf{n}(t) \quad \text{and} \quad \dot{\mathbf{n}}(t) = -\kappa(t) \cdot v(t) \cdot \mathbf{t}(t) \tag{8.10}$$

These two equations are called the *Frenet formulae*. If we use the arc length parameterization with unit speed $v(l) = 1$, the formulae simplify to the following:

$$\dot{\mathbf{t}}(l) = \kappa(l) \cdot \mathbf{n}(l) \quad \text{and} \quad \dot{\mathbf{t}}(l) = -\kappa(l) \cdot \mathbf{t}(l)$$

For example, for a circle, we have $\gamma(t) = (r\cos t, r\sin t)$, $v(t) = r$, and $\mathbf{n} = \dot{\mathbf{t}}$ so that $\kappa(t) = 1/r$ by the first Frenet formula.

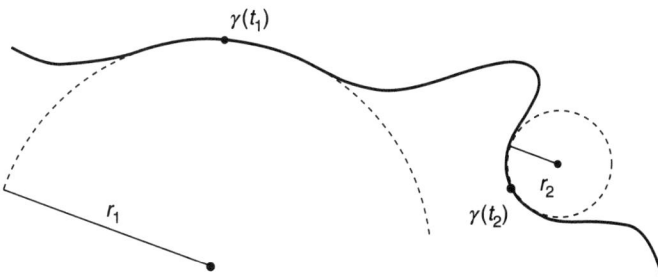

FIGURE 8.5 The osculating circles at $p_i = \gamma(t_i)$, where $i = 1, 2$; the circle at p_i has radius $r(t_i)$.

The absolute value of the curvature at $p = \gamma(t)$ is equal to the inverse of the radius $r(t)$ of the *osculating circle* at p, which is the largest circle tangent to γ on the concave side of p; see Figure 8.5. Note that we cannot have $r(t) = 0$, but r is infinite when p is on a straight line segment. In general, we have the following:

$$|\kappa(t)| = \frac{1}{r(t)}, \quad \text{if } r(t) < \infty \tag{8.11}$$

The sign of κ depends on whether γ is locally convex or concave at $\gamma(t)$.

The curvature of a digital curve can be estimated using approximations to the tangent vector, estimated derivatives along the curve, or approximations to the osculating circle; see Chapter 10.

8.1.4 Angle

We saw in Section 3.1.3 that the angle η between two vectors can be expressed in terms of the scalar product of the vectors. In n-dimensional Euclidean space ($n \geq 2$), we have the following:

$$\cos \eta = \frac{\langle p, q \rangle_e}{\|p\|_2 \cdot \|q\|_2}$$

Let two smooth curves $\gamma_1 = (x_1, y_1)$ and $\gamma_2 = (x_2, y_2)$ intersect at $p = \gamma_1(t_1) = \gamma_2(t_2)$. Let the unit tangent vectors of γ_1 at $t = t_1$ and of γ_2 at $t = t_2$ be $\mathbf{t}_1$ and $\mathbf{t}_2$. The angle η between p_1 and p_2 at p satisfies the following, because $\|\mathbf{t}_1\|_2 = \|\mathbf{t}_2\|_2 = 1$:

$$\cos \eta = \langle \mathbf{t}_1, \mathbf{t}_2 \rangle_e \tag{8.12}$$

Equation 3.7 shows that angular values can also be defined by weak scalar products; this generalizes the Euclidean space approach. See Section 3.3.3 for a discussion of angular values for grid-based metrics. Weak scalar products can also be defined for graph metrics on adjacency graphs.

Angles between digital curves can be estimated using approximations to their tangent vectors; see Chapter 10.

8.1.5 Area

A *planar region* R is a compact set in $\mathbb{R}^2$. Every compact is closed (and hence *measurable*, which means that it has a well-defined content) and bounded (so that its measure is finite); thus it is *integrable*. The *area* of R is given by the following:

$$\mathcal{A}(R) = \int_R dx dy \tag{8.13}$$

8.1 Planar Curves and Arcs

Area is additive[2]; if R_1 and R_2 are planar regions that have disjoint interiors, we have $\mathcal{A}(R_1 \cup R_2) = \mathcal{A}(R_1) + \mathcal{A}(R_2)$. This allows us to measure the area of a set by partitioning the set into (e.g., convex) subsets and adding the areas of these subsets.

Let T be a triangle pqr where $p = (x_1, y_1)$, $q = (x_2, y_2)$, and $r = (x_3, y_3)$. Then we have the following:

$$\mathcal{A}(T) = \frac{1}{2} \cdot |D(p, q, r)| \tag{8.14}$$

where $D(p, q, r)$ is the *determinant*:

$$\begin{vmatrix} x_1 & y_1 & 1 \\ x_2 & y_2 & 1 \\ x_3 & y_3 & 1 \end{vmatrix} = x_1 y_2 + x_3 y_1 + x_2 y_3 - x_3 y_2 - x_2 y_1 - x_1 y_3$$

Note that $D(p, q, r)$ can be positive or negative; this can be used to define the orientation of the ordered triple (p, q, r). More generally, let $P = <p_1, p_2, \ldots, p_n>$ be a simple polygon, $n \geq 3$, and $p_i = (x_i, y_i)$. Then the following is true,

$$\mathcal{A}(P) = \frac{1}{2} \cdot \left| \sum_{i=1}^{n} x_i (y_{i+1} - y_{i-1}) \right| \tag{8.15}$$

where $p_0 = p_n$ and $p_{n+1} = p_1$. Let the coordinate transformation $u = u(x, y)$, $v = v(x, y)$ map an integrable set R in the xy coordinate system into S in the uv-coordinate system, and let u_x, u_y, v_x, v_y be the partial derivatives of u and v with respect to x and y. Then the following is called the *Jacobian matrix* of the transformation:

$$\mathcal{A}(R) = \int_S |J| \, du dv \quad \text{where } J = \begin{pmatrix} u_x(x, y) & v_x(x, y) \\ u_y(x, y) & v_y(x, y) \end{pmatrix} \tag{8.16}$$

Note that $|J| = |u_x v_y - u_y v_x|$. For example, for $u = 5x^2 + 3y$ and $v = 6x^2 y + 2y^2$, we have $u_x(x, y) = 10x$, $u_y(x, y) = 3$, $v_x(x, y) = 12xy$, and $v_y(x, y) = 6x^2 + 4y$.

The inner and outer Jordan digitizations can be used to estimate the area of a planar region. If S is a compact subset of $\mathbb{R}^2$, the areas $\mathcal{A}(J_h^-(S))$ and $\mathcal{A}(J_h^+(S))$ converge to the area $\mathcal{A}(S)$ as the grid resolution $h \to \infty$.

In Figure 8.6, S (upper left) is an unknown set; G (upper middle) is its Gauss digitization; and the upper right shows its inner and outer Jordan digitizations. The area of the relative convex hull (lower right) of the inner digitization with respect to the outer digitization (see Section 1.2.9) can also be used as an estimate of $\mathcal{A}(S)$. The lower row of the figure shows three other polygons with areas that could be used to estimate $\mathcal{A}(S)$. Theorems 2.2, 2.3, and 2.4 provide theoretic justifications of the multigrid convergence of these area estimators; their measurement bias is also

---

2. Area, as a measure, is even countably additive.

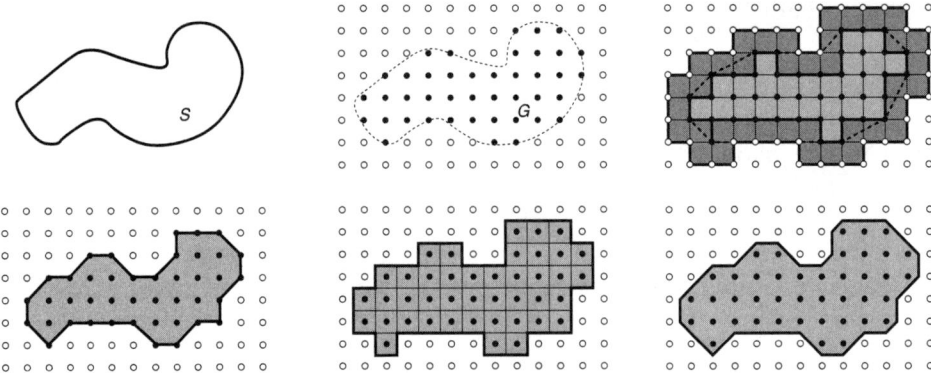

FIGURE 8.6 Upper row: unknown set S; Gauss digitization of S; and inner and outer approximations to S, which also show the convex hull of the inner approximation relative to the outer approximation. Lower row: 8-curve passing through the 4-border pixels; union of the 2-cells; and curve passing through the midpoints of the invalid edges.

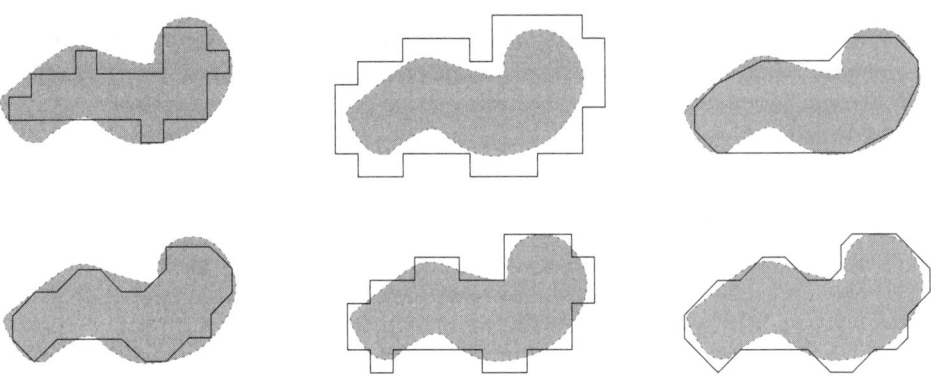

FIGURE 8.7 Visual comparison of an unknown set S with polygons that could be used to estimate its area, starting with inner and outer Jordan digitizations in the upper row and then following Figure 8.6.

important. Figure 8.7 shows how they differ from S for a fixed grid resolution. It can be shown experimentally that the inner Jordan digitization (lower left) leads to underestimation of the area, but the estimates based on the relative convex hull, the cardinality of the set of grid points, and the polygonal frontier defined by the midpoints of the invalid edges are "less biased."

8.1.6 Isothetic grid polygons

In this section, we discuss geometric properties of simple isothetic polygons Π with vertices that are grid points. Let f be the number of grid squares contained in Π, α_0 the number of grid points in Π, and l the number of grid points on the frontier $\delta \Pi$. By *Pick's formula* (see the comments following Theorem 4.10), we have the following:

$$f = \alpha_0 - \frac{l}{2} - 1 \tag{8.17}$$

When we trace the frontier (see Chapter 5) of a region in the frontier grid, the resulting polygonal chain does not necessarily circumscribe a single simple grid polygon. For example, in Figure 8.8 (right), the frontier of an 8-connected region in the picture grid circumscribes three simple polygons. In general, it circumscribes a sequence of simple isothetic grid polygons $\Pi_1, \ldots, \Pi_n$ ($n \geq 1$). Because the area is additive, from Pick's formula (Equation 8.17) we have the following,

$$\begin{aligned} f &= \mathcal{A}(\bigcup_{k=1}^{n} \Pi_k) \\ &= (\alpha_0^{(1)} - \frac{l_1 - 1}{2} - 1) + \sum_{k=1}^{n-1} \left(\alpha_0^{(k)} - \frac{l_k - 2}{2} - 1 \right) + (\alpha_0^{(n)} - \frac{l_n - 1}{2} - 1) \\ &= \alpha_0 - \frac{L}{2} - 1 \end{aligned}$$

where L is the total length of the frontier (i.e., $L = \sum_{k=1}^{n} l_k - 2(n-1)$) and $\alpha_0 = \sum_{k=1}^{n} \alpha_0^{(k)} - (n-1)$ is the number of grid points in $\bigcup_{i=1}^{n} \Pi_k$. To count α_0, we can use *discrete column-wise integration*: start with $\alpha_0 := 0$; $\alpha_0 := \alpha_0 + y$ for all $p = (x, y)$ at the upper end of a run of object pixels in the y direction; and $\alpha_0 := \alpha_0 - y + 1$ for all $p = (x, y)$ at the bottom of such a run. Figure 8.8 shows all of the local patterns that can occur in the frontier tracing, each with its corresponding increment, and also illustrates how the α_0 values change during frontier tracing when we start at

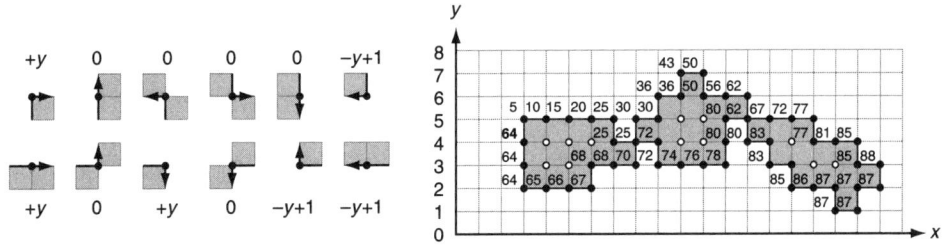

FIGURE 8.8 Left: local patterns used in discrete column-wise integration; right: an example of counting α_0.

the uppermost-leftmost vertex and trace the frontier clockwise. In this example, we have $\alpha_0 = 64$ and $L = 54$ so that $f = 36$.

Finally, we prove a proposition (compare with Theorem 4.9) about the numbers of convex and concave vertex angles of a frontier of a simple isothetic grid polygon Π. We will use the same method in Section 8.3.7 to prove a theorem about the angles of isothetic simple grid polyhedra.

In the following discussion, we consider only the outer frontier of Π, which we assume is traversed clockwise. Inner frontiers (see Figure 8.9), which we assume are traversed counterclockwise, can be treated similarly. If three consecutive vertices pqr on the frontier are not collinear, we call the vertex at q *convex* if its vertex angle is $\pi/2$ and *concave* if it is $3\pi/2$.

Proposition 8.1 Let Π_A and Π_C be the numbers of convex and concave angles of Π; then we have the following:

$$\Pi_A - \Pi_C = 4$$

Proof For a single rectangle, we have $\Pi_A = 4$ and $\Pi_C = 0$. Any Π can be partitioned into a finite number of isothetic rectangles. Hence Π can be constructed by starting with a single isothetic rectangle and joining isothetic rectangles one at a time to an isothetic simple polygon. The four ways of performing the joins are shown in Figure 8.10.

In case (1), Π_A and Π_C remain unchanged; in case (2), both Π_A and Π_C increase by 1; and in cases (3) and (4), both Π_A and Π_C increase by 2. Hence Π_A through Π_C remain 4 throughout the process. ∎

For an inner frontier, we similarly have $\Pi_A - \Pi_C = -4$.

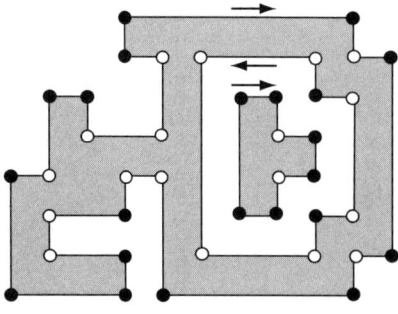

FIGURE 8.9 Three examples of isothetic polygonal circuits.

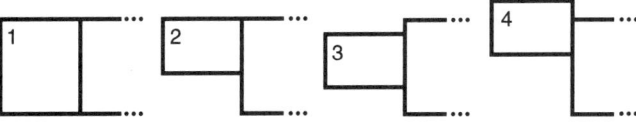

FIGURE 8.10 Four possible joins.

8.2 Space Curves and Arcs

A curve in $\mathbb{R}^n$ is a one-dimensional continuum; see Definition 7.2. For $n = 3$, we can use either of two analytic representations for a curve: two equations define a curve as the intersection of two surfaces, or a parameterization $\gamma(t) = (x(t), y(t), z(t))$, which also provides a direction and speed $v(t)$. For example, the curve $\gamma(t) = (a\cos t, a\sin t, bt)$, which is a *circular helix* that "winds around" a cylindric surface, is characterized by a diameter $2a$ and a vertical distance $2\pi b$; here a second surface that defines such a helix—together with the cylindric surface—is not uniquely defined. Changing b to $-b$ changes the helix from "right-handed" to "left-handed" or vice versa.

When we use a parametric representation, we must also specify the domain of t, for example $a \leq t \leq b$. γ is *smooth* iff $x(t)$, $y(t)$, and $z(t)$ are continuously differentiable. Analogously with Equation 8.1, the *arc length* of γ between the starting point $\gamma(a)$ and the general point $\gamma(t)$ is as follows:

$$\mathcal{L}(t) = \int_a^t \sqrt{\dot{x}^2 + \dot{y}^2 + \dot{z}^2}\, \mathrm{d}s = \int_a^t v(s)\, \mathrm{d}s \tag{8.18}$$

If $v(t) = 0$, $\gamma(t)$ is called a *singular point*. A point can be singular with respect to one parameterization but not with respect to another. A parameterization is called *regular* if it has no singular points. Unit speed $v(t) = 1$ can be achieved by using the arc length parameterization, which is regular.

For any regular parameterization of γ, we can define a unit tangent vector $\mathbf{t}$ at each point of γ. The *curvature* of γ is the absolute rate of change of direction of its tangent vector, where $l = \mathcal{L}(t)$ is arc length:

$$\kappa(t) = \left\| \frac{\mathrm{d}\mathbf{t}(t)}{\mathrm{d}l} \right\|_2 = \left\| \frac{\mathrm{d}\mathbf{t}(t)}{\mathrm{d}t} \frac{\mathrm{d}t}{\mathrm{d}\mathcal{L}(t)} \right\|_2 = \frac{\|\dot{\mathbf{t}}(t)\|_2}{v(t)} \tag{8.19}$$

For the arc length parameterization $\gamma(l)$, it follows that $\kappa(l) = \|\dot{\mathbf{t}}(l)\|_2$. $1/\kappa(t)$ is the *radius of (the circle of) curvature*. As in the planar case, this is also called the radius of the *osculating circle*. On a straight segment of γ, we have $\kappa = 0$, and the radius of curvature becomes undefined (or infinite).

Any direction in the plane perpendicular to the tangent vector at a point p of γ is a normal direction to γ at p. A natural choice is $\dot{\mathbf{t}}$, because $\dot{\mathbf{t}} \cdot \mathbf{t} = 0$, so $\dot{\mathbf{t}}$ is in this plane. The unit vector $\mathbf{n} = \dot{\mathbf{t}}/\|\dot{\mathbf{t}}\|_2$ is called the *principal normal*. In accordance with Equation 8.19, we have the following:

$$\dot{\mathbf{t}}(t) = \kappa(t)v(t)\mathbf{n} \tag{8.20}$$

The *cross product (vector product)* $\mathbf{b} = \mathbf{t} \times \mathbf{n}$ defines a third vector $\mathbf{b}$ called the *binormal*, which is orthogonal to both $\mathbf{t}$ and $\mathbf{n}$. Let $\mathbf{t} = (t_1, t_2, t_3)$ and $\mathbf{n} = (n_1, n_2, n_3)$, and let the unit vectors in the coordinate directions be $\mathbf{e}_1 = (1,0,0)$, $\mathbf{e}_2 = (0,1,0)$, and $\mathbf{e}_3 = (0,0,1)$; then the following is given:

$$\mathbf{b} = (t_2 n_3 - t_3 n_2)\mathbf{e}_1 + (t_3 n_1 - t_1 n_3)\mathbf{e}_2 + (t_1 n_2 - t_2 n_1)\mathbf{e}_3$$

$\mathbf{t}(t)$, $\mathbf{n}(t)$, and $\mathbf{b}(t)$ define an orthogonal coordinate system at $\gamma(t)$ called the *3D Frenet frame*.

The *torsion* of γ is defined by the following, where $l = \mathcal{L}(t)$ is arc length:

$$\tau(t) = \left\| \frac{\mathrm{d}\mathbf{b}(t)}{\mathrm{d}l} \right\|_2 \tag{8.21}$$

$1/\tau(t)$ is the *radius of (the circle of) torsion*. A curve is planar iff its torsion is identically zero; hence the torsion of γ can be thought of as measuring the departure of γ from planarity.

Estimation of arc length, curvature, and angle for 3D digital curves will be discussed in Chapter 10.

8.3 Surfaces and Solids

In Section 7.4, we gave a topologic definition and classification of surfaces. In this section, we discuss geometric properties of surfaces and of "solid" objects.

8.3.1 Analytic representations

A surface can be represented analytically either by an equation $f(x,y,z) = 0$ or in parametric form. In this section, we will use a parametric representation: $\Gamma(u,v) = (x(u,v), y(u,v), z(u,v))$. We will assume that the functions x, y, and z have partial derivatives as follows and similarly for y and z:

$$x_u(u,v) = \frac{\partial x(u,v)}{\partial u}, \quad x_v(u,v) = \frac{\partial x(u,v)}{\partial v}$$

Definition 8.2 A smooth surface patch Γ is defined by a simply connected compact set $B \subseteq \mathbb{R}^2$ and three functions $x(u,v)$, $y(u,v)$, and $z(u,v)$, each of which is continuously differentiable for all $(u,v) \in B$, such that the following is true:

$$\Gamma = \{(x,y,z): x = x(u,v) \wedge y = y(u,v) \wedge z = z(u,v) \wedge (u,v) \in B\}$$

8.3 Surfaces and Solids

We assume that two points $(u,v) \in B$ never define the same point (x,y,z) of Γ (injectivity condition) and that the matrix of first derivatives has rank 2 for all $(u,v) \in B$:

$$\begin{pmatrix} x_u & y_u & z_u \\ x_v & y_v & z_v \end{pmatrix}$$

If the other conditions are satisfied but the rank of the matrix is less than 2 at some points $(u,v) \in B$, we say that γ has singularities at those points.

A smooth surface patch Γ cannot be a Jordan surface, which must be homeomorphic to the surface of a sphere. The $C^{(1)}$ property of the functions $x(u,v)$, $y(u,v)$, and $z(u,v)$ also allows no discontinuities in the derivatives; a polyhedral surface patch is not smooth.

As an example, consider the surface Γ of the sphere $x^2 + y^2 + z^2 - r^2 = 0$. A possible parameterization for this surface is as follows:

$$\Gamma(u,v) = (r\cos u \cos v, r\cos u \sin v, r\sin u) \quad \text{where } 0 < u < \pi \text{ and } 0 < v < 2\pi$$

Here u is called the latitude angle and v is called the longitude angle. The sin and cos functions are continuously differentiable for all (u,v), and the matrix of first derivatives has rank 2. Note that this parameterization does not represent the entire surface; one semicircle joining the poles is not included. Definition 8.2 requires B to be compact; this can be achieved by using $B = \{(u,v) : \varepsilon \leq u \leq \pi - \varepsilon \wedge \varepsilon \leq v \leq 2\pi - \varepsilon\}$ for some small $\varepsilon > 0$. In accordance with the wording latitude and longitude for u and v, the bounds on u,v mean that the Greenwich meridian (including the north and south poles) is removed. This is necessary in order to guarantee the injectivity condition of Definition 8.2.

As a second example, let B be a simply connected compact region in a plane Π. We can define a smooth surface patch Γ parallel to Π by taking $x(u,v) = $ constant, $y(u,v) = u$, and $z(u,v) = v$. The matrix of partial derivatives has rank 2:

$$\begin{pmatrix} 0 & 1 & 0 \\ 0 & 0 & 1 \end{pmatrix}$$

Such a Γ is called *planar*.

As a third example, let Γ be defined by the equation $z = f(x,y)$ for $(x,y) \in B \subseteq \mathbb{R}^2$. Then Γ has the parameterization $\{(x,y,f(x,y)) : (x,y) \in B\}$. Such a Γ is called a *Monge patch*; it is smooth iff f is continuously differentiable for all $(x,y) \in B$.

8.3.2 Surface area

Let Γ have no singularities so that at least one of the subdeterminants given here is nonzero at each point of B:

$$D_1 = \begin{vmatrix} y_u & z_u \\ y_v & z_v \end{vmatrix}, D_2 = \begin{vmatrix} z_u & x_u \\ z_v & x_v \end{vmatrix}, D_3 = \begin{vmatrix} x_u & y_u \\ x_v & y_v \end{vmatrix} \quad (8.22)$$

The *area* of Γ is then defined as follows,

$$A(\Gamma) = \int_B \sqrt{D_1^2 + D_2^2 + D_3^2}\, du dv \tag{8.23}$$

provided Γ is measurable (see Definition 8.4 and Theorem 8.3 on p. 286).

If Γ is a planar surface patch, we have the following,

$$A(\Gamma) = \int_B d(u,v) = A(B)$$

so that the surface area measurement problem is reduced to a planar area measurement problem.

The definition of measurability of Γ is based on the triangulation of a bounded set B_1 such that $B \subseteq B_1^\circ$ and such that the first derivatives of $x(u,v)$, $y(u,v)$, and $z(u,v)$ exist and are continuous in B_1° (formally, $x(u,v)$, $y(u,v)$, and $z(u,v)$ are functions in $C^{(1)}(B_1^\circ)$).

In No. 108 of [696], it is shown that the angles η of the triangles in such a triangulation must satisfy $\eta < 2\pi/3$. This was proved independently by O. Hölder (p. 29 of [441]) in 1882, by G. Peano [806] in 1890, and by H.A. Schwarz [968] in 1890. We will now review the example given by H.A. Schwarz.

In 3D picture analysis, objects are often acquired slice by slice. The question arises whether surface area estimates can be based on triangulations that involve different slices (e.g., calculated by a marching cubes algorithm; see Section 8.4.2). It is also important that refining the acquisition process should result in more accurate estimates of surface area; see the concept of multigrid convergence in Section 2.4.3.

Example 8.1 Let a right circular cylinder of radius r and height h be cut by $k-1$ planes ($k \geq 2$) parallel to the bases of the cylinder, which segment the cylinder into k congruent parts. Construct a regular n-gon ($n \geq 3$) in each of the $k+1$ cross-sections (including the bases); see Figure 8.11, where $n = 6$. Rotate the n-gon by π/n between successive slices. Connect each edge of each n-gon to the vertex of each neighboring n-gon that is closest to the edge. This results in a triangulation $Z_{k,n}$ of the surface of the cylinder into $2kn$ congruent triangles that have the following total area:

$$A(Z_{k,n}) = 2\pi r \cdot \frac{\sin(\pi/n)}{\pi/n} \sqrt{\frac{1}{4}\pi^4 r^2 \left(\frac{\sin(\pi/2n)}{\pi/2n}\right)^4 \left(\frac{k}{n^2}\right)^2 + h^2}$$

As k and n go to infinity, the lengths of the edges of the triangles go to zero. However, the area of $Z_{k,n}$ does not converge to the surface area $A(L) = 2\pi rh$ of the cylinder unless k/n^2 goes to zero. If k/n^2 converges to $g > 0$, $A(Z_{k,n})$ converges to the following

$$2\pi r \cdot \frac{\sin(\pi/n)}{\pi/n} \sqrt{\frac{1}{4}\pi^4 r^2 g^2 + h^2}$$

8.3 Surfaces and Solids

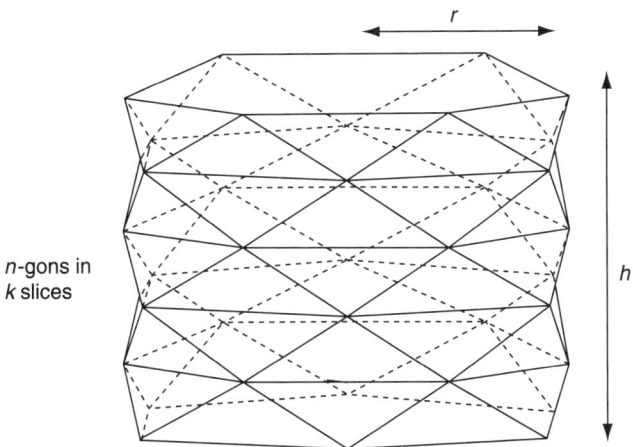

n-gons in k slices

FIGURE 8.11 Triangulation of the lateral surface of a right circular cylinder.

k/n^2 can even go to infinity (e.g., when $k = n^3$); in that case, $\mathcal{A}(Z_{k,n})$ goes to infinity as well.

Note that this example uses sample points on the surface; such points are usually not available in digital geometry, so surface area estimates in digital geometry may be even less accurate. The example also shows that a method (approximation by n-gons) can work in one case (see Section 1.2.7 about the estimation of π) but not in another.

We now define a class of triangulations that satisfy the constraint $\eta < 2\pi/3$:

Definition 8.3 Let $B_1 \subseteq \mathbb{R}^2$ be a simply connected compact set such that $B \subseteq B_1^\circ$, and let $0 < \omega < \pi/3$. A network Z of triangles that completely covers B_1 is called a triangular subdivision of B_1 with respect to B if it satisfies *the following conditions:*

(i) The angles of the triangles in Z do not exceed $\pi - \omega$.

(ii) If a triangle in Z has a nonempty intersection with B, its vertices are all in B_1.

A triangular subdivision Z of B_1 with respect to B defines a polyhedral approximation $\Gamma(Z)$ of Γ by orthogonal projection of the vertices of Z onto[3] Γ, and the adjacency relation between the vertices of Z defines an adjacency relation between these vertices on Γ. The surface area $\mathcal{A}(\Gamma(Z))$ is defined to be the sum of the areas of the triangular faces of $\Gamma(Z)$.

---

3. Note that the vertices of $\Gamma(Z)$ are **on** Γ and not merely "close to" Γ, as they are in a digitization. Also, in picture analysis, Γ is "unknown," so we have no way of constructing $\Gamma(Z)$.

Definition 8.4 Suppose there exists a sequence $Z_1, Z_2, Z_3, \ldots$ of triangular subdivisions of B_1 with respect to B such that $a_t \to 0$, where a_t is the maximum length of any side of any triangle in Z_t. This sequence defines a sequence of polyhedral approximations $\Gamma(Z_1), \Gamma(Z_2), \Gamma(Z_3), \ldots$ of Γ that have well-defined surface areas $\mathcal{A}\Gamma(Z_1)), \mathcal{A}(\Gamma(Z_2)), \mathcal{A}(\Gamma(Z_3))$, ... We say that Γ is measurable if the following is bounded:

$$\mathcal{A}(\Gamma) = \sup_t \mathcal{A}(\Gamma(Z_t))$$

Theorem 8.3 If Γ is a measurable smooth surface patch, we have the following,

$$\mathcal{A}(\Gamma) = \int_B \sqrt{D_1^2 + D_2^2 + D_3^2} \, \mathrm{d}(u,v)$$

which is independent of the parameterization of Γ, where D_1, D_2, and D_3 are the subdeterminants defined in Equation 8.22.

Let Γ be a Monge surface patch defined by $z = f(x,y)$ for (x,y) in a closed bounded measurable set $B \subset \mathbb{R}^2$, where the first-order partial derivatives of f exist and are continuous on a set B_1 such that $B \subseteq B_1^\circ$. By Theorem 8.3, the area of $\Gamma = \{(x,y,f(x,y)) : (x,y) \in B\}$ is as follows:

$$\mathcal{A}(\Gamma) = \int_B \sqrt{1 + (f_x(x,y))^2 + (f_y(x,y))^2} \, \mathrm{d}(x,y) \qquad (8.24)$$

The vector $(f_x(x,y), f_y(x,y))$ is called the *gradient* of Γ at $(x,y) \in B$. Let $\mathbf{n}_p(x,y) = (-f_x(x,y), -f_y(x,y), 1)$ and $\mathbf{n}_n(x,y) = (f_x(x,y), f_y(x,y), -1)$; these vectors are the *normals* to Γ at (x,y). Then we have the following:

$$\mathcal{A}(\Gamma) = \int_B \left\| \mathbf{n}_p(x,y) \right\|_2 \mathrm{d}(x,y) = \int_B \left\| \mathbf{n}_n(x,y) \right\|_2 \mathrm{d}(x,y) \qquad (8.25)$$

This formula can be used to design surface area estimators for digital sets using estimated surface normals.

The formula in Theorem 8.3 can be rewritten in terms of the following expressions, which define the *first fundamental form* (E,F,G) of Γ:

$$E = x_u^2 + y_u^2 + z_u^2, \quad F = x_u x_v + y_u y_v + z_u z_v, \quad G = x_v^2 + y_v^2 + z_v^2$$

This form satisfies both of the following conditions:

$$D_1^2 + D_2^2 + D_3^2 = EG - F^2$$

and

$$\mathcal{A}(\Gamma) = \int_B \sqrt{EG - F^2} \, \mathrm{d}(u,v) \qquad (8.26)$$

8.3 Surfaces and Solids

Chapter 11 will discuss the estimation of the surface area of a 3D digital object based on approximating its surface by, for example, planar surface patches. A 3D picture can be regarded as a sequence of 2D pictures, each of which is a digitization of a 2D cross-section of the scene. The inference of 3D properties from 2D cross-sections is studied in *stereology* using probability-theoretic models for the sampling processes involved; see Section 1.2.11. A general formula of stereology in traditional "stereologic notation" is as follows,

$$\mathcal{V}_V = \mathcal{A}_A = \mathcal{L}_L = \mathcal{P}_P \tag{8.27}$$

where $\mathcal{V}_V$ is the volume of a solid K (in unit test volumes), $\mathcal{A}_A$ is the area of K in test planes per unit test area, $\mathcal{L}_L$ is the length of line intercepts with K in test lines per unit test line length, and $\mathcal{P}_P$ is the ratio between the number of points in K and the total number of test points. Equation 8.27 assumes that these measurements are statistically uniform. To estimate surface area, we can use the following identities,

$$\mathcal{S}_V = \frac{4}{\pi}\mathcal{L}_A = 2\mathcal{P}_L = 2\frac{\mathcal{P}_V}{\mathcal{L}_V} = 4\frac{\mathcal{P}_A \mathcal{P}_L}{\mathcal{L}_V} = \frac{\mathcal{P}_V}{\mathcal{P}_A} \tag{8.28}$$

where $\mathcal{S}_V$ is surface area per unit test volume, $\mathcal{L}_A$ is the length of line elements per unit test area, $\mathcal{P}_L$ is the number of points per unit test line length, $\mathcal{P}_V$ is the number of points per unit test volume, $\mathcal{L}_V$ is the length of line segments per unit test volume, and $\mathcal{P}_A$ is the number of points per unit test area.

8.3.3 Example: an ellipsoid

To illustrate the difficulty of calculating surface area, we consider the example of an ellipsoid. Gauss, or outer and inner Jordan, digitizations (see Figure 8.12) of ellipsoids or combinations of ellipsoids (e.g., subtracting smaller ellipsoids from larger ones,

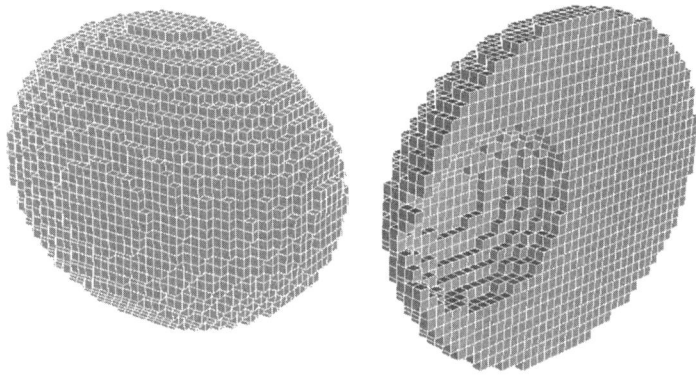

FIGURE 8.12 Left: a digitized ellipsoid. Right: a slice through a digitized ellipsoid from which a smaller ellipsoid that touches the larger ellipsoid at one surface point has been subtracted.

thereby defining holes or cavities within the larger ellipsoids) provide useful data sets for testing 3D algorithms (e.g., for estimating surface area and comparing it with the correct value).

An ellipsoid is certainly an "elementary" solid. Its surface area can be expressed by elementary analytic functions if two of its radii are equal (i.e., if it is an ellipsoid of revolution [216]). It was asserted in 1979 [494] that, "Except for the special cases of the sphere, the prolate spheroid, and the oblate spheroid, no closed form expression exists for the surface area of the ellipsoid. This situation arises because of the fact that it is impossible to carry out the integration in the expression for the surface area in closed form for the most general case of three unequal axes. In spite of the widespread use of the ellipsoid as a mathematical model and the extensive knowledge about the theory of curved surfaces and numeric integration, the problem of approximating the surface area of a triaxial ellipsoid does not appear to have been addressed."

In fact, in 1825, A.M. Legendre (p. 352–359 in [640]) expressed the surface area of a general ellipsoid in terms of incomplete elliptical integrals (for an accessible reference, see [628]). If the ellipsoid is nearly spherical, Legendre's explicit formula is unsuitable for numerical computation, but a rapidly convergent series can be used.[4] A simpler computation of the surface area of a general ellipsoid is due to G. Tee [498]; it is not limited to be an ellipsoid of revolution. This computation is given here next; it allows accurate area estimation independent of the parameters of the ellipsoid.

An ellipsoid $E_{a,b,c}$ with semiaxes a, b, and c centered at the origin and with axes of symmetry along the coordinate axes has the following equation:

$$\frac{x^2}{a^2} + \frac{y^2}{b^2} + \frac{z^2}{c^2} = 1$$

If two semiaxes are equal (e.g., $b = c$), $E_{a,b,c}$ is generated by rotation around the x-axis of the halfellipse, with $y \geq 0$:

$$\frac{x^2}{a^2} + \frac{y^2}{b^2} = 1$$

The surface area $\mathcal{A}(E_{a,b,b})$ is as follows, where $u = x/a$ and $q = 1 - b^2/a^2$:

$$4\pi ab \int_0^1 \sqrt{1 - qu^2}\, du$$

Therefore, we have the following:

$$\mathcal{A}(E_{a,b,b}) = \begin{cases} 2\pi b\left(a \times \frac{\arcsin\sqrt{q}}{\sqrt{q}} + b\right) & \text{if } q > 0 \\ 2\pi b(a+b) & \text{if } q = 0 \\ 2\pi b\left(a \times \frac{\operatorname{arcsinh}\sqrt{-q}}{\sqrt{-q}} + b\right) & \text{if } q < 0 \end{cases}$$

---

4. The website http://documents.wolfram.com/v4/MainBook/G.1.7.html gives an obscure statement, without proof or references, of an incorrect version of Legendre's formula for the surface area of a general ellipsoid; see also [1130] (p. 976). Corrections have been reported to the administrator of that website, but without effect. A corrected version of that formula (also without proof or references) is given at http://home.att.net/numericana/answer/ellipsoid.htm in lieu of a correction, which should have appeared on the first website.

8.3 Surfaces and Solids

The truncated power series shown here can be used if $|q| \ll 1$:

$$A(E_{a,b,b}) = 2\pi b \left(a \left[1 + \tfrac{1}{6}q + \tfrac{3}{40}q^2 + \tfrac{5}{112}q^3\right] + b \right)$$

$A(E_{a,b,c})$ can be calculated using Equation 8.24. Without loss of generality, the coordinate axes can be chosen so that $a \geq b \geq c$. The surface $z = f(x,y)$ of $E_{a,b,c}$ satisfies the following conditions:

$$\frac{\partial f}{\partial x} = \frac{-c^2 x}{a^2 z}, \qquad \frac{\partial f}{\partial y} = \frac{-c^2 y}{b^2 z}$$

In the octant in which x, y, and z are all nonnegative, the surface area is as follows:

$$\int_0^a \int_0^{b\sqrt{1-x^2/a^2}} \sqrt{1 + \frac{c^4 x^2}{a^4 z^2} + \frac{c^4 y^2}{b^4 z^2}}\, dy\, dx$$

$$= \int_0^a \int_0^{b\sqrt{1-x^2/a^2}} \sqrt{\frac{1 - \frac{x^2}{a^2} - \frac{y^2}{b^2} + \frac{c^2}{a^2}\frac{x^2}{a^2} + \frac{c^2}{b^2}\frac{y^2}{b^2}}{1 - \frac{x^2}{a^2} - \frac{y^2}{b^2}}}\, dy\, dx$$

The integral with respect to y is as follows,

$$\int_0^{b\sqrt{1-x^2/a^2}} \sqrt{\frac{1 - \left(1 - \frac{c^2}{a^2}\right)\frac{x^2}{a^2} - \left(1 - \frac{c^2}{b^2}\right)\frac{y^2}{b^2}}{1 - \frac{x^2}{a^2} - \frac{y^2}{b^2}}}\, dy = b\sqrt{1 - \left(1 - \frac{c^2}{a^2}\right)\frac{x^2}{a^2}}\, h(m)$$

where $s = x/a$, $t = y / \left[b\sqrt{1-s^2}\right]$,

$$m = \frac{\left(1 - \frac{c^2}{b^2}\right)(1 - s^2)}{1 - \left(1 - \frac{c^2}{a^2}\right)s^2}, \tag{8.29}$$

and

$$h(m) = \int_0^1 \sqrt{\frac{1 - mt^2}{1 - t^2}}\, dt$$

is Legendre's complete elliptical integral of the second kind.[5] Hence, we have the following:

$$A(E_{a,b,c}) = 8b \int_0^a \sqrt{1 - \left(1 - \frac{c^2}{a^2}\right)\frac{x^2}{a^2}}\, h(m)\, dx$$

$$= 8ab \int_0^1 \sqrt{1 - \left(1 - \frac{c^2}{a^2}\right)s^2}\, h(m)\, ds$$

---

5. $E(m)$ in Milne-Thomson's notation in Chapter 17 of [3].

This expression can be evaluated numerically. Archimedes showed that, for $a = b = c$, the integrand is constant at $\pi/2$; see [1083]. As $c/a \searrow 0$, the integrand converges to $\sqrt{1-s^2} = \sqrt{(1-s)(1+s)}$, which has a singular derivative at $s=1$. To get a smoother integrand, let $u^2 = 1-s$; then the following is true,

$$\mathcal{A}(E_{a,b,c}) = 16ab \int_0^1 \sqrt{1 - \left(1 - \frac{c^2}{a^2}\right) s^2} \, h(m) \, u \, du \qquad (8.30)$$

where $s = 1 - u^2$ and m ($0 \le m \le 1$) is as defined in Equation 8.29. The integral $h(m)$ decreases from $\pi/2$ to 1 as m increases. The Arithmetic-Geometric Mean of C.F. Gauss is a very efficient tool for evaluating both $h(m)$ and Legendre's complete elliptical integral $k(m)$ of the first kind (see Milne-Thomson's Section 17.6 in [3]):

$$k(m) = \int_0^1 \frac{dt}{\sqrt{(1-t^2)(1-mt^2)}}$$

However, if m is very close to 1, the power series in $m_1 = 1 - m$ (p. 54 in [168]) should be used to evaluate $k(m)$ for $m < 1$ and to evaluate $h(m)$ for $m \le 1$.

The integrand with respect to u in Equation 8.30 is smooth enough for the integral to be evaluated by Romberg integration for $c > 0$. As $c/a \searrow 0$, the ellipsoid converges to a two-sided elliptical lamina with surface area $2\pi ab$. Hence, for $c/a \ll 1$, the area should be $2\pi ab$ plus some terms in c/a.

8.3.4 Gauss' definition of surface curvature

Let Γ be a smooth surface patch such that $\Gamma(u,v) = (x(u,v), y(u,v), z(u,v))$ and $(u,v) \in B$ is not a singular point. Then $\Gamma_u(u,v) = (x_u(u,v), y_u(u,v), z_u(u,v))$ and $\Gamma_v(u,v) = (x_v(u,v), y_v(u,v), z_v(u,v))$ are *tangent vectors* to Γ at $\Gamma(u,v)$. These vectors are not parallel; they therefore define a plane called the *tangent plane* to Γ at $\Gamma(u,v)$ that contains all of the tangent vectors to Γ at $\Gamma(u,v)$.

The cross product $\Gamma_u(u,v) \times \Gamma_v(u,v)$ defines a normal direction perpendicular to the tangent plane. The unit *surface normal* at $\Gamma(u,v)$ is as follows:

$$\mathbf{n}(u,v) = \frac{\Gamma_u(u,v) \times \Gamma_v(u,v)}{\|\Gamma_u(u,v) \times \Gamma_v(u,v)\|_2}$$

The normal points inward or outward, depending on whether its sign is positive or negative.

The vector emanating from the origin in direction $\mathbf{n}(u,v)$ intersects the surface of the unit sphere at a point. This defines a mapping of Γ onto the unit sphere called the *Gauss map* of Γ. C.F. Gauss defined surface curvature based on this mapping; the unit sphere is therefore sometimes called the *Gaussian sphere*.

Let $U_\varepsilon(p)$ be a small neighborhood of a point $p \in \Gamma$ defined by intersecting Γ with a sphere of radius $\varepsilon > 0$ centered at p. Then the following is a smooth surface patch on the unit sphere:

$$\mathbf{n}(U_\varepsilon(p)) = \{\mathbf{n}(u,v) : \Gamma(u,v) \in U_\varepsilon(p)\}$$

8.3 Surfaces and Solids

Gauss originally defined the *curvature* of Γ at p as the ratio between the area of this patch and the area of $U_\varepsilon(p)$ as $\varepsilon \to 0$:

Definition 8.5 (C.F. Gauss, 1828) The (historic) *Gaussian curvature* of a smooth surface patch Γ at $p = \Gamma(u,v)$ is as follows:

$$\kappa_G(u,v) = \lim_{\varepsilon \to 0} \frac{\mathcal{A}\big(\mathbf{n}(U_\varepsilon(p))\big)}{\mathcal{A}(U_\varepsilon(p))}$$

For numeric calculations, it is usually easier to replace the circular U-neighborhoods with rectangular neighborhoods. A point on the Gaussian sphere is specified by its longitude ξ and latitude η. Let δR be the ("rectangular") surface patch on the Gaussian sphere between ξ and $\xi + \delta\xi$ and between η and $\eta + \delta\eta$; thus $\mathcal{A}(\delta R) = \delta\xi\,\delta\eta\,\cos\eta$. Let $\delta\Gamma_{(u,v)}$ be the patch of Γ around $\Gamma(u,v)$ that defines δR. Then we have the following:

$$\kappa_G(u,v) = \lim_{\delta\eta \to 0} \lim_{\delta\xi \to 0} \frac{\delta\xi\,\delta\eta\,\cos\eta}{\mathcal{A}(\delta\Gamma_{(u,v)})}$$

For example, a sphere of radius $r > 0$ has constant Gaussian curvature $1/r^2$.

In computer vision, $G(\xi,\eta) = 1/\kappa_G(u,v)$ is called the *Gaussian image* of Γ; it is a "labeled" Gauss map. If $\kappa_G(u,v) = 0$ (e.g., if Γ is locally planar at $\Gamma(u,v)$), G has a point impulse at (ξ,η) (defined by the surface normal at $\Gamma(u,v)$); the impulse is "labeled" by the surface area that contributes to the impulse.

A finite convex polyhedron that has $n > 0$ faces has a Gaussian image that consists of n point impulses, each labeled by the area of the corresponding face. H. Minkowski has shown that convex polyhedra are determined up to translation by their face areas and normals; hence the labeled Gaussian image determines a convex polyhedron up to translation.

As an example of the calculation of Gaussian curvature, consider again the ellipsoid $E_{a,b,c}$ defined by the following:

$$\Gamma(\phi,\psi) = (a\cos\psi\cos\phi, b\sin\psi\cos\phi, c\sin\phi)$$

(see Figure 8.13 for the definitions of the angles ϕ and ψ). Here the Gaussian curvature is as follows,

$$\kappa_G(u,v) = \left(\frac{abc}{(bc\cos\psi\cos\phi)^2 + (ac\sin\psi\cos\phi)^2 + (ab\sin\phi)^2}\right)^2$$

and the Gaussian image is as follows, where ξ and η are longitude and latitude on the Gaussian sphere:

$$G(\xi,\eta) = \left(\frac{abc}{(a\cos\xi\cos\eta)^2 + (b\sin\xi\cos\eta)^2 + (c\sin\eta)^2}\right)^2$$

Two smooth surface patches Γ_1 and Γ_2 are called *isometric* iff there is a one-to-one mapping φ from Γ_1 onto Γ_2 such that φ and φ^{-1} are both differentiable and φ

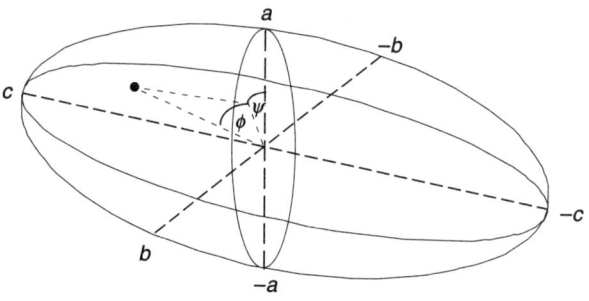

FIGURE 8.13 The angles ϕ and ψ in a general ellipsoid.

maps any smooth arc in Γ_1 onto a smooth arc in Γ_2 of the same length. It is proved in differential geometry that two smooth surface patches are isometric iff they have identical first fundamental forms (E, F, G).

A *local isometry* φ from Γ_1 onto Γ_2 is a one-to-one differentiable function such that, for all $p \in \Gamma_1$, there exist neighborhoods $U(p)$ in Γ_1 and $U(\varphi(p))$ in Γ_2 such that φ is an isometry between the neighborhoods. For example, the surface of an infinite cylinder is locally isometric to a plane.

Theorem 8.4 (Teorema Egregium C.F. Gauss, 1828) If a smooth surface patch can be mapped locally isometrically onto another such patch, the values of their (historic) Gaussian curvatures are the same at corresponding points.

This theorem points out a limitation of the historic Gaussian curvature. We are usually interested in distinguishing between the local curvature of a plane (a *planar point*) and that of a cylinder (a *parabolic point*), but they are not distinguishable when the historic definition is used. Also, the historic Gaussian curvature is always nonnegative; it does not allow us to distinguish between the local curvature of an elliptical surface patch (an *elliptic point*) and that of a hyperbolic surface patch (a *hyperbolic point*). To overcome these limitations, differential geometry has introduced other measures of surface curvature (e.g., based on pairs of curves lying on the surface patch and intersecting at the given point). (The curvature of a space curve was defined in Section 8.2.) These measures will be defined in the next section.

8.3.5 Principal, Gaussian, and mean surface curvature

Let Γ be a Monge patch defined by $z = f(x_1, x_2)$ on a planar region B, where f is in $C^{(2)}(B)$, so that it is twice continuously differentiable for all $(x_1, x_2) \in B$ and therefore satisfies the *integrability condition*:

$$f_{x_1 x_2} = \frac{\partial^2 f(x_1, x_2)}{\partial x_1 \partial x_2} = \frac{\partial^2 f(x_1, x_2)}{\partial x_2 \partial x_1} = f_{x_2 x_1}$$

8.3 Surfaces and Solids

We calculate the surface curvature at $p \in \Gamma$ using arcs Γ_1 and Γ_2 that are contained in Γ, pass through p, and are not parallel in a neighborhood of p. Let $\mathbf{t}_1$ and $\mathbf{t}_2$ be the tangent vectors to Γ_1 and Γ_2 at p. These vectors span the tangent plane Π_p to Γ at p. We assume angular orientations $0 \leq \eta < \pi$ in Π_p (a halfcircle centered at p).

The surface normal $\mathbf{n}_p$ at p is orthogonal to Π_p and collinear with the cross product $\mathbf{t}_1 \times \mathbf{t}_2$. Let Π_η be a plane that contains $\mathbf{n}_p$ and has orientation η; see Figure 8.14. Π_η makes a dihedral angle η with Π_0 and cuts Γ in an arc γ_η. (Π_η may cut Γ in several arcs or curves, but we consider only the one that contains p.) Let $\mathbf{t}_\eta$, $\mathbf{n}_\eta$, and κ_η be the tangent, normal, and curvature of γ_η at p. κ_η is the *normal curvature* of any arc $\gamma \subset \Gamma \cap \Pi_\eta$ at p that is incident with p. For example, take f to be the constant function with a value of 0 everywhere. Then the Monge patch is part of the horizontal plane through $o = (0,0,0)$ in $\mathbb{R}^3$. Any γ_η is a straight line segment in the plane that is incident with o; we have $\kappa_\eta = 0$ and $\mathbf{t}_\eta = \gamma_\eta$. If we take f to be a "cap" of a sphere centered at its north pole p, γ_η is a segment of a great circle on the sphere, $\mathbf{n}_\eta$ is incident with the straight line passing through p and the center of the sphere, and $\kappa_\eta = 1/r$ where r is the radius of the sphere.

The *characteristic polynomial* p of an $n \times n$ matrix $\mathbf{A}$ is defined as $p(\lambda) = \det(\mathbf{A} - \lambda \mathbf{I}) = (-\lambda)^n + \cdots + \det(\mathbf{A})$, where $\mathbf{I}$ is the $n \times n$ identity matrix. According to the Cayley–Hamilton Theorem [311], $p(\mathbf{A}) = \mathbf{0}$. The *eigenvalues* λ_i of an $n \times n$ matrix $\mathbf{A}$ are the n roots of its characteristic equation $\det\left(\mathbf{A} - \lambda \mathbf{I}\right) = 0$.

Let $\mathbf{v}$ be a unit vector in Π_p. The negative derivative $-D_\mathbf{v} \mathbf{n}$ of the unit normal vector field $\mathbf{n}$ of a surface, which is regarded as a linear map from Π_p to itself, is called the *shape operator* (or *Weingarten map* or *second fundamental tensor*) of the surface. Let $\mathbf{M}_p$ be the Weingarten map in matrix representation at p (with respect to any orthonormal basis in Π_p), and let λ_1 and λ_2 be the eigenvalues of the 2×2 matrix $\mathbf{M}_p$. Note that these eigenvalues do not depend on the choice of orthonormal basis in Π_p.

Definition 8.6 λ_1 and λ_2 are called the *principal curvatures* or *main curvatures* of Γ at p. The product $\lambda_1 \lambda_2$ is called the *Gaussian curvature*, and the mean $(\lambda_1 + \lambda_2)/2$ is called the *mean curvature*.

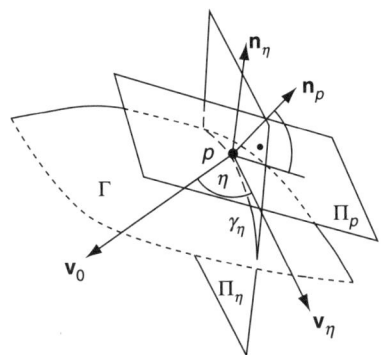

FIGURE 8.14 A surface Γ cut by a plane Π_η.

It is proved in differential geometry [1014] that the absolute value of the Gaussian curvature is equal to the historic Gaussian curvature of Definition 8.5. The mean curvature is equal to $(\kappa_\eta(p)+\kappa_{\eta+\pi/2}(p))/2$ for any $\eta \in [0,\pi)$.

In the case of a Monge patch, it is natural to choose $\mathbf{b}_1 = (1,0,f_{x_1})$ and $\mathbf{b}_2 = (0,1,f_{x_2})$ as basis elements for Π_p; they are the images of the usual basis $\mathbf{e}_1 = (1,0)$ and $\mathbf{e}_2 = (0,1)$ of $\mathbb{R}^2$ under differentiation of the Monge patch. We then calculate the following matrix of inner products,

$$\mathbf{P}_p = \begin{pmatrix} \langle -D_{\mathbf{b}_1}\mathbf{n}_p, \mathbf{b}_1 \rangle & \langle -D_{\mathbf{b}_1}\mathbf{n}_p, \mathbf{b}_2 \rangle \\ \langle -D_{\mathbf{b}_2}\mathbf{n}_p, \mathbf{b}_1 \rangle & \langle -D_{\mathbf{b}_2}\mathbf{n}_p, \mathbf{b}_2 \rangle \end{pmatrix}$$

which unfortunately is *not* the matrix $\mathbf{M}_p$ of the Weingarten map with respect to the basis $\{\mathbf{b}_1, \mathbf{b}_2\}$, because the latter is not necessarily orthonormal. The relationship between $\mathbf{P}_p$ and $\mathbf{M}_p$ is explained in [1014] (at the end of p. 50 of Volume 3),

$$\mathbf{M}_p = \mathbf{G}_p^{-1}\mathbf{P}_p$$

where $\mathbf{G}_p$ is the metric tensor of Γ at p and we use the fact that $\mathbf{G}_p$ and $\mathbf{P}_p$ are symmetric matrices:

$$\mathbf{G}_p = \begin{pmatrix} \langle \mathbf{b}_1, \mathbf{b}_1 \rangle = 1+f_{x_1}^2 & \langle \mathbf{b}_1, \mathbf{b}_2 \rangle = f_{x_1}f_{x_2} \\ \langle \mathbf{b}_2, \mathbf{b}_1 \rangle = f_{x_2}f_{x_1} & \langle \mathbf{b}_2, \mathbf{b}_2 \rangle = 1+f_{x_2}^2 \end{pmatrix}$$

The following 2×2 matrix is called the *Hessian matrix* of Γ[6]:

$$\mathbf{H}(x_1, x_2) = ((a_{ij}))_{1 \le i,j \le 2} \quad \text{where } a_{ij} = \frac{\partial^2 f(x_1, x_2)}{\partial x_i \partial x_j}$$

The Monge patch satisfies the integrability condition $a_{12} = a_{21}$ (i.e., its Hessian matrix is also symmetric).

Theorem 8.5 The Gaussian curvature of the surface defined by $z = f(x_1, x_2)$ is as follows:

$$\frac{f_{x_1 x_1} f_{x_2 x_2} - f_{x_1 x_2}^2}{\left(1 + f_{x_1}^2 + f_{x_2}^2\right)^2} = \frac{\det \mathbf{H}}{\left(1 + f_{x_1}^2 + f_{x_2}^2\right)^2}$$

$\mathbf{M}_p$ has a quadratic characteristic equation with two eigenvalues λ_1 and λ_2 at any surface point p. These eigenvalues define a pair of orthogonal eigenvectors $\mathbf{v}_1$ and $\mathbf{v}_2$ that satisfy the equations $\mathbf{M}_p \mathbf{v}_1 = \lambda_1 \mathbf{v}_1$ and $\mathbf{M}_p \mathbf{v}_2 = \lambda_2 \mathbf{v}_2$. By the symmetry of

---

6. Named after the German mathematician L.O. Hesse (1811–1874).

8.3 Surfaces and Solids

$\mathbf{M}_p$ (i.e., $\mathbf{M}_p = \mathbf{M}_p^T$), we have the following:

$$\begin{aligned}\lambda_1 \langle \mathbf{v}_1, \mathbf{v}_2 \rangle &= \langle \lambda_1 \mathbf{v}_1, \mathbf{v}_2 \rangle = \langle \mathbf{M}_p \mathbf{v}_1, \mathbf{v}_2 \rangle \\ &= \langle \mathbf{v}_1, \mathbf{M}_p^T \mathbf{v}_2 \rangle = \langle \mathbf{v}_1, \mathbf{M}_p \mathbf{v}_2 \rangle = \langle \mathbf{v}_1, \lambda_2 \mathbf{v}_2 \rangle \\ &= \lambda_2 \langle \mathbf{v}_1, \mathbf{v}_2 \rangle\end{aligned}$$

It follows that $\lambda_1 \neq \lambda_2$ implies $\langle \mathbf{v}_1, \mathbf{v}_2 \rangle = 0$, so $\mathbf{v}_1$ and $\mathbf{v}_2$ are orthogonal. (Sometimes, $\lambda_1 = \lambda_2$ [e.g., for a planar patch or a patch on a sphere]. In such cases, $\mathbf{v}_1$ and $\mathbf{v}_2$ are chosen as an arbitrary orthonormal pair of basis elements of Π_p.) Let $\eta = 0$ be the direction of $\mathbf{v}_1$ or $\mathbf{v}_2$ in Π_p (i.e., either λ_1 or λ_2 is $\kappa_0(p)$).

Now let $\mathbf{w}_1$ and $\mathbf{w}_2$ be any two orthogonal vectors that span the tangent plane Π_p to Γ at p (i.e., they are tangent vectors that define normal curvatures in directions η and $\eta + \pi/2$). Then the *Euler formula*

$$\kappa_\eta(p) = \lambda_1 \cdot \cos(\eta)^2 + \lambda_2 \cdot \sin(\eta)^2 \tag{8.31}$$

allows us to calculate the normal curvature $\kappa_\eta(p)$ in any direction η at p from the principal curvatures λ_1 and λ_2 and the angle η.

Chapter 11 will discuss the estimation of mean surface curvature based on digital approximations to the surface. We can locally estimate the tangent plane Π_p and thus the normal $\mathbf{n}_p$ at any point p on Γ (see Figure 8.14). We can also cut Γ at p by a plane Π_c in some direction (e.g., parallel to the xz-plane at the y-coordinate of p). The intersection of Π_c with Γ defines an arc γ_c in the neighborhood of p, and we can estimate the curvature κ_c of γ_c at p. However, we cannot assume that Π_c is incident with the surface normal $\mathbf{n}_p$ at p. Let $\mathbf{n}_c$ be the principal normal of γ_c at p. A theorem of Meusnier[7] [727] tells us that the normal curvature κ_η in any direction η is related to the curvature κ_c and the normals $\mathbf{n}_p$ and $\mathbf{n}_c$ by the following:

$$\kappa_\eta = \kappa_c \cdot \cos(\mathbf{n}_p, \mathbf{n}_c)$$

By estimating two normal curvatures κ_η and $\kappa_{\eta+\pi/2}$, we can estimate the mean curvature.

8.3.6 Volume

Let $K \subset \mathbb{R}^3$ be a compact set and $f(x,y,z)$ a function such that $0 \leq f(x,y,z) \leq 1$ for all $(x,y,z) \in K$. $f(x,y,z)$ can be regarded as the "density" of K at $p = (x,y,z) \in K$. The following integral defines the *volume* of f on K:

$$I = \iiint_K f(x,y,z)\, \mathrm{d}x\mathrm{d}y\mathrm{d}z$$

---

7. French aeronautic theorist and military general J.B.M. Meusnier (1754–1793).

If $f(x,y,z) = 1$ iff $(x,y,z) \in K$, I is the *volume* $\mathcal{V}(K)$. K is *integrable* iff it has a well-defined and finite volume. Every compact set is integrable.

Section 2.3.2 described Jordan's method of estimating the volume of a set using its inner and outer digitizations. It estimates the (Jordan) volume $\mathcal{V}(S)$ of S by its inner or outer volume, provided that these volumes converge to the same limit as the grid constant goes to zero.

Grid-based studies by Jordan and Peano initiated the field of combinatorial geometry (the geometry of numbers; see Section 1.2.8). A fundamental theorem of H. Minkowski says that a convex subset of $\mathbb{R}^n$ that is symmetric about $o = (0,\ldots,0)$ (i.e., if p is in the set so is $-p$) and that has a volume of at least 2^n contains at least two grid points in addition to o.

Cavalieri's Principle[8] [1031] is often used for area or volume estimation. Let R_1 and R_2 be two 2D regions that lie between two parallel lines Γ_1 and Γ_2. Suppose any line between Γ_1 and Γ_2 and parallel to them cuts R_1 and R_2 into straight line segments of equal (total) length; then R_1 and R_2 have equal area. Similarly, let R_1 and R_2 be two solids that lie between two parallel planes. If any plane between the two planes and parallel to them cuts the solids in regions of equal (total) area, then the solids have equal volume.

Estimating the volume of a digital set by counting voxels is justified by Cavalieri's Principle and by Jordan's method of volume measurement. Volume estimation can also be based on polyhedral approximation (as in the use of polygonal approximation for area estimation that was discussed at the end of Section 8.1.5) or on integral-geometry–based methods. For the latter approach, we briefly state a few basic equations.

Let $B_r = \{p \in \mathbb{R}^n : \|p\|_e \leq r\}$ be the n-dimensional ball of radius $r > 0$ around the origin, let $M \subset \mathbb{R}^n$, and let $M_\rho = M \oplus B_r$ be the Minkowski sum of M and B_r. An *oval* is a bounded closed convex subset of $\mathbb{R}^n$. For example, if $n = 1$, M is a straight segment, and $\mathcal{L}(M_r) = \mathcal{L}(M) + 2r$. Let $M \subset \mathbb{R}^2$ be an oval with $\mathcal{A}(M) > 0$; then we have the following:

$$\mathcal{A}(M_r) = \mathcal{A}(M) + \mathcal{P}(M)r + \pi r^2 \tag{8.32}$$

If $M \subset \mathbb{R}^3$ is an oval with $\mathcal{V}(M) > 0$, then the following is true, where $\overline{\mathcal{W}}(M)$ is the *mean width* of the oval M:

$$\mathcal{V}(M_r) = \mathcal{V}(M) + \mathcal{A}(M)r + 2\pi\overline{\mathcal{W}}(M)r^2 + \tfrac{4}{3}\pi r^3 \tag{8.33}$$

Both equations follow from *Steiner's formula*; see page 220 in [958]. The volume determines the surface area:

$$\mathcal{A}(M) = \lim_{r \to 0} \frac{\mathcal{V}(M_r) - \mathcal{V}(M)}{r}$$

---

8. Named after the Italian mathematician B. Cavalieri (1598–1647), a pupil of G. Galilei.

8.3 Surfaces and Solids

The mean width of a simple convex polyhedron M that has α_1 edges is given by the following,

$$\overline{\mathcal{W}}(M) = \frac{1}{4\pi} \sum_{i=1}^{\alpha_1} l_i \eta_i \tag{8.34}$$

where l_i is the length of the ith edge e_i and α_i is the angle between the normals of the faces that intersect in e_i. For the general case (i.e., the mean width of an arbitrary oval), see [958].

$2\pi\overline{\mathcal{W}}(M)$ is also called the *mean curvature* $\mathcal{M}(M)$ of the oval M. The *total curvature* $\mathcal{C}(M)$ is 4π for all ovals of nonzero volume. In general, these properties are defined by the following:

$$\mathcal{M}(M) = \tfrac{1}{2} \lim_{r \to 0} \frac{\mathcal{A}(M_r) - \mathcal{A}(M)}{r} \quad \text{and} \quad \mathcal{C}(M) = \lim_{r \to 0} \frac{\mathcal{M}(M_r) - \mathcal{M}(M)}{r}$$

Equation 8.33 becomes the following:

$$\mathcal{V}(M_r) = \mathcal{V}(M) + \mathcal{A}(M)r + \mathcal{M}(M)r^2 + \tfrac{1}{3}\mathcal{C}(M)r^3$$

Steiner's formula also allows us to conclude the following (for an oval $M \subset \mathbb{R}^3$):

$$\begin{aligned}
\mathcal{A}(M_r) &= \mathcal{A}(M) + 2\mathcal{M}(M)r + \mathcal{C}(M)r^2 \\
\mathcal{M}(M_r) &= \mathcal{M}(M) + \mathcal{C}(M)r \\
\mathcal{C}(M_r) &= \mathcal{C}(M)
\end{aligned}$$

Equations 1.10 and 1.11 also deal with ovals in $\mathbb{R}^3$.

8.3.7 Isothetic polyhedra

Sections 5.2.2 and 5.2.3 discussed combinatorial topology for incidence pseudographs, and Section 5.3 discussed it in detail for regular infinite incidence pseudographs. The results can be interpreted as counting formulas for properties of isothetic grid polygons or polyhedra.

Let Π be the simply connected union of a finite set of 6-connected simple polyhedra. The volume α_3 of Π can be calculated using formulas given in Section 5.3.2 (e.g., on the values α_0 and b_{03} and the constants $a_{03} = a_{30} = 8$). b_{03} can be calculated during frontier tracing, and α_0 can be calculated using discrete column-wise integration as in Section 8.1.6. Section 5.3.2 also provides ways of calculating α_3 based on the values α_1 and b_{13} and the constants $a_{13} = 4$ and $a_{31} = 12$ or on the values α_2 and b_{23} and the constants $a_{23} = 2$ and $a_{32} = 6$.

We now derive a result about the angles of isothetic simple grid polyhedra. The counting formulas in Section 5.3 are formulas for isothetic simple grid polyhedra defined by unions of cells. The angle formula given in this section is based

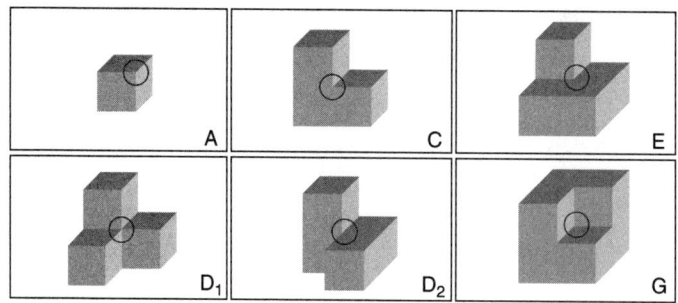

FIGURE 8.15 The six kinds of angles in an isothetic simple polyhedron.

on a geometric construction, unlike the purely combinatorial approach used in Section 5.3. It also provides a method of defining topologic invariants that distinguish inner surfaces from outer surfaces.

There are six kinds of angles in an isothetic simple polyhedron; see Figure 8.15. These angles will be referred to as being of types A, C, D_1, D_2, E, and G, respectively.

Theorem 8.6 Let Π be an isothetic simple polyhedron, and let Π_A, Π_C, Π_{D_1}, Π_{D_2}, Π_E, and Π_G be the numbers of angles of Π of types A, C, D_1, D_2, E, and G. Then we have the following:

$$T = (\Pi_A + \Pi_G) - (\Pi_C + \Pi_E) - 2(\Pi_{D_1} + \Pi_{D_2}) = 8$$

Proof Π can be obtained by a finite sequence of joins of a simple isothetic polyhedron Π_1 to a simple isothetic polyhedron Π_2 in which a face P_2 of Π_2 is brought into coincidence with a subset of a face P_1 of Π_1. We can start the process with two isothetic rectangular parallelepipeds for which $\Pi_A = 8$ and all the other Πs are zero; thus these parallelepipeds satisfy the formula. At each step of the joining process Π_1 and Π_2 can have only A and C angles on their merging faces P_1 and P_2. By Proposition 8.1, the number of A-angles on P_2 is always four greater than the number of C-angles. (P_2 is a subset of P_1; this excludes joining a C-angle on P_2 with an A-angle on P_1 or joining a C-angle on P_2 with a point on an edge of P_1.) Hence there are only six types of joinings, as shown in Figure 8.16:

i. an A-angle on P_2 is joined with an A-angle on P_1;

ii. an A-angle on P_2 is joined with a C-angle on P_1;

iii. an A-angle on P_2 is joined with a point on an edge of P_1;

iv. an A-angle on P_2 is joined with an interior point of P_1;

8.3 Surfaces and Solids

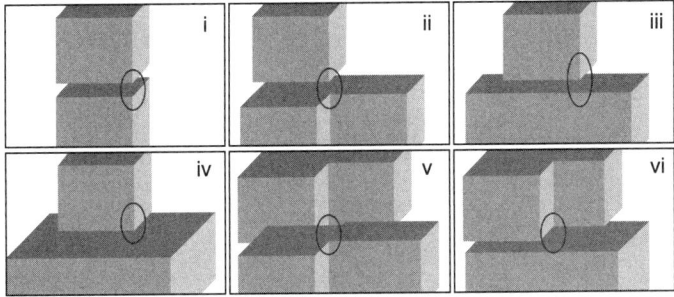

FIGURE 8.16 The six possible joining operations.

v. a C-angle on P_2 is joined with a C-angle on P_1; and

vi. a C-angle on P_2 is joined with an interior point on P_1.

We can assume that Theorem 8.6 is valid for Π_1 and Π_2. Thus, before the joining operation, the total count for the two polyhedra is $T = 16$. We will now prove that $T = 8$ after the joining operation.

In case (i), an A-angle of Π_1 and an A-angle of Π_2 are lost; this decreases T by 2. In case (ii), a C-angle of Π_1 and an A-angle of Π_2 are lost, but the union gains a D_1- or D_2-angle, so T decreases by 2. In case (iii) an A-angle of Π_2 is lost, but the union gains a C-angle, so T decreases by 2. In case (iv), an A-angle of Π_2 is lost, but the union gains an E-angle, so T decreases by 2. In case (v), a C-angle of Π_1 and a C-angle of Π_2 are lost; this increases T by 2. In case (vi), a C-angle of Π_2 is lost, but the union gains a G-angle so T increases by 2.

In summary, the four joinings of A-angles decrease T by 2, and the two joinings of C-angles increase T by 2. The number of A-angles on P_2 is always four more than the number of C-angles; hence, after we perform the joining operations that involve all of the A- and C-angles of P_2, we have $T = 8$. ∎

The formulas in Proposition 8.1 and Theorem 7.13 are both useful for analyzing isothetic frontiers. In a 2D binary picture, an isothetic frontier is inner (the frontier of a union of proper holes) or outer, depending on whether $\Pi_V - \Pi_C < 0$ or > 0. In a 3D binary picture, there are dualities between angles of types A and G, C and E, D_1 and D_1, and D_2 and D_2. Thus $T = 8$ whether we consider the surface as being the outer frontier of a simple polyhedron or an inner frontier of a proper polyhedral hole. However, we can use other quantities to distinguish between inner and outer frontiers; for example, $\Pi_A - \Pi_G$ is negative for an inner frontier and positive for an outer frontier. The quantity T is also of practical interest; for example, it provides a test for whether the entire surface of an isothetic polyhedron has been traced. The six angle counts can also be used as shape descriptors.

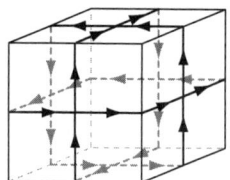

FIGURE 8.17 Three circuits in the graph that represents the faces of a 3-cell.

8.4 Surface Tracing and Approximation

In Section 5.4.4, we briefly discussed the use of the FILL procedure to trace the 3D frontier of a simply connected region. In Section 8.4.1, we will describe a more efficient frontier tracing algorithm. In Section 8.4.2, we will describe a "marching cubes" algorithm that approximates the frontier of a region by constructing a triangulation "between" the border and coborder of the region.

8.4.1 The Artzy-Herman algorithm

The frontier of a 1- or 0-region of voxels splits in general into frontiers of a finite number of 2-regions; we will consider only 2-regions here. The FILL procedure discussed in Section 5.4.4 does not attempt to minimize the number of accesses to frontier faces to determine whether they have already been labeled; it is therefore also applicable to 0- or 1-regions. In a 2-region, any frontier face is edge-adjacent to exactly four[9] other frontier faces; hence FILL makes at most four visits to each frontier face, including the first visit, when the face is labeled. The algorithm described in this section requires at most two visits to each face, including the first visit.

Let $[F, A]$ be the undirected graph in which each node represents a frontier face of a closed 2-region M of voxels in the 3D incidence grid $\mathbb{C}_3$. We call two nodes f_1 and f_2 adjacent ($f_1 A f_2$) iff they are 1-adjacent in $\mathbb{C}_3$. Each $f \in M$ is incident with one 3-cell $I(f)$ in M and with one 3-cell $O(f)$ in $\overline{M}$. Evidently, $f_1 A f_2$ iff $O(f_1) = O(f_2)$ (so that $I(f_1)$ and $I(f_2)$ are 1-adjacent) or $O(f_1)$ and $O(f_2)$ are 2-adjacent (so that $I(f_1)$ and $I(f_2)$ are also 2-adjacent) or $O(f_1)$ and $O(f_2)$ are both 2-adjacent to the same 3-cell in $\overline{M}$ (i.e., $I(f_1) = I(f_2)$). It follows that each node in $[F, A]$ has degree 4.

Let $I(f)$ be the 3-cell shown in Figure 8.17. In the graph representing the faces of $I(f)$, there are three circuits, each of which is parallel to one of the coordinate planes. Each face has two outgoing and two incoming edges; each of these edges goes from one face to another by "crossing" a 1-cell (grid edge). Each grid edge in the frontier of M is incident with exactly two frontier faces of M. If we start at any face f in the frontier of M, the two outgoing edges of f on $I(f)$ point to two frontier

---

9. See Section 3.1.3 in [1107] for a proof, based on invalid edges in the 6-adjacency grid, that $\nu(p) = 4$ for any frontier face p in this adjacency graph.

> 1. Create a list F of faces and push f_0 into F; create a queue Q of faces, and push f_0 into Q; create a list L of labeled faces, and push f_0 twice into L.
> 2. While the queue Q is not empty:
>   a) pop face f out of Q (and delete it from Q),
>   b) for both out-faces g of f, do the following:
>   i. if g is on L, delete it from L,
>   ii. otherwise, push g into F, push g into Q, and push g into L.

ALGORITHM 8.1 The Artzy-Herman algorithm for tracing all faces of a 3D frontier.

faces called the *out-faces* of f, and the two incoming edges are outgoing edges of two *in-faces* of f. The frontier tracing algorithm makes use of this directed graph structure (Algorithm 8.1). We accumulate all of the frontier faces of M in the list F. The queue Q implies a breadth-first access order; using a stack instead implies a depth-first access order. We can return to a face at most twice; the second copy of f_0 on L is removed when we enter Step 2.a.i. for the first time.

8.4.2 Marching cubes

A set of voxels in a 3D picture P is supposed to be a representation (e.g., a Gauss digitization) of an unknown 3D object with an unknown surface. We choose a threshold T ($0 < T \leq G_{\max}$) to define object voxels ($P(p) \geq T$) and nonobject voxels ($P(p) < T$). The voxels of P at each z-coordinate define a 2D picture called a *layer*. We scan each layer of P to detect border voxels, which can be combined in successive layers to define triangular faces of the border [1141]. A *marching cubes algorithm* [667] uses eight voxels (a $2 \times 2 \times 2$ block) on two successive layers of P to define triangles that approximate the unknown surface. These triangles are chosen by analyzing how the unknown surface could intersect the $2 \times 2 \times 2$ block. The surface is assumed to intersect each grid edge (between two neighboring voxels) at most once. It follows that it can intersect a $2 \times 2 \times 2$ block in $2^8 = 256$ ways. The differences between T and the voxel values can be used to estimate the intersection points with grid edges. As a default, we assume that all edges are intersected at their midpoints.

Figure 8.18 illustrates situations that do not lead to unique triangulations. A marching cubes algorithm assumes that the $2 \times 2 \times 2$ voxel configurations are mapped into possible triangulations by a *look-up table*. The 256 eight-voxel configurations can be reduced modulo rotations and symmetries. This leads to one homogeneous configuration (all eight voxels are inside or outside the object), which produces no triangle, and 14 other cases. Figure 8.19 shows a look-up table that specifies one triangulation for each of the 14 configurations. The union of the triangulations defines an *isosurface* for P that depends on the threshold T and the look-up table.

Look-up tables occasionally produce isosurfaces that have holes. This happens if adjacent eight-voxel configurations produce "nonmatching" triangles. The

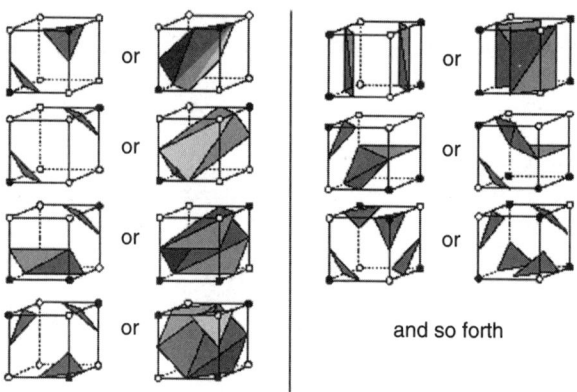

FIGURE 8.18   Local eight-voxel configurations do not lead to a unique triangulation.

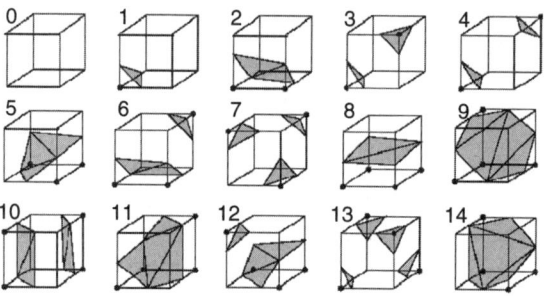

FIGURE 8.19   The 15 marching cubes configurations of [667].

resulting isosurface has an acyclic vertex with a star (the set of triangles incident with it) that is not a cycle of triangles. Figure 8.20 shows an example that gives rise to about 1% acyclic vertices, depending on the threshold used. In this example, a 23-case look-up table was used; there were two choices for 8 of the 14 nontrivial cases.

In Figure 8.20, a large number of triangles were generated. Reduction algorithms[10] have been designed to decrease the number of triangles (e.g., by merging triangles that have almost identical normals).

8.4.3   Local and global polyhedrization

A solid Θ is a set in $\mathbb{E}^3$ with a frontier that is a closed, bounded, measurable, hole-free surface (see Chapter 7). A digitization of Θ in a grid with resolution h is a finite union

---

10. These are called "decimation algorithms" in the literature; in ancient Roman armies, decimation was a severe punishment in which every tenth person was killed.

8.4 Surface Tracing and Approximation

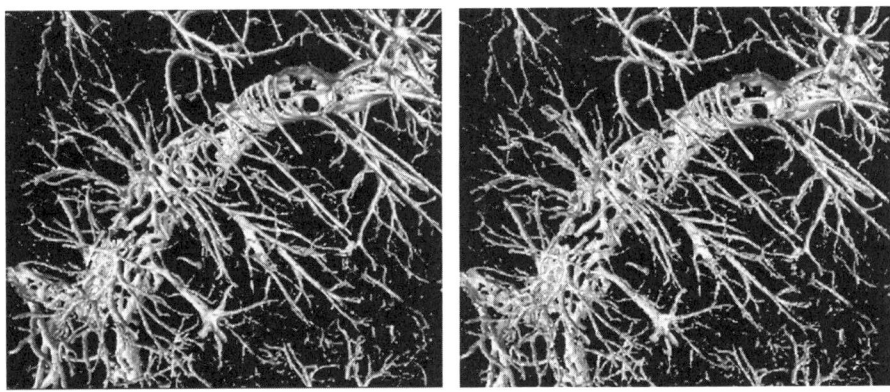

FIGURE 8.20 Left: isosurface generated by a marching cubes algorithm (astrocytes in brain tissue). A 3D picture ($256 \times 256 \times 51$ voxels) was mapped into an isosurface defined by 407,216 vertices of generated triangles. Right: the isosurface after vertex reduction; it has only about 40% as many vertices as the isosurface shown on the left.

$dig_h(\Theta)$ of 3-cells. A *digital surface* S is the frontier $\vartheta(dig_h(\Theta))$ of such a set; it can be regarded as a 2D complex of 0-, 1-, and 2-cells or as a set of frontier faces. Note that S is an isothetic polyhedral surface.

Θ can be magnified by a factor $h > 0$ (see Section 2.3.2) and digitization can be limited to a grid that has grid constant 1. Alternatively, Θ can be kept at its original size, but digitization can be done on a grid of resolution $h > 0$ (grid constant $\theta = 1/h$). The first approach was preferred by Jordan and Minkowski; it has the advantage that the calculation of surface area involves only integer arithmetic and is therefore preferable in implementations. The second approach is common in numeric analysis; we will use it in multigrid convergence studies (following Definition 2.10).

Polyhedrization maps a digital surface S into a finite set of polygonal faces in $\mathbb{E}^3$ but does not always provide a simple polyhedral approximation of the (unknown) surface $\vartheta_h(\Theta)$. For example, a triangulation produced by a marching cubes algorithm may not be hole-free and so may not be the surface of a simple polyhedron.

For each polygon Π of a polyhedrization of S, there exists a subset $A_h(\Pi)$ of S such that Π depends only on $A_h(\Pi)$; if $S \setminus A_h(\Pi)$ is modified in any way into another digital surface, its polyhedrization still contains Π. $A_h(\Pi)$ is called the *ball of influence* of Π. Because a polyhedrization of S contains only finitely many Πs, the set of radii of these balls of influence has a maximum value $R(h, S)$.

Definition 8.7 A polyhedrization method is called *local* iff there exists a constant $R_0 > 0$ such that $R(h, S) \leq R_0/h$ for any digital surface S and any grid resolution h. A polyhedrization method that is not local is called global.

The frontier faces of a simply 2-connected region (e.g., visited by the Artzy-Herman traversal algorithm) define a polyhedrization with constant $R_0 = \sqrt{3}/2$. The marching cubes algorithm is a polyhedrization method with constant $R_0 = \sqrt{3}$.

Definition 8.7 can be applied to other methods or properties that involve measurements in a metric space. Digital straight line segment (DSS) methods (as discussed in Chapters 9 and 10) are global methods. All of the methods of digitization defined in Chapter 2 are local methods.

8.5 Exercises

1. Give a parametric representation $x = x(t)$, $y = y(t)$ for the parabola defined by the equation $y^2 = 4ax$. What is the domain of the parameter t? Do the same for the ellipse $x^2/a^2 + y^2/b^2 = 1$.

2. Calculate the curvature $\kappa(t)$ of the parabola and ellipse defined in Exercise 8.1. (Hint: Apply one of the Frenet formulae.)

3. Let $< p_1, \ldots, p_n >$ be a simple polygon, and let $\mathbf{n}_i$ be the unit normal of the side $p_i p_{i+1}$ $(1 \leq i \leq n)$. Let $\mathbf{v}_i = \|p_i p_{i+1}\|_2 \mathbf{n}_i$ (where $p_{n+1} = p_1$) be the normal weighted by the length of the side. Prove that $\mathbf{v}_1 + \ldots + \mathbf{v}_n$ is the zero vector $(0,0)$.

4. Prove that, if the vectors defined in Exercise 3 are the same for two convex simple polygons, they must differ by a translation.

5. Show that the determinant $D(p,q,r)$ in Equation 8.14 can be calculated using only two multiplications (and several additions/subtractions).

6. The following figure shows a unit square S_0 in the upper left corner. Delete an "open cross" from S_0 as shown in the upper right corner; this results in a disconnected set S_1 of four squares. Repeat this process for each resulting square to obtain a set S_2 composed of 16 squares, a set S_3 composed of 64 squares, and so on. Note that each S_n is a closed set and consists of a finite number of squares. Continue the process to infinity; recall that the family of closed sets is closed under arbitrary intersections. What is the area of the *limiting set* $S_0 \cap S_1 \cap S_2 \ldots$?

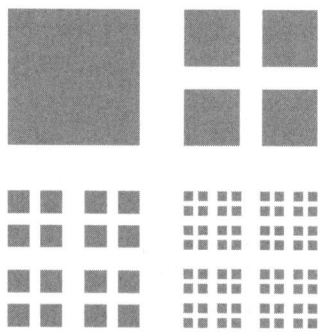

7. Prove the area formula in Equation 8.15 for simple polygons.

8. Let D be a planar disk of radius $a > 0$ centered at the origin. What is the equation of D in polar coordinates $x = r\cos\phi$ and $y = r\sin\phi$? What are the Jacobian matrix and Jacobians of this coordinate transformation?

9. Prove that a sphere of radius $r > 0$ has constant Gaussian curvature $1/r^2$.

10. The vector representation in Exercises 3 and 4 can be generalized to simple polyhedra: $\mathbf{v}_i = \mathcal{A}(F_i)\mathbf{n}_i$ where the F_is are the faces of the polyhedron and $\mathbf{n}_i$ is the unit normal to F_i. Prove that $\mathbf{v}_1 + \ldots + \mathbf{v}_n$ is the zero vector $(0,0,0)$.

11. Define cylindric coordinates $x = r\cos\phi$, $y = r\sin\phi$, and $z = z$ and spheric coordinates $x = r\sin\psi\cos\phi$, $y = r\sin\psi\sin\phi$, and $z = r\cos\psi$. What are the Jacobian matrices and Jacobians of these coordinate transformations?

12. The characteristic polynomial of a general 2×2 matrix $\mathbf{A}$ is $p(\lambda) = \det(\mathbf{A} - \lambda\mathbf{I})$ $= \lambda^2 - (a_{11} + a_{22})\lambda + (a_{11}a_{22} - a_{12}a_{21})$. Prove that $p(\mathbf{A}) = \mathbf{A}^2 - (a_{11} + a_{22})\mathbf{A} + (a_{11}a_{22} - a_{12}a_{21})\mathbf{I} = \mathbf{0}$, where $\mathbf{I}$ and $\mathbf{0}$ are the 2×2 identity and null matrices. (This proves the case $n = 2$ of the Cayley-Hamilton Theorem.)

13. Let S be a connected polygonal region that has b grid points on its n borders and i grid points in its interior. Prove that the area of S is $\frac{b}{2} + i - n - 1$.

8.6 Commented Bibliography

The geometry of curves and surfaces is treated in many textbooks, for example, [249, 781]. The Steinhaus method was proposed in [1022]. For parameter estimation of digital curves, see [1156]; for estimation of normals, see [751]. Higher-order approximations of curves (Theorems 8.1 and 8.2) were used for length estimation in [778, 779]. [704] gives a comprehensive review of methods of estimating curvature, including methods using vector algebra, which were not discussed in Section 8.1.3; see also [595].

The material in Sections 8.3.1 and 8.3.3 partially follows [498, 696]. Example 8.1 was published in [968] in 1890 and is also given in [696]. Theorem 8.3 was proved in [696]. For the study of surfaces in materials science, see [72, 147].

The *Teorema Egregium* was published by C.F. Gauss in [357]. Examples of the calculation of Gaussian curvature, including the formulas for the ellipsoid given in Section 8.3.4, are given in [971]. Theorem 8.4 is proved in [781, 1014]. For a proof of the Theorem of Meusnier, see [1014]. The application of basic theorems of differential geometry to digital geometry is discussed in [630].

For results about volumes of solids in the geometry of numbers, see [733, 842, 960]. The fundamental formulas about the content of "parallel bodies" M_r (see the

2D and 3D formulas, Equations 8.32 and 8.33, and Section 1.2.11) are due to J. Steiner [1020]; due to Minkowski, they are of central importance in the theory of convex bodies today. G. Pick gave a formula for the area of not necessarily simply connected grid polygons in [816]. For applications of his formula to simple grid polygons in digital geometry, see, for example, [605, 877, 952, 1107]. The effect of ratio-based digitization (i.e., taking the percentage of the 2-cell occupied by the set into account) on area estimation using pixel counts was studied in [398]. For stereology, see [1079]. Cavalieri's Principle was published in 1635 in [166]. Equation 8.34 is from [964]. For integral geometry and geometric or grid inequalities for measures such as volume or surface area, see [98, 152, 393, 413, 746, 788, 963, 965, 1021].

Simpler and more efficient algorithms often exist for isothetic objects; see, for example, [1128]. Such objects can be regarded as boundaries of cell complexes [1005], and they often allow combinatorial approaches (see, for example, [226] about the visibility of isothetic rectangles). Isothetic polygons and polyhedra also have applications to VLSI (Very Large Scale Integration) layout, database design, computational morphology, stock cutting, and partitioning problems [227, 801]. Sections 8.1.6 and 8.3.7 review [1149].

The Artzy-Herman algorithm was published in 1981 [48]; see the preface of [430] for its history. For a proof that this algorithm always generates a 1-cycle (see Section 6.4.5) of faces, see [429]; [433] is an earlier version that contains a shorter proof. A modified algorithm for surface traversal of 26-connected regions is discussed in [809]. For other surface traversal algorithms, see [143, 582, 828, 1077].

Topologic ambiguities in local 3D surface approximation techniques such as the marching cubes look-up tables are discussed, for example, in [775]. The 23-case look-up table in Figure 8.20 was published in [1126]. For vertex reduction, see, for example, [776, 967]. Marching cubes algorithms are sometimes (inaccurately) used as a generic name for methods of level surface extraction from gridded data. A marching cubes algorithm is easy to implement and will be used in Chapter 12 to illustrate a failure of local methods to correctly estimate surface area. Bilinear interpolation on frontier faces, or trilinear interpolation on border voxels, provides topologically correct surface approximations. Recent algorithms (e.g., *adaptive skeleton climbing* [821], *triangulation by ear clipping* [292]), are based on digital geometry, topology, and multiresolution concepts and avoid [291] expensive postprocessing for triangle reduction, which is typically necessary with marching cubes algorithms.

[495] discusses simplicial surface constructions on the boundary (i.e., "between the border and coborder") of a 3D set of voxels. The surface of the resulting combinatorial polyhedron provides a triangulation without holes (compare Section 8.4.2).

For Theorem 8.3 and its generalizations to polyhedra with "tunnels," "cavities," and "poles," see [461, 462]. The proof of Theorem 8.3 is from [1149]. For the recovery of n-dimensional isothetic polyhedra from their sets of corner vertices and on definitions of vertex types, see [629].

The classification of polyhedrization schemes based on balls of influence $B(p)$ was proposed in [498].

Local polyhedrization techniques include variants of the marching cubes algorithm [499, 667, 741, 1125, 1141], the marching tetrahedron algorithm [386, 861]

8.6 Commented Bibliography

(which constructs triangular surface patches using look-up tables for four-voxel configurations that form a tetrahedron that is obtained by subdividing a $2 \times 2 \times 2$ cube), the dividing cube algorithm [949, 1125] (which differs from the marching cubes algorithm in its use of a hierarchical octree data structure), and the algorithm for generating discrete combinatorial polyhedra [497] (where it is shown that the resulting polyhedra do not have topologic ambiguities like those obtained from marching cubes algorithms).

Global polyhedrization techniques can use convex hulls, relative convex hulls [1005], or discrete standard polyhedra [340]. For 3D convex hull algorithms, see, for example, the textbook [97] or Chapter 19 (by R. Seidel) in [371]. The resulting polyhedra can also be studied with respect to reversibility, compression [593], and surface smoothness.

For Exercise 12 (the Cayley-Hamilton Theorem), see [1047], which also discusses basics of eigenvalue and eigenvector calculations. The extension of Pick's Theorem in Exercise 13 is from [952]. ([816] is usually cited only for the area of a grid polygon that has only one border, but it actually includes the formula for grid polygons of any genus.)

CHAPTER 9

2D Straightness

This chapter discusses digital straightness in the grid point and grid cell models. We consider its relationships with other disciplines such as number theory and the theory of words as well as its role in picture analysis. We also discuss algorithms for recognizing digital straight line segments (DSSs) and partitioning digital arcs into such segments.

9.1 Basics

We consider the grid-intersection digitization (see Section 2.3.3) or outer Jordan digitization (see Section 2.3.2) of a ray

$$\gamma_{\alpha,\beta} = \{(x, \alpha x + \beta) : 0 \leq x < +\infty\}$$

in the set $\mathbb{N}^2 = \{(i,j) : i,j \in \mathbb{N}\}$ of grid points with nonnegative integer coordinates or in the set of 2-cells that have centers in $\mathbb{N}^2$. Because of the symmetry of the grid, we can assume that $0 \leq \alpha \leq 1$.

$\gamma_{\alpha,\beta}$ has a sequence of intersection points $p_0, p_1, p_2, \ldots$ with the vertical grid lines at $n \geq 0$. Let $(n, I_n) \in \mathbb{Z}^2$ be the grid point closest to p_n, and let the following be true:

$$I_{\alpha,\beta} = \{(n, I_n) : n \geq 0 \wedge I_n = \lfloor \alpha n + \beta + 0{\cdot}5 \rfloor\}$$

If there are two closest grid points, we choose the upper one; see Section 2.3.3. $I_{\alpha,\beta}$ is the set of centers of a set of grid squares $R(\gamma_{\alpha,\beta})$. The differences between successive I_ns define the following *chain codes*:

$$i_{\alpha,\beta}(n) = I_{n+1} - I_n = \begin{cases} 0 & \text{if } I_n = I_{n+1} \\ 1 & \text{if } I_n = I_{n+1} - 1 \\ & \text{for } n \geq 0 \end{cases}$$

In accordance with our assumption that $0 \leq \alpha \leq 1$, we need to use only the codes 0 and 1. We recall that code 0 is a horizontal increment and code 1 is a diagonal increment; see Figure 9.1.

Definition 9.1 $i_{\alpha,\beta} = i_{\alpha,\beta}(0) i_{\alpha,\beta}(1) i_{\alpha,\beta}(2) \ldots$ is a *digital ray* (in the grid point model) with *slope* α and *intercept* β.

This definition can easily be adapted to handle straight lines instead of rays. The code sequence of a *digital straight line* (DSL) is infinite in both directions.

If we use the grid cell model, we can use sequences of grid squares to define digital rays or straight lines:

Definition 9.2 A *cellular straight line* is a set M of 2-cells such that each cell in M has a nonempty intersection with a straight line γ, and γ is contained in the union of the cells.

This definition uses the outer Jordan digitization of γ. A *cellular straight line segment* is defined by a straight line segment γ in the same way. An example is shown in Figure 9.2.

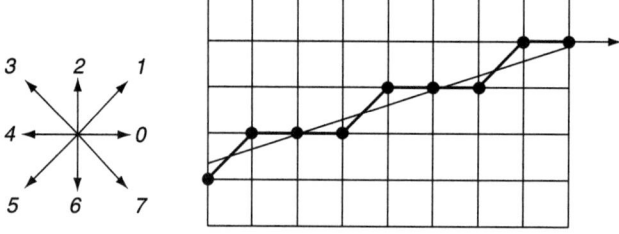

FIGURE 9.1 Segment of a digital ray defined by grid-intersection digitization.

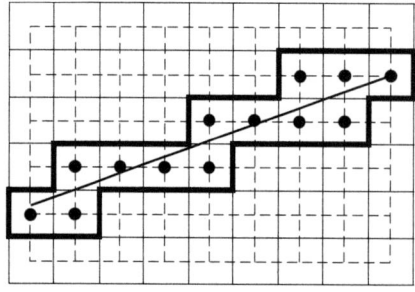

FIGURE 9.2 A cellular straight line segment.

9.1 Basics

If we translate a cellular straight line by (0·5,0·5) so that its 2-cells are in grid point positions, its frontier consists of two infinite 4-arcs. These 4-arcs can be used to define "upper" and "lower" digital straight lines.

In this introductory section, we present three basic theorems. Theorem 9.1 is about connectivity, which will be treated in Section 9.2; Theorem 9.2 is about self-similarity, which will be treated in Section 9.3; and Theorem 9.3 is about periodicity, which will be treated in Section 9.4.

A finite or infinite 8-arc is called *irreducible* iff its set of grid points does not remain 8-connected if a nonendpoint is removed from it.

Theorem 9.1 A digital ray is an irreducible 8-arc.

Proof A ray $\gamma_{\alpha,\beta}$ ($0 \leq \alpha \leq 1$) intersects the grid lines $x = n$ once each. Its intercepts with any two successive grid lines $x = n$ and $x = n+1$ differ vertically by α; hence the digitizations of these intersections differ vertically by ≤ 1. Thus successive grid points of the digital ray are 8-neighbors. Removing the grid point at any $x = n$ would leave the grid points at $x = n-1$ and $x = n+1$ 8-disconnected. ∎

The ray $\gamma_{\alpha,\beta}$ *generates* the digital ray $i_{\alpha,\beta}$. If $\beta - \beta'$ is an integer, we have $i_{\alpha,\beta} = i_{\alpha,\beta'}$. Thus we can assume without loss of generality that the βs are limited to $0 \leq \beta \leq 1$. Evidently, $i_{0,\beta} = 000\ldots$ and $i_{1,\beta} = 111\ldots$.

Theorem 9.2 If α is irrational, $I_{\alpha,\beta}$ uniquely determines both α and β. If α is rational, $I_{\alpha,\beta}$ uniquely determines α and determines β up to an interval.

Proof $I_{\alpha,\beta} = I_{\alpha',\beta'}$ implies $\alpha = \alpha'$, because otherwise the vertical distances between $\alpha x + \beta$ and $\alpha' x + \beta'$ would not be bounded as x goes to infinity, so the I_n values would differ beyond some large enough n.

If α is irrational, the values of $\alpha x + \beta$ modulo 1 for all $x \geq 0$ are dense in $[0,1]$. Hence, for every $\varepsilon > 0$, there exist n_0 and m_0 such that the following are true, and changing β by ε would result in a change in $I_{\alpha,\beta}$:

$$\alpha n_0 + \beta - \lfloor \alpha n_0 + \beta \rfloor < \varepsilon$$

and

$$\alpha m_0 + \beta - \lfloor \alpha m_0 + \beta \rfloor > 1 - \varepsilon$$

Thus $I_{\alpha,\beta}$ uniquely determines β when α is irrational.

If α is rational, the set of values of $\alpha x + \beta$ modulo 1 is finite for $x \geq 0$. Hence β is determined only up to an interval with a length that depends on α. ∎

According to Theorem 9.2, $i_{\alpha,\beta}$ always determines α uniquely. A digital ray is called *rational* if its slope is rational and *irrational* if its slope is irrational. For more about the intercepts β, see Section 9.5.

Digital rays are (right) infinite words over $\{0,1\}$. We recall a few basic definitions from the theory of words. A (finite) *word* defined on (or "over") an alphabet A is a finite sequence of elements of A. The *length* $|u|$ of the word $u = a_1 a_2 \ldots a_n$ (where each $a_i \in A$) is the number n of *letters* a_i in u. The *empty word* ε has length zero. The set of all words defined on A is denoted by $A^\star$.[1] A word v is a *factor* of a word u iff there exist words v_1 and v_2 such that $u = v_1 v v_2$. v is a *subword* of u iff $v = a_1 a_2 \ldots a_n$ and there exist words $v_0, v_1, \ldots, v_n$ such that $u = v_0 a_1 v_1 a_2 \ldots a_n v_n$.

Let $X \subset A^\star$. The set of all *infinite words* $w = u_0 u_1 u_2 \ldots$ (where each $u_i \in X - \{\varepsilon\}$) is denoted by X^ω. If all of the u_is are equal, for example to v, we write $w = v^\omega$. For all $v \in A^\star$ and $w \in A^\omega$, v is a *prefix* and w a *suffix* of the concatenation vw.

An integer $k \geq 1$ is a *period* of a word $u = a_1 a_2 \ldots a_n$ if $a_i = a_{i+k}$ ($i = 1, \ldots, n-k$). The smallest period of u is called *the* period of u. An infinite word $w \in A^\omega$ is called *periodic* if it is of the form $w = v^\omega$ for some nonempty word $v \in A^\star$. A word $w \in A^\omega$ is *eventually periodic* if it is of the form $w = uv^\omega$ for some $u \in A^\star$ and some nonempty $v \in A^\star$. A word $w \in A^\omega$ is called *aperiodic* if it is not eventually periodic.

The digitization of a ray $\gamma_{\alpha,\beta}$ in the grid point model is periodic if α is rational and aperiodic if it is irrational:

Theorem 9.3 Rational digital rays are periodic, and irrational digital rays are aperiodic.

A shortest word v such that $w = v^\omega$ is called a *basic segment* of w, and $|v|$ is *the* period of w. If α is an irreducible rational fraction, the length of the basic segment is the denominator of α. For example, if $\alpha = 3/7$, the basic segment may be 1010100 or any of its cyclic permutations. A rational slope does not uniquely specify a basic segment, but a rational slope α together with an intercept β uniquely specify it.

All rational DSLs that have the same slope α can be transformed into one another by translation in the x-direction (i.e., they are all *equivalent up to translation* [675]). This implies that the intercepts β do not influence the translation-invariant properties of rational DSLs.

9.2 Supporting Lines

An alternative way of defining a digital ray is as the 4-border of the upper or lower dichotomy of $\mathbb{N}^2$ defined by a (real) ray. Let the following be true,

$$U_{\alpha,\beta} = \{(n, U_n) : n \geq 0 \land U_n = \lceil \alpha n + \beta \rceil\}$$
$$L_{\alpha,\beta} = \{(n, L_n) : n \geq 0 \land L_n = \lfloor \alpha n + \beta \rfloor\}$$

---

1. $A^\star$ was also used in the definition of dimension, but the context will always clarify the meaning.

9.2 Supporting Lines

and let $u_{\alpha,\beta}(n) = U_{n+1} - U_n$ and $l_{\alpha,\beta}(n) = L_{n+1} - L_n$. The chain codes $u_{\alpha,\beta}$ and $l_{\alpha,\beta}$ are the *upper digital ray* and *lower digital ray* generated by $\gamma_{\alpha,\beta}$. These upper and lower digital rays are not the same as the grid-intersection digitization of $l_{\alpha,\beta}$, but they are also irreducible 8-arcs.

Because $L_{\alpha,\beta} = I_{\alpha,\beta-0.5}$, any lower digital ray is also a digital ray and vice versa. If $\alpha n + \beta$ is not an integer, then $U_n = L_n + 1$. Otherwise, $U_n = L_n$; the digital rays $u_{\alpha,\beta}$ and $l_{\alpha,\beta}$ differ in this case, and $\gamma_{\alpha,\beta}$ has an *integral point* at n. If $\gamma_{\alpha,\beta}$ has no integral points, then $u_{\alpha,\beta} = i_{\alpha,\beta-0.5} = l_{\alpha,\beta}$. If $\gamma_{\alpha,\beta}$ has integral points and α is rational, there exists β' such that $U_{\alpha,\beta} = I_{\alpha,\beta'}$. If $\gamma_{\alpha,\beta}$ has integral points and α is irrational, $U_{\alpha,\beta}$ and $L_{\alpha,\beta}$ differ by subsequences of length 2 only.

The grid points of a rational ray are the integer solutions of a finite set of linear equations with rational coefficients. *Arithmetic geometry*[2] specifies n-dimensional digital hyperplanes by pairs of linear Diophantine inequalities. In the 2D case, let a and b be relatively prime integers, let μ and ω be integers, and let the following be true:

$$D_{a,b,\mu,\omega} = \{(i,j) \in \mathbb{Z}^2 : \mu \leq ai + bj < \mu + \omega\}$$

$D_{a,b,\mu,\omega}$ is called an *arithmetic line* with *slope* a/b, *approximate intercept* μ, and *arithmetic width* ω [848].

Theorem 9.4 Any set of grid points $D_{a,b,\mu,\max\{|a|,|b|\}}$ is the set of grid points of a DSL. Conversely, for any rational DSL, there exist a, b, and μ such that the set of grid points of the given DSL is $D_{a,b,\mu,\max\{|a|,|b|\}}$.

This theorem also implies that $\omega = \max\{|a|,|b|\}$ defines an irreducible 8-arc. DSLs are called *naive lines* in arithmetic geometry, and $\omega = |a| + |b|$ defines a *standard line*. See Exercise 9.12 for characterizations of gaps in arithmetic lines. (See Definition 7.13 for definitions of gaps, separation, and gap-freeness.) An arithmetic line is gap-free (8-gap-free) iff it is 4-connected and 4-gap-free iff it is 8-connected. A naive line is 8-connected and 4-separating in $\mathbb{Z}^2$, and a standard line is 4-connected and 8-separating in $\mathbb{Z}^2$.

Because we are considering only lines with slope $0 \leq a/(-b) \leq 1$, we have $0 \leq a \leq -b$. For a naive line, we have $\omega = -b$, and all the grid points in $D_{a,b,\mu,\omega}$ lie between or on two lines $ax + by = \mu$ and $ax + by = \mu - b - 1$ (i.e., $y = \alpha x + \beta$ and $y = \alpha x + \beta - (1 - \frac{1}{-b})$ where $\alpha = a/-b$ and $\beta = \mu/b$). (This proves Corollary 9.1; see p. 314.) These lines are called *supporting lines* $D_{a,b,\mu,\omega}$. (As we will see, supporting lines are used in definitions of DSLs; they are called *leaning lines* in arithmetic geometry.)

Digital 4-rays are 4-arcs, where code 2 is a vertical increment:

$$i^{\circ}_{\alpha,\beta}(n) = \begin{cases} 0 & \text{if } I_n = I_{n+1} \\ 02 & \text{if } I_n = I_{n+1} - 1 \end{cases}$$

---

2. See also Chapter 11.

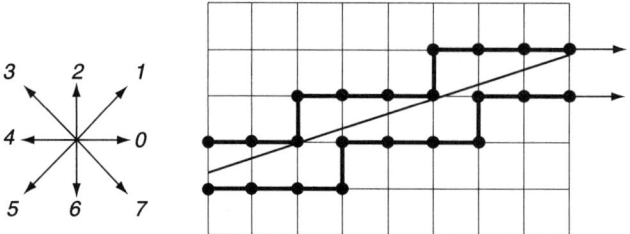

FIGURE 9.3 Segments of lower and upper digital 4-rays that follow the boundaries of the upper and lower dichotomies, which are linearly separated by a ray.

We can also define *upper digital 4-rays* $u^\circ_{\alpha,\beta}(n)$ and *lower digital 4-rays* $l^\circ_{\alpha,\beta}(n)$; examples are shown in Figure 9.3. We still have $i^\circ_{0,\beta} = 000\ldots$, but we now have $i^\circ_{1,\beta} = 020202\ldots$.

We can define a morphism that maps digital rays into digital 4-rays. A *morphism* or *substitution* $\varphi : A^\star \to B^\star$ is a function such that $\varphi(xy) = \varphi(x)\varphi(y)$ for all $x, y \in A^\star$. φ is uniquely determined by its values for all of the letters in A. φ is called *nonerasing* if it never maps a letter into the empty word. A nonerasing morphism $\varphi : A^\star \to B^\star$ defines a function (also called a morphism) from A^ω to B^ω by $\varphi(a(0)a(1)\ldots a(n)\ldots) = \varphi(a(0))\varphi(a(1))\ldots\varphi(a(n))\ldots$. Digital 4-rays can be defined by the following morphism, which maps digital rays into digital 4-rays:

$$\varphi : \begin{array}{l} 0 \mapsto 0 \\ 1 \mapsto 02 \end{array}$$

The theory of words studies morphisms on infinite words.

A chain code is a word over the alphabet $\{0, \ldots, 7\}$ (in our case, $\{0, 1\}$ or $\{0, 2\}$). Hence, digital rays and DSSs can be regarded as words. In terms of this interpretation, we have the following:

Definition 9.3 A *digital straight line segment* (*DSS* for short) is a nonempty factor of a digital ray. A *digital 4-straight line segment* (*4-DSS* for short) is a nonempty factor of a digital 4-ray.

A DSS u *connects* two points $p = (m_p, n_p)$ and $q = (m_q, n_q)$ of N^2 ($m_p < m_q$) iff the geometric interpretation of $u = u(1)\cdots u(m_q - m_p + 1)$ defines a sequence of horizontal and diagonal steps from p to q. Let $u = u(1)u(2)\cdots u(n)$ be an 8-arc of length n, and let $G(u) = \{p_0, p_1, \ldots, p_{n-1}\}$ be the *assigned set of grid points* such that $p_0 = (0,0)$ and u connects p_0 with p_{n-1} via a sequence of horizontal and diagonal steps through $p_1, \ldots, p_{n-2}$. Theorem 9.4 implies the following:

Corollary 9.1 A word $u \in \{0,1\}^\star$ is a DSS iff $G(u)$ lies between or on two parallel lines with a distance apart (in the y direction) that is less than 1.

9.2 Supporting Lines

There are four possible oriented diagonals in a grid square. The (oriented) *main diagonal* for a pair of parallel lines is the one that maximizes the dot product with the normal to the lines.[3]

Theorem 9.5 A finite 4-arc $u \in \{0,2\}^*$ is a 4-DSS iff $G(u)$ lies on or between a pair of parallel lines with a distance apart in the main diagonal direction that is less than $\sqrt{2}$.

Proof Let μ be the mapping from $\{0,1,2\}^*$ into $\{0,1,2\}^*$ that is defined by replacing any factor 02 by 1. $u \in \{0,1,2\}^*$ is a 4-DSS iff $\mu(u)$ is a DSS.

Let γ_1 and γ_2 be parallel lines with a main diagonal distance that is less than $\sqrt{2}$. Consider a finite 4-arc $u \in \{0,2\}^*$ such that $G(u)$ lies between or on this pair of parallel lines. If the slope α of these lines is either 0 or 1, u is either 0^n or $(02)^n$ so that it is a 4-DSS. If $0 < \alpha < 1$, we shift the lower line (e.g., γ_2) parallel to itself so it moves closer to γ_1 and becomes the line ζ. Then $G(\mu(u))$ lies between or on γ_1 and ζ, and the distance between them in the y-direction is less than 1, so $\mu(u)$ is a DSS and u is a 4-DSS.

On the other hand, suppose $u \in \{0,2\}^*$ is such that the minimum main diagonal distance between a pair of parallel lines that have $G(u)$ between or on them is at least $\sqrt{2}$. Then u must contain at least one subword 22, so $\mu(u)$ is not a DSS and u is not a 4-DSS. ∎

Two parallel lines at minimum diagonal distance that have $G(u)$ between or on them are called a *pair of supporting lines* of u. A finite 4-arc is also a finite 8-arc, but lying between or on a pair of parallel lines with a main diagonal distance that is less than $\sqrt{2}$ does not imply that the 4-arc is a DSS because it may not be an irreducible 8-arc. A pair of supporting lines with respect to a set $D_{a,b,c,b}$ of grid points has intercepts that differ by $0 < 1 - \frac{1}{b} < 1$ so that it has a main diagonal distance of less than $\sqrt{2}$.

The distance between a pair of parallel lines is measured in the direction of the normal to the lines. Let M be a bounded set in the plane and θ a direction $(0 \leq \theta < 2\pi)$. The *width* $w_\theta(M)$ is the minimum distance between a pair of parallel lines such that θ is the direction of the normal to the lines, and M lies between or on them. Let $R_{2\times 2}$ be a square formed by four 2-cells.

Theorem 9.6 An edge-connected set M of 2-cells is cellularly straight iff there exists a direction θ such that $w_\theta(\bigcup M) \leq w_\theta(R_{2\times 2})$.

---

3. The main diagonal direction makes angle $135°$ with the positive x-axis if the digitized line has a slope in $[0,1)$.

9.3 Self-Similarity

Self-similarity properties of digital rays and DSSs have been studied in picture analysis. An initial formulation of necessary conditions for self-similarity of DSLs was given by H. Freeman in [344]:

> "To summarize, we thus have the following three specific properties which all chains of straight lines must possess [342]:
>
> **(F1)** at most two types of elements can be present, and these can differ only by unity, modulo eight;
>
> **(F2)** one of the two element values always occurs singly;
>
> **(F3)** successive occurrences of the element occurring singly are as uniformly spaced as possible."

These properties (listed as (1), (2), and (3) in the historic source) were based on heuristic insights and illustrated by examples. Note that property (**F3**) is not precisely formulated.

9.3.1 The chord property

A. Rosenfeld in [883] gave a first formal characterization of DSLs, which led to a better specification of property (**F3**).

> **Definition 9.4** A set M of grid points satisfies the *chord property* iff, for any two distinct p and q in M and any point r on the (real) line segment pq, there exists a grid point $t \in M$ such that $L_\infty(r,t) \equiv \max(|x_r - x_t|, |y_r - y_t|) < 1$.

> **Theorem 9.7** A finite irreducible 8-arc $u \in \{0,1\}^*$ is a DSS iff $G(u)$ satisfies the chord property.
>
> *Proof* First, we show that $G(u)$ satisfies the chord property if u is a DSS (Theorem 1 in [883]). Let p and q be points of $G(u)$. The line segment pq intersects the grid lines $x = n$ that lie between p and q. Thus, for any point $r = (x,y)$ of pq, we have $|n - x| \leq \frac{1}{2}$ for some point $(n,m) \in G(u)$. It suffices to show that, whenever pq crosses a line $x = n$, the point $t = (n,m)$ of $G(u)$ on that line lies at vertical distance $|y - m| < 1$ above or below the crossing point $r = (n,y)$.
>
> Let u be a nonempty factor of a digitization of $\gamma_{\alpha,\beta}$; then neither p nor q can be more than $\frac{1}{2}$ vertically above $\gamma_{\alpha,\beta}$ or $\frac{1}{2}$ or more vertically below $\gamma_{\alpha,\beta}$. Let $r = (n,y)$ be $a_r \geq 0$ vertically above $\gamma_{\alpha,\beta}$ or $b_r \geq 0$ vertically below $\gamma_{\alpha,\beta}$. It follows that $0 \leq a_r \leq \frac{1}{2}$ or $0 \leq b_r < \frac{1}{2}$. If r is above t, $\gamma_{\alpha,\beta}$ intersects $x = n$ at vertical distance $0 \leq a_t < \frac{1}{2}$ above (or at) t, and we have $y - m \leq a_r + a_t < 1$. If

9.3 Self-Similarity

r is below t, $\gamma_{\alpha,\beta}$ intersects $x = n$ at vertical distance $0 \le b_t \le \frac{1}{2}$ below (or at) t, and we have $m - y \le b_r + b_t < 1$.

Now we prove that u is a DSS if $G(u)$ satisfies the chord property. To do this, we will use the *Transversal Theorem* of L.A. Santaló (see Section 1.2.8).

Let the 8-arc u join the grid points (n, y_0) and $(n+m, y_m)$ where $m > 0$ and $y_m - y_0 \le m$. If $y_m - y_0 = m$, the 8-arc is a diagonal line segment, and the chord property implies that $G(u)$ consists of grid points along this diagonal so that it is a DSS.

Suppose without loss of generality that $y_m - y_0 \le m - 1$. Let $T_i (0 \le i \le m)$ be the set of grid points of $G(u)$ on grid line $x = i$. The chord property implies that $T_i \ne \emptyset$ and that there are two integers l_i and u_i such that T_i is the set of all grid points $(n+i, y)$ where $l_i \le y \le u_i$. Assign a (real) straight line segment $L(p)$ to any grid point $p = (x, y)$, where the following is true:

$$L(p) = \{(x, v) : \quad y - 0.5 < v \le y + 0.5\}$$

Let L_i be the union of the $L(p)$s assigned to the grid points of T_i:

$$L_i = \{(n+i, v) : \quad l_i - 0.5 < v \le u_i + 0.5\}$$

Let $\mathcal{F} = \{L_0, \ldots, L_m\}$.

Clearly, $L_0, \ldots, L_m$ are parallel. Choose any L_i, L_j, and L_k ($0 \le i < j < k \le m$), and consider two grid points $p \in L_i$ and $q \in L_k$. The line segment pq intersects the grid line $x = j$ at a point $r = (j, y_r)$. By the chord property, there is a grid point $t = (x_t, y_t) \in G(u)$ such that $L_\infty(r, t) < 1$. Thus, t is also on $x = j$ (i.e., $x_t = j$). Let s be the midpoint of rt, and let $\varepsilon = |y_t - y_r|/2$. Let γ be the straight line through s parallel to pq; then γ intersects $x = i$ at $x_p \pm \varepsilon$ and $x = k$ at $x_q \pm \varepsilon$. Because $\varepsilon < 0.5$, it follows that γ intersects $L(p), L(t)$, and $L(q)$, so it intersects L_i, L_j, and L_k. By the Transversal Theorem, it follows that there exists a straight line γ that intersects all of the L_i. It remains only to show that γ generates all of the grid points in $G(u)$ by grid-intersection digitization.

Each T_i contains a grid point p_i such that γ intersects $L(p_i)$. We have $p_0 = (n, y_0)$ and $p_m = (n+m, y_m)$. Let q_0 and q_m be the intersection points of γ with $L(p_0)$ and $L(p_m)$ so that $q_0 = (n, y_0 + \lambda)$ and $q_m = (n+m, y_m + \mu)$ where $-0.5 < \lambda, \mu \le 0.5$. The horizontal distance between q_0 and q_m is m, and the vertical distance is $|y_0 + \lambda - y_m - \mu| \le |y_0 - y_m| + |\lambda - \mu| \le m - 1 + |\lambda - \mu| < m$. The line segment $q_0 q_m$ makes an angle smaller than $45°$ with the horizontal direction, so its grid-intersection digitization is specified by intersections with the vertical grid lines $x = n+i, 0 \le i \le m$. The grid points produced by these intersections are in $G(u)$, because γ is a transversal of all of the L_is, and $G(u)$ contains only these grid points, because u is an irreducible 8-arc. ∎

Note that infinitely many irreducible two-sided infinite 8-arcs satisfy the chord property but are not digital straight lines (e.g., arcs defined by "sparse" occurrences of $1s$ in 0^ω).

Theorem 9.7 was used in [883] to derive the following necessary conditions on the chain codes of DSSs. These conditions are stated in terms of the runs[4] in the chain code.

- **(R1)** "The runs have at most two directions, differing by $45°$, and for one of these directions the run length must be 1.
- **(R2)** The runs can have only two lengths, which are consecutive integers.
- **(R3)** One of the runs can occur only once at a time.
- **(R4)** ..., for the run length that occurs in runs, these runs can themselves have only two lengths, which are consecutive integers; and so on."

These properties, which were listed as 1), 2), 3) and 4) in the historic source, do not provide sufficient conditions, but they provide a recursive formulation of (**F3**).

The chord property is equivalent to a *compact chord property* that uses the real polygonal arc joining the points of the DSS rather than the real line segment joining its endpoints and the metric L_1 rather than L_∞.

The property of *evenness* ("the slope must be the same everywhere") of a DSS is equivalent to the chord property. "Balance" of words (see Section 9.4) is related to evenness.

9.3.2 Syntactic characterization

Let $s = (s(i))_{i \in I}$ ($I \subseteq \mathbb{Z}$) be a finite or infinite word over $\mathbb{N}$. A letter (in our case, a digit) k is called *singular in s* iff the following are true:

- it appears in s; and
- for all $i \in I$ such that $i-1$ and $i+1$ are in I, if $s(i) = k$, then $s(i-1) \neq k$ and $s(i+1) \neq k$.

k is *nonsingular in s* iff it appears in s and is not singular in s. Word s is *reducible* iff s contains no singular letter or any factor of s that contains only nonsingular letters is of finite length.

Let s be reducible, and let $R(s)$ be the following:

(1) the length of s if s is finite and contains no singular letter;

(2) the word that results from s by replacing by their run lengths all factors of nonsingular letters in s that are between two singular letters in s and deleting all other letters in s; or

(3) the letter a if $s = a^\omega$.

---

4. A *run* is a maximum-length factor a^n of a word.

9.3 Self-Similarity

Recursive application of this *reduction operation* R produces a sequence of words $s_0, s_1, \ldots$ where $s_0 = s$ and $s_{n+1} = R(s_n)$. The sequence u_0, u_1, u_2 in Algorithm 9.1 is an example.

Definition 9.5 The chain code u of a two-sided infinite 8-arc has the *DSL property* iff $u_0, u_1, \ldots$ (where $u_{n+1} = R(u_n)$) are reducible words and satisfy the following two conditions:

(**L1**) There are at most two different letters a and b in u_n, and, if there are two, then $|a - b| = 1$ (modulo 8 in the case of u_0).

(**L2**) If there are two different letters in u_n, at least one of them is singular.

Using the DSL property, it is possible to define necessary and sufficient conditions for chain codes of DSSs. Words of nonsingular letters at the ends of the code require special attention. Let s be a finite word, and let $l(s)$ and $r(s)$ be the run lengths of nonsingular letters to the left of the first singular letter and to the right of the last singular letter in s.

Definition 9.6 A finite chain code u has the *DSS property* iff $u = u_0$ satisfies (**L1**) and (**L2**), and any sequence $u_n = R(u_{n-1})$ satisfies (**L1**) and (**L2**) as well as the following two conditions:

(**S1**) If u_n contains only one letter a or two different letters a and $a+1$, then $l(u_{n-1}) \leq a+1$ and $r(u_{n-1}) \leq a+1$.

(**S2**) If u_n contains two different letters a and $a+1$ and a is nonsingular in u_n, then u_n starts with a if $l(u_{n-1}) = a+1$ and ends with a if $r(u_{n-1}) = a+1$.

L.D. Wu [1137] proved that a chain code has the DSS property iff the corresponding arc is irreducible and has the chord property. This concluded, in 1982, the process of formalizing Freeman's constraints (**F1-F3**) and provided a set of constraints on the design of DSS recognition procedures:

Theorem 9.8 A finite 8-arc is a DSS iff its chain code satisfies the DSS property.

(We give no proof here; see the discussion of continued fractions in Section 9.3.3.)
A finite factor of a two-sided infinite chain code u satisfies the DSS property iff exactly one straight line defines u by grid-intersection digitization. This implies the following:

Theorem 9.9 A two-sided infinite 8-arc is a digital straight line iff its chain code satisfies the DSL property.

Wu's proof of Theorem 9.8 is based on number theory and consists of many case discussions. Researchers therefore tried to find shorter, "more elegant" proofs of Wu's theorem (e.g., proofs based on continued fractions).

9.3.3 Continued fractions

A rational number a_1/a_0 ($a_0 > a_1 > 0$) can be represented by a finite continued fraction with integer coefficients $q_i > 0$ ($1 \le i < n$) and $q_n > 1$:

$$\frac{a_1}{a_0} = [q_1, q_2, \ldots, q_n] = \cfrac{1}{q_1 + \cfrac{1}{q_2 + \cfrac{\ddots}{q_{n-1} + \cfrac{1}{q_n}}}}$$

The Euclidean algorithm can be used to derive such continued fractions:

$$\frac{a_0}{a_1} = q_1 + \frac{a_2}{a_1} \quad \text{with} \quad 0 < \frac{a_2}{a_1} < 1$$

$$\frac{a_1}{a_2} = q_2 + \frac{a_3}{a_2} \quad \text{with} \quad 0 < \frac{a_3}{a_2} < 1$$

$$\cdots\cdots\cdots\cdots\cdots$$

$$\frac{a_{n-2}}{a_{n-1}} = q_{n-1} + \frac{a_n}{a_{n-1}} \quad \text{with} \quad 0 < \frac{a_n}{a_{n-1}} < 1$$

$$\frac{a_{n-1}}{a_n} = q_n \quad \text{with} \quad a_{n+1} = 0$$

Irrational numbers can be represented by infinite continued fractions.

The value of a continued fraction can be expressed in the following form, where $\alpha_n, \beta_n, \gamma_n, \delta_n$ depend on the q_is:

$$\frac{a_1}{a_0} = [q_1, q_2, \ldots, q_n] = \frac{\alpha_n q_n + \beta_n}{\gamma_n q_n + \delta_n}$$

For $n \ge 1$, we have $\alpha_n \delta_n - \beta_n \gamma_n = (-1)^n$ and the following:

$$[q_1, q_2, \ldots, q_n, q_{n+1}] = \frac{\alpha_{n+1} q_{n+1} + \beta_{n+1}}{\gamma_{n+1} q_{n+1} + \delta_{n+1}}$$

$$= \left[q_1, q_2, \ldots, q_{n-1}, q_n + \frac{1}{q_{n+1}}\right] = \frac{\alpha_n \left(q_n + \frac{1}{q_{n+1}}\right) + \beta_n}{\gamma_n \left(q_n + \frac{1}{q_{n+1}}\right) + \delta_n}$$

$$= \frac{\alpha_n (q_n q_{n+1} + 1) + \beta_n q_{n+1}}{\gamma_n (q_n q_{n+1} + 1) + \delta_n q_{n+1}}$$

9.4 Periodicity

so that the following is true:

$$\frac{\alpha_{n+1}q_{n+1}+\beta_{n+1}}{\gamma_{n+1}q_{n+1}+\delta_{n+1}} = \frac{(\alpha_n q_n + \beta_n)q_{n+1}+\alpha_n}{(\gamma_n q_n + \delta_n)q_{n+1}+\gamma_n} \tag{9.1}$$

Continued fractions can be manipulated using a "concatenation" operator $\otimes$; for simple fractions, this operator is defined by $(a/b) \otimes (c/d) = (a+b)/(c+d)$. In terms of $\otimes$, we have the following:

$$[q_1, q_2, \ldots, q_n] = \begin{cases} [q_1, q_2, \ldots, q_{n-1}+1] \otimes (q_n-1)[q_1, q_2, \ldots, q_{n-1}] & \text{if } n \text{ is even} \\ (q_n-1)[q_1, q_2, \ldots, q_{n-1}] \otimes [q_1, q_2, \ldots, q_{n-1}+1] & \text{if } n \text{ is odd} \end{cases}$$

This expression is called the *splitting formula* in [1107]. The splitting process can continue until only *atomic slopes* $[q] = 1/q$ remain.

An atomic slope $[q]$ can be encoded by $q-1$ 0s and one 1. Alternating the splitting formulas for odd and even values of n yields a balanced code sequence. For example, $46/87 = [1,1,8,5]$ gives the following:

$$\begin{aligned} [1,1,8,5] &= [1,1,9] \otimes 4 \cdot [1,1,8] \\ &= (8 \cdot [1,1] \otimes [1,2]) \otimes 4 \cdot (7 \cdot [1,1] \otimes [1,2]) \\ &= (8 \cdot [2] \otimes ([2] \otimes [1])) \otimes 4 \cdot (7 \cdot [2] \otimes ([2] \otimes [1])) \end{aligned}$$

This yields the following code sequence, which has length 87 and contains 46 1s:

(01010101010101)(011)
((01010101010101)(011))
((01010101010101)(011))
((01010101010101)(011))
((01010101010101)(011))

The splitting formula allows us to express Freeman's conjecture and Rosenfeld's recursive process in a very concise way by applying the splitting formula twice, first for n and then for $n-1$:

$$\begin{aligned} [q_1, q_2, \ldots, q_n] &= (q_{n-1} \cdot [q_1, q_2, \ldots, q_{n-2}] \otimes [q_1, q_2, \ldots, q_{n-2}+1]) \\ &\otimes (q_n - 1) \cdot ((q_{n-1}-1) \cdot [q_1, q_2, \ldots, q_{n-2}] \otimes [q_1, q_2, \ldots, q_{n-2}+1]) \end{aligned}$$

(This example assumes that n is even.) This approach handles only DSSs that are factors of rational rays, but we will see in Corollary 9.2 that all DSSs have this property.

9.4 **Periodicity**

Self-similarity studies have a long history in number theory. The theory of words is a relatively recent discipline that contains many interesting results about self-similarity, often with a focus on irrational straight rays. Rational digital rays are periodic infinite words (Theorem 9.3), and irrational digital rays are aperiodic infinite words that are studied under the name of *Sturmian words*.[5] This section gives some basic definitions and results as well as a few proofs.

Let w be a finite or infinite word over $A = \{0,1\}$. Let $F(w)$ be the set of all factors of w and $F_n(w)$ the set of factors of w of length $n \geq 0$. The *complexity function* of w is as follows:

$$P(w,n) = \mathrm{card}(F_n(w))$$

$P(w,0) = 1$ (the empty word is always a factor), and $P(w,1)$ is the number of distinct letters in w. For an infinite word w, we have $P(w,n) \leq P(w,n+1)$, because every factor of length n can be extended to the right by at least one letter. Furthermore, $F_{m+n}(w) \subseteq F_m(w)F_n(w)$, so $P(w,m+n) \leq P(w,m)P(w,n)$.

Let w be an infinite periodic word with period k. Then $P(w,n) \leq k$ for all $n \geq 0$ (i.e., the complexity of a periodic word is limited by its period). The following theorem shows that the converse is also true and generalizes these statements to eventually periodic words. (For example, 10^ω is not periodic, but it is eventually periodic, and it is not a rational digital ray.)

Theorem 9.10 The following statements about an infinite word w are equivalent:

(i) w is eventually periodic;

(ii) $P(w,n) = P(w,n+1)$ for some $n \geq 0$;

(iii) $P(w,n) < n+k-1$ for some $n \geq 1$, where k is the number of distinct letters in w; and

(iv) $P(w,n)$ is bounded.

Proof

(i) $\Rightarrow$ (iv): Let $w = uv^\omega$. Then $P(w,n) \leq |uv|$ for all $n \geq 0$.

(iv) $\Rightarrow$ (iii): Let $P(w,n) < p$ for all $n \geq 0$. If k is the number of distinct letters in w, we have $P(w,1) = k < p$, so $p \geq k+1$. Hence $P(w, p-k+1) < p$.

(iii) $\Rightarrow$ (ii): Suppose (ii) is not true, so $P(w,m-1) < P(w,m)$ for all $m \geq 0$. Then $n+k-1 > P(w,n) \geq P(w,1) + n - 1 = k+n-1$ for some $n \geq 1$, which is impossible.

(ii) $\Rightarrow$ (i): Consider the *factor graph* $G_n(w)$, which has node set $F_n(w)$ and edge set $E = \{(bu,a,ua) : a,b \in A \wedge bua \in F_{n+1}(w)\}$. At least one edge is incident with each node, because every factor of length n is a prefix of a factor of length $n+1$. Because $P(w,n) = P(w,n+1)$, exactly one edge is incident with each

---

5. [748] introduced the term *Sturmian trajectories* after the mathematician C.F. Sturm (1803–1855), who is famous for his rule for computing the roots of an algebraic equation. [748] defined *Sturmian words* as zeros of solutions of linear homogeneous second-order differential equations.

9.4 Periodicity

node; thus any strongly connected component of $G_n(w)$ is a simple circuit. Hence w defines an infinite path in $G_n(w)$. Because $F_n(w)$ is finite, the path must eventually repeat and become a fixed circuit after some initial prefix; thus w is eventually periodic. ∎

A sequence $(v_n)_{n \geq 0}$ of finite words over an alphabet A *converges* to an infinite word w if every prefix of w is a prefix of all but a finite number of v_ns. For example, the sequence $0^n 1^n$ converges to 0^ω.

Let $f_1 = 1$, $f_2 = 0$, and $f_{n+1} = f_n f_{n-1}$ for $n \geq 2$. The sequence of lengths $|f_n|$ is the Fibonacci sequence[6] $F_1 = 1, F_2 = 1, F_3 = 2, F_4 = 3, F_5 = 5, \ldots$. The sequence $(f_n)_{n \geq 0}$ converges to the *Fibonacci word*:

$$f = 010010100100101001010010010100100 1 \ldots$$

The Fibonacci word can also be defined by a morphism:

$$\varphi: \begin{array}{l} 0 \mapsto 01 \\ 1 \mapsto 0 \end{array}$$

Indeed, $f = \varphi^\omega(0)$.

Definition 9.7 A *Sturmian word* is an infinite word $w = a_1 a_2 a_3 \ldots$ over $\{0, 1\}$ that has exactly $n+1$ factors of length n for every $n \geq 0$.

Any suffix of a Sturmian word is Sturmian. The Fibonacci word is Sturmian; so is the *Thue-Morse word* $t = \mu^\omega(0) = 0110100110010110\ldots$, where the following are true:

$$\mu: \begin{array}{l} 0 \mapsto 01 \\ 1 \mapsto 10 \end{array}$$

According to Theorem 9.10, any aperiodic infinite word has complexity $P(w, n) \geq n+1$ for $n \geq 0$; hence Sturmian words have the least possible complexity. Because a Sturmian word has $P(w, 1) = 2$, it must be defined on a binary alphabet.

A *right special factor* of an infinite word w is a finite word u such that $u0$ and $u1$ are factors of w. A word w is Sturmian iff it has exactly one right special factor of each length $n \geq 0$. (The empty word is the right special factor of length 0.) For the Fibonacci word, 11 is not a factor, so 0 is the only right special factor of length 1; 000 and 011 are not factors, so 10 is the only factor of length 2; and so on.

The *height* $h(w)$ of a word over $\{0, 1\}$ is the number of 1s in w. If v and w have the same length, $\delta(v, w) = |h(v) - h(w)|$ is called their *balance*. A set X of words is *balanced* iff $|v| = |w|$ implies $\delta(v, w) \leq 1$ for all pairs of words $v, w \in X$. An infinite word w is *balanced* if its set of factors is balanced.

The *slope* of a nonempty word w is $\pi(w) = h(w)/|w|$. We have the following:

$$\pi(uv) = \frac{|u|}{|uv|} \pi(u) + \frac{|v|}{|uv|} \pi(v)$$

---

6. Standard notation as used in *The Fibonacci Quarterly*.

It can be shown that an infinite word w is balanced iff, for all nonempty factors u and v of w, we have the following:

$$|\pi(u) - \pi(v)| < \frac{1}{|u|} + \frac{1}{|v|} \tag{9.2}$$

This shows that the sequence of slopes of w is a Cauchy sequence. Thus a balanced infinite word has a uniquely defined slope that is the limit of the slopes of its finite prefixes. Let w be an infinite balanced word and let w_n be the prefix of w of length $n \geq 1$. Then the sequence $(\pi(w_n))_{n \geq 1}$ converges as $n \to \infty$. For example, for the Fibonacci word f, we have $h(f_n) = F_{n-2}$ and $|f_n| = F_n$, and F_{n-2}/F_n converges to $\pi(f) = 1/\tau^2$ where $\tau = (1 + \sqrt{5})/2$ is the golden ratio.

A digital ray is an infinite word that is periodic (aperiodic) iff it is the digitization of a ray of rational (irrational) slope. In the theory of words, digital rays are called *mechanical words*.

Theorem 9.11 (M. Morse and G.A. Hedlund, 1940) The digitization of a ray of slope α is a balanced word of slope α.

Proof Let w be a lower digital ray. The height of a factor $u = w(n) \ldots w(n + p - 1)$ is $h(u) = \lfloor \alpha(n+p) + \beta \rfloor - \lfloor \alpha n + \beta \rfloor$. Hence $\alpha \cdot |u| - 1 < h(u) < \alpha \cdot |u| + 1$ so that $\lfloor \alpha \cdot |u| \rfloor \leq h(u) \leq 1 + \lfloor \alpha \cdot |u| \rfloor$. Thus $h(u)$ takes on only two consecutive values when u ranges over factors of w of fixed length, so w is balanced. Moreover, the following is true,

$$\left| \frac{h(u)}{|u|} - \alpha \right| = |\pi(u) - \alpha| < \frac{1}{|u|}$$

so that $\pi(u) \to \alpha$ for $|u| \to \infty$. ∎

The inequality $|\pi(u) - \alpha| < 1/|u|$ provides a criterion for evaluating the accuracy of a slope estimated from a finite DSS. Another method of evaluating the accuracy of an estimated slope will be discussed at the end of Section 9.5.

Corollary 9.2 Any DSS is a factor of a rational digital ray.

Proof An interval in $[0, 1)$ of width $1/|u|$ that contains an irrational number α also contains rational numbers α' that satisfy $|\pi(u) - \alpha'| < 1/|u|$. ∎

9.5 Number-Theoretic Properties

Finally, we state the following:

Theorem 9.12 (M. Morse and G.A. Hedlund, 1940) *The following statements about an infinite word w are equivalent:*

(i) *w is Sturmian;*

(ii) *w is balanced and aperiodic; and*

(iii) *w is an irrational digital ray.*

Note that a balanced infinite word is not always a digital ray if its slope is rational. For example, 01^ω has slope 1, but it is not a digital ray. Only periodic infinite balanced words are rational digital rays.

9.5 Number-Theoretic Properties

9.5.1 Counting segments and partitions

The following theorem is from the theory of words:

Theorem 9.13 *The number of balanced words of length n is as follows,*

$$1 + \sum_{i=1}^{n} (n+1-i)\phi(i)$$

where ϕ is Euler's totient function.

It can be shown that a finite word u is balanced iff it is a factor of an irrational digital ray. In accordance with Corollary 9.2, any finite balanced word is also a factor of a rational digital ray (i.e., Theorem 9.13 gives the number of DSSs of length n starting at the origin $(0,0)$). It is the set of segments u of lower digital rays defined by $0 \leq x \leq n$, $0 \leq \alpha \leq 1$, and $0 \leq \beta < 1$; the first grid point in $G(u)$ is $(0,0)$, and $G(u)$ contains $n+1$ grid points. The number of such DSSs that pass through the origin is given as follows:

$$\frac{1}{\pi^2} \cdot n^3 + \mathcal{O}\left(n^2 \cdot \log n\right) \tag{9.3}$$

The Euler function, ϕ, satisfies the following:

$$\sum_{i=1}^{n} \phi(i) \approx \frac{3}{\pi^2} \cdot n^2 \quad \text{and} \quad \sum_{i=1}^{n} i \cdot \phi(i) \approx \frac{2}{\pi^2} \cdot n^3$$

Hence the formula in Theorem 9.13 implies Equation 9.3.

The *Farey series* $F(n)$ of order $n \geq 1$ is the increasing sequence of irreducible fractions between 0 and 1 with denominators that do not exceed n. For example, $F(5)$ is given as follows:

$$\frac{0}{1}, \frac{1}{5}, \frac{1}{4}, \frac{1}{3}, \frac{2}{5}, \frac{1}{2}, \frac{3}{5}, \frac{2}{3}, \frac{3}{4}, \frac{4}{5}, \frac{1}{1}$$

The DSSs of length n that pass through the origin are in one-to-one correspondence with $F(n)$.

There is an obvious one-to-one correspondence between the set of DSSs that start at $(0,0)$ and the set of linear partitions of an $n \times n$ grid. (A *linear partition* of a set S divides it into two parts that are contained in the two halfplanes bounded by a line.) Evidently, any DSS that has $n+1$ points and begins at $(0,0)$ defines a linear partition of the $n \times n$ grid, but there also exist linear partitions that do not correspond to such DSSs.

The number of linear partitions of an $m \times n$ grid ($m \leq n$) is given as follows:

$$\frac{3}{\pi^2} \cdot m^2 \cdot n^2 + \mathcal{O}(m^2 \cdot n \cdot \log n + m \cdot n^2 \cdot \log \log n) \tag{9.4}$$

This shows how many different digital rays exist on an $m \times n$ grid. The asymptotic formulas for the numbers of DSSs and linear partitions can be derived from formulas for average values of number-theoretic functions.

9.5.2 Spirographs

L. Dorst and R. P. W. Duin developed in [278] a theory of *spirographs* that establishes additional links between digital rays and number theory. Figure 9.4 (left) shows some parallel shifts of a ray with slope α ($0 < \alpha < 1$) that have y-intercepts in the interval $[0, 1)$. For any grid line $x = n$, there is exactly one grid point (n, y_n) such that the ray $y = \alpha x + \beta_n$ passes through (n, y_n) and intersects the y-axis in the interval $[0, 1)$. Spirographs[7] are diagrams that exhibit the distribution of these y-intercepts. As shown in Figure 9.4 (right), the y-intercepts corresponding to rays that pass through grid points (n, y_n), $n = 0, 1, 2, \ldots$, are represented by successive vertices of a (non-simple) polygon inscribed in a circle of perimeter 1:

Definition 9.8 A *spirograph* $S(\alpha, n)$ is a set of n points marked on a circle of unit perimeter. The points are the y-intercepts of rays with slope α that intersect the grid lines $x = 0, x = 1, \ldots, x = n-1$ at grid point positions.

If α is rational, there are only finitely many such rays, thereby resulting in a finite set of y-intercepts in $[0, 1)$ that repeat periodically. The *signature* of $S(\alpha, n)$ is the order modulo n of the marked points on the circle. (If α is rational, $y = \alpha x + \beta$ generates the same lower digital ray for all β in an interval; see Theorem 9.2.)

---

7. This was named after a children's toy that is used for drawing curves.

9.5 Number-Theoretic Properties

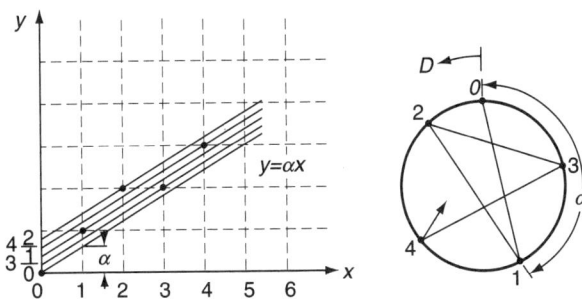

FIGURE 9.4 Left: grid points on grid lines $x = 0, 1, \ldots, 4$. Right: y-intercepts of the rays through these grid points, mapped into a spirograph.

The distance $D_\alpha(i,j)$ between two points $i, j \in S(\alpha, n)$, where $0 \le i, j < n$, is the length of the anticlockwise arc from i to j:

$$D_\alpha(i,j) = (i-j)\alpha - \lfloor (i-j)\alpha \rfloor$$

The smallest clockwise distance from $0 \in S(\alpha, n)$ is $D_{\text{right}} = \min\{D_\alpha(i, 0) : i \ne 0 \wedge i \in S(\alpha, n)\}$. Let the following be the point at this minimum distance:

$$i_{\text{right}} = \min\{k \ne 0 : k \in S(\alpha, n) \wedge D_\alpha(k, 0) = D_{\text{right}}\}$$

Similarly, let $D_{\text{left}} = \min\{D_\alpha(0, i) : i \ne 0 \wedge i \in S(\alpha, n) \wedge D_\alpha(0, i) \ne 0\}$ and the following be true:

$$i_{\text{left}} = \max\{k \ne 0 : k \in S(\alpha, n) \wedge D_\alpha(0, k) = D_{\text{left}}\}$$

By definition, $D_{\text{right}} = \alpha i_{\text{right}} - \lfloor \alpha i_{\text{right}} \rfloor$ and $D_{\text{left}} = \alpha i_{\text{left}} - \lfloor \alpha i_{\text{left}} \rfloor$. Therefore, the bounds on α that preserve the signature of the spirograph are as follows:

$$\frac{\lfloor \alpha i_{\text{right}} \rfloor}{i_{\text{right}}} \le \alpha < \frac{\lfloor \alpha i_{\text{left}} \rfloor}{i_{\text{left}}}$$

These bounds are the best rational approximations to α using fractions with denominators that do not exceed $n-1$. This can be proved from the fact that $\lfloor \alpha i_{\text{right}} \rfloor / i_{\text{right}}$ and $\lfloor \alpha i_{\text{left}} \rfloor / i_{\text{left}}$ are two successive fractions in the Farey series $F(n-1)$.

For every α ($0 \le \alpha < 1$) and n ($n \ge 1$), there is an interval of possible intercepts β ($0 \le \beta < 1$) such that the given lower straight line segment of length n is a digitization of the ray $\alpha x + \beta$. The width of this interval defines the maximal possible error $E_{\max}(\alpha, n)$; see Figure 9.5. $E_{\max}(\alpha, n)$ is the maximum arc length in the spirograph $S(\alpha, n)$:

Theorem 9.14 $E_{\max}(\alpha, n) = D_{\text{right}} + D_{\text{left}}$ in the spirograph $S(\alpha, n+1)$.

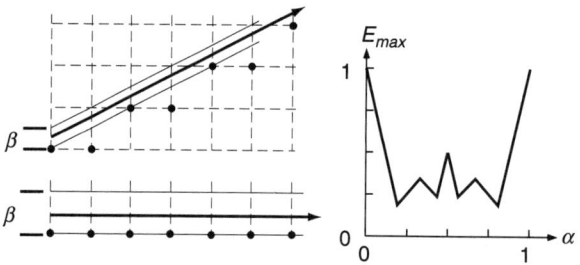

FIGURE 9.5 Left: the maximum error in β as a function of the estimated α value for $n = 6$ [278]. Right: the maximum error in β is 1 for $\alpha = 0$ and 0·5 for $\alpha = 0·5$.

The formula $D_{\text{right}} + D_{\text{left}} = \lceil \alpha i_{\text{left}} \rceil - \lfloor \alpha i_{\text{right}} \rfloor + \alpha(i_{\text{right}} - i_{\text{left}})$ using the values from $S(\alpha, n+1)$ provides a simple method of calculating the errors $E_{\max}(\alpha, n)$. If α is a fraction a/b in the Farey series $F(n)$, then $E_{\max}(a/b, n) = 1/b$.

9.6 Algorithms

Many efficient DSS recognition algorithms have been published. The computational problem is as follows: the input is a sequence of chain codes $i(0), i(1), \ldots$ where $i(k) \in \{0, 1\}$ and $k \geq 0$. An *offline DSS recognition algorithm* decides whether a finite word $u \in \{0, 1\}^\star$ is a DSS. An *online DSS recognition algorithm* reads the successive chain codes $i(0), i(1), \ldots$ and determines the maximum $k \geq 0$ such that $i(0), i(1), \ldots, i(k)$ is a DSS but $i(0), i(1), \ldots, i(k), i(k+1)$ is not. A recognition algorithm has linear run time behavior (is a *linear algorithm*) if it runs in $\mathcal{O}(n)$ time (i.e., it performs at most $\mathcal{O}(|u|)$ computation steps for any finite input word $u \in \{0, 1\}^\star$). Analogous definitions can be given for 4-DSS recognition algorithms. An online algorithm is linear if it uses *on the average* a constant number of operations for each input chain code symbol.

9.6.1 Design paradigms

The decomposition of a 4- or 8-arc into a sequence of DSSs or 4-DSSs is a more general computational problem that includes DSS or 4-DSS recognition as a subproblem. Obviously a linear online DSS recognition algorithm supports a linear decomposition algorithm, but linear offline algorithms allow only quadratic runtime behavior.

The design of a DSS recognition algorithm may be based on a particular characterization of DSSs, such as the following:

(C1) the original definition of a DSS based on grid-intersection digitization;

9.6 Algorithms

(C2) a characterization by pairs of supporting lines (e.g., **(C2.1a)** Theorem 9.4, **(C2.1b)** Corollary 9.1, **(C2.2)** Theorem 9.5, **(C2.3)** Theorem 9.6);

(C3) the equivalence with the chord property (see Theorem 9.7); and

(C4) the DSS property (see Theorem 9.8).

Algorithms that use **(C4)** are often called *linguistic techniques*.

9.6.2 A linear online DSS recognition algorithm

We review in detail one of the first linear online DSS recognition algorithms, published in 1982 in [223], which uses **(C4)**.

Algorithm CHW1982a

The input sequence is $u = i(0)i(1)i(2)\ldots i(n)$, $i(k) \in A = \{0, 1, \ldots, 7\}$, $0 \le k \le n$. Let $u_0 = u$, and, if $u_{k-1} \ne \varepsilon$ (the empty word), let $u_k = R(u_{k-1})$ where R is the reduction operation used in Section 9.3 to define the DSL and DSS properties. Let $l(k)$ and $r(k)$ be the run lengths of nonsingular letters to the left of the first singular letter in u_k or to the right of the last singular letter; see Definition 9.6. Let $s(k)$ be the singular letter in u_k if there is one or -1 otherwise. Let $n(k)$ be the second letter in u_k if there is one or -1 otherwise. Algorithm 9.1 shows an example. The input chain code u_0 is now represented by a *syntactic code*, which is as follows for the given example:

	s	n	l	r
0	0	1	2	3
1	4	3	3	0
2	3	2	2	1

In general, the code consists of integers in four columns s, n, l, and r. The DSS property (see Definition 9.6) imposes constraints on these integers so that the given

---

$u_0 = $ 11011101110111011101110111011101110
1111011101110111011101110111011110111
$s(0) = 0, \quad n(0) = 1, \quad l(0) = 2, \quad r(0) = 3$
$u_1 = $ 33343343343334334
$s(1) = 4, \quad n(1) = 3, \quad l(1) = 3, \quad r(1) = 0$
$u_2 = $ 2232
$s(2) = 3, \quad n(2) = 2, \quad l(2) = 2, \quad r(2) = 1$
$u_3 = \varepsilon$

ALGORITHM 9.1 Input example for algorithm **CHW1982a** [223].

word $u = i(0)i(1)i(2)\ldots i(n)$ can be classified as being a DSS or not. Before starting to read a word, initialize the values in columns s and n to -1 and the values in columns l and r to 0. Suppose the syntactic code has been calculated for an input sequence of length ≥ 0 and the next input chain code is d. Let $\mathbf{N}(k,a,b)$ be true iff $|a-b| = 1$ for $k \geq 1$ and $|a-b|$ (modulo 8) $= 1$ for $k = 0$. The algorithm uses tests that follow straightforwardly from the DSS property:

$T_1(k,d):$ $\quad n(k) = -1 \wedge s(k) = -1 \wedge$
$\quad\quad\quad\quad [k > 0 \rightarrow l(k-1) \leq d+1 \wedge r(k-1) \leq d+1]$

$T_2(k,d):$ $\quad n(k) \neq -1 \wedge s(k) = -1 \wedge T_{2.1}(k,d) \wedge T_{2.2}(k,d)$

$T_{2.1}(k,d):$ $\quad d = n(k)$

$T_{2.2}(k,d):$ $\quad \mathbf{N}(k,d,n(k)) \wedge [k > 0 \rightarrow$
$\quad\quad\quad\quad \{l(k-1) \leq n(k) \vee l(k-1) = d \wedge l(k) \neq 0\} \wedge$
$\quad\quad\quad\quad \{r(k-1) \leq n(k) \vee r(k-1) = d \wedge r(k) \neq 0\}]$

$T_3(k,d):$ $\quad d = s(k) \wedge r(k) = 0 \wedge$
$\quad\quad\quad\quad l(k) = 1 \wedge s(k+1) = -1 \wedge n(k+1) \leq 1 \wedge [k > 0 \rightarrow$
$\quad\quad\quad\quad l(k-1) \leq s(k) \wedge \{r(k-1) \leq s(k) \vee r(k-1) = n(k)\}]$

$T_4(k,d):$ $\quad d = n(k) \wedge [s(k+1) = -1 \rightarrow r(k) \leq n(k+1)] \wedge$
$\quad\quad\quad\quad [s(k+1) \neq -1 \rightarrow r(k) + 1 \leq n(k+1) \vee$
$\quad\quad\quad\quad \{r(k)+1 = s(k+1) \wedge r(k+1) \neq 0\}]$

$T_5(k,d):$ $\quad d = s(k) \wedge r(k) \neq 0$

The algorithm inserts new elements d into the code of Algorithm 9.1 as long as the incoming chain code satisfies the DSS property.

Algorithm **CHW1982a** runs in linear time: $|u_{k+1}| \leq 1/2 \cdot |u_k|$ for all $k \geq 0$ and any input DSS chain code. There is only one loop in the algorithm in the case in which a new element needs to be added to one of the u_ks. Therefore its run time $t(n)$ for inputs of length $n = |u_0|$ is on the order of the following:

$$\mathcal{O}(|u_0| + |u_1| + \ldots + |u_{\log n}|) = \mathcal{O}\left(\sum_{k=0}^{\log_2 n} \frac{n}{2^k}\right) = \mathcal{O}(n)$$

It also follows that the number of relevant integers in the syntactic code is at most $\mathcal{O}(\log n)$, because the index m of the last nonempty word u_m satisfies $m \leq \log_2 n$. A stronger inequality is as follows:

$$n \geq (\frac{1}{2} + \frac{1}{4}\sqrt{2})(1+\sqrt{2})^m - 2$$

For example, $n = 2377\ldots 5739$ requires only reduced chain code words u_k for $k \leq m = 9$. Of course, representing a DSS by the endpoints of one of its possible preimages is an even shorter representation.

9.6 Algorithms

```
            k = 0
    1       if  T₁(k,d)    go to 10
            if  T₂(k,d)    go to 20
            if  T₃(k,d)    go to 30
            if  T₄(k,d)    go to 40
            if  T₅(k,d)    go to 50
            go to 100
   10       n(k) = d,  l(k) = 1,   return "yes"
   20       if  T₂.₁(k,d)   go to 21
            if  T₂.₂(k,d)   go to 22
            go to 100
   21       l(k) = l(k) + 1,   return "yes"
   22       s(k) = d,   return "yes"
   30       s(k) = n(k),  n(k) = d,  l(k) = 0,  r(k) = 2,
            return "yes"
   40       r(k) = r(k) + 1,   return "yes"
   50       d = r(k),  r(k) = 0,  k = k + 1,   go to 1
  100       for m = 0 until k − 2 do r(m) = s(m + 1)
            if k ≠ 0 then r(k − 1) = d,
            return "no"
```

ALGORITHM 9.2   DSS recognition algorithm **CHW1982a** using syntactic codes.

9.6.3 Review of other algorithms

This section briefly reviews some other DSS recognition algorithms and details an algorithm that segments an 8-arc into maximum-length DSSs.

Algorithm CHW1982b

Our second linear online algorithm, published in [223], uses the set of possible preimages (see approach **(C1)**); as long as this set is nonempty, we can continue to read chain code symbols.

Let the DSS $u \in \{0,1\}^*$ of length n connect grid point $p_0 = (0,0)$ with grid point p_n through grid points $p_1, \ldots, p_{n-1}$. Let $-0.5 \leq l_x \leq u_x < n + 0.5$ be segments of the grid lines $x = 0, x = 1, \ldots, x = n$, which are the unions of all intersections of these grid lines, with possible straight line segment preimages of u with respect to grid-intersection digitization, so that $x - 0.5 \leq l_x \leq u_x < x + 0.5$. Note that segment $(x, l_x)(x, u_x)$ may degenerate into a single point (i.e., $l_x = u_x$). Segment $(x, l_x)(x, u_x)$ must not contain the grid point p_x ($0 \leq x \leq n$); see Figure 9.6. The sequence of points $(0, u_0), (1, u_1), \ldots, (n, u_n), (n, l_n), (n-1, l_{n-1}), \ldots, (0, l_0)$ is called the *digitization polygon* of u. Because $(x, l_x)(x, u_x)$ may degenerate into a single point, the digitization

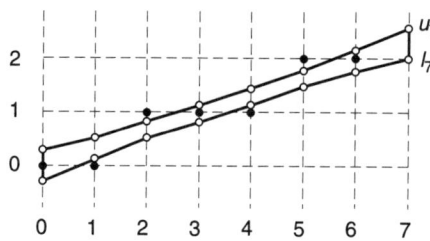

FIGURE 9.6 Digitization polygon for $u = 0100100$.

polygon need not be simple. Note that the segments $(0, u_0)(n, l_n)$ and $(0, l_0)(n, u_n)$ are contained in the digitization polygon.

Let u be extended by adding another chain code $a \in \{0, 1\}$. The 8-arc ua is a DSS iff it has a digitization polygon. The linear online algorithm **CHW1982b**, described in detail in [223] and also published in [222], uses the digitization polygon of u to construct the polygon for ua if possible or returns "no" if there is no polygon for ua.

The digitization polygon of a DSS is the union of all of the line segments with digitization that is the given DSS. Digitization polygons were also studied in [279]. Any DSS is uniquely characterized by a quadruple of integers: its length, its shortest period, its lowest-terms slope, and its phase. From this quadruple, we can calculate the digitization polygon.

Algorithm S1983

[988] gives a linguistic technique (see **(C4)**) for segmenting an 8-arc into DSSs. As in **CHW1982a**, this algorithm involves only integer operations using the syntactic rules specified in the DSS property. A parser checks the rules related to each layer k and (eventually) activates a parser for the next layer $k + 1$. Several parsers at different layers may be active simultaneously. The maximum number of layers is bounded by $4 \cdot 785 \cdot \log_{10} n + 1 \cdot 672$ and is taken on for digital rays of slopes a/b where a and b are consecutive Fibonacci numbers [566]. The average number of layers is less than half of this value [566].

[988] reports on experiments that compared polygons with vertices that are the *break points* of segmented 8-arcs with the polygonal preimages of these 8-arcs with respect to grid-intersection digitization. It describes an ambiguity in defining maximum-length DSSs defined by these break points.

Algorithm AK1985

[20] has already been cited in connection with pairs of supporting lines for 8-arcs; it gives a DSS recognition algorithm that follows approach **(C2.1b)**. Let $u \in \{0, 1\}^*$ be an 8-arc of length n connecting grid point $p_0 = (0,0)$ with grid point p_n through grid points $p_1, \ldots, p_{n-1}$. *Critical points* are a minimal subset of $G(u) = \{p_0, p_1, \ldots, p_n\}$ that defines a pair of supporting lines that have a minimum distance in the y direction

9.6 Algorithms

and that have $G(u)$ between or on them. An 8-arc u is a DSS iff this distance is < 1; see Corollary 9.1.

Without loss of generality, let u have four critical points $q_1, q_2, r_1, r_2 \in G(u)$ where $q_1 q_2$ specify a *nearest support below u* and $r_1 r_2$ a *nearest support above u*. Then u is uniquely specified either by n and q_1, q_2 or by n and r_1, r_2. [20] describes a linear offline algorithm for calculating the nearest support below and/or above u. A final test (Corollary 9.1) decides whether or not u is a DSS. This algorithm has also been used to define a linear offline algorithm for calculating the digitization polygon (see algorithm **CHW1982b**). [20] also discusses the calculation of digitization polyhedra for DSSs in 3D space.

Algorithm CHS1988a

[222] gives three linear online DSS recognition algorithms. The first is a slightly improved version of algorithm **CHW1982b**. The second also uses approach **(C1)**; however, this time the definition of grid-intersection digitization is used to perform DSS recognition by solving a separability problem for a monotonic polygon.

Let $u \in \{0, 1\}^*$ be an 8-arc of length n connecting grid point $p_0 = (0, 0)$ with grid point p_n through grid points $p_1, \ldots, p_{n-1}$. Let $p_k = (k, I_k)$, $k = 0, 1, \ldots, n$. The *weak digitization polygon* of u is defined by vertices $(0, I_0 + 0.5), (1, I_1 + 0.5), \ldots, (n, I_n + 0.5), (n, I_n - 0.5), (n-1, I_{n-1} - 0.5), \ldots, (0, I_0 - 0.5)$. This polygon is monotonic in the x direction. The separability problem is as follows: u is a DSS iff the upper polygonal chain $(0, I_0 + 0.5), (1, I_1 + 0.5), \ldots, (n, I_n + 0.5)$ of its weak digitization polygon can be separated from its lower polygonal chain $(n, I_n - 0.5), (n-1, I_{n-1} - 0.5), \ldots, (0, I_0 - 0.5)$ by a straight line that does not intersect either polygonal chain. [222] gives a linear online algorithm for solving this separability problem for ua based on a solution for u. The separability problem can also be stated as the problem of determining the visibility of $(0, I_0 - 0.5)(0, I_0 + 0.5)$ from $(n, I_n - 0.5)(n, I_n + 0.5)$ or vice versa.

Algorithm CHS1988b

The third linear online DSS recognition algorithm in [222] uses **(C2.1b)**. It is similar to (but independent of) the linear offline algorithm **AK1985**. It uses the critical points calculated for u to calculate critical points for ua if possible and returns "no" otherwise. The algorithm is quite short and allows a quick implementation. [222] also contains a geometric analysis of possible and impossible locations of critical points. If a critical point of u is cancelled in an extended uv, it cannot become a critical point in extensions of uv.

Algorithm SD1991

[1007] discusses a linear offline DSS recognition algorithm that uses the linguistic approach **(C4)**. It begins with Wu's linear offline algorithm [1137] and corrects the flaw described in [454].

Algorithm DR1995

[256] describes a linear online DSS recognition algorithm that uses the **(C2.1a)** approach; the naive line in [256] is the same as a DSL (see the comment following Theorem 9.4). The algorithm is based on updating two linear Diophantine inequalities, which amount to a test of whether $G(u)$ is in a strip (see algorithm **K1990**) of width $\max\{|a|,|b|\}$. For details of this algorithm, see Chapter 11, where it is used for 3D DSS recognition.

9.6.4 A linear online 4-DSS algorithm

[592] discusses the recognition of 4-DSSs on the frontier of a region in a 2D incidence grid using approach **(C2.3)**. Because this algorithm is one of the simplest and most efficient linear online 4-DSS recognition algorithms, we will give it in full detail. It is based on the calculation of a *narrowest strip* defined by the nearest support below and above (see Theorem 9.5). It resembles the linear offline algorithm **AK1985** and the linear online algorithm **CHS1988b** for 8-arcs. For the notation, see Figure 9.7; for examples, see Figure 9.6.4.

Algorithm K1990

This algorithm follows a digital 4-curve and extends a 4-DSS as long as it has at most two directions and all of its grid points lie between or on a pair of parallel lines that have a main diagonal distance of less than $\sqrt{2}$. On the parallel line to the left of the digital curve, we define a *negative base* between the grid points p_N=*StartN* and q_N=*EndN*, and, on the parallel line to the right of the digital curve, we define a *positive base* between the grid points p_P=*StartP* and q_P=*EndP*.

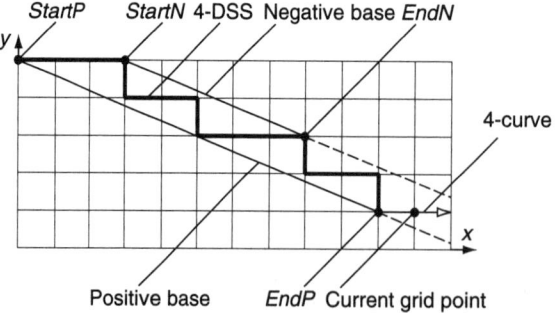

FIGURE 9.7 [592] Notation for algorithm **K1990**.

9.6 Algorithms

A sequence of grid points (x,y) on the digital 4-curve is a 4-DSS iff the following is true,

$$0 \leq bx - ay + c \leq |a| + |b| - 1$$

where $(a,b)^T$ is a vector $\mathbf{v}$ parallel to the negative (or positive) base of the 4-DSS that has relatively prime integer coordinates and c is an integer (constant for the 4-DSS) such that $c = ay - bx$ for any grid point (x,y) on the negative base. Let $h(x,y) = bx - ay + c$, and suppose the inequalities are true for $n-1$ accepted grid points of the 4-DSS. When a new 4-DSS is initialized, let the first step be from $p = (x_1, y_1)$ to the 4-adjacent $q = (x_2, y_2)$; let $p_N := q_P := p$, $q_N := p_P := q$, $a := x_2 - x_1$, $b := y_2 - y_1$, and note the direction of the step from p to q. Whenever a new step is not in one of the (at most two) directions in the current 4-DSS, we start a new DSS. When there has been only one direction so far, we continue the current DSS. If there have been two directions, we consider the cases shown in Algorithm 9.3.

Note that in case (iii), we have a new vector $\mathbf{v}$ that defines new values a, b, and c. In cases (i) and (ii), we have to move either q_N or q_P forward into position r_n. Figure 9.8 illustrates a clockwise and an anticlockwise traversal of the frontier of a digital region that produce different segmentations into maximum-length 4-DSSs.

(i) $h(x_n, y_n) = 0$: (x_n, y_n) is on the negative base, and the n vertices form a 4-DSS;

(ii) $h(x_n, y_n) = |a| + |b| - 1$: (x_n, y_n) is on the positive base, and the n vertices form a 4-DSS;

(iii) $h(x_n, y_n) = -1$ or $h(x_n, y_n) = |a| + |b|$: the n vertices form a 4-DSS, because the new grid point r_n is still within the distance limits from the points between the two supporting lines but the values a, b, and c need to be updated:

    if $h(x_n, y_n) = -1$ then
       begin
          $q_N := r_n$; $p_P := q_P$; $\mathbf{v} := r_n - p_N$;
       end

    if $h(x_n, y_n) = |a| + |b|$ then
       begin
          $q_P := r_n$; $p_N := q_N$; $\mathbf{v} := r_n - p_P$;
       end

(iv) otherwise, the n vertices do not form a 4-DSS; stop at the previous vertex r_{n-1}, and initialize a new 4-DSS.

ALGORITHM 9.3 Cases in algorithm **K1990**.

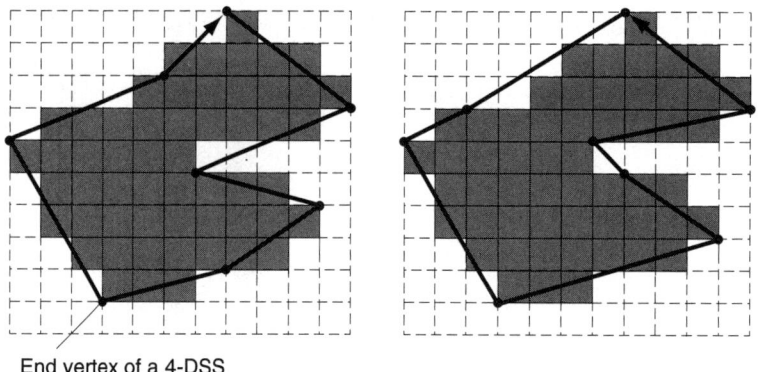

End vertex of a 4-DSS

FIGURE 9.8 Examples of algorithm **K1990**.

9.7 Exercises

1. Segment the infinite word $w = 0^1 10^2 10^3 \ldots 0^n 10^{n+1} \ldots$ into a sequence of consecutive DSSs of maximum lengths.

2. Consider the "spiral" in $\mathbb{Z}^2$ that starts at the origin $o = (0,0)$ and is represented by the infinite chain code $w = 0^1 1^2 3^4 \ldots n^{n+1} \ldots = 0112223333\ldots$. Let $u_n = 0^1 1^2 3^4 \ldots n^{n+1}$ be a prefix of w that defines a path from o to p_n. Characterize the asymptotic growth of the Euclidean distance $d_e(o, p_n)$. Is this growth linear, quadratic, polynomial of order m, or exponential?

3. Prove Equation 9.2 for infinite words w.

4. Let $F(i_{\alpha,\beta})$ be the set of all factors of the DSL $i_{\alpha,\beta}$. Prove that $F(i_{\alpha,\beta}) = F(i_{\alpha,0})$ for any irrational α.

5. Let i_{α,β_1} and i_{α,β_2} be DSLs with the same rational slope α. Prove that there exists an $m \in \mathbb{N}$ such that $i_{\alpha,\beta_1}(n) = i_{\alpha,\beta_2}(n+m)$ for all $n \in \mathbb{Z}$.

6. Let G be the set of grid points of a DSS, and let $C(G)$ be the Euclidean convex hull of G (see Chapter 13). Prove that the points of G are the only grid points in $C(G)$.

7. Two DSSs are called *parallel* (*perpendicular*) iff they are grid-intersection digitizations of parallel (perpendicular) straight line segments. Show that there exists a simply connected 4-region G that is not the set of grid points contained in a rectangle but with a 4-border that consists of two mutually perpendicular pairs of parallel DSSs.

8. Let $U_m(p) = \{q \in \mathbb{Z}^2 : d_8(q,p) \leq m\}$ where $m \geq 0$. We say that an 8-arc ρ has the *skeleton property* iff, for any distinct p and q in $G(\rho)$ and any point r on the (real) line segment pq, there exists a grid point $t \in G(\rho)$ such that $\max\{d_8(r,s) : s \in U_m(t)\} < m+1$. Prove that an 8-arc has the chord property iff it has the skeleton property.

9. Let $G_{p,q}$ be the set of all grid points that are on DSSs that connect p with q. Show that $G_{p,q}$ is an 8-region and that the 4-border of $G_{p,q}$ can be segmented into at most two DSSs.

10. Algorithms **CHW1982a** and **K1990** both have $\mathcal{O}(n)$ run time complexity (say $f(n) = an + b$ with $a, b > 0$) on DSSs of length n ($n_L \leq n \leq n_U$). Estimate a and b for each algorithm.

11. Give an example of a simple 8-curve and an initial point such that the numbers of DSSs in a segmentation of the curve into maximum-length DSSs are different for a clockwise and a counterclockwise traversal of the curve.

12. Let $D_{a,b,\mu,\omega} = \{(i,j) \in \mathbb{Z}^2 : \mu \leq ai + bj < \mu + \omega\}$ be an arithmetic line, where a and b are relatively prime integers such that $0 \leq a \leq b$ and μ and ω are integers. Prove the following:

    (i) D is 8-disconnected (i.e., D has 4-gaps; see Definition 7.13 if $\omega < b$).

    (ii) D is 8-connected and has 8-gaps if $b \leq \omega < a+b$.

    (iii) D is 4-connected and gap-free if $a+b \leq \omega$.

9.8 Commented Bibliography

The computer representation of lines and curves has been an active subject of research for nearly half a century [121, 342, 664, 883]. Related work even earlier on the theory of words [748] (specifically on mechanical or Sturmian words) remained unnoticed in the pattern recognition community. The material on the theory of words cited in this chapter follows [667, 668].

Theorem 9.1 is from [883]. For Theorem 9.2, see [140]. It has been known since [137] that grid-intersection digitization of rays $\gamma_{\alpha,\beta}$ produces periodic digital rays if α is rational and aperiodic digital rays if it is irrational (Theorem 9.3). An algorithm for "symmetric grid-intersection digitization" that does not depend on the clockwise or counterclockwise orientation of the curve is given in [1052]. [137] gives an algorithm for calculating the basic segment of a rational digital ray for $\beta = 0$. [1137] gives an algorithm for calculating the basic segment of such a ray using α and β as inputs.

The grid points of a rational ray are the integer solutions of a finite set of linear equations with rational coefficients [101]. This property was of basic importance for the establishment of arithmetic geometry [332, 848]. Theorem 9.4 and Exercise 12 are from [848].

Corollary 9.1 was proved in [30] using the chord property of Theorem 9.7; see also [20, 222].

[140, 312, 592] used digital 4-rays instead of 8-rays. [312] introduced digital rays in the grid cell model; sequences of bordering 1-cells define 4-rays. Digital 4-rays can also be generated by global mappings defined on digital 8-rays. The geometric characterization of 4-DSSs is discussed in [592] based on results in [20] about the "nearest support below or above" of a DSS.

Theorem 9.5 is an unproved statement in [592]. For Theorem 9.6, see [312].

An early algorithm for generating a DSS that connects two arbitrary grid points p and q was published in [843]. [137] proposed grammars for chain code generation of rational digital rays based on Freeman's criteria (**F1**), (**F2**), and (**F3**).

Theorem 9.7 was proved in [883]. The proof given in Section 9.3.1 that u is a DSS if $G(u)$ satisfies the chord property is from [864]. Alternative proofs of some of the results in [883] (at most two run lengths that are successive integers and one of which occurs only as singletons) are given in [348]. The compact chord property is discussed in [978]. A variant of the chord property for the outer Jordan digitization of a straight line segment was given in [867]. The chord property and its generalization to a chordal triangle property for $\mathbb{Z}^3$ are discussed in [866]. The property of evenness is studied in [454]. In [456], it is proved that the absence of runs that differ by more than 1 is equivalent to the chord property. [454] calls nonbalanced words *uneven* and proves that an infinite 8-arc has the chord property iff it has no uneven finite factors.

[528] gives three necessary and sufficient conditions (detailed definitions omitted here) for a digital arc to be a DSS: (i) its total absolute curvature is 0; (ii) its width in some direction is 0; and (iii) its length in some direction is less than half of the perimeter of its convex hull.

It has also been proved that point sequences generated by (a version of) Brons' parallel algorithm have the chord property [30, 31]. The formal language L of DSSs is context-sensitive; see [316] and later publications [802, 931, 1138]. This implies that linear-bounded or cellular automata can recognize DSSs using "string rewriting rules." A result in the theory of words [288] says that the complement $\{0,1\}^* \setminus L$ of the set of all DSSs is a context-free language.

Definition 9.5 is a precise formulation of criteria (**F1–F3**); it is used in [451] to define a DSS algorithm. The given formulation follows [449]. Definition 9.6 and Theorem 9.9 are also from [449].

[1136] does not contain the important algorithm published in [1137]. Regarding Theorem 9.8, [1137] does not contain a theorem but only statements about an algorithm specified by a flow chart. However, it is easily seen that this algorithm is actually an implementation of the DSS property as described above, so [1137] actually contains a proof of Theorem 9.8, including the generation of straight lines that have rational or irrational slopes. [1137] also considers the case of an infinite code sequence and shows that any finite factor of a two-sided infinite chain code c has the DSS property iff exactly one straight line with slope α and intercept β defines c by grid-intersection digitization.

9.8 Commented Bibliography

Material for a concise proof of Wu's theorem based on properties of Farey series was published in [278] in the form of an algorithm.[8] Proofs of Wu's theorem based on continued fractions (see Section 9.3.3) were published in 1991 in [140, 1105]; see also [1107]. The use of continued fractions to model digital rays was discussed in [137]. Related results in number theory [464] have been useful in these studies. We reviewed related definitions from number theory and reported results given by K. Voss in [1107].

For the theory of words, see [667, 668]. The definition of "Sturmian words" follows [668]. Some authors use the name "Sturmian words" for lower DSLs; see, for example, [74]. The proof of Theorem 9.10 (from [219]) is a citation of the proof of Theorem 1.3.13 in [668] as given by J. Berstel and P. Séébold. Theorems 9.11 and 9.12 are from [748]. The proof of Theorem 9.11 is a citation of the proof of Lemma 2.1.14 in [668] as given by J. Berstel and P. Séébold. Periodicity studies of digital rays can also be based on signal-theoretic (Fourier transform) methods (see [635]); this allows characterization of *approximate periodicity*.

Theorem 9.13 is from [728]. Asymptotic estimates of the number of DSSs of length n are given in [73]. [74] gives another proof of Theorem 9.13 and also an algorithm for random generation of lower DSSs of length n. Equation 3 is from [585]. [656] contains earlier related work. [653] uses the author's results on the number of DSSs in an $n \times n$ grid to show that piecewise DSS coding of digital curves requires $\mathcal{O}(n^4)$ table entries.

Suggestions about using Farey series to model digitized lines were made in [137, 274, 332, 738, 931]. In [931], it is shown that DSSs of length n through the origin are in one-to-one correspondence with the nth Farey series. The number of linear partitions of an $m \times n$ orthogonal grid is considered in [4] (Equation 4). Theorem 9.14 is from [279]. See [1099] for a more recent discussion about Farey numbers for characterizing DSLs.

Linear offline algorithms for DSS recognition based on the DSS property (see Definition 9.6) were published in 1981 in [451] and in 1982 in [1137].[9] A linear offline algorithm for cellular straight segment recognition based on convex hull construction is briefly sketched in [517, 509]. In [516], it is shown that a digital set is digitally convex iff the border digital arcs between the vertices of its convex hull are DSSs. A finite set of lattice points that lie between two lines at unit min (horizontal, vertical) distance is a DSS. A digital arc is a DSS iff it is convex. Convexity will be treated in Chapter 13.

The extended abstract [515] discusses digital arcs and digital convexity: a digital arc is a DSS iff it has the chord property; a digital set is digitally convex iff the convex hull of its set of corner points contains no corner point of its complement; and a digital arc is a DSS iff it is digitally convex (this is proved for several definitions of digitization in [516]). These conditions can be checked in linear offline time using run length coding. Algorithms in [516] deal with determining whether a digital region is a digital convex polygon.

---

8. The DSL property is called "linearity conditions" in [278]. See also our discussion of [278] in Section 9.5.

9. [454] discusses a flaw in Wu's algorithm.

Two linear online algorithms for DSS recognition were published in 1982 by E. Creutzburg, A. Hübler, and V. Wedler [223]; one of them is an online version of the offline algorithm published in [451]. For an early version of a linguistic DSS recognition algorithm, see [931]. (It was not based on the correct DSS property, which became known later.)

The general problem of decomposing a 4- or 8-arc into a sequence of 4-DSSs or DSSs, which includes 4-DSS or DSS recognition as a subproblem, is discussed in, among others, [256, 542, 592, 1007].

Many DSS recognition algorithms have not been reviewed here due to space limitations (e.g., [175, 341, 601, 650, 654, 655, 912, 990, 1155]). The original source of algorithm **K1990** is [592]; see also [542, 593] for discussions of this algorithm. A method of segmentation into a minimum number of DSSs in linear time based on calculating a tangential representation is given in [324]. Segmentation into "fuzzy DSSs" (defined using arithmetic geometry) is discussed in [254]. Scattering of points away from a "true" DSS is considered in [159].

Section 10.3 will describe performance evaluations of a few DSS recognition algorithms. Still lacking is a comprehensive and comparative evaluation of such algorithms. A statistic analysis of empiric time complexities would also be of interest. The random DSS generation algorithm of [74] could be used to create input data.

The segmentation of an 8-curve into maximum-length DSSs (see [223]) depends on the starting point and orientation of the traversal of the curve. It would be of interest to analyze the possible variation in such segmentations.

Different adjacency definitions may be worth studying in greater detail in connection with DSS algorithms; see, for example, [699] about digitizations in a 16-neighborhood. Straightness in 3D or higher-dimensional digital spaces would also be worth studying. (See, for example, [1028]: a set of grid points is an n-dimensional DSS iff $n-1$ of its projections onto the coordinate planes are 2D DSSs.) Finally, straightness can be discussed in multivalued digital pictures. (In [529], positional errors are estimated for straight edges between regions that have given constant values as a function of the size of the picture and its number of values.)

Exercises 4 and 5 are from [675]. For Exercise 6, see [516]. Exercise 7 is from [599]. Parallel DSSs are studied in [812]. Exercise 8 is from [979], and Exercise 9 from [811]. For other references about digitized straight lines, see [174, 280, 382, 722].

CHAPTER 10

2D Arc Length; Curvature and Corners

This chapter discusses ways of estimating the length or curvature of a 2D digital arc or curve using geometric constructions such as local or global polygonal approximations. We evaluate these methods in terms of theoretic criteria such as multigrid convergence as well as by experimental comparisons. Digitization and arc length are also defined for 3D curves in the first section of this chapter; for further discussion of 3D curves, see Chapter 11.

10.1 The Length of a Digital Curve

This section discusses methods of estimating the length of a 2D digital arc or curve. These methods can also be used to measure the perimeter of a simply connected region.

We first define curve digitization and the length of a digital curve for both 2D and 3D curves.

In this section, $\mathbb{Z}_h^2$ and $\mathbb{Z}_h^3$ are grids (in the grid point model), with grid constant $0 < \theta \leq 1$ and grid resolution $h = 1/\theta$ (see Section 2.1.2). $\mathbb{Z}_h^2$ consists of grid points with coordinates that are $(\theta \cdot i, \theta \cdot j)$ where $i, j \in \mathbb{Z}$, and $\mathbb{Z}_h^3$ consists of grid points with coordinates that are $(\theta \cdot i, \theta \cdot j, \theta \cdot k)$ where $i, j, k \in \mathbb{Z}$.

10.1.1 Curve digitizations

The topologic frontier of a simply connected compact set S in the Euclidean plane is a simple curve $\gamma : [0, 1] \to \mathbb{R}^2$. We assume that γ is rectifiable. In a grid, γ and S are represented in digitized form. We are interested in estimating the length and curvature of γ from its digital representation. In accordance with Definition 2.10, let $dig_h(\gamma)$ be a digitization of γ in $\mathbb{Z}_h^2$. Three possible digitizations of γ are as follows:

(i) a cyclic 4-path $\rho_{h,4}(\gamma)$ or 8-path $\rho_{h,8}(\gamma)$ of grid points derived from grid-intersection digitization of γ in $\mathbb{Z}_h^2$;

(ii) a cyclic 4- or 8-path of vertices of 2-cells on the frontier of the Gauss digitization $G_h(S)$ of S; and

(iii) the closed difference set (in $\mathbb{E}^2$) between the outer and inner Jordan digitizations $A = J_h^+(S)$ and $B = J_h^-(S)$ (i.e., $A \setminus B° = (A \setminus B)^\bullet$).

These digitizations define 2D digital curves in either the grid point or grid cell model. Option (iii) is, in general, the same as the outer Jordan digitization of γ (e.g., if γ does not follow any grid edge).

The *h-frontier* $\vartheta_h(S) = \vartheta(G_h(S))$ of S may consist of several nonconnected curves, even if S is convex; see Figure 10.1. Figure 10.1 also shows how this situation can be resolved; if h is sufficiently large, the dark shaded rectangle in Figure 10.1 provides a connection between the components. However, situations in which components cannot be connected for any grid resolution can arise at a vertex of a polygon that has a small interior angle $0 < \eta \ll \pi/2$ (see Figure 10.2). The vertex of the

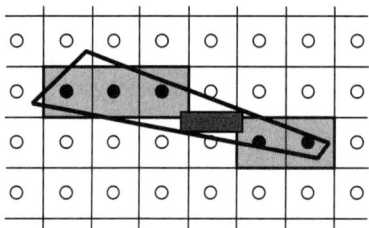

FIGURE 10.1 A simple polygon S for which $G_h(S)$ splits into two components. The dark shaded rectangle shows how the components can be reconnected if a higher grid resolution is used.

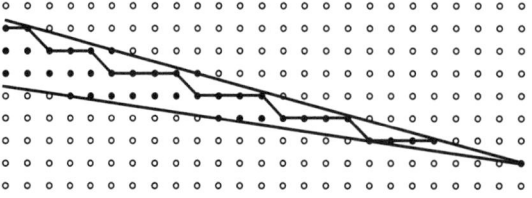

FIGURE 10.2 A DSS segment that does not continue to the vertex of an angle. The angular sector contains a ray with slope 1/5 [560].

angular sector (assumed here to be a grid point in $\mathbb{Z}^2$) cannot be connected with the other points in $G_h(S)$ (shown as filled dots) by any 8-path. In this example, there is a ray with rational slope $1/5$ "inside" the sector. Any grid resolution $5nh$ ($n \geq 1$) leads to the same situation.

S is called *h-compact* iff there is an $h_0 > 0$ such that $\vartheta_h(S)$ is a single (connected) curve for any $h \geq h_0$. When we deal with Gauss digitizations, we may sometimes have to require h-compactness.

In 3D, we will consider two possible digitizations of an arc or curve $\gamma : [0, 1] \to \mathbb{R}^3$:

(i) a (cyclic) α-path $\rho_{h,\alpha}(\gamma)$ of grid points ($\alpha \in \{6, 18, 26\}$) derived from grid-intersection digitization of γ in $\mathbb{Z}_h^3$; and

(ii) the outer Jordan digitization of γ.

If γ does not pass through any 1-cells, its outer Jordan digitization is a 2-connected sequence of voxels.

In either 2D or 3D, the input can be given as a sequence of pixels or voxels in an adjacency grid and can be encoded by a chain code $i(0), i(1), \ldots$, where $i(k) \in A = \{0, \ldots, 7\}$ in 2D and $i(k) \in A = \{0, \ldots, 26\}$ in 3D ($k \geq 0$).

10.1.2 Local and global estimation

Arc length estimates can be based on local metrics such as weighted or chamfer distances; see the end of Section 3.2.3. For example, isothetic steps in the 2D grid can be weighted by θ and diagonal steps by $\sqrt{2} \cdot \theta$, or we can use weights θ, $\sqrt{2} \cdot \theta$, and $\sqrt{3} \cdot \theta$ in the 3D grid. An advantage of such estimates is that they have linear online algorithms. The same is true for local curvature estimates, which take only fixed numbers of neighboring pixels or voxels into account.

A local estimator makes use of a polygonization of the digital arc or curve obtained by connecting successive grid points or grid vertices in a neighborhood of constant size. Global estimators, on the other hand, do not use fixed neighborhoods. ("Local" and "global" are formally defined in Section 8.4.3.) They may be based on maximum-length digital straight segments (DSSs), on *minimum-length polygons* (MLPs), or on approximations of normals or tangents.

10.1.3 DSS-based estimation

A DSS segmentation of an α-path ρ consists of DSSs of maximum length that define a polygonization of ρ; see Figure 10.3 (left) for a 2D example. The DSSs of the polygon are of potentially unlimited length. The sum of the lengths of the DSSs defines a *DSS-based length estimator*. (No weights are needed; the lengths of the DSSs are the Euclidean lengths of the corresponding line segments.) This length may depend on the starting point and on whether ρ consists of grid points or of vertices of grid cells.

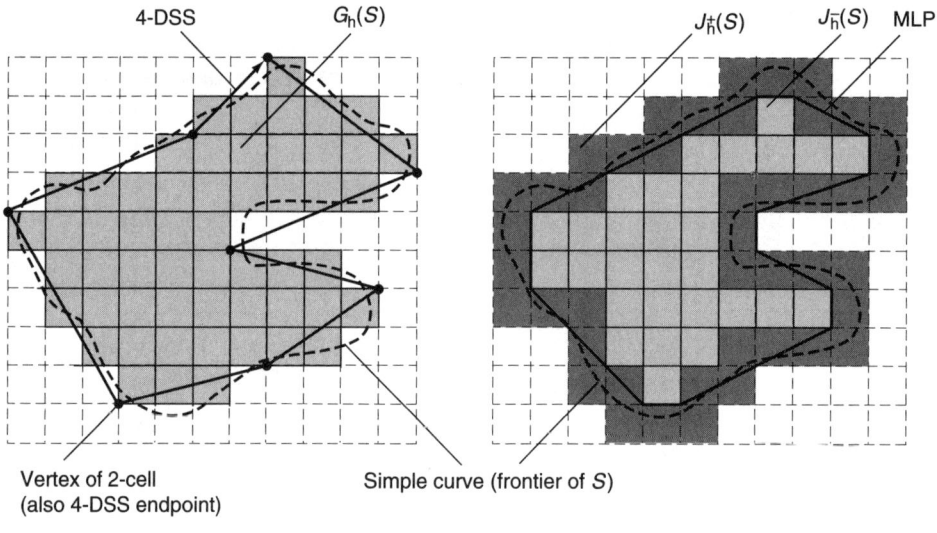

FIGURE 10.3 Left: segmentation of a 4-path into a sequence of maximum-length 4-DSSs. Right: MLP between two polygonal frontiers [555].

A DSS centered at a pixel or voxel p of ρ is also a global approximation of a tangential line segment[1] to ρ and can be used in the estimation of the curvature of ρ at p.

10.1.4 MLP-based estimation

A second class of global arc length estimators are MLP-based; see Figure 10.3 (right) for a 2D example. In 2D, an MLP is a minimum-length polygon that circumscribes the inner frontier of S and is in the interior of its outer frontier. As we saw in Section 1.2.9, the 2D MLP is the convex hull of the inner frontier relative to the outer frontier and is uniquely defined.

The *intrinsic distance* (see Section 3.2.4) between two points in a simple polygon is the length of a shortest arc that connects the points and is contained in the polygon. It is not hard to see that this shortest arc is polygonal and that the polygon is a metric space under intrinsic distance. The *intrinsic diameter* of the polygon is the maximum intrinsic distance between any two of its points. It is not hard to see that this diameter is the intrinsic distance between two vertices of the polygon. Analogous definitions can be given for a simple polyhedron.

---

1. ρ is assumed to be a digitization of a smooth arc γ; the tangential line segment at a point on γ is approximated by a DSS that is a subset of a digitized supporting line of a subsequence of ρ.

10.1 The Length of a Digital Curve

Let M_θ be the union of the cells in a simple 1-curve in $[\mathbb{C}^2, \leq]$ (see Section 7.2.2). ϑM_θ can be partitioned into two curves γ_1 and γ_2 that are the frontiers of two isothetic polygons P_1 and P_2 such that $P_1 \subset P_2^\circ$. The *length of M_θ* is defined as the length of the minimum-length curve in M_θ that encircles γ_1 (see Figure 10.4, left). Similarly, let M_θ be the union of all of the cells in a simple 1-arc in $[\mathbb{C}^2, \leq]$. The length of M_θ is defined as its intrinsic diameter (see Figure 10.4, right). It is not hard to prove that M_θ is a DSS iff its intrinsic diameter is defined by a straight line segment.

Similarly, let $M_\theta \subset \mathbb{R}^3$ be the union of all of the cells in a simple 2-arc or 2-curve in $[\mathbb{C}^3, \leq]$. If M_θ is a simple 2-curve, its *length* is the length of a minimum-length curve that is contained in M_θ and is not contractible into a single point in M_θ (see Figure 10.5, left). If M_θ is a simple 2-arc, its *length* is its intrinsic diameter (see Figure 10.5, right), and M_θ is a DSS iff its intrinsic diameter is defined by a straight line segment. M_θ is called *planar* iff the centers of all of its cells are coplanar. It can be shown that the intrinsic diameter of a nonplanar simple 2-arc is defined by a unique polygonal arc. For a given M_θ, these *MLP-based length estimates* are uniquely defined, but they may depend on the value of θ.

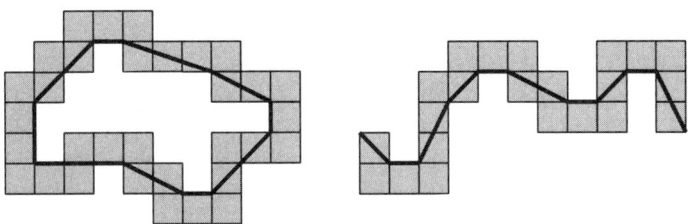

FIGURE 10.4 The length of a simple 1-curve (left) and of a simple 1-arc (right) in the 2D grid cell model [1005].

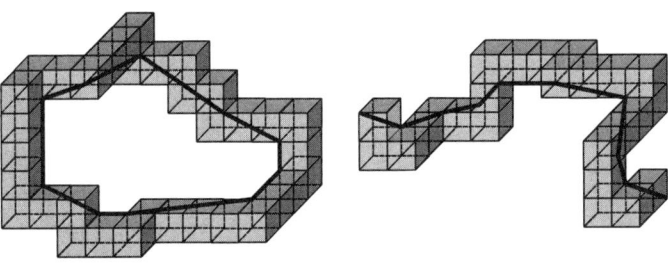

FIGURE 10.5 The length of a simple 2-curve (left) and of a simple 2-arc (right) in the 3D grid cell model [1005].

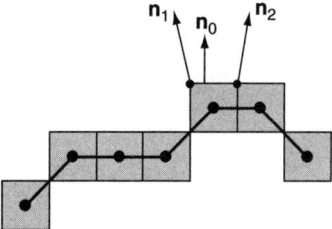

FIGURE 10.6 An 8-path in the grid cell model, and three normals at a cell.

10.1.5 Tangent-based estimation

Arc length estimates can also be based on estimated tangents; see Equations 8.1 and 8.18 for the 2D and 3D cases. Figure 10.6 shows an example. An 8-path is traced along, for example, the "upper" alternating sequence of 0-cells and 1-cells of its grid cell representation. For each 1-cell, we estimate normals $\mathbf{n}_1$ and $\mathbf{n}_2$ at its endpoint 0-cells, using a DSS or MLP method to find straight line approximations to the 8-path on both sides of the 1-cell. Let $\mathbf{n}_0$ be the unit normal to the 1-cell, and let $\mathbf{n} = (\mathbf{n}_1 + \mathbf{n}_2)/2$. We use $\langle \mathbf{n}, \mathbf{n}_0 \rangle = \mathbf{n} \cdot \mathbf{n}_0$ as an approximation to $\|(\dot{x}, \dot{y})\|$. Evidently, such a *tangent-based length estimate* depends on how the normals are estimated and combined. The normals can also be used to estimate the curvature of the path at the 1-cell.

In this chapter, we will study estimates of the arc lengths or curvatures of 2D digital arcs or curves, particularly with respect to their multigrid convergence; for 3D curves, see Chapter 11. Statistic properties of digital arcs and curves can also be studied; see Section 1.2.11.

10.2 Definitions of 2D Arc Length Estimators

In this section, we define the methods of 2D arc length estimation that will be evaluated in the next section.

10.2.1 Local estimators

Local estimations were historically the first ones used for arc length estimation in picture analysis. These estimators were applied to digital curves defined by digitization methods (i) or (ii) in Section 10.1.1.

Local estimators are based on shortest paths in a weighted adjacency grid of pixels. The weights are chosen to approximate Euclidean distance; see the discussion of chamfer metrics at the end of Section 3.2.3.

10.2 Definitions of 2D Arc Length Estimators

To make arc length estimates as accurate as possible, it has been suggested that statistic analysis be used to find weights that minimize the mean square error between the estimated and true length of a straight line segment. This leads to the definition of a *best linear unbiased estimator* (BLUE, for short) for straight line segments. One such estimator [281] is as follows,

$$E_{\text{chm}}(\rho_{h,8}(\gamma)) = \frac{1}{h} \cdot (0 \cdot 948 \cdot n_i + 1 \cdot 343 \cdot n_d) \qquad (10.1)$$

where n_i is the number of isothetic steps and n_d the number of diagonal steps in the digital arc or curve. (The subscript "chm" refers to the chessboard metric d_8.) Similar estimators have been proposed in [104] for chamfer distances (e.g., coefficients 0·95509 and 1·33693 for isothetic and diagonal steps for chamfer distances using a 3×3 neighborhood). Another local estimator [1115] is the *cornercount estimator* (coc, for short), where n_c is the number of odd-even transitions in the chain code of the digital arc or curve:

$$E_{\text{coc}}(\rho_{h,8}(\gamma)) = \frac{1}{h} \cdot (0 \cdot 980 \cdot n_i + 1 \cdot 406 \cdot n_d - 0 \cdot 091 \cdot n_c) \qquad (10.2)$$

Algorithms for calculating these local estimators are straightforward: trace a digital curve and locally update the estimator's value.

10.2.2 DSS estimators

Section 9.6 reviewed algorithms for DSS recognition. These algorithms decide whether a given sequence of grid points is a DSS; some of them also segment a digital arc or curve into a sequence of maximum-length DSSs. A length estimator E_{DSS} is then defined by the length of the resulting polygon or polygonal arc. For example, algorithm **DR1995** (for 8-curves) defines the E_{8ss} estimator, and algorithm **K1990** (for 4-curves) defines the E_{4ss} estimator. Note that the segmentation into DSSs is not uniquely defined; it depends on the method, the chosen starting point, and the direction in which the arc or curve is traced.

[283] defined a most probable original (mpo) length estimation method for DSSs that has superlinear convergence $O(r^{-1 \cdot 5})$; it is multigrid convergent for grid-intersection digitization and for the class of all straight line segments γ. Let n be the length of the DSS $\rho_{h,8}(\gamma) = i(0), \ldots, i(n-1)$, and let a/b be the best possible rational estimate of its slope. (Formally, b is the length of the shortest period, which is the smallest $k \in \{1, \ldots, n\}$ such that $k = n$ or $i(m+k) = i(m)$ for $0 \leq m \leq n-k-1$. a is the height difference in one period; for example, if $i(m) \in \{0,1\}$ for all $0 \leq m \leq n-1$, then $a = i(0) + \ldots + i(q-1)$.) Then we have the following:

$$E_{\text{mpo}}(\rho_{h,8}(\gamma)) = \frac{1}{h} \cdot (n\sqrt{1 + (a/b)^2}) \qquad (10.3)$$

In [283], this estimator was limited to single straight line segments, but it also defines an 8-DSS–based arc length estimator 8mp: apply the 8-DSS segmentation algorithm **DR1995** and sum the mpo length estimates of the 8-DSSs.

10.2.3 MLP estimators

In MLP-based length estimators, a simple 1-curve ρ is described by two isothetic polygons γ_1 and γ_2 that are the frontiers of sets S_1 and S_2 such that S_2 is contained in the interior S_1° of S_1 and ρ is contained in $B = S_1 \setminus S_2^\circ$. Here γ is defined by digitization method (iii) in Section 10.1.1, and S_1 and S_2 are $J_h^+(S)$ and $J_h^-(S)$, with the constraint that γ_1 and γ_2 are at Hausdorff distance 1 with respect to L_∞. We find an MLP that is contained in B and circumscribes γ_2; the E_{mlp} length estimate is then the length of this (uniquely defined) MLP.

We will describe two MLP-based length estimators. The standard MLP estimator E_{mlp} defines B to be the difference set between the outer and inner Jordan digitizations of S. The *approximating-sausage* estimator E_{aps} also involves a parameter δ ($0 < \delta \leq 0.5/h$).

10.2.4 MLPs of simple 1-curves

We assume that γ_1 and γ_2 are traced in the positive orientation (i.e., counterclockwise in a righthand coordinate system). A vertex v_i on γ_1 or γ_2 is called a *convex vertex* if the frontier makes a positive turn at v_i (i.e., $D(v_{i-1}, v_i, v_{i+1}) > 0$ [see Equation 8.14]). Similarly, v_i is called a *concave vertex* if the frontier makes a negative turn ($D(v_{i-1}, v_i, v_{i+1}) < 0$) and a *collinear vertex* if $D(v_{i-1}, v_i, v_{i+1}) = 0$.

Our algorithm traces γ_1 (or γ_2), detects convex and concave vertices, puts their coordinates into a list L, and marks them as convex or concave. For simplicity, assume that the coordinates are integers. The coordinates of two successive vertices with indices i and $i+1$ satisfy $|x_{i+1} - x_i| + |y_{i+1} - y_i| = 1$.

It is easy to show that only convex vertices of the "inner" curve γ_2 and only concave vertices of the "outer" curve γ_1 can be vertices of the MLP. There exists a mapping from the set of all concave vertices of γ_2 onto the set of all concave vertices of γ_1 such that each concave vertex of γ_1 corresponds to at least one concave vertex of γ_2. See Figure 10.7 for an example. The numbers denote successive vertices in the list L; 1 is a start vertex (i.e., an already known MLP vertex); 3 and 5 are successive convex vertices on γ_2; 2, 4, and 6 are successive concave vertices on γ_1. Vertex 7 is not between the negative (black line) and positive (white line) sides of sector (6,1,5); therefore 5 is the next MLP vertex and is a new start vertex.

The algorithm starts by putting all of the vertices of γ_2 into L. We then replace each concave vertex of γ_2 in L with its corresponding concave vertex of γ_1 by modifying its coordinates by ± 1, where the sign depends on the orientations of the incident edges. L now contains all of the vertices that will form the MLP. Note that the vertices in L are still in the original vertex order on γ_2.

Suppose we already know that vertex v_i of L is an MLP vertex. Then v_j ($j > i$) can be an MLP vertex only if all convex vertices v_k^+ such that $i < k < j$ lie on the positive side of (v_i, v_j) or are collinear with it (i.e., $D(v_i, v_k^+, v_j) \geq 0$). Similarly, all concave vertices v_l^- such that $i < l < j$ must lie on the negative side of (v_i, v_j) or be

10.2 Definitions of 2D Arc Length Estimators

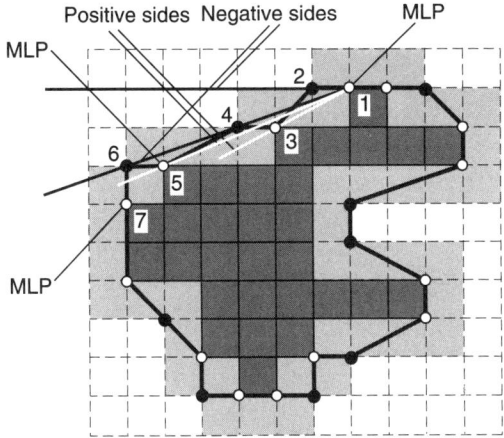

FIGURE 10.7 Finding the next MLP vertex [555].

collinear with it. (Otherwise (v_i, v_j) would cross either γ_1 or γ_2, which is not allowed to happen.) Suppose a convex vertex v^+ and a concave vertex v^- both satisfy these conditions. When we consider a vertex v as a candidate MLP vertex, the following situations can occur:

1. v lies on the positive side of (v_i, v^+) (i.e., $D(v_i, v^+, v) \geq 0$);

2. v lies on the negative side of (v_i, v^+) or is collinear with it and also lies on the positive side of (v_i, v^-) or is collinear with it; or

3. v lies on the negative side of (v_i, v^-).

In case (1), v^+ becomes the next MLP vertex. In case (2), v becomes a candidate for the MLP and must replace either v^+ or v^- depending on the sign of v. In case (3), v^- becomes the next MLP vertex. This is also correct in the trivial case where v^+, v^-, or both coincide with v_i. Thus we can start with a vertex v_1 that is known to be an MLP vertex (e.g., the uppermost-leftmost vertex of γ_2); set v^+ and v^- equal to v_1; and then test all subsequent vertices as just described. Whenever the next MLP vertex is detected, it becomes a new start vertex.

The algorithm is given in Algorithm 10.1; it also provides a length estimate $\mathcal{L}$.

If we are given a simply connected 4-component in the grid point model, the 4-path through all of the 8-border points of S defines the inner frontier γ_2. We can expand γ_2 into an outer frontier γ_1 at Hausdorff distance 1 from γ_2 (with respect to L_∞) and use γ_1 and γ_2 as inputs for the MLP algorithm. Figure 10.8 shows that the algorithm can handle "narrow concavities." The resulting MLP intersects all of the invalid edges of S.

Figure 10.8 also illustrates how the algorithm proceeds. The top left shows an example of the set between γ_1 and γ_2. The top right shows the original sequence

1. Initialize list $L = (v_1, \ldots, v_n)$ as described in the text; it contains all of the vertices on γ_2 except the concave vertices, which are replaced by concave vertices on γ_1. Each v_i in L is labeled by the sign of $D(v_{i-1}, v_i, v_{i+1})$.
2. Let $k := 1$, $a := 1$, $b := 1$ and $i := 2$. Let $\mathcal{L} := 0$ and $p_1 := v_1$.
   //v_1 is the first MLP vertex//
3. If $i > n+1$, stop.
4. If $i \leq n$, then $j := i$; else $j := 1$. //go back to v_1//
5. If $D(p_k, v_b, v_j) > 0$, then //v_j lies on the positive side//
   $\{k := k+1, p_k := v_b, i := b, a := b,$ and $\mathcal{L} := \mathcal{L} + d_e(p_{k-1}, p_k)\}$;
   else
   a) If $D(p_k, v_a, v_j) \geq 0$, then //v_j is in the sector//;
      if v_j has a positive label, then $b := j$, else $a := j$;
      else //v_j lies on the negative side//
   b) $\{k := k+1, p_k := v_a, i := a, b := a,$ and $\mathcal{L} := \mathcal{L} + d_e(p_{k-1}, p_k)\}$
6. Go to Step 3.

ALGORITHM 10.1 MLP calculation and length estimation.

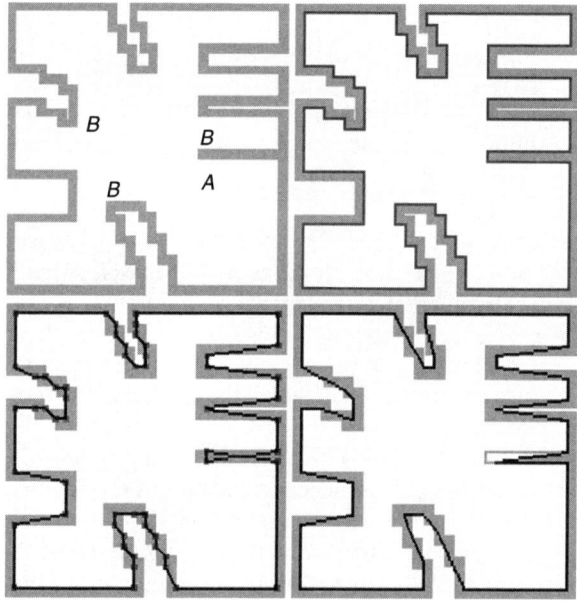

FIGURE 10.8 Successive steps of the MLP algorithm. In this example, γ_1 and γ_2 cross one another at A, and different segments of γ_1 coincide at the three Bs [555].

10.2 Definitions of 2D Arc Length Estimators

of 8-border points as a dark curve (γ_2); the bottom left shows the curve that connects all convex vertices on γ_2 and concave vertices on γ_1 in the order tested in the algorithm; and the bottom right shows the resulting MLP. The algorithm has linear time complexity.

10.2.5 The approximating sausage approach

Let the h-frontier of S be represented by $\Pi = \langle v_0, v_1, \ldots, v_{n-1} \rangle$ where the vertices are in clockwise order and the interior of S lies to the right. We define the *forward shift* $f(v_i)$ of v_i as the point on the edge (v_i, v_{i+1}) at distance δ from v_i, and the *backward shift* $b(v_i)$ of v_i as the point on the edge (v_{i-1}, v_i) at distance δ from v_i.

To generate the approximating sausage "around" Π, for each edge (v_i, v_{i+1}), we define the line segment $(v_i, f(v_{i+1}))$ that joins v_i to the forward shift of v_{i+1}; this is referred to as the *forward approximating segment* and is denoted by $L_f(v_i)$. The *backward approximating segment* $(v_i, b(v_{i-1}))$ is defined similarly and denoted by $L_b(v_i)$. We now have three sets of edges: the original edges of the h-frontier and the forward and backward approximating segments. Using these edges, we define a connected region $A_h^\delta(S)$ that is homeomorphic to an annulus as follows:

Given a polygonal circuit Π that describes an h-frontier in clockwise orientation, by reversing Π, we obtain a polygonal circuit Π^{-1} in counterclockwise orientation. In the initialization step of our approximation procedure, we consider Π and Π^{-1} as the *external* and *internal* bounding polygons of a polygon Π_B that is homeomorphic to the annulus. This initial polygon Π_B has area 0, and its points coincide with those of $\vartheta_h(S)$.

We now "move" the external polygon Π "away" from $G_h(S)$ and the internal polygon Π^{-1} "into" $G_h(S)$, as described below. (The Hausdorff distance between Π and Π^{-1} is nonzero when the sets are not identical.) This process expands Π_B step by step into a final polygon that contains $\vartheta_h(S)$, and such that the Hausdorff distance between Π and Π^{-1} is always less than or equal to $1/h$. For this purpose, we add forward and backward approximating segments to Π and Π^{-1} to increase the area of Π_B.

In the "moving" process, for any forward or backward approximating segment $L_f(v_i)$, or $L_b(v_i)$, we first remove the part lying in the interior of the current polygon Π_B and update Π_B by adding the remaining part of the segment as a new frontier edge. The orientation of this edge is chosen so that the interior of Π_B lies to the right of it. The resulting polygon Π_B^δ is referred to as the *approximating sausage* of the h-frontier and is denoted by $A_h^\delta(S)$ (see Figure 10.9, left, for $\delta = 1/2$). Note that the width of the approximating sausage depends on δ. An aps approximation of the frontier of S is a shortest closed curve $\gamma_h^\delta(S)$ that lies entirely in the interior of $A_h^\delta(S)$ and encircles the internal frontier of $A_h^\delta(S)$ (see Figure 10.9, right). It follows that $\gamma_h^\delta(S)$ is uniquely defined and is a polygonal curve defined by finitely many straight line segments. Note that $\gamma_h^\delta(S)$ depends on the choice of δ; we have used $\delta = 0.5/h$ in our experiments.

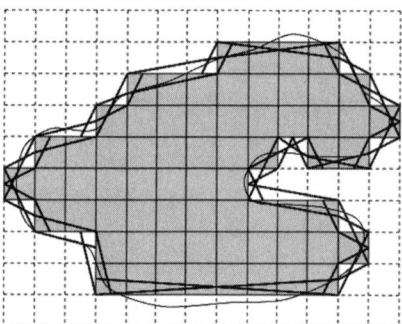 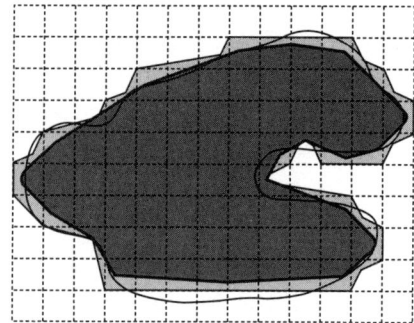

FIGURE 10.9 Left: construction of the approximating sausage. Right: resulting aps curve [50].

10.2.6 Tangent-based estimators

We can use a tangent vector integration process to estimate the length of a curve; see Equation 8.1. Let $\|(\dot{x},\dot{y})\| : [a,b] \to \mathbb{R}^2$ be the length of the tangent vector associated with $\gamma(t)$. Then the following can be approximated by using discrete estimates of the products $\|(\dot{x},\dot{y})\|$ dt:

$$l(\gamma) = \int_a^b \|(\dot{x},\dot{y})\| \, dt \qquad (10.4)$$

To estimate the normals $\mathbf{n}_1$ and $\mathbf{n}_2$ (see Figures 10.6 and 10.10), we regard the 0-cell as the center point p of a maximum-length DSS of the 8-curve that is obtained by tracing one frontier of the digitized (in the grid cell model) curve γ. A centered maximum-length DSS always has even length. The estimates can also be based on 4-DSSs that approximate the 4-curve of the chosen frontier. Figure 10.10 shows a DSS approximation on the left and a 4-DSS approximation on the right.

The DSS (or 4-DSS) approximates the tangent line to γ, and the normal is perpendicular to this line. This method can make use of any DSS definition and recognition algorithm. Straightforward application of a linear online DSS recognition algorithm starting at p and proceeding alternatingly in both directions along the curve leads to an $\mathcal{O}(n^2)$ solution, where n is the number of pixels on the curve.

An optimization method proposed in [323] allows us to compute all of the discrete tangents in linear time by using the discrete tangent computed at each vertex to initialize the discrete tangent at the next vertex; on average, this requires only a constant number of changes. We will use DSS approximation of the outer frontiers of simple 0-curves. The discrete version of Equation 10.4 is as follows, where S is the set of all pixels in $dig_h(\gamma)$:

$$E_{\tan}(dig_h(\gamma)) = \sum_{p \in S} \mathbf{n}(p) \cdot \mathbf{n}_0(p) \qquad (10.5)$$

10.3 Evaluation of 2D Arc Length Estimators

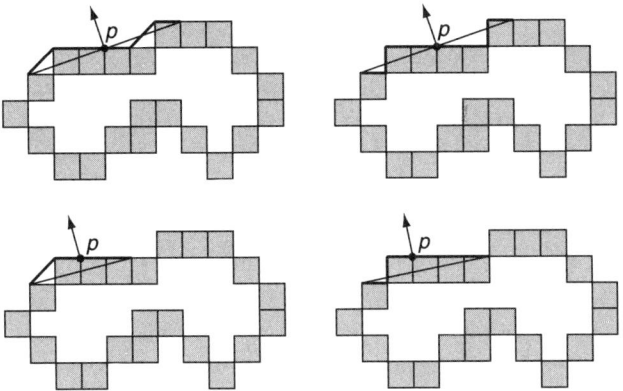

FIGURE 10.10 Left column: DSS centered at p. Right column: 4-DSS centered at p. Upper row: estimation of $\mathbf{n}_1$; both estimated tangential lines, and hence both estimated normals, are identical. Lower row: estimation of $\mathbf{n}_2$; different estimated tangential lines result in different estimated normals, depending on how the estimator is defined.

For each 1-cell c in the alternating sequence of 0-cells and 1-cells along the chosen frontier, we calculate $\mathbf{n}(c)$ and $\mathbf{n}_0(c)$.

10.3 Evaluation of 2D Arc Length Estimators

This section presents both experimental and theoretic comparisons of the 2D arc length estimators that were described in Section 10.2.

10.3.1 Online and offline algorithms and time complexity

The notions of online, offline, and linear time apply to estimators of all four classes; we describe them here for DSS estimators (see Section 9.6). An offline DSS algorithm takes an entire arc or curve (defined by a finite word $u \in A^*$) as input and decides whether it is a DSS. An online DSS algorithm reads successive chain codes $i(0), i(1), \ldots$; for each n, it decides whether $i(0), i(1), \ldots, i(n)$ is a DSS; if not, it initializes a new DSS with $i(n-1)i(n)$. After a maximum-length DSS has been found, its length estimate (usually the Euclidean distance between its end vertices) is added to the length estimate of the arc or curve. An offline algorithm is *linear* iff it runs in $\mathcal{O}(n)$ time (i.e., it performs at most $\mathcal{O}(|u|)$ basic computation steps for any input word $u \in A^*$). An online algorithm is linear iff it performs *on the average* a constant number of operations for each chain code symbol.

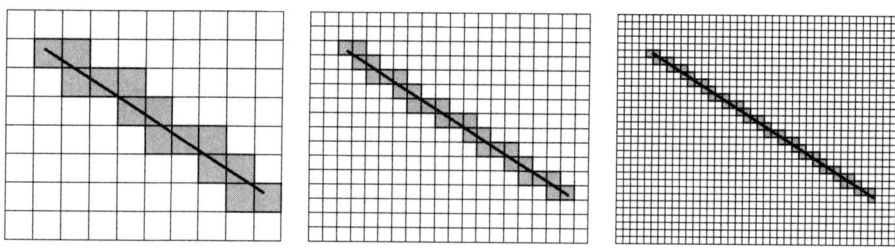

FIGURE 10.11 Outer Jordan digitization of a straight line segment in grids of varying resolution.

10.3.2 Multigrid convergence theorems

We apply the general definition of multigrid convergence (Definition 2.10) to the problem of estimating the length $\mathcal{L}(\gamma)$ of an arc or curve γ. Assume that the estimate E is defined for all curves γ in a given class (e.g., the class of all simple curves in the Euclidean plane) and for all digitizations $dig_h(\gamma)$ where $h > 0$. E is multigrid convergent to $\mathcal{L}$ with respect to digitization model dig_h iff $E(dig_h(\gamma))$ converges to $\mathcal{L}(\gamma)$ as $h \to \infty$ for any curve γ in the class of interest. More formally, we have the following:

$$|E(dig_h(\gamma)) - \mathcal{L}(\gamma)| \leq \kappa(h) \tag{10.6}$$

where $\lim_{h \to \infty} \kappa(h) = 0$; the speed of convergence is $\mathcal{O}(1/\kappa(h))$.

In general, we expect that an increase in grid resolution will result in an increase in accuracy. Suppose the digitization of a straight line segment consists of all cells that have nonempty intersections with the segment (i.e., we use outer Jordan digitization; see Figure 10.11). If the grid resolution is $h > 0$, the Hausdorff distance between the segment and the frontier of its outer Jordan digitization is $\sqrt{2} \cdot 1/h$. Hence the frontier converges to the segment with respect to the Hausdorff metric generated by the Euclidean metric. However, the perimeter of the union of the cells remains constant as h increases; this shows that this perimeter cannot be used for length estimation.

The local estimators chm and coc of the lengths of digitized arcs or curves (see Section 10.3) are not multigrid convergent. No proof has yet been published that local estimators cannot achieve multigrid convergence for "sufficiently complex" input data (e.g., not just for isothetic rectangles).[2] Given an algorithm for constructing a DSS approximation of the h-frontier $\vartheta_h(S)$ of a simply connected digital set $G_h(S)$, we define $\varepsilon_{\text{DSS}}/h$ as the maximum Hausdorff distance between $\vartheta_h(S)$ and γ_h:

$$d_e(\vartheta_h(S), \gamma_h) \leq \frac{\varepsilon_{\text{DSS}}}{h} \tag{10.7}$$

---

[2]. In a presentation at Dagstuhl/Wadern in April 2002, M. Tajine showed that local estimators are not multigrid convergent in some situations. In [1042], it is shown that any local length estimator is not multigrid convergent for almost any line segments in $\mathbb{R}^2$.

10.3 Evaluation of 2D Arc Length Estimators

Theorem 10.1 Let S be a convex h-compact polygonal set in $\mathbb{R}^2$. Then there exists a grid resolution h_0 such that, for all $h \geq h_0$, any DSS approximation of the h-frontier $\vartheta_h(S)$ is a connected polygon with perimeter $\mathcal{P}_h$ that satisfies the following inequality:

$$|\mathcal{L}(\vartheta(S)) - \mathcal{P}_h| \leq \frac{2\pi}{h}\left(\varepsilon_{\text{DSS}}(h) + \frac{1}{\sqrt{2}}\right) \tag{10.8}$$

(This theorem is from [560]; for a proof, see Section 10.3.3.) The value of h_0 depends on the given set. If $\varepsilon_{\text{DSS}}(h) = 1/h$, it follows from Equation 10.8 that the upper error bound for DSS approximations is as follows[3]:

$$\frac{2\pi}{h^2} + \frac{2\pi}{h \cdot \sqrt{2}} \approx \frac{4.5}{h} \quad \text{if } h \text{ is large} \quad (h \gg 1) \tag{10.9}$$

In [883], a DSS is assumed to be a finite 8-path. When we use cell complexes, it is appropriate to consider a finite 4-path to be a DSS iff its main diagonal width is less than $\sqrt{2}$; see [20, 597, 848].

Both of the MLP-based arc length estimators described in Section 10.2 (mlp and aps) are known to be multigrid convergent for convex Jordan curves γ.

Theorem 10.2 Let S be a bounded h-compact convex polygonal set. Then there exists a grid resolution h_0 such that, for all $h \geq h_0$, any aps approximation of the h-frontier $\vartheta_h(S)$ ($0 < \delta \leq 0.5/h$) is a connected polygon, for example, with perimeter $\mathcal{P}_h$ and the following:

$$|\mathcal{L}(\vartheta S) - \mathcal{P}_h| \leq (4\sqrt{2} + 8 * 0.0234)/h \approx 5.844/h \tag{10.10}$$

There are several convergence theorems for MLP approximations in [1005]. The perimeter of such an approximation is a convergent estimator of the perimeter of a bounded convex smooth or polygonal set in the Euclidean plane. The following theorem is basically from [1005]; it specifies the asymptotic constant for mlp perimeter estimates.

Theorem 10.3 Let γ be a convex planar curve that is contained in a simple 1-curve ρ in $[\mathbb{C}^2, \leq]$ for grid resolution $h \geq 1$. Then the MLP approximation of ρ is a connected polygonal curve of length $\mathcal{P}_h$ that satisfies the following:

$$\mathcal{P}_h \leq \mathcal{L}(\gamma) < \mathcal{P}_h + \frac{8}{h} \tag{10.11}$$

---

3. Let $\kappa(h) = 2\pi/h^2 + 2\pi/h \cdot \sqrt{2}$; then $\kappa(h) \approx \pi\sqrt{2}/h$ as $h \to \infty$.

From Theorems 10.1, 10.2, and 10.3, we see that the DSS upper error bound $4.5/h$ is smaller than the aps upper bound $5.844/h$, which is in turn smaller than the mlp upper bound $8/h$. However, this does not imply anything about the relative size of these errors. The determination of optimum error bounds remains an open problem.

Theorem 10.4 Let γ be a simple $C^{(2)}$ curve with bounded curvature. Then both the estimated discrete tangent direction and the tangent-based length estimate $E_{\tan}(dig_h(\gamma))$ are multigrid convergent.

The speed of convergence and the maximum error bound for these estimates have not yet been determined.

10.3.3 Proof of Theorem 10.1

To prove Theorem 10.1, we use two lemmas from integral geometry [958].

Lemma 10.1 If a convex planar polygonal set S is contained in a convex planar set C, the perimeter of S is at most equal to the perimeter of C.

We recall that an ε-*sausage* of a curve γ (as originally discussed by H. Minkowski [732]; see Figure 7.2) is the set of all points p such that $d_e(\{p\},\gamma) \leq \varepsilon$.

Lemma 10.2 The length of the outer frontier of the ε-sausage of the frontier of a convex planar polygon S is $\mathcal{P}(S) + 2\pi\varepsilon$, where $\mathcal{P}(S)$ is the perimeter of S.

We now prove Theorem 10.1. We assume that S is h-compact for $h \geq h_0$, so $G_h(S)$ is connected, and the DSS approximation of $\vartheta_h(S)$ is a single (connected) polygonal curve. We will prove that there exists a constant $h_1 \geq 1$ such that the following is true for all $h \geq h_1$:

$$d_e(\vartheta S, \vartheta_h(S)) \leq \frac{1}{h \cdot \sqrt{2}} \qquad (10.12)$$

Suppose this Hausdorff distance was greater than $(h \cdot \sqrt{2})^{-1}$. Then there would exist either (A) at least one point p on $\vartheta_h(S)$ with a minimum Euclidean distance to ϑS that is greater than $(h \cdot \sqrt{2})^{-1}$ or (B) at least one point q on ϑS with a minimum Euclidean distance to $\vartheta_h(S)$ that is greater than $(h \cdot \sqrt{2})^{-1}$.

In case (A), the circle with center p and radius $(h \cdot \sqrt{2})^{-1}$ would not contain any point of ϑS; hence this circle would be either (A1) disjoint from S or (A2) completely

10.3 Evaluation of 2D Arc Length Estimators

inside of S. Let p be on the frontier of a grid square with midpoint g_{ij}^h. The grid point g_{ij}^h is inside the circle. In case (A1), it follows that g_{ij}^h cannot be in S (i.e., g_{ij}^h is not in $G_h(S)$). It follows that p cannot be on $\vartheta_h(S)$, which contradicts our assumption. In case (A2), p is on an h-edge incident with two h-squares with midpoints that are both in the circle and thus in S; hence, in this case, too, p cannot be on $\vartheta_h(S)$. Thus case (A) is impossible.

In case (B), because S is h-compact for $h \geq h_0$, the distance between q and the nearest grid point can become arbitrarily small as $h \to \infty$, while $G_h(S)$ still remains connected. Thus, in case (B), we must increase h so that $h \geq h_1 \geq h_0$ for some h_1, which represents the situation in which the minimum Euclidean distance from $q \in \vartheta S$ to $\vartheta_h(S)$ is less than or equal to $(h \cdot \sqrt{2})^{-1}$. Only a finite number of vertices on the frontier of S can require such increases in h_0. This concludes the proof of Equation 10.12.

Equations 10.12 and 10.7 and the triangle inequality for Hausdorff distance imply the following:

$$d_e(\vartheta S, \gamma_h) \leq \frac{\varepsilon_{\text{DSS}}}{h} + \frac{1}{h \cdot \sqrt{2}}$$

Let $\varepsilon = \varepsilon_{\text{DSS}}/h + 1/(h \cdot \sqrt{2})$. Then the perimeter of S and the length of γ_h differ by at most $2\pi\varepsilon$. To show this, let γ_h be the frontier of a convex (polygonal) set C so that the following is true:

$$d_e(\vartheta S, \vartheta C) \leq \varepsilon \qquad (10.13)$$

We will show that the following is true, which will complete the proof of the theorem:

$$|\mathcal{P}(S) - \mathcal{P}(C)| \leq 2\pi\varepsilon \qquad (10.14)$$

The frontier ϑC lies in the ε-sausage of the frontier ϑS. Let $\vartheta_\varepsilon S$ be the outer frontier of the ε-sausage of ϑS. By Lemma 10.1, we have the following:

$$\mathcal{P}(C) \leq |\vartheta_\varepsilon S|$$

By Lemma 10.2, we have the following:

$$|\vartheta_\varepsilon S| = \mathcal{P}(S) + 2\pi\varepsilon$$

Hence, the following is true:

$$\mathcal{P}(C) \leq \mathcal{P}(S) + 2\pi\varepsilon \qquad (10.15)$$

ϑS lies in the ε-sausage of ϑC, because Hausdorff distance is symmetric. Let $\vartheta_\varepsilon C$ be the outer frontier of the ε-sausage of ϑC. By Lemma 10.1, we have the following:

$$\mathcal{P}(S) \leq |\vartheta_\varepsilon C|$$

By Lemma 10.2, we have the following:

$$|\vartheta_\varepsilon C| = \mathcal{P}(C) + 2\pi\varepsilon$$

Hence, the following is true:

$$\mathcal{P}(S) - 2\pi\varepsilon \leq \mathcal{P}(C) \tag{10.16}$$

From Equations 10.15 and 10.16, we have the following,

$$\mathcal{P}(S) - 2\pi\varepsilon \leq \mathcal{P}(C) \leq \mathcal{P}(S) + 2\pi\varepsilon$$

which proves Equation 10.14 and thus the theorem.

10.3.4 Experimental evaluation

Comparative evaluations of length estimators should investigate their accuracy and stability both on convex and nonconvex curves. In the experiments reported in [206], we used the test curves shown in Figure 10.12 digitized on grids of sizes between 30×30 and 1000×1000 using any of the three digitization methods in Section 10.1.1.

Two performance measures were used: (i) the relative error (in percent) between the estimated and true curve length; and (ii) for the DSS and MLP methods, a *tradeoff measure* defined as the product of the relative error and the number of generated segments (the *efficiency of convergence*).

In the plots in Figure 10.13, convergence is evident for all of the estimators. However, for estimators chm and coc, the convergence is to a false value! The errors are calculated for each curve, combined into a mean error for each grid size, and then the plots are generated by taking sliding means over 30 grid sizes. We show both linear and logarithmic plots of the errors. 8mp slightly overestimates the length as compared with 8ss.

The tradeoff measure is plotted in Figure 10.14 for the DSS and MLP estimators. Here, too, the errors are calculated for each curve and combined into a mean error for a given grid size, and the plots are generated by taking sliding means over 30 grid sizes.

As an additional test, a square of fixed size was rotated in a grid of resolution 128. Figure 10.15 shows its estimated perimeter as a function of rotation angle. Except for chm and coc, the estimates are relatively orientation-independent.

Figure 10.16 shows the runtimes of the DSS and MLP estimators as compared with local estimators. Obviously, the local estimators are fastest. MLP as implemented in [555] (as described in Section 10.2.3) is the fastest global estimator, but 4ss and 8ss come close to it. The aps estimator has not yet been optimized;

FIGURE 10.12 Test curves [555].

10.3 Evaluation of 2D Arc Length Estimators

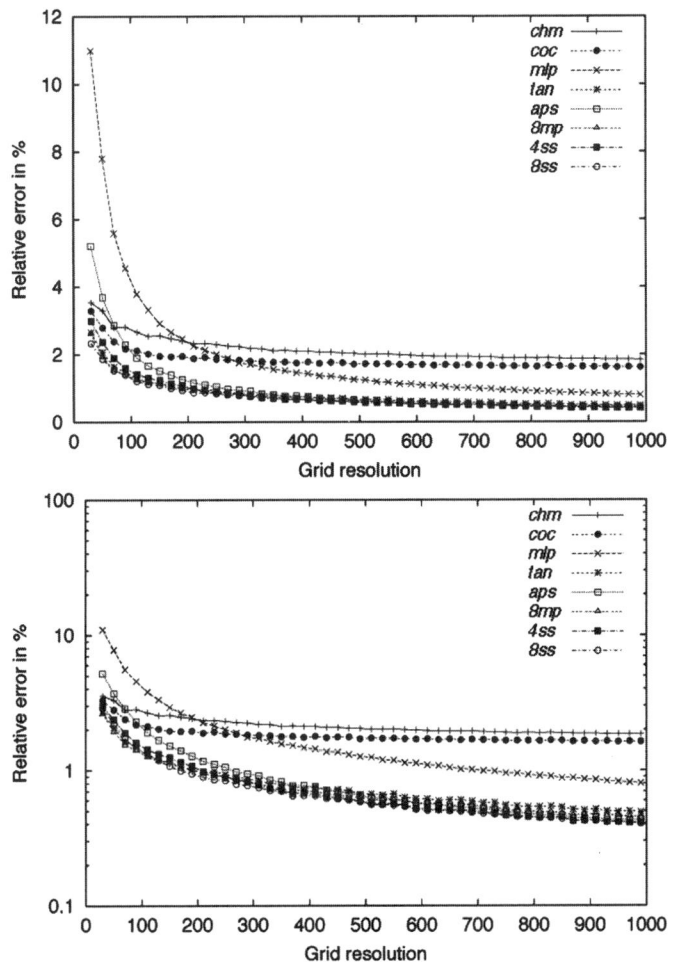

FIGURE 10.13 Multigrid convergence of length estimators: plots of sliding means of the relative errors (above); logarithmic plots (below) [206].

faster implementations may be possible. Theoretically, the tan estimator has a linear asymptotic runtime implementation [323], but the estimator that was tested had quadratic runtime; optimization is needed here, too.

These experiments show that the local estimators are not multigrid convergent even for our simple test data set. For the five global estimators, we obtained experimental confirmations of known theoretic convergence results.

Interestingly, the runtimes of the polygonal DSS and MLP estimators were only slightly greater than those of the local estimators; hence the use of local estimators is not justified by a runtime argument. The choice of a global estimator may depend

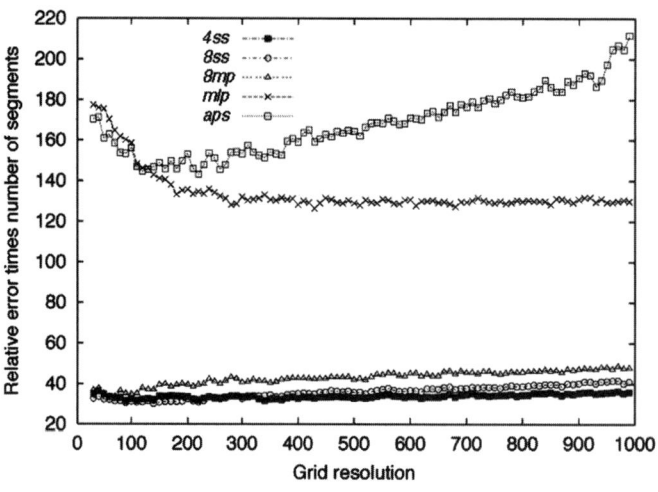

FIGURE 10.14 Trade off plots for the DSS and MLP estimators [206].

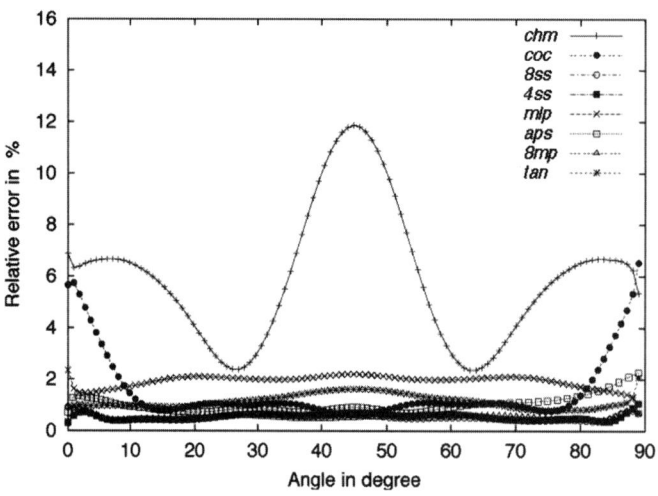

FIGURE 10.15 Relative errors for a rotating square [206].

on the available software. Studies involving more extensive test data might be useful in the selection of the most efficient estimator for a given application.

All of the tested estimators have linear online complexity. Of all of the possible MOP extensions of the DSS and MLP estimators we have reported only on 8mp. Our theoretic studies answered (in most cases) the following questions:

10.3 Evaluation of 2D Arc Length Estimators

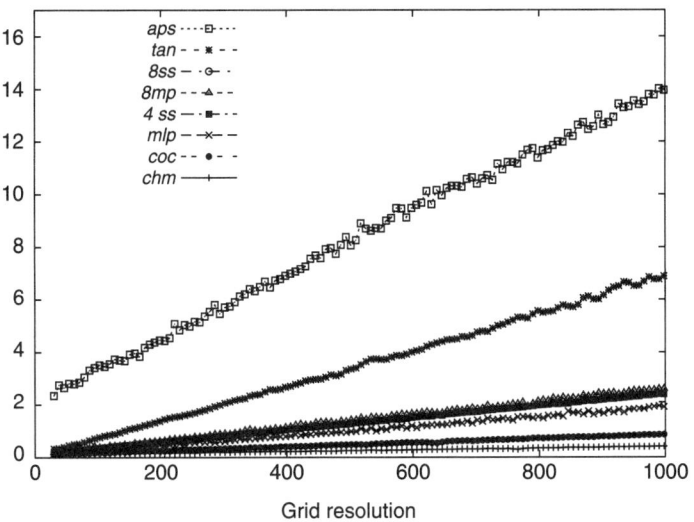

FIGURE 10.16 Runtimes of the DSS- and MLP-based estimators on an Ultra 10 Sparc workstation [206].

TABLE 10.1 Tested arc length estimators [206].

Method	Multigrid	Discrete	Unique	References	3D extension
8ss	Yes	Yes	No	[203]	[256]
8mp	Yes	Yes	No	-	[256]+[283]
4ss	Yes	Yes	No	-	[592]
mlp	Yes	Yes	Yes	[148]	[739, 1000]
aps	Yes	No	No	-	[49, 50]
tan	Yes	Yes	Yes	-	[305], [206]
local	No	Possibly	Yes	[481]	[104, 281]

multigrid convergence: Is the estimator multigrid convergent at least for convex curves? (If so, we are also interested in its convergence speed.)

discrete: Does the core of the estimation algorithm deal only with integers?

unique: Is the result independent of initialization (starting point, tracing orientation, and so forth)?

3D extension: Can the estimator be extended to digital curves in 3D space?

Table 10.1 provides a summary of the answers to these questions. Convergence speed is known to be linear for 8ss, 4ss, mlp and aps; see Theorems 10.1, 10.2, 10.3, and 10.4.

10.4 The Curvature of a Planar Digital Curve

In this section, we discuss methods of estimating the curvature of a planar curve and detecting "corners" (high-curvature pixels) on such curves. Let $\rho = \langle p_1, \ldots, p_m \rangle$ be an α-curve (usually an 8-curve) in $\mathbb{Z}^2$. Let $p_i = (x_i, y_i)$, where $i = 1, \ldots, m$, and let $1 \leq k \leq m$. For each pixel p_i on ρ, we define a *forward vector* $\mathbf{f}_{i,k} = p_i - p_{i+k}$ and a *backward vector* $\mathbf{b}_{i,k} = p_i - p_{i-k}$, where the indices are modulo m. Let $\mathbf{f}_{i,k} = (x_{i,k}^+, y_{i,k}^+)$ and $\mathbf{b}_{i,k} = (x_{i,k}^-, y_{i,k}^-)$.

10.4.1 Corner detectors

Curvature analysis is often oriented toward detecting high-curvature pixels ("corners"). We first review two early methods that detect such pixels without calculating curvature estimates.

Algorithm RJ1973

This algorithm [910] detects a corner at p_i based on analysis of the cosines $c_{i,k}$ of the angles between the forward vector $\mathbf{f}_{i,k}$ and the backward vector $\mathbf{b}_{i,k}$; $c_{i,k}$ is called the *k-cosine angle measure*. Let $0 < a < 1$ (e.g., $a = 0.05$) and $k_0 = \lfloor am \rfloor$. We start at $k = k_0$ and decrement k as long as $c_{i,k}$ increases. Suppose this occurs at $k = k_i$ so that $c_{i,k_i} \geq c_{i,k_i-1}$. In a subsequence of ρ of length k_i centered at p_i

a corner is detected at p_i iff $c_{i,k_i} > c_{i,k_j}$ for all $j \neq i$ such that $|i - j| \leq k_i/2$ (modulo m).

Note that the algorithm depends on only one parameter a. Because it does not use a constant initial value k_0, it is adaptive to the length of the 8-curve and is therefore global.

Algorithm RW1975

This algorithm [924] uses averaged cosine values given here:

$$\bar{c}_{i,k} = \begin{cases} \frac{2}{k+2} \sum_{l=k/2}^{k} c_{i,l} & \text{if } k \text{ is even} \\ \frac{2}{k+3} \sum_{l=(k-1)/2}^{k} c_{i,l} & \text{if } k \text{ is odd} \end{cases}$$

Otherwise, it follows algorithm **RJ1973**.

The k-cosine angle measure is an example of a *significance measure* defined in a sliding interval of width $2k_0 + 1$ and evaluated in a reduced interval of width $k_0 + 1$.

10.4 The Curvature of a Planar Digital Curve

The parameter k_0 depends on the number m of points on the given digital curve. The sliding interval defines a *region of support* (ROS) for a detected corner, which is normally considered to be at the center of its ROS. Evaluation of the measure is limited to a subset of the ROS. In the aforementioned examples, the size $\lfloor am \rfloor$ of the ROS is uniformly determined (i.e., it is the same at every point of the curve).

[703] reviews more than 100 corner detection and curvature estimation algorithms for 2D digital curves. Curvature estimators **(C1)** will be reviewed in the next subsection. Other approaches are based on the following:

- **(C2)** Estimation of curve properties in a uniformly determined ROS defined by input parameters. Examples of relevant properties are **(C2.1)** angles between forward and backward vectors (see above); **(C2.2)** approximation errors (e.g., defined by distances between arcs and chords [see Figure 10.17]); and **(C2.3)** directional changes between forward and backward ROSs.

- **(C3)** Estimation of curve properties in an individual determined ROS.

In the following paragraphs, we give a few examples of such approaches.

[251] uses approach **(C2.1)** at various scales to polygonally approximate the curve at different levels of detail. [257] defines *multiscale curvature* based on minima and maxima of the k-cosine at different scales, where k varies with the scale. In [194], corners are locations at which a triangle of specified size and angle can be inscribed in the curve.

Two of the corner detectors proposed in [932] are based on approach **(C2.2)**, specifically on distances between arcs and chords (see Figure 10.17). In [327], points of the curve are classified into the categories "on smooth interval," "on noisy interval," and "corner" by analyzing deviations of arcs of the curve from their chords. A more efficient implementation of this method was presented in [815]. Arc-chord distance is combined with Gaussian scale-space techniques in [407].

As an example of approach **(C2.3)**, [295] defines a *curvature chain* $c_i = \mathrm{mod}_8(s_i - s_{i-1} + 11) - 3$ where the s_is are chain codes. This curvature chain is "smoothed" by convolution with a filter. [46] uses comparisons of forward and backward chain code histograms based on a correlation coefficient measure. Chain code histograms on a sliding arc are used in [47].

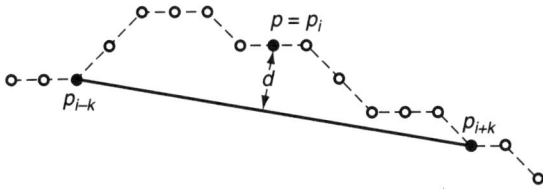

FIGURE 10.17 The *arc-chord distance measure* is defined by the distance d between a point p on the curve and a symmetric chord defined by a parameter k.

Other approaches based on curve properties include analysis of the chord lengths between pairs of points k steps apart along the curve [603] or of the number of DSSs centered at a point of the curve [586]. Corners can also be defined by local maxima of a measure of local symmetry [782]. [1065] describes a corner detection method based on eigenvalues of covariance matrices of neighborhoods of curve points.

As an example of approach **(C3)**, studies of the visual perception of shapes [618] motivated [879] to assign individually sized neighborhoods to border points of convex regions to detect "dominant points." [837] gives an algorithm for the calculation of backward and forward vectors and also allows asymmetric ROSs at a point. The backward and forward vectors point to p_{i-k} and p_{i+l}, respectively, where k and l are determined by studying the angular changes that take place when moving away from p_i along the curve. A (k,l)-cosine measure is used to detect corners; see [838] for further studies along these lines. For a combination of multiscale and individually sized ROSs, see [770].

Finally, we mention two examples of other approaches **(C4)**. In [455], chain codes are transformed into differential chain codes $c_i = s_i - s_{i-1}$. The maximum lengths of nonzero sequences in the differential code — and sums of consecutive nonzero differential codes (starting with *pair sums* and generalizing to *group sums*) — are used to detect "critical points." See [32] for other applications of differential codes. [421] derives a hierarchical, polygonal approximation to a curve by detecting candidate "dominant points" and then iteratively removing "least significant points" until a stable polygonal approximation is reached.

10.4.2 Curvature estimators

Curvature can be estimated from **(C1.1)** the change in the slope angle of the tangent line (e.g., relative to the x-axis); **(C1.2)** derivatives along the curve; or **(C1.3)** the radius of the osculating circle (circle of curvature); see Definition 8.1, Equation 8.9, and Equation 8.11, respectively.

Algorithm FD1977

This algorithm [346] (in category **(C1.1)**) estimates changes in the slope angles θ_i of the tangent lines at points p_i. Let the following be the angle estimate based solely on backward vectors of "length" k:

$$\theta_{i,k} = \begin{cases} \tan^{-1}(y^-_{i,k} x^-_{i,k}) & \text{if } |x^-_{i,k}| \geq |y^-_{i,k}| \\ \cot^{-1}(x^-_{i,k} y^-_{i,k}) & \text{otherwise} \end{cases}$$

Let the following be the centered difference,

$$\delta_{i,k} = \theta_{i+1,k} - \theta_{i-1,k}$$

10.4 The Curvature of a Planar Digital Curve

which is "accumulated" by the following using $\Delta = \tan^{-1}(1/(k-1))$:

$$E_{i,k} = \ln t_1 \cdot \ln t_2 \cdot \sum_{j=i}^{i+k} \delta_{j,k}$$

Let the following also be true:

$$t_1 = \max\{t: \forall s\,(1 \le s \le t \to -\Delta \le \delta_{i-s,k} \le \Delta)\}$$
$$t_2 = \max\{t: \forall s\,(1 \le s \le t \to -\Delta \le \delta_{i+k+s,k} \le \Delta)\}$$

These are the maximum lengths of the arcs preceding p_i and following p_{i+k} in which the differences $\delta_{j,k}$ remain close to zero (in an interval $[-\Delta, +\Delta]$) and which define the "legs" of a corner.

A corner is detected at p_i iff $E_{i,k} > T$ and the previous corner is at a distance of at least k from p_i on ρ.

Note that this procedure depends on a parameter k (e.g., $k = \lfloor am \rfloor$ as in algorithm **RJ1973**) and on a threshold T. Because t in the definition of t_1 and t_2 can be arbitrarily large, this is a global method.

Algorithm BT1987

This algorithm [84] modifies algorithm **FD1977** as follows: t_1 and t_2 are upper-bounded by $\lfloor bm \rfloor$ ($0 < b < 1$), and the curvature estimates $E_{i,k}$ are calculated and averaged over a range of values of k ($k_L \le k \le k_U$):

$$E_i = \frac{1}{k_U - k_L + 1} \sum_{k=k_L}^{k_U} E_{i,k}$$

This method involves four parameters: k_L, k_U, b, and the threshold T. Because the upper bound is $\lfloor bm \rfloor$, it is a global method.

Algorithm HK2003

This algorithm [434], which is also in category **(C1.1)**, estimates changes in the slope angles of tangents. It calculates a maximum-length 8-DSS $p_{i-b}p_i$ of the following Euclidean length

$$l_b = \left((x_{i-b} - x_i)^2 + (y_{i-b} - y_i)^2\right)^{-1/2}$$

that goes "backward" from p_i on ρ and a maximum-length 8-DSS $p_i p_{i+f}$ of the following length

$$l_f = \left((x_{i+f} - x_i)^2 + (y_{i+f} - y_i)^2\right)^{-1/2}$$

that goes "forward" from p_i on ρ. It then calculates the following angles,

$$\theta_b = \tan^{-1}\left(\frac{|x_{i-b} - x_i|}{|y_{i-b} - y_i|}\right) \quad \text{and} \quad \theta_f = \tan^{-1}\left(\frac{|x_{i+f} - x_i|}{|y_{i+f} - y_i|}\right)$$

the mean $\theta_i = \theta_b/2 + \theta_f/2$, and the angular differences $\delta_1 = |\theta_f - \theta_i|$ and $\delta_2 = |\theta_b - \theta_i|$. Finally, the curvature estimate at p_i is as follows:

$$\frac{\delta_1}{2l_f} + \frac{\delta_2}{2l_b}$$

Algorithm M2003

[736] is an early example of an algorithm in category **(C1.2)**. More recently, [703] assumes that the given 8-curve $\langle p_1, \ldots, p_m \rangle$, where $p_j = (x_j, y_j)$ for $1 \leq j \leq m$, is sampled along a parameterized curve $\gamma(t) = (x(t), y(t))$ where $t \in [0, m]$. At point p_i, we assume that $\gamma(0) = \gamma(m) = p_i$ and $\gamma(j) = p_{i+j}$ for $1 \leq j \leq m-1$. Functions $x(t)$ and $y(t)$ are locally interpolated at p_i by the following second-order polynomials,

$$x(t) = a_0 + a_1 t + a_2 t^2$$
$$y(t) = b_0 + b_1 t + b_2 t^2$$

and curvature is calculated using Equation 8.9. We have $x(0) = x_i$, $x(1) = x_{i-k}$, and $x(2) = x_{i+k}$ with integer parameter $k \geq 1$; this is analogous for $y(t)$. The curvature at p_i is then defined by the following:

$$E_i = \frac{2(a_1 b_2 - b_1 a_2)}{(a_1^2 + b_1^2)^{1.5}}$$

Algorithm CMT2001

This algorithm [208], which is in category **(C1.3)**, involves approximation of the radius of the osculating circle. At each point p_i, we calculate a maximum-length DSS centered at p_i. This DSS is used as an approximate segment of the tangent at p_i, and its length l_i corresponds to the radius of the osculating circle. The algorithm calculates an *inner radius* $I_i = \lceil (l_i - 1/2)^2 - 1/4 \rceil$ and an *outer radius* $O_i = \lfloor (l+1/2)^2 - 1/4 \rfloor$ and returns $2/(I_i + O_i)$ as an estimate of the radius of curvature at p_i.

10.4.3 Experimental evaluation

[661] contains an experimental comparison of several curvature estimation methods (see Sections 10.4.1 and 10.4.2) that were published before 1990 as well as some other local methods (see the references in Section 10.6). The results obtained from algorithm **BT1987** were described as being "in best correspondence to human perception."

We illustrate comparative evaluations for only two algorithms, **CMT2001** and **HK2003**. The curvature of a circle of radius R is $\kappa = 1/R$. Figures 10.18 and 10.19 show the error $|1/R - \tilde{\kappa}|$ averaged over all pixels on 8-curves that are the borders of Gauss digitizations of disks $x^2 + y^2 = r^2$ or $x^2 + y^2 = r^2 + r$ ($r = 100, \ldots, 1000$). The plots are sliding means, which eliminate "digitization noise" by symmetrically averaging 19 values before and 19 values after the current value r.

10.4 The Curvature of a Planar Digital Curve

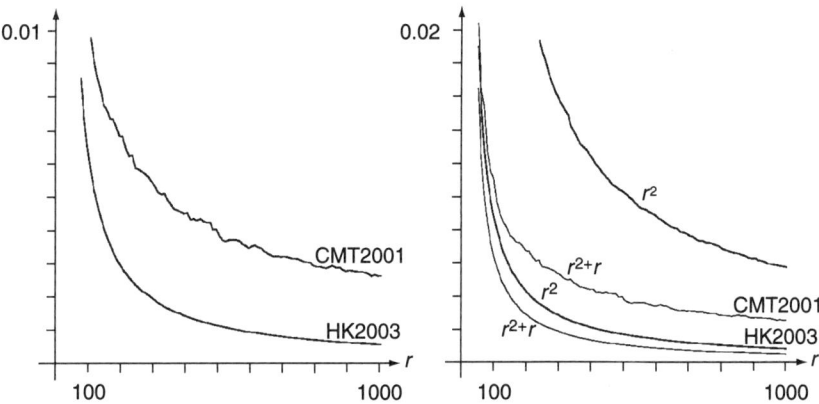

FIGURE 10.18 Left: sliding means of errors in the interval $[0,\ldots,0{\cdot}01]$ for algorithms **CMT2001** (upper curve) and **HK2003** (lower curve) for the digitized disk $x^2+y^2 = r^2+r$. Right: sliding means of errors in the interval $[0\ldots0{\cdot}02]$ for $x^2+y^2 = r^2$ and $x^2+y^2 = r^2+r$; the upper pair of curves are for **CMT2001** and the lower pair for **HK2003**. In both cases, the (larger!) disk $x^2+y^2 = r^2+r$ produces smaller errors.

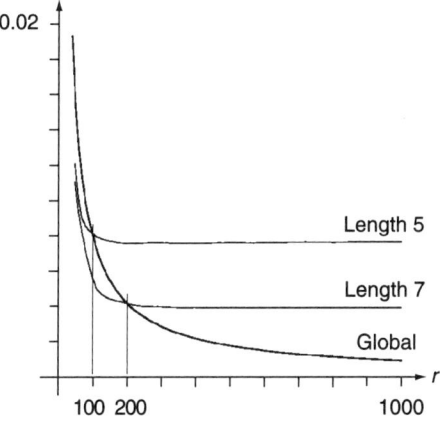

FIGURE 10.19 Sliding means of errors for algorithm **HK2003** (lower curve) for the disk $x^2+y^2 = r^2$ digitized at grid resolutions $h = 100,\ldots,1000$, as compared with local restrictions of the algorithm in which the length of the approximating line segments is limited to 5 (upper curve) or 7 pixels. The curves intersect between $r = 100$ and $r = 101$ and between $r = 203$ and $r = 204$.

Figure 10.18 shows total errors for algorithms **CMT2001** and **HK2003**, on the left for the digitized disk $x^2 + y^2 = r^2 + r$ and on the right also for the disk $x^2 + y^2 = r^2$. In the latter case, the digital disk has small "peaks" on its border due to the fact that its radius is an integer. These "unsmooth" situations slightly increase the error for both algorithms.

Figure 10.19 shows how the errors behave if the unlimited-length DSS approximation in **HK2003** is replaced by backward $b_{i,k}$ and forward vectors $f_{i,k}$ of constant length (5 and 7 in the figure). The errors are smaller than for the global method up to about $r = 100$ for $k = 5$ and up to about $r = 200$ for $k = 7$.

[703] contains an extensive experimental comparison of corner detectors and curvature estimation methods; it demonstrates good performance of algorithm **M2003** as compared with 28 other methods. We illustrate the experimental comparison in [703] by giving results obtained by five methods on two 8-curves (4-borders of 4-regions); Figure 10.20 shows these 8-curves.

Figure 10.21 shows corner or curvature values for the symmetric curve (which is of length 222) calculated by methods **RJ1973** (with $a = 0.1$), **FD1977** (with $k = 9$), **BT1987** (with $k_L = 9$, $k_U = 14$, and $b = 0.1$), **M2003** (with $k = 10$), and **HK2003**.

RJ1973 produces a "plateau" between border points 115 and 124. The original algorithm would label all of these points as being corners. In the resulting segmentation, which is shown in Figure 10.22 on the upper left, only the midpoint of this "plateau" is taken as a corner.

FD1977 reproduces the curve's symmetry in the calculated values $E_{i,k}$. The original algorithm maps these into asymmetric corner positions; see the segmentation shown in Figure 10.22, which uses $k = 9$. The following modification would produce perfectly symmetric corners: select p_i as a corner iff $E_{i,k} > E_{j,k}$ for all j such that $|i - j| \leq k$.

A maximum of 14 corners was used for **BT1987**. No corners were detected by **FD1977** at any of pixels 63 through 93 or 146 through 176.

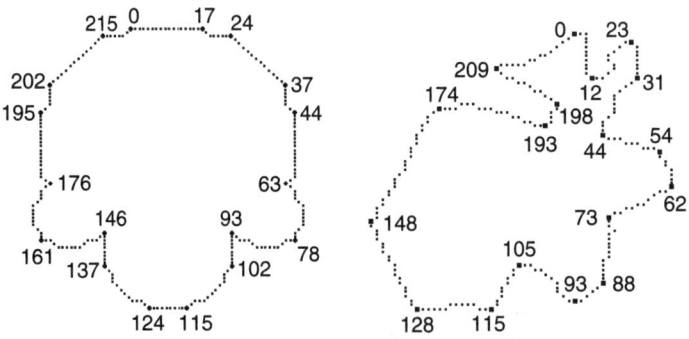

FIGURE 10.20 A symmetric 8-curve (left) and an asymmetric 8-curve (right) that were used in comparisons of corner detectors (using only the grid resolution shown in the figure).

10.4 The Curvature of a Planar Digital Curve

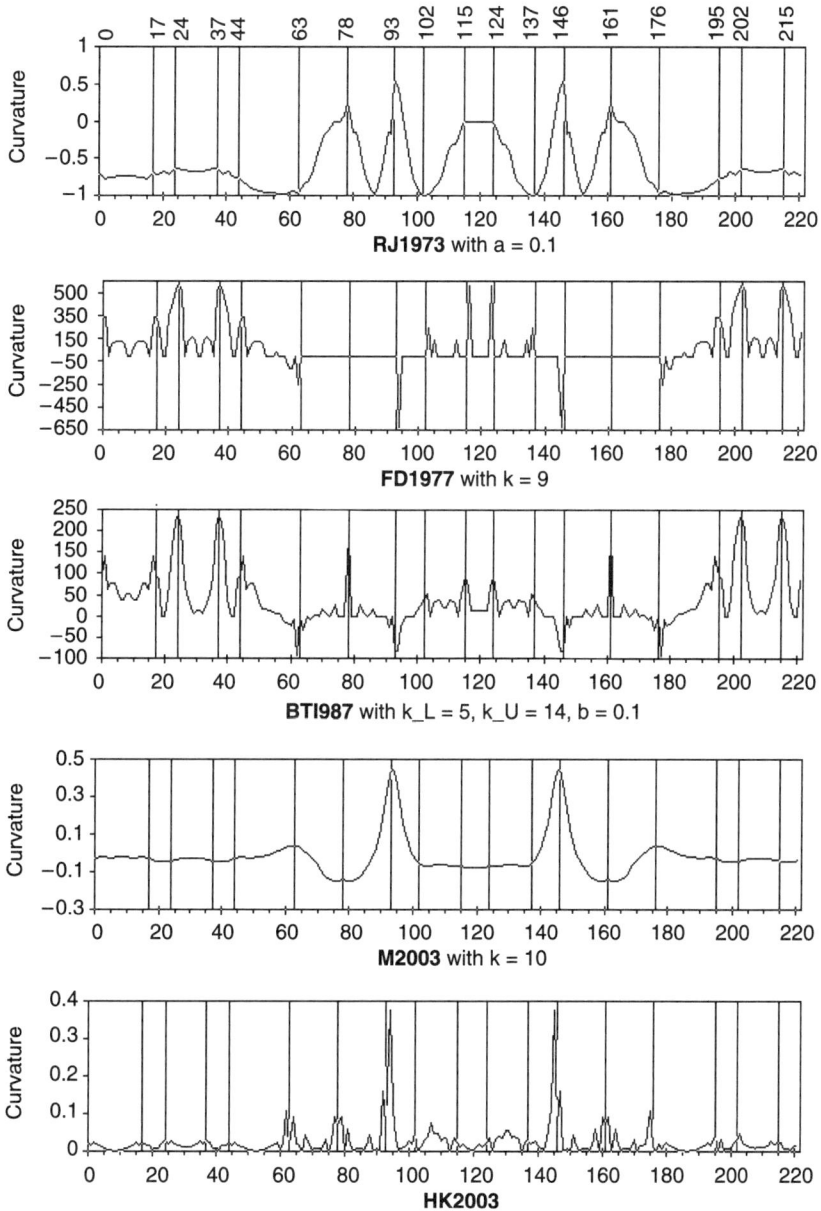

FIGURE 10.21 Corner or curvature values measured by five methods. The parameters used are specified in the figure. The x-coordinate gives border point numbers as shown in Figure 10.20 for the symmetric curve.

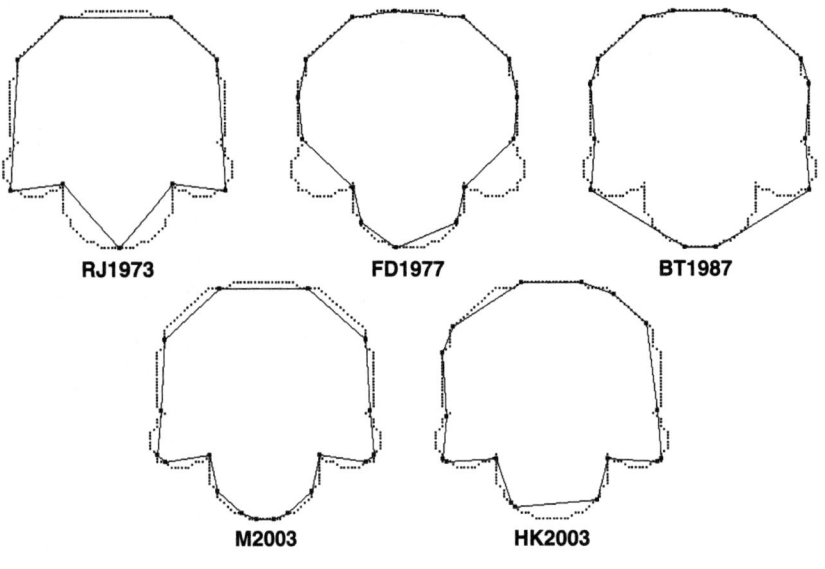

FIGURE 10.22   Resulting segmentations for the symmetric curve.

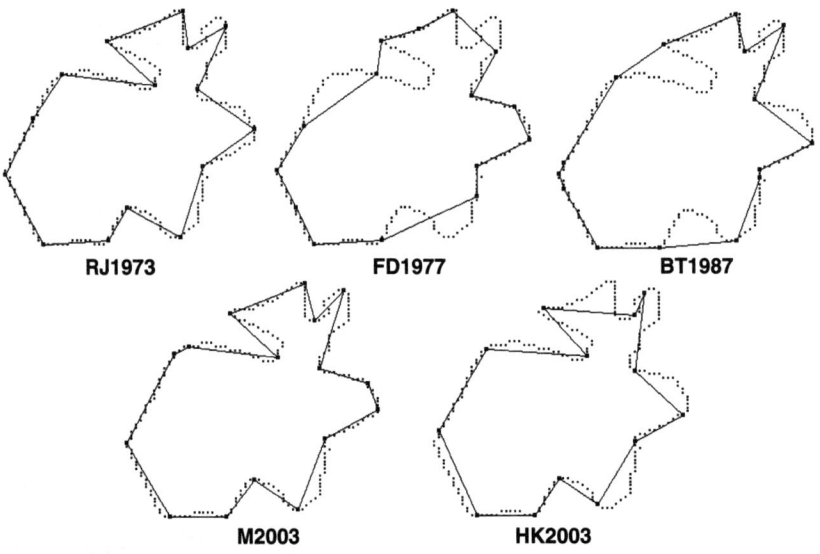

FIGURE 10.23   Resulting segmentations for the asymmetric curve.

10.4 The Curvature of a Planar Digital Curve

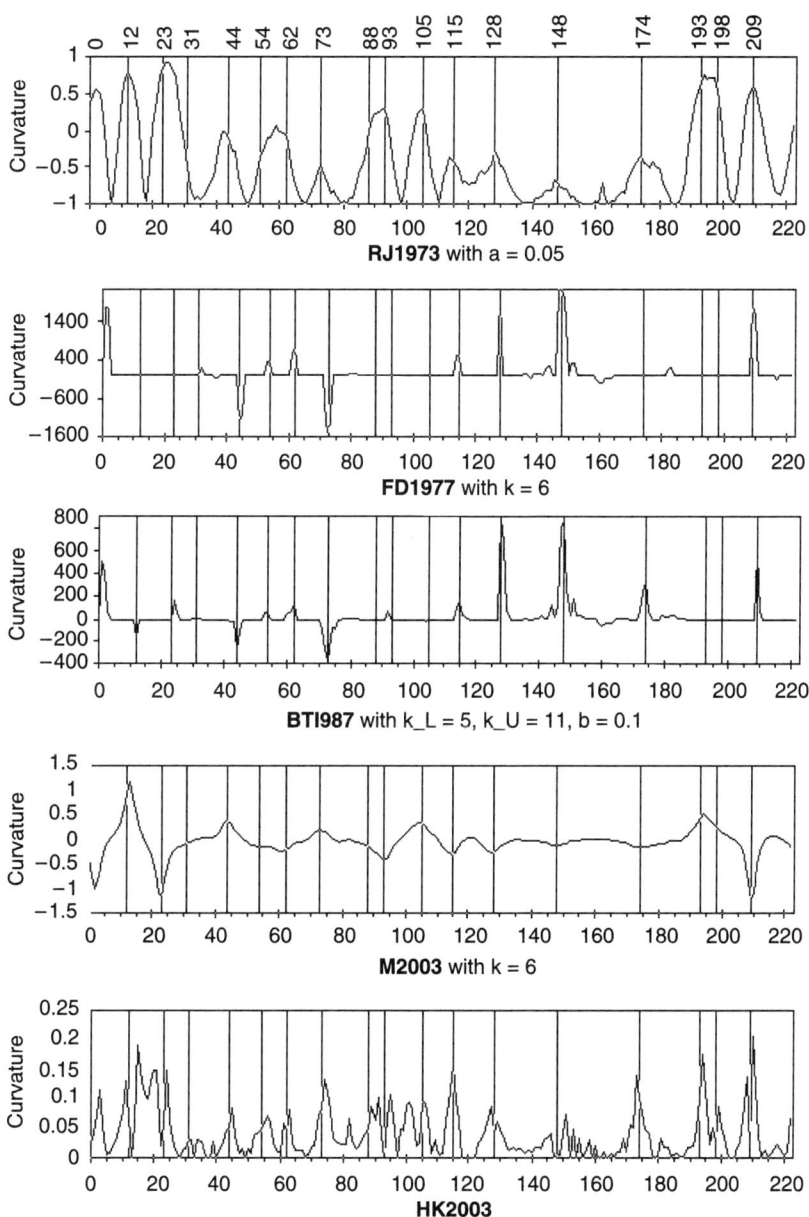

FIGURE 10.24 Corner or curvature values measured by five methods. The parameters used are specified in the figure. The x-coordinate gives border point numbers as shown in Figure 10.20 for the asymmetric curve.

Threshold 0.08 was used for corner detection in **M2003**. This algorithm also detects subsequences of corners that are "nearly collinear"; see the three corners on the left and on the right in Figure 10.22. Elimination of corners that are nearly collinear with the preceding and following corners would eliminate this type of redundancy.

Method **HK2003** does not require any specification of parameters, but it depends on the orientation in which the border is traversed. The curvature values are not symmetric because a maximum-length DSS from p_i to p_j is not necessarily a maximum-length DSS from p_j to p_i. At each p_i, a "backward interval" and a "forward interval" are defined by backward and forward DSSs. A corner is detected iff the curvature at p_i is at least equal to the curvatures at every p_j in p_is backward and forward intervals.

Figure 10.24 shows corner or curvature values for the asymmetric curve in Figure 10.20 (which has length 223) calculated by the same five methods. The resulting segmentations are shown in Figure 10.23.

10.5 Exercises

1. The *Archimedes-Hui constant* of a length estimation algorithm is the minimum grid resolution h_0 such that the algorithm initially estimates π within the error interval defined by Equation 1.6 when a circular region centered at the origin is digitized in a grid with edge length $1/h$. Note that the estimates "oscillate," so later estimates may be outside of this interval. Using a sliding mean is critical because the behavior of the errors is uncertain. Can you give a "better" definition of such a constant? Calculate the constants (using the above definition or an improved definition) for DSS and MLP length estimation methods.

2. Consider the inner and outer Jordan digitizations of a bounded connected planar set S and the MLP contained in the difference set of the two digitizations and circumscribing the inner digitization. Identify cases in which the grid-intersection digitizations of this MLP and of the frontier of S are (are not) the same.

3. In cases A and B in Figure 10.8, two border vertices are at distance 1 or 2 in the grid but at distance greater than 2 on the border 4-curve. Prove that the Algorithm 10.1 produces a grid polygon that circumscribes the inner polygon, even in cases A and B.

4. If two grid cells that share an edge e in a 2D multivalued picture P have values u and v, we say that e has *strength* $|u-v|$. The sum of the strengths of all of the grid edges in P is called the *total frontier strength* of P. Prove that, if P is binary, its total frontier strength is equal to the sum of the lengths (in grid edge units) of all of the frontier of P along which 0s meet 1s.

5. Implement algorithm **M2003** and at least one other curvature estimation algorithm of your choice, and evaluate their performance as was done in Section 10.4.3.

6. Let ρ be a 4-path with pixels $p_1, \ldots, p_n$ that are all distinct. At each pixel p_i ($1 < i < n$), the path is either straight or makes a right or left turn, depending on whether the determinant $D(p_{i-1}, p_i, p_{i+1})$ is zero, positive, or negative. In these cases, we say that the turn of ρ at p_i is 0, $\pi/2$, or $-\pi/2$, respectively; θ_i is the turn of ρ at p_i. If p_1 is a 4-neighbor of p_n (so that the path defines a closed curve), we can also define its turn at p_n (modulo n). Prove that the *total turn* $\sum_{i=1}^{n} \theta_i$ of a closed curve is 2π.

10.6 Commented Bibliography

The length-related sections in this chapter review material presented in [50, 51, 149, 206, 543, 555, 560, 1005, 1042]. [604] discusses local and global methods from the viewpoint of parallel computation. [928] compares 23 algorithms for polygonal approximation of 8-curves.

For length estimators based on chamfer metrics, see [104, 281, 293, 826, 1093]. Estimator chm was proposed in [281] and estimator coc in [1115].

DSS algorithms are based on characterizations of digital lines using (see Chapter 9) syntactic properties [451, 1137], arithmetic properties defining tangential lines [256], or properties of feasible regions in the (dual) parameter space [283, 584] or for linear programming methods such as the Fourier-Motzkin algorithm [341]. Linear offline and online algorithms for DSS recognition are discussed in Chapter 9.

The general problem of decomposing a digital curve into a sequence of DSSs, which includes DSS recognition as a subproblem, is discussed in, for example, [256, 283, 555, 592]. The implementation of **DR1995** used in the experiments in [206] follows [256]. The implementation of **K1990**, which was originally defined in [592], is described in [555]. The multigrid convergence of DSS-based length estimators is studied in [203, 560, 597] for frontiers of bounded convex sets.

For the minimum-perimeter polygon, its relation to the relative convex hull, and its application to perimeter estimation of digitized sets, see [739, 1000]. Relative convex hulls (which, in the case of inner and outer polygons, are MLPs) have been studied in computational geometry and robotics. The MLP algorithm tested here is the one reported in [555]. See [657] for an alternative algorithm and [973] for an application of MLPs to characterizing "star-shapedness." For the approximating-sausage (aps) approach, see [49, 50]. The algorithm tested here for this method was provided by the authors of [49, 50]. Both MLP-based length estimators are multigrid convergent [49, 50, 1002] for convex curves.

For the perimeter of a fuzzy set, see [909].

Gauss digitizations of convex polygons S are analyzed in [597] with respect to the effects of small interior angles on $G_h(S)$. Theorem 10.1 was proved in [560]; the proof is based to a large extent on material in [597]. Lemmas 10.1 and 10.2 correspond to statements in [958]; they were proved independently in [597].

The tradeoff measure of efficiency of convergence was proposed in [1005] and was used in [555]. The test data (see Figure 10.12) were proposed in [555]. In future comparisons, stereologic methods [1079] of length or total curvature estimation could also be included.

[661] is a survey of corner detectors and curvature estimators covering the methods from Section 10.4 published before 1990 as well as algorithms from [190, 715], which are characterized in [661] as being very noise-sensitive, and a weighted k-curvature corner detector from [932], which requires many weight assignments. [207, 1131] also contain review sections on curvature estimators. [1164] evaluates several detectors of "critical points." [703] gives a comprehensive review of corner detectors ("dominant point detectors") and curvature estimators. Approaches to corner detection were discussed in Section 10.4.1; we give a few additional references in the following.

[1131] proposes a curvature estimator based on the variation of the angles between fitted lines and an axis, using optimization in a $k \times k$ window of constant size. For a modification that uses a purely discrete line fitting process, see [1096]; this "centered tangential DSS" method was modified in [207] (see algorithm **CMT2001**). A linear-time algorithm for calculating centered tangential DSSs is given in [323]. Theorem 10.4 was proved in [208]. The normal vector-based method was proposed in [305]. Discrete tangents to a digital curve are discussed in [1096]. [462] studies principal normals of 8-curves in the context of shape deformation and defines the digital curvature flow that occurs when the 8-curve changes its shape.

Digital curves can be transformed into smooth curves in Euclidean space (e.g., using B-splines), and their curvature can then be calculated using numeric methods [396]; this approach deserves more attention (see algorithm **M2003**). [90] uses quadratic Bezier approximations through "key pixels" for border representation.

Estimation of the osculating circle is the basic principle in **CMT2001** [208]. See also [397] and [950] for methods that estimate radii of osculating circles.

[71, 174] estimate the curvature at p_i by the angle between p_{i-k}, p_i, and p_{i+k} divided by $d_e(p_{i-k}, p_i) + d_e(p_{i+k}, p_i)$. Optimization of distances between smoothed curve data and Euclidean circles was proposed in [1131]. [407] accumulates the distance from a pixel p_i on an 8-curve to a chord specified by moving endpoints.

For curvature estimation based on dual Voronoi diagram construction, see [448]. Other methods of polygonal approximation (i.e., detection of "critical points") of 8-curves are based on genetic algorithms [588], on principles of perceptual organization [445], and on the Haar transform [252].

Section 10.4.3 reviews multigrid convergence studies of **HK2003** and **CMT2001** by S. Hermann; algorithm **CMT2001** was provided by D. Coeurjolly, and it also reviews curvature estimation and corner detection experiments by M. Marji using five different methods. Method **HK2003** was not covered in [703].

CHAPTER 11

3D Straightness and Planarity

This chapter discusses digital straightness in 3D space, thereby generalizing the DSS- and MLP-based concepts, models, and algorithms that were studied in Chapters 9 and 10. It also discusses digital planarity in the 3D grid adjacency and incidence models, including relationships with other disciplines. Algorithms for recognizing digital planar segments are briefly reviewed, and one algorithm for partitioning a digital surface into such segments is discussed in detail.

11.1 3D Straightness

A digital straight line (DSL) in $\mathbb{Z}^3$ can be defined by 3D grid-plane intersection digitization, arithmetic geometry, or outer 3D Jordan digitization of a straight line $\gamma \subset \mathbb{R}^3$. It can be treated in 3D grid adjacency models or in the 3D grid incidence model, which is based on 0-, 1-, 2-, and 3-cells (see Section 11.1.4).

11.1.1 Grid-plane intersection digitization

Using the approach in Section 2.3.5, the 3D grid-plane intersection digitization of γ in $\mathbb{C}_3^{(3)}$ is $dig_\sigma^+(\gamma)$ where the following is true:

$$\Pi_\sigma = \{(x_1, x_2, x_3) : \exists i (1 \leq i \leq 3 \wedge x_i = 0) \wedge \max_{1 \leq i \leq 3} |x_i| \leq \tfrac{1}{2}\}$$

γ can intersect a grid plane in such a way that up to four grid points are closest to the intersection point. As is the case in 2D grid-intersection digitization, we exclude such cases by selecting one of these grid points using some decision rule. The resulting 3D DSL is the doubly infinite sequence $\ldots, q_{-1}, q_0, q_1, \ldots$ of center points (grid points in $\mathbb{Z}^3$) of the 3-cells in $dig_\sigma^+(\gamma)$.

A (26-) DSL segment (3D DSS or 26-DSS) is a finite subsequence of a 3D DSL. Evidently, a 3D DSS is a simple 26-arc. A *rational* 3D DSL is the digitization of a *rational straight line* $\gamma = \{(\beta_x + \alpha_x t, \beta_y + \alpha_y t, \beta_z + \alpha_z t) : t \in \mathbb{R}\}$; here $\beta = (\beta_x, \beta_y, \beta_z) \in \mathbb{R}^3$ can be arbitrary, but the slopes α_x, α_y, and α_z must be rational.

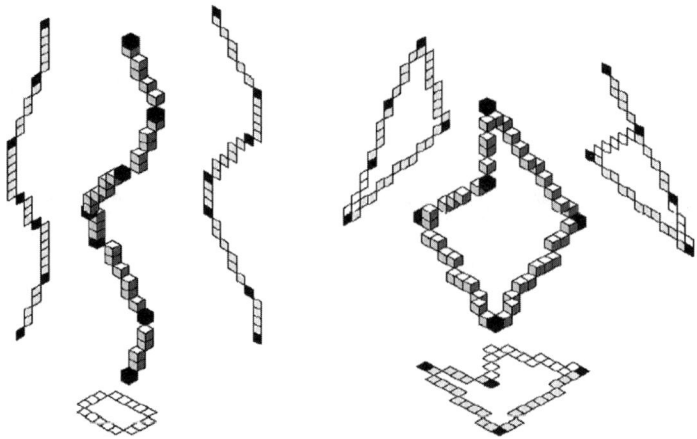

FIGURE 11.1 Segmentations of a 26-arc (left) and a 26-curve (right) into maximum-length 26-DSSs (ending at black voxels), together with their projections into the $(x=0)$-, $(y=0)$-, and $(z=0)$-planes [203].

The conditions for 2D straightness discussed in Chapter 9 can be extended to 3D. For example, $S \subseteq \mathbb{Z}^3$ has the chord property, iff for any $p, q \in S$, every point on the straight line segment $pq \subset \mathbb{R}^3$ is at L_∞-distance < 1 from some point of S. (We recall that the Minkowski metric L_∞ coincides with d_{26} on $\mathbb{Z}^3$.) It can be shown [511] that a simple 26-arc is a 3D DSS iff it has the chord property. In [511], the following theorem is also proved:

Theorem 11.1 A simple 26-arc is a 26-DSS iff two of its projections onto the $(x=0)$-, $(y=0)$-, and $(z=0)$-planes are 8-DSSs.

Either of these projections may be a single grid point; the third projection may not even be a simple 8-arc. Figure 11.1 illustrates projections of 26-DSSs.

Theorem 11.1 provides a straightforward method of recognizing 26-DSSs and, therefore, of segmenting a 26-arc into a sequence of maximum-length 26-DSSs. We project the given 26-arc onto the three planes and apply 8-DSS recognition to the projections; we continue as long as 8-DSS recognition is successful for at least two of the projections.

11.1.2 Arithmetic geometry

3D DSSs are defined in arithmetic geometry by linear Diophantine inequalities; see Section 9.2. A 3D digital line is characterized by relatively prime integers a, b, and c (where $0 \leq c \leq b \leq a$) that correspond to the coordinates x, y, and z in that order. For other orders of a, b, and c, we can permute the coordinates.

11.1 3D Straightness

Definition 11.1 $G \subset \mathbb{Z}^3$ is a *3D arithmetic line* defined by integers $a, b, c, \mu_1, \mu_2, \omega_1$, and ω_2 iff the following is true:

$$G = \{(x,y,z) \in \mathbb{Z}^3 : \mu_1 \leq cx - az < \mu_1 + \omega_1 \wedge \mu_2 \leq bx - ay < \mu_2 + \omega_2\}$$

The parameters μ_1 and μ_2 are called the *lower bounds* of G, and the parameters ω_1 and ω_2 define its *arithmetic thickness*.

Let G be a 3D arithmetic line defined by $a, b, c, \mu_1, \mu_2, \omega_1, \omega_2 \in \mathbb{Z}$ where $0 \leq c < b < a$. Then the following are true:

$$\text{if } a+c \leq \omega_1 \text{ and } a+b \leq \omega_2, G \text{ is 6-connected;} \quad (11.1)$$

$$\text{if } a+c \leq \omega_1 \text{ and } a \leq \omega_2 < a+b$$
$$\text{or } a+b \leq \omega_2 \text{ and } a \leq \omega_1 < a+c, G \text{ is 18-connected;} \quad (11.2)$$

$$\text{if } a \leq \omega_1 < a+c \text{ and } a \leq \omega_2 < a+b, G \text{ is 26-connected;} \quad (11.3)$$

$$\text{if } \omega_1 < a \text{ or } \omega_2 < a, G \text{ is 26-disconnected.} \quad (11.4)$$

G is called a *3D naive line* iff $\omega_1 = \omega_2 = \max\{|a|, |b|, |c|\} = a$. When $c = 0$, G is a (2D) naive line, as in Chapter 9. If $c > 0$, in accordance with Equation 11.4, G is 26-connected. $c \leq b \leq a$ corresponds to the following parameterization of G, using parameter x only:

$$z = \lfloor \frac{cx - \mu_1}{a} \rfloor \text{ and } y = \lfloor \frac{bx - \mu_2}{a} \rfloor$$

Evidently, a 3D naive line is an unbounded simple 26-arc. It is provided in [203] that the following is true:

Theorem 11.2 Any rational 3D digital line defined by grid-plane intersection digitization is a 3D naive line and vice versa.

This theorem and Theorem 11.1 imply the following:

Corollary 11.1 A simple 26-arc is a 3D naive line iff two of its projections onto the $(x = 0)$-, $(y = 0)$-, and $(z = 0)$-planes are DSSs iff two of these projections are (2D) naive lines.

11.1.3 A linear online 3D DSS segmentation algorithm

Corollary 11.1 justifies the following 3D DSS segmentation algorithm, which uses the **(C2.1a)** approach that was defined in Chapter 9 (Theorem 9.4) and that was originally published in [256].

We consider a projection of the given 26-curve into one of the $(x=0)$-, $(y=0)$-, or $(z=0)$-planes. Without loss of generality, assume an 8-curve in the $(z=0)$-plane; this allows us to suppress the z-coordinate. The grid points (x,y) on a DSS must satisfy the following condition,

$$D: \mu \leq ax - by < \mu + \max\{|a|,|b|\}$$

where a and b are relatively prime integers and μ is an integer. We describe the calculation of a, b, and μ.

Algorithm DR1995

The initial grid point q_1 of a new 8-DSS can be chosen as the origin. It is sufficient to implement the algorithm for the first octant (i.e., slope in $[0,1)$); other cases can be mapped into this case by reflection. At $q_1 = (0,0)$, we start with condition D_1: $0 \leq -y < 1$ (i.e., $a = 0$, $b = 1$, and $\mu = 0$). We assume that the given sequence of grid points involves moves in at most three directions: $(1,0)$, $(1,1)$, or $(1,-1)$; thus we proceed in the x-direction in the first octant.

The points $q_i = (a_i, b_i)$, where $1 \leq i \leq n$, form an 8-DSS or a segment of a naive line iff (see Theorem 9.4) there are relatively prime integers a and b with $a < b$ (recall that we are assuming the first octant) and an integer ν such that all n points are between or on a lower supporting line $ax - by = \mu$ and an upper supporting line $ax - by = \mu + \max\{|a|,|b|\} - 1 = \mu + b - 1$ (note again that we are in the first octant). For a new point $q_{n+1} \in A_8(q_n)$, where $x_{n+1} > x_n$, we have three cases: q_{n+1} is between or on these two lines (i.e., no update is needed) or it is ("just") above the upper or ("just") below the lower supporting line.

[253, 256] give simple decision and updating criteria for these cases. Let u_1, u_2 and l_1, l_2 be the points on the upper and lower supporting line, respectively, where index 1 denotes the point q_i ($1 \leq i \leq n$) with the smallest x-coordinate and index 2 denotes the point with the largest x-coordinate (compare points *StartN*, *EndN* and

---

1. Let $r = ax_{n+1} - by_{n+1}$ be the remainder of the new point q_{n+1}.
2. If $\mu \leq r < \mu + b$, then $u_2 = q_{n+1}$ (if $r = \mu$) or $l_2 = q_{n+1}$ (if $r = \mu + b - 1$), and stop; otherwise, go to Step 3.
3. If $r = \mu - 1$, then $l_1 = l_2$, $u_2 = q_{n+1}$, $a = |y_{n+1} - u_{12}|$, $b = |x_{n+1} - u_{11}|$ (where $u_1 = (u_{11}, u_{12})$), $\mu = ax_{n+1} - by_{n+1}$, and stop; otherwise, go to Step 4.
4. If $r = \mu + b$, then $u_1 = u_2$, $l_2 = q_{n+1}$, $a = |y_{n+1} - l_{12}|$, $b = |x_{n+1} - l_{11}|$ (where $l_1 = (l_{11}, l_{12})$), $\mu = ax_{n+1} - by_{n+1} - b + 1$ (or $\mu = au_{21} - bu_{22}$), and stop; otherwise, go to Step 5.
5. The new point does not form an 8-DSS with the previous n points; initialize a new 8-DSS at q_n.

ALGORITHM 11.1 Algorithm **DR1995** for the first octant; $\max\{|a|,|b|\} = b$.

11.1 3D Straightness

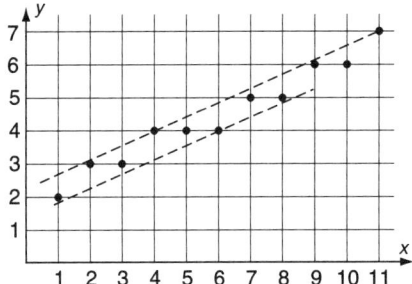

FIGURE 11.2 An 8-DSS with $p_1 = (1,2), p_2 = (2,3), \ldots, p_{11} = (11,7)$.

StartP, EndP in Figure 9.7). Let $r = ax_{n+1} - by_{n+1}$ be the *remainder* of point q_{n+1} with respect to the slope a/b of the given naive line segment $\{q_1, \ldots, q_n\} \subset D_{a,b,\mu,-b}$.

Theorem 11.3 If $\mu \leq r < \mu + \max\{|a|,|b|\}$, then $q_{n+1} \in D_{a,b,\mu,\max\{|a|,|b|\}}$. If $r = \mu - 1$, then $\{q_1, \ldots, q_n, q_{n+1}\}$ is a segment of a naive line with a slope that is defined by vector $u_1 q_{n+1}$. If $r = \mu + \max\{|a|,|b|\}$, then $\{q_1, \ldots, q_n, q_{n+1}\}$ is a segment of a naive line with a slope that is defined by vector $l_1 q_{n+1}$. If $r < \mu - 1$ or $r > \mu + \max\{|a|,|b|\}$, then $\{q_1, \ldots, q_n, q_{n+1}\}$ is not a segment of a naive line.

The theorem translates into an 8-DSS algorithm, which is shown in Algorithm 11.1.

Figure 11.2 shows a sequence $q_i = p_i - p_1$ where, in the first octant, $p_1 = (1,2)$ and $i = 1, \ldots, 11$. q_1 and q_2 satisfy condition D_2: $0 \leq x - y < 1$ with $\mu = 0$; the upper supporting lines are both $y = x$. $q_1, q_2,$ and q_3 satisfy condition D_3: $-1 \leq x - 2y < 1$; here, the lower supporting line is $2y = x$, and $\mu = -1$ is defined by q_2. At $q_6 = (5,2)$, we have condition D_6: $-4 \leq 2x - 5y < 1$; $\mu = -4$, $a = 2$, and $b = 5$, and the upper supporting line is defined by q_4. At q_{10}, we still have a naive line segment $q_1, \ldots, q_{10}$ with $l_1 = q_1, l_2 = q_6, u_1 = q_4,$ and $u_2 = q_9$. The remainder r at q_{11} is -5 (i.e., we have the second case in Theorem 11.3). The new slope is defined by vector $u_1 q_{11} = (7,3)$, and $\{q_1, \ldots, q_{11}\}$ is a segment of the naive line D_{11}: $-5 \leq 3x - 7y < 2$. We now have $l_1 = l_2 = q_6, u_1 = q_4,$ and $u_2 = q_{11}$.

Figure 11.3 shows a circle and an ellipse in general 3D positions (in the grid cell model) and their DSS segmentations.

Suppose we successfully apply a DSS recognition algorithm in two of the coordinate planes, for example, by obtaining supporting lines in the $(y = 0)$-plane with normal $\mathbf{n}_y = (y_1, y_2)$ and in the $(z = 0)$-plane with normal $\mathbf{n}_z = (z_1, z_2)$; then these two 8-DSSs are projections of a 26-DSS that has normal $\mathbf{n} = (y_2 z_2, y_2 z_1, y_1 z_2)$. Note that, if (y_1, y_2) and (z_1, z_2) are 2D grid points with relatively prime integer coordinates, the coordinates of $\mathbf{n}$ need not be relatively prime.

Figure 11.4 illustrates the linear run time of the 3D DSS segmentation algorithm and shows two successful 2D DSS segmentations for each segment in 3D space.

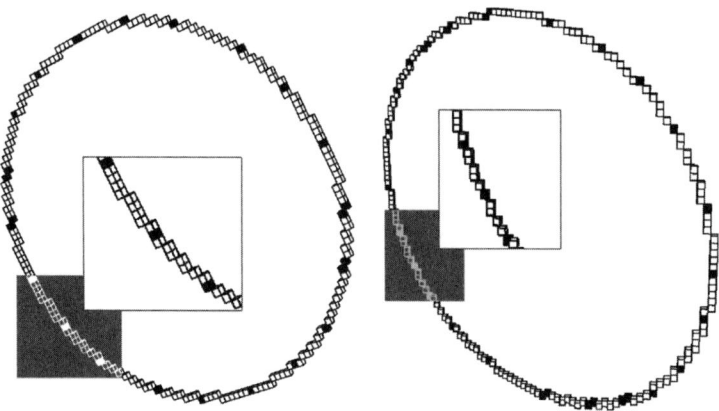

FIGURE 11.3 Examples of DSS segmentations; the windows (gray) are magnified to show details. The curves do not lie in planes parallel to the $(x = 0)$-, $(y = 0)$-, or $(z = 0)$-plane. Left: a circle. Right: an ellipse [203].

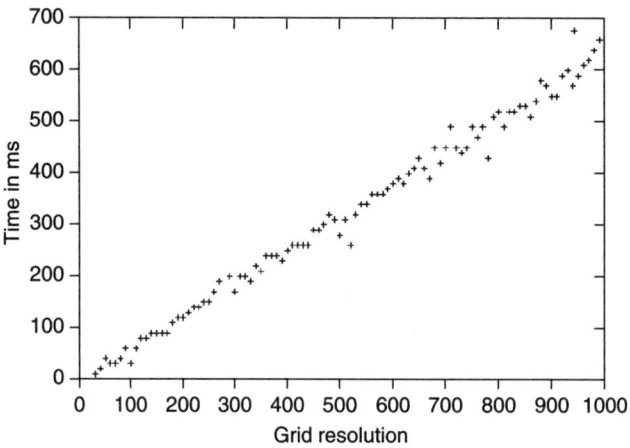

FIGURE 11.4 Run time (in ms) of the 3D DSS segmentation procedure on a Sun Ultra SparcStation [203] for a digitized circle in 3D space; see the left side of Figure 11.3.

11.1.4 MLPs of simple 2-curves

In this section, we consider DSLs in the 3D incidence grid. The outer 3D Jordan digitization of a straight line $\gamma \subset \mathbb{R}^3$ is a finite set of 3-cells. We also use grid vertices, edges, and faces in the following discussion. We assume that γ is not incident with any grid vertex; it follows that the DSL is a 2-arc in the 3D grid cell model.

11.1 3D Straightness

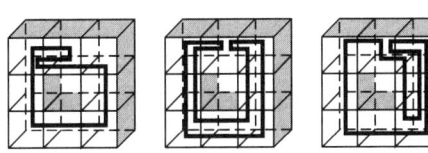

FIGURE 11.5 Curves complete in a tube [546].

A 2-curve in $\mathbb{C}^3$ can be represented as an alternating sequence $\rho = (f_0, c_0, f_1, c_1, \ldots, f_n, c_n)$ of faces f_i and cubes c_i ($0 \leq i \leq n$) such that f_i and f_{i+1} are sides of c_i and $f_{n+1} = f_0$. A 2-arc ξ is a 2-connected subsequence of 3-cells starting at a face f_i and ending at a face $f_j \neq f_i$. Evidently, ξ is a DSL (with respect to outer 3D Jordan digitization) iff there is a straight line segment in $\mathbb{R}^3$ that has endpoints in f_i and f_j and that is contained in the union of the 3-cells of ξ, so f_i is visible from f_j (and vice versa) in ξ.

The 3D DSL segmentation problem is defined as follows: starting at one 3-cell of a 2-curve ρ, segment the 2-curve into consecutive maximum-length 2-arcs that are DSSs, where the endvoxel of each DSS is the startvoxel of the next DSS. In the rest of this section, we discuss how to solve this problem by calculating an MLP "within ρ."

We assume from now on that ρ is simple. This is equivalent to assuming that $n \geq 4$, and, for any two cubes c_i and c_k in ρ such that $|i - k| \geq 2 \pmod{n+1}$, if $c_i \cap c_k \neq \emptyset$, then either $|i - k| = 2 \pmod{n+1}$ and $c_i \cap c_k$ is an edge or $|i - k| = 3 \pmod{n+1}$ and $c_i \cap c_k$ is a vertex.

The union M_ρ of all of the cubes in ρ is called the *tube* of ρ; it is a compact polyhedron and is homeomorphic to a torus if ρ is simple.

A simple 2-curve is called *complete in* M_ρ iff it has a nonempty intersection with any cube of ρ. A nonplanar simple 2-curve in $\mathbb{R}^3$ determines exactly one MLP that is complete and contained in its tube. For planar simple 2-curves, the MLP is not uniquely determined, but such curves can be treated like 1-curves in $[\mathbb{C}^2, \leq]$.

There is no straightforward way to extend 2D MLP algorithms to 3D. One reason is that the vertices of 2D MLPs are vertices of the given polygons, whereas a 3D MLP can have vertices with irrational coordinates, even if it is contained in a tube that has only grid point vertices.

Let ρ be a simple 2-curve, and let $\Pi = \langle p_0, p_1, \ldots, p_m \rangle$ (where $p_0 = p_m$) be a polygon that is complete and contained in M_ρ.

Lemma 11.1 If $\Pi = \langle p_0, p_1, \ldots, p_m \rangle$ is a polygon that is complete and contained in the tube of a simple 2-curve M_θ, then $m \geq 3$. Two line segments cannot be complete in the tube of any simple 2-curve.

Proof If $m \leq 2$, Π is contained in a straight line segment; this is impossible, because a simple 2-curve is homeomorphic to a torus. $m = 3$ (a triangle) is possible (e.g., for the simple 2-curve shown in Figure 11.5), but, in this minimal case, no side of the triangle can be contained in one of the cubes. ∎

The curves on the left and right in Figure 11.5 are not contractible into single points in M_ρ, but the curve in the middle is contractible.

A simple curve γ *passes through* a face f iff there exist t_1, t_2, and T such that $\{\gamma(t): t_1 \leq t \leq t_2\} \subseteq f$, $\gamma(t_1-\varepsilon) \notin f$ and $\gamma(t_2+\varepsilon) \notin f$ for all ε such that $0 < \varepsilon \leq T$. During a traversal of γ, we *enter* a cube c at $\gamma(t_1) \in c$ if $\gamma(t_1 - \varepsilon) \notin c$, and we *leave* c at $\gamma(t_2) \in c$ if $\gamma(t_2+\varepsilon) \notin c$ for all ε such that $0 < \varepsilon \leq T$. A traversal is defined by a starting vertex p_0 of γ and a direction.

Let $C_\rho = (c_0, c_1, ..., c_n)$ be the sequence of cubes of ρ in the order in which they are entered during curve traversal. If a polygon Π is complete and contained in M_ρ, C_ρ contains all of the cubes of ρ and no other cubes.

Lemma 11.2 If Π is an MLP of a simple 2-curve ρ, C_ρ contains each cube of ρ just once.

Proof Suppose Π enters the same cube c of ρ twice, for example, first at q_1 and then at q_2. q_1 and q_2 may be on the same face of c (see Figure 11.5, left and right) or on different faces of c (see Figure 11.5, middle).

Suppose first that both q_1 and q_2 are on the face f common to cubes c and c'. Suppose the number of times Π passes through f is odd. We insert q_1 and q_2 into Π as new vertices that split it into two polygonal chains $\Pi_1 = \langle q_2, ..., q_1 \rangle$ and $\Pi_2 = \langle q_1, ..., q_2 \rangle$, which have a union of Π. The lengths of Π_1 and Π_2 exceed the length of the straight line segment $q_1 q_2$. Without loss of generality, let Π_1 be the chain that does not pass through f; thus Π_1 is complete in M_ρ. Because c is convex, it contains the straight line segment $q_1 q_2$. Replace Π_2 with $q_1 q_2$ (i.e., replace Π with $Q = (q_1, q_2, ..., q_1)$). Q is still complete and contained in M_ρ, but it is shorter than Π, which contradicts the assumption that Π is an MLP of ρ.

Now suppose the number of passes of Π through f is even. Suppose Π enters c at q_1, then passes through f and enters c' at r_1, then passes through f again and enters c at q_2, then passes through f again and enters c' at r_2. (Π may make an even number of other passes through f before it returns to q_1.) We insert q_1, r_1, q_2, and r_2 into Π as new vertices that split it into four polygonal chains $\Pi_1 = (q_1, ..., r_1)$, $\Pi_2 = (r_1, ..., q_2)$, $\Pi_3 = (q_2, ..., r_2)$, and $\Pi_4 = (r_2, ..., q_1)$ with a union of Π. It follows that the following is true,

$$C_{\Pi_1} \subseteq C_{\Pi_3} \vee C_{\Pi_3} \subseteq C_{\Pi_1}$$

and analogously for Π_2 and Π_4. Without loss of generality, let $C_{\Pi_1} \subseteq C_{\Pi_3}$. We replace Π_1 with the straight line segment $q_1 r_1$; that segment is in f, and the length of Π_1 exceeds the length of $q_1 r_1$. Thus the resulting polygonal curve is still complete and contained in M_ρ, but, it is shorter than Π, which contradicts the assumption that Π is an MLP of ρ.

11.1 3D Straightness

Next we consider the case where q_1 and q_2 are on different faces of c, for example, q_1 on f_1 and q_2 on f_2. Because q_2 is a point of reentry into c, there must be a point q_{ex} on f_2 where we leave c before reentering it at q_2. If there is another reentry point on f_2, we are back to the first case. Hence, we can assume that Π leaves c once and enters c once. Let f_2 be a face of $c' \neq c$. If Π does not intersect the second face of c' that is contained in ρ, we replace $(q_{ex},...,q_2)$ (which is contained in c' but not in f_2) with the shorter straight line segment $q_{ex}q_2$, which is contained in f_2 and thus in c'. The resulting polygon would be shorter than Π and still complete and contained in ρ, which is a contradiction. It follows that Π must leave c' through its second face, which is contained in ρ. When we trace around ρ, we arrive at the cube $c'' \neq c$, which is incident with f_1; we leave c'' (and enter c) at a point that may be the same as q_1; and we enter c'' again through f_1. Thus Π contains two polygonal subsequences that are both complete and contained in ρ; this contradicts the shortest-length assumption. ∎

Let $\Pi = \langle p_0, p_1, ..., p_m \rangle$ be a polygon contained in a tube M_ρ. A polygon Q is called an M_ρ-*transform* of Π iff Q can be obtained from Π by a finite number of steps, each of which is a replacement of a triple a,b,c of vertices by a polygonal arc $a, b_1, ..., b_k, c$ contained in the same set of cubes of ρ as a,b,c. ($k=0$ means deletion of b; $k=1$ means a move of b within M_ρ; and $k \geq 2$ means replacement of two straight line segments with a sequence of $k+1$ straight line segments that are all contained in M_ρ.)

Lemma 11.3 Let Π be a polygon that is complete and contained in the tube M_ρ of a simple 2-curve ρ such that $\mathcal{C}_\rho$ has no repetitions of cubes. Then any M_ρ-transform of Π is also complete and contained in M_ρ.

Proof By definition of the M_ρ-transform, it is contained in M_ρ. Because $\mathcal{C}_\rho$ has no repetitions of cubes, Π traces M_ρ cell by cell. From Lemma 11.1, we know that Π has at least three vertices (i.e., at least three line segments) and that $m \geq 3$ (two line segments cannot be complete in ρ; i.e., at least one cube is not intersected by the two line segments). Thus a replacement of the two line segments within the same set of cells of ρ cannot transform Π into a curve contractible in M_ρ; hence the curve remains complete in M_ρ. ∎

An edge contained in a tube M_ρ is called *critical* iff the edge is the intersection of three cubes contained in ρ. Figure 11.6 illustrates critical edges of two 2-curves; only the left one is simple.

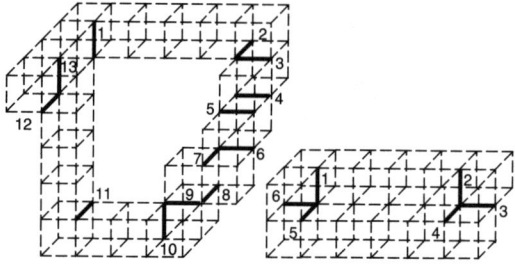

FIGURE 11.6 Critical edges of two curves [546].

Note that simple 2-curves have edges that are contained in at most three cubes. For example, the curve consisting of four cubes only (an edge is contained in four cubes in this case) was excluded by the constraint $n \geq 4$.

Theorem 11.4 Let ρ be a simple 2-curve. The vertices of a shortest simple polygon that is complete and contained in the tube M_ρ must be located on critical edges.

Proof We consider both planar and nonplanar simple 2-curves ρ (i.e., the MLP may not be uniquely defined).

Let $\Pi = \langle p_0, p_1, ..., p_m \rangle$ be a shortest simple polygon that is complete and contained in M_ρ, where $p_0 = p_m$ and $m \geq 3$. Without loss of generality, we consider the subsequence (p_0, p_1, p_2) of Π and show that p_1 is on a critical edge. By Lemma 11.2, C_ρ has no repetitions; thus we can apply Lemma 11.3 to Π and M_ρ.

We can exclude the case in which p_1 is collinear with p_0 and p_2, because such a p_1 would not be a vertex. Three noncollinear points p_0, p_1, and p_2 define a triangular region $\triangle(p_0, p_1, p_2)$ in a plane $\mathcal{E}$ in $\mathbb{R}^3$. The following discussion deals with geometric configurations in $\mathcal{E}$. *Frontier points* are points on the frontier ϑM_ρ.

First, we ask whether p_1 can be moved toward $p_0 p_2$ in $\triangle(p_0, p_1, p_2)$ so that the resulting polygonal arc $(p_0, ..., p_{new}, ..., p_2)$ is still contained in M_ρ. This would be an M_ρ-transform of Π, and the resulting curve would be complete and contained in M_ρ. If the intersection of an ε-neighborhood of p_1 with $\triangle(p_0, p_1, p_2)$ were in M_ρ for some $\varepsilon 0$, the resulting curve would be shorter, which is impossible; hence, for any $\varepsilon > 0$, at least one frontier point q in an ε-neighborhood of p_1 must be on one of the line segments $p_0 p_1$ or $p_1 p_2$, so p_1 itself is a frontier point.

11.1 3D Straightness

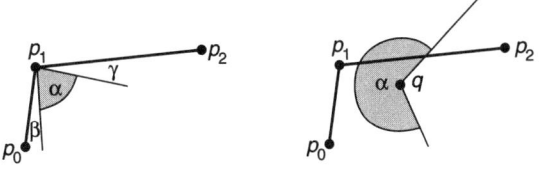

FIGURE 11.7 Neighborhood of p_1 (left). Intersection with an uncritical edge (right) [546].

A neighborhood of p_1 is illustrated in Figure 11.7 (left). The angular sector α represents the region not in M_ρ. Because α is bounded by an interior angle of $\triangle(p_0, p_1, p_2)$, we have $\alpha < \pi$.

A frontier point can be in a face or on an edge. Suppose first that p_1 is in a face f. Either $\mathcal{E}$ and f intersect in a straight line segment or f is contained in $\mathcal{E}$. Intersecting in a straight line segment would contradict the fact that $\alpha < \pi$ in the ε_0-neighborhood of p_1, and $f \subset \mathcal{E}$ would allow p_1 to move toward $p_0 p_2$ in $\triangle(p_0, p_1, p_2)$, which contradicts our MLP assumption.

There are three possibilities for an edge contained in M_ρ. We call it *uncritical* if it is in only one cube contained in ρ; *ineffective* if it is in exactly two such cubes; and *critical* (as previously defined) if it is in three such cubes. p_1 cannot be on an ineffective edge, and it cannot be on a critical or an uncritical edge, because this corresponds to it being within a face, as discussed before. p_1 also cannot be on an uncritical edge or a critical edge.

Figure 11.7 (right) illustrates an intersection q in $\mathcal{E}$ with an uncritical edge that is not coplanar with $\mathcal{E}$. The angular sector $\alpha > \pi$ (the region not in M_ρ in an ε-neighborhood of q) does not allow p_1 to be such a point. If the uncritical edge were in $\mathcal{E}$, α would be equal to π, which is also impossible for p_1. Thus only one option remains: p_1 must be on a critical edge; indeed, $\alpha < \pi$ for such an edge. ∎

Note that this theorem also applies to planar simple 1-curves in $[\mathbb{C}^2, \leq]$. It follows that MLP vertices on such a 1-curve can only be convex vertices of the inner frontier or concave vertices of the outer frontier, because these are the only vertices incident with three squares of the 1-curve.

11.1.5 The rubber band algorithm

This algorithm, which is from [148], is based on the following physical model. Suppose a rubber band passes through the tube M_ρ. If it can move freely, it will contract to

TABLE 11.1 Calculated points on edges [149].

Critical edge	1	2/3	4	5	6/7	8	9	10	11	12/13
First run	a	b	c	d	e	f	g	h	i	j
Second run	a	b	D	D	e	D	D	h	i	j

the MLP, which is complete and contained in M_ρ, assuming it is smooth enough to slide over the critical edges of the tube.

The algorithm consists of two subprocesses. The first is an initialization process that defines a simple polygonal curve Π_0 that is complete and contained in M_ρ and such that C_{Π_0} contains each cube of ρ just once (see Lemma 11.2). The second is an iterative process (a M_ρ-transform; see Lemma 11.3) in which each iteration transforms Π_t into Π_{t+1} ($t \geq 0$) such that $l(\Pi_t) \geq l(\Pi_{t+1})$. Thus the resulting polygon is also complete and contained in ρ.

The following are three methods of initializing the polygon. Init I: The vertices of the initial polygon are at the centers of the cubes that constitute the 2-curve. Init II: The vertices are at the midpoints of critical edges. Init III: The initial polygon connects only vertices that are endpoints of consecutive critical edges.

For Init III, we scan the curve until we find a pair (e_0, e_1) of consecutive critical edges that are not parallel or (if parallel) that are not in the same grid layer. (Figure 11.6 (right) shows a nonsimple 2-curve; searching for a pair of noncoplanar edges would be insufficient in this case.) For such a pair (e_0, e_1), we choose vertices (p_0, p_1), where p_0 bounds e_0 and p_1 bounds e_1 such that the line segment $p_0 p_1$ has minimum length; note that (p_0, p_1) is not always uniquely defined. This is the first line segment of the initial polygon Π_0.

Suppose $p_{i-1} p_i$ is the last line segment on the Π_0 specified so far, where p_i bounds e_i. Then there is a unique vertex p_{i+1} on the next critical edge e_{i+1} such that $p_i p_{i+1}$ has minimum length. (Length zero is possible if $p_{i+1} = p_i$; in this case, we skip p_{i+1}, which means that we do not increase i.) Note that this $p_i p_{i+1}$ will always be included in the tube, because the centers of all cubes between two consecutive critical edges are collinear. The process stops when we connect p_n (on e_n) with p_0. (Note that a minimum-distance criterion for this final step might prefer a line segment between p_n and the second vertex bounding e_0 [i.e., not p_0]). Table 11.1

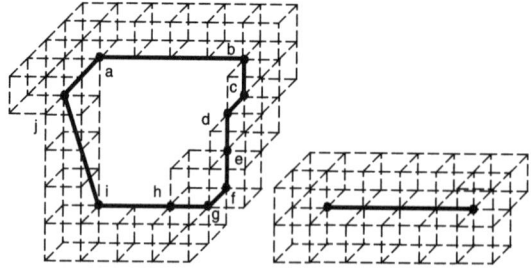

FIGURE 11.8 Curve initializations (Init III; "clockwise") [149].

11.1 3D Straightness

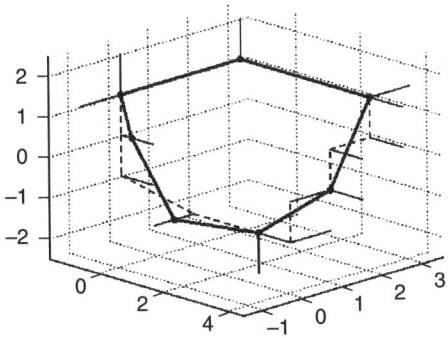

FIGURE 11.9 If the initialization starts below on the left, the final step of the initialization process would prefer the second vertex of the first edge [149].

shows the list of vertices for the curve on the left in Figures 11.6 and 11.8. The first row lists all of the critical edges shown in Figure 11.6. The second row contains the vertices of the initial polygon shown in Figure 11.8 (initialization = first iteration of the algorithm). Vertex b is on edge 2 and also on edge 3, so there is only one column (2/3) for these edges.

Initialization methods Init I through Init III construct polygons Π_0 that are always complete and contained in the given tube. Note that traversals in opposite directions or that start at different critical edges may lead to different initial polygons if Init III is used. For example, a "counterclockwise" traversal of the curve shown in Figure 11.6 (left) starting at edge 1 selects edges 11 and 10 as the first pair of consecutive critical edges; the resulting "counterclockwise" polygon differs from the one shown in Figure 11.8. Figure 11.9 shows a curve for which the final step does not return to the starting vertex.

The results of Init III are shown in Figure 11.8. The curve on the right is already an MLP for this nonsimple 2-curve. For a planar curve, the process may fail to find the first pair of critical edges; in this case, a 2D algorithm can be used to calculate the MLP.

In the iteration process, we move pointers (addressing three consecutive vertices of the polygonal curve found so far) around the curve until a complete iteration leads only to an improvement that is below a given threshold τ (i.e., $l(\Pi_t) - \tau < l(\Pi_{t+1})$). We cannot wait until there is no change at all, because this may never happen. In all of the experiments reported in [149], the algorithm terminated quickly for a reasonable value of τ.

Let $\Pi_t = (p_0, p_1, ..., p_m)$ be a polygon with three pointers addressing the vertices at positions $i-1$, i, and $i+1$. Three cases can occur that define specific M_ρ-transforms.

($\mathbf{O}_1$) p_i can be deleted iff $p_{i-1}p_{i+1}$ is a line segment within the tube. Triple (p_{i-1}, p_i, p_{i+1}) is then replaced by (p_{i-1}, p_{i+1}), and we continue with vertices p_{i-1}, p_{i+1}, and p_{i+2}.

($\mathbf{O}_2$) The closed triangular region $\triangle(p_{i-1}p_ip_{i+1})$ intersects more than just the three critical edges of p_{i-1}, p_i, and p_{i+1} (see Figure 11.10) (i.e., simple deletion of p_i

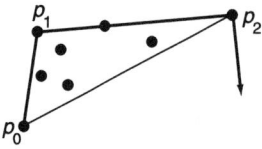

FIGURE 11.10   Intersection points with edges [149].

would not be sufficient). This situation is handled by constructing a convex arc (the shortest curve that surrounds a given finite set of planar points is a convex polygon [155]) and replacing p_i with the sequence of vertices $q_1,\ldots,q_k$ between p_{i-1} and p_{i+1} on this arc iff the sequence of line segments $p_{i-1}q_1,\ldots,q_kp_{i+1}$ lies inside the tube. Because the vertices are ordered, we can use a linear-time convex hull algorithm. Barycentric coordinates with basis $\{p_{i-1}, p_i, p_{i+1}\}$ can be used to decide which of the intersection points are inside the triangle.[1] In this case, we continue with a triple of vertices that starts with q_k.

($\mathbf{O_3}$) p_i can be moved on its critical edge to a position p_{new} that minimizes the total length of $p_{i-1}p_{\text{new}}$ and $p_{\text{new}}p_{i+1}$. An $\mathcal{O}(1)$ algorithm for this move is given below. (p_{i-1}, p_i, p_{i+1}) is then replaced by $(p_{i-1}, p_{\text{new}}, p_{i+1})$, and we continue with the triple $p_{\text{new}}, p_{i+1}, p_{i+2}$.

In situation ($\mathbf{O_3}$), suppose p_i lies on critical edge e and is not collinear with $p_{i-1}p_{i+1}$. Let l_e be the line containing e. We first find the point $p_{\text{opt}} \in l_e$ such that the following is true,

$$|p_{\text{opt}} - p_{i-1}| + |p_{i+1} - p_{\text{opt}}| = \min_{p \in l_e} L(p)$$

and where we also have the following:

$$L(p) = (|p - p_{i-1}| + |p_{i+1} - p|)$$

If p_{opt} lies on the closed critical edge e, we simply replace p_i with p_{opt}; if not, we replace p_i with the vertex bounding e that lies closest to p_{opt}.

The following is a slightly simpler method of finding p_{opt} than the one described in [148]. Without loss of generality assume that l_e is parallel to the x-axis, where y_e and z_e are constants:

$$l_e = \{(t, y_e, z_e)^T | t \in \mathbb{R}\}$$

If $x_{i-1} = x_{i+1}$, we take $p_{\text{opt}} = (x_{i-1}, y_e, z_e)^T$. Otherwise, the following

$$\frac{\partial L}{\partial x}(p_{\text{opt}}) = 0$$

leads to a quadratic equation in x_{opt}, where $\alpha_i = (y_e - y_i)^2 + (z_e - z_i)^2$:

$$(\alpha_{i+1} - \alpha_{i-1})x_{\text{opt}}^2 + 2(\alpha_{i-1}x_{i+1} - \alpha_{i+1}x_{i-1})x_{\text{opt}} + \alpha_{i+1}x_{i-1}^2 - \alpha_{i-1}x_{i+1}^2 = 0$$

---

1. In the majority of cases, we found that $k = 1$ (i.e., p_i is replaced by q_1).

11.1 3D Straightness

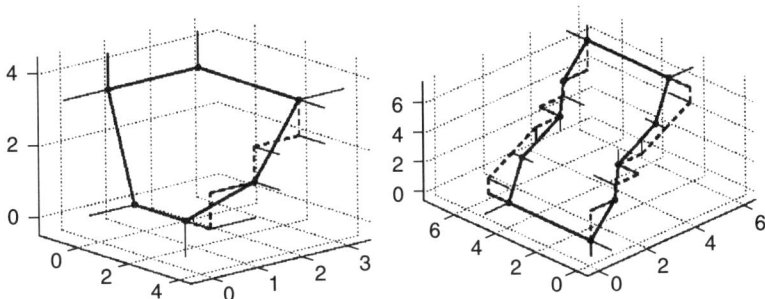

FIGURE 11.11 Two examples of initial polygons (dashed) and MLPs (solid). Critical edges are shown as short line segments. The rest of the tube is not shown [149].

Table 11.1 shows that the second run starts with the polygonal curve $\Pi_1 = \langle a,b,c,d,e,f,g,h,i,j \rangle$. For triple a,b,c, none of $(\mathbf{O}_1)$ through $(\mathbf{O}_3)$ leads to a better location of b. For triple b,c,d, we use $(\mathbf{O}_1)$ to delete c (symbol 'D'). Triple b,d,e then leads to the deletion of d, and so on, and finally triple j,a,b does not delete or move a. In the following run, nothing changes.

The process stops if an entire cycle does not lead to a "significant modification" as defined by a threshold τ.[2] Experiments in [149] used τs between $l(\Pi_t) \cdot 10^{-5}$ and $l(\Pi_t) \cdot 10^{-7}$. In Figure 11.11, the initial polygon Π_0 is dashed, and the solid line is the final polygon. The short line segments are the critical edges of the tube.

We conclude this section by giving an estimate of the complexity of the rubber-band algorithm as a function of the number n of cubes of ρ.

The algorithm completes each run in $\mathcal{O}(n)$ time. A move of p_i on a critical edge requires constant time. In the experiments, τ was set to $l(\Pi_0) \cdot 10^{-6}$; this ensured that the measured time complexity was $\mathcal{O}(n)$ (i.e., the number of runs did not depend on n).

Figure 11.12 shows the time needed for the algorithm to stop as a function of the number of cubes n and the number of critical edges. The test set contained 70 randomly generated simple 2-curves. In Figure 11.12 (left), each error bar shows the mean convergence time and its standard deviation for a set of 10 digital curves.

The algorithm is iterative. Because $l(\Pi_{t+1}) < l(\Pi_t)$ (if they are equal, the algorithm stops) and there is a lower bound on the length $l(\Pi_t)$ of the MLP, the algorithm converges. However, it is not certain to what polygon it converges. Lemmas 11.2 and 11.3 give a partial answer: it always converges to a polygon that is complete and contained in M_ρ.

The algorithm provides a method of polygonal approximation and length measurement of simple 2-curves that has run time $\mathcal{O}(n)$. The method has been successfully used for a wider class of curves, including cases such as the curve on the right in

---

2. We distinguish between *elementary iteration steps* $(\mathbf{O}_1)$, $(\mathbf{O}_2)$, or $(\mathbf{O}_3)$ and *cycles*, which consist of sequences of iteration steps that go completely around M_ρ.

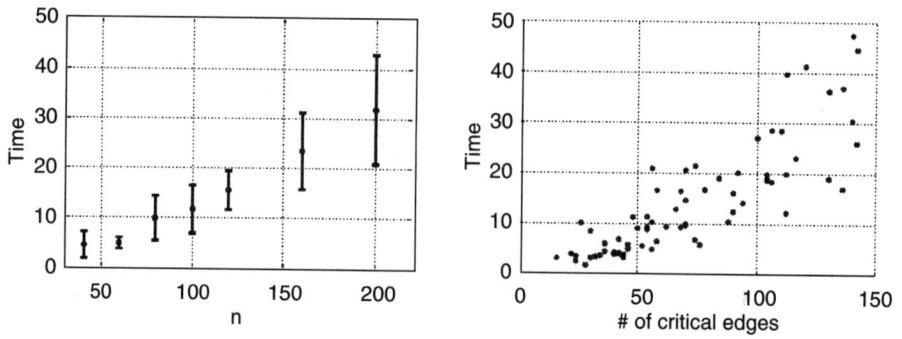

FIGURE 11.12   Central Process Utility (CPU) time in seconds as a function of the number n of 3-cells (left) and the number of critical edges (right) [149].

Figure 11.6 in which each cube in ρ has exactly two bounding faces in ρ. This simpler definition also allows for the simpler generation of test examples.

It would be desirable to prove that the time complexity of the algorithm is always $\mathcal{O}(n)$ and that it always converges to the MLP. Both of these statements can be conjectured from the experiments: the number of cycles was independent of the resolution of the 2-curve, and the minima found by the algorithm were independent of the initialization method.

11.2 Digital Planes in 3D Adjacency Grids

A plane in $\mathbb{E}^3$ that has a z-coefficient that is not 0 is defined by an expression of the following form, where $\alpha_1, \alpha_2, \beta \in \mathbb{R}$:

$$\Gamma(\alpha_1, \alpha_2, \beta) = \{(x, y, z) \in \mathbb{R}^3 : z = \alpha_1 x + \alpha_2 y + \beta\}$$

The symmetry of the grid allows us to assume $0 \leq \alpha_1 \leq 1$ and $0 \leq \alpha_2 \leq 1$ in the rest of this chapter; we also assume $0 \leq \beta < 1$ for convenience, using an argument similar to that for DSLs.

A digital plane can be obtained from a plane in $\mathbb{E}^3$ by outer 3D Jordan digitization, 3D grid-line intersection digitization (using the grid points nearest to the intersections of the plane with grid lines; compare the 3D grid-plane intersection digitization in Section 11.1.1), or simply by applying the floor or ceiling function to the coordinates of the points in $\Gamma(\alpha_1, \alpha_2, \beta)$.

11.2.1 3D grid-line intersection digitization

If we use 3D grid-line intersection digitization, under the above assumptions about $\alpha_1, \alpha_2, \beta$, we need to consider only grid lines parallel to the z-axis.

11.2 Digital Planes in 3D Adjacency Grids

Let $\Gamma(\alpha_1, \alpha_2, \beta)$ intersect the vertical grid line $(x = m, y = n)$ at $p_{m,n}$ where $m, n \geq 0$. Then the grid point closest to $p_{m,n}$ is $(m, n, I_{m,n})$ where the following is true:

$$I_{\alpha_1,\alpha_2,\beta} = \{(m, n, I_{m,n}) : m, n \geq 0 \wedge I_{m,n} = \lfloor \alpha_1 m + \alpha_2 n + \beta + 0.5 \rfloor\}$$

If there are two closest grid points, we choose the lower one. The set $I_{\alpha_1,\alpha_2,\beta}$ uniquely determines both the slopes α_1 and α_2 and the intercept β if α_1 or α_2 is irrational. If both α_1 and α_2 are rational, $I_{\alpha_1,\alpha_2,\beta}$ uniquely determines α_1 and α_2 but determines β only up to an interval. This can be proved by a straightforward generalization of the proof of Theorem 9.2.

In analogy with the chain codes in Chapter 9, we define *step codes*, starting with $i_{\alpha_1,\alpha_2,\beta}(0,0) = I_{0,0} \in \{0,1\}$:

$$i_{\alpha_1,\alpha_2,\beta}(0, n+1) = I_{0,n+1} - I_{0,n} = \begin{cases} 0 & \text{if } I_{0,n+1} = I_{0,n} \\ 1 & \text{if } I_{0,n+1} = I_{0,n} + 1 \end{cases} \quad \text{for } n \geq 0$$

$$i_{\alpha_1,\alpha_2,\beta}(m+1, 0) = I_{m+1,0} - I_{m+1,0} = \begin{cases} 0 & \text{if } I_{m+1,0} = I_{m,0} \\ 1 & \text{if } I_{m+1,0} = I_{m,0} + 1 \end{cases} \quad \text{for } m \geq 0$$

In addition to these "initial values," we define column-wise step codes:

$$i^{(c)}_{\alpha_1,\alpha_2,\beta}(m, n+1) = I_{m,n+1} - I_{m,n} = \begin{cases} 0 & \text{if } I_{m,n+1} = I_{m,n} \\ 1 & \text{if } I_{m,n+1} = I_{m,n} + 1 \end{cases} \quad \text{for } m \geq 1$$

We also define row-wise step codes:

$$i^{(r)}_{\alpha_1,\alpha_2,\beta}(m+1, n) = I_{m+1,n} - I_{m,n} = \begin{cases} 0 & \text{if } I_{m+1,n} = I_{m,n} \\ 1 & \text{if } I_{m+1,n} = I_{m,n} + 1 \end{cases} \quad \text{for } n \geq 1$$

The values in the 0th row and 0th column are used in both the column-wise and row-wise step codes; see Figure 11.13. The assumptions $0 \leq \alpha_1 \leq 1$ and $0 \leq \alpha_2 \leq 1$ guarantee that codes 0 and 1 are sufficient. It follows that $i^{(c)}_{\alpha_1,\alpha_2,\beta}(m,n) = i^{(r)}_{\alpha_2,\alpha_1,\beta}(m,n)$ where $m, n \geq 0$. On the basis of the additional assumption $\alpha_1 \leq \alpha_2$, we will use only row-wise step codes in the sequel, and we will omit the superscript (r).

Definition 11.2 $i^+_{\alpha_1,\alpha_2,\beta} = \{(m, n, i_{\alpha_1,\alpha_2,\beta}(m,n)) : m, n \geq 0\}$ is a *digital plane quadrant* (in the grid point model) with slopes α_1 and α_2 and intercept β.

If we do not require m and n to be nonnegative integers, we obtain *digital planes* $i_{\alpha_1,\alpha_2,\beta}$. For $D \subseteq \mathbb{R}^2$, let $i^D_{\alpha_1,\alpha_2,\beta} = \{(m, n, i_{\alpha_1,\alpha_2,\beta}(m,n)) : (m,n) \in D \cap \mathbb{Z}^2\}$. If α_1 or α_2 is irrational, we speak about *irrational digital planes* and otherwise about *rational digital planes*.

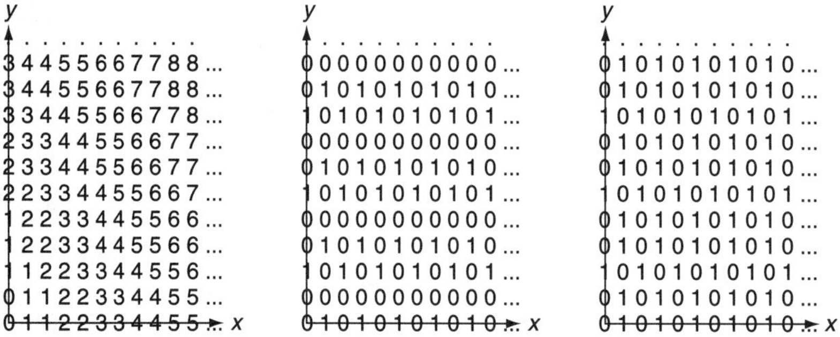

```
y                          y                          y
↑                          ↑                          ↑
. . . . . . . . . . . .    . . . . . . . . . . . .    . . . . . . . . . . . .
3 4 4 5 5 6 6 7 7 8 8 ...  0 0 0 0 0 0 0 0 0 0 0 ...  0 1 0 1 0 1 0 1 0 1 0 ...
3 4 4 5 5 6 6 7 7 8 8 ...  0 1 0 1 0 1 0 1 0 1 0 ...  0 1 0 1 0 1 0 1 0 1 0 ...
3 3 4 4 5 5 6 6 7 7 8 ...  1 0 1 0 1 0 1 0 1 0 1 ...  1 0 1 0 1 0 1 0 1 0 1 ...
2 3 3 4 4 5 5 6 6 7 7 ...  0 0 0 0 0 0 0 0 0 0 0 ...  0 1 0 1 0 1 0 1 0 1 0 ...
2 3 3 4 4 5 5 6 6 7 7 ...  0 1 0 1 0 1 0 1 0 1 0 ...  0 1 0 1 0 1 0 1 0 1 0 ...
2 2 3 3 4 4 5 5 6 6 7 ...  1 0 1 0 1 0 1 0 1 0 1 ...  1 0 1 0 1 0 1 0 1 0 1 ...
1 2 2 3 3 4 4 5 5 6 6 ...  0 0 0 0 0 0 0 0 0 0 0 ...  0 1 0 1 0 1 0 1 0 1 0 ...
1 2 2 3 3 4 4 5 5 6 6 ...  0 1 0 1 0 1 0 1 0 1 0 ...  0 1 0 1 0 1 0 1 0 1 0 ...
1 1 2 2 3 3 4 4 5 5 6 ...  1 0 1 0 1 0 1 0 1 0 1 ...  1 0 1 0 1 0 1 0 1 0 1 ...
0 1 1 2 2 3 3 4 4 5 5 ...  0 0 0 0 0 0 0 0 0 0 0 ...  0 1 0 1 0 1 0 1 0 1 0 ...
0̶ 1̶ 1̶ 2̶ 2̶ 3̶ 3̶ 4̶ 4̶ 5̶ 5̶ → x   0̶ 1̶ 0̶ 1̶ 0̶ 1̶ 0̶ 1̶ 0̶ 1̶ 0̶ → x   0̶ 1̶ 0̶ 1̶ 0̶ 1̶ 0̶ 1̶ 0̶ 1̶ 0̶ → x
```

FIGURE 11.13 Left: $I_{\frac{1}{2},\frac{1}{3},0}(m,n)$. Middle: $i^{(r)}_{\frac{1}{2},\frac{1}{3},0}(m,n)$. Right: $i^{(c)}_{\frac{1}{2},\frac{1}{3},0}(m,n)$ [131].

Analogously to the translation-equivalence of rational DSLSs with the same slope α [675], we have translation-equivalence of rational digital planes with the same slopes α_1 and α_2 [134]. This implies that the intercepts β do not influence translation-invariant properties of rational digital straight lines; we can therefore study translation-equivalence classes i_{α_1,α_2} of rational digital planes.

Using Definition 7.5 of digital surfaces (and digital surface patches) in adjacency grids, we have the following:

Theorem 11.5 *A digital plane is an unbounded digital surface.*

Proof Let $p = (i,j,k)$ be a point of digital plane $I_{\alpha_1,\alpha_2,\beta}$, and consider $I_{\alpha_1,\alpha_2,\beta} \cap \{(x,y,z) \in S : x = i\}$. Let $p' = (i, j-1, k')$ and $p'' = (i, j+1, k'')$ be the only two points of $I_{\alpha_1,\alpha_2,\beta}$ on the vertical lines $x = i$ and $y = j-1$ and $x = i$ and $y = j+1$, respectively. Because $\alpha_1 \le \alpha_2$, we have $0 \le |k-k'|, |k-k''| \le 1$; thus $(j-1, k')$ and $(j+1, k'')$ are the only two points, defined by p and $x = i$, which are 8-adjacent to (j,k), and they are not mutually 8-adjacent. Similarly, p and $y = j$ define only two 8-adjacent points in $I_{\alpha_1,\alpha_2,\beta} \cap \{(x,y,z) \in S : y = j\}$, which are not mutually 8-adjacent. In $I_{\alpha_1,\alpha_2,\beta} \cap \{(x,y,z) \in S : z = k\}$, p and $z = k$ may define more than two 8-adjacent points, but $(i, j, k-1)$ and $(i, j, k+1)$ are not both in $I_{\alpha_1,\alpha_2,\beta}$, because $p = (i,j,k)$ is the only point of $I_{\alpha_1,\alpha_2,\beta}$ on the vertical grid line $x = i$, $y = j$. Thus $I_{\alpha_1,\alpha_2,\beta}$ is a digital surface. ∎

Corollary 11.2 *Let $D \subset \mathbb{Z}^2$ be a 4-region; then $i^D_{\alpha_1,\alpha_2,\beta}$ is a digital surface patch.*

Such a patch is called a *digital plane segment* (DPS), which is defined in the grid point model with respect to grid-line intersection digitization.

11.2.2 Self-similarity

Definition 11.3 $S \subseteq \mathbb{Z}^3$ is said to have the *chordal triangle property* iff, for any $p_1, p_2, p_3 \in S$, every point on the triangle $p_1 p_2 p_3 \subset \mathbb{R}^3$ is at L_∞-distance < 1 from some point of S.

Obviously, a simple digital surface that satisfies the chordal property cannot be bounded.

Theorem 11.6 A simple digital surface is a digital plane iff it has the chordal triangle property.

This theorem is from [513], where it is also shown that, for a bounded digital plane segment, the chordal triangle property is neither necessary nor sufficient. Additional conditions (defined by distances to two lines) that characterize digital planarity are discussed by C. Ronse in [867, 871]. These papers study the use of Helly-type theorems (see Helly's First Theorem and the Transversal Theorem in Section 1.2.8) for approximating affine functions. Like the proof of Theorem 9.7 regarding the digitization of straight lines, these theorems can also be used to characterize digitizations of planes in $\mathbb{R}^3$.

For any $p = (p_x, p_y, p_z) \in \mathbb{Z}^3$, let $p_{z=0} = (p_x, p_y, 0)$ be the projection of p into the xy-plane.

Definition 11.4 $S \subseteq \mathbb{Z}^3$ is called *even* iff its projection into the xy-plane $\{(x, y, 0) : (x, y) \in \mathbb{Z}^2\}$ is one-to-one and, for every quadruple (p, q, r, s) of points in S such that $p_{z=0} - q_{z=0} = r_{z=0} - s_{z=0}$, we have $|(p_z - q_z) - (r_z - s_z)| \le 1$.

Defining evenness with respect to the xy-plane is consistent with our previous assumptions about digital planes. By requiring a one-to-one mapping into the xy-plane, we consider only unbounded sets $S \subseteq \mathbb{Z}^3$ as being even. The following theorem does not make use of the assumption $\alpha_1 \le \alpha_2$ about a digital plane.

Theorem 11.7 $S \subseteq \mathbb{Z}^3$ is a digital plane iff it is even.

11.2.3 Supporting and separating planes

A *supporting plane* of $S \subseteq \mathbb{Z}^3$ divides $\mathbb{R}^3$ into two (closed) halfspaces such that S is completely contained in one of them.

Theorem 11.8 $S \subseteq \mathbb{Z}^3$ is a digital plane iff it has a supporting plane Γ such that the L_∞-Hausdorff distance between S and Γ is < 1.

In [513], it was claimed that if $S \subseteq \mathbb{Z}^3$ is a (finite) digital plane segment, the points of S are at L_∞-Hausdorff distance < 1 from at least one plane incident with one of the faces of the convex hull of S; thus one of these planes is a supporting plane in the sense of Theorem 11.8. However, [253] gave a counterexample: if $D = [0,6] \times [0,7]$, the L_∞-Hausdorff distance between $i^D_{5/29,9/29,1/2}$ and any plane incident with one of the faces of the convex hull of $i^D_{5/29,9/29,1/2}$ is greater than 1.

Let $S \subset \mathbb{Z}^3$, and let $S_{z+1} = \{(x,y,z+1) : (x,y,z) \in S\}$. A plane $\Gamma \subset \mathbb{R}^3$ *separates* the sets $S_1, S_2 \subset \mathbb{Z}^3$ iff S_1 and S_2 are in opposite open halfspaces defined by Γ.

Theorem 11.9 A finite set $S \subset \mathbb{Z}^3$ is a subset of a digital plane iff there exists a plane that separates S from S_{z+1}.

11.2.4 Arithmetic planes

Arithmetic geometry as suggested by S. Forchhammer in [332] and developed by J.-P. Reveillès in [848] provides a uniform approach to the study of digitized hyperplanes in n dimensions. We discuss here only the 3D case. Let a, b, and c be relatively prime integers, and let μ and ω be integers.

Definition 11.5 $D_{a,b,c,\mu,\omega} = \{(i,j,k) \in \mathbb{Z}^3 : \mu \le ai + bj + ck < \mu + \omega\}$ is called an *arithmetic plane* with *normal* $\mathbf{n} = (a,b,c)^T$, *approximate intercept* μ, and *arithmetic thickness* ω.

Arithmetic planes are a generalization of arithmetic lines $D_{a,b,\mu,\omega} = \{(i,j) \in \mathbb{Z}^2 : \mu \le ai + bj < \mu + \omega\}$. From Theorem 9.4, we know that naive lines ($\omega = \max\{|a|,|b|\}$) are the same as rational DSSs and that standard lines ($\omega = |a| + |b|$) are the same as rational 4-DSSs. If $\omega = \max\{|a|,|b|,|c|\}$, the arithmetic plane $D_{a,b,c,\mu,\omega}$ is called a *naive digital plane*; if $\omega = |a| + |b| + |c|$, it is called a *standard plane*.

Theorem 11.10 Every digital plane with rational slopes is a naive plane and vice versa.

In other words, for any digital plane $i_{\alpha_1,\alpha_2,\beta}$ with rational α_1 and α_2, there exist relatively prime integers a, b, and c and an integer μ such that $i_{\alpha_1,\alpha_2,\beta} = D_{a,b,c,\mu,\max\{|a|,|b|,|c|\}}$. In addition, for any $D_{a,b,c,\mu,\max\{|a|,|b|,|c|\}}$, there exist rational slopes α_1 and α_2 and an intercept β such that $D_{a,b,c,\mu,\max\{|a|,|b|,|c|\}} = i_{\alpha_1,\alpha_2,\beta}$.

Now assume $0 < a \le b \le c$, and consider digitizations of Euclidean planes that are incident with the origin (e.g., by assuming $\mu = 0$). If $D_{a,b,c,0,\omega}$ is a naive plane, each voxel $(x,y,z) \in D_{a,b,c,0,c}$ projects into exactly one pixel (x,y) in the xy-plane. From the assumption $0 < a \le b \le c$, it follows that there is at least one voxel $(x,y,z) \in D_{a,b,c,0,\omega}$ for every $(x,y) \in \mathbb{Z}^2$. The *height map* $H_{a,b,c,\omega}$ is defined on $\mathbb{Z}^2$ by assigning the maximum value z to (x,y) such that $(x,y,z) \in D_{a,b,c,0,\omega}$.

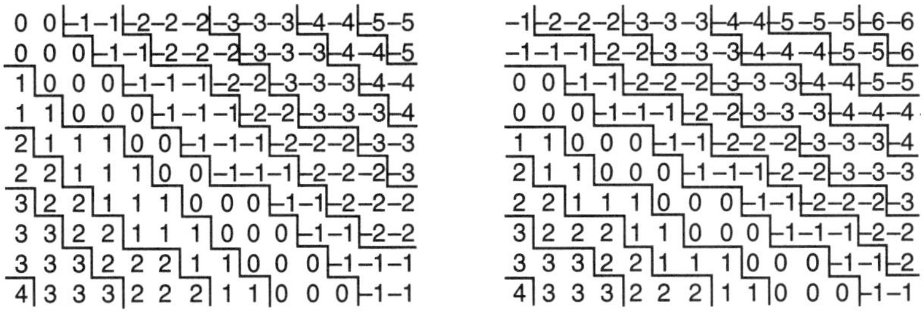

11.2 Digital Planes in 3D Adjacency Grids

```
 0  0 -1 -1 -2 -2 -2 -3 -3 -3 -4 -4 -5 -5      -1 -2 -2 -2 -3 -3 -3 -4 -4 -5 -5 -5 -6 -6
 0  0  0 -1 -1 -2 -2 -2 -3 -3 -3 -4 -4 -5      -1 -1 -1 -2 -2 -3 -3 -3 -4 -4 -4 -5 -5 -6
 1  0  0  0 -1 -1 -1 -2 -2 -3 -3 -3 -4 -4       0  0 -1 -1 -2 -2 -2 -3 -3 -3 -4 -4 -5 -5
 1  1  0  0  0 -1 -1 -1 -2 -2 -3 -3 -3 -4       0  0  0 -1 -1 -1 -2 -2 -3 -3 -3 -4 -4 -4
 2  1  1  1  0  0 -1 -1 -1 -2 -2 -2 -3 -3       1  1  0  0  0 -1 -1 -2 -2 -2 -3 -3 -3 -4
 2  2  1  1  1  0  0 -1 -1 -1 -2 -2 -2 -3       2  1  1  0  0  0 -1 -1 -1 -2 -2 -3 -3 -3
 3  2  2  1  1  1  0  0  0 -1 -1 -2 -2 -2       2  2  1  1  1  0  0  0 -1 -1 -2 -2 -2 -3
 3  3  2  2  1  1  1  0  0  0 -1 -1 -2 -2       3  2  2  2  1  1  0  0  0 -1 -1 -1 -2 -2
 3  3  3  2  2  2  1  1  0  0  0 -1 -1 -1       3  3  3  2  2  1  1  1  0  0  0 -1 -1 -2
 4  3  3  3  2  2  2  1  1  1  0  0  0 -1 -1    4  3  3  3  2  2  2  1  1  1  0  0  0 -1 -1
```

FIGURE 11.14 Height maps: $D_{6,7,16,0,16}$ on the left; $D_{6,9,16,0,16}$ on the right [134].

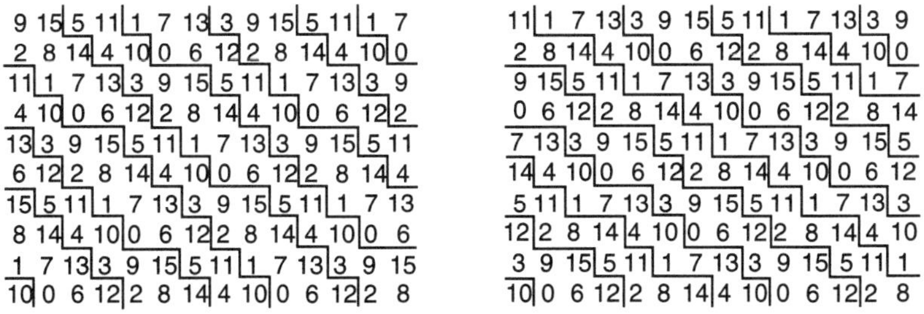

```
 9 15  5 11  1  7 13  3  9 15  5 11  1  7      11  1  7 13  3  9 15  5 11  1  7 13  3  9
 2  8 14  4 10  0  6 12  2  8 14  4 10  0       2  8 14  4 10  0  6 12  2  8 14  4 10  0
11  1  7 13  3  9 15  5 11  1  7 13  3  9       9 15  5 11  1  7 13  3  9 15  5 11  1  7
 4 10  0  6 12  2  8 14  4 10  0  6 12  2       0  6 12  2  8 14  4 10  0  6 12  2  8 14
13  3  9 15  5 11  1  7 13  3  9 15  5 11       7 13  3  9 15  5 11  1  7 13  3  9 15  5
 6 12  2  8 14  4 10  0  6 12  2  8 14  4      14  4 10  0  6 12  2  8 14  4 10  0  6 12
15  5 11  1  7 13  3  9 15  5 11  1  7 13       5 11  1  7 13  3  9 15  5 11  1  7 13  3
 8 14  4 10  0  6 12  2  8 14  4 10  0  6      12  2  8 14  4 10  0  6 12  2  8 14  4 10
 1  7 13  3  9 15  5 11  1  7 13  3  9 15       3  9 15  5 11  1  7 13  3  9 15  5 11  1
10  0  6 12  2  8 14  4 10  0  6 12  2  8      10  0  6 12  2  8 14  4 10  0  6 12  2  8
```

FIGURE 11.15 Remainder maps for the naive planes shown in Figure 11.14 [134].

Figure 11.14 illustrates two height maps of naive planes $D_{a,b,c,0,c}$. Let $L_{a,b,c}(z_0) = \{(x,y) \in \mathbb{Z}^2 : (x,y,z_0) \in D_{a,b,c,0,c}\}$ where $z_0 \in \mathbb{Z}$. Then $L_{a,b,c}(z_0)$ is an arithmetic line $D(a,b,\mu,\omega)$ with $\mu = -cz_0$ and $\omega = c$. $D(a,b,\mu,\omega)$ is standard if $c = a+b$, "thicker than standard" if $c > a+b$, and "thinner than standard" but "thicker than naive" if $c < a+b$. The arithmetic lines $L_{a,b,c}(z_0)$ with $z_0 \in \mathbb{Z}$ partition $\mathbb{Z}^2$ into equivalence classes that are all translation equivalent[3] iff a and b are relatively prime [134]. Figure 11.14 shows height maps for a case in which a and b are relatively prime (left) and a case in which they are not relatively prime (right).

$0 < a \leq b \leq c$ implies that the projections $L_{a,b,c}^{(x)}(x_0) = \{(y,z) \in \mathbb{Z}^2 : (x_0,y,z) \in D_{a,b,c,0,c}\}$ and $L_{a,b,c}^{(y)}(y_0) = \{(x,z) \in \mathbb{Z}^2 : (x,y_0,z) \in D_{a,b,c,0,c}\}$ (where $x_0, y_0 \in \mathbb{Z}$) are naive lines with approximate intercepts $\mu = -ax_0$ and $\mu = -by_0$, respectively. The arithmetic lines $L_{a,b,c}^{(x)}(x_0)$, where $x_0 \in \mathbb{Z}$, partition $\mathbb{Z}^2$ into translation-equivalent equivalence classes. The same is true for the arithmetic lines $L_{a,b,c}^{(y)}(y_0)$ for $y_0 \in \mathbb{Z}$; see [253, 255].

Naive planes can also be represented by arrays of remainders [253]. Let $(x,y,z) \in D_{a,b,c,0,c}$. We assign value $ax + by + cz$ to grid point (x,y) (i.e., its remainder modulo c). This results in a *remainder map* $R_{a,b,c}$. Figure 11.15 shows two examples. On the left, we have $a = 6$ and $b = 7$ (i.e., the integers are relatively prime, which results in remainders in the entire range $0, \ldots, 15$ for $c = 16$). On the right, we

---

3. $A, B \subset \mathbb{Z}^n$ are *translation equivalent* iff there is a translation vector $\mathbf{t} \in \mathbb{Z}^n$ such that $A = \mathbf{t} \oplus B$.

have $a=6$ and $b=9$ (i.e., the remainders in each equivalence class of the depth map are all equal modulo $\gcd(6,9) = 3$).

Proposition 11.1 $R_{a,b,c} = R_{c-a,b,c} = R_{a,c-b,c} = R_{c-a,c-b,c}$ $(0 < a \leq b \leq c)$.

This is called the *Symmetry Lemma* in [134], which defines a special type of symmetry between naive planes $D_{a,b,c,0,c}$, $D_{c-a,b,c,0,c}$, $D_{a,c-b,c,0,c}$, and $D_{c-a,c-b,c,0,c}$. If a or b is larger than $c/2$, the Symmetry Lemma allows us to consider without loss of generality symmetric naive planes $D_{a,c-b,c,0,c}$ or $D_{c-a,c-b,c,0,c}$ for which the first two parameters do not exceed $c/2$.

11.2.5 Periodicity

A *position* (i,j) in an array $X = (X(i,j))_{0 \leq i, 0 \leq j}$ is defined by a row i and a column j; $X(i,j)$ is the *element* of X at position (i,j). The elements of X are letters in an alphabet A. We continue to assume $0 \leq \alpha_1, \alpha_2 \leq 1$; hence, in a digital plane quadrant, we have $A = \{0,1\}$.

Let $S \subseteq \mathbb{Z}_+^2 = \{(i,j) \in \mathbb{Z}^2 : i, j \geq 0\}$. The restriction $X[S]$ of X to positions in S is called a *factor of X on S*.

> **Definition 11.6** A vector $\mathbf{v}$ in $\mathbb{Z}^2$ is called a *symmetry vector* for X and S iff $X(i,j) = X(\mathbf{v}+(i,j))$ for all $(i,j) \in S$ such that $\mathbf{v}+(i,j) \in S$. $\mathbf{v}$ is called a *periodicity vector* or a *period* for X and S iff, for any integer k, the vector $k\mathbf{v}$ is a symmetry vector for S.

An infinite array X on $\mathbb{Z}_+^2$ is called *2D-periodic* iff there are two linearly independent vectors $\mathbf{u}$ and $\mathbf{v}$ in $\mathbb{Z}^2$ such that $\mathbf{w} = i\mathbf{u} + j\mathbf{v}$ is a period for X for any $(i,j) \in \mathbb{Z}^2$ and $\mathbf{w} \in \mathbb{Z}_+^2$. X is called *1D-periodic* iff all periods of X are parallel vectors.

Let X be a 2D-periodic infinite array on $\mathbb{Z}_+^2$. The set of symmetry vectors of X defines (by additive closure) a subgrid Λ of $\mathbb{Z}^2$. Any basis of Λ is a *basis* of X.

We say that an infinite array X on $\mathbb{Z}_+^2$ is *tiled* by a (finite) rectangular factor W if X is a pairwise disjoint repetition of W. Evidently, any 2D-periodic array on $\mathbb{Z}_+^2$ can be tiled.

> **Theorem 11.11** Any rational digital plane quadrant is 2D-periodic. Any irrational digital plane quadrant is either 1D-periodic or aperiodic.

Any basis of a rational digital plane quadrant defines a lattice with cells that are parallelograms. Let $ax + by + cz = d$ be a rational plane in which a, b, c, and d are integers and a, b, and c are relatively prime.

11.2 Digital Planes in 3D Adjacency Grids

Theorem 11.12 The lattice cells of all bases of a rational digital plane quadrant have constant area $\max\{|a|,|b|,|c|\}$.

Let X be an array on $\mathbb{Z}_+^2$. An $m \times n$ $S \subset \mathbb{Z}_+^2$ defines an $m \times n$-factor of X. Any rectangular factor of an irrational digital plane is also a factor of a rational plane.

Let $P_X(m,n)$ be the number of $m \times n$-factors of X. For example, $P_X(0,0) = 1$ for any X, and $P_X(1,1)$ is the number of distinct letters in X. (In our case, $A = \{0,1\}$.) P_X generalizes the complexity function $P(w,n)$ defined in Chapter 9.

An array X on $\mathbb{Z}_+^2$ is called *Sturmian* iff it is an infinite array of an irrational digital plane where both slopes, α_1 and α_2, are irrational numbers. The *height* $h(W)$ of an $m \times n$ array W is the number of 1s in W. If V and W are of the same size, $\delta(V,W) = |h(V) - h(W)|$ is called their *balance*. A set of arrays X of identical size is called *balanced* iff $\delta(V,W) \leq 1$ for all pairs V,W of the arrays. An infinite array X on $\mathbb{Z}_+^2$ is called *array!balanced* iff its set of $m \times n$-factors is balanced. An array X on $\mathbb{Z}_+^2$ is called *eventually periodic* iff there exist integers k and l such that the array $(X(i,j))_{k \leq i, l \leq j}$ (which is a *suffix* of X) is periodic.

Theorem 11.13 Let X be a digital plane quadrant. If $P_X(m,n) \leq mn$ for some $m,n \geq 0$, X has at least one periodicity vector.

In summary, a digital plane quadrant is 2D-periodic iff it is rational iff P_X is bounded; it is Sturmian iff it is irrational and not 1D-periodic.

11.2.6 Connectivity of arithmetic planes

An arithmetic line becomes 8-disconnected iff $\omega < \max\{|a|,|b|\}$. Similarly, an arithmetic plane $D_{a,b,c,\mu,\omega}$ no longer has grid points on all of the vertical grid lines iff $\omega < \max\{|a|,|b|,|c|\}$.

A standard arithmetic plane is 26-separating and gapfree; it has no 6-, 18-, or 26-gaps. A naive arithmetic plane is 6-separating but not necessarily 18- or 26-separating; it can have 18- or 26-gaps. Note that, if S is not α-connected, any of its subsets is α-separating in S.

Theorem 11.14 Let $D_{a,b,c,\mu,\omega}$ be an analytic plane with $0 \leq a \leq b \leq c$ and $0 \leq \mu$. If $\omega < c$, the plane has 6-gaps; if $c \leq \omega < b+c$, it has 18-gaps and is 6-separating in $\mathbb{Z}^3$; if $b+c \leq \omega < a+b+c$, it has 26-gaps and is 18-separating in $\mathbb{Z}^3$; and if $a+b+c \leq \omega$, it is 26-gapfree.

In Chapter 9, we formulated equivalences between 8-gap-freeness and 4-connectedness (4-gap-freeness and 8-connectedness) for arithmetic lines; this cannot be done for arithmetic planes. Connectivity is a translation-invariant property. Without loss

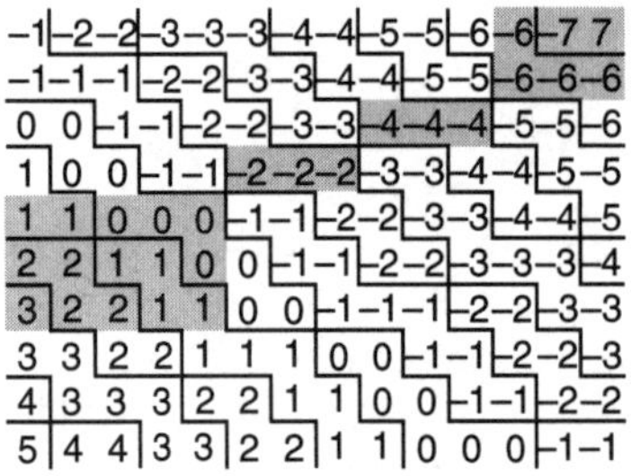

```
-1|-2-2|-3-3-3|-4-4|-5-5|-6-6|-7 7
-1-1-1|-2-2|-3-3|-4-4|-5-5|-6-6-6
 0 0|-1-1|-2-2|-3-3|-4 4-4|-5-5|-6
 1|0 0|-1-1|-2-2-2|-3-3|-4-4|-5-5
 1 1|0 0 0|-1-1|-2-2|-3-3|-4-4|-5
 2 2|1 1|0 0|-1-1|-2-2|-3-3-3|-4
 3|2 2|1 1|0 0|-1-1-1|-2-2|-3-3
 3 3|2 2|1 1 1|0 0|-1-1|-2-2|-3
 4|3 3 3|2 2|1 1|0 0|-1-1|-2-2
 5|4 4|3 3|2 2|1 1|0 0 0|-1-1
```

FIGURE 11.16 Height map of the naive plane $D_{5,7,11,0,11}$. The 8-connected set of pixels (shown in gray) is a projection of a 26-disconnected set of voxels of this plane [134].

---

of generality, we consider grid-line intersection digitizations of rational planes $ax + by + cz = 0$ that are incident with the origin. Let $D_{a,b,c,\omega}$ be the corresponding arithmetic plane with thickness $\omega \in \mathbb{Z}_+$ where $a, b, c \in \mathbb{Z}$ and $\gcd(a,b,c) = 1$. In the case of a naive plane ($\omega = \max\{|a|, |b|, |c|\}$), we simply write $D_{a,b,c}$.

Definition 11.7 $\Omega_\alpha(a,b,c) = \max\{\omega : D_{a,b,c,0,\omega} \text{ is } \alpha\text{-disconnected}\}$ is called the α-*connectivity number* of the class of all arithmetic planes $D_{a,b,c,0,\omega}$ such that $\omega \in \mathbb{Z}_+$ where $\alpha \in \{6,18,26\}$ and $a, b, c \in \mathbb{Z}$.

$\omega = \Omega_\alpha(a,b,c) + 1$ is the smallest integer such that $D_{a,b,c,0,\omega}$ is α-connected. Evidently, $\alpha \geq \beta$ ($\alpha, \beta \in \{6,18,26\}$) implies $\Omega_\alpha(a,b,c) \leq \Omega_\beta(a,b,c)$. Naive planes are always 26-connected (i.e., $\Omega_{26}(a,b,c) \leq \max\{|a|,|b|,|c|\}$), and standard planes are always 6-connected (i.e., $\Omega_6(a,b,c) \leq |a|+|b|+|c|$). Connectivity numbers remain unchanged when a, b, and c are permuted (e.g., $\Omega_\alpha(a,b,c) = \Omega_\alpha(b,c,a)$).

Exercise 7 in Section 11.5 defines "jumps" and shows that they exist in naive planes if $c < a + b$. Figure 11.16 illustrates such a naive plane in which 8-connected sets of pixels in the height map may be projections of 26-disconnected sets of voxels in the plane. The Symmetry Lemma (Proposition 11.1) allows us to transform such naive planes into symmetric (in the sense of the Lemma) naive planes for which $c < a + b$ is no longer true. This implies the following:

Proposition 11.2 $\Omega_{26}(a,b,c) = \Omega_{26}(c-a,b,c) = \Omega_{26}(a,c-b,c) = \Omega_{26}(c-a,c-b,c)$ if a, b, and c are relatively prime integers such that $0 < a \leq b \leq c$.

A *graceful plane* is a naive plane for which $c = a + b$; see [133]. Proposition 11.2 and Exercise 12 in Section 11.5 imply that $\Omega_{26}(a,b,a+b) = b - 1$. The main result in [134] is as follows:

Theorem 11.15 Let a, b, and c be relatively prime integers such that $c \geq a + 2b$ and $a > 0$. Then $\Omega_{26}(a,b,c) = c - a - b + \gcd(a,b) - 1$.

This theorem, combined with Proposition 11.2, allows us to derive other solutions, such as the following:

$$\Omega_{26}(a,b,c) = b-a+\gcd(a,c-b)-1 \text{ if } c < 2b-a$$
$$\Omega_{26}(a,b,c) = b+a-c+\gcd(c-b,c-a)-1 \text{ if } c < a+b/2$$
$$\Omega_{26}(a,b,c) \geq c-a-b+\gcd(a,b)-1 \text{ if } c < a+2b \text{ and } a \neq b$$

Theorem 11.16 If $a+b < c < a+2b$, then $a-1 \leq \Omega_{26}(a,b,c) \leq b-1$.

A more instructive understanding of the the plane connectivity issue can be obtained by generating remainder maps $M_{a,b,c}$ for thicknesses ω in the range from 0 to c. The patterns in Figure 11.17 visualize the pixels that belong to $M_{11,19,35}$ for $\omega = 1, 2, 6, 7, 11, 12, 13, 20,$ and 35. We have $\Omega_{26}(11,19,35) = 12$. By varying the parameters a, b, c, and ω, we can calculate a family of patterns, some of which have interesting (local or recursive) configurations. Studying their properties might be of interest for studying tilings of the plane.

11.3 Digital Planes in the 3D Incidence Grid

A *cellular digital plane* can be defined by considering a digital plane (or naive plane) in the grid cell model. It can also be defined by outer Jordan digitization of a plane Γ in the grid cell model. However, if Γ passes through a grid vertex or contains a grid edge, outer Jordan digitization produces "locally thicker" cellular planes.

The frontier of a cellular digital plane consists of two "parallel layers" of frontier faces that define an *upper* and a *lower digital frontier plane* in the incidence grid $\mathbb{C}_3$; they are analogous to the lower and upper digital rays or lines defined in Chapter 9.

Each 0-cell of a 3-cell c is incident with three 2-cells of c. The normals to these 2-cells form a *tripod*, and there are eight different tripods. All of the face normals of any upper or lower digital frontier plane belong to one tripod.

Definition 11.8 $S \subset \mathbb{C}_3^{(2)}$ is called a *digital plane of 2-cells* in the incidence grid iff it is an upper or lower digital frontier plane defined by a cellular digital plane.

A finite 1-connected subset of a digital plane of 2-cells is called a *digital plane segment* (DPS) in the 3D incidence grid.

In the plane (see Theorem 9.5), a 4-path is a 4-DSS iff it is contained between or on a pair of supporting lines with a main diagonal distance of less than $\sqrt{2}$. Let G be a finite 1-connected set of faces of grid cubes. Let the faces be contained between or on a pair of parallel planes. The *main diagonal* $\mathbf{v} = (\pm 1, \pm 1, \pm 1)$ of a pair of parallel planes is the diagonal direction (one of the eight possible directed diagonals in the 3D grid with length $\|\mathbf{v}\| = \sqrt{3}$) that has the greatest dot product (inner product),

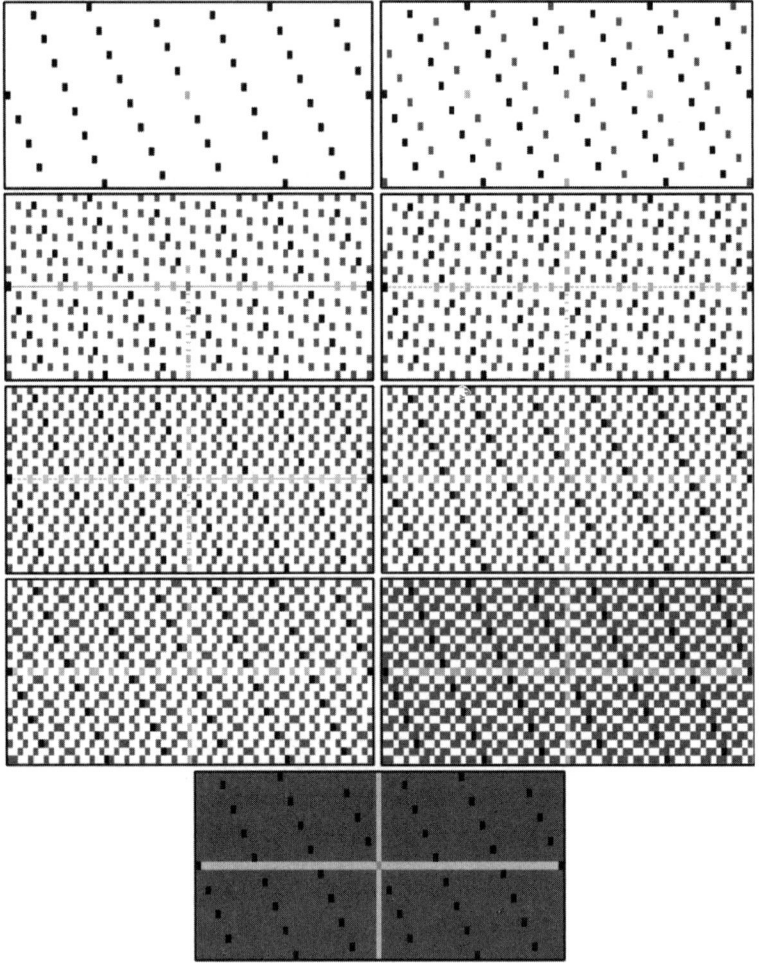

FIGURE 11.17 Remainder maps $M_{11,19,35}$ for (from top left to bottom) $\omega = 1, 2, 6, 7, 11, 12, 13, 20,$ and 35. (Courtesy of Valentin Brimkov, SUNY Fredonia.)

with the outward pointing normal **n** to the planes. If there is more than one such direction, we can choose one of them arbitrarily. The distance between the planes in the main diagonal direction is called their *main diagonal distance*.

> **Proposition 11.3** A finite 1-connected set of frontier faces of a set of 3-cells is a DPS iff all of the face normals belong to one tripod and the faces are contained between or on a pair of parallel planes with a main diagonal distance of less than $\sqrt{3}$.

11.3 Digital Planes in the 3D Incidence Grid

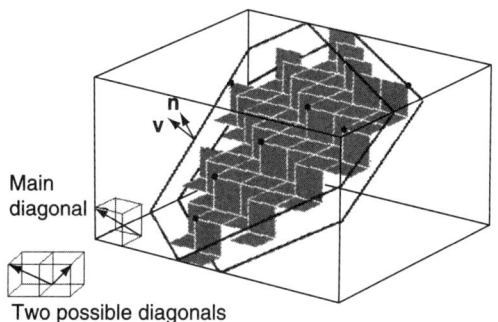

FIGURE 11.18 A DPS; the main diagonal distance between the two parallel planes is less than $\sqrt{3}$.

In other words, a DPS in the incidence grid can be assumed to be a 1-connected set of 2-cells in the frontier of a 6-region of voxels. If considered together with its incident 0- and 1-cells, it is a 2D Euclidean cell complex. A *simply connected DPS* consists of faces that have a union that is homeomorphic to the unit disk (i.e., it is a 1- simply connected set of 2-cells). Figure 11.18 shows a DPS; **n** is its normal, and **v** is the vector in the main diagonal direction.

If we are given the frontier of the projection of the DPS onto one of the two parallel planes, it is possible to reconstruct the DPS in 3D space (up to a translation in the normal direction to the planes).

Let **v** be the vector of length $\sqrt{3}$ in main diagonal direction, and let **n** be an outward pointing normal to the pair of parallel planes. Let **p** be a grid vertex incident with the DPS, and let $\mathbf{v} \cdot \mathbf{p} = d_p$ be the equation of a plane incident with **p** that has normal **v**. In accordance with Proposition 11.3, the vertices **p** of the grid faces of a DPS must satisfy the following:

$$0 \leq \mathbf{n} \cdot \mathbf{p} - d_p < \mathbf{n} \cdot \mathbf{v} \tag{11.5}$$

Let $\mathbf{n} = (a,b,c)$. The scalars a, b, and c can have different signs, but, because **n** and **v** must point in the same direction "modulo a directed diagonal," we can assume without loss of generality that $a, b, c > 0$. Equation 11.5 then becomes the following:

$$0 \leq ax + by + cz - d_p < a + b + c \tag{11.6}$$

Hence, a DPS in the grid-cell model is equivalent (by mapping vertices into grid points) to a finite 6-connected set of grid points in a standard digital plane (see Definition 11.5) for which $\nu = d_p$ and $\omega = a+b+c$.

In addition to checking the tripod condition (which is easy), DPS recognition (in the grid cell model) can be performed by answering the following question: Given n vertices $\{p_1, p_2, \ldots, p_n\}$, does each p_i such that $d_i = \mathbf{v} \cdot p_i$ satisfy Equation 11.5? Or, to put it another way, do we have the following?

$$0 \leq \mathbf{n} \cdot p_i - d_i < \mathbf{n} \cdot \mathbf{v} \quad \text{for } i = 1, \ldots, n \tag{11.7}$$

11.4 DPS Recognition and Generation

Theorem 11.9 was used in [1028] to suggest a DPS recognition algorithm based on convex hull separability. The recognition of DPSs in grid-adjacency models (i.e., DPSs regarded as subsets of $\mathbb{Z}^3$) is also discussed in [1089] (using the characterization by evenness given previously), [549] (using least-square optimization), and [157, 716, 823, 1100] (using linear programming). [255] discusses the use of arithmetic "limits." $ax + by + cz = \mu$ and $ax + by + cz = \mu + \omega$ are called the lower and upper *supporting planes* of an arithmetic plane; a test for the existence of these planes provides another method of designing algorithms for DPS recognition. Supporting planes are called *leaning planes* in arithmetic geometry.

[341] suggests a method of recognizing DPSs in the incidence grid based on conversion of Equation 11.5 into a system of linear inequalities by eliminating the d_ps:

$$\mathbf{n} \cdot p_i - \mathbf{n} \cdot p_j < \mathbf{n} \cdot \mathbf{v} \quad \text{for } i, j = 1, \ldots, n \tag{11.8}$$

This system of n^2 inequalities can be solved in various ways. [341, 340] uses the Fourier-Motzkin algorithm. We can also use Fukuda's *cdd algorithm*[4] for solving systems of linear inequalities by successive intersection of halfspaces defined by the inequalities.

11.4.1 An incremental DPS algorithm

We will next describe an incremental algorithm [550]. Typical timing results for these three algorithms are shown in Figure 11.19 for a polyhedrized digital ellipsoid at grid resolutions ranging from 10 to 100.

Algorithm KS2001

Π is called a *supporting plane* of a finite set of faces if the faces are all in one of the closed halfspaces defined by Π and their diagonal distances to Π are all less than $\sqrt{3}$. If the set of faces has $n \geq 4$ vertices, Π must be incident with three noncollinear vertices, and all the other vertices must lie on or on one side of Π. A set of faces can have more than one supporting plane.

The incremental algorithm repeatedly updates a list of supporting planes; if this list is empty, the set of points is not a DPS. The updating step is as follows: if we have $n \geq 0$ points, we add an $(n+1)$st point iff the list of supporting planes remains nonempty. To test this, we first check the new point against each of the listed supporting planes to see if it is on the same side of the plane as the other points and within the allowed diagonal distance. We delete the plane from the list if these conditions are not satisfied. We then construct new supporting planes by combining the new point with pairs of existing points. A new supporting plane is added to the

---

4. http://www.cs.mcgill.ca/~fukuda/soft/cdd_home/cdd.html

11.4 DPS Recognition and Generation

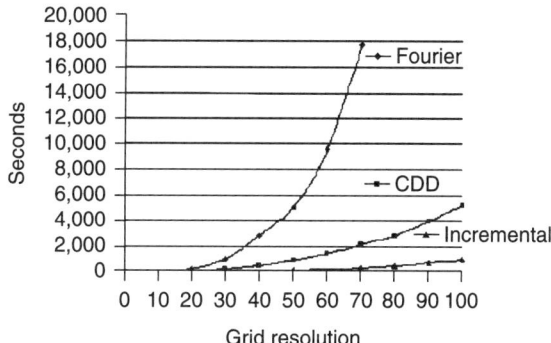

FIGURE 11.19 Running times of three DPS recognition algorithms on a PIII 450 running Linux. (Laurent Papier provided the Fourier-Motzkin program.)

list if all $n+1$ points satisfy the conditions. The set of points is accepted as a DPS iff the final list of planes is not empty. The updating step is time-efficient, because we can restrict the tests to points that have extreme positions in any of the eight diagonal directions.

A surface consists of edge-connected faces. These faces can be represented by a face graph that has nodes that are the faces and in which each node has four pointers to its edge-adjacent faces. The face graph can be constructed using, for example, the Artzy-Herman surface tracing algorithm.

We can perform a breadth-first search of the face graph to agglomerate the faces into DPSs. This process is implemented using two queues. The first is called a *seeds queue*; it contains all of the faces found by the search that do not belong to any yet recognized DPS. A face is inserted into the seeds queue if it cannot be added to the current DPS. The next DPS starts from a face chosen from the seeds queue; the choice of this face determines how the DPS "grows." The second queue is used to maintain the breadth-first search. "Growing a DPS" looks like propagating a "circular wave" on the surface from a center at the original seed face.

We try to add an adjacent face to the current DPS by testing each vertex of the face that is not yet on the DPS. If all four vertices pass the test, the face is added to the DPS and deleted from the seeds queue (if it was on that queue). Otherwise, we insert the face into the seeds queue and try another adjacent face. If no more adjacent faces can be added, we start a new DPS from a face on the seeds queue.

A list of the frontier vertices of each DPS is maintained during the agglomeration process, not only to simplify the tests for whether a new vertex can be added but also to maintain the topologic equivalence of the DPS to a unit disk. This ensures that the frontier always remains a simple polygon so that the algorithm constructs only simply connected DPSs. (This condition can be removed, if desired.)

Figure 11.20 illustrates results of the agglomeration process for a digitized sphere and for an ellipsoid with semiaxes 20, 16, and 12. Faces that have the same gray level belong to the same DPS. The numbers of faces of the digital surfaces of

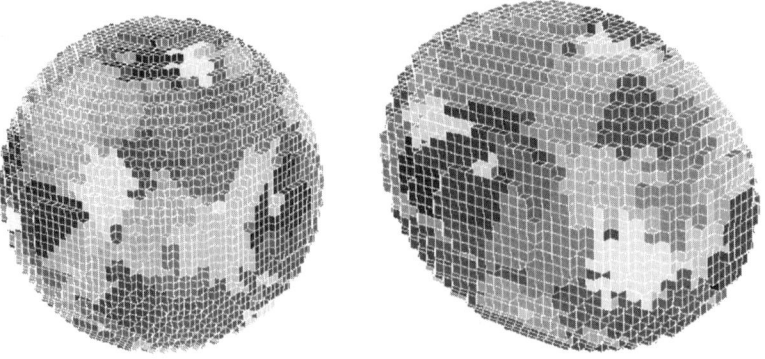

FIGURE 11.20 Agglomeration into DPSs of the faces of a sphere and an ellipsoid (with a grid resolution of $h = 40$).

FIGURE 11.21 A polyhedrized sphere and ellipsoid.

the sphere and ellipsoid are 7584 and 4744, respectively. The numbers of DPSs are 285 and 197; the average sizes of these DPSs are 27 and 24 (faces).

To complete the polyhedrization process we take all of the face vertices that are incident with at least three of the DPSs as the vertices of the polyhedron. Figure 11.21 shows the final polyhedra for the sphere and ellipsoid. Note that these polyhedra are not simple; their surfaces are not hole-free.

Restricting the depth of the breadth-first search changes the polyhedrization from global to local and results in "more uniform" polyhedra. Figure 11.22 shows results when the depth is restricted to 7. The number of small DPSs is reduced, and the sizes of the DPSs are more evenly distributed. The numbers of DPSs are 282 and 180, and their average sizes are 27 and 26; note that these are nearly the same as in the unrestricted case.

11.5 Exercises

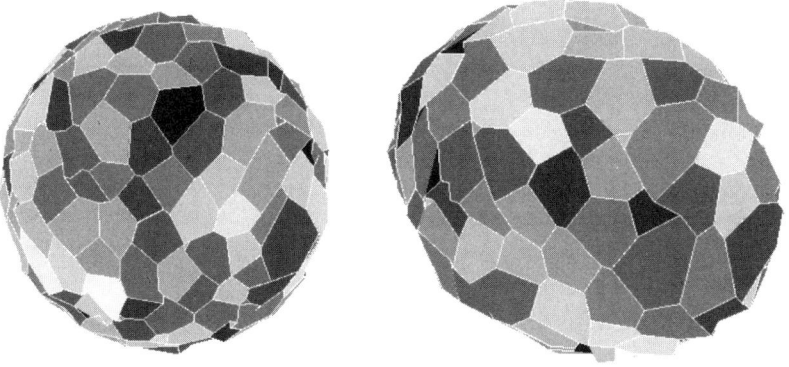

FIGURE 11.22 The polyhedrized sphere and ellipsoid when the breadth-first search depth is restricted to 7.

11.4.2 DPS generation algorithms

Consider the task of obtaining a gap-free digitization of the surface of a simple polyhedron "face-by-face." If every polygonal patch on the surface of the polyhedron is approximated (digitized) individually (!) by a naive plane, gaps may result "near the edges" of the polyhedron.

A method from [62] for solving this problem is based on reducing the 3D problem to a 2D one by projecting the polygonal patches onto suitable coordinate planes, digitizing the resulting 2D polygons (in the thinnest way possible, in the terminology of arithmetic geometry), and then finally calculating the 3D digital polygons as subsets of naive planes.

Another method [135] approximates every 3D polygon by a digital polygon, again in the thinnest way possible, and approximating the edges of the polygons by 3D DSSs. The resulting digitization is "optimally thin" in the sense that removing any voxel from the digital surface produces a gap. A third method [133] is based on using graceful planes and lines, respectively, to approximate the surface polygons and their edges. The algorithms in the cited references run in times that are linear in the number of generated surface voxels.

11.5 Exercises

1. Prove statements 11.1 through 11.4.

2. If the projections of a 26-DSS are an 8-DSS with normal (x_1, x_2) in the $(x=0)$-plane and an 8-DSS with normal (y_1, y_2) in the $(y=0)$-plane, what is the normal of the 26-DSS?

3. Generate the following circles $\gamma(t)$ in $\mathbb{R}^3$, where R_x and R_z are 3×3 matrices of rotations around the x- and z-axes, respectively:

$$\gamma(t) = R_x(\eta_1) R_z(\eta_2) R_x(\eta_3) \begin{pmatrix} r \cdot \cos t \\ r \cdot \sin t \\ 0 \end{pmatrix} \quad \text{where } 0 \leq t < 2\pi$$

The angles η_1, η_2, and η_3 and the radius r are randomly chosen from uniform distributions in $[0, 2\pi)$ and $[0 \cdot 5, 1]$, respectively. Perform grid-plane intersection digitization of each generated circle using varying grid sizes (e.g., between 100^3 and 1000^3). Implement 3D DSS segmentation and 3D MLP approximation, and compare their run times.

4. (Open problem) Is there a simple 2-curve such that none of the vertices of its 3D MLP is a grid vertex (i.e., none of these vertices is an endpoint of a critical edge)?

5. Give an example of a 2-curve with a 3D MLP that has at least one vertex that has an irrational coordinate.

6. A straight line in 3D space can be represented as the intersection of two planes $a_i x + b_i y + c_i c + d_1 = 0$ $(i = 1, 2)$. Give an example in which the intersection of two digital planes (i.e., two naive planes) obtained by grid-line intersection digitization of two distinct planes is 26-disconnected.

7. Two voxels $p = (i, j, k)$ and $q = (i+1, j+1, k+2)$ in the grid cell model (see the following figure) define a *jump*. Show that a naive plane $D_{a,b,c,\mu,c}$ (where $c = \max\{a, b, c\}$) has a jump iff $c < a + b$.

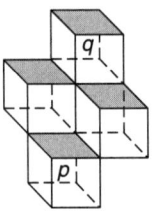

8. There are four possible factors of size 2×2 of a digital plane quadrant for which $0 \leq \alpha_1, \alpha_2 \leq 1$, and $\alpha_1 \leq \alpha_2$. Which pairs of these factors can be contained in the same rational digital plane quadrant?

9. Prove that a rational digital plane quadrant has at most mn distinct factors of size $m \times n$.

10. Implement algorithm **KS2001** and test it on digitized ellipsoids with varying semi-axes $0 < a, b, c \leq 1$ for different grid resolutions, for example, $h = 100, \ldots, 1000$.

Study the impact of different search strategies and search depth thresholds on the run time and on the resulting DPS segmentation.

11. Design a DPS approximation algorithm along the following lines: incrementally calculate the 3D convex hull of a set of grid points to which one 6-connected grid point is added at a time, and incrementally update the minimum width of the convex hull as long as the minimum width is below a predefined threshold (e.g., $\sqrt{3}$ or larger, allowing for "minor variations").

12. Prove that the connectivity number Ω_{26} satisfies (i) $\Omega_{26}(a,a,a) = 0$, (ii) $\Omega_{26}(0,b,c) = c-1$, (iii) $\Omega_{26}(a,b,b) = b-1$, (iv) $\Omega_{26}(a,a,c) = c-a-1$, and (v) $\Omega_{26}(a,b,2b) = b-1$, where a, b, and c are relatively prime integers such that $0 < a \leq b \leq c$.

11.6 Commented Bibliography

For reviews of 3D straight lines, see [210] (in computer graphics) and [480] (in 3D picture analysis).

Theorem 11.1 is from [511]; for a generalization to straight lines in $\mathbb{Z}^n$, see [540]. The digitization of hyperplanes in arbitrary dimensions is discussed in [540].

[325] defines 3D DSSs in arithmetic geometry using Diophantine inequalities. A more specific definition is given in [203] for the purpose of deriving an efficient 3D DSS recognition algorithm. Sections 11.1.2 and 11.1.3 review parts of [203]. Theorem 11.3 is proved in [253]; see also [256].

Algorithm **DR1995** is used in [203] to recognize 26-DSSs (see Section 11.1) and to estimate the lengths of 3D curves; see Chapter 12 for estimation results.

A brief description of the iterative 3D MLP algorithm and some preliminary experimental results were presented in [148]. The difficulty of the subject is illustrated by the fact that the *Euclidean shortest path problem* (given a finite collection of polyhedral obstacles in 3D space and a source and a target point, find a shortest obstacle-avoiding path from source to target) is known to be NP-hard [162]. However, there are polynomial-time algorithms for the *approximate Euclidean shortest path problem*; see [198]. The convergence behavior of the algorithm in [148] is not yet known; it may converge to the exact 3D MLP, or it may approximate it up to some error. Experiments performed so far suggest that the algorithm always converges to the correct 3D MLP, and time measurements support the hypothesis that its runtime behavior is asymptotically linear in the number of input 3-cells, even if a very small threshold is used for termination.

[171] describes an algorithm for estimating 3D straight lines based on projections onto a plane.

Row- and column-wise step codes of digital planes were introduced in [129]. For the recognition of digital planes, see [519]; for the recognition of digital naive planes, see [1099, 1100]. For the approximation of linear and affine functions, see [866, 871]. For arithmetic straight lines and planes, see [341].

Theorems 11.5 and 11.6 are from [513]. Definition 11.3 is from [517]. Evenness of digital arcs was studied in [454] and generalized in [1088] to evenness of sets in $\mathbb{Z}^n$. Theorem 11.7 is proved in [1088].

For the definition of supporting planes and Theorem 11.8, see [513]. Theorem 11.9 is from [1028], where it is proved for digitized hyperplanes of any dimension.

An arithmetic representation of digital planes was introduced in [332]. For a study of digitized hyperplanes of arbitrary dimension based on pairs of Diophantine inequalities, see [23]. Definition 11.5 and Theorem 11.10 are from [848]. For Theorem 11.14, see [23], which actually discusses the general n-dimensional case:

Theorem 11.17 Let $P = P(b, a_1, a_2, \ldots, a_n, \omega) = \{x \in \mathbb{Z}^n : 0 \leq b + \sum_{i=1}^{n} a_i x_i < \omega\}$ be a digital hyperplane, where $b \geq 0$, $a_i \geq 0$ for all i and $a_i \leq a_{i+1}$ for $1 \leq i \leq n-1$. Then, if $\omega < a_n$, the digital hyperplane has $(n-1)$-gaps. For $0 < k < n$, if $\sum_{i=k+1}^{n} a_i \leq \omega < \sum_{i=k}^{n} a_i$, the digital hyperplane has $(k-1)$-gaps and is k-separating. If $\omega \geq \sum_{i=1}^{n} a_i$, the digital hyperplane is gapfree.

This theorem answers the question about the maximal thickness ω for which α-gaps appear. Analytic definitions of digital hyperplanes can also be based on real parameters $b, a_1, \ldots, a_n$; this allows us to characterize digital planes that have irrational slopes (compare Theorem 11.10) with arithmetic planes defined by at least one irrational parameter.

In [129], V.E. Brimkov extended periodicity studies in the theory of words (Chapter 9) to 2D words based on [18, 351]. Section 11.2.5 briefly reviews his work. [849] proves that $P_X(m, n) \leq mn$ if X is a rational digital plane quadrant (see Exercise 9). Section 11.2.6 reviews [134]. This report also gives an algorithm for computing $\Omega_{26}(a, b, c)$, which has a runtime (measured in arithmetic operations) that is $\mathcal{O}(a \log b)$.

The incremental algorithm **KS2001** in Section 11.4, including its derivation and experimental results, was published in [550]. See also [796] for polyhedrization algorithms.

Exercise 3 describes an experiment reported in [149]. For Exercise 6, see [62, 255]. Exercise 7 is from [133]. For Exercise 12, see [134].

CHAPTER 12

3D Arc Length, Surface Area, and Curvature

Length and curvature estimation for arcs in 2D spaces were treated in Chapter 10. This chapter begins by discussing length, curvature, and torsion estimation for arcs in 3D spaces. It then discusses how the area or curvature of a surface can be estimated using either polyhedrization of the surface or estimation of the surface normal. Estimation methods can be classified as local or global; such a classification can also be used in other contexts. Estimators can be evaluated in terms of their multigrid convergence and their computational complexity. Local methods are fast, but their accuracy is limited. Global methods are more complex, but they can potentially provide accurate estimates.

12.1 3D Arcs

This section discusses estimation of arc length, curvature, and torsion for 3D arcs from their grid-plane intersection or outer Jordan digitizations. Grid-plane methods use 3D chain code representations or DSS approximation; Jordan methods use MLP approximations (see Chapter 11).

12.1.1 Arc length estimation

Grid-plane intersection digitization of a curve or arc $\gamma(t)$ in $\mathbb{R}^3$ using a grid of resolution $h > 0$ results in a 26-curve $\rho_{h,26}(\gamma)$ of grid points in $\mathbb{Z}_h^3$ generated by incrementing t. This 26-curve has the following length:

$$\mathcal{L}_{ssl}(\rho_{h,26}(\gamma)) = \frac{1}{h} \cdot \left(n_1 + \sqrt{2}n_2 + \sqrt{3}n_3 \right) \tag{12.1}$$

Here n_1 counts isothetic steps of length 1, n_2 counts diagonal steps of length $\sqrt{2}$, and n_3 counts diagonal steps of length $\sqrt{3}$. This *sum of step lengths* $\mathcal{L}_{ssl}$ is a local length estimator for γ.

Statistic analysis of step distributions was used in [523] to design an unbiased length estimator based on step counts. Minimization of the rms error (the square root of the mean squared error) led to the following estimator:

$$E_{rms}(\rho_{h,26}(\gamma)) = \frac{1}{h} \cdot (0 \cdot 9016 \cdot n_1 + 1 \cdot 289 \cdot n_2 + 1 \cdot 615 \cdot n_3) \qquad (12.2)$$

Comparison of this local length estimator with the two global length estimators described below is analogous to comparison of E_{chm} (see Section 10.2.1) with E_{8ss} and E_{mlp} in the planar case. E_{rms} is advantageous for grid resolutions of up to about 200 for digitized curves of diameter 1; see Exercise 1 in Section 12.4.

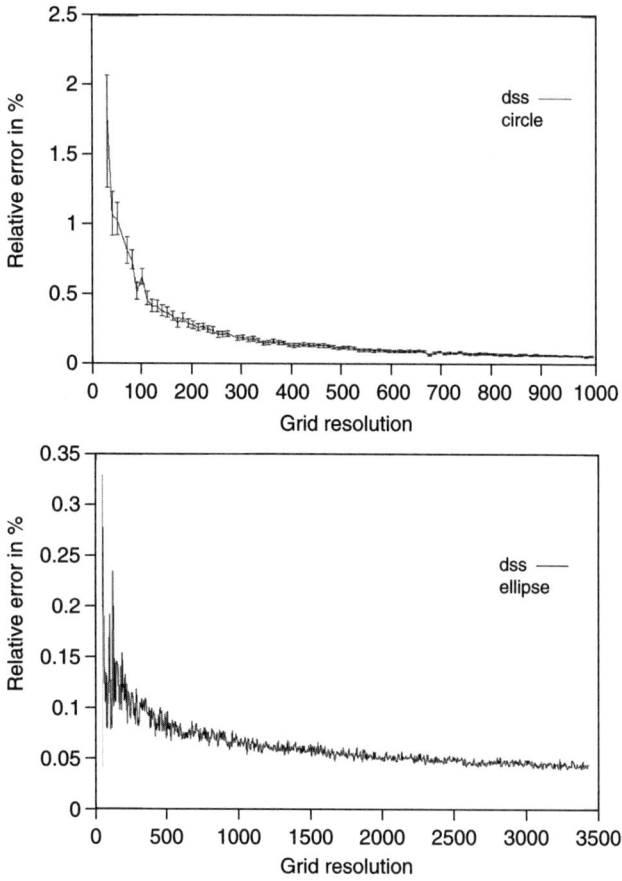

FIGURE 12.1 Relative errors of E_{dss} for digitized 3D curves [203]. Top: circle; see Figure 11.3 (left). Bottom: ellipse with semiaxes 1 and $\sqrt{2}$; see Figure 11.3 (right). Note that the scales are different.

12.1 3D Arcs

The linear online 3D DSS segmentation algorithm **DR1995** (see Section 11.1.3) provides a global length estimator E_{dss}: the length of the polygonal arc or curve where vertices are the endpoints of the resulting DSSs. [203] proves multigrid convergence of E_{dss} to the true length for a class of 3D curves when grid-plane intersection digitization is used. Speed of convergence and maximum error bounds for E_{dss} have not yet been determined, but experimental studies indicate linear convergence speed; see Figure 12.1. [203] illustrates the linear decrease of the relative error in the case of a digitized ellipse. A DSS segmentation depends on the choice of starting point; this creates some variations in the length estimates. The upper plot in Figure 12.1 shows 95% confidence intervals (vertical bars) for all possible DSS segmentations of the digital circle.

Another global length estimator, E_{rba}, is provided by applying the rubber band algorithm of Section 11.1.5 to the outer Jordan digitization of γ. This algorithm approximates the minimum-length polygonal arc or polygon inscribed in γ. The length of this polygonal arc is the E_{rba} estimate of the length of γ. This estimate depends on the initialization method used in the rubber band algorithm and on the threshold τ.

[149] reports on experiments using the threshold $\tau = \mathcal{L}(\Pi_0) \cdot 10^{-6}$ for all three initialization methods discussed in Section 11.1.5. The experiments used outer Jordan digitizations of circles in R^3 as defined in Exercise 3 in Chapter 11. In each experiment (for each initialization method), the lengths of 25 digitized randomly generated circles were estimated relative to the true circumference π. Figure 12.2 (left) shows the resulting estimates of π. The error bars are centered at the medians of the estimated lengths of the 25 circles; the lengths of the bars are twice the variance of the estimates. Figure 12.2 (right) shows the differences in the estimated lengths when the

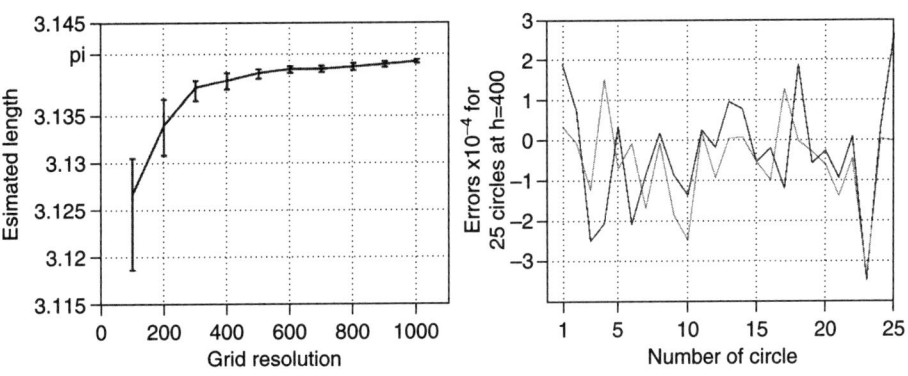

FIGURE 12.2 Left: E_{rba} length estimates of digitized circles in 3D for grid resolutions 100^3 to 1000^3. Right: differences between the estimated lengths for 25 randomly generated circles at resolution 400^3. The black plot shows the differences between initialization methods I and III; the gray curve shows the differences between Init II and III.

TABLE 12.1 Means of the numbers of cycles of the rubber band algorithm that were needed to achieve the threshold for different resolutions and initialization methods.

	Grid resolution				
	100	200	300	400	500
Init I	9·20	9·60	9·64	9·12	9·15
Init II	8·92	9·08	8·92	8·12	8·85
Init III	7·96	8·56	8·80	8·68	8·85

	Grid resolution				
	600	700	800	900	1000
Init I	9·30	9·32	8·50	8·86	8·85
Init II	8·48	8·40	10·00	8·32	8·28
Init III	8·61	8·64	8·50	8·73	8·57

three initialization methods were used. These differences are on the order of 10^{-4}; this indicates convergence of the algorithm to a global minimum. The experiments also showed differences on the order of 10^{-4} in the positions of the vertices of the approximated MLP.

The average numbers of iteration cycles of the rubber band algorithm in the experiments are shown in Table 12.1 and Figure 12.3. Only a small gain results from applying an initialization method that is more sophisticated than just placing the vertices at grid points (i.e., centers of 3-cells) or at midpoints of critical edges. The results also show that the number of cycles needed until the threshold was reached did not depend on the grid resolution or on the type of curve that was digitized.

The rubber band algorithm had $\mathcal{O}(n)$ runtime for a class of 2-curves ρ in which each cube in ρ has exactly two bounding faces in ρ (see Figure 11.6, right). Test

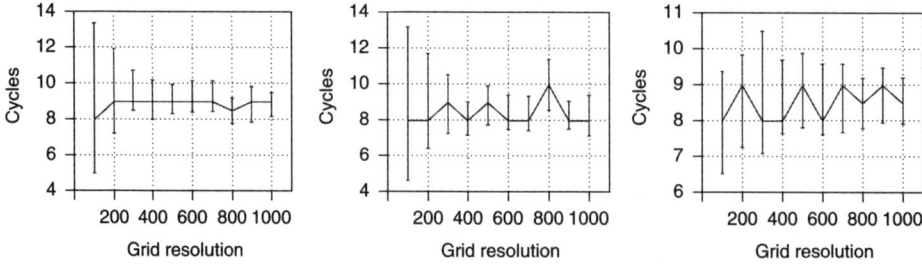

FIGURE 12.3 Numbers of cycles needed to achieve the threshold for circles digitized at resolutions 100^3 to 1000^3: Init I (left), Init II (middle), and Init III (right).

12.1 3D Arcs

examples of such curves are very simple to generate. However, two questions remained unanswered in [149]: Is the time complexity of the algorithm always $\mathcal{O}(n)$? Does the resulting polygon always converge to the MLP?

12.1.2 Estimation of curvature and torsion

Many of the methods of estimating the curvature of a digitized planar curve can be adapted to curvature estimation for digitized curves in $\mathbb{R}^3$. Curvature can then be used to estimate torsion; see Section 8.2.

One way to estimate the curvature and torsion of a 26-curve $\rho = \langle p_0, \ldots, p_n \rangle$ is as follows: the symmetric maximum-length DSS $p_{i-k}p_{i+k}$ (e.g., in the direction from p_{i-k} to p_{i+k}, where the subscripts are added modulo $n+1$) centered at p_i is an approximation $\mathbf{t}_i$ to the tangent vector at p_i. The changes in direction between $\mathbf{t}_i$ and its predecessor $\mathbf{t}_{i-1}$ and successor $\mathbf{t}_{i+1}$ are approximations to $\dot{\mathbf{t}}_i$, and the magnitude of $\dot{\mathbf{t}}_i$ is an estimate of the curvature κ_i of ρ at p_i. $\mathbf{n}_i = \dot{\mathbf{t}}_i / \kappa_i$ approximates the principal normal at p_i, and $\mathbf{b}_i = \mathbf{t}_i \times \mathbf{n}_i$ approximates the binormal. The magnitude of $\dot{\mathbf{b}}_i$ approximates the torsion τ_i of ρ at ρ_i.

Such approximations can also be obtained using two maximum-length DSSs that begin and end at p_i; see algorithm **HK2003** in Section 10.4.2 for curvature estimation of digitized planar curves. Estimates of the derivative $\dot{t}_i$ or $\dot{b}_i$ can be based on weighted sums of local differences.

As an example, consider the circular helix $\gamma(t) = (a\cos t, a\sin t, bt)$ shown in Figure 12.4. Its curvature and torsion are as follows:

$$\kappa(t) = \frac{a}{a^2 + b^2} \text{ and } \tau(t) = \frac{\pm b}{a^2 + b^2}.$$

The torsion is positive if the helix is turning counterclockwise (as seen from above) and negative if it is turning clockwise. Values estimated by the methods described above can be compared with these true values.

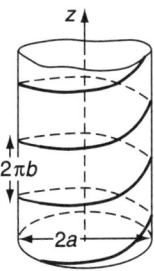

FIGURE 12.4 A circular helix of diameter $2a$ and vertical spacing $2\pi b$.

12.2 Surface Area Estimation

In this section, we consider digital surfaces defined by sets of voxels in an adjacency grid or by frontier faces of sets of voxels in an incidence grid.

12.2.1 Local methods

Let $\vartheta_h(S)$ be the frontier of a solid S. We can use local methods such as marching cubes to approximate $\vartheta_h(S)$ by a set of polygonal faces. The sum of the areas of these faces is an estimate of the surface area of S.

A marching cubes algorithm produces a set of triangles that form an isosurface of S. Let $p_i = (x_i, y_i, z_i)$ where $1 \leq i \leq 3$. Then the triangle $p_1 p_2 p_3$ has the following area,

$$\tfrac{1}{2} |\vec{p_1 p_2} \times \vec{p_1 p_3}| = \tfrac{1}{2} \left((a_2 b_3 - a_3 b_2)^2 + (a_3 b_1 - a_1 b_3)^2 + (a_1 b_2 - a_2 b_1)^2 \right)^{1/2}$$

where $(a_1, a_2, a_3) = \vec{p_1 p_2} = (x_2 - x_1, y_2 - y_1, z_2 - z_1)$ and $(b_1, b_2, b_3) = \vec{p_1 p_3} = (x_3 - x_1, y_3 - y_1, z_3 - z_1)$. Figure 12.5 shows a triangle for which $p_1 = (0, 0, 0.5)$, $p_2 = (0.5, 0, 1)$, and $p_3 = (0, 0.5, 1)$, so $\mathbf{a} = (0.5, 0, 0.5)$, $\mathbf{b} = (0, 0.5, 0.5)$, $\mathbf{a} \times \mathbf{b} = (-0.25, -0.25, 0.25)$, and the area of the triangle is $\frac{1}{8}\sqrt{3}$. The sum of the areas of the triangles is a surface area estimate. Evidently, this estimate depends on the threshold T used to compare the voxel values and on how the vertices of the triangles are chosen (at edge midpoints or based on differences between T and the voxel values; see Section 8.4.2).

Let $E_{\mathrm{MCU}}(S)$ be the sum of the areas of the triangles generated by the marching cubes algorithm that uses the 23-case look-up table of [1126], where the vertices of the triangles are at the midpoints of the grid edges. Because local length estimators such as E_{chm} and E_{coc} (see Chapter 10) do not provide multigrid convergence to the correct length, it is not surprising that $E_{\mathrm{MCU}}(S)$ also fails to provide multigrid convergence to the correct surface area.

The accuracy of surface area (and curvature) estimates can be improved by replacing the linear approximation used in marching cubes algorithms with quadratic or cubic approximations of $2 \times 2 \times 2$ voxel configurations. The size of the voxel

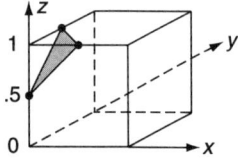

FIGURE 12.5 Example of a triangle in a polyhedrization produced by a marching cubes algorithm.

12.2 Surface Area Estimation

configurations can also be expanded to $m \times m \times m$ ($m \geq 2$). Note that the radius of the ball of influence of the marching cubes algorithm is $R_0 = \frac{m}{2}\sqrt{3}$.

Surface area estimates can be compared with the surface area $\mathcal{A}(S)$ of the isothetic polyhedral surface S. For grid constant 1, this is simply the number of frontier faces in S.

It is of theoretic interest to study worstcase deviations in the surface area estimates of digitizations of simple solids Θ. We calculate the Gauss digitization $G_h(\Theta)$ for different rotational orientations of Θ in a 3D grid of resolution $h > 1$. Let S_h be the digital surface of Θ in one of these orientations. Figure 12.6 shows, at the top, the behavior of surface area $\mathcal{A}(S_h)$ when Θ is a cube and, at the bottom, the behavior of the estimated values $E_{\mathrm{MCU}}(S_h)$ when Θ is a cube, sphere, or cylinder. The figure shows "obvious" convergence (as $h \to \infty$) in all cases. The (estimated) surface area values depend on the rotational orientation of the cube. The deviation d is 0 for a isothetic cube; there are only minor deviations due to the corners of the cube. For

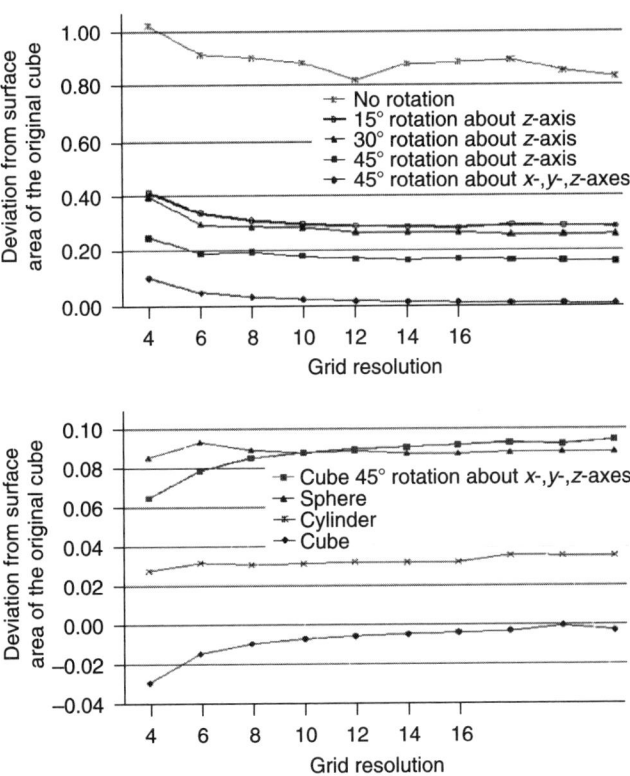

FIGURE 12.6 Top: "obvious" convergence of estimated surface area $\mathcal{A}(S_h)$ for a cube Θ in different rotational orientations; the convergence is toward the value $(1+d)s$, where s is the surface area of the original cube and the scale is up to 100%. Bottom: surface area estimates $E_{\mathrm{MCU}}(S_h)$; here the scale is only up to 10%.

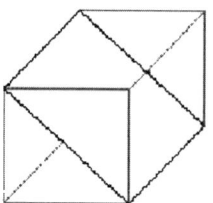 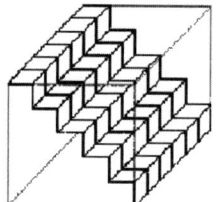 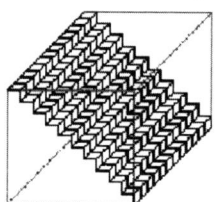

FIGURE 12.7 The digitization of a rectangular surface patch with slope $(45°, 45°, 0°)$ is a regular staircase for any grid resolution h. The surface area of this staircase is constant and is different from that of the original patch.

a cube that makes angles of $45°$ with the x-, y-, and z-axes, d is as great as 85% for $\mathcal{A}(S_n)$ and is about 9.5% for E_{MCU}. The sphere also produces a deviation of about 9%.

The "staircase" example in Figure 12.7 illustrates the "repeated self-similarity" of the digital surface of a regular solid. The isosurface produced by a marching cubes algorithm for a digitized sphere shows similar patterns of repeated self-similarity.

Differences between isothetic (after digitization) and original surface areas are analyzed in [643]. For a planar surface patch, the estimated surface area may be as much as $\sqrt{3}$ times the true value; its deviation from the true value does not depend on the grid resolution h but only on the position and orientation of the patch. If the patch has normal $(1,1,1)$ and passes through a grid point, its estimated surface area is $\sqrt{3} \cdot s$. The value 0.85 reported above for the digitized cube that makes angles of $45°$ with the x-, y-, and z-axes would decrease toward $\sqrt{3} - 1 = 0.7320508$ for larger values of h.

These results hold for any local polyhedrization technique; the deviation between the estimated surface area and the true value depends on position and orientation and not on grid resolution h. The deviation is lower if the number of normal vectors to the surface is greater; for example, it is lower for the triangular patches in an isosurface produced by a marching cubes algorithm as compared with the case of an isothetic cube [550], which has only six normals.

12.2.2 RCH methods

Let $J_h^-(\Theta)$ and $J_h^+(\Theta)$ be the inner and outer Jordan digitizations of a solid Θ in a grid of resolution h, and suppose $J_h^-(\Theta)$ is simply 6-connected. The relative convex hull (RCH) of $J_h^-(\Theta)$ with respect to $J_h^+(\Theta)$ is a polyhedron $\Pi_h^{\text{RCH}}(\Theta)$ that is contained in $J_h^+(\Theta)$ and contains $J_h^-(\Theta)$ (see Figure 12.8). The surface area of this polyhedron is an estimate of the surface area of Θ.

Theorem 12.1 The surface area of $\Pi_h^{\text{RCH}}(\Theta)$ is a multigrid-convergent estimator for the surface area of a solid Θ that has a smooth surface $\vartheta\Theta$.

12.2 Surface Area Estimation

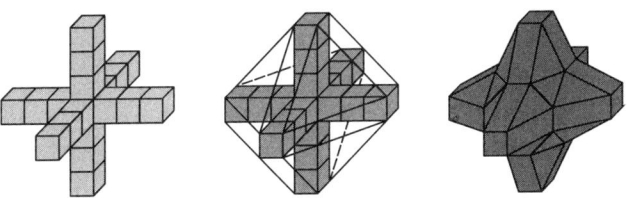

FIGURE 12.8 Left: inner polyhedron. Middle: convex hull. Right: convex hull relative to the outer polyhedron [1005].

There is no uniquely defined polyhedron of minimum surface area that is contained in $J_h^+(\Theta)$ and contains $J_h^-(\Theta)$. The introduction of additional vertices that are not at grid vertex positions allows the surface area be reduced (see Exercise 2 Section 12.4). The MLP and the RCH are the same in 2D, but the minimum area surface and the RCH are not the same in 3D.

Inner and outer polyhedral surfaces for a given digital surface S can be defined in different ways. (i) We can let S be the frontier of the inner polyhedron, and we can obtain the frontier of the outer polyhedron by Minkowski addition of a cube $[-1, +1]^3$. (ii) We can obtain the frontiers of the inner and outer polyhedra by Minkowski subtraction and addition of a cube $[-0.5, +0.5]^3$. Method (i) was used in the experiments reported subsequently.

The efficient computation of RCHs for isothetic grid polyhedra is still an open problem.

The RCH is the same as the convex hull if the inner and outer polyhedra are the inner and outer Jordan digitizations of a convex set Θ, such as an ellipsoid (see Section 8.3.3). An ellipsoid is defined by a triple a, b, c of radii and a rotational orientation. The Gauss digitization of the ellipsoid at resolution $h > 1$, where the unit of length is the largest radius of the ellipsoid, defines a digital surface S_h. Figure 12.9 shows examples of the convex hull $C_e(S_h)$.

The surface area of an approximating convex polyhedron defines an estimator E_{CH}. The planar patches of this polyhedron are not limited in size; hence,

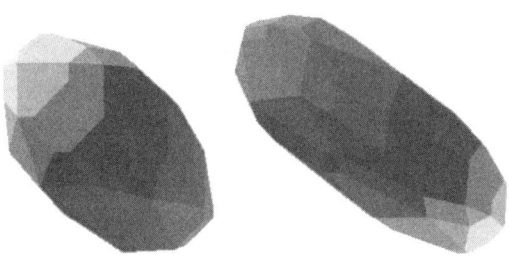

FIGURE 12.9 Examples of the convex hull of a rotated and Gauss-digitized ellipsoid.

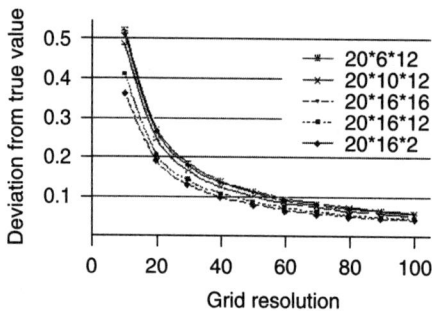

FIGURE 12.10 Relative errors in the surface area estimates E_{CH} for ellipsoids of different sizes rotated by $45°$ around the z- and y-axes.

this estimator is global. Figure 12.10 shows examples of relative deviations in E_{CH} that "obviously" go to zero.

Convex-hull–based surface area estimates E_{CH} of digitized ellipsoids have relative errors lower than 1% for grid resolutions close to $h = 50$, but marching-cubes–based estimates E_{MCU} have relative errors higher than 2% [550]. Figure 12.11 shows relative errors for four grid resolutions. The convergence to the true value is slightly slower for ellipsoids that have greater surface curvature (e.g., smaller minimum radius).

12.2.3 NOR methods

Surface area can be calculated by the integration of surface normals (see Equation 8.25) or surface gradients (see Equation 8.24). Normals and gradients can

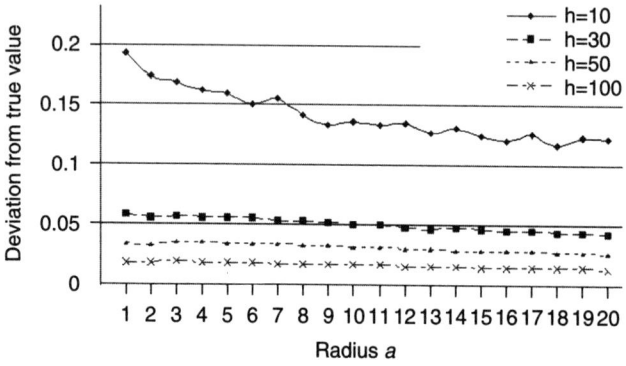

FIGURE 12.11 The relative errors in E_{CH} for ellipsoids of sizes $20 \times 20 \times a$ ($a = 1, 2, \ldots, 20$) digitized at grid resolutions $h = 10, 30, 50, 100$ show that the convergence is slightly faster if the ellipsoid is more spheric and thus has lower curvature.

12.2 Surface Area Estimation

be estimated for digital surfaces by calculating tangents or tangent planes. Note that a normal is orthogonal to a tangent plane and has one of two possible directions.

We generalize the method reported in Section 10.1.5 from length estimation to surface area estimation. Let $\mathbf{n}_0(f)$ be the unit normal to a frontier face f of a digital surface S, and let $\mathbf{n}(f)$ be an estimated normal for f. $\mathbf{n}(f)$ can be estimated using an approximating plane estimated by a global method (see Section 8.4.3). Care must be taken that $\mathbf{n}(f)$ always points "outward" relative to the interior of S. The discretization of Equation 8.25 leads to the following estimate:

$$E_{\text{NOR}}(S) = \sum_{f \in S} \langle \mathbf{n}_0(f), \mathbf{n}(f) \rangle \qquad (12.3)$$

Let Θ be a solid that has a smooth ($C^{(1)}$) surface $\vartheta\Theta$. Let dig_h be a digitization that maps Θ into a subset $\text{dig}_h(\Theta)$ of $\mathbb{C}^{(3)}_{3,h}$, and let Θ_h be the union of the 3-cells in $\text{dig}_h(\Theta)$. The vertices of Θ_h converge to $\vartheta\Theta$ as $h \to \infty$. We expect that the estimated normals to $\vartheta(\Theta_h)$ should also converge to normals of $\vartheta\Theta$, but this depends on how the normals are estimated and on the "microstructure" of the surface.

Theorem 12.2 Let Θ be a solid that has a smooth surface. The surface area estimates $E_{\text{NOR}}(\vartheta(\Theta_h))$ are multigrid convergent to the area $\mathcal{A}\vartheta(\Theta)$ iff the estimated normals on $\vartheta(\Theta_h)$ converge to the normals of $\vartheta(\Theta)$.

The following normal estimation algorithm [204, 331] ensures this convergence:

1. Calculate a 3D distance transform (see Section 3.4.2) for all voxels in Θ_h. ([204] suggests the use of a chamfer metric because of its simplicity; for chamfer metrics, see Section 3.2.3.)
2. Calculate a gradient map of the 3D distance map using differences in a $3 \times 3 \times 3$ neighborhood to approximate the spatial derivatives. In particular, calculate gradient vectors at all of the frontier vertices or faces of Θ_h.
3. For each frontier vertex or face, estimate a normal by taking the mean of selected gradient vectors in a spheric neighborhood in the frontier complex. Use the following selection criteria:
    (a) Gradients with directions that differ from that of the mean by more than an a priori threshold ("outliers") are excluded. This threshold defines the "smallest details" that are taken into account.
    (b) The spheric neighborhood must remain symmetric around the estimated normal.

Note that the selection criteria depend on the mean normal direction, so the algorithm is iterative. The initial value is the gradient at the voxel. The selective averaging process is repeated for a fixed number of iterations or until the results at successive iterations differ by less than a threshold.

This estimation procedure converges to the normal field on $\vartheta(\Theta)$ if Θ has a smooth surface, the radius of the spheric neighborhood is $\mathcal{O}(\sqrt{h})$, and the iterative averaging process converges [204]. Figure 12.12 shows experimental comparisons between E_{NOR}, E_{CH}, and E_{MCU} for digitized spheres and right circular cylinders (height = 2· radius +1). The estimates are also compared with theoretic results of normal vector integration for the digitized spheres.

Calculating gradients at all voxel positions and performing iterations in (large) spheric neighborhoods is not computationally efficient. As an alternative, an algorithm for estimating tangents to digital curves [1096] can be used to estimate tangents

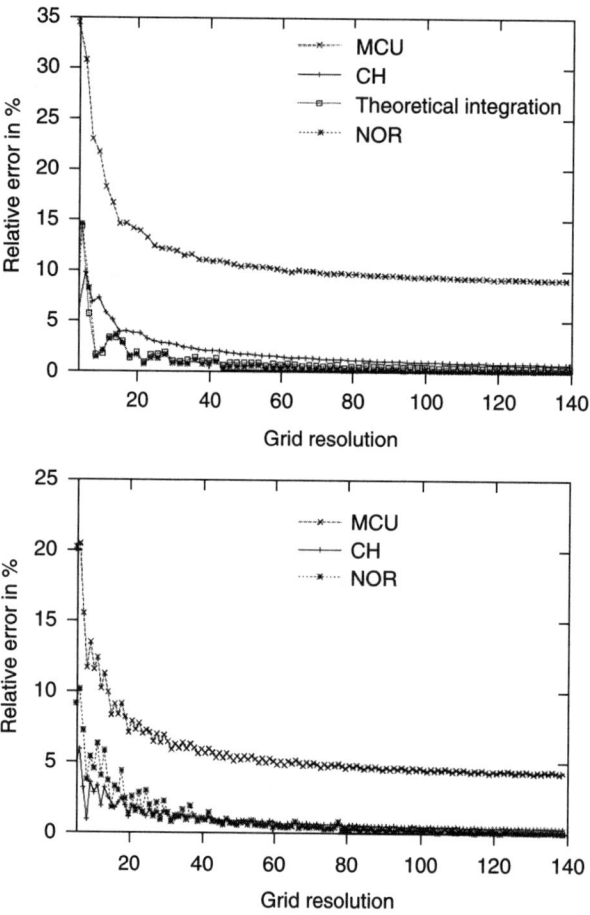

FIGURE 12.12 Comparisons of E_{MCU}, E_{CH}, and E_{NOR} for a sphere (above, also including a comparison with theoretic results of normal vector integration) and a right circular cylinder (below) [204].

12.2 Surface Area Estimation

to curves that are cross-sections (e.g., in the x-, y-, and $x+y$-directions) of a surface. These tangents can also be used to estimate surface curvature (see Section 12.3).

12.2.4 DPS methods

A DPS-based surface area estimator is based on a partition of the given digital surface S into a finite set of DPSs. We assume that the surface is defined by frontier faces in the incidence grid. Two DPSs in the partition cannot share a 2-cell but can share 0- or 1-cells on their frontiers. The union of the 2-cells in the DPSs is the set of all 2-cells in S.

A DPS is a finite 1-connected set of 2-cells in $\mathbb{C}_3$ that have a union G that is contained between or on two parallel planes Γ_1 and Γ_2; see Figure 11.18. A projection of a DPS G in the normal direction to Γ_1 and Γ_2 is a polygonal region of Γ_1. The area a_G of this region can be calculated using Equation 8.15. The sum of the a_Gs defines an estimate $E_{\text{DPS}}(S)$ of $\mathcal{A}(S)$. The experiments reported in this section are based on an implementation of algorithm **KS2001** (see Section 11.4.1), which defines the estimator E_{DPS}.

Figure 12.13 compares the true surface area of an ellipsoid to its DPS-estimated surface area for three rotational orientations. The search depth was restricted to 10. For these ellipsoids, the E_{CH} and E_{MCU} estimates have relative errors of 3.22% and 10.80% for $h = 100$, whereas the relative error of the E_{DPS} estimate is less than 0.8%. All three estimates show a good tendency to converge as $h \to \infty$.

Unlike E_{CH}, E_{DPS} is applicable to nonconvex objects. Interestingly, the relative error in the estimated surface area is smaller for nonconvex objects. Figure 12.14 shows (on the left) half of an ellipsoid that has an inner tangential ellipsoidal hole. E_{DPS} surface area estimates for this nonconvex solid in three rotational orientations

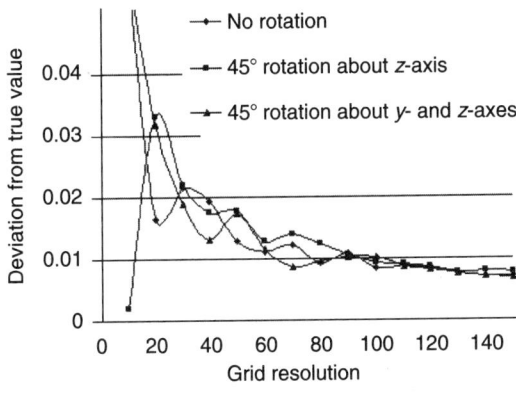

FIGURE 12.13 Relative errors in E_{DPS} surface area estimates for an ellipsoid in three rotational orientations and for grid resolutions from $h = 10$ to $h = 150$.

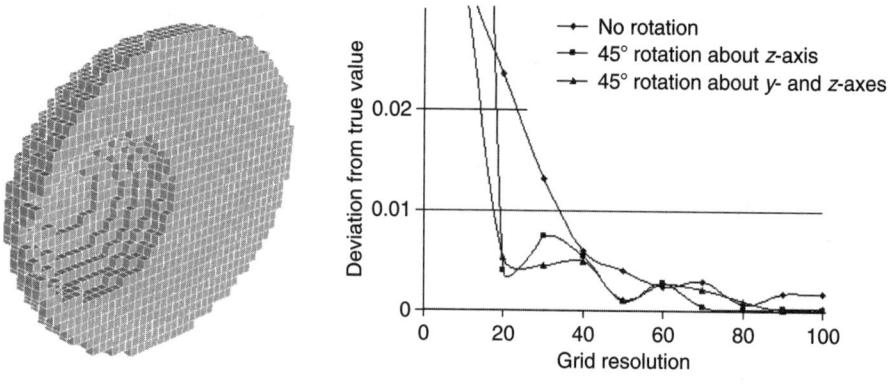

FIGURE 12.14 Left: A nonconvex solid. Right: relative errors of E_{DPS} surface area estimates for this solid in three rotational orientations.

are shown on the right. (In this experiment, the DPSs were not required to be simply connected.)

Because the DPS estimator involves several processes (search, possibly with restricted depth, and selection of first and subsequent "seed vertices"), it would be difficult to prove theorems about its convergence behavior.

12.3 Surface Curvature Estimation

Curvature estimation is often based on polyhedrization. A local method might estimate curvature using differences in the orientations of frontier faces that are incident with a vertex p of S. A global method might search S to find a maximum-size planar polygon (in a plane Γ_p) that approximates S "around" p; changes in the orientation of Γ_p between adjacent vertices of S then provide estimates of curvature. Alternatively, it might integrate the (oriented) difference between Γ_p and S; see Figure 8.4 for the 2D case.

Equation 8.25 also provides a method of estimating surface area based on surface normals. Thus, estimating planes Γ_p around vertices p of a surface allows us to estimate both curvature and surface area; see Section 12.2.3.

Estimation of the curvature of S at a point p is less computationally complex if it is based on estimates of the curvature along a few curves that pass through p and are contained in S; this allows us to apply the curvature estimation methods for curves that were discussed in Chapter 10. The values of these curvatures can be used to estimate the coefficients of the Hessian (see the end of Section 8.3.5); the eigenvalues of the Hessian provide estimates of the mean, principal, and Gaussian curvature of S.

12.3 Surface Curvature Estimation

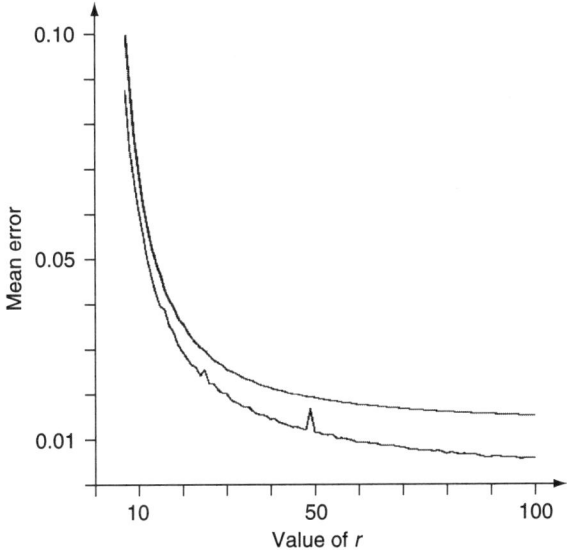

FIGURE 12.15 Curvature estimation errors for circles (lower curve) and spheres (upper curve) [434]. Sliding means were not used (e.g., the upper curve shows mean errors in the mean curvature estimates for all of the frontier faces of the sphere).

Curvature estimates of S at p in a few directions (e.g., the positive x- and y-directions) can also be used to calculate a mean value; this is not theoretically justified (see the discussion in Section 8.3.5), but it provides a simple and efficient method of detecting points of high curvature. We illustrate this approach by using algorithm **HK2003** (see Section 10.4.2) for surface curvature estimation [434].

Algorithm HK2003a

This algorithm allows us to estimate curvature along (planar) 8-curves. We apply it to 8-paths of frontier vertices obtained by "slicing" the digital spheres by $(x = i)$-, $(y = j)$-, or $(z = k)$-planes $(i, j, k \in \mathbb{Z})$. Each frontier face is cut by exactly two of these planes; this gives us two 8-paths that pass through the face. Suppose the $(x = i)$-plane passes through the center point p of the face. The resulting tangent $\mathbf{t}_x$ at p is $(0, \cos\theta, \sin\theta)$, where θ is the mean angle θ_i defined in algorithm **HK2003**. Analogously, a cut by a $(y = j)$-plane results in $\mathbf{t}_y = (\cos\theta, 0, \sin\theta)$, and a cut by a $(z = k)$-plane results in $\mathbf{t}_z = (\cos\theta, \sin\theta, 0)$.

Without loss of generality, assume that cuts by $(x = i)$- and $(y = j)$-planes were used, which leads to curvature estimates κ_x and κ_y. A $90°$ rotation of the tangent vectors $\mathbf{t}_x$ and $\mathbf{t}_y$ (e.g., in the positive direction, because below we are interested in angles between normals) produces the (estimated) principal normals of the two 8-paths at p.

For example, $\mathbf{t}_y = (\cos\theta, 0, \sin\theta)$ produces $\mathbf{n}_y = (-\sin\theta, 0, \cos\theta)$. The cross product of $\mathbf{t}_x$ and $\mathbf{t}_y$ is an estimated unit normal $\mathbf{n}_p$ of the estimated tangent plane Π_p. We can then compute the normal curvatures κ_{η_1} and κ_{η_2} of the two 8-paths ($\eta = 0$ corresponds to one of the principal curvatures λ_1 or λ_2 at p; the angles η_1 and η_2 remain unknown in general) using Meusnier's theorem (see Section 8.3.5):

$$\kappa_{\eta_1} = \kappa_x \cos(\mathbf{n}_p, \mathbf{n}_x) \text{ and } \kappa_{\eta_2} = \kappa_y \cos(\mathbf{n}_p, \mathbf{n}_y)$$

The mean of κ_{η_1} and κ_{η_2} is used as an estimate of the mean curvature and based on the assumption that η_1 and η_2 differ by about $90°$.

Figure 12.15 compares curvature estimation errors for planar curves (applying algorithm **HK2003** to circles $x^2 + y^2 \leq r^2 + r$) with curvature estimation errors (using the method described here) for spheres $x^2 + y^2 + z^2 \leq r^2 + r$ ($r = 1, \ldots, 100$). The digital circles and spheres were obtained by Gauss digitization of disks and balls. A sliding mean was not used; the curves show exact values. The error decreases faster in the 2D case. The 8-paths used in the 3D case were digitizations of circles of radius $\leq \sqrt{r^2 + r}$. Smaller circles yield larger errors in the 2D case.

Curvature measurement can also make use of a statistic approach that was invented by physicists [72, 147]. To estimate the mean curvature at a point p of the surface of a solid, measure the volume of the intersection of the solid with a small sphere centered about p. There is a relation between that volume and an approximation to the mean curvature. The approach [99, 800] also works in 2D for estimating the curvature of the frontier of a planar set.

12.4 Exercises

1. Perform a comparative accuracy analysis (estimated length vs true length as a function of grid resolution) for the two programs for the global estimators implemented in Exercise 3 in Section 11.5.

2. What is the outer Jordan digitization of $\gamma_h(t) = (t, h^2 \cos(t/h))$ in $\mathbb{Z}_h^2$, where $t \in \mathbb{R}$ and $h > 0$?

3. The line through the points (x_1, y_1, z_1) and (x_2, y_2, z_2) has the following equations:

$$\frac{x - x_1}{x_2 - x_1} = \frac{y - y_1}{y_2 - y_1} = \frac{z - z_1}{z_2 - z_1}$$

What are the projections of this line into the ($x = 0$)-, ($y = 0$)-, and ($z = 0$)-planes? (Use the standard form $z = a_x y + b_x$, $z = a_y x + b_y$, and $y = a_z x + b_z$.) Design a random line generator based on the random generation of (some of) the parameters $a_x, \ldots, b_z$.

12.4 Exercises

4. Give an example of a solid Θ (in some rotational orientation) for which the estimation of surface area by adding the areas of the frontier faces of $G_n(\Theta)$ results in values that converge to $\sqrt{3} \cdot \mathcal{A}(S)$ as $h \to \infty$.

5. The figure below shows the inner Jordan digitization I of a solid. Let its outer Jordan digitization O be identical to the Minkowski sum of I and $[-1,+1]^3$ (one layer shown by dashed cubes). Find a minimum-area polyhedral surface that is contained in $O \setminus I°$ and that has exactly two vertices on each of the bold edges of the "tower": one on the top (a grid vertex) and one somewhere below on the edge. (Hint: Two vertices are shown on one of the bold edges.)

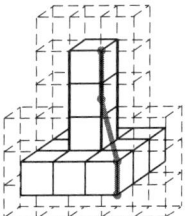

6. Let $C(S)$ be the convex hull of S, and let $C_B(S)$ be the B-convex hull of S, where $S \subseteq B \subset \mathbb{R}^3$ (see Definition 1.4; replace $\mathbb{R}^2$ with $\mathbb{R}^3$). Prove that (i) $S \subseteq C_B(S) \subseteq B \cap C(S)$; (ii) $S = C_B(S)$ iff S is B-convex; and (iii) $C_B(S) = C(S)$ iff $C(S) \subseteq B$.

7. Prove that, if B and S are simple polyhedra and $S \subseteq B \subset \mathbb{R}^3$, then $C_B(S)$ is a simply connected polyhedron.

8. Suppose we approximate the normal at each convex corner of a digital surface by the mean of the three vectors of the tripod. Show that this method does not result in convergence of the normals (see Theorem 12.2).

9. Replace the breadth-first strategy in the incremental DPS recognition algorithm with a depth-first strategy. Discuss how this affects the "shapes" of the resulting DPSs when the input surface is a digital sphere.

10. Perform curvature estimation experiments (generalizing those illustrated in Figure 12.15) for ellipsoids with varying radii a, b, and c, and include "very flat" ellipsoids. (Use algorithm **HK2003a** or another 3D curvature estimation algorithm of your choice.)

12.5 Commented Bibliography

Length estimation for simple digital curves in 3D adjacency grids based on weighted local moves is also treated in [16, 481, 522]. The weights were optimized using the BLUE approach, as discussed in Chapter 10 for the 2D case.

[203] proved a multigrid convergence theorem for the estimator E_{dss} specified by algorithm **DR1995**. For a generalization of the global estimator E_{mpo} (see Equation 10.3) to lines in $\mathbb{R}^3$, see [177].

The problem of defining multigrid convergent methods of surface area measurement has been studied for more than 100 years; see, for example, [695]. For combinatorial surfaces, see [339, 849].

The areas of the isosurfaces generated by local polyhedrization techniques (see Section 8.6) have behavior similar to those in the marching cubes case shown at the bottom of Figure 12.6. The use of local tilings for surface area estimation is studied in [55]. Global polyhedrization techniques (see Section 8.6) are potential candidates for multigrid convergent surface area estimators. Theorem 12.1 is from the unpublished manuscript [1004]. [1154] describes approximative calculation of MLPs: the frontier of a 6-region is "sliced" in the x- or y-direction, and a triangulation of the surface is created by connecting vertices of adjacent 2D MLPs of the resulting curves.

Normal-based surface area calculation is discussed in [294, 330]. Discrete integration of vector fields (see Section 12.2.3) was proposed in [641, 643], thereby generalizing the tangent method of arc length estimation [305] to surface area estimation. For multigrid convergence results for this method (including Theorem 12.2), see [204]; for an earlier version of the normal estimator, see [331]. For a review of normal estimation, see [1144]; for weighted averaging of normals of frontier faces, see [185, 797]. Normals of "2D slices" of 3D surfaces were used in [643, 1048].

Curvature estimation is discussed in [145, 330] for digital surfaces, and in [56, 713, 714, 1091] for curved 2D manifolds. For curvature estimation in the frontier grid (incidence grid), see [642]. Algorithm **HK2003a** is from [434]. Exercise 2 is an example from [203].

CHAPTER 13

Hulls and Diagrams

The convex hull of a set can be regarded as a "domain of influence" of the set. Similarly, the Voronoi diagram of a set of points defines "domains of influence" of the points. This chapter discusses hulls and diagrams in Euclidean space or in a grid, with emphasis on 2D. We discuss definitions of digital convexity and digital Voronoi diagrams and give algorithms based on adjacency grid models. We also discuss "domains of influence" in pictures.

13.1 Hulls

Let **S** be a class of subsets of a set S. A function H that takes sets in **S** into sets in **S** is called a *hull* operator iff it has the following properties:

H1: $M \subseteq H(M)$ for all $M \in$ **S**.

H2: $M_1 \subseteq M_2$ implies $H(M_1) \subseteq H(M_2)$ for all $M_1, M_2 \in$ **S**.

H3: $H(H(M)) \subseteq H(M)$ for all $M \in$ **S**.

H1 and **H3** imply that $H(H(M)) = H(M)$ (i.e., H is an idempotent operator). A hull operator is also called a *closure operator* in algebra. The identity operator $I(M) = M$ and the topologic closure are examples of hull operators. In the following examples, S is Euclidean space $\mathbb{R}^n$.

1) Let **S** $= \wp(S)$ be the class of all subsets of S. $M \subseteq S$ is called *convex* if, for any distinct $p, q \in M$, the straight line segment pq is contained in M. It is easy to see that any intersection of convex subsets of S is convex. The intersection $C(M)$ of all of the convex subsets of S that contain M is called the *convex hull* of M. It is not hard to show that C is a hull operator.

2) Let **S** be the class of all bounded subsets of S. For any $M \in$ **S**, let $D_e(M)$ be a disk of smallest radius (defined by metric d_e) that contains M. Such a disk is not necessarily uniquely defined, but the smallest radius is uniquely defined, because the radius is a continuous function on the compact set defined by the

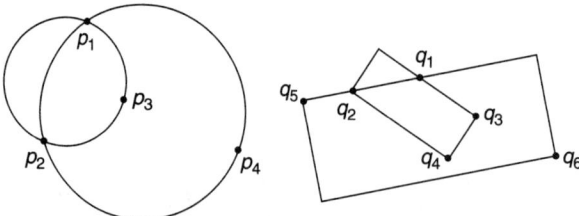

FIGURE 13.1 Left: $D_e(\{p_1,p_2,p_3\})$ is not contained in $D_e(\{p_1,\ldots,p_4\})$. Right: $R(\{q_1,\ldots,q_4\})$ is not contained in $R(\{q_1,\ldots,q_6\})$.

closed convex hull of the set. It is not hard to see that D_e satisfies **H1** and **H3** but not **H2** (see Figure 13.1). Such an operator is called a *pseudohull*. Similar remarks apply to the operator R for which $R(M)$ is the rectangle of smallest area that contains M. (Note that $R(M)$ may also not be unique, for example, if M is a disk; to make it unique, we can also require that one of its sides make the smallest possible angle with the positive x-axis. As in the previous argument for the radius, the smallest area is also uniquely defined.) It can be shown that these operators also satisfy the following:

H4: $M_1 \subseteq M_2$ implies $\mathcal{A}(H(M_1)) \leq \mathcal{A}(H(M_2))$ for all sets $M_1, M_2 \in \mathbf{S}$ where $\mathcal{A}(S)$ is the area of S.

Note that this axiom also implies that $H(M)$ is measurable for $M \in \mathbf{S}$. An operator that satisfies **H1**, **H3**, and **H4** is called a *near-hull*. **H2** implies **H4** for any family $\mathbf{S}$ of sets S such that $H(S)$ is always measurable.

3) Let $\mathbf{S}$ be the class of all finite subsets of S that contain at least two points. For any $M \in S$ and any $p \in M$, let $D_e^M(p)$ be a disk of smallest radius centered at p that contains another point q of M. Let $E_e(M) = \bigcup_{p \in M} D_e^M(p)$. It is not hard to show that E_e is a near-hull. Figure 13.2 (left) shows (in gray) the E_e-hull of a set of grid points (the filled dots). Figure 13.2 (right) shows the E_4-hull of the same set of points defined using d_4 "disks" (diamonds centered at the points).

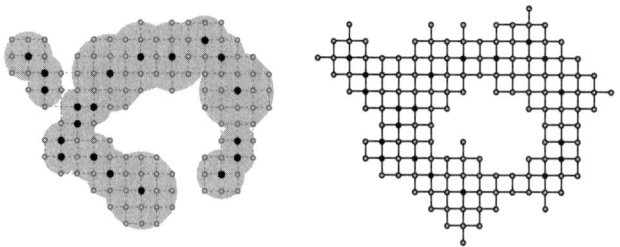

FIGURE 13.2 Left: the E_e-hull of the set of dark grid points using Euclidean disks. Right: the E_4-hull of the same set of grid points using city block "disks."

13.1 Hulls

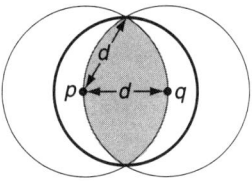

FIGURE 13.3 Reduction of diameter from d to $d\sqrt{3}/2$.

H.W.E. Jung [485] analyzed balls $B_e(M)$ with smallest radius in $\mathbb{R}^n$ (i.e., the generalization of disks of smallest radius to n dimensions [$n \geq 2$]). He showed that, if $M \subset \mathbb{R}^n$ is finite and has diameter d, the radius of $B_e(M)$ is upper bounded by the following:

$$d\sqrt{\frac{n}{2(n+1)}}$$

An example for $n = 2$ is shown in Figure 13.3. A disk of radius d centered at any of the points of M must contain M. Let $p, q \in M$ be two points such that $d_e(p,q) = d$. Draw two disks of radius d centered at p and q. Because M lies in each disk, it lies in their intersection, which is shaded in the figure. The bold circle contains the shaded region and has radius $d\sqrt{3/2} \approx 0.866 \cdot d$. Jung's theorem shows that this bound can be reduced to $d\sqrt{3}/3 \approx 0.577 \cdot d$. Jung's upper bound is the best possible; there are finite sets $M \subset R^2$ for which $D_e(M)$ has exactly this radius. The radius of $D_e(M)$ also has the trivial lower bound $d/2$.

13.1.1 Convex hull computation in the Euclidean plane

Let Π be a simple polygon in the Euclidean plane together with its interior, and let $\langle p_1, \ldots, p_n \rangle$ ($n \geq 3$) be the vertices of Π. It is not hard to show (see Section 1.2.9) that the convex hull of Π is a polygon and has $m \leq n$ vertices. In this section, we show how to compute the vertices $q_1, \ldots, q_m$ of $C(\Pi)$.

We use the determinant $D(p,q,r)$ (see Equation 8.14) to classify triples of successive vertices p,q,r of Π as negative if p,q,r is a right turn; zero if p, q, and r are collinear; and positive if p,q,r is a left turn. Figure 13.4 shows an example of a left turn for which $D(p,q,r) = 15$. In convex hull algorithms, we use the notation $D(p,q,r) = L$ and $D(p,q,r) = R$ for left and right turns.

We use a stack for the vertices of $C(\Pi)$. At a given step of a convex hull algorithm, the stack contains, for example, $q_0, q_1, \ldots, q_t$ so that its depth is $t+1$. The operation PUSH(p) means that t becomes $t+1$ and p becomes the last element of the stack. The operation POP means that t becomes $t-1$ by removal of q_t from the stack. $p \leftarrow p+1$ denotes a move to the next vertex of Π.

We assume that p_1 is known to be a vertex of $C(\Pi)$; for example, we can choose p_1 to be the uppermost of the leftmost vertices of Π. *Sklansky's algorithm* [999] for calculating $C(\Pi)$ is shown in Algorithm 13.1. This algorithm has computation time

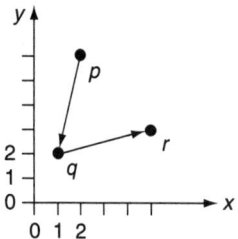

FIGURE 13.4 A right-handed xy-coordinate system and a left turn.

linear in the number of vertices of Π. It is correct provided that the frontier of Π is completely visible from the outside (i.e., through any point q on the frontier of Π there is a ray that intersects Π only at q). Step 4 of Graham's Scan (see Algorithm 1.1) also assumes that the simple polygon is completely visible from the outside.

Figure 13.5 shows an example of a simple polygon that is not completely visible from the outside. Sklansky's algorithm fails for this polygon. The algorithm starts at p_1 and proceeds correctly up to p_5 but cannot deal properly with p_6. (A *concavity* in Π is the closure of a connected component of $C(\Pi) \setminus \Pi$. It is not hard to see that a concavity of a polygon is a polygon. An edge of a concavity *closes* the concavity iff it is not an edge of Π. When the tracing of the frontier of Π goes from p_6 to p_7, it crosses $p_2 p_5$, which is an edge that closes a concavity in Π.)

One way to correct the behavior of Sklansky's algorithm is to continue tracing the frontier (doing nothing to the stack) until $p_2 p_5$ is crossed again, and then resume with the algorithm until another concavity is found. An algorithm that uses this strategy is shown in Algorithm 13.2. Here we use the Boolean variable $flag$ to indicate whether the vertex tracing is outside a concavity ($flag = 0$) or inside a concavity ($flag = 1$). $p^\star$ denotes the vertex immediately preceding p on Π. This algorithm has linear time complexity.

The situation is more complex if the input is not an ordered sequence of vertices of a simple polygon but rather a set of n points in the plane. In this situation, the worst-case time complexity of any convex hull algorithm is $\Omega(n \log n)$. In fact, there exist optimal algorithms that have a worst-case complexity of $\mathcal{O}(n \log n)$; Graham's Scan (Section 1.2.9) is such an algorithm.

1. Let q_0 and q_1 be the first two vertices of Π, and let $t := 1$. Let p be the next vertex (p_3) of Π.
2. If $p = q_0$, stop.
3. As long as $t > 0$ and $D(q_{t-1}, q_t, p) \neq R$, POP.
4. PUSH(p), $p \leftarrow p+1$, and go to Step 2.

ALGORITHM 13.1 Sklansky's algorithm for calculating the convex hull of a "visible" simple polygon.

13.1 Hulls

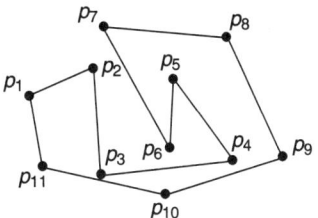

FIGURE 13.5 A simple polygon with a frontier that is not completely visible from the outside.

Evidently, the convex hull of a set of n points is a polygon that can have as many as n vertices. It can be shown [847] that the expected number of vertices of the convex hull of a set of n points is $\frac{2r}{3}\log_2 n + \mathcal{O}(r)$ if the n points are independently and uniformly distributed in a bounded convex r-gon; it is $\mathcal{O}(n^{1/3})$ if they are independently and uniformly distributed in a bounded, simply connected set that has a smooth frontier; and it is $\mathcal{O}(\sqrt{\log n})$ if their distribution is Gaussian.

13.1.2 Convex hull computation in the (2D) grid

Knowing that the vertices of a simple polygon have integer coordinates (i.e., that it is a grid polygon) results in no asymptotic time benefit for convex hull computation, but methods of convex hull computation exist that are applicable only to grid polygons. For example, the leftmost (downward) and rightmost (upward) 1s in each row of Figure 13.6 can be used as an input sequence for Sklansky's algorithm; they form a

---

1. Let q_0 and q_1 be the first two vertices of Π, and let $t := 1$. Let p be the next vertex of Π. Set *flag* to 0.
2. If $p = q_0$ or $p = q_1$, POP as long as $t > 0$ and $D(q_{t-1}, q_t, p) \neq R$, and stop; otherwise, go to Step 3.
3. If *flag* = 0,
   a) As long as $t > 0$ and $D(q_{t-1}, q_t, p) \neq R$, POP.
   b) If $t = 0$ or $D(p^\star, q_t, p) = R$, PUSH(p); otherwise, set *flag* to 1 and go to Step 5. // The vertex sequence encounters a concavity.//
4. Otherwise,
   a) $p \leftarrow p+1$ until $D(q_{t-1}, q_t, p) = L$.
   b) Set *flag* to 0 and POP until $t = 0$ or $D(q_{t-1}, q_t, p) = R$.
   c) PUSH(p).
5. $p \leftarrow p+1$, and go to Step 2.

ALGORITHM 13.2 A convex hull algorithm for arbitrary simple polygons.

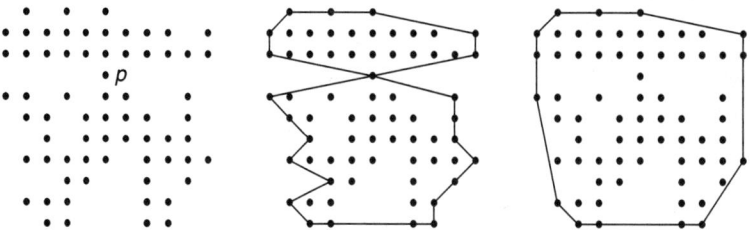

FIGURE 13.6 An 8-connected set in which we first identify extreme pixels in each row (pixel p is extreme in both directions) and then apply Sklansky's convex hull algorithm.

polygon that is completely visible from the outside. Note that this polygon may not be simple (see pixel p).

This algorithm can be applied even if the set of 1s is 8-disconnected. Let x_{max}, x_{min}, y_{max}, and y_{min} be (estimates of) the greatest and least coordinates of all the 1s. (Exact values of these quantities can be calculated in time linear in the number of 4-border pixels of a region by adding four counters to the border tracing algorithm; see Algorithm 4.3.) We have to scan $m = y_{max} - y_{min} + 1$ rows of pixels each of length up to $n = x_{max} - x_{min} + 1$ to identify the leftmost and rightmost 1s in each row; this takes $\mathcal{O}(mn)$ time. We then apply Sklansky's algorithm to the sequence of leftmost (downward) and rightmost (upward) 1s; this takes $\mathcal{O}(\max\{m, n\})$ time, which is less than $\mathcal{O}(mn)$, so the asymptotic upper time bound is $\mathcal{O}(mn)$. On the other hand, if we take all of the 1s in the rectangle as unsorted points and use an optimum-time convex hull algorithm (e.g., Graham's Scan), we obtain an asymptotic upper time bound of $\mathcal{O}(mn \log mn)$, because there may be up to mn 1s in the rectangle. This shows that the optimal-time bound for unrestricted input may not be optimal when the input consists only of grid points.

A convex hull that is a grid polygon and that is contained in the grid $\mathbb{G}_{m+1,m+1}$ can have only a limited number of vertices. Conversely, let $e(m)$ be the maximum number of grid vertices. Let $m = s(n)$ be the minimal side length of a square with vertices that are grid points and that contains a convex grid polygon that has n vertices. It can be shown that the following is true:

$$s(n) = m \text{ implies } e(m) \geq n \text{ and } e(m) = n \text{ implies } s(n) \leq m \qquad (13.1)$$

Figure 13.7 shows that $e(7) = 13$, $e(8) = 14$, and $e(9) = 16$; $s(13) = 7$, $s(14) = 8$, $s(15) = 9$, and $s(16) = 9$. These examples also show that, in convex hulls that have maximal number of vertices, the edges tend to be defined by pairs (a, b) of relatively prime integers, where a and b are the increments in the x and y directions. For example, for $n = 8$, we have the edges $(0, 1)$, $(1, 1)$, $(2, 1)$, $(1, 0)$, $(2, -1)$, $(1, -2)$, $(0, -1)$, $(-1, -2)$, $(-1, -1)$, $(-1, 0)$, $(-2, 1)$, $(-1, 1)$, and $(-1, 2)$, which constitute a traversal around the convex hull.

13.1 Hulls

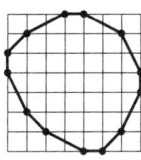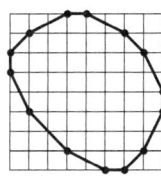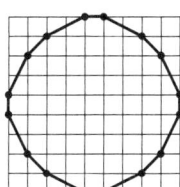

FIGURE 13.7 Three examples of maximal numbers of vertices on a convex hull [1110]. Left: $m = 7$. Middle: $m = 8$. Right: $m = 9$.

In the following discussion, we use *Euler's totient function* ϕ; $\phi(t)$ is the number of natural numbers $r \leq t$ that are relatively prime to t. For example, $\phi(10) = 4$ ($r = 1, 3, 7, 9$). Note that $\phi(1) = 1$.

We now describe a method of constructing convex hulls that have at most t vertices. Sort all of the pairs (a, b) of relatively prime integers such that $a + b \leq t$ in decreasing order of the slope of the line segment defined by (a, b) starting at $(0, 1)$ and ending at $(t-1, 1)$. There are $\sum_{j=1}^{t} \phi(j)$ pairs (a, b) in this sequence. The polygonal arc defined by this sequence is a quadrant of the convex hull. The example $m = 9$ in Figure 13.7 illustrates this construction for $t = 3$. The resulting convex grid polygon has $n_t = e(m_t) = 4 \cdot \sum_{j=1}^{t} \phi(j)$ vertices and is contained in a square with vertices that are grid points and that has side length $m_t = s(n_t) = \sum_{j=1}^{t} j \cdot \phi(j)$. Based on this discussion, we have the following:

Proposition 13.1 There exist strictly monotonically increasing sequences of natural numbers $\{m_t\}$ and $\{n_t\}$ ($t = 1, 2, 3, \ldots$) such that the following is true:

$$m_t = s(n_t) = \sum_{j=1}^{t} j \cdot \phi(j) \text{ and } n_t = e(m_t) = 4 \cdot \sum_{j=1}^{t} \phi(j)$$

In number theory, it is known [991] that, for sufficiently large t, $e(m_t)/t^2 \approx 12/\pi^2 \approx 1 \cdot 2159$, $s(n_t)/t^3 \approx 2/\pi^2 = 0 \cdot 2026$, and $s(n_t)/e(m_t)^{3/2} \approx \pi/12 \cdot 3^{1/2} \approx 0 \cdot 1511$. Table 13.1 illustrates the speed of convergence of these approximations. Because s and e are monotonically increasing functions (see Exercise 2 in Section 13.4), we can conclude that $s(n) \in \mathcal{O}(n^{3/2})$ and $e(m) \in \mathcal{O}(m^{2/3})$. More precisely, it can be shown that *the following is true:*

Theorem 13.1 $e(m) = \frac{12}{(2\pi)^{2/3}} m^{2/3} + \mathcal{O}(m^{1/3} \log m)$ and $s(n) = \frac{2\pi}{12^{3/2}} n^{3/2} + \mathcal{O}(n \log n)$.

Table 13.1 shows that e and s intersect between $t = 37$ and $t = 48$.

The exponent $2/3$ in the estimate of $e(m)$ also occurs in the estimation of the maximal number of grid points on a convex Jordan curve. I.V. Jarnik showed in 1925 [475] that a convex Jordan curve of length l is incident with at most

TABLE 13.1 Values of $s(n_t)$, $e(m_t)$, and related quotients [1110].

t	$s(n_t)$	$e(m_t)$	$\dfrac{e(m_t)}{t^2}$	$\dfrac{s(n_t)}{t^3}$	$\dfrac{s(n_t)}{e(m_t)^{3/2}}$
1	1	4	4.0000	1.0000	0.1250
2	3	8	2.0000	0.3750	0.1326
3	9	16	1.7778	0.3333	0.1406
4	17	24	1.5000	0.2656	0.1446
5	37	40	1.6000	0.2960	0.1463
6	49	48	1.3333	0.2269	0.1473
7	91	72	1.4694	0.2653	0.1490
8	123	88	1.3750	0.2402	0.1490
9	177	112	1.3827	0.2428	0.1493
10	217	128	1.2800	0.2170	0.1498
20	1745	512	1.2800	0.2181	0.1506
30	5601	1112	1.2356	0.2074	0.1510
40	13,111	1960	1.2250	0.2049	0.1511
50	26,021	3096	1.2384	0.2082	0.1511
60	44,231	4408	1.2344	0.2048	0.1511
70	69,821	5976	1.2196	0.2036	0.1511
80	105,389	7864	1.2287	0.2058	0.1511
90	149,301	9920	1.2247	0.2048	0.1511
100	203,085	12,176	1.2176	0.2031	0.1512
110	273,901	14,864	1.2284	0.2058	0.1511
120	351,177	17,544	1.2183	0.2032	0.1511
130	447,393	20,616	1.2199	0.2036	0.1511
140	561,941	24,000	1.2245	0.2048	0.1511
150	686,733	27,432	1.2192	0.2035	0.1511
160	833,935	31,224	1.2107	0.2036	0.1511
170	1,003,303	35,320	1.2221	0.2042	0.1511

$$\frac{3}{(2\pi)^{1/3}} l^{2/3} + \mathcal{O}(l^{1/3})$$

grid points and that the first term is the best possible upper bound (i.e., there exist curves that achieve this upper bound).

13.1.3 Near-hull computation in the Euclidean plane

In this section, we discuss algorithms for computing the rectangle $R(M)$ with the smallest possible area that contains a finite set of points or a simple polygon M in the Euclidean plane. It can be shown [347] that the following is true:

13.1 Hulls

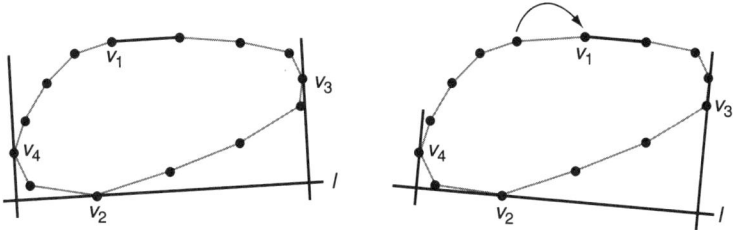

FIGURE 13.8 Vertex v_1 moves to the next position and triggers a move of pointer v_2; v_3 and v_4 need not move.

Proposition 13.2 *A side of $R(M)$ is incident with an edge of $C(M)$.*

On the basis of this, we can compute $R(M)$ as follows. Let v_1 be the first vertex of $C(M)$. Examine the vertices in sequence until the first vertex v_2 (not on the first edge) is found such that a line l through v_2 parallel to the first edge does not intersect the interior of $C(M)$; see Figure 13.8 (left). Choose vertices v_3 between v_1 and v_2 and v_4 between v_2 and v_1 that define a circumscribing rectangle of M. After this initialization step, move v_1 to successive vertices of $C(M)$ and move pointers v_2, v_3, and v_4 (if necessary) so that the four vertices continue to define a circumscribing rectangle. Figure 13.8 (right) illustrates a move of v_1. Compare the areas of the resulting rectangles, and choose the smallest one.

Let n be the number of vertices of $C(M)$. The initialization step takes $\mathcal{O}(n)$ computation time; each of the vertices $v_1, \ldots, v_4$ moves into at most n positions, and each move takes constant time. Hence the algorithm is linear in the number of vertices on the convex hull (if the initial calculation of $C(M)$ is not taken into account).

Let **S** be the class of subsets of $\mathbb{E}^2$ with frontiers that are Jordan curves so that they are homeomorphic to the unit disk. It can be shown that any $M \in \mathbf{S}$ has a circumscribing square (i.e., a square with all four edges intersecting M in at least one point). Let $Q(M)$ be the circumscribing square that has smallest area; we can make $Q(M)$ unique by minimizing the angle that one of its sides makes with the positive x-axis. As was the case in Section 13.1, $Q(M)$ defines a near-hull.

Proposition 13.3 *Any plane Jordan curve γ has a circumscribing square (i.e., a square that contains γ, with all four sides having a nonempty intersection with γ).*

Proof Let l be a line that does not intersect γ and l' a line parallel to l (e.g., with slope θ) such that γ is in the stripe between l and l'. Move both lines toward γ by parallel translation until they intersect γ for the first time. The resulting lines are *lines of support* of γ. Similarly, construct two other lines of support

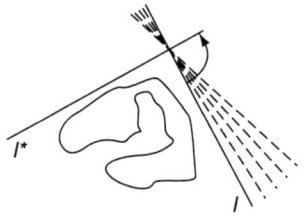

that are orthogonal to l. These four lines of support define a *circumscribing rectangle* of γ. Let the side lengths of this rectangle be $a(l)$ and $b(l)$. If we perform this construction for lines l with slopes between θ and $\theta + \pi/2$, we obtain circumscribing rectangles with side lengths that change from $(a(l), b(l))$ to $(b(l), a(l))$ so that, during the rotation, $a(l) - b(l)$ changes sign. Evidently, $a(l) - b(l)$ is a continuous function of θ.[1] Hence $a(l) - b(l)$ must be zero for some slope between θ and $\theta + \pi/2$ so that the circumscribing rectangle for this slope is a square. ∎

13.2 2D Digital Convexity

Digitally convex sets were defined in Section 2.3.4 with respect to a digitization model. This section assumes the 2D grid point model (i.e., digitizations are subsets of $\mathbb{Z}^2$). A *digitization by dilation* is an outer σ-digitization using domains of influence $\Pi_\sigma(q)$, where Π_σ contains the origin. For example, Gauss, outer Jordan, and grid-intersection digitization are digitizations by dilation. For any digitization by dilation, a finite $S \subset \mathbb{Z}^2$ is digitally convex iff S is the digitization of its Euclidean convex hull $C(S)$.

The Gauss digitization $G(S)$ of a convex set $S \subseteq \mathbb{R}^2$ may not be 8-connected (see Exercise 1 in Section 2.5), but its cross digitization (defined in Section 2.3.5 by generalizing grid intersection digitization to arbitrary planar sets) must be simply 8-connected (see Exercise 4 in Section 2.5). In this section, we use cross digitization in the 8-adjacency grid. A set $\emptyset \neq S \subseteq \mathbb{Z}^2$ will be called *digitally convex* (with respect to cross digitization) iff there exists a convex set $M \subseteq \mathbb{R}^2$ such that $S = \text{dig}_{\text{cross}}(M)$.

13.2.1 Digital convex hulls

Evidently, any digital straight line (DSL) or straight line segment (DSS) is digitally convex. Moreover, a finite irreducible 8-arc is digitally convex iff it is an (8-)DSS.

---

1. Let $f : \mathbb{R} \to \mathbb{R}$ be continuous on $[a,b]$ ($a < b$); let $x_1, x_2 \in [a,b]$ be such that $f(x_1) > 0$ and $f(x_2) < 0$. Then $f(x_0) = 0$ for some $x_0 \in [a,b]$.

13.2 2D Digital Convexity

Theorem 13.2 A finite set $M \subseteq \mathbb{Z}^2$ is digitally convex iff any one of the following is true:

1. M satisfies the chord property.
2. For all $p, q \in M$, at least one DSS that has p and q as endpixels is contained in M.
3. For all $p, q, r \in M$, all of the grid points in the (closed) triangle pqr are in M.
4. Any grid point on the (real) line segment between two grid points of M is also in M.
5. Let $[M]$ be the union of the grid cells that have centers in M. For any two grid points p and q in M, let M_{pq} be the region surrounded by the frontier of $[M]$ and the real line segment pq; then any grid point in M_{pq} is in M.

There exist infinite sets $M \subseteq \mathbb{Z}^2$ that satisfy the chord property but that are not digitally convex. For example, let H be a halfplane defined by a straight line γ with irrational slope, and let M be the cross digitization of M. If $p \in \overline{M}$ is 4-adjacent to two 4-border pixels of M, $M \cup \{p\}$ still satisfies the chord property.

The Euclidean convex hull $C(M)$ of any $M \subseteq \mathbb{Z}^2$ is a convex polygon with vertices that are all 4-border pixels of M. If M is simply 8-connected, 4-border tracing visits all of the 4-border pixels of M in sequence; see the border tracing algorithm in Section 4.3.4. The vertices of $C(M)$ partition this sequence into finite or (at most two) infinite subsequences that start and end at successive vertices of $C(M)$. These subsequences will be called *border segments* of M.

Theorem 13.3 An 8-connected $M \subseteq \mathbb{Z}^2$ is digitally convex iff $G(C(M)) \subseteq M$. If $M \neq \emptyset$ and $M \neq \mathbb{Z}^2$, M is digitally convex iff every border segment of M is a DSS.

This leads to a convexity test for finite sets M of grid points. Calculate $C(M)$, and test whether each border segment of M is a DSS.

S is called a *digitally convex completion* of M iff S contains M and is digitally convex, and $S \setminus U$ is not digitally convex for any $U \subseteq S \setminus M$. For example, let $M = \{p, q\}$; then any 8-DSS that contains p and q is an 8-connected digitally convex completion of M. $G(C(M))$ is a digitally convex completion of M and is 8-connected if M is 8-connected. There may also exist grid points p that are not in $G(C(M))$ but such that $G(C(M)) \cup \{p\}$ is digitally convex.

Proposition 13.4 If $M \subseteq \mathbb{Z}^2$ is 8-connected, it has a unique digitally convex completion.

Definition 13.1 The digitally convex completion of an 8-connected $M \subseteq \mathbb{Z}^2$ is called the *digital 8-convex hull* of M and is denoted by $C_8(M)$.

Proposition 13.5 Let $M \subseteq \mathbb{Z}^2$ be 8-connected, and let $S \subseteq \mathbb{Z}^2$ be digitally convex. The following statements are equivalent:

1. $S = C_8(M)$.
2. $M \subseteq S$, and, if $M \subseteq U$ where $U \subseteq \mathbb{Z}^2$ is digitally convex, then $S \subseteq U$.
3. $S = \bigcap \{U : M \subseteq U \subseteq \mathbb{Z}^2 \wedge U \text{ is digitally convex}\}$.

Using Proposition 13.5 and Theorem 13.3, we can prove the following:

Proposition 13.6 If $M \subseteq \mathbb{Z}^2$ is 8-connected, $C_8(M) = G(C(M))$.

It follows that the border of $C_8(M)$ can be calculated in $\mathcal{O}(n)$ time where n is the cardinality of the border of M.

13.2.2 Row and column convexity

$S \subseteq \mathbb{Z}^2$ is called *row-convex (column-convex)* iff each row (column) of the grid contains at most one run of pixels of S. Let $M \subseteq \mathbb{E}^2$ be convex, and let $G(M)$ be nonempty and 4-connected. It is not hard to see that $G(M)$ is row- and column-convex. It can be shown [530] that the number $b(n)$ of row-convex (or column-convex) n-ominoes satisfies the following:

$$\lim_{n \to \infty} b(n)^{1/n} = a \text{ where } 3 \cdot 20 < a < 3 \cdot 21 \tag{13.2}$$

(See Equation 2.4 for the number of all n-ominoes.) It can further be shown [69] that the number $c(n)$ of row-column–convex n-ominoes satisfies the following:

$$\lim_{n \to \infty} c(n) = b \cdot a^n \text{ where } b = 2 \cdot 6756 \ldots \text{ and } a = 2 \cdot 3091 \ldots \tag{13.3}$$

A *staircase* is a 4-arc that has x- and y-coordinates that are monotonically nonincreasing or nondecreasing. The 8-border of a row-column–convex 4-connected set can be partitioned into at most four staircases; see Figure 13.9 for an example. We see from Figure 13.9 (left) that such a set is not necessarily digitally convex.

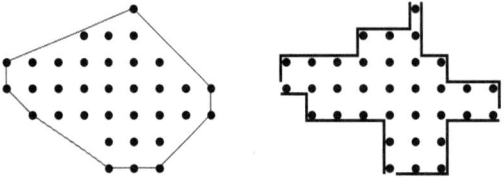

FIGURE 13.9 Left: a row-column–convex 4-connected set. Right: the 8-border of the set can be partitioned into four staircases.

13.2.3 Fuzzy digital convexity

Let the values of the pixels in a picture P be in the range [0,1] so that P defines a fuzzy subset μ of the grid (see Section 1.2.10). P is called μ-*convex* iff, for all pixels p, q, there exists a DSS with endpixels p and q such that $\mu(t) \geq \min(\mu(p), \mu(q))$ for all pixels on the DSS.

P is called μ-(8-)connected (see Section 13.3.3) iff, for all pixels p, q, there exists an (8-)path ρ from p to q such that $\mu(t) \geq \min(\mu(p), \mu(q))$ for all pixels on ρ. Thus μ-convexity can be regarded as a special case of μ-connectedness in which ρ is required to be a DSS.

In Proposition 13.11, we will see that P is μ-connected iff all of its level sets are connected. Similarly, by Theorem 13.2, we have the following:

Proposition 13.7 P is μ-convex iff all of its level sets are digitally convex.

13.3 Diagrams

Diagrams are defined in metric spaces for countable sets S of "simple" geometric objects such as points, line segments, polygons, polyhedra, and so on. One type of diagram divides the metric space into cells such that each element of S is contained in exactly one cell. Another type connects some pairs of elements of S by line segments.

Let $S = \{p_1, \ldots, p_n\}$ be a set of points in the Euclidean plane $\mathbb{E}^2$. The *Voronoi cell*[2] of $p_i \in S$ ($i = 1, \ldots, n$) is the closure of its *zone of influence*, which is the set of all points in E^2 that are closer (with respect to metric d_e) to p_i than to any other point of S:

$$V_e(p_i) = \{q : q \in \mathbb{E}^2 \land d_e(q, p_i) \leq d_e(q, p_j) \text{ for } j = 1, \ldots, n\} \quad (13.4)$$

Euclidean distance was used in this definition [1101], but Voronoi cells can be defined in any metric space. (The closure of the zone of influence of p_i is not necessarily identical to $V_e(p_i)$ if d_e is replaced by another metric on $\mathbb{R}^2$, such as a chamfer metric.)

Definition 13.2 The *Voronoi diagram* of S is the union of the frontiers of the Voronoi cells $V_e(p_1), \ldots, V_e(p_n)$.

The Voronoi diagram is called the *skeleton by influence zones* (SKIZ) in mathematic morphology. p_i and p_k are called *Voronoi neighbors* iff $V_e(p_i)$ and $V_e(p_k)$ share an edge (i.e., have more than a single point in common). Distinct Voronoi

---

2. This is named after the Ukrainian mathematician G.F. Voronoi (1868–1908). An n-dimensional generalization was studied by the German mathematician J.P.G.L. Dirichlet (1805–1859) [273]. The U.S. officer A.H. Thiessen also used such polygonal cells in the discussion of meteorologic data; see, for example, [1050]. Voronoi cells are therefore also called *Thiessen polygons* and are the 2D case of *Dirichlet polyhedra* or *Dirichlet cells*.

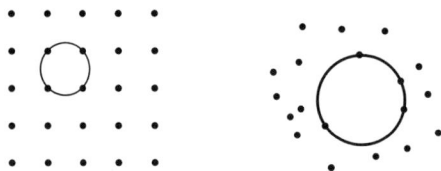

FIGURE 13.10 Two sets of points that are not circle-free.

neighbors are called *Voronoi adjacent*. The *dual Voronoi diagram* is obtained by joining the pairs of Voronoi adjacent points with straight line segments. It can be shown that the bounded faces in a dual Voronoi diagram are all triangles iff S is *circle-free*: whenever more than three points of S lie on a circle, another point of S lies inside of the circle. (Figure 13.10 shows two examples of sets that are not circle-free.) If S is circle-free, its dual Voronoi diagram is called its *Delaunay triangulation* [259][3]; Figure 13.11 shows an example. A dual Voronoi diagram that has edges added to complete a triangulation if S is not circle-free is called a *Delaunay diagram*.

Voronoi and Delaunay diagrams are well-known examples of diagrams in $\mathbb{E}^2$. The Voronoi diagram has a straightforward generalization to $\mathbb{E}^n$. The set S can be a family of pairwise disjoint simply connected bounded sets rather than a set of points.

Let S be a finite set of points in $\mathbb{E}^n$. A *nearest neighbor diagram* of S is defined by the set of (undirected) line segments pq such that p is a nearest neighbor of q in S or vice versa. (In such a diagram, each $p \in S$ is joined by a line segment to only one of its nearest neighbors.) A *minimum spanning tree* diagram is a tree with nodes that are the points of S and with edges that have minimum total length. Note that these diagrams are not necessarily unique; the same is true for Delaunay diagrams of non–circle-free sets.

Proposition 13.8 Any nearest neighbor diagram of $S \subset \mathbb{E}^2$ is a subdiagram of a minimum spanning tree diagram of S, which is in turn a subdiagram of a Delaunay diagram of S.

13.3.1 Diagram computation in the Euclidean plane

Let $S = \{p_1, \ldots, p_n\}$ be a set of points in the Euclidean plane. The Voronoi cells of the points of S are convex. The Voronoi cell $V_e(p)$ is bounded iff p is in the interior of the convex hull $C(S)$. The frontier of an unbounded Voronoi cell consists of at most two straight lines or rays and finitely many straight line segments. A bounded Voronoi cell is a simple polygon. The line segments, rays, or lines on the frontier of a

---

3. This is named after the Russian mathematician B.N. Delaunay (also called Delone; 1890–1980).

13.3 Diagrams

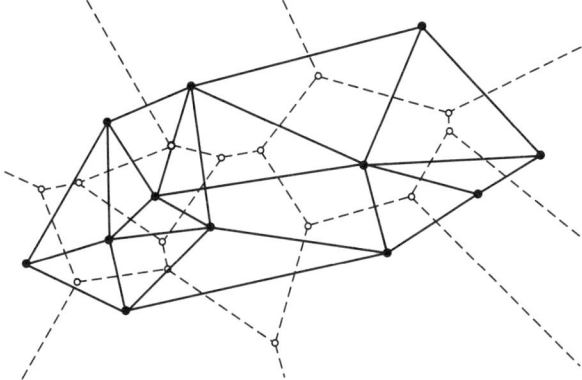

FIGURE 13.11 A Voronoi diagram (dashed) and the corresponding Delaunay triangulation (bold).

Voronoi cell are contained in the perpendicular bisectors of pairs of points $p, q \in S$; see Figure 13.12.

The construction of Voronoi diagrams and Delaunay triangulations are dual processes; either diagram can be derived from the other. We will describe a simple Delaunay diagram construction algorithm that has asymptotic time complexity $\mathcal{O}(n^2)$ and a small asymptotic constant; it supports time-efficient constructions for (about) $n \leq 1000$. For larger sets of points, it is preferable to use optimized algorithms that run in $\mathcal{O}(n \log n)$ time. The worst-case time complexity of Delaunay diagram construction has lower bound $\Omega(n \log n)$. If the set of points is not circle-free, the algorithm chooses one of the possible diagrams.

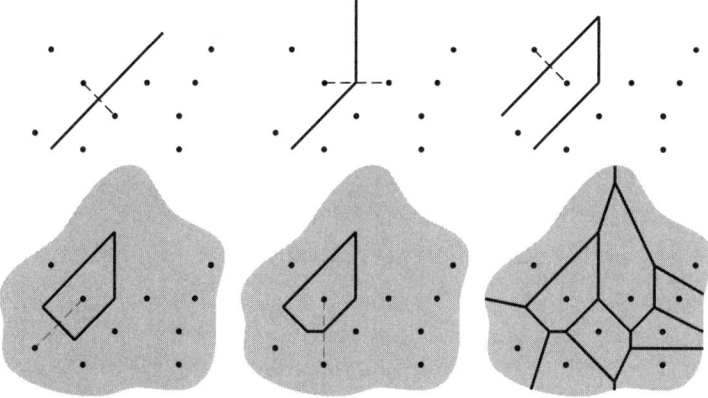

FIGURE 13.12 The Voronoi diagram is composed of (segments or rays of) perpendicular bisectors. Five "construction steps" for one of the cells are shown; the final Voronoi diagram is at the lower right.

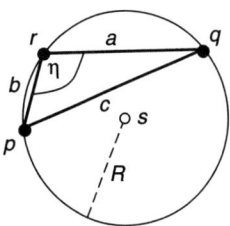

FIGURE 13.13 Points p, q, and r define a circle of radius R.

According to Proposition 13.8, a Delaunay diagram is a subdiagram of a nearest neighbor diagram. Given a finite set S of points, we pick a point p in S and search in S for one of the nearest neighbors q of p. We start the construction with pq, which is an edge of one of the Delaunay triangles.

In Figure 13.13, we have the following, where A is the area of triangle pqr:

$$\cos \eta = \frac{a^2 + b^2 - c^2}{2ab} \quad \text{and} \quad R = \frac{abc}{4A}$$

The points of S are all collinear, for example, on line l, iff $A = 0$ for all $r \in S \setminus \{p, q\}$; in this case, the Delaunay diagram connects the points along l in sorted order. From now on, we assume that the points of S are not all collinear.

Points p, q, and r of S form a Delaunay triangle if[4] the following is maximal for r among all points of $S \setminus \{p, q\}$:

$$R \cdot \cos \eta = \frac{c(a^2 + b^2 - c^2)}{8 \cdot A}$$

Because c is constant for a given p and q, we need only maximize $(a^2 + b^2 - c^2)/A$. We recall that $A = \frac{1}{2}|D(p,q,r)|$ where D is the determinant; see Equation 8.14. This gives us the following test for identifying a third point of a Delaunay triangle:

Let $r \in S \setminus \{p, q\}$ be the first point for which $D(p,q,r) \neq 0$. If $D(p,q,r) > 0$, we identify third points (throughout the procedure) by maximizing as follows:

$$K(p,q,r) := \frac{a^2 + b^2 - c^2}{D(p,q,r)}$$

If $D(p,q,r) < 0$, we identify third points by minimizing $K(p,q,r)$.

The sign of $D(p,q,r)$ also tells us whether the ordered triple (p,q,r) describes a left turn or a right turn. The oriented straight line segment from p to q defines two halfplanes; r is in the left (right) halfplane iff (p,q,r) is a left turn (right turn). pq can also be incident with a second Delaunay triangle that has a third point r' that is in the right (left) halfplane. This is why we put oriented line segments (vectors)

---

4. Not "iff," the line containing p and q defines two halfplanes, and there can be a Delaunay triangle in each of the halfplanes. The maximization can be performed for each halfplane.

13.3 Diagrams

---

1. Choose $p \in S$, and find a $q \in S$ that is a nearest neighbor of p. Choose a point $t \in S \setminus \{p,q\}$; depending on the sign of $D(p,q,t)$, we use maximization or minimization in this run of the algorithm (see text).
2. Find a third point $r \in S \setminus \{p,q\}$ that maximizes (minimizes) $K(p,q,r)$. Put the vectors $\vec{qp}$, $\vec{pr}$, and $\vec{rq}$ on a list.
3. As long as the list is not empty, go to Step 4; otherwise, stop.
4. Take $\vec{pq}$ off of the list. Set *flag* := 0, set pointer *ThirdPoint* := VOID, and set $K := -\infty$ (for maximization) or ∞ (for minimization).
5. For all $r \in S \setminus \{p,q\}$ (go to Step 6 when finished):
   a) If $K \leq K(p,q,r)$ for maximization or $K \geq K(p,q,r)$ for minimization, go to Step 5.c.
   b) Set *flag* := 1, $K := K(p,q,r)$, and *ThirdPoint* := r.
   c) Go back to Step 5 for the next point $r \in S \setminus \{p,q\}$.
6. If *flag* = 0, go to Step 3. Otherwise, let $r := ThirdPoint$.
   //A new Delaunay triangle (p,q,r) has been detected; take appropriate action.//
   a) If $\vec{pr}$ is on the list, delete it; otherwise, put $\vec{pr}$ on the list.
   b) If $\vec{rq}$ is on the list, delete it; otherwise, put $\vec{rq}$ on the list.
7. Go to Step 3.

ALGORITHM 13.3 Delaunay diagram algorithm.

rather than line segments on the list in the algorithm; if we arrive at the same vector a second time, we delete it from the list. The algorithm is given in Algorithm 13.3, and Figure 13.14 shows three examples of Delaunay triangulations.

13.3.2 Diagram computation in the digital plane

If S contains only grid points, only integer arithmetic is needed for Algorithm 13.3. Let $G(p,q,r) = a^2 + b^2 - c^2$. When we use maximization, the test $K(p,q,r_1) < K(p,q,r_2)$? can be replaced with $G(p,q,r_1) \cdot D(p,q,r_2) < G(p,q,r_2) \cdot D(p,q,r_1)$? at the cost of using two variables G and D instead of K.

Sets of grid points are often not circle-free; Figure 13.15 (left) shows an example. However, the Delaunay diagram algorithm (Algorithm 13.3) always produces a triangulation if the points in S are not all collinear. By keeping track of points with equal (maximal or minimal) values of K in Steps 5.a and 5.b, we can also detect points that are incident with more than three Voronoi cells.

In the grid, we are typically interested only in assigning grid points to Voronoi cells or Delaunay triangles. The Euclidean metric was used in Equation 13.4; the resulting Voronoi diagram is called the d_e-*Voronoi diagram*; see Figure 13.15, (right). If we use a grid metric d_α instead of d_e in Equation 13.4, we obtain a d_α-*Voronoi*

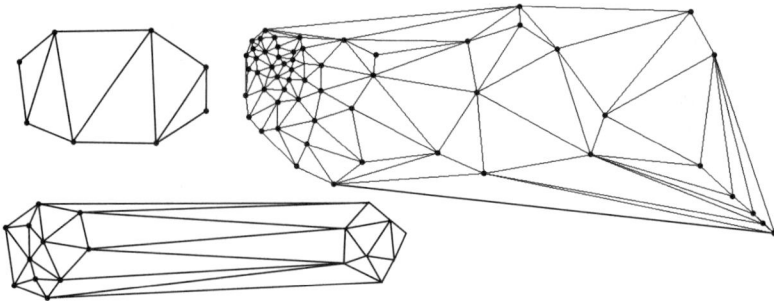

FIGURE 13.14 Three examples of Delaunay triangulations. The union of the Delaunay triangles is the convex hull of the set of points.

diagram. Figure 13.16 shows d_4- and d_8-Voronoi diagrams for the set of points used in Figure 13.15.

d_α-Voronoi diagrams can be constructed as follows. Initially, we give each point of S a different *Voronoi cell label*. In each run of the algorithm, we label all of the grid points q that have not yet been labeled but that are α-adjacent to already labeled grid points. If q is α-adjacent only to points that have the same label, give q that label; if q is adjacent to points that have different labels, q is a *Voronoi diagram point*. Continue this process for as long as unlabeled grid points remain.

For a chamfer metric, the Voronoi diagram can be computed using the Rosenfeld-Pfaltz two-pass distance transform algorithm.

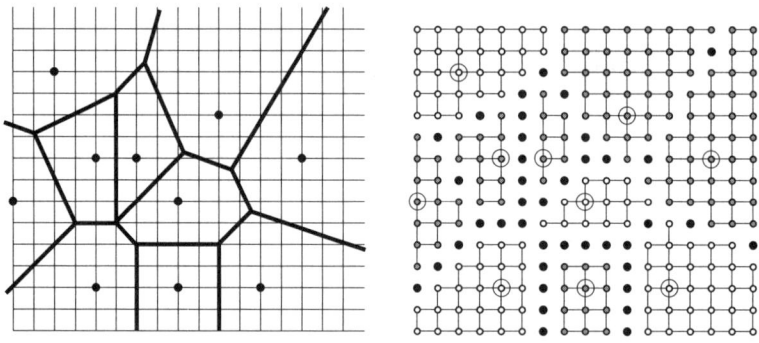

FIGURE 13.15 Left: Voronoi diagram for a set of grid points. The result of the Delaunay diagram algorithm depends on the order in which the points of S are visited, because S is not circle-free (see the Voronoi point that is incident with four Voronoi cells). Right: digital representation of the diagram; grid points on the frontiers of Voronoi cells are black.

13.3 Diagrams

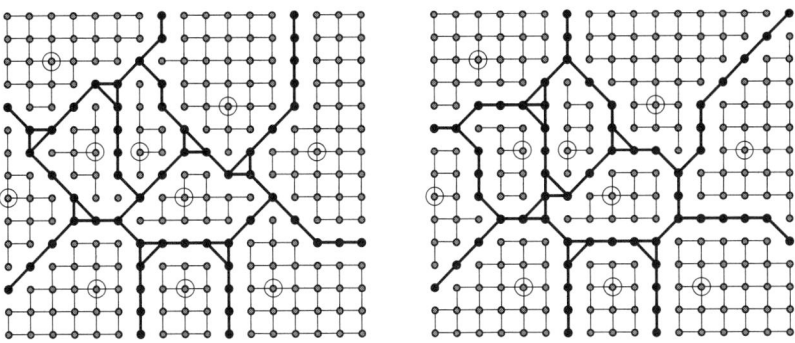

FIGURE 13.16 Left: d_4-Voronoi diagram. Right: d_8-Voronoi diagram.

13.3.3 Diagrams in pictures

Let A be an adjacency relation on a grid $\mathbb{G}_{m,n}$. An A-path is a sequence of pixels $\rho = (p_0, \ldots, p_h)$ such that p_i is A-adjacent to p_{i-1} $(0 < i \leq h)$. Let $\{S_1, \ldots, S_h\}$ be a set of disjoint subsets of $\mathbb{G}_{m,n}$. A-paths can be used to define "domains of influence" $D(S_i)$ of the S_is in various ways. For example, we can say that p belongs to $D(S_i)$ iff the length of a shortest path from p to S_i is less than the length of a shortest A-path from p to any other S_j (i.e., if p's "A-distance" to S_i is less than its A-distance to any other S_j). If we use this definition, $D(S_i)$ is the Voronoi cell of S_i with respect to A-distance. (We will usually drop the A in what follows.)

If P is a picture defined on $\mathbb{G}_{m,n}$, we can restrict the class of allowable paths or assign weights to paths using the values of the pixels on the paths. For example:

a) We can define the value-weighted length of a path ρ as the sum of the values of the pixels of ρ. The value-weighted Voronoi cell of S_i then consists of pixels that have a value-weighted distance to S_i (see Section 3.4) that is less than their value-weighted distance to any other S_j.

b) We can call ρ allowable iff its pixels all have values greater (or less) than some threshold t. This leads to a Voronoi cell concept based on intrinsic A-distance in the set of pixels that have values that are greater (or less) than t.

c) We can define the slope-weighted length of ρ as the sum of the differences (or absolute differences) of the values of successive pairs of pixels of ρ. (We can think of "steep" paths as taking longer to "climb.")

d) We can call ρ allowable if its pixel differences all have values greater (or less) than some threshold t. Of particular interest are *monotonic* paths in which the differences are all nonnegative or all nonpositive.

In the remainder of this section, we will consider only domains of influence defined by monotonic paths, and we will assume that $A = A_4$ or A_8.

A maximal (4- or 8-) connected set Π of pixels that has values in P that are constant will be called a *plateau*[5]; the constant value will be denoted by $P(\Pi)$. Evidently, every pixel p belongs to exactly one plateau, which we denote with $\Pi(p)$.

Two plateaus Π and Π' are called (4- or 8-) *adjacent* if some pixel of Π is adjacent to some pixel of Π'. Evidently, if Π and Π' are adjacent, we must have $P(\Pi) \neq P(\Pi')$. Π is called a *top* (*bottom*) if its value is higher (lower) than that of any plateau adjacent to it. A top is also called a *peak* or *summit* and a bottom is also called a *pit* or *sink*.

A path $\rho = (p_0, \ldots, p_n)$ is called *nondescending* if $P(p_i) \geq P(p_{i-1})$ and *nonascending* if $P(p_i) \leq P(p_{i-1})$ for all $1 \leq i \leq n$.

Proposition 13.9 For any pixel p, there exists a nondescending (nonascending) path $(p_0, \ldots, p_n)$ such that $p = p_0$, $q = p_n$, and q is a pixel of a top (bottom).

In the following paragraphs, we assume that the pixel values in P are in the range [0,1]. As we saw in Section 1.2.10, such a picture defines a fuzzy subset μ of $\mathbb{G}_{m,n}$.

We define the *strength* $\sigma(\rho)$ of a path $\rho = (p_0, \ldots, p_n)$ as $\min_{0 \leq i \leq n} \mu(p_i)$. We define the *connectedness* $c(p,q)$ of pixels p and q as $\max_\rho \sigma(\rho)$, where the max is taken over all paths $\rho = (p_0, \ldots, p_n)$ such that $p_0 = p$ and $p_n = q$. Evidently, $c(p,q) \leq \min(\mu(p), \mu(q))$. We say that p and q are μ-*connected* iff $c(p,q) = \min(\mu(p), \mu(q))$. We say that a set S of pixels is μ-connected iff any two pixels of S are μ-connected; in particular, we say that P is μ-connected iff any two pixels of $\mathbb{G}_{m,n}$ are μ-connected.

Proposition 13.10 Let p and q be two pixels of P, and let $\lambda(p,q) = \min(\mu(p), \mu(q))$. Let $\mu_{\lambda(p,q)}$ be the $\lambda(p,q)$-level set of P (see Section 1.2.10). Then p and q are μ-connected iff p and q are connected in $\mu_{\lambda(p,q)}$.

Proof Evidently, p and q are in $\mu_{\lambda(p,q)}$, and there is a path between them in that set iff there is a path $\rho = (p_0, \ldots, p_k)$ between them such that $\mu(p_i) \geq \lambda(p,q) = \min(\mu(p), \mu(q))$ for all p_i on ρ. ∎

Proposition 13.11 P is μ-connected iff μ_λ is connected for all $0 \leq \lambda \leq 1$.

Proof p and q are in μ_λ iff $\lambda \leq \min(\mu(p), \mu(q))$; hence every p and q are μ-connected iff every μ_λ is connected. ∎

Proposition 13.12 For any $0 \leq \lambda \leq 1$, any component C of μ_λ contains a top.

---

5. This terminology is suggested by regarding the picture as defining an isothetic surface in which the value of a pixel defines the height of the surface above the pixel. Plateaus are (4- or 8-) components of P-equivalence classes.

13.3 Diagrams

Proof According to Proposition 13.9, from any pixel of C, there is a nondescending path ρ to a top Π, and all of the pixels of ρ and Π must be in C. ∎

Proposition 13.13 If P has a unique top, it is μ-connected.

Proof If P were not μ-connected, according to Proposition 13.11, some μ_λ would not be connected. Thus μ_λ would have at least two components C and D. According to Proposition 13.12, each of these components would contain a top, and these tops could not be the same. ∎

We now prove the converse: if P is μ-connected, it must have a unique top. Suppose P had two tops Π and Π'. Let $\Pi_0, \ldots, \Pi_n$ be a sequence of plateaus such that $\Pi_0 = \Pi, \Pi_n = \Pi'$, and Π_i is adjacent to Π_{i-1} ($1 \leq i \leq n$). Choose such a sequence for which $\min(\mu(\Pi_i)) = v$ is as large as possible and the value v is taken on as few times as possible. Let $P(\Pi_j) = v$; then, evidently, $0 < j < n$ and $\mu(\Pi_{j-1}), \mu(\Pi_{j+1})$ are greater than v. Let Π_{j-1} and Π_{j+1} be in μ_w where $w > v$. Because Π is μ-connected, μ_w must be connected, according to Proposition 13.11. Hence the sequence of Πs can be diverted around Π_j through plateaus higher than v; the diverted sequence either has a minimum value higher than v or takes on the value v fewer times, which is a contradiction. We have thus proved the following:

Theorem 13.4 P is μ-connected iff it has a unique top.

Evidently, μ-connectedness is reflexive and symmetric but not transitive. For example, let P consist of three pixels p, q, and r such that $\mu(q) < \min(\mu(p), \mu(r))$; then p and q, and q and r are μ-connected, but p and r are not μ-connected. Thus, we cannot partition P into "μ-components." However, we will now show that the sets of pixels that are μ-connected to a given top have partition-like properties.

Let Π be a top, and let $[\Pi]$ be the set of pixels that are μ-connected to pixels of Π. If Π' is another top, $[\Pi]$ and $[\Pi']$ can overlap; indeed, in the three-pixel example in the preceding paragraph, $\{p\}$ and $\{r\}$ are tops, and q belongs to both $[\{p\}]$ and $[\{r\}]$. However, we will now see that $[\Pi]$ and Π' must be disjoint.

Proposition 13.14 Let $p \notin \Pi$ be μ-connected to a pixel of Π; then $\mu(p) < \mu(\Pi)$.

Proof Suppose $\mu(p) \geq \mu(\Pi)$. If p were μ-connected to a pixel q of Π, there would be a path $\rho = (p_0, \ldots, p_h)$ such that $p_0 = p, p_n = q$, and $\mu(p_i) \geq \min(\mu(p), \mu(q))$ for all $0 \leq i \leq h$. Because $\mu(p) \geq \mu(\Pi) = \mu(q)$, we would thus have $\mu(p_i) \geq \mu(\Pi)$ for all i. However, because Π is a top, the pixel just preceding Π on the path must have a value of less than $\mu(\Pi)$, which is a contradiction. ∎

Theorem 13.5 If Π and Π' are distinct tops, $[\Pi]$ and Π' must be disjoint.

Proof Suppose Π' and $[\Pi]$ were not disjoint; then some pixel p of Π' would be μ-connected to some pixel q of Π, so there would exist a path $\rho = (p_0, \ldots, p_n)$ such that $p = p_0, q = p_n$, and $\mu(p_i) \geq \min(\mu(p), \mu(q))$ for all i. Because Π' and Π are distinct tops, they must be disjoint; thus $p \notin \Pi$. Hence, according to Proposition 13.14, $\mu(p) < \mu(\Pi)$, so $\min(\mu(p), \mu(q)) = \mu(p)$. However, because Π and Π' are distinct tops, they cannot be adjacent, and the first pixel on ρ that is not in Π' must have a value of less than $\mu(\Pi') = \mu(p)$, which is a contradiction. ∎

Theorem 13.6 p and q are μ-connected iff there exists a top Π such that p and q are both in $[\Pi]$.

Proof If p and q are both in $[\Pi]$, then $\mu(p)$ and $\mu(q)$ are both $\leq \mu(\Pi)$, according to Proposition 13.4. Hence there is a path ρ from p to a pixel p' of Π such that, for all pixels t on ρ, we have $\mu(t) \geq \min(\mu(p), \mu(\Pi)) = \mu(p)$, and there is a path ρ' from q to a pixel q' of Π such that, for all pixels t on ρ', we have $\mu(t) \geq \min(\mu(q), \mu(\Pi)) = \mu(q)$. Also, because Π is connected, there is a path ρ^* from p' to q' such that, for all pixels t on ρ^*, we have $\mu(t) = \mu(\Pi) \geq \min(\mu(p), \mu(q))$. Combining these paths yields a path from p to q on which $\mu \geq \min(\mu(p), \mu(q))$ so that p and q are μ-connected. Conversely, let p and q be μ-connected; let, for example, $\mu(p) \leq \mu(q)$; and let ρ be a nondescending path from q to some pixel q' of a top Π (see Proposition 13.9). For all pixels t on ρ, we have $\mu(t) \geq \mu(q) \geq \mu(p)$; because p and q are μ-connected, there is a path ρ' between them such that, for all pixels t on ρ', we have $\mu(t) \geq \min(\mu(p), \mu(q)) = \mu(p)$. Hence there is a path $\rho'\rho$ from p to q' such that, for all pixels t on $\rho'\rho$, we have $\mu(t) \geq \mu(p) = \min(\mu(p), \mu(q'))$ so that $p \in [\Pi]$. ∎

Thus the μ-connected sets of P can be regarded as "domains of influence" of the tops of P.

We now return to the viewpoint from which we regard the pixel values as representing surface heights, and we no longer assume that they are in the range [0,1]. We say that p is *directly upstream* from q and q *directly downstream* from p iff $P(p) > P(q)$ and $\Pi(p)$ is adjacent to $\Pi(q)$. Evidently, if $\Pi(p)$ is a top, no q can be directly upstream from p, and, if $\Pi(p)$ is a bottom, no q can be directly downstream from p. We say that p is *upstream* from q and q is *downstream* from p if there exists a sequence of pixels $(p_0, p_1, \ldots, p_n)$ such that $p = p_0, q = p_n$, and p_i is directly upstream from $p_{i+1} (0 \leq i < n)$. We can also apply the terms (directly) *upstream* and *downstream* to the plateaus themselves.[6]

---

6. This terminology too is motivated by regarding $P(p)$ as the height of a surface above p. If we pour water on the surface at the point above p, it will flow to the point above q iff either $q \in \Pi(p)$ or q is downstream from p.

13.3 Diagrams

Evidently, no pixel can be upstream from a top or downstream from a bottom. On the other hand, according to Proposition 13.9, every pixel either belongs to a top or is downstream from a top, and every pixel either belongs to a bottom or is upstream from a bottom. Note that a pixel may be downstream from more than one top or upstream from more than one bottom. If P is downstream from only one top Π, we say that it belongs to the *runoff* of Π; if it is upstream from only one bottom Π, we say that it belongs to the *watershed* (sometimes called the *catchment basin*) of Π. Let G be the directed graph with nodes that are the plateaus of P and in which (Π, Π') is a directed edge of G iff Π' is directly downstream from Π. If Π^* is a bottom, p is in the watershed of Π^* iff the plateau containing p is connected to Π^* in G.

The watersheds of a picture can be identified by assigning a label to each bottom and propagating that label to all of the plateaus that are upstream from that bottom. A pixel p is in the watershed of a bottom Π^* if $\Pi(p)$ receives only the label of Π^*. A simple picture with pixel values $0, \ldots, 4$ and the watershed of one of its bottoms (the singleton 0 in the last row) are shown in Figure 13.17. The runoffs of a picture can be identified analogously by assigning labels to the tops.

In our discussion of runoffs and watersheds, we have ignored the rate at which the surface elevation changes. If q is adjacent to p and $P(q) < P(p)$, the rate at which water flows from p to q depends on how much smaller $P(q)$ is than $P(p)$. If p is in the watershed of a bottom Π, there are paths of steepest descent from p to Π. If the edges of G are weighted by their slopes and Π is in the watershed of Π^*, we can find strongest paths from Π to Π^* (i.e., paths along which the rate of flow of water from Π to Π^* is maximal).

A plateau (or pixel) that is upstream from more than one bottom is said to belong to a *divide of a surface*, because the water that flows from such a plateau is "divided": it does not all flow to the same bottom. Pixels that lie on the *ridges* of a surface are usually divide pixels, because the water that flows down the two sides of the ridge usually reaches two or more bottoms. Similarly, pixels that lie in *ravines* are usually downstream from more than one top, because the water flowing into a ravine from its two sides usually comes from two or more tops. A divide pixel (or plateau) that is directly upstream from pixels (or plateaus) that belong to two different watersheds belongs to the *divide line* (sometimes called a *watershed line*)

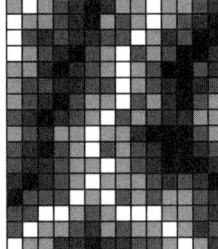  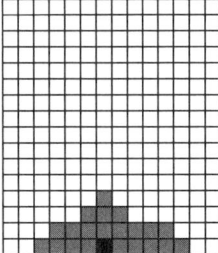

FIGURE 13.17 A watershed: Left, middle: a picture with pixel values $0, \ldots, 4$. Right: the watershed of the 0 in the last row.

between the two watersheds. If water flowed upward from the bottoms, water from different bottoms would meet at divide lines. Thus there is an analogy between a divide line and a medial axis (see Section 3.4.4) at which the grassfire from the boundary of a set meets itself; however, note that a "divide line" can be very thick, because its pixels can belong to arbitrarily large plateaus. Divide lines that are (usually) not more than two pixels thick can be obtained by using a "viscous" water flow that has a rate that depends on both distance and elevation difference.

13.4 Exercises

1. A *shape* is a simply connected measurable compact set in the Euclidean plane. For any $M \subseteq \mathbb{E}^2$, let $S(M)$ be the smallest shape that is similar to a given shape and that contains M. Which of the properties **H1** through **H4** are satisfied by the operator S?

2. Let S be the class of all compact subsets of the plane, let $M \subseteq S$, and let $I(M)$ be the isothetic rectangle of smallest area that contains M. Prove that I is a hull operator.

3. Give algorithms for constructing the smallest circumscribing square or disk of a finite set of points of $\mathbb{Z}^2$. (Hint: It is simpler to construct an approximate solution.)

4. Let $M \subset \mathbb{R}^2$ be finite, let d be its diameter, and let its smallest circumscribing rectangle $R(M)$ have sides of lengths a and b. Find lower and upper bounds for a and b in terms of d.

5. Give examples that show that properties **H1**, **H2**, and **H3** are independent (e.g., a class **S** and a mapping H that satisfy **H1** but not **H2** and not **H3**).

6. Let Π_0 be a simple polygon. We recall (see Section 13.1.1) that a (first order) *concavity* in Π_0 is the closure of a connected component of $C(\Pi_0) \setminus \Pi_0$. Evidently, Π_0 is convex iff it has no first-order concavities. Define a kth-order concavity $(k = 2, 3, \ldots)$ in Π_0 as a concavity in a $(k-1)$st-order concavity in Π_0. Prove the following:

   a) For any k, kth-order concavities in Π_0 are simple polygons with vertices that are vertices of Π_0.

   b) If k is odd, kth-order concavities in Π are subsets of $C(\Pi_0) \setminus \Pi_0$; if k is even, kth-order concavities in Π_0 are subsets of Π_0.

   c) If Π_0 has n vertices, it cannot have kth-order concavities for $k > n$. Can you give a better bound on k?

13.4 Exercises

7. Design an algorithm for constructing the convex hull of a finite set S of points in the Euclidean plane using the following strategy. Identify in S the up to eight points that are extreme in the x-, y-, $(x+y)$-, and $(x-y)$-directions. Points of S in the interior of the octagon defined by these eight points cannot be on the frontier of the convex hull. The remaining points of S — in addition to the eight points — must lie in eight triangles outside of the octagon. Construct the frontier of the convex hull using only these remaining points.

8. Let $e(m)$ and $s(n)$ be the functions defined in Section 13.1.2. Prove that $e(m) \leq 2m+2$ for $m \geq 1$; $s(n) \geq \frac{n}{2} - 1$ for $n \geq 3$; and s and e are monotonically increasing functions.

9. Show that an 8-disconnected set M can have more than one digitally convex completion.

10. Let dig be a digitization function that maps Euclidean line segments $pq \subset \mathbb{R}^2$ ($p,q \in \mathbb{Z}^2$) into 8-regions $dig(p,q) \subset \mathbb{Z}^2$ such that $p,q \in dig(p,q)$. dig is called *close* iff, for all $r \in dig(p,q)$, there is a point $v \in pq$ such that $d_8(r,v) < 1$, and, for all $v \in pq$, there is a grid point $r \in dig(p,q)$ such that $d_8(r,v) < 1$. dig is called *convex* iff, for all $r,s \in dig(p,q)$, we have $dig(r,s) \subseteq dig(p,q)$ for all $p,q \in \mathbb{Z}^2$. Show that closeness and convexity are mutually exclusive properties of dig.

11. Let M be a finite set of points in $\mathbb{R}^2$. If a convex hull edge pq does not intersect "its" Voronoi diagram edge $V(p) \cap V(q)$, replace pq with two new edges, one of which joins p at the point where the "old" edge pq crosses the frontier of $V(p)$. Continue this process until there is no further change. This process "shrinks" the original convex hull $C(M)$ into a polygon $G(M)$; see Figure 13.18 (left).

   (i) Show that G is a pseudohull; it is called the *Gabriel pseudohull* of M.

   (ii) Give an algorithm for calculating the Gabriel pseudohull.

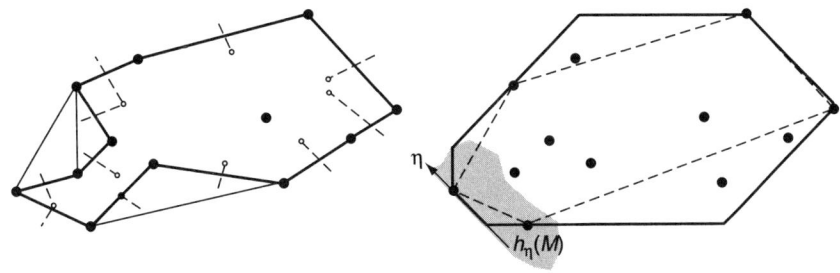

FIGURE 13.18 Left: the bold polygonal line is the frontier of the Gabriel pseudohull; see Figure 13.11 for the Voronoi and Delaunay diagrams. Right: the 8-hull (bold) and 8-approximation (dashed).

12. Let $M \subset \mathbb{R}^2$ be finite. The frontier of the *8-approximation* of M is defined by the cyclic sequence of points $p = (x, y) \in M$ that maximize x, $x + y$, y, $-x + y$, $-x$, $-x - y$, $-y$, and $x - y$, in that order; see Figure 13.18 (right). The *8-hull* is defined by intersecting the eight halfplanes defined by tangent lines in directions $\eta = i2\pi/8$ ($i = 0, \ldots, 7$) that contain M.

   (i) Which of the properties **H1** through **H4** are satisfied by the 8-approximation and the 8-hull?

   (ii) Implement the 8-approximation and the 8-hull, generate h points in a square, and compare the mean difference between the areas of their 8-approximation, their 8-hull, and their convex hull (e.g., calculated on the basis of Graham's algorithm) for $h = 100, \ldots, 1000$.

13. Let $\eta \in [0, 2\pi)$ be a direction, let $p \in \mathbb{Z}^2$, and let $\gamma_\eta(p)$ be the oriented straight line through $p = (x_0, y_0)$ in direction $\eta + \pi/2$ if $\eta \neq 0$ and $\eta \neq \pi$:

$$\gamma_\eta(p) = \{(x, y) : y = ax + b \wedge a = \tan(\eta + \pi/2) \wedge b = y_0 - ax_0\}$$

   Let it otherwise be as follows:

$$\gamma_\eta(p) = \{(x, y) : x = x_0\}$$

   Let $M \subset \mathbb{Z}^2$ be finite. $p \in M$ is an extreme point of M in direction η (notation: $p \in X_\eta(M)$) iff M is contained in the closed halfplane $h_\eta(M)$ to the right of $\gamma_\eta(p)$. Let $\mathrm{dir}(n) = \{i2\pi/n : i = 0, \ldots, n - 1\}$ be a set of n directions, and let the following be true:

$$H_n(M) = \bigcap_{\eta \in \mathrm{dir}(n)} h_\eta(M) \text{ and } A_n(M) = C\left(\bigcup_{\eta \in \mathrm{dir}(n)} X_\eta(M)\right)$$

   $H_n(M)$ is the *n-hull* of M (see Exercise 12 for $n = 8$), and $A_n(M)$ is the *n-approximation* of M.

   (i) Extend the experiments of Exercise 12 by comparing the area of the n-hull and n-approximation with the area of the convex hull for $n = 4, \ldots, 64$.

   (ii) Let $f(n)$ be the largest integer m such that, for all finite sets $M \subset \mathbb{Z}^2$ such that $\max\{|x_1 - x_2|, |y_1 - y_2| : (x_1, y_1), (x_2, y_2) \in M\} \leq m$, the n-approximation $A_n(M)$ is always equal to the convex hull $C(M)$. For example, $f(1) = 0$, $f(2) = f(3) = 1$, and $f(4) = 2$. Give lower and upper bounds on $f(n)$ when n is a multiple of 4.

14. Show that a set $S \subseteq \mathbb{Z}^2$ is digitally convex iff there is a supporting (real) line through every point of its 4-border. (Note: *Supporting lines* of a planar set have the set on one side of them.)

13.5 Commented Bibliography

For a discussion of hulls, pseudohulls, and near-hulls, see [535], which also discusses the Gabriel pseudohull [1062] (see Exercise 10) and the 8-hull [537] (see Exercise 11). The case $n = 2$ of Jung's upper bound for the radius of the smallest disk that contains a finite set $M \subset \mathbb{R}^2$ is also proved in [830]. Relative convex hulls (see Section 1.2.9) define hulls of "inner sets" with respect to a fixed "outer set." In the context of this chapter, relative convex hulls have applications to analyzing convexity in the grid cell model; see [508]. "Approximate convexity" based on concepts of geometric probability and shape analysis is proposed in [420]. [973] discusses digital "star-shapedness."

Sklansky's algorithm was published in [999]. The correct convex hull algorithm for simple polygons (Algorithm 13.1) appeared in [539], thereby correcting an erroneous algorithm in [451]. See the literature about computational geometry (e.g., the textbooks [97, 823] or Chapter 19 [by R. Seidel] in [371]) for other 2D and 3D convex hull algorithms (for simple polygons, finite sets of points, simple polyhedra, and so on). The algorithm sketched in Exercise 6 has linear expected time assuming a uniform distribution of the points in a rectangle [265]. [302] gives expected values of the areas and diameters of the convex hulls of sets of n points.

See [451] for convex hull computation for sets of grid points and [180, 865] for the computation of digital convex hulls. The discussion of maximal numbers of vertices of convex grid polygons in Section 13.1.2 and Exercise 7 follows [1110], which also contains empiric formulas for estimating the functions e and s. See also [465] for asymptotic estimates of e and s. Theorem 13.1 is from [5].

[449] contains a review of digital convexity (see Section 13.2). See [1165] for number-theoretic studies of digital convex polygons. The results in this section are from [348, 451, 510, 515, 516, 854, 911, 952, 998, 999]. See [508, 518] for convexity in the grid cell model. [473] and [1157] discuss fuzzy convexity, and [906] discusses convexity on graphs.

[902] discusses measures of concavity. For more about convex digital regions of specific shapes, see [333, 765] (squares) and [276, 328, 512, 606] (disks). See [905] about fuzzy orthoconvexity and rectangles and [906] about fuzzy triangles.

The Delaunay diagram algorithm in Algorithm 13.3 (including the "integer arithmetic only" optimization for sets of grid points) was published in [453]. For a review of Voronoi and Delaunay diagram algorithms (Dirichlet cell algorithms in the general case) and related geometric properties, see Chapter 20 (by S. Fortune) in [371]. For performance analysis of Voronoi tessellation algorithms, see [819]. For Proposition 13.8, see [1062].

An algorithm for Euclidean Dirichlet labeling on $\mathbb{Z}^n$ ($n \geq 2$), assuming a finite set of grid points as input, is given in [373]. The algorithm is based on space-filling curves in $\mathbb{Z}^n$; it generalizes examples given in Section 1.1.3 for $\mathbb{Z}^2$.

[588] applies dual Voronoi diagrams to the recognition of circular arcs. [119] approximates the skeleton of a set $S \subset \mathbb{R}^2$ from the Voronoi diagram of points sampled along the frontier of S. [217] studies topologic connectedness in the plane defined by Voronoi cells. For other references regarding Voronoi diagrams, see [8, 38, 310, 385].

For fuzzy connectedness in 2D pictures, see [891, 897]. For classic references about monotonicity-based segmentation of surfaces, see [167, 710]. For watershed segmentation, see [83, 87, 727, 735, 824, 1098]. For Exercise 2 (smallest circumscribing disk), see [1120]. Exercise 9 is from [683], and Exercise 12 follows [537].

CHAPTER **14**

Transformations

Subfields of *geometry* are sometimes defined by systems of axioms and sometimes by groups of transformations under which certain geometric properties remain invariant. This chapter describes the transformation-based approach and presents a set of axioms for digital geometry. It discusses transformation groups and symmetries, neighborhood-preserving transformations, applying transformations to pictures, magnification and demagnification, and digital tomography.

14.1 Geometries

A wide variety of geometries have been developed, motivated by a wide variety of applications: *Euclidean* (Thales of Miletus, Hippocrates of Chios, the secret society of the Pythagoreans, Euclid, Archimedes); *analytic* (Descartes, also known as Cartesius); *perspective* (Alberti, da Vinci, della Francesca, Dürer); *projective* (Desargues, Pascal); *descriptive* (Monge); *non-Euclidean*, such as *elliptic* and *hyperbolic* (Lobachevsky, Bólyai, Riemann); and *combinatorial* (Helly, Borsuk, Erdős).

The Norwegian mathematician S. Lie and the German mathematician F. Klein formulated a classification system for all of the geometries known at their time:

> Geometric properties of objects are those which are invariant with respect to a specified group of transformations.

This classification scheme is known as the 1872 *Erlangener Programm* of F. Klein. Let **B** be a base set and **G** a group of transformations defined on **B**. The theory of invariants with respect to **B** and **G**, which studies quantities that can be measured in **B** that have values that are invariant under the transformations in **G**, defines a geometry in which a nonempty family of subsets ("objects" or "figures") $\mathcal{F} \subseteq \wp(\mathbf{B})$ is specified as the class of objects of interest. We assume that these objects and the quantities ("properties") that are measured for them are in fact of practical or theoretic interest.

For example, **B** might be the 3D Euclidean space $\mathbb{E}^3 = [\mathbb{R}^3, d_e]$ defined by the manifold $\mathbb{R}^3$ of triples of real numbers and the Euclidean metric d_e; $\mathcal{F}$ might be the family of all bounded polyhedra; and **G** might be the group of (i) all similarity transformations, (ii) all affine transformations, or (iii) all projective transformations of **B** into the Euclidean plane. This defines three different geometries of polyhedra in the sense of the Erlangener Programm.

The study of invariants also requires some relationship among the elements of **B** (i.e., **B** must have a *structure* defined by a metric [see Chapter 3] or a topology [see Chapter 6]). For example, for the invariance of cross-ratios in projective geometry, we make use of the Euclidean metric; for the invariance of simple connectedness, we usually make use of the Euclidean topology.

If at least one of **B**, **G**, or $\mathcal{F}$ is finite or discrete, the geometry is called a *discrete geometry*. A set $A \subseteq \mathbf{B}$ is called *discrete* in **B** iff any $p \in \mathbf{B}$ has a neighborhood $U(p) \subseteq \mathbf{B}$ such that $U(p) \cap A$ is finite. For example, any set of grid points is discrete in $\mathbb{R}^n$. A family of sets $\mathcal{G} \subseteq \wp(\mathbf{B})$ is *discrete* in **B** iff any $p \in \mathbf{B}$ has a neighborhood $U(p) \subseteq \mathbf{B}$ that has nonempty intersections with only a finite number of the sets in $\mathcal{G}$. For example, any set of cells of a Euclidean complex (see Section 6.4.2) is discrete in $\mathbb{R}^n$.

Digital geometry, using either the grid point or grid cell model, is a discrete geometry. The structure in the base set **B** of a discrete geometry is specified by (at least) a system of *algebraic neighborhoods* $U(p) \subseteq \mathbf{B}$ for all $p \in \mathbf{B}$, where U is reflexive, symmetric, and nontransitive. An element q is a *proper neighbor* of p *(adjacent to p)* iff $q \in U(p)$ and $p \neq q$. Adjacency structures $[S, A]$ (see Chapter 4) allow us to introduce such neighborhoods; $q \in U(p)$ iff $p = q$ or $q \in A(p)$.

Packings, tessellations, polyhedra, and Euclidean complexes all involve discrete families of sets in Euclidean space. Geometry on graphs and finite geometries are other examples of discrete geometries. Some of them can be regarded as generalizations of digital geometry if their base set **B** has an adjacency grid or incidence grid (or cell complex) as an interpretation (i.e., as a model in the sense of logic).

14.2 Axiomatic Digital Geometry

Many geometries can be defined by systems of axioms. Eukleides (also known as Euclid; see Section 1.2.2) also formulated in 13 books an *axiom system* for the geometry known today as Euclidean geometry. Such a system should be complete, nonredundant, and consistent [384].[1] This book has introduced theories based on axioms (e.g., the theory of oriented adjacency graphs based on properties **A1** through **A4**). Consistency is ensured by the existence of nontrivial models.

At the end of the 19th century, there were still many open problems related to the axiomatic foundations of geometries. Poincaré, Pasch, and Hilbert were among those who studied such problems. At the end of the 19th century, mathematicians recognized that Euclid's axiom system was incomplete. For example, Euclid did not define precise concepts of "between," "inside," or "outside"; his reasoning about these concepts was based on pictures. In 1899, D. Hilbert [436] proposed a modified

---

1. A set of axioms is *consistent* iff it is impossible to deduce a contradiction from it; it is *nonredundant* iff none of the axioms can be deduced from the others; and it is *complete* iff it can be used to prove or disprove every proposition.

14.2 Axiomatic Digital Geometry

axiom system for Euclidean geometry that removed some of these defects. His work and the work of H.G. Forder [334] formalized Euclidean geometry and proved the consistency of its axioms.

Removing redundancy from a set of axioms system is not critical (in principle), but constructing a complete and consistent set of axioms is. K. Gödel [363] proved that no set of axioms can be both consistent and complete (Gödel's Incompleteness Theorem). It follows that a consistent set of axioms cannot be complete.

The following axiomatic definition of digital geometry (based on an adaptation of a subset of Hilbert's axiom system) was presented in a 1989 habilitation thesis [449] by A. Hübler.

Let **B** be a set of *points* and $\mathcal{G} \subseteq \wp(\mathbf{B})$ a nonempty set of *lines*. We begin with two *incidence axioms*:

G1: For any two distinct points $p, q \in \mathbf{B}$, exactly one line in $\mathcal{G}$ contains p and q.

G2: For any line $\gamma \in \mathcal{G}$, there exist pairwise distinct points $p, q, r \in \mathbf{B}$ such that $p \in \gamma$ and $q \in \gamma$ but $r \notin \gamma$.

According to axiom **G2**, there are at least three points in **B**. Using axiom **G1**, it then follows that there are at least three pairwise distinct lines in $\mathcal{G}$. It also follows that any two distinct lines can have at most one point in common. (As we saw in Figure 7.11, "crossings" of lines that have no common point are also possible.)

Let $\parallel$ be a *relation of parallelism* on $\mathcal{G}$ that satisfies the following:

G3: $\parallel$ is an equivalence relation on $\mathcal{G}$, and, for any line $\gamma \in \mathcal{G}$ and any point $p \in \mathbf{B}$, there is exactly one line γ' such that $p \in \gamma'$ and $\gamma \parallel \gamma'$.

The equivalence relation $\parallel$ defines equivalence classes in $\mathcal{G}$ that are called *directions*.

Let $\gamma(p, q)$ be the line uniquely defined by points p and q (see axiom **G1**). A one-to-one mapping Φ from **B** onto **B** is called a *translation* on **B** iff it is either the identity mapping I or it has the following three properties:

1. $\Phi(\gamma) = \{\Phi(p) : p \in \gamma\} \parallel \gamma$ for all lines γ.
2. $\Phi(p) \neq p$ for all $p \in \mathbf{B}$.
3. $\{\gamma(p, \Phi(p)) : p \in \mathbf{B}\}$ is a direction.

The following axiom guarantees that there are translations that are different from I:

G4: For any two distinct points $p, q \in \mathbf{B}$, there is a translation Φ such that $\Phi(p) = q$.

From this and the previous axioms, it follows that there is exactly one translation Φ such that $\Phi(p) = q$.

Proposition 14.1 Two distinct lines γ_1 and γ_2 are parallel iff there is a translation $\Phi_1 \neq I$ such that $\Phi_1(\gamma_1) = \gamma_1$ and $\Phi_1(\gamma_2) = \gamma_2$, iff there is a translation $\Phi_2 \neq I$ such that $\Phi_2(\gamma_1) = \gamma_2$.

Let **T** be the set of all translations on **B**.

Theorem 14.1 *T* is a commutative group under the operation of composition and has *I* as its identity element.

A translation $\Phi \neq \mathbf{I}$ is called *cyclic* iff there is an integer $n > 0$ such that $\Phi^n(p) = p$ for all $p \in \mathbf{B}$. A translation is cyclic iff $\Phi^n(p) = p$ for some $n > 0$ and some $p \in \mathbf{B}$. The smallest such integer is called the *cycle* of Φ. [449] shows that, if there exists a cyclic translation with cycle $m > 0$, all translations in **T** that are different from **I** are also cyclic with the same cycle m, and m is a prime number. [449] provided models of axioms **G1** through **G4** with cyclic translations, for finite or countable infinite sets of points **B**. Nonparallel lines need not intersect.

Next we introduce axioms of order. Let < and > be two opposite total orders on a line γ (i.e., for all $p, q \in \gamma$, we have $p < q$ iff $q > p$). Let $[\gamma, <]$ and $[\gamma, >]$ denote these *oriented lines*. Let $\mathbf{B} \subseteq \mathbf{B}^3$ be the *betweenness relation* on **B** for any three pairwise distinct points p, q, r on $[\gamma, <]$:

$$B(p,q,r) \text{ iff } p < q < r \text{ or } r < q < p$$

G5: For any oriented line $[\gamma, <]$ and any point $p \in \gamma$, there exist points $q, r \in \gamma$ such that $B(q, p, r)$.

(Possibly this axiom already excludes models with cyclic translations; see the statement following **G7**.) It follows that any line contains (countably) infinitely many points. Axioms **G1** through **G5** no longer have finite models.

The following axiom excludes nondiscrete models such as Euclidean geometry:

G6: For any two points $p, q \in \gamma$, there are only finitely many points r such that $B(p, r, q)$.

This requires lines to be countable and isomorphic (with respect to the order) to $[\mathbb{Z}, <]$. Oriented lines can be identified with ordered point sequences $\{p_i\}_{i \in \mathbb{Z}}$.

G7: Let $\gamma_1, \gamma_2,$ and γ_3 be three pairwise distinct parallel lines. Let γ and γ' be two lines that intersect $\gamma_1, \gamma_2,$ and γ_3, and let $\{p_i\} = \gamma \cap \gamma_i$ and $\{p'_i\} = \gamma' \cap \gamma_i$ ($i = 1, 2, 3$). Then $B(p_1, p_2, p_3)$ iff $B(p'_1, p'_2, p'_3)$.

This axiom implies that $B(p, q, r)$ implies $B(\Phi(p), \Phi(q), \Phi(r))$ for any translation Φ. It also follows [449] that a model of axioms **G1** through **G7** cannot have a cyclic translation.

For any line γ and any point $p \in \gamma$, there is a translation Φ such that $\gamma = \{\Phi^i(p) : i \in \mathbb{Z}\}$. Such a translation Φ is called a *generator* of γ. If Φ is a generator of γ, so is Φ^{-1}. Any line γ has exactly two generators, and two lines are parallel iff they have the same generators.

A translation Φ is called *atomic* iff there is no translation Ψ such that $\Phi = \Psi^n$ and $n \neq 1$. A translation is atomic iff it is a generator of a line.

$S \subseteq \mathbf{B}$ is called *complete* with respect to B iff, for all $p, q \in S$ and all $r \in \mathbf{B}$, $B(p, r, q)$ implies $r \in S$. $S \subseteq \mathbf{B}$ is called *convex* iff it is complete (with respect to B). The intersection of any (finite or infinite) family of convex sets is convex.

The *convex hull* $C(S)$ of $S \subseteq \mathbf{B}$ is the smallest (with respect to set inclusion) convex set that contains S. It follows that the following is true for all $S \subseteq \mathbf{B}$:

$$C(S) = \bigcap \{A : A \subseteq \mathbf{B} \wedge A \text{ is convex}\}$$

Any line is convex. A *line segment* is a finite complete subset S of a line that contains at least two points. An infinite complete proper subset of a line is called a *ray*. The points of any line segment can be ordered (e.g., $(p_1, \ldots, p_k)$); p_1 and p_k are called the *endpoints* of the segment.

Let $p \in \mathbf{B}$, and let $\Phi, \Psi \in \mathbf{T}$ be two translations in different directions. $\mathbb{G}_{p,\Phi,\Psi} = \{\Phi^i \Psi^k(p) : i, k \in \mathbb{Z}\}$ is called the *2D grid defined by p, Φ, and Ψ*. The triple p, Φ, Ψ defines a *coordinate system* on $\mathbb{G}_{p,\Phi,\Psi}$. Any 2D grid is isomorphic to the grid $\mathbb{Z}^2$ (i.e., they differ by a one-to-one mapping that is betweenness-invariant). Thus, all 2D grids are isomorphic to each other.

A *basis* of a set $\mathbf{T}$ of translations is a set $\mathcal{B} \subseteq \mathbf{T}$ such that any translation in $\mathbf{T}$ is a finite composition of translations in $\mathcal{B}$. The *dimension!of a set of translations* of $\mathbf{T}$ is the smallest cardinality of any basis of $\mathbf{T}$.

Let γ_1, γ_2, and γ_3 be pairwise distinct parallel lines. We define $B(\gamma_1, \gamma_2, \gamma_3)$ (i.e., γ_2 is between γ_1 and γ_3) iff there is a translation Φ such that $\gamma_2 = \Phi^i(\gamma_1)$ and $\gamma_3 = \Phi^k(\gamma_1)$ for some natural numbers i and k such that $i < k$.

G8: For any two parallel lines γ_1 and γ_2, only a finite number of lines γ satisfy $B(\gamma_1, \gamma, \gamma_2)$.

Every 2D grid satisfies axiom **G8**. Models of axioms **G1** through **G8** are called *digital grid geometries*. Axiom **G6** can be deduced from axioms **G1** through **G5** and **G7** and **G8**. (We can also possibly deduce **G8** from **G1** through **G7**.) There are models of this axiom system in orthogonal grids $\mathbb{Z}^n$ that have any dimension $n > 0$ [228]; hence the system is consistent. The axioms also allow us to define other geometric concepts such as planes, halfplanes, and coplanarity. They do not allow us to define the topologic concepts of adjacency sets or neighborhoods, but the order relations on lines determine adjacencies between their points.

Theorem 14.2 *Every digital grid geometry is uniquely determined by a set $\mathbf{B}$ of points and a set $\mathbf{T}$ of translations.*

14.3 Transformation Groups and Symmetries

The study of symmetry has a long history in the arts, architecture, crystallography, and mathematics. H. Weyl wrote the following in [1122]:

> "Symmetry, as widely or as narrowly as you may define its meaning, is an idea by which man through the ages has tried to comprehend and create order, beauty, and perfection."

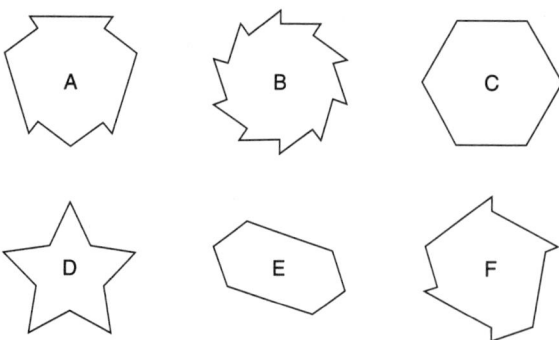

FIGURE 14.1 Symmetry examples in the Euclidean plane.

Symmetries are defined with respect to geometric transformations. A *symmetry operation* maps a set into itself. Polygon A in Figure 14.1 is reflection-symmetric with respect to the vertical axis, but it is not rotation-symmetric. Polygon B is rotation-symmetric (in multiples of $36°$) but not reflection-symmetric. Polygon F is neither rotation- nor reflection-symmetric. Polygons C, D, and E are both reflection- and rotation-symmetric, with varying numbers of axes and rotation angles.

Any planar set is symmetric with respect to the identity. Let us denote geometric transformations with capital letters (e.g., I for the identity and R for a counterclockwise rotation by $90°$). We have $R^4 = I$, and R has period 4, because 4 is the smallest power of R that is equal to I. I is the only transformation that has period 1. We can also consider inverses of transformations; the inverse $R^{-1} = R^3$ satisfies $RR^{-1} = I$. A nonempty set G of transformations is a group iff, for any transformations $U, V \in G$, we have $UV \in G$ and $U^{-1} \in G$. The cardinality of a group (which may be finite or infinite) is called its *order*. For example, $\{I, R, R^2, R^3\}$ is a group of order 4. The set of all symmetry operations on a set S is a group, which is called the *symmetry group* of S. Two groups are said to be isomorphic iff there is a one-to-one mapping f between them that preserves the group operations: For all U, V, we have $f(UV) = f(U)f(V)$ and $f(U^{-1})f(V)^{-1}$. In the sequel, we regard isomorphic groups as being identical.

Polygon F in Figure 14.1 has the trivial symmetry group that contains only the identity I. A symmetry group that contains a translation must also contain all integer multiples of that translation. It follows that a bounded set cannot be translation-symmetric.

The group-theoretic approach to symmetry arose at the end of the 19th century. The classifications of symmetry groups in the Euclidean plane and in the Euclidean 3D space[2] are well-known. We review the planar case. A discrete symmetry group in the plane may contain the following:

---

2. Parallel publications in 1891 by E.S. Fedorov [317], a director of mines in the Ural region, and A. Schönflies [966], a professor at Göttingen University, classified all 230 symmetry groups of crystal in 3D space. Their results were in close agreement, and Schönflies granted Fedorov priority for the discovery [153].

14.3 Transformation Groups and Symmetries

(i) no translations, and either of the following:

(i.1) no reflections (these are the *circle groups* C_n generated by a rotation about an angle $2\pi/n$ for some $n \geq 2$; see polygon B in Figure 14.1 for $n = 10$); or

(i.2) at least one reflection (these are the *dihedral groups* D_n generated by reflection about an axis through the origin at an angle of $2\pi/n$ for some $n \geq 2$; see polygons $A, C, D,$ and E in Figure 14.1 for $n = 2, n = 6, n = 5,$ and $n = 4$);

(ii) translations, but only in one direction; or

(iii) translations in more than one direction.

The symmetry group of a bounded set is either a circle group C_n or a dihedral group D_n. The group D_n contains n rotations and n reflections. Translation-symmetric (unbounded) sets are periodic in one or several directions (cases (ii) and (iii) above, see [221]). (A picture can be regarded as a "finite window" of an unbounded periodic set.) Case (ii) contains only seven groups, which are called the *frieze groups*. Case (iii) contains 17 groups, called the *wallpaper groups*, which were fully classified in [318]. Drawings by the Dutch artist M.C. Escher[3] illustrate at least 16 of the 17 wallpaper groups.

[486] proposes using naive lines (DSLs) to define reflections in the digital plane. The line of reflection $\gamma_0 = \{(x,y) \in \mathbb{Z}^2 : 0 \leq ax - by < b\}$ where $0 < a \leq b$ passes through the origin and has slope a/b. γ_0 defines a direction (i.e., an equivalence class Γ of lines in the sense of Section 14.2). Γ consists of all of the following lines:

$$\gamma_i = \{(x,y) \in \mathbb{Z}^2 : ib \leq ax - by < (i+1)b\} \quad (i \in \mathbb{Z})$$

The direction $\Gamma^\perp$ perpendicular to Γ is defined by slope $b/-a$ and consists of all of the following lines:

$$\gamma_k^\perp = \{(x,y) \in \mathbb{Z}^2 : kb \leq bx + ay < (k+1)b\} \quad (k \in \mathbb{Z})$$

Any direction defines a partition of $\mathbb{Z}^2$ into parallel DSLs. A grid point $p = (x,y) \in \mathbb{Z}^2$ determines the following values,

$$i = \lfloor \frac{bx + ay}{b} \rfloor$$

and

$$k = \lfloor \frac{ax - by}{b} \rfloor$$

which identify the line $\gamma_k^\perp$ such that $p \in \gamma_k^\perp$ is at 8-distance i from γ_0 "along $\gamma_k^\perp$". p is in one of the two digital halfplanes defined by γ_0. We reflect p in γ_0, mapping p into the opposite digital halfplane; the reflected p is also on $\gamma_k^\perp$ and also at 8-distance i from γ_0. Reflection in γ_0 defines a one-to-one mapping on the grid

---
3. See the picture gallery "Symmetry" at http://www.mcescher.com/.

points of $\mathbb{Z}^2$. Note that this is not simply a mapping of a point with "γ-coordinates" (i, k) into a point with γ-coordinates $(-i, k)$.

These mappings are digital analogs of reflections in the Euclidean plane. A DSL is digitally symmetric with respect to reflection in itself, but it is not necessarily a symmetric set in the Euclidean plane. Translations defined by vectors $\vec{op}$ are also one-to-one mappings of $\mathbb{Z}^2$ into itself. In the case of rotations, however, we can only deal with rotations by multiples of $90°$; this greatly limits the study of digital rotational symmetry.

14.4 Neighborhood-Preserving Transformations

In this section, we discuss transformations f that take $\mathbb{Z}^2$ into $\mathbb{Z}^2$. We define a concept of "continuity" for such an f and show that f is continuous iff it preserves 4-connectedness. We also show that f is continuous and one-to-one iff it is a translation (possibly combined with a reflection or with a rotation by a multiple of $90°$).

Let d be a metric on $\mathbb{Z}^2$ that has the following properties:

a) $p \neq q$ implies $d(p, q) \geq 1$.

b) $d(p, q) = 1$ iff p and q are 4-adjacent.

Note that d_4 and d_e have these properties but that d_8 does not.

Let f be a function from $\mathbb{Z}^2$ into $\mathbb{Z}^2$. We call f *d-continuous at* p if, for all $\epsilon \geq 1$, there exists a $\delta \geq 1$ such that $d(p, q) \leq \delta$ implies $d(f(p), f(q)) \leq \epsilon$. Note that this definition is analogous to the familiar epsilon-delta definition of continuity in the Euclidean plane, because $d > 0$ is equivalent to $d \geq 1$.

Proposition 14.2 f is d-continuous at p iff $d(p, q) \leq 1$ implies $d(f(p), f(q)) \leq 1$.

Proof If f is d-continuous at p, let $\epsilon = 1$; then there exists a $\delta \geq 1$ such that $d(p, q) \leq \delta$ implies $d(f(p)f(q)) \leq 1$, and, in particular, this must be true for $\delta = 1$. Conversely, if $d(p, q) \leq 1$ implies $d(f(p), f(q)) \leq 1$, then $d(p, q) = 1$ implies $d(f(p), f(q)) \leq 1 \leq \epsilon$ for any $\epsilon \geq 1$, so f is d-continuous at p with $\delta = 1$. ∎

Note that, if f is d-continuous at every pixel in a set $S \subseteq \mathbb{Z}^2$, it is "uniformly" d-continuous on S. We will drop the phrase "at p" from now on.

Proposition 14.3 f is d-continuous iff it takes 4-connected sets of pixels into 4-connected sets of pixels.

14.4 Neighborhood-Preserving Transformations

Proof If f preserves 4-connectedness and $d(p,q) \leq 1$, the set $\{p,q\}$ is 4-connected; hence the set $\{f(p),f(q)\}$ must be 4-connected, so $d(f(p),f(q)) \leq 1$. This proves that f is d-continuous by Proposition 14.2. Conversely, let f be d-continuous, let S be a 4-connected set of pixels, and let $f(p), f(q)$ be any two pixels in $f(S)$. Because S is 4-connected, there exists a 4-path $p_0, p_1, \ldots, p_n$ from $p = p_0$ to $q = p_n$ in S, so $d(p_i, p_{i+1}) = 1$ for all $0 \leq i < n$. According to Proposition 14.2, this implies $d(f(p_i), f(p_{i+1})) \leq 1$ for all $0 \leq i < n$; hence $f(p_0), f(p_1), \ldots, f(p_n)$, omitting repetitions if necessary, is a 4-path in $f(S)$ from $f(p)$ to $f(q)$ so that $f(S)$ is 4-connected. ∎

Proposition 14.4 If $d(f(p), f(q)) \leq d(p,q)$ for all p and q, f is d-continuous. If $d = d_4$, the converse is also true.

Proof The first part is an easy consequence of Proposition 14.2. Conversely, let f be d_4-continuous, and let $d_4(p,q) = a$; then there exists a 4-path $p_0, p_1, \ldots, p_a$ of length a from $p = p_0$ to $q = p_a$. Because $d_4(p_i, p_{i+1}) = 1$, by Proposition 14.2, we have $d_4(f(p_i), f(p_{i+1})) \leq 1$; hence, by the triangle inequality and induction, we have $d_4(f(p), f(q)) \leq a = d_4(p,q)$. ∎

$f(x,y) = (x+y, 0)$ is an example of a d_e-continuous function for which we have $d_e(f(p), f(q)) > d_e(p,q)$. In the Euclidean plane, the property of f used in Proposition 14.4 is called a *Lipschitz condition*; it implies continuity but not conversely.

If f is d-continuous and one-to-one and q is a 4-neighbor of p, $f(q)$ must be a 4-neighbor of $f(p)$. Moreover, if q_1 and q_2 are consecutive 4-neighbors of p (i.e., they have a common 4-neighbor [a diagonal neighbor of p]), $f(q_1)$ and $f(q_2)$ must also be consecutive. It is not hard to show that one of the following must be true:

a) The neighbors of p are mapped into the corresponding neighbors of $f(p)$.

b) The neighbors are mapped into their reflections in a vertical, horizontal, or diagonal line (through p).

c) The neighbors are mapped into their rotations (around p) by $\pm 90°$ or $(\pm)180°$.

Moreover, if any of these conditions are true for some p, it is not hard to show that it must be true for every p. Thus we have the following:

Theorem 14.3 If f is d-continuous and one-to-one, it must be a translation, possibly combined with a symmetry of the square.

(A symmetry of the square is a vertical, horizontal, or diagonal reflection or a rotation by $\pm 90°$ or $(\pm)180°$.)

14.5 Applying Transformations to Pictures

In this section, we discuss how to apply a geometric transformation of the plane to a picture in such a way as to ensure that the result of the transformation is also a picture.

A geometric transformation of the plane is defined by a pair of equations of the following form, which specify the new coordinates of each point as functions of the old coordinates:

$$x' = h_1(x,y) \qquad y' = h_2(x,y) \tag{14.1}$$

If we want to apply such a transformation to a picture, we must deal with the fact that, even if (x,y) are integers, (x',y') may not be integers. In this section, we discuss how to apply a geometric transformation to a picture in such a way that the result of the transformation is still a picture.

A naive approach to applying the geometric transformation of Equation 14.1 to a picture P would be as follows. For each pixel of P (e.g., at location (x,y)), compute its new coordinates (x',y') using Equation 14.1; round these coordinates to the nearest integers $[x'], [y']$; and construct a new picture P' in which the pixel at location $([x'], [y'])$ has value $P(x,y)$. Unfortunately, this naive approach would lead to unacceptable results; it would assign no values to some pixels of P' and more than one value to other pixels. To see this, consider a simple example in which we apply a 45° counterclockwise rotation to the 3 × 3 picture P that has pixel values that are shown in Figure 14.2. This rotation takes (x,y) into the point that has the following coordinates:

$$x' = (x-y)\sqrt{2}/2 \qquad y' = (x+y)\sqrt{2}/2$$

Let the center of rotation and the origin be at the pixel that has value g; then the new coordinates of the nine pixels and their rounded values are as shown in Table 14.1. Thus the pixels at locations $(-1, 1)$ and $(1,1)$ each get two values (a and d, h and i), and the pixel at location $(0,2)$ gets no value, even though it is surrounded by pixels that do get values. Schematically, the new picture looks like Figure 14.3.

Evidently, the rounded 45° rotation is not a one-to-one correspondence of the grid with itself. For example, when we apply it to a DSS consisting of five diagonally consecutive pixels, we obtain five pixels that lie on a horizontal (or vertical) line, where they occupy an interval of length $5\sqrt{2} \approx 7$; thus the five original pixels rotate

a	b	c
d	e	f
g	h	i

FIGURE 14.2 A 3 × 3 picture P.

14.5 Applying Transformations to Pictures

TABLE 14.1 Rotated coordinates of the nine pixels of P when the naive approach is used.

Pixel of P	Original coordinates	New coordinates	Rounded new coordinates
a	(0,2)	$(-\sqrt{2},\sqrt{2})$	$(-1,1)$
b	(1,2)	$(-\sqrt{2}/2, 3\sqrt{2}/2)$	$(-1,2)$
c	(2,2)	$(0, 2\sqrt{2})$	$(0,3)$
d	(0,1)	$(-\sqrt{2}/2, \sqrt{2}/2)$	$(-1,1)$
e	(1,1)	$(0, \sqrt{2})$	$(0,1)$
f	(2,1)	$(\sqrt{2}/2, 3\sqrt{2}/2)$	$(1,2)$
g	(0,0)	$(0,0)$	$(0,0)$
h	(1,0)	$(\sqrt{2}/2, \sqrt{2}/2)$	$(1,1)$
i	(2,0)	$(\sqrt{2}, \sqrt{2})$	$(1,1)$

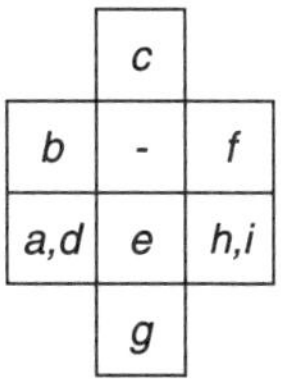

		c	
b	-	f	
a,d	e	h,i	
	g		

FIGURE 14.3 The result of rotating P by 45° using the naive approach.

into five out of seven horizontally consecutive pixels. Two of these seven pixels will have no values assigned to them, so the rotated line segment will have gaps. Conversely, when we rotate a horizontal DSS consisting of seven horizontally consecutive pixels, the rotated pixels occupy an interval of length $\approx 5\sqrt{2}$ on a diagonal; thus the values of the seven original pixels must be assigned to five pixels, so two of these pixels will have more than one value assigned to them.

We can avoid these difficulties by using the inverse of the transformation in Equation 14.1 to map each pixel in the new picture (e.g., with coordinates (x',y')), into a (real) point (x,y) in the plane that contains the original picture P. (We are assuming that the transformation in Equation 14.1 is invertible.) Let the inverse of the transformation in Equation 14.1 be as follows:

$$x = H_1(x',y'), \qquad y = H_2(x',y') \tag{14.2}$$

It specifies the old coordinates of a point as functions of the point's new coordinates. Equation 14.2 maps (x',y') into a point (x'',y'') in the plane of the old picture P.

We assign a value to the pixel of the new picture at location (x',y') by interpolating between the values of the pixels of P that lie near (x'',y''). Alternative methods of performing this interpolation will be discussed shortly; for the moment,

TABLE 14.2 Inverse-transformed coordinates of the pixels in P'.

Pixel of P'	Inverse transformed coordinates	Closest pixel of P	Value
(0,0)	(0,0)	(0,0)	g
(0,1)	($\sqrt{2}/2, \sqrt{2}/2$)	(1,1)	e
(0,2)	($\sqrt{2}, \sqrt{2}$)	(1,1)	e
(0,3)	($3\sqrt{2}/2, 3\sqrt{2}/2$)	(2,2)	c
(1,1)	($\sqrt{2}, 0$)	(1,0)	h
(1,2)	($3\sqrt{2}/2, \sqrt{2}/2$)	(2,1)	f
(−1,1)	($0, \sqrt{2}$)	(0,1)	d
(−1,2)	($\sqrt{2}/2, 3\sqrt{2}/2$)	(1,2)	b

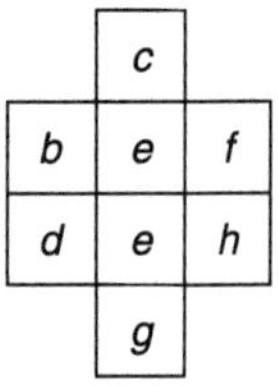

	c	
b	e	f
d	e	h
	g	

FIGURE 14.4 The result of rotating P by 45° using nearest-neighbor interpolation.

we will use *nearest-neighbor* (sometimes called *zero-order*) *interpolation*, in which (x', y') is given the value of the pixel of P that lies closest to (x'', y'').

When we use the inverse transformation and nearest-neighbor interpolation, our rotation example works as follows. The inverse transformation is a clockwise 45° rotation that takes (x', y') into the point that has the following coordinates:

$$x = (y' + x')\sqrt{2}/2, \qquad y = (y' - x')\sqrt{2}/2$$

When we apply this transformation, the nearest-neighbor values for the pixels of P' are as shown in Table 14.2. Note that the same value is assigned to more than one pixel (both (0,1) and (0,2) get e), and some of the values (a and i) are not assigned to any pixel; however, every pixel in P' gets a unique value, as shown in Figure 14.4.

As we see from Figure 14.4, the rotated picture is no longer an upright square; it is approximately a 45° rotated square. In general, if we want to display a geometrically transformed picture on a finite square grid, we may have to leave parts of that grid (such as its corners) blank, or we may have to discard the values of parts of the original picture. If the transformation is not continuous, the transformed picture may not even be connected!

We now consider alternative methods of performing the interpolation. The simplest such method is *bilinear interpolation*, which is defined as follows. Let the integer parts $\lfloor x'' \rfloor, \lfloor y'' \rfloor$ of x'' and y'' be i and j so that (x'', y'') is surrounded by the four grid points having coordinates that are shown in Figure 14.5. Let the

14.5 Applying Transformations to Pictures

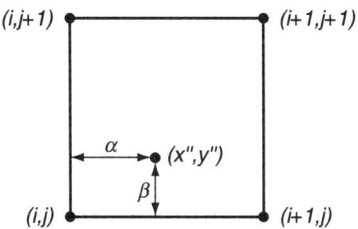

FIGURE 14.5 A point and its four surrounding grid points.

TABLE 14.3 Bilinear interpolation of the pixel values of P.

Pixel location in P'	Inverse transformed coordinates	α, β (rounded to one decimal place)	Values of surrounding pixels in P	Interpolated value (rounded to one decimal place)
(0,0)	(0,0)	0, 0	g	g
(0,1)	$(\sqrt{2}/2, \sqrt{2}/2)$	0.7, 0.7	d, e, g, h	$0.1g + 0.2(d+h) + 0.5e$
(0,2)	$(\sqrt{2}, \sqrt{2})$	0.4, 0.4	b, c, e, f	$0.4e + 0.2(b+f) + 0.2c$
(0,3)	$(3\sqrt{2}/2, 3\sqrt{2}/2)$	0.1, 0.1	$c, \text{-}, \text{-}, \text{-}$	$0.8c$
(1,1)	$(\sqrt{2}, 0)$	0.4, 0	h, i	$0.6h + 0.4i$
(1,2)	$(3\sqrt{2}/2, \sqrt{2}/2)$	0.1, 0.7	$f, i, \text{-}, \text{-}$	$0.3i + 0.6f$
(−1,1)	$(0, \sqrt{2})$	0, 0.4	d, a	$0.6d + 0.4a$
(−1,2)	$(\sqrt{2}/2, 3\sqrt{2}/2)$	0.7, 0.1	$a, b, \text{-}, \text{-}$	$0.3a + 0.6b$

fractional parts of x'' and y'' be $\alpha = x'' - \lfloor x'' \rfloor$ and $\beta = y'' - \lfloor y'' \rfloor$; thus $0 \leq \alpha, \beta < 1$. Then the value that we assign to the pixel at location (x', y') is as follows:

$$(1-\alpha)(1-\beta)P(i,j) + (1-\alpha)\beta P(i,j+1) + \alpha(1-\beta)P(i+1,j) + \alpha\beta P(i+1,j+1)$$

Note that, if x'' is an integer (i.e., $\alpha = 0$), (x'', y'') is on the line segment between pixels (i, j) and $(i, j+1)$, and a value is assigned to (x', y') by linear interpolation between their values: $(1-\beta)P(i,j) + \beta P(i,j+1)$. Similarly, if y'' is an integer, we have $\beta = 0$; here (x'', y'') is collinear with (i, j) and $(i+1, j)$, and (x', y') gets value $(1-\alpha)P(i,j) + \alpha P(i+1,j)$. Finally, if x'' and y'' are both integers, we have $\alpha = \beta = 0$ and $(x'', y'') = (i, j)$ so that (x', y') gets value $P(i, j)$, as would be expected.

To illustrate this method, we again consider our rotation example. Here the values to be assigned to the pixels of the new picture are determined as shown in Table 14.3. All nine of the original values now contribute to the values of the new pixels, although two of them (a and i) do not make the largest contribution to the value of any pixel, and e makes the largest contribution to the values of two different pixels. No values get entirely discarded, but only one of the old values (g) is preserved exactly, and the other values are blurred or attenuated. Note that the inverse transforms of (0,3), (1,2), and (−1, 2) lie slightly outside of the old 3 × 3 grid, so some of the pixels that surround their preimages (x'', y'') are "blank" (indicated

a	a	a	b	b	b	c	c	c
a	a	a	b	b	b	c	c	c
a	a	a	b	b	b	c	c	c
d	d	d	e	e	e	f	f	f
d	d	d	e	e	e	f	f	f
d	d	d	e	e	e	f	f	f
g	g	g	h	h	h	i	i	i
g	g	g	h	h	h	i	i	i
g	g	g	h	h	h	i	i	i

FIGURE 14.6 $3\times$ magnification of P using nearest-neighbor interpolation.

in Table 14.3 by hyphens). When computing the values for the pixels at (0,3), (1,2), and (−1, 2), we have treated the blanks as having value 0.

Higher-order interpolation schemes can also be used; these generally yield better-appearing results. For example, we can use *bicubic spline interpolation*, in which the pixel values are approximated by a linear combination of products of cubic polynomials $\Sigma\Sigma c_{ij}g_i(x)g_j(y)$. The coefficients of these polynomials can be chosen so that the approximation is an exact fit to the values at the pixel locations. The details will not be given here.

The preferred interpolation scheme depends on the nature of the transformation and on the type of picture that is being transformed. For example, if P is a binary picture that has only values 0 and 1 (representing black and white), we might prefer to use nearest-neighbor interpolation, because the higher-order interpolation methods introduce intermediate values, which may be of no interest. As another example, suppose we are magnifying a picture using a transformation of the following form, where $k \gg 1$:

$$x' = kx \qquad y' = ky$$

Nearest-neighbor interpolation would assign the same values to large blocks of pixels in the magnified picture (e.g., magnification of our 3×3 picture using $k = 3$ would give the 9×9 array shown in Figure 14.6); this blocky appearance would usually be objectionable. If we use bilinear interpolation, the pixel values of the magnified picture would be as shown (in part) in Figure 14.7; thus the picture would have a smoother appearance. A real picture that has been magnified by a factor of 10 using nearest-neighbor, bilinear, and bicubic interpolation is shown in Figure 14.8; note that very little improvement is obtained by using bicubic interpolation.

Conversely, if we demagnify a picture (i.e., $k \ll 1$) using low-order interpolation, the pixel values in the demagnified picture depend only on a few of the old pixel values in the vicinities of the preimages of the new pixels; many of the old pixel values will have no influence on the demagnified picture. If we do not want to discard these old values completely, we should use an interpolation scheme in which the pixels in a large neighborhood of the preimage of a new pixel contribute to the value of that pixel.

14.5 Applying Transformations to Pictures

...			
0.3g + 0.7d	...		
0.7g + 0.3d	0.4g+ 0.2d+ 0.2h+ 0.1e	...	
g	0.7g + 0.3h	0.3g + 0.7h	...

FIGURE 14.7 $3\times$ magnification of a portion of P using bilinear interpolation.

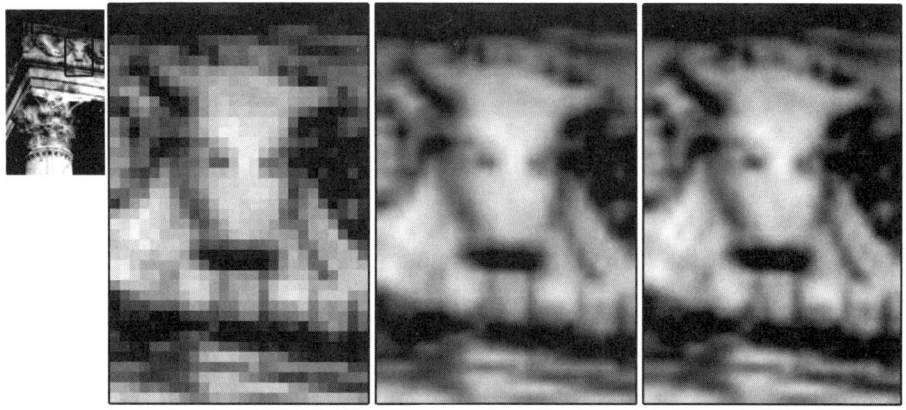

FIGURE 14.8 $10\times$ magnification of part of a picture using (left) nearest-neighbor interpolation, (middle) bilinear interpolation, (right) bicubic interpolation.

Interpolation may be necessary even if the geometric transformation is a translation $x' = x + \alpha, y' = y + \beta$. If the translation is by an integer amount, no interpolation is needed; otherwise, bilinear interpolation is appropriate. Interpolation is also unnecessary when a picture is rotated around a grid point by a multiple of $90°$ or reflected in a grid line; linear interpolation is appropriate when a picture is reflected in a (non-grid) horizontal or vertical line.

When we use nearest-neighbor interpolation, a geometric transformation takes binary pictures into binary pictures but may not preserve geometric properties of the pictures. For example, we saw in Figure 14.4 that the result of rotating the nine-pixel picture by $45°$ had only eight pixels; thus the rotation did not preserve area. Similarly, $45°$ rotation of a five-pixel diagonal DSS yields a seven-pixel horizontal or vertical DSS (and vice versa). Moreover, the horizontal or vertical segment is 4-connected, but the diagonal segment is only 8-connected; thus the rotation does not even preserve topology. Evidently, translation does preserve geometric properties, and magnification or demagnification evidently does not preserve area. Magnification preserves topology (a magnified connected or disconnected set is evidently

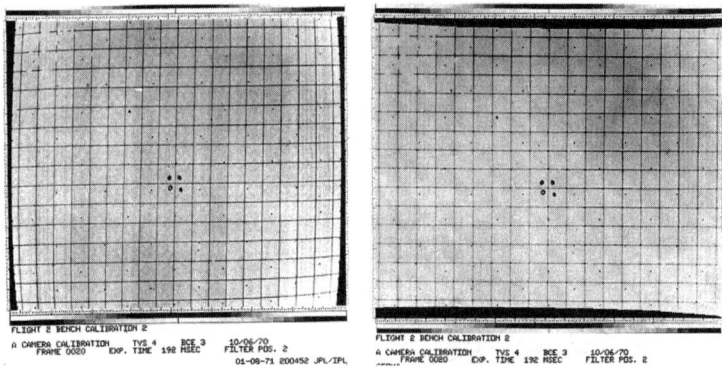

FIGURE 14.9 Conversion of a distorted grid of lines (left) into an undistorted grid (right) using a piecewise linear geometric transformation (from [784]).

still connected or disconnected), but demagnification apparently cannot preserve disconnectedness, because distant pixels may become neighbors.

A general geometric transformation of the plane can be approximated by, for example, linear transformations that are defined on small pieces of the plane. For example, we can divide the plane into small squares and use a linear transformation on each square; this maps the square into a quadrilateral. If the transformations of adjacent squares agree at the common vertices of the squares, these quadrilaterals will "fit together" to tessellate the plane. An historic example of a piecewise linear geometric transformation that converts a distorted grid of lines into an undistorted grid is shown in Figure 14.9.

14.6 Magnification and Demagnification

As we saw in Section 14.5, magnification of a picture P by an integer factor $k \neq 1$ replaces each pixel p of P with a $k \times k$ block of pixels. We can think of the new pixels (in the grid cell model) as having the same size as the original pixels; this implies that the height and width of the new picture are k times those of the original picture and that the area of the new picture is k^2 times that of the original picture. Alternatively, we can think of the new picture as being obtained by subdividing each pixel of the original picture into a $k \times k$ block of "mini-pixels," each of which has height and width $1/k$ and area $1/k^2$ (if the original pixels were of unit size). k-fold magnification using this alternative approach is equivalent to redigitizing P on a grid of resolution k (see Chapter 2) and using nearest-neighbor interpolation.

In the grid cell model, a (nonextremal) 0-cell is contained in four 1-cells and four 2-cells; a 1-cell contains two 0-cells and (if it is nonextremal) is contained in two 2-cells; and a 2-cell contains four 0-cells and four 1-cells. When we perform

14.6 Magnification and Demagnification

k-fold magnification, a magnified 1-cell contains $(k+1)$ 0-cells and k unit 1-cells; a magnified 2-cell contains $(k+1)^2$ 0-cells, $2k(k+1)$ unit 1-cells, and k^2 unit 2-cells.

If we use outer Jordan digitization (in the grid cell model), an α-arc ρ in P is a sequence of cells $p_0, p_1, \ldots, p_n$ such that p_i is α-adjacent to $p_{i-1}(1 \leq i \leq n)$. k-fold magnification converts ρ into a sequence of $k \times k$ blocks of cells $p_0, p_1, \ldots, p_n$ such that p_i is α-adjacent to $p_{i-1}(1 \leq i \leq n)$; we can think of the magnified ρ as a "thick" arc. If we use grid-intersection digitization, ρ is a sequence of grid points. The vector $v_i = p_{i-1}p_i$ is either an isothetic unit vector or a diagonal vector of length $\sqrt{2}$. Let v_i make angle $45i°$ with the positive x-axis; then v_i can be represented by the *chain code* $i \in \{0, \ldots, 7\}$, and ρ can be represented by a chain code sequence $i_0, i_1, \ldots i_n$. It is not hard to see that the k-fold magnification of ρ has the chain code $i_0^k, i_1^k, \ldots i_n^k$, where i^k denotes k-fold repetition of i [342].

Demagnifying a picture cannot preserve all of its pixel values, but it provides a "summary" of the picture that may be useful for many purposes. Even when a picture is substantially demagnified, its gross structure usually remains quite recognizable; the demagnified picture is a "miniature" (sometimes called a "thumbnail") of the original. Figure 14.10 shows three pictures and the results of demagnifying them by a factor of 8. The demagnified pictures are also shown remagnified (by factors of slightly less than 8, using nearest-neighbor interpolation) to illustrate the information that is still available after eightfold demagnification.

Different degrees of demagnification may be useful for different purposes, depending on what information about the original picture needs to be preserved. It may therefore be desirable to use a set of demagnifications (e.g., by powers of 2) so that any desired demagnification is approximated by one of them. Figure 14.11 shows demagnifications of a 512×512 picture by successive powers of 2 (from 2 to 512).

If the size of the original picture (in pixels) is $2^n \times 2^n$, its demagnifications by powers of 2 have sizes $2^{n-1} \times 2^{n-1}, 2^{n-2} \times 2^{n-2}$, and so on. The total number of pixels in all of these demagnified pictures is less than the following (i.e., less than $\frac{1}{3}$ more than in the original picture):

$$2^{2n}(1 + \frac{1}{4} + \frac{1}{16} + \cdots) = 2^{2n} \cdot 1\frac{1}{3}$$

This succession of demagnified versions of a picture can be thought of as a *"pyramid"* of pictures; note that this pyramid "tapers" exponentially rather than linearly.

From a picture of size $2^n \times 2^n$, we can construct an $n+1$-level pyramid in which level k is a picture of size $2^k \times 2^k (0 \leq k \leq n)$. In this pyramid, the value of a pixel p on any level $k < n$ is the average of the values of a 2×2 block of pixels on level $k+1$. The value of the single pixel on level 0 is the average of the values of all of the pixels of the original picture, the values of the four pixels on level 1 are the averages of the values of the pixels in the four quadrants of the original picture, and so on.

We can construct a rooted tree of degree 4 in which the pixel on level 0 is the root and each pixel p on any level $k < n$ is linked to its four "children" (a 2×2 block of pixels) on level $k+1$. The leaves of this tree are the pixels of the full-size picture at the base of the pyramid. Using this pyramid structure, we can explore the

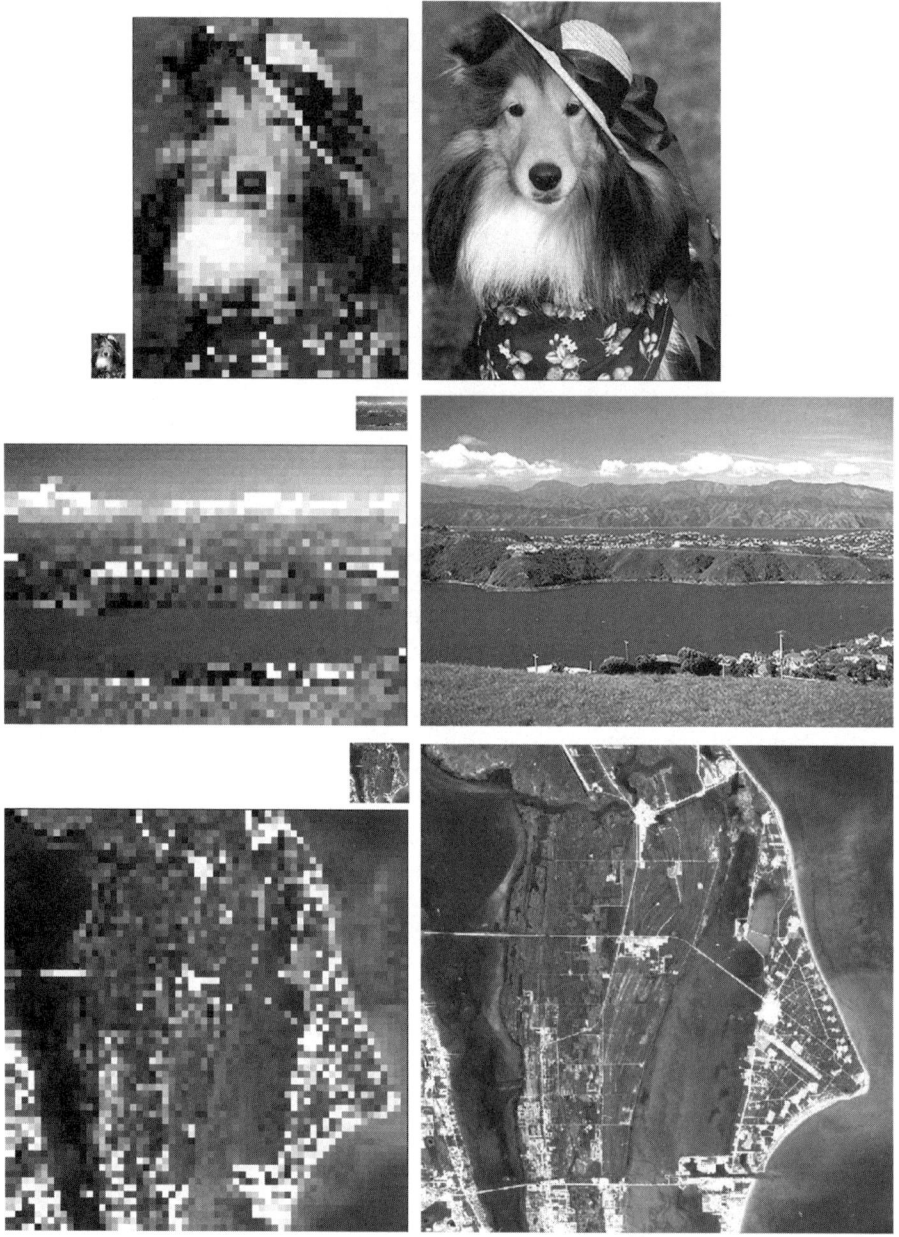

FIGURE 14.10 Three examples of 8× demagnification and slightly less than 8× remagnification.

14.6 Magnification and Demagnification

FIGURE 14.11 Demagnification of a 512×512 picture (of J.B. Listing) by factors of 2, 4, 8, 16, 32, 64, 128, 256, and 512.

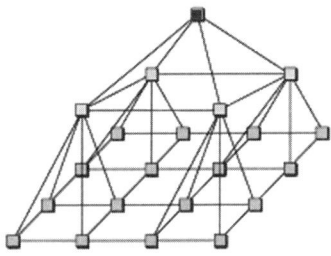

FIGURE 14.12 A three-level pyramid.

picture at any given scale by moving between adjacent pixels on a given level, and we can explore it in (discrete) scale-space by moving between parents and children on consecutive levels. The structure of a three-level pyramid is shown in Figure 14.12.

Demagnification of a picture is not a translation-invariant operation; the results depend on the positions of the blocks of pixels that are represented by single pixels in the demagnified picture. For example, consider the one-dimensional binary picture with the following pixel values:

$$\ldots, 1, 1, 1, 1, 0, 0, 0, 0, \ldots$$

Suppose we demagnify this picture by a factor of 2 so that the pixels at locations x and $x+1$ ($x = \ldots, -2, 0, 2, 4, \ldots$) are represented by a single pixel. If the coordinate of the rightmost 1 is odd, the demagnification yields $\ldots, 1, 1, 0, 0, \ldots$; however, if this coordinate is even, the demagnification yields $\ldots 1, 1, \frac{1}{2}, 0, 0, \ldots$. As an extreme example, consider the following picture:

$$\ldots, 1, 1, 0, 0, 1, 1, 0, 0, 1, 1, \ldots$$

If the righthand 1s have odd coordinates, demagnification by a factor of 2 yields $\ldots, 1, 0, 1, 0, 1, \ldots$; if they have even coordinates, it yields $\ldots, \frac{1}{2}, \frac{1}{2}, \frac{1}{2}, \frac{1}{2}, \frac{1}{2}, \ldots$.

The position-dependence of the effects of demagnification can be reduced by smoothing the picture before demagnifying it. Smoothing is accomplished by replacing the value of each pixel with a weighted average of the values of the pixel and some of its neighbors. If the pixel at coordinate x in a one-dimensional picture has value $P(x)$ the weighted averaging gives it the following new value:

$$\overline{P}(x) = \Sigma_{k=-m}^{m} w_k P(x+k)$$

(This is a weighted average of the values of an odd number of pixels; averaging neighborhoods of even sizes can also be used.) The weights should satisfy the following conditions:

(1) $\Sigma_{k=-m}^{m} w_k = 1$: This ensures that the weighting process does not result in an increase or decrease in the average value of the pixels. To further ensure that each pixel of P contributes the same total weight to each pixel of $\overline{P}$, we require that the sums of the ws in even-numbered and odd-numbered positions both be $\frac{1}{2}$.

(2) $w_{-k} = w_k$ for all $1 \leq k \leq m$: The weights are symmetric.

(3) $w_k \geq w'_k$ for all $0 \leq k < k' \leq m$: The center weight is the highest, and the weights decrease as they get farther from the center.

For $m = 2$, there are five weights; because they are symmetric, we can denote them by $\gamma, \beta, \alpha, \beta, \gamma$ where $\gamma \leq \beta \leq \alpha$. Condition (1) implies that $\alpha + 2\beta + 2\gamma = 1$ and $\beta = \frac{1}{4}$, so $\gamma = \frac{1}{4} - \frac{\alpha}{2}$. Condition (3) requires that $\alpha \geq \frac{1}{4}$, so $\gamma \leq \frac{1}{8}$. If we require the weights to be nonnegative, we must have $\alpha \leq \frac{1}{2}$, but we are still free to choose α in the range $[\frac{1}{4}, \frac{1}{2}]$. In our extreme example,

$$\ldots, 1, 0, 1, 0, 1, 0, 1, \ldots$$

if we smooth using weights for which $\beta = \frac{1}{4}$ and $\gamma = \frac{1}{4} - \frac{\alpha}{2}$, the value of the smoothed picture at the 1s is $\alpha + (\frac{1}{2} - \alpha) = \frac{1}{2}$, and its value at the 0s is $2\beta = \frac{1}{2}$; thus the smoothed picture is constant, so its demagnification is no longer position-dependent. In 2D, it is usual to require that the weights w_{jk} be separable (i.e., $w_{jk} = w_j w_k$).

14.7 Digital Tomography

Section 1.2.11 defined digital tomography in very general terms. To be more specific, assume (see Section 14.2) a digital geometry on $\mathbb{Z}^n$ ($n \geq 2$) and a set **T** of translations on $\mathbb{Z}^n$ (e.g., translations in isothetic directions).

Let $D = \{\gamma^{(1)}, \ldots, \gamma^{(m)}\}$ ($m \geq 1$) be a set of pairwise nonparallel (see Proposition 14.1) lines in $\mathbb{Z}^n$. Each line is uniquely determined by a grid point $p \in \mathbb{Z}^n$ and a generator $\Phi \in$ **T**.

Let Γ_i be the set of lines in $\mathbb{Z}^n$ that are parallel to $\gamma^{(i)}$ ($1 \leq i \leq m$). The equivalence class Γ_i is a direction in the set of all lines. $\mathbb{Z}^n$ and **T** satisfy axiom **G8** (i.e., we can enumerate the lines in $\Gamma_i = \{\gamma_j^{(i)} : j \in \mathbb{Z}\}$, where $\gamma^{(i)} = \gamma_k^{(i)}$ for some $k \in \mathbb{Z}$).

Definition 14.1 Let $S \subset \mathbb{Z}^n$ be finite, let $j \in \mathbb{Z}$, and let $\mathcal{P}_i(j) = \text{card}\left(S \cap \gamma_j^{(i)}\right)$. $\mathcal{P}_i$ is called the *projection* of S in direction Γ_i.

Figure 14.13 shows an example in which $n = 2$ and **T** is the set of translations along the x- or y-axis. For a finite set S, we have nonzero projections $\mathcal{P}_i(j)$ only in finite intervals of j-values. Let $I_1 \times \ldots \times I_m$ be the smallest cuboidal subset of $\mathbb{Z}^n$ that covers all of the nonzero projections. The *reconstruction problem* in digital tomography is as follows: given a set $\mathcal{P} = (\mathcal{P}_1, \ldots, \mathcal{P}_m)$ of projections in specified directions, find a finite set $S \subset \mathbb{Z}^n$ ($n > 2$) that has the given projections.

The reconstruction problem for the set shown in Figure 14.13 has a unique solution; the corresponding system of linear equations (which identifies each pixel in $I_1 \times \ldots \times I_m$ with a variable $x_k \in \{0,1\}$), where $k = 1, \ldots, 16$, has only one solution.

A possible method of solving the reconstruction problem of digital tomography is to apply algorithms of *computed tomography* (CT) [215] (e.g., algorithms based on Fourier transforms [implementing the inverse Radon transform] or iterative reconstruction algorithms). However, these algorithms calculate real values at pixel positions. The *algebraic reconstruction technique* (ART) of [372] can be modified to

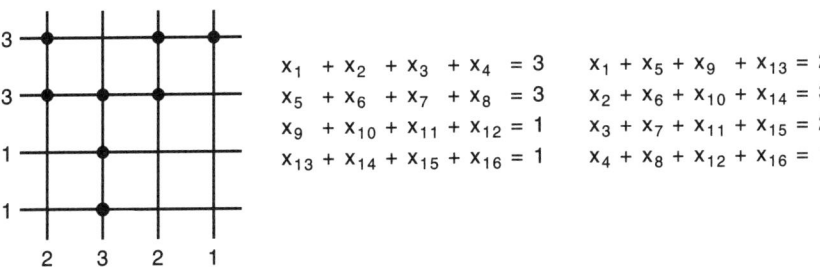

FIGURE 14.13 Left: two projections of a column-convex polyomino. Middle and right: corresponding system of linear equations.

yield a binary solution [423, 170]. Linear programming provides other alternatives (simplex method, interior point methods); see [381].

We conclude this section by discussing the reconstruction of a binary picture (matrix) from two projections, as illustrated in Figure 14.13. The input $\mathcal{P} = \{R, S\}$ is a pair of projections in the row and column directions. In Figure 14.13, we have $R = (3, 3, 1, 1)$ and $S = (2, 3, 2, 1)$. Two vectors $R = (r_1, \ldots, r_m)$ and $S = (s_1, \ldots, s_n)$ are *compatible* iff they contain only nonnegative integers $r_i \leq n$ ($i = 1, \ldots, m$) and $s_j \leq m$, ($j = 1, \ldots, n$) and the sum of the r_is equals the sum of the s_js.

The solvability of a reconstruction problem defined by a pair of compatible vectors $\{R, S\}$ is characterized by the following theorem, independently discovered by H.J. Ryser [934] and D. Gale [350] in 1957. We permute the vector S into $S' = (s'_1, \ldots, s'_n)$ such that $s'_1 \geq s'_2 \geq \ldots s'_n$. We define an $m \times n$ matrix $\overline{A}$ in which row i consists of r_i 1s followed by $n - r_i$ 0s. Let $\overline{S} = (\bar{s}_1, \ldots, \bar{s}_n)$ be the projection vector of $\overline{A}$ in the column direction.

Theorem 14.4 The reconstruction problem for $R = (r_1, \ldots, r_m)$ and $S = (s_1, \ldots, s_n)$ has at least one solution (a binary picture of size $n \times m$) iff the following is true:

$$\sum_{i=k}^{n} s'_i \geq \sum_{i=k}^{n} \bar{s}_i \text{ for } 2 \leq k \leq n$$

The case $k = 1$ is already covered by the compatibility of the vectors R and S. In our example, we have $S' = (3, 2, 2, 1)$ and $\overline{S} = (4, 2, 2, 0)$, where $5 \geq 4$ for $k = 2$, $3 \geq 2$ for $k = 3$, and $1 \geq 0$ for $k = 4$.

A *switching component* of a binary picture is a 2×2 block of pixels of the following form:

$$\begin{bmatrix} 1 & 0 \\ 0 & 1 \end{bmatrix} \text{ or } \begin{bmatrix} 0 & 1 \\ 1 & 0 \end{bmatrix}$$

If a picture has a switching component, it is evidently not uniquely reconstructible from its row and column projections. In [934], Ryser proved the following:

Theorem 14.5 A binary picture is uniquely reconstructible from its row and column projections iff it has no switching component; if P_1 and P_2 are two solutions, P_1 is transformable into P_2 by a finite number of swaps of switching components.

Note that additional knowledge about the given binary picture (e.g., that it contains only one row-convex, column-convex, or convex polyomino) may allow a unique reconstruction that would not otherwise be possible.

14.8 Exercises

1. The set of all geometric transformations of the plane is closed under composition of functions; if $f : (x' = h_1(x,y),\ y' = h_2(x,y))$ and $g : (x' = k_1(x,y),\ y' = k_2(x,y))$ are geometric transformations, then $g \circ f\ (x'' = k_1(h_1(x,y), h_2(x,y)),\ y'' = k_2(h_1(x,y), h_2(x,y)))$ is also a geometric transformation. Composition of functions is *associative* (i.e., for all f, g, h, we have $(f \circ g) \circ h = f \circ (g \circ h)$). The identity transformation $i : (x' = x,\ y' = y)$ is an *identity* element for the composition of geometric transformations (i.e., for all f, we have $i \circ f = f \circ i = f$). g is called an *inverse* of f if $g \circ f = f \circ g = i$. A set G of geometric transformations of the plane is called a *group* if G contains i and any $f \in G$ has an inverse in G. Prove that the following sets of transformations are groups:

    (1) $\{i\}$

    (2) $\{i, r_x\}, \{i, r_y\}$, and $\{i, r_o\}$ where r_x, r_y, and r_o are the reflections in the x-axis, the y-axis, and the origin

    (3) $\{i, r_x, r_y, r_1, r_2\}$ where r_1 and r_2 are the reflections in the two diagonals

    (4) $\{i, r_\pi\}$ where r_π is the rotation around the origin by $180°$

    (5) $\{i, r_{\pi/2}, r_{-\pi/2}, r_\pi\}$ where $r_{\pi/2}$ and $r_{-\pi/2}$ are the rotations around the origin by $\pm 90°$

    (6) $\{i, r_{\pi/2}, r_{-\pi/2}, r_\pi, r_x, r_y, r_1, r_2\}$

    (7) All of the rotations around the origin

    (8) All of the translations

    (9) All of the scale changes

    (10) All of the rigid motions

    (11) All of the affine transformations

2. Prove that a rotation by angle θ around the origin $(0,0)$ is defined by the following pair of equations:
$$x' = x\cos\theta + y\sin\theta;\quad y' = -x\sin\theta + y\cos\theta$$

3. The reflections in the x-axis, the y-axis, and the origin are defined (respectively) by the following pairs of equations:
$$x = -x,\ y' = y;\ x' = x,\ y' = -y;\ x' = -x,\ y' = -y$$

    Prove that these reflections are not the same as rotations through any angle around the origin.

4. Write the equations for reflection in a line L through the origin that makes angle θ with the x-axis using the fact that this reflection can be accomplished by rotating L (through angle $-\theta$) until it coincides with the x-axis, then reflecting in the x-axis, and then rotating back through angle θ.

5. An *affine transformation* is defined by the following equations, where $a_1 b_2 \neq a_2 b_1$:

$$x' = a_1 x + b_1 y + c_1; \; y' = a_2 x + b_2 y + c_2$$

Prove that translations, rotations, reflections, and scale changes (magnifications or demagnifications) are all affine transformations.

6. Characterize affine transformations that consist of a translation followed by a rotation (or vice versa). Such a transformation is called a *rigid motion*. Two pictures that differ by a rigid motion are called *congruent*. (Note that most rigid motions do not take pixels into pixels. Thus mapping a picture into a picture by a rigid motion requires interpolation; the pixel values in two pictures that differ by a rigid motion may not be the same.) Two pictures that differ by a rigid motion followed by a scale change (or vice versa) are called *similar*.

7. A *shear* is a transformation defined by the equations $x' = x + ay, y' = y$ or $x' = x$, $y' = y + ax$; thus a shear takes pixels into pixels iff a is an integer. Prove that a shear cannot be d-continuous unless it is the identity ($a = 0$).

8. Let R be a bounded region of the plane, and let f be a d-continuous function from $\mathbb{Z}^2$ into $\mathbb{Z}^2$ that takes pixels in R into pixels in R. Prove that there must exist a pixel p in R such that $d_8(p, f(p)) \leq 1$.

9. Show that the converse of Proposition 14.4 is not always true for d_e.

10. Generalize the results of Section 14.4 to 3D.

11. The group of one-to-one geometric transformations of pictures P defined on $\mathbb{G}_{n,n}$ ($n \geq 3$) consists of eight transformations: the identity *id*; the vertical, horizontal and diagonal reflections *ver*, *hor*, and *dia*; the $90°$, $180°$, and $270°$ (clockwise) rotations *rot*, *rot²*, and *rot³*; and a transformation *mir* defined by $mir(P) = rot(hor(P))$ that satisfies $mir(P) = ver(rot(P)) = dia(ver(hor(P)))$. Construct a table of all compositions $op_1(op_2(P))$ of pairs of these transformations, and show that this table proves that these transformations form a group.

12. In the 3D grid cell model, a nonextremal 0-cell is contained in six 1-cells, 12 2-cells, and eight 3-cells; a 1-cell contains two 0-cells and (if it is nonextremal) is contained in four 2-cells and four 3-cells; a 2-cell contains four 0-cells and four 1-cells and is contained in two 3-cells; and a 3-cell contains eight 0-cells, 12 1-cells, and six 2-cells. How many 0-cells, unit 1-cells, unit 2-cells, and unit 3-cells are contained in a k-fold magnified 3-cell?

14.9 Commented Bibliography

The introductory remarks about the history of geometry follow [544]. Geometry is one of the best axiomatized disciplines in mathematics [334, 384]. For definitions of discrete geometry, see [96, 319, 394]. The conferences on "Discrete Geometry for Computer Imagery," of which [111] was the first held outside France, actually focus on digital geometry.

The habilitation [449] of A. Hübler, which unfortunately has remained unpublished except for some small notes such as [450], contains valuable contributions to the definition and analysis of the translation group T of digital geometry and the axiomatic foundations of digital geometry on the regular orthogonal grid in $\mathbb{R}^n$. See also the essay [1009] about the axiomatic foundations of convexity and linearity. Hübler's axiomatic theory was recently studied in [228].

Symmetry groups in geometry are discussed, for example, in [221, 704, 1008, 1122]. Digital symmetry, as discussed in Section 14.3, is further detailed in [486]; see also [863]. For accumulator space-based symmetry analysis of "noisy" planar polygons, see [463]. [267] discusses symmetry analysis in pyramidal picture representations.

For Exercise 11, see [533], which also studied the cardinalities of families of geometric transformations defined on $\mathbb{G}_{n,n}$ ($n \geq 2$), including magnifications, demagnifications, shifts, and cyclic shifts, as well as the cardinalities of (see the definition in Section 17.7) families of local operations of order $k \geq 1$. For geometric constructions in the digital plane, see [1090].

d-continuous functions on $\mathbb{Z}^2$ were introduced in [903]. For "continuous" functions on nonbinary pictures, see [766]. For digital versions of homeomorphism, retraction, and homotopy, see [114]. A "calculus" for d-continuous functions is discussed in [767].

Methods of applying geometric transformations to pictures were discussed in [477] and were used for the geometric correction of pictures in [784]. For another method of digital rotation, see [21].

For magnification by an integer factor, see [1107]. [524] proves that Hough transforms based on real spaces are superior (with respect to the size of the accumulator array) to "digital Hough transforms."

The pyramid data structure was introduced in [1045]. For a collection of papers about multiresolution picture representation and processing, see [899]. For methods of smoothing a picture prior to reducing its resolution, see [154]. Uses of pyramids in picture analysis and computer vision are discussed in [479]. A pyramid is a discrete version of a scale-space in which scale varies continuously; see [652].

In the pyramids described in Section 14.6, the value of each pixel on level k is the average of a 2×2 block of pixels on level $k+1$. More generally, we can construct pyramids in which each pixel value on level k is a weighted average of a block of pixel values on level $k+1$, where the blocks can overlap and can be of any desired size (or shape). Still more generally, for any graph G, we can construct a pyramid of graphs $G_0, G_1, \ldots, G_n$ in which $G_n = G$, each node of G_k ($k < n$) is linked to the nodes of a subgraph of G_{k+1}, and every node of G_{k+1} belongs to one of these subgraphs. [551]

discusses repeated derivation of region-adjacency graphs, layer by layer, starting with a general adjacency graph. If values are associated with the nodes of G, we can give each node p of each G_k a value that is a function of the values of the nodes of the subgraph of G_{k+1} to which p is linked. We can also identify each node of G_k with a "representative" node of G_{k+1} (i.e., we can regard G_k as being obtained by the "reduction" of G_{k+1}). For a recent review of such general methods of constructing pyramids and the uses of such pyramids in picture analysis, see [478]. Pyramidal approaches are also closely related to multiscale representations of shapes. [70] discusses representations of borders of regions at different scales in terms of DSSs, circular arcs, corners, and points that delimit the arcs.

Section 14.7 reviews parts of Chapter 1 (by A. Kuba and G.T. Herman) of [431], which is a collection of papers about discrete (particularly, digital) tomography; see also [380]. Theorem 14.4 is from [350, 934]. The proof of this theorem involves an algorithm that constructs a solution in $\mathcal{O}\left(n\left(m+\log n\right)\right)$ time. For the reconstruction of pictures from multiple projections (a combinatorial subject), see [58, 60, 141, 142, 261, 353, 354, 355]. The reconstruction of isothetic polygons from labeled (convex or concave) vertices is discussed in [629].

CHAPTER **15**

Morphologic Operations

Mathematic morphology deals with operations that replace the value of a pixel p with the max or min of the values of a set of neighbors of p. This chapter introduces *dilation* and *erosion* operations as well as combinations of such operations known as *hit-and-miss transforms* and *opening* and *closing* operations. It illustrates how such operations can be used to simplify a picture (e.g., to remove "noise" from it) or to decompose or "segment" it into parts.

The primary subject of this chapter is the study of operations on a picture that replace the value of each pixel p with the maximum (or the minimum) of the values of a set of neighbors of p. (Such operations were briefly introduced in Section 1.2.12 and were also used in Sections 4.1 and 12.2.3.) If the picture is binary (its pixel values are all 0s and 1s), such a "local maximum" operation has the effect of "dilating" the 1s; p's value becomes (or remains) 1 if any of its neighbors (that belong to the specified set) had value 1. Similarly, a local minimum operation has the effect of dilating the 0s. Note that dilating the 1s is equivalent to "eroding" the 0s, and dilating the 0s is equivalent to eroding the 1s. Evidently, dilation is quite different from magnification (which "expands" both the 1s and the 0s), and erosion is quite different from demagnification (which "contracts" both the 1s and the 0s.)

Dilation and erosion (and the other operations discussed in this chapter) are defined for multivalued pictures, but they have simple geometric interpretations when they are applied to binary pictures. These operations are therefore usually called *morphologic operations; morphology* is the study of form and pattern (i.e., of geometric properties of binary pictures). We will define the concepts in this chapter only for 2D pictures, but they generalize straightforwardly to higher dimensions.

15.1 Dilation

Let σ be a nonempty set of pixels at locations specified relative to an origin o. For any pixel p, the σ-*neighborhood* $\sigma(p)$ of p is the set of pixels that coincide with the pixels of σ when they are translated so that o coincides with p. Evidently, $\sigma(p) = \{p+s : s \in \sigma\}$

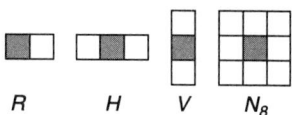

FIGURE 15.1 Four structuring elements; the origin is shaded.

where + denotes coordinate-wise addition $((x,y)+(u,v) = (x+u, y+v))$. Note that o is not necessarily in σ, so p may not be in its own σ-neighborhood. We will assume from now on that σ is finite. In mathematic morphology (the standard name for the subject of this chapter), σ is usually called a *structuring element*.

Definition 15.1 Let $P(p)$ be the value of the pixel p in the picture P. The σ-dilation $P^{(\sigma)}$ of P is the picture in which the following is true:

$$P^{(\sigma)}(p) = \max_{q \in \sigma(p)} P(q)$$

If $o \in \sigma$, we have $p \in \sigma(p)$ so that $\max_{q \in \sigma(p)} P(q) \geq P(p)$. Thus, if $o \in \sigma$, we have $P^{(\sigma)} \geq P$ at every pixel p; however, this is not true if $o \notin \sigma$.

If P is a binary picture, $P^{(\sigma)}$ is evidently also a binary picture. Moreover, $P^{(\sigma)}(p) = 1$ iff $P(q) = 1$ for some pixel q of $\sigma(p)$. Let $\langle P \rangle$ denote the set of pixels of P that have value 1; then $\langle P^{(\sigma)} \rangle = \{p : \sigma(p) \cap \langle P \rangle \neq \emptyset\}$. Using the notation of set-theoretic mathematic morphology (see Section 1.2.12), we have $\langle P^{(\sigma)} \rangle = \langle P \rangle \oplus \sigma$.

Evidently, if $\sigma = \{o\}$, we have $P^{(\sigma)} = P$; in general, if σ is a single pixel in some location (x,y) relative to o, then $P^{(\sigma)}$ is the result of translating P by $(-x,-y)$. The term "dilation" is more appropriate if σ is a set of two or more pixels that contains o. For example (see Figure 15.1), if $\sigma = R$ consists of the pixels in locations $(0,0)$ (the location of o) and $(0,1)$, $P^{(\sigma)}$ is obtained by "smearing" the values of P toward the left; the value of p in $P^{(\sigma)}$ is the max of the values of p and its righthand neighbor. Similarly, if $\sigma = H$ consists of o and its two horizontal neighbors, $P^{(\sigma)}$ is obtained by "smearing" the values of P to the right and left; if $\sigma = V$ consists of o and its two vertical neighbors, $P^{(\sigma)}$ is obtained by "smearing" P up and down; and if $\sigma = N_8$ is the 8-neighborhood of o (the set consisting of o and its horizontal, vertical, and diagonal neighbors), $P^{(\sigma)}$ is obtained by "smearing" P horizontally, vertically, and diagonally. Note that, in these last three cases, σ is symmetric about o.

The results of dilating some simple binary pictures (1s = black, 0s = white)[1] using the four σs in Figure 15.1 are shown in Figure 15.2. Analogous results for a picture of some text (in which dark pixels have high values) are shown in Figure 15.3.

---

1. When displaying a picture, high pixel values are usually represented by light shades of gray so that 0 corresponds to black and G_{max} to white. Using this convention, the 0s in a binary picture should be black and the 1s should be white. However, it is also common to regard the 1s in a binary picture as "object" pixels and the 0s as "background" pixels, and, in many situations (e.g., a picture of a document page), the "objects" (characters) are black and the "background" is white. In this chapter (and sometimes in the following chapters), dark pixels in a multivalued picture have high pixel values, and the 1s and 0s in a binary picture represent black and white pixels, respectively.

15.2 Erosion

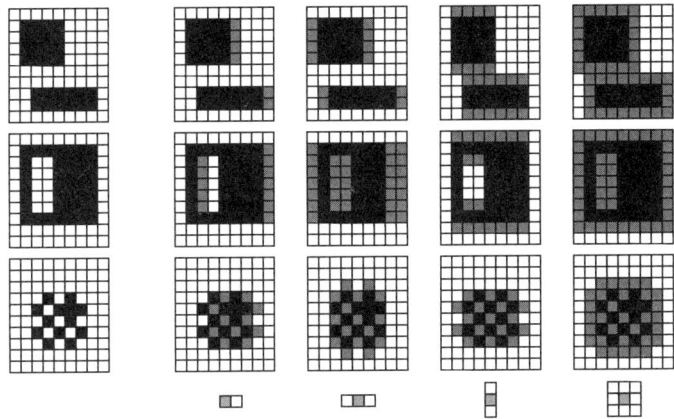

FIGURE 15.2 Dilations of a set of binary pictures (shown in the left column) using the four structuring elements shown in Figure 15.1. Note that "object" pixels (which have value 1) are displayed as black. The pixels added by the dilations are shown in gray.

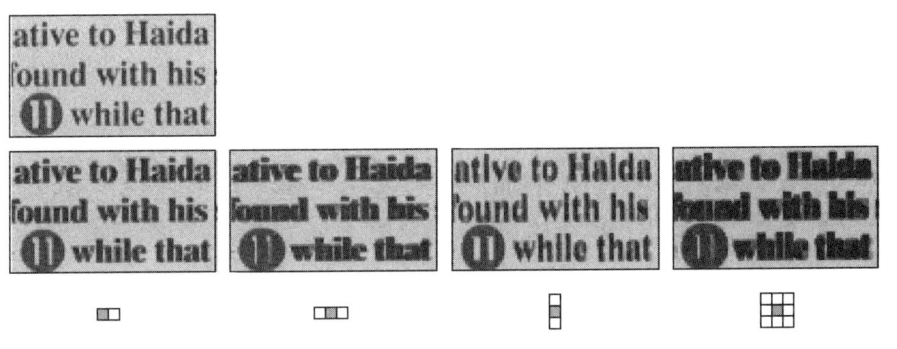

FIGURE 15.3 Dilation of a picture of some text using the same four structuring elements. Note that, in this example, dark pixels have high values, so dilation makes the characters thicker.

15.2 Erosion

Definition 15.2 The σ-erosion $P_{(\sigma)}$ of P is the picture in which the following is true:

$$P_{(\sigma)}(p) = \min_{q \in \sigma(p)} P(q)$$

If $o \in \sigma$, we have $p \in \sigma(p)$, so $\min_{q \in \sigma(p)} P(q) \leq P(p)$ (i.e., $P_{(\sigma)} \leq P$).

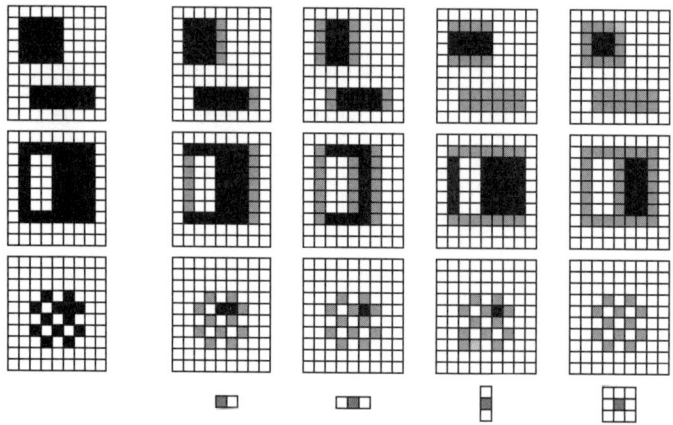

FIGURE 15.4 Erosions of the same set of binary pictures as shown in Figure 15.2 using the same four structuring elements. Note that "object" pixels (value 1) are displayed as black. The pixels deleted by the erosions are shown in light gray.

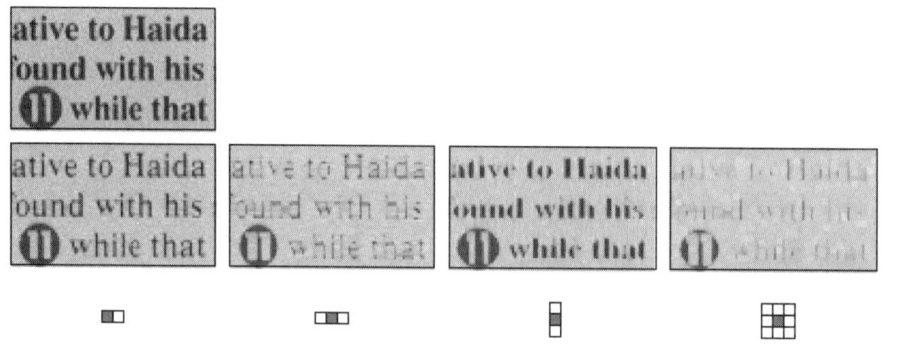

FIGURE 15.5 Erosion of the text picture (the same one as was used in Figure 15.3) using the same four structuring elements. Note that dark pixels have high values.

If P is a binary picture, so is $P_{(\sigma)}$. Moreover, $P_{(\sigma)}(p) = 1$ iff $P(q) = 1$ for every pixel q of $\sigma(p)$; in other words, $\langle P_{(\sigma)} \rangle = \{p : \sigma(p) \subseteq \langle P \rangle\}$ and $\langle P_{(\sigma)} \rangle = \langle P \rangle \ominus \sigma$ in the notation of Section 1.2.12.

Evidently, if σ is a single pixel in location (x, y), $P_{(\sigma)}$ is the same as $P^{(\sigma)}$ (i.e., it is the result of translating P by $(-x, -y)$). The term "erosion" is more appropriate if σ is a set of two or more pixels that contains o.

The results of eroding the binary pictures shown in Figure 15.2 using the four structuring elements in Figure 15.1 are shown in Figure 15.4, and the results of eroding the text picture of Figure 15.3 using these structuring elements are shown in Figure 15.5. Note that, in these pictures, dark pixels have high values (in the binary pictures, object pixels have value 1), so erosion removes object pixels.

Dilations and erosions commute with monotonic transformations of the pixel values of a picture P. Indeed, let P^* be a picture obtained from P by applying a monotonic function f to the pixel values of P so that, for any pixel p, we have $P^*(p) = f(P(p))$. p has value $\max_{q \in \sigma(p)} P(q)$ in $P^{(\sigma)}$ and value $\max_{q \in \sigma(p)} P^*(q) = \max_{q \in \sigma(p)} f(P(q))$ in $P^{*(\sigma)}$. Because f is monotonic, both maxima are taken on at the same pixel(s) q; hence, the second maximum is f of the first maximum. Thus, dilating a picture and then applying a monotonic transformation to its pixel values gives the same result as applying the monotonic transformations and then dilating the picture; this is similar for erosion.

15.3 Combining Dilations and Erosions

In this section, we discuss two methods of combining dilations and erosions:

a) hit-and-miss transforms, which dilate and erode a picture using disjoint structuring elements; and

b) opening and closing operations, which erode a picture and then dilate it (or vice versa) using the same structuring element (which we assume to be symmetric and to contain o).

15.3.1 Hit-and-miss transforms and templates

Let P be a binary picture and let σ and τ be two structuring elements. We have seen that $P_{(\sigma)}(p) = 1$ iff $P = 1$ at every pixel of $\sigma(p)$. Similarly, $\overline{P}_{(\tau)}(p) = 1$ iff $\overline{P} = 1$ (i.e., $P = 0$) at every pixel of $\tau(p)$.

Definition 15.3 $\min(P_{(\sigma)}, \overline{P}_{(\tau)})$ is called the *hit-and-miss transform* of P by the pair of structuring elements (σ, τ).

Hit-and-miss transforms can be used to identify pixels of P with neighborhoods that have 1s in specified locations (defined by σ) and 0s in specified locations (defined by τ). For example, if $\sigma = \{(0,0)\}$ and $\tau = \{(0,1)\}$, then $p = 1$ in $\min(P_{(\sigma)}, \overline{P}_{(\tau)})$ iff, in P, we have $p = 1$ and $q = 0$ where q is the upper neighbor of p. Evidently, if σ and τ intersect, $\min(P_{(\sigma)}, \overline{P}_{(\tau)})$ must be identically zero, because we can never have both $P_{(\sigma)}(p) = 1$ and $\overline{P}_{(\tau)}(p) = 1$; hence, we can assume that σ and τ are disjoint. We can think of σ and τ as defining a "template" in which the pixels of σ are 1s and the pixels of τ are 0s; evidently, we have $\min(P_{(\sigma)}(p), \overline{P}_{(\tau)}(p)) = 1$ iff the neighborhood of p "matches" this template. Mathematic morphologists also use the notation $\langle P \rangle * (\sigma, \tau)$ for the hit-and-miss transform $(\langle P \rangle \ominus \sigma) \cap (\langle \overline{P} \rangle \ominus \tau)$ of $\langle P \rangle$. It follows that $\langle P \rangle * (\sigma, \tau) = (\langle P \rangle \ominus \sigma) \setminus (\langle P \rangle \ominus \underline{\tau})$.

If P is multivalued, a hit-and-miss transform can be used to identify pixels with σ-neighbors that have high values and with τ-neighbors that have low values. Let $\overline{P}$ be the picture in which $\overline{P}(p) = G_{max} - P(p)$ for all p; then $\min(P_{(\sigma)}, \overline{P}_{(\tau)})$ has a high value iff the σ-neighbors of p have high values and its τ-neighbors have low values.

In Section 15.5, we will give examples of the use of binary hit-and-miss transforms to detect local features in a picture. Unlike dilation, erosion, opening, and closing, the hit-and-miss transform is uniquely defined for binary pictures only. See [874, 1012] for alternative definitions of hit-and-miss transforms for multivalued pictures. [1012] applies structuring elements (σ, τ) to different "layers" (binary pictures defined by thresholds) of a multivalued picture.

15.3.2 Opening and closing

Let σ be a structuring element that is not necessarily symmetric. (We recall that it is symmetric iff $\sigma = \underline{\sigma}$, where $\underline{\sigma}$ denotes the mirror set.)

Definition 15.4 Dilating a picture P by σ and then eroding it by $\underline{\sigma}$ (i.e., constructing the picture $(P^{(\sigma)})_{(\underline{\sigma})}$) is called σ-*closing*. Similarly, eroding P by σ and then dilating it by $\underline{\sigma}$ (i.e., constructing $(P_{(\sigma)})^{(\underline{\sigma})}$) is called σ-*opening*.

For binary pictures, opening by σ transforms $\langle P \rangle$ into the union of all translates of σ that are contained in $\langle P \rangle$. Closing of $\langle P \rangle$ by σ is the same as opening $\langle \overline{P} \rangle$ by σ.

Proposition 15.1 $(P^{(\sigma)})_{(\underline{\sigma})} \geq P$ and $(P_{(\sigma)})^{(\underline{\sigma})} \leq P$.

Proof We prove the inequality for opening; the proof for closing is similar. Eroding $\langle P \rangle$ by σ and then dilating by $\underline{\sigma}$ gives the following:

$$(P_{(\sigma)})^{(\underline{\sigma})}(p) = \max\{P_{(\sigma)}(q) : q \in \underline{\sigma}(p)\} = \max\{\min\{P(r) : r \in \sigma(q)\} : q \in \underline{\sigma}(p)\}$$

For every q in $\underline{\sigma}(p)$, p is in $\sigma(q)$, so $\min\{P(r) : r \in \sigma(q)\}$ is $\leq P(p)$, and, because this is true for all such q, its maximum over $q \in \underline{\sigma}(p)$ is $\leq P(p)$. Hence, $P_{(\sigma)})^{(\underline{\sigma})}(p) \leq P(p)$. ∎

If P is a binary picture, we recall that $\langle P^{(\underline{\sigma})} \rangle$ is the union of $\sigma(p)$ for all $p \in \langle P \rangle$, and $\langle P_{(\sigma)} \rangle$ is the set of all p such that $\sigma(p) \subseteq \langle P \rangle$. It follows that $\langle (P_{(\sigma)})^{(\underline{\sigma})} \rangle$ is the union of the $\sigma(p)$s that are contained in $\langle P \rangle$. The result of σ-closing a binary picture can be characterized similarly. In Sections 15.4 and 15.6, we will show how opening and closing operations can be used to simplify or "segment" a (not necessarily binary) picture.

15.4 Simplification

In this section, we will show how opening and closing operations can be used to "simplify" or "smooth" a picture by eliminating minor irregularities in its pixel

15.4 Simplification

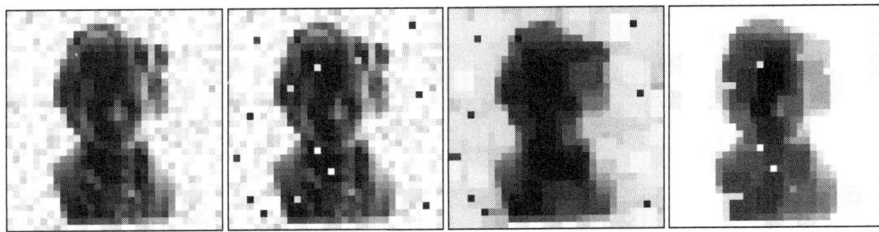

FIGURE 15.6 Removing "salt-and-pepper" noise by opening and closing. From left to right: original picture (note that dark pixels have high values); picture with noise; closing with N_8; opening with N_8.

values—in particular, by eliminating small (more precisely, thin) groups of high-valued pixels from a region of low-valued pixels or vice versa. Such irregularities may result if random changes are made in the pixel values.[2] In a *smooth* region of a picture (i.e., a region in which the values of adjacent pixels are the same or differ only slightly; see Section 14.4), random changes will produce many pixels with values that differ significantly from those of their neighbors, because the changes are unlikely to have the same effect on the pixel and its neighbors. Similarly, along an *edge* in a picture (a smooth locus of large changes in pixel values; e.g., when a high-valued region is adjacent to a low-valued region along a smooth frontier), random changes of pixel values from high to low and vice versa will produce irregularities in the edge.

A picture is said to contain *salt-and-pepper noise* if low-valued pixels occasionally occur in regions of high-valued pixels and vice versa. (Recall that, in this chapter, low values are light and high values are dark.) "Salt" can be eliminated by dilating the picture and "pepper" by eroding it, provided we use a structuring element σ such that the exceptional pixels always have nonexceptional σ-neighbors. However, dilation enlarges dark regions and erosion shrinks them (and vice versa for light regions), so dilation or erosion distorts the picture. We will now show how to eliminate salt or pepper but avoid enlarging the dark or light regions by eroding and then dilating (i.e., opening) or dilating and then eroding (i.e., closing).

Figure 15.6 shows, on the left, a picture P that contains a dark object on a light background. The values of some of the pixels of P have been complemented (i.e., value w has been occasionally replaced by $G_{max} - w$); this results in salt-and-pepper noise. Closing P (using the structuring element N_8; see Figure 15.1, right) eliminates the "salt" but does not enlarge the object; similarly, opening P using N_8 eliminates the "pepper" but does not shrink the object. These effects are also illustrated in Figure 15.7 for a larger (binary) picture; note that clusters of noise pixels are often not eliminated.

---

2. In picture processing, undesired fluctuations in pixel values are referred to as *noise*. The engineers who developed video communication systems borrowed this "acoustic" terminology from audio communications, where fluctuations in a signal may result in unpleasant sounds. Elimination of such fluctuations is called *noise cleaning* or *noise removal*.

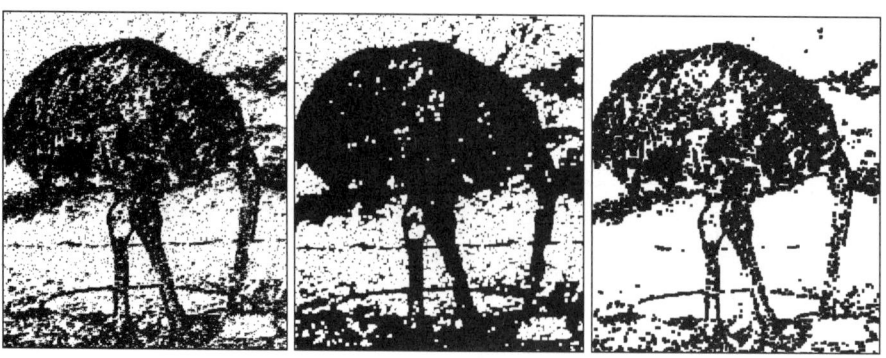

FIGURE 15.7 Left: noisy binary picture P; note that black pixels have value 1. Center: result of closing P using N_8. Right: result of opening P using N_8.

Opening eliminates not only isolated high-valued pixels but also "thin" sets of high-valued pixels, and closing eliminates not only isolated low-valued pixels but also "thin" sets of low-valued pixels. Specifically, we call a set S of pixels σ-*thin* if every pixel of S has σ-neighbors in the complement $\overline{S}$ of S. If S consists of light pixels and $\overline{S}$ of dark pixels and S is σ-thin, σ-erosion eliminates all of the light pixels that were in S. Subsequent σ-dilation cannot replace these light pixels, because the pixels of $\overline{S}$ are all dark; hence, σ-opening eliminates S. Similarly, if S consists of dark pixels and $\overline{S}$ of light pixels and S is σ-thin, σ-closing eliminates S. Thus, opening or closing should not be applied to a picture if it contains thin light or dark regions such as lines or curves that are significant (i.e., that are not merely "noise").

To illustrate how opening eliminates thin sets of high-valued pixels, let P be a binary picture and let σ be N_8. Suppose $\langle P \rangle = U \cup V$ where U is a square of size of at least 3×3 and V is a rectangle of width 2 so that V is N_8-thin. $\langle (P_{(\sigma)})^{(\sigma)} \rangle$ contains U, because every pixel of U is contained in a 3×3 square that is contained in $\langle P \rangle$; however, it does not contain any pixel of V, because no pixel of V is contained in a 3×3 square that is contained in $\langle P \rangle$. On the left side of Figure 15.8, the top picture is P; in the middle picture, the pixels of $\langle P_{(N_8)} \rangle$ are black, and the remaining pixels of $\langle P \rangle$ are light gray; in the bottom picture, the pixels of $\langle (P_{(N_8)})^{(N_8)} \rangle$ are black or dark gray. We see that N_8-opening of P eliminates V but leaves U intact. A less trivial example is shown on the right side of Figure 15.8. We see that the "thick" parts of the object have survived but the thin parts have been eliminated; as a result, the shapes of the surviving thick parts have been simplified.

Similarly, closing adjoins to $\langle P \rangle$ thin parts of the complement of $\langle P \rangle$. For example, let $\langle P \rangle = U - V$, where U is a square and V a rectangle within U of sizes as in the preceding paragraph, as illustrated on the left of Figure 15.9. Then $\langle (P^{(N_8)})_{(N_8)} \rangle$ contains V, because every pixel of V is contained in a 3×3 square that is centered at a pixel of $U - V$. A less trivial example is shown on the right of Figure 15.9.

In these examples, the thin regions are N_8-thin (i.e., they are not more than two pixels thick). In general, "thin" regions can be arbitrarily thick, but they can be

15.4 Simplification

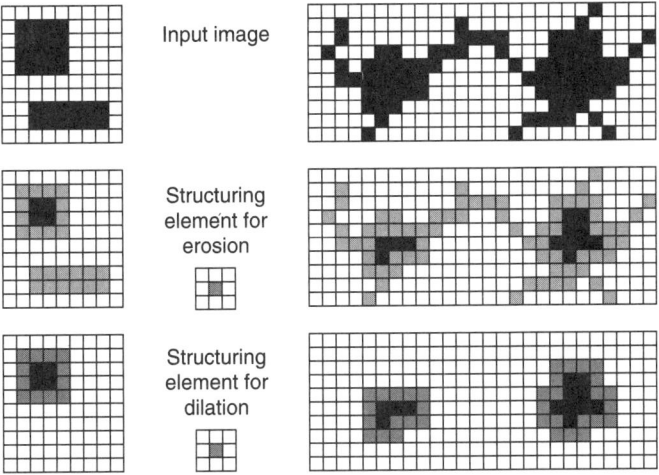

FIGURE 15.8 Two examples of how opening can be used to eliminate thin parts of an object. Note that object pixels are black. In the second row, pixels of $\langle P_{(\sigma)} \rangle$ are black, and the remaining pixels of $\langle P \rangle$ are light gray. In the third row, pixels of $\langle (P_{(\sigma)})^{(\sigma)} \rangle$ are black or dark gray.

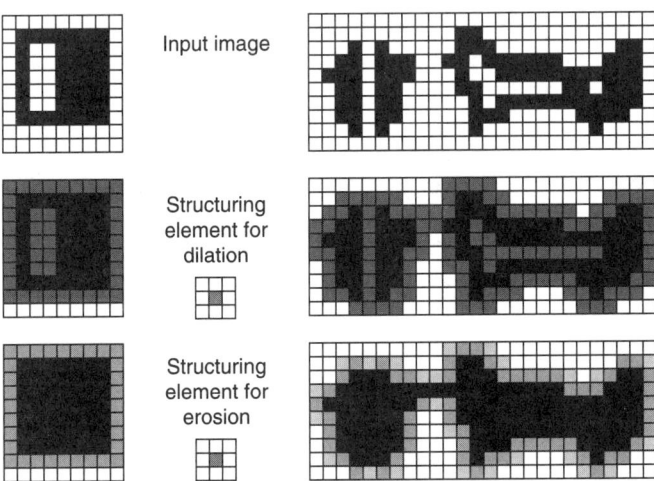

FIGURE 15.9 Two examples of how closing can be used to eliminate thin parts of the complement of an object: (top) P; (bottom) $(P^{(\sigma)})_{(\sigma)}$. Note that object pixels are black.

eliminated from or adjoined to a set by performing an opening (or closing) operation that uses a sufficiently large structuring element; see Section 15.6. The pictures in these examples are binary, but the same methods can be applied to multivalued pictures to eliminate thin parts of dark (or light) objects or to adjoin to them thin parts of their background.

15.5 Segmentation

This section discusses ways of defining distinctive sets of pixels in a picture. Partitioning a picture into distinctive subsets is called *segmentation*. We briefly describe some basic methods of segmentation, with emphasis on their relationship to morphologic operations. Segmentation is treated extensively in books about picture processing and computer vision (e.g., segmentation of a picture into "watersheds" was discussed in Section 13.3.3).

15.5.1 Thresholding

If a subset S of the pixels in a picture P has values that differ significantly from those of (most of) the other pixels of P, we call S *distinctive*. For example, if the pixels of S have significantly higher values than the other pixels, we can choose a *threshold* t such that pixels with values of t or greater (almost all) belong to S. We can then define a binary picture P_t in which $P_t(p) = 1$ iff $P(p) \geq t$. This process of converting a multivalued picture into a binary picture by comparing its pixel values to a threshold is called *thresholding*. Some simple examples of thresholding are shown in Figure 15.10.

Evidently, thresholding a picture P is a monotonic transformation of the pixel values of P. It follows (see Section 15.2) that dilations and erosions commute with thresholding. Thus, dilating or eroding a picture and then thresholding it has the same effect as thresholding the picture and then dilating or eroding the resulting binary picture. Similarly, because opening is erosion followed by dilation and closing is

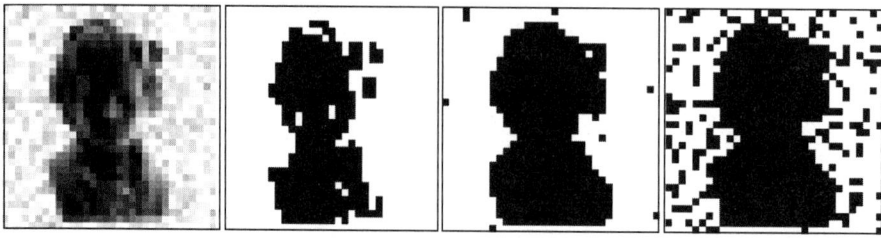

FIGURE 15.10 Original picture (as seen in Figure 15.6) and results of thresholding it using thresholds 115, 204, and 232 (from left to right).

15.5 Segmentation

dilation followed by erosion, these operations also commute with monotonic transformations of the pixel values and, in particular, with thresholding; thus, opening or closing a picture and then thresholding it has the same effect as thresholding the picture and then opening or closing the resulting binary picture.

More generally, a picture can be segmented using two (or more) thresholds $[l, h]$ ($0 \leq l < h \leq G_{\max}$); all pixels with values u that are between l and h are kept, and all those with values that are $< l$ or $> h$ are rejected. See [557] for other methods of binarizing multivalued pictures.

15.5.2 Local features

A pixel p of P is said to belong to a *local feature in P* if the neighbors of p have a distinctive pattern of values. For example, p (e.g., at (x, y)) lies on a vertical edge if a set of its neighbors on the right (e.g., $(x+1, y+1), (x+1, y)$, and $(x+1, y-1)$) has high values and a set of its neighbors on the left (e.g., $(x-1, y+1), (x-1, y)$, and $(x-1, y-1)$) has low values, or vice versa. Similarly, the pixel at (x, y) lies on a vertical line if the pixels at $(x, y+1), (x, y)$, and $(x, y-1)$ have high values and the pixels at $(x-1, y+1), (x-1, y), (x-1, y-1), (x+1, y+1), (x+1, y)$, and $(x+1, y-1)$ have low values or vice versa. Edges and lines in other directions are defined analogously. "Spots" are pixels with values that are higher (or lower) than those of all of their adjacent pixels.

Pixels that belong to local features can be identified using hit-and-miss transforms. For example, pixels that lie on a vertical edge in a binary picture P can be identified by eroding P using $\sigma = \{(1, 1), (1, 0), (1, -1)\}$ and dilating P using $\sigma = \{(-1, 1), (-1, 0), (-1, -1)\}$ (or vice versa) and taking the min of the results. Even if P is not binary, the value of this min will be high at p iff p lies on a vertical edge. Examples of the use of hit-or-miss transforms to identify spots and "notches" in horizontal lines are shown in Figure 15.11.

15.5.3 Texture

The texture of a region in a picture can be characterized by the presence of many pixels that belong to local features of particular types. For example, the texture is "spotted" if the region contains many spots, "busy" if it contains many edges, "striped" if it contains many lines, and "directional" if the edges or lines all have similar directions.

In general, when such hit-and-miss transforms are applied to a picture, regions of the picture that contained many occurrences of specific local features will contain many high-valued pixels. In Section 15.6, we will describe a method of finding regions of a picture that contain many high-valued pixels.

Texture is discussed extensively in books about picture analysis and computer vision. In this book, we do not discuss signal-theoretic, statistic, or perceptual approaches to texture characterization.

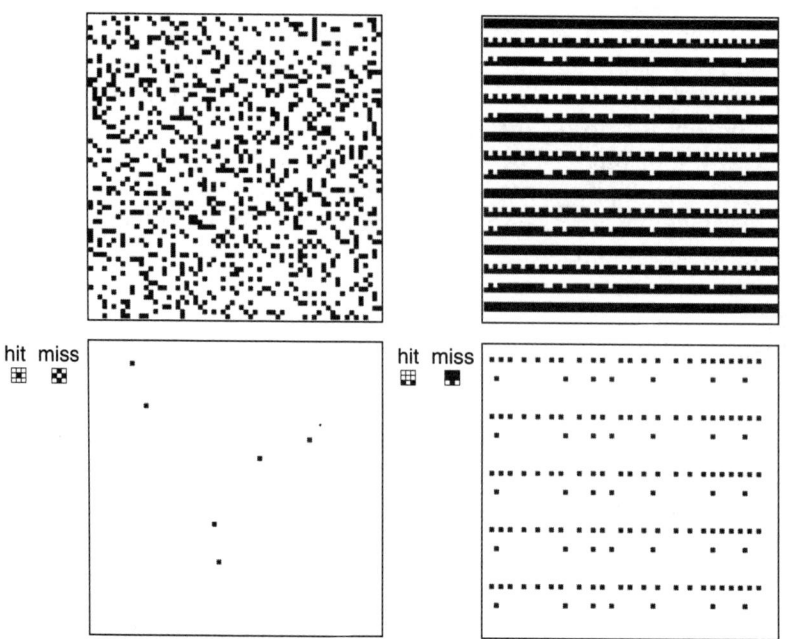

FIGURE 15.11 Detection of local features using hit-and-miss transforms. Upper row: two binary pictures containing "pepper" noise (left) and "notched" horizontal lines (right). Bottom row: results of applying hit-and-miss transforms using the structuring elements shown on the left of the pictures.

15.6 Decomposition

The segmentation methods described in Section 15.5 used individual pixel values or local patterns of pixel values to identify distinctive pixels. In this section, we describe methods of segmenting a picture into parts that are characterized by geometric properties. These methods make use of morphologic operations in which the structuring elements can be of arbitrary size. To distinguish these methods from segmentation methods based on individual or local pixel values, we call them *decomposition* methods.

Let σ_h be the structuring element that consists of the pixels that have distances from the origin (using any desired metric) that are at most $h > 0$. For example, if we use the metric d_8, σ_1 is N_8. Note that σ_h is symmetric and contains o. We will abbreviate σ_h by h in superscripts and subscripts (e.g., the σ_h-dilation of P will be denoted by $P^{(h)}$).

15.6.1 Clusters

We first show how to use closing to "fuse" a cluster of high-valued pixels (i.e., a large number of such pixels in a region R made up of other pixels that have low

15.6 Decomposition

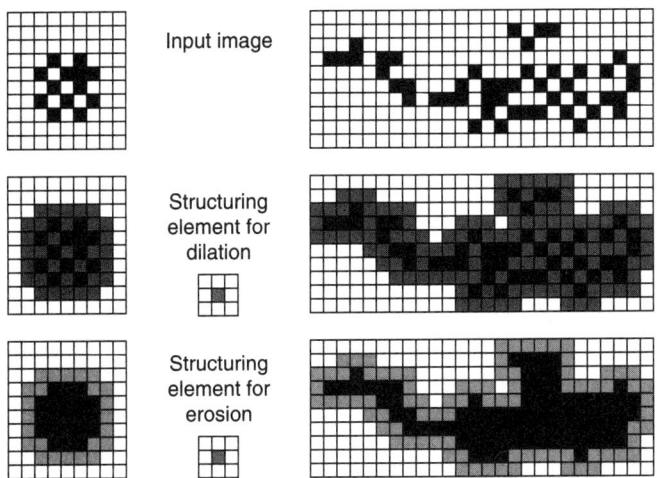

FIGURE 15.12 The use of closing to fuse a cluster: (top) P; (bottom) $(P^{(1)})_{(1)}$. Note that cluster pixels are black.

values) so that, after the closing, all of the pixels of R have high values. The basic idea is that, if the pixels in the cluster are at distances less than $2h$ from each other, the sets of low-valued pixels between the cluster pixels are σ_h-thin. As we saw in Section 15.4, closing can be used to adjoin to a set S thin parts of the complement of S. Hence, closing can be used to adjoin to the cluster pixels the thin sets of low-valued pixels between the cluster pixels so that all of the pixels of R are adjoined to the cluster.

Figure 15.12 shows simple examples of how eliminating thin parts of $\langle \overline{P} \rangle$ can be used to fuse a cluster of dark pixels of $\langle P \rangle$. If S is a cluster, any pixel of $\overline{S}$ that is "surrounded" by pixels of S belongs to a thin part of the complement $\overline{S}$. The closing adjoins these thin parts to S and thus fuses S into a solid region. The picture in this example is binary, but the example generalizes straightforwardly to multivalued pictures and to sparser clusters (i.e., those that use larger structuring elements).

15.6.2 Elongated object parts

Closing with structuring elements of different sizes can be used to segment a picture into clusters (of high-valued pixels on a low-valued background) that have different densities. Similarly, opening with structuring elements of different sizes can be used to eliminate thin high-valued regions that have different thicknesses.

Figure 15.13 shows a binary picture P that contains sets of 1s that have different thicknesses and the results of opening P using the structuring elements $\sigma_1, \sigma_2,$ and σ_4. (These structuring elements are squares of sizes $3 \times 3, 5 \times 5,$ and 9×9; they are the sets of pixels with d_8-distances from o that are $\leq 1, \leq 2,$ and ≤ 4. Such

FIGURE 15.13 An example of how opening can be used to detect and eliminate sets of object pixels that have various thicknesses. Note that object pixels are black. Upper left: original object. Remaining pictures: results of opening using square structuring elements of sizes 3×3, 5×5, and 9×9.

a sequence of openings by balls of increasing diameter is called a *granulometry* in mathematic morphology [706, 969].) The σ_1-opening eliminates the thinnest sets of 1s; the σ_2-opening eliminates thicker sets of 1s; and the σ_4-opening eliminates all but the thickest sets of 1s. Another example is shown in Figure 15.14; here, the sets of high-valued pixels do not have constant thicknesses, but they can be almost entirely eliminated by opening using a structuring element of the appropriate size.

Opening using structuring elements of increasing sizes can be used to identify elongated sets of high-valued pixels in a picture. Let P be a binary picture, and let $\langle P \rangle_h$ be the set of 1s of P that remain after opening $\langle P \rangle$ using an $h \times h$ structuring element. Thus, $\langle P \rangle_h$ consists of "h-thick" parts of $\langle P \rangle$ (i.e., parts of which every pixel is contained in an $h \times h$ square that is contained in $\langle P \rangle$). This fact can be used to identify elongated parts of $\langle P \rangle$ that have thickness of at most h. Suppose we call an $h \times k$ rectangle "elongated" if $k \geq 3h$. Then a connected component of $\langle P \rangle - \langle P \rangle_h$ that contains at least $3h^2$ pixels must have been an elongated part of $\langle P \rangle$ having a thickness of less than h. For example, the sets of pixels in Figure 15.13 that are

15.6 Decomposition

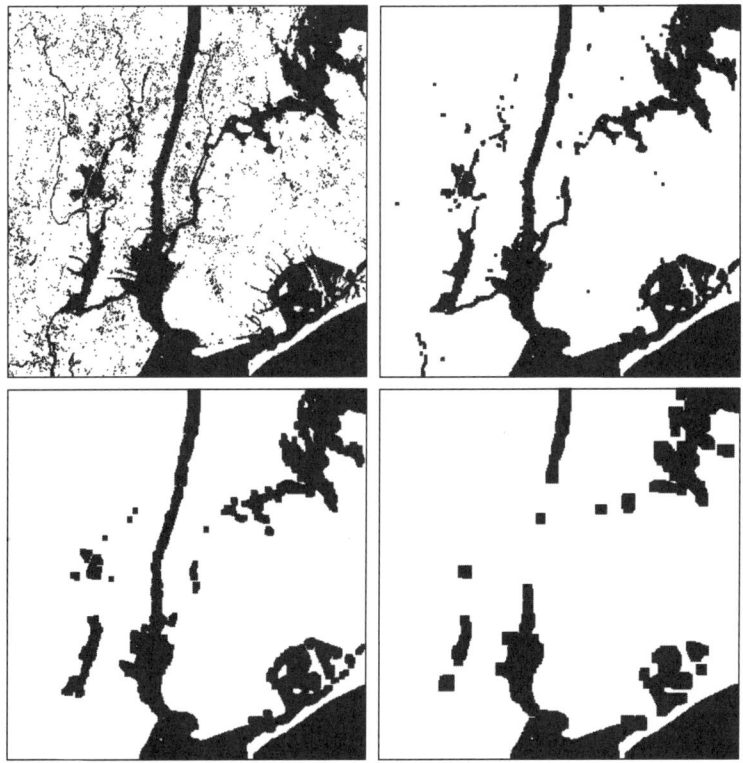

FIGURE 15.14 Opening of a thresholded aerial picture (upper left) using structuring elements of different sizes.

eliminated by σ_i-opening ($i = 1, 2, 4$) have 819, 2829, and 9681 pixels, respectively, so they are all elongated. Similarly, in Figure 15.14, the sets of pixels that are eliminated by σ_i-opening ($i = 1, 2, 3$) have 7824, 10,020, and 14,042 pixels, respectively, and so are all elongated.

An alternative method of detecting elongated sets is to use the union of openings by elongated structuring elements in all (main) directions.

15.6.3 Distance transforms and medial axes

We conclude this section by showing how morphologic operations can be used to compute distance transforms (see Section 3.4.2) and medial axes (see Section 3.4).

Proposition 15.2 Let P be a binary picture, let $P_{(0)} = P$, and let $P_{(i+1)}$ be the result of eroding $P_{(i)}$ ($i = 0, 1, \ldots, D-1$) using the structuring element $H \cup V$ or N_8 (see Figure 15.1), where D is the diameter of P. Then $\Sigma_{i=0}^{D} P_{(i)}$ is the distance transform of P; using $H \cup V$ gives the d_4 transform, and using N_8 gives the d_8 transform.

Proposition 15.3 Let $P_{(i)}$ be as in Proposition 15.2, and let $P_{(i)*}$ be the result of redilating $P_{(i)}$ (once) using the same structuring element that was used in the erosions. Thus $P_{(i)*}$ is an opening of $P_{(i-1)}$ so that, in accordance with Proposition 15.1, $P_{(i)*} \subseteq P_{(i-1)}$. Then $\bigcup_{i=1}^{D}(P_{(i-1)} \setminus P_{i*}) = M(P)$.

Note that this method of constructing the medial axis can also be applied to multi-valued pictures.

15.7 Exercises

1. If $\sigma = \{(x_1, y_1), \ldots, (x_n, y_n)\}$, prove that $P^{(\sigma)}$ is the max of the translates of P by $(-x_1, -y_1), \ldots, (-x_n, -y_n)$.

2. Let $H = \{(-1,0), (0,0), (1,0)\}$ and $V = \{(0,-1), (0,0), (0,1)\}$, and let $E = N_8$ consist of $(0,0)$ and its eight horizontal, vertical, and diagonal neighbors. Prove the following:
$$P^{(E)} = (P^{(H)})^{(V)} = (P^{(V)})^{(H)}$$

3. Let $\underline{\sigma}$ denote the reflection of σ in the origin (i.e., $\{(-x,-y) : (x,y) \in \sigma\}$). If σ is symmetric, we have $\sigma = \underline{\sigma}$. If P is a binary picture, prove that $\langle P^{(\sigma)} \rangle$ is the union of $\underline{\sigma}(p)$ for all $p \in \langle P \rangle$.

4. If $\sigma = \{(x_1, y_1), \ldots, (x_n, y_n)\}$, prove that $P_{(\sigma)}$ is the min of the translates of P by $(-x_1, -y_1), \ldots, (-x_k, -y_k)$.

5. Let P be a binary picture, and let $\overline{P}$ be the complement of P so that $\overline{P}$ has 0s where P has 1s and vice versa. Prove that, for any structuring element σ, we have the following:
$$\overline{P^{(\sigma)}} = \overline{P}_{(\underline{\sigma})} \text{ and } \overline{P_{(\sigma)}} = \overline{P}^{(\underline{\sigma})}$$

6. Generalize the concepts in this chapter to 3D pictures.

15.8 Commented Bibliography

The theoretic study of morphologic operations was initiated nearly 50 years ago [392] in connection with the problem of estimating the sizes of objects. For the

15.8 Commented Bibliography

origins of "mathematic morphology," see [708]. Early work on morphologic analysis of micrographs is described in [744], where dilation and erosion are called "plating" and "etching," and only simple symmetric structuring elements are used. See also Section 1.2.12 for historic references.

Systematic treatments of morphologic operations can be found in [417, 706, 969, 1012] (see also Chapter 9 in [370] and [418, 874]). For local minimum and maximum operations, see also [762]. The method of computing the distance transform and the medial axis described in Propositions 15.2 and 15.3 is described in [749]; for the multivalued case, see [808].

CHAPTER **16**

Deformations

In this chapter, we discuss operations that "deform" a picture to derive new pictures from it while preserving geometric properties of the original picture, particularly topologic properties. We emphasize deformations of 2D binary pictures, but we also briefly discuss deformations of 3D and multivalued pictures.

16.1 Topology-Preserving Deformations and Simple Pixels

In this section, we assume that P is a 2D binary picture. We will usually use 8-adjacency for the 1s and 4-adjacency for the 0s, and we will assume that the set $\langle P \rangle$ of 1s is finite, but our results also hold (with appropriate modifications) under the opposite assumptions. The 3D case will be discussed in Section 16.6 and the multivalued case in Section 16.7.

We have seen (see Section 4.2.2) that the (8,4) region-adjacency graph of a binary picture is a tree. If we take the infinite background component B of 0s as the root of this tree, then the region corresponding to any node of the tree surrounds the regions corresponding to the children of that node, and the leaves of the tree correspond to simply connected regions. Two binary pictures are topologically equivalent (isotopic) iff there is an isomorphism f between their rooted region-adjacency trees such that T is a child of S iff $f(T)$ is a child of $f(S)$ (see Proposition 6.4).

Let $A_8(p)$ be the set of pixels that are 8-adjacent to p, and let $N_8(p) = A_8(p) \cup \{p\}$. Evidently, p is (4- or 8-) adjacent to a (4- or 8-) component S of $\langle P \rangle$ or $\langle \overline{P} \rangle$ iff it is (4- or 8-) adjacent to $S \cap A_8(p)$. It is easily verified that at most four 4- or 8-components can intersect $A_8(p)$. These components are all 8-adjacent to p, and they are 4-adjacent to p iff they contain a 4-neighbor of p.

> **Definition 16.1** A pixel p is called (8,4) *simple* iff it is 8-adjacent to exactly one 8-component of 1s in $A_8(p)$ and 4-adjacent to exactly one 4-component of 0s in $A_8(p)$.

Some examples of simple and nonsimple pixels are shown in Figure 16.1.

If we use the grid cell model, we can give a criterion for the simplicity of p in terms of the *attachment set* of the grid square p. If $p \in \langle P \rangle$, we define the *P-attachment*

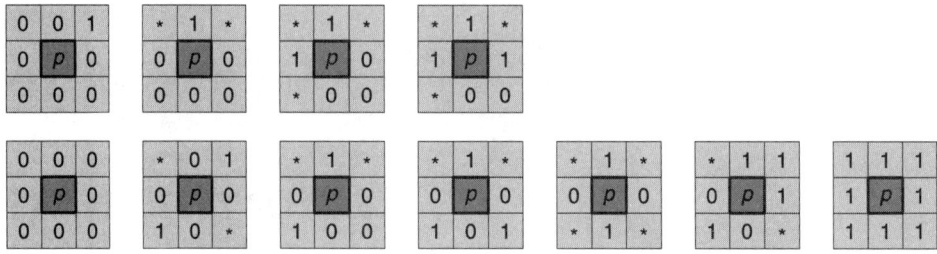

FIGURE 16.1 Examples of (upper row) simple pixels and (lower row) nonsimple pixels in the (8,4)-adjacency grid; p and the *s can be either 0 or 1.

set p_P as the union of intersections of the frontier ϑp (see Definition 5.11) with the frontier of some other pixel q of $\langle P \rangle$; in other words, $p_P = \bigcup_{q \in \langle P \rangle, q \neq p}(\vartheta p \cap \vartheta q)$. It can be shown that p is (8,4)-simple iff p_P is a connected proper subset of ϑp.

Theorem 16.1 Changing the value of a simple pixel of a picture P from 1 to 0 or from 0 to 1 results in a picture P' that is topologically equivalent to P.

Proof We make use of Proposition 6.4, in which topologic equivalence (isotopy of geometric representations in the grid cell model) is characterized by isomorphy of rooted region-adjacency graphs. Let p be 8-adjacent to the 8-component C of $\langle P \rangle$ and 4-adjacent to the 4-component D of $\langle \overline{P} \rangle$. If $p = 1$, let $C^* = C \cup \{p\}$, and, if $p = 0$, let $D^* = D \cup \{p\}$. It follows that C^* is 4-adjacent to D and that C is 8-adjacent to D^*; in the latter case, if a black and a white component are 8-adjacent, then they must also be 4-adjacent, because, for any pair of diagonally adjacent black or white pixels, the two "intermediate" 4-adjacent pixels must be either black or white. Hence each of them is either in the black or the white component, so C^* and D (C and D^*) are 4-adjacent.

If we change p from 1 to 0 (or vice versa), C^* and D become C and D^* (or vice versa); all other components of $\langle P \rangle$ and $\langle \overline{P} \rangle$ remain the same and

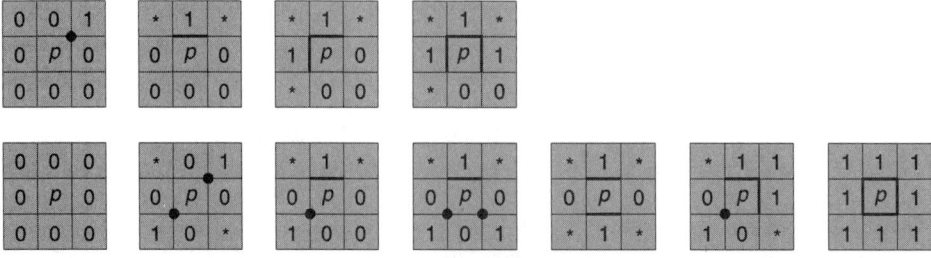

FIGURE 16.2 Attachment sets for the examples shown in Figure 16.1.

16.1 Topology-Preserving Deformations and Simple Pixels

adjacency relationships between pairs of components remain the same (with C^*, D replacing C, D^* or vice versa). Thus, the components of P' are in one-to-one correspondence with those of P, and this correspondence preserves adjacency. Hence the region-adjacency graphs of P and P' are isomorphic, and this isomorphism evidently preserves (parent, child) relationships in the rooted trees. ∎

Definition 16.2 Two pictures differ by a *simple deformation* (for brevity, SD) iff each of them can be obtained from the other by repeatedly changing simple pixels from 1 to 0 (or vice versa).

According to Theorem 16.1, SD is *topology-preserving*. It is easily verified that, when the value of a simple pixel is changed, the pixel remains simple; hence, SD is reversible. In the grid cell model, the geometric representations of two binary pictures (before and after removing a simple pixel) are isotopic; "squeezing" the simple pixel (a square) into its frontier is a continuous deformation from one representation to the other.

In the next two sections, we will discuss "one-way" SD processes that change simple 1s to 0s (but not vice versa). Such processes are used to "shrink" or "thin" components of 1s. In this section, we will establish some important properties of general ("two-way") SD processes. Specifically, we will show that SD can be used to magnify a binary picture or to translate it, and we will prove (see [916]) that, if two pictures are topologically equivalent, they differ by SD.

Let the portion of P that contains 1s have n rows that we number $1, \ldots, n$, starting from the bottom row; the row of 0s below the bottom row is numbered 0. Each column of P consists of alternating runs (maximal sequences) of 1s (if any) and 0s; a run with its uppermost pixel in row i will be denoted by r_i. (Note that there may be several r_is for a given i.) Let $h(n) \geq h(n-1) \geq \ldots \geq h(1) \geq h(0)$ be a monotonically nonincreasing sequence of nonnegative integers. We will dilate each r_i upward by the amount $h(i)$. A run r_i will be dilated upward only after the runs r_j (for all $j > i$) have been dilated upward. For each i, the runs r_i of 0s will all be dilated first (in any order), and then the runs r_i of 1s will be dilated (in any order). (If we were using 4-adjacency for the 0s and 8-adjacency for the 1s, we would dilate the 1s before the 0s.) Note that when the run (e.g., r_k) just below r_i is (later) dilated upward, r_i is re-eroded (shortened, from the bottom). We will show in the following paragraphs that any such sequence of upward dilations involves only changes in the values of simple pixels.

There are no runs r_n of 0s, because vertical runs of 0s that intersect the top row extend infinitely far into the background component B. Thus the dilation process begins by dilating each run r_n of 1s upward (into B) by the amount $h(n)$. Let t be the top pixel of any such run; then the pixel p just above t (see Figure 16.3) is in B, and so is its upper neighbor x. If $u = 0$ or $u = 1$ and u's run has not yet been dilated, we have $q = y = 0$; if $u = 1$ and u's run has already been dilated, we have $q = 1$; this is also similar for v, r, and z. In any case, p is simple; it is 4-adjacent to

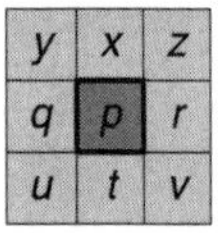

FIGURE 16.3 p is the pixel above the top pixel t of a run (see text).

only one 4-component of 1s (the component containing t) and 8-adjacent to only one 8-component of 0s (the component containing x). If we change p from 0 to 1 and replace t with p and p with x, the same argument shows that x is simple, and so on. Hence the upward dilation of r_n by $h(n)$ involves only changes of simple 0s to 1s.

The argument is similar (but slightly more complicated) regarding the runs r_i for $i < n$. We first consider the runs of 0s. Let r_i be a run of 0s that is about to be dilated (by $h(i)$), and let P^* be the picture just before dilation of r_i. (We can assume that $h(i) > 0$; otherwise no dilation is needed.) Let t be the uppermost pixel of r_i. The run of 1s in P just above t has its uppermost pixel in a row higher than i; hence, it has already been dilated by at least $h(i+1)$ so there are more than $h(i)$ 1s above t in P^*. Let u be t's west neighbor in P (as in Figure 16.3). If u is not the uppermost pixel of its run in P, that run has already been dilated by at least $h(i+1)$ (or is an infinite run of 0s and needs no dilation); hence, in P^*, there are more than $h(i)$ pixels above u that have the same value as u. If $u = 1$ is the uppermost pixel of its run in P, that run has not yet been dilated, but u's north neighbor x in P belongs to a run of 0s in P that has already been dilated by at least $h(i+1)$, so there are more than $h(i)$ 0s above u in P^*. Finally, if $u = 0$ is the uppermost pixel of its run in P, that run may or may not have been dilated yet. If it has not, the run of 1s above it in P has already been dilated by at least $h(i+1)$, so there are more than $h(i)$ 1s above u in P^*; however, if the run of 0s whose uppermost pixel in P being u has already been dilated, there are exactly $h(i)$ 0s above u in P^*. Similar remarks apply to v (and y). Thus the pixel values in P^* in the columns containing u and v are constant for at least $h(i)$ rows above t's row. Moreover, if q or r is 1 in P^*, there are more than $h(i)$ 1s above u or v in P^*. It follows easily from this that p is 4-adjacent to exactly one 4-component of 1s (which contains the pixel x above p; this is because t is 0 and, if q or r is 1, so is y or z, so q or r is 4-connected to x), and p is 8-adjacent to exactly one 8-component of 0s (which contains the pixel t; this is because, if y or z is 0, so is q or r, so y or z is 8-connected to t). Hence p is simple. If $h(i) > 1$, when we change p from 1 to 0, the pixel above it is still a 1, and it is still simple for the same reason; this continues to be true for at least $h(i)$ pixels above t. Thus, the run of 0s that has as its top pixel t can be dilated upward by at least the amount $h(i)$ by repeatedly changing simple 1s to 0s.

A similar argument applies to the runs r_i of 1s. Here the run of 0s in P beginning just above t has already been dilated by at least $h(i+1)$, so there are more than $h(i)$ 0s above t in P^*. The case in which u is not the uppermost pixel of its run in P is the same as in the previous paragraph. If $u = 0$ is the uppermost pixel of its run in P, that run has already been dilated by $h(i)$. If $u = 1$ is the uppermost pixel of its run in P, that run too may have already been dilated by $h(i)$, and, if it has not, the run

16.1 Topology-Preserving Deformations and Simple Pixels

of 0s above it has already been dilated by at least $h(i+1)$. Similar remarks apply to v (and y). Thus, for at least $h(i)$ rows above t, the pixel values in P^* in the columns containing u and v are constant. It follows easily that p can be 8-adjacent to only one 8-component of 0s (if u or $v = 0$, q or r must also be 0, so u or v is in the same 8-component of 0s as x) and can be 4-adjacent to only one 4-component of 1s (if q or $r = 1$, u or v must also be 1, so q or r is in the same 4-component of 1s as t); thus, p is simple. If $h(i) > 1$, when we change p from 0 to 1, the same argument holds, with q, r replacing u, v and y, z replacing q, r, and this continues to be true for at least $h(i)$ pixels above t. Thus, the run that has as its top pixel t can be dilated upward by at least the amount $h(i)$ by repeatedly changing simple 0s to 1s.

We can now prove the following:

Theorem 16.2 SD can be used to magnify a picture by any integer factor.

Proof We first use SD to dilate the vertical runs upward as described in the previous paragraphs using $h(i) = m \cdot i$. Let r_i be a run with its uppermost pixel in row i and with its lowest pixel in row $k+1 \leq i$ so that the length of r_i is $i-k$. Evidently, none of the dilations affect r_i except the dilation of r_i by $m \cdot i$ and the dilation of the run r_k below r_i by $m \cdot k$. The first dilation lengthens r_i by $m \cdot i$, and the second dilation shortens r_i by $m \cdot k$. Thus, the length of r_i becomes $(i-k) + mi - mk = (i-k) + (i-k)m = (i-k)(m+1)$, so the length of r_i is magnified by the factor $m+1$. Horizontal magnification is then achieved in an exactly analogous way using dilations of horizontal runs. ∎

Proposition 16.1 A picture can be translated horizontally or vertically by one step using SD.

Proof We give the proof for upward translation; the proofs for the other three directions are exactly analogous. P' is the upward translation of P by one step iff, for all (x,y), the value of P' at (x,y) is the same as the value of P at $(x, y-1)$. We again use SD to dilate the vertical runs upward using $h(i) = 1$ for $i = n, n-1, \ldots, 1, 0$. Let r be a run with its top pixel in row i and its bottom pixel in row $k+1$. When the process reaches row i, r is dilated upward by one pixel; when the process reaches row k, r is eroded from below by one pixel. The process does not affect r at any other stage, so its net effect is to translate r upward by one pixel. ∎

Proposition 16.2 A picture can be translated in a diagonal direction by one step using SD.

Proof A one-step diagonal translation can be achieved by a one-step horizontal translation followed by a one-step vertical translation (or vice versa). ∎

Theorem 16.3 A picture can be translated along any 4-path or 8-path using SD.

Proof The required translation consists of a sequence of one-step translations in isothetic or diagonal directions. ∎

Note that translation "along an 8-path" includes one-step horizontal or vertical translations that "complete" the 8-path into a 4-path.

According to Theorem 16.1, if two pictures differ by SD, they are topologically equivalent. We can now prove the converse:

Theorem 16.4 If two pictures are topologically equivalent, they differ by SD.

Proof Let the adjacency graph of the pictures be a rooted tree of height h; we recall that the root node corresponds with the infinite background component B of 0s. The children of the root correspond with components of 1s, which we call "top components." Note that each top component C corresponds with the root of a subtree of height at most $h-1$. We will now show how SD can be used to move the top components of any binary picture (together with all of the components that they surround) far apart from one another and to put each of them into a standard form.

Let C be a top component. Because C is adjacent to B, there is an 8-path ρ from some pixel of C to a distant pixel of B; all but the first pixel of ρ is in B. We can assume that ρ is a shortest such path; thus, ρ does not touch or cross itself and is evidently simply connected. According to Theorem 16.2, we can use SD to magnify the picture by an integer factor $t > 2h+1$. This expands each pixel into a $t \times t$ square; thus, ρ becomes a succession of such squares of 0s that either share a side or touch at a corner. Suppose two successive squares H and K of 0s touch at a corner; then two squares M and N (of 1s or 0s) also share that corner. We can thus, use SD to dilate H downward by any amount less than t. Suitable dilations of this kind, together with the magnification, convert ρ into a "thick 4-path" P^* of 0s of thickness of at least $2h+1$. Evidently, if ρ is simply connected, so is P^*. Because the far end of ρ (hence, of P^*) is distant from all of the components of the picture other than B, we can assume that the last part of ρ (and P^*) is straight, because once it gets far enough away from the other components, it need not bend to avoid them.

We now use SD to dilate C so as to create a simply 4-connected protrusion P of 1s of thickness $2h-1$ interior to P^*. To do this, we use SD to create a

4-path of 1s that extends along the midline of P^* by repeatedly changing 0s to 1s just beyond the endpoint of the path. We then thicken this path by creating another path alongside it (on each side) that is still interior to P^*; this can be done by repeatedly changing corner 0s to 1s. This process can be repeated until the desired thickness is reached, with P remaining interior to P^*. Note that, because the last part of P^* is straight, so is the last part of P.

Let D be a hole in C. Because C is 4-connected, the magnification guarantees that there is a thick 4-path Q^* of 1s from D to the beginning of P. We now use SD to dilate D so as to create a 4-connected protrusion Q of thickness $2h - 3$ interior to Q^*. When Q^* reaches the beginning of P, Q continues along P and fills the interior of P with 0s.

Let E be an island in D (a component of 1s adjacent to and surrounded by D). Because D is 8-connected, the magnification (and subsequent dilations) guarantees that there is a thick 4-path R^* of 0s from E to the beginning of Q. We now use SD to dilate E so as to create a 4-connected protrusion R of thickness $2h - 5$ interior to R^*. When R^* reaches the beginning of Q, R continues along Q and fills the interior of Q with 1s.

This process continues, using SD to create a nested collection of protrusions from $C, D, E, \ldots$ that correspond to successively lower nodes of C's subtree. Because the height of this subtree is at most $h - 1$, even when we reach a leaf node of C's subtree, there is still room for its protrusion in the nest. Because the component L that corresponds with the leaf node is simply connected (see Exercise 2 in Section 16.8), it and its protrusion can then be eroded down to the last row (or column) of pixels in the straight part of the protrusion so that it becomes a horizontal or vertical straight line segment.

If the parent K of L has other children, we can do the same thing to them; each protrusion stops eroding just before it reaches the place where the previous protrusion stopped, so the eroded protrusions remain nonadjacent. After the children have been eroded down to line segments in this way, there are no longer any holes in K, except far away along its protrusion. We can then erode K (see Exercise 3 in Section 16.8) until it is near the end of the straight part of its protrusion. We erode K until just before the point to which its children were eroded. This converts it into a hollow rectangle that just surrounds the line segments that resulted from the erosions of the children.

This process continues. When all of the children in the subtree of a region H have been eroded, H itself is eroded. Its siblings are dilated to create protrusions, the children of the siblings are dilated to create nested protrusions, and so on until leaves are reached; these can then be eroded. This process continues until C itself has been eroded.

The result of the entire process is a concatenation of nests of hollow rectangles or line segments; each rectangle surrounds the rectangles or segments obtained from its children and is adjacent to the rectangles or segments obtained from its siblings. The sequence in which this concatenation is created depends on the order in which we process the children of each component; however, if two Cs have isomorphic subtrees, there exist processing orders for the Cs that result in congruent concatenations.

Using this process, we can transform each top component C to a canonic form (a concatenation of nests) that depends only on C's subtree and that is located far away from the other top components (or from their canonic forms). If two pictures P and P' are topologically equivalent, they can give rise to the same set of canonic forms. SD can then be used (see Theorem 16.3) to translate the canonic forms; because they are far apart, they have room to translate. Hence SD can be used to rearrange the canonic forms obtained from P so that they occupy the same relative positions as the canonic forms obtained from P'. By reversing the process that was used to create P''s canonic form, we can then reconstruct P'. Thus SD can be used to transform P to a canonic form, rearrange that form (if necessary), and transform it into P'. ∎

16.2 Shrinking

We saw in Section 16.1 that changing the value of a simple pixel from 1 to 0 is topology-preserving. It is known that, if a simply connected component of $\langle P \rangle$ contains at least two pixels, it contains at least two simple pixels (Exercise 2 in Section 16.8). It follows that, if we repeatedly change the values of simple pixels from 1 to 0 until no further change is possible, every simply connected component of $\langle P \rangle$ shrinks to a single pixel. Note that, at every step of this shrinking process, the components remain simply connected.

Algorithms can also be defined that shrink simply connected components of $\langle P \rangle$ to single pixels more rapidly by repeatedly changing the values of many simple pixels from 1 to 0 simultaneously ("in parallel") while preserving topology. Note that, if we changed the values of all simple pixels from 1 to 0 simultaneously, topology would sometimes not be preserved; for example, if $\langle P \rangle$ is a 2×2 block of 1s, all of its pixels are simple, so changing their values from 1 to 0 destroys a component of 1s. To preserve topology while shrinking in parallel, it is necessary to be selective about which pixels have their values changed from 1 to 0. The following are some examples of algorithms that perform selective parallel shrinking. (In this section, we use 4-adjacency for 1s and 8-adjacency for 0s.)

a) Repeatedly change the values of all pixels p from 1 to 0 if they satisfy both of the following conditions (where (x,y) are the coordinates of p and where "$(i,j) = 0$" means "the pixel with the coordinates (i,j) has value 0"):

(1) p is simple and $(x, y-1) = 0$, or $(x, y+1) = (x+1, y) = (x-1, y) = 0$, but $(x, y-1) = 1$.

16.2 Shrinking

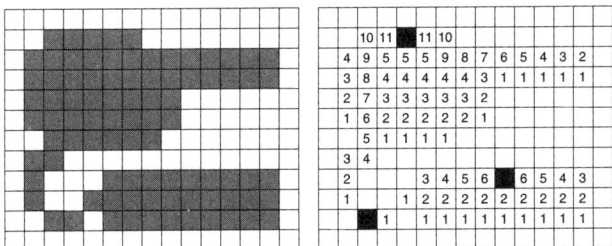

FIGURE 16.4 Left: picture (1s = black, 0s = white). Right: result of applying algorithm (a). The number in each cell is the step at which the value of that pixel is changed from 1 to 0.

(2) p's neighborhood is neither $\begin{smallmatrix} & 0 & 0 & \\ 0 & p & 1 & 0 \\ & 0 & 0 & \end{smallmatrix}$ nor $\begin{smallmatrix} & 0 & & \\ 0 & 1 & 0 \\ 0 & p & 0 \\ & 0 & & \end{smallmatrix}$.

An example of shrinking using this algorithm is shown in Figure 16.4.

b) A variant of algorithm (a) alternates between two criteria for changing pixels from 1 to 0: At odd-numbered steps, it uses criterion (1); at even-numbered steps, it changes all pixels p from 1 to 0 if they are simple and either $(x, y-1) = (x-1, y) = 0$ or $(x, y+1) = (x+1, y) = 0$. An algorithm that uses alternating criteria for changing 1s to 0s is called a *subiteration* algorithm. The application of this algorithm to the same picture is shown in Figure 16.5.

c) Another type of algorithm is based on partitioning the pixels of P into "subfields"; for example, the evenness or oddness of the x- and y-coordinates of a pixel define four subfields (both odd, both even, x odd and y even, or vice versa). The algorithm operates on one subfield at a time; at each iteration, it changes simple pixels from 1 to 0 iff they belong to one of the subfields. Note that two 8-adjacent pixels cannot belong to the same subfield; hence, topology preservation is assured, even if all of the simple pixels in a subfield are changed

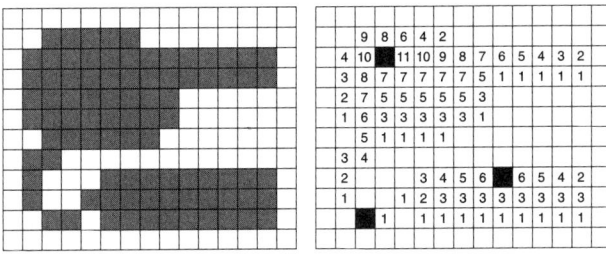

FIGURE 16.5 Picture and result of applying algorithm (b) The notation is the same as in Figure 16.4.

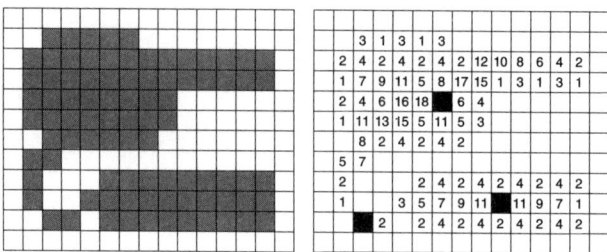

FIGURE 16.6   Picture and result of applying approach (c). The top row and left column have odd coordinates.

from 1 to 0 simultaneously. The application of this approach to the same picture is shown in Figure 16.6. (It is also common to use two "checkerboard" subfields, which are defined by the parity of $x+y$.)

In general, simultaneously changing the values of a set S of pixels of $\langle P \rangle$ from 1 to 0 preserves topology iff the pixels of S can be arranged in a sequence $p_1, p_2, \ldots$ such that, if the values of the p_is are successively changed from 1 to 0, p_i is simple just before its value is changed. Let $\mathcal{A}$ be a shrinking algorithm that consists of a sequence of iterations $\mathcal{A}_j$. We call a pixel p $\mathcal{A}_j$-simple if p satisfies the conditions used in $\mathcal{A}_j$ for changing the value of a pixel from 1 to 0. The following result is from [872]:

Proposition 16.3 $\mathcal{A}$ preserves topology if, for every j, the following are true:

a) Every $\mathcal{A}_j$-simple 1 is simple.

b8) For every pair of 8-adjacent $\mathcal{A}_j$-simple 1s, each of the 1s in the pair remains $\mathcal{A}_j$-simple when the value of the other 1 in the pair is changed to 0.

If we use 8-adjacency for $\langle P \rangle$ and 4-adjacency for $\langle \overline{P} \rangle$, then $\mathcal{A}$ preserves topology if, for every j, the following are true:

a) Every $\mathcal{A}_j$-simple 1 is simple.

b4) For every pair of 4-adjacent $\mathcal{A}_j$-simple 1s, each of the 1s in the pair remains $\mathcal{A}_j$-simple when the value of the other 1 in the pair is changed to 0.

c) In any set of mutually 8-adjacent 1s, at least one of them is not $\mathcal{A}_j$-simple.

[75] introduced "P-simple pixels," which allow a necessary and sufficient characterization of topology preservation.

Many topology-preserving shrinking algorithms have been defined. Most of them are used for thinning (see Section 16.3) (i.e., they are designed not only to preserve topology but also to have no effect on "thin" subsets of $\langle P \rangle$). (However, a few paragraphs below, "thinning" and "skeletonization" are used synonymously. Section 16.3 deals particularly with skeletonization, which is a class of locally defined topology-preserving shrinking methods that preserves arcs [881].) Most of these algorithms are of the subiteration type, a few are of the subfield type, and some are even "fully parallel."

The shrinking algorithms described so far are capable of shrinking simply connected components of $\langle P \rangle$ to single pixels; however, because these algorithms can change only simple 1s to 0s, they cannot shrink non–simply connected components to single pixels. (A connected component that has a hole will shrink into a simple closed curve.) For example, no pixel of a simple curve C is simple; hence, if C is a component of $\langle P \rangle$, an algorithm that changes only simple 1s to 0s has no effect on C.

We conclude this section by describing an algorithm that shrinks every component of $\langle P \rangle$ or $\langle \overline{P} \rangle$, except for the background component, to a single pixel. The algorithm also removes single-pixel components (note that this violates topology preservation) so that these components cannot interfere with the shrinking of components that surround them. The following version of this algorithm assumes that we use 4-adjacency for $\langle P \rangle$ and 8-adjacency for $\langle \overline{P} \rangle$. Let p and its east, south, and southeast neighbors have the following values, where z, u, v, and w are 0 or 1:

Then p becomes (or remains) 1 iff $z+u=2$, $z+v=2$, or $u+v+w=3$. An example of the operation of this algorithm is shown in Figure 16.7; some of its properties are described in Exercises 3 and 4 in Section 16.8.

16.3 Thinning

In this section, we discuss algorithms that shrink each component K of $\langle P \rangle$ into a connected *skeleton* S: a union of arcs or curves that are centrally located in K. (If we use 4- (8-)adjacency for $\langle P \rangle$, these are 4- (8-)arcs and 4- (8-)curves. Most thinning algorithms use 8-adjacency for $\langle P \rangle$, because this yields thinner skeletons.[1]) If K consists of elongated parts, there will ideally be an arc or curve of S centrally located in each of these parts, as illustrated schematically in Figure 16.8. A process that shrinks K into a skeleton is called *thinning* or *skeletonization*. We require that the process be topology-preserving; S must be connected, and, if H is a hole in K, there must be a unique hole H' in S that contains H (H' is the region surrounded by a closed curve that is contained in S). However, we require more than just topology preservation [868]; "drilling a hole" at one place and connecting the hole

---

[1]. For example, when we use 8-adjacency for $\langle P \rangle$, the skeleton of a diagonally oriented rectangle is a diagonal line segment; however, when we use 4-adjacency, the skeleton is a diagonal staircase.

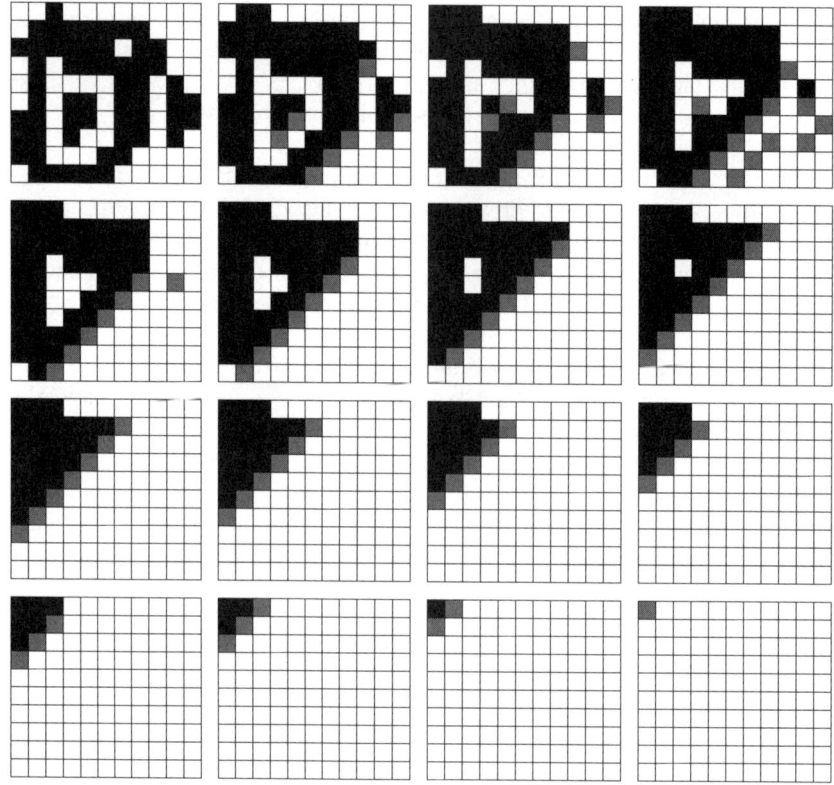

FIGURE 16.7 Sixteen steps in the operation of an algorithm that shrinks every (non-background) component to an isolated pixel (starting with the picture on the upper left). Gray squares show black pixels that are removed at that step.

to a complementary component in another place may preserve topology, but it does not correspond to our idea of "adequate shape representation."

Because S must be centrally located in K, the pixels of S must be as far away as possible from the complement $\overline{K}$ of K. This implies that (most of) these pixels belong to the medial axis of K (see Section 3.4.4). In particular, if A is an arc (a "branch") of S contained in an elongated part L of K, an endpoint of A must be near the border of L at a point where the border has high curvature (see Section 10.4).

The geometric requirement of a centrally located skeleton can also be specified in topologic terms; the inclusion relations between the components of a skeleton S and the components of the given set K and between the holes of S and the holes of K must both be bijections (one-to-one and onto). In other words, a component cannot split or disappear, and a hole cannot be newly created or merged with another complementary component. [868] shows that this topologic constraint is equivalent to a skeletonization process defined by a sequence of removals of simple pixels. To ensure topology preservation and these two bijections, most thinning algorithms

16.3 Thinning

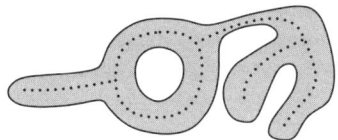

FIGURE 16.8 Schematic illustration of the desired result of thinning. A connected object (gray) that has a hole and protrusions; its skeleton (dotted) has a loop that surrounds the hole and branches that are centrally located in the protrusions.

operate by removing simple pixels from K by changing their values from 1 to 0. As we saw in Section 16.2, topology preservation and these two bijections can be guaranteed, even if sets of simple pixels are removed simultaneously. To ensure that the skeleton is centrally located, a thinning algorithm that removes simple pixels from K must remove them from all sides of K. (A pixel p at location (x, y) is called a north border pixel if $(x, y + 1)$ has value 0; a south border pixel if $(x, y - 1)$ has value 0; an east border pixel if $(x + 1, y)$ has value 0; and a west border pixel if $(x - 1, y)$ has value 0.) We cannot remove simple pixels from all sides of K at once; as we saw in Section 16.2, this would completely destroy Ks that are only two pixels thick. However, we can remove them from one side at a time, and we can obtain a centrally located skeleton by removing them from opposite sides alternatingly (e.g., successively removing sets of simple pixels that are north, east, south, west, north, and so on border pixels). Four subiterations can be reduced to two by removing both north and east pixels in one subiteration and both south and west pixels in the second.

It should be pointed out that central location of the skeleton can be achieved only approximately. For example, the skeleton of an upright rectangle R of height 2 and length n cannot lie exactly on the midline of R, because the midline passes midway between two rows of pixels. In general, if K has even width and S is thin (i.e., has unit width), then either the position of S must be biased (it must lie on one side of the midline of K), or S must zigzag from one side of the midline to the other. Zigzags in skeletons of sets of even width are especially relevant in algorithms that use subfields. Algorithms that use directional subiterations do not create zigzags.

A thinning algorithm must have no effect on parts of K that are already thin. To ensure this, we must not remove a simple pixel from K if it might be an arc endpoint. We regard a pixel p of a 4-connected set K as an arc endpoint if it is 4-adjacent to exactly one pixel q of K; in order to allow an endpoint of a diagonal staircase to be an arc endpoint, we also allow one diagonal neighbor of p that is 4-adjacent to q to be in K. Thus we call p an arc endpoint if only one 4-component of 1s in $A_8(p)$ is 4-adjacent to p and this component has either one or two pixels. Note that, when this type of local definition of an arc endpoint is used, minor irregularities on the border of K may qualify as arc endpoints, so S may have noisy "spurs." Less noisy results can be obtained by estimating the curvature of the border of K and regarding a simple pixel p (which evidently must be a border pixel) as an arc endpoint only if the curvature of the border at p is sufficiently high.

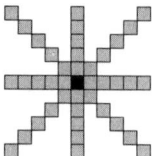

FIGURE 16.9 This set of gray pixels has no simple pixels except for arc endpoints, but it is not entirely thin, because it has an interior pixel (black).

It should be pointed out that, even if S has no simple pixels except for arc endpoints, it can still have interior pixels, so it may not be entirely thin. An 8-connected example of this situation is shown in Figure 16.9.

An example of thinning by successively removing simple north, south, east, west, north, and so on border pixels [881] that are not arc endpoints is shown in Figure 16.10. This algorithm determines whether to remove a pixel p at a given iteration by examining $N_8(p)$ only. Algorithms can be defined that require fewer iterations but that apply different tests to the $N_8(p)$s at alternating iterations or examine larger neighborhoods of the ps. Algorithms based on subfields (see Section 16.2) can also be defined. Another type of thinning algorithm removes border pixels of K that are visited only once when tracing the border and that are not adjacent to interior pixels of K.

Definition 16.3 A maximal subset A of a skeleton S that is an arc and such that no non-endpoint of A is adjacent to any pixel of $S \setminus A$ is called a *branch* of S.

It is not hard to see that S can be decomposed into a set of curves and branches (junctions can be decomposed into branches); note that the endpoints of a branch can be adjacent to curve points or to endpoints of other branches. (In the S shown in Figure 16.9, the interior pixel is a singleton branch.) If a branch endpoint has only one neighbor in S, it is called an *end*; if it has more than two neighbors in S, it is called a *junction*; otherwise, it is called *normal*.

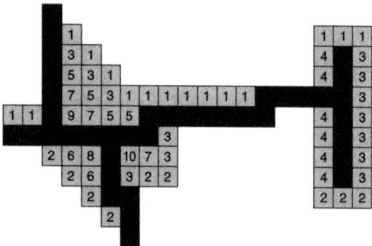

FIGURE 16.10 A (4-,8-) skeleton. The number in each cell is the step at which that 1 was changed to 0. The black pixels are 1s that were never changed; they belong to the skeleton.

16.4 Deformations of Curves

In general, the parts of K that give rise to branches of S will not all be equally significant. One way to evaluate the significance of a branch A is to examine the values of the distance transform of K (see Section 3.4.2) at the pixels of A. If the values are high, A must have arisen from a thick elongated part of K; if the values decrease as an endpoint of A is approached, A must have arisen from a tapering part of K; if they increase as the endpoint is approached, A must have arisen from a bulbous part of K; and so on. Criteria based on the distance values can be defined for pruning (eliminating or shortening) branches of S so as to retain only branches that should arise from perceptually significant parts of K.

16.4 **Deformations of Curves**

We recall (see Definition 7.7) that a *simple (4-,8-) curve* C (from now on we omit "simple") is a finite nonempty (4-,8-) connected set of pixels, each of which has exactly two (4-,8-) neighbors in C. Figure 16.11 shows some examples in the grid cell model (where we should use the terms *1-curve* and *0-curve*). We usually assume that a 4-curve has at least eight pixels and an 8-curve has at least four pixels, but it will be convenient in this section to regard a (4-,8-) isolated pixel as a *trivial (4-,8-) curve*.

We could say that two (4-,8-) curves C and D differ by a *local deformation* iff every pixel of C is a neighbor of a pixel of D (and vice versa). (If C and D are 4-(8-) curves, "neighbor" means 8-(4-) neighbor.) We could also say that D is a deformation of C (or vice versa) iff there exists a sequence of (4-,8-) curves $C_0, C_1, \ldots, C_n$ such that $C_0 = C$, $C_n = D$, and C_i and C_{i-1} differ by a local deformation ($1 \leq i \leq n$). Unfortunately, these definitions are not restrictive enough. Let C be a subset of a connected set S in the Euclidean plane that has a hole H. If C surrounds (does not surround) H and C is continuously deformed while remaining a subset of S, the deformed C still surrounds (does not surround) H. Hence, if C surrounds H and D does not, they cannot be continuous deformations "in S" (i.e., isotopic with respect to the topologic space defined by base set S) of one another. On the other hand, Figure 16.12 shows two 4-curves (8-curves) C and D such that C surrounds the pixel p but D does not, and C and D differ by a local deformation. Because p could be a one-pixel hole, the examples in Figure 16.12 show that two (4-,8-) curves can differ by a deformation even if one of them surrounds a hole and the other does not; this can be true even if the hole is large (see Figure 16.13).

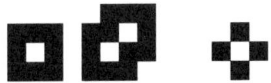

FIGURE 16.11 Left: two 4-curves. Right: an 8-curve.

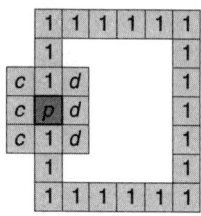 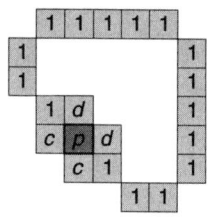

FIGURE 16.12 Two curves that differ by a local deformation and a pixel p that is inside of one curve but outside of the other. Left: 4-curves. Right: 8-curves. The cs are pixels of C, the ds are pixels of D, and the 1s are pixels of both C and D.

FIGURE 16.13 C and D differ by a local deformation, but a large hole (the ps) is inside of C and outside of D. (The cs and ds are as in Figure 16.12.)

A better way to define deformations of a (4-,8-) curve C is to make use of the fact (see Section 7.3.2) that, if C is nontrivial, its complement $\overline{C}$ consists of two (8-,4-) components C_I and C_O (the "inside" and "outside" of C). As we saw in Section 7.3.2, C (8-,4-) separates C_I from C_O; it (8-,4-) surrounds C_I (but surrounds no pixel of C_O) and is (8-,4-) surrounded by C_O; every pixel of C is (8-,4-) adjacent to both C_I and C_O. (The prefixes 4- and 8- will usually be omitted in what follows.)

Definition 16.4 We say that two curves C and D differ by a *strong local deformation* if C_I and D_O are disjoint and C_O and D_I are disjoint.

Proposition 16.4 If C and D differ by a strong local deformation, they differ by a local deformation.

Proof Because the definition of a (strong) local deformation is symmetric in C and D, we need to prove only one side of the conclusion (e.g., that any pixel of C coincides with or is adjacent to a pixel of D). Let p be a pixel of C that is not on D; then p is either in D_I or D_O (e.g., the former). Let q be a neighbor

16.4 Deformations of Curves

of p that lies in C_O. Because D separates D_I from D_O, q cannot be in D_O, and, because D_I and C_O are disjoint, q cannot be in D_I. Hence q must be on D so that p is adjacent to a pixel of D. A symmetric argument shows that, if p is in D_O, it must be adjacent to a pixel of D. ∎

Note that the pairs of curves in Figures 16.12 and 16.13 do not differ by a strong local deformation; they violate the first part of the definition. We say that C and D differ by a *strong deformation* if there exists a sequence of curves $C_0, C_1, \ldots, C_n$ such that $C_0 = C$, $C_n = D$, and C_i and C_{i-1} differ by a strong local deformation ($1 \leq i \leq n$).

We will now study the topology preservation properties of strong (local) deformations. We assume that the curves $(C, D, \ldots)$ are subsets of a connected set T and that the deformations take place "in T." We will usually assume that T is 4-connected and that the curves are 4-curves; it can be shown that similar results hold in the 8-case.

Proposition 16.5 Let H be a hole in T. C and D cannot differ by a strong local deformation in T if C surrounds H but D does not.

Proof If C surrounds H, H must be contained in C_I, and if D does not surround H, H must be contained in D_O; however, if C and D differ by a strong local deformation, C_I and D_O must be disjoint. ∎

Corollary 16.1 C and D cannot differ by a strong deformation in T if C surrounds H but D does not.

Proof Let C_j ($1 \leq j \leq n$) be the first C_i that does not surround H; then C_{j-1} and C_j violate Proposition 16.5. ∎

Proposition 16.6 If neither C nor D surrounds a hole in T, they differ by a strong deformation in T.

Proof Clearly any two trivial (one-pixel) "curves" $C, D \subseteq T$ differ by a strong deformation. Indeed, because T is connected, we can define a sequence of one-pixel translations (which are evidently strong local deformations) that move C

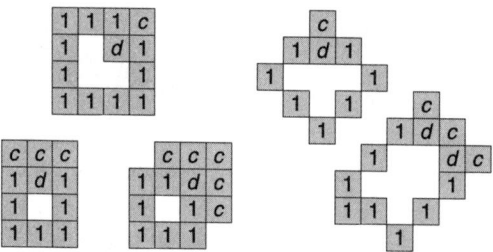

FIGURE 16.14 Examples of strong local deformations in the neighborhoods of simple pixels. Left: 4-curves. Right: 8-curves.

through T until it becomes D. To complete the proof, we will show that there exists a sequence of strong local deformations that "shrinks" C (or D) into a trivial curve.

It is well known (Exercise 2 in Section 16.8) that, if C_I is simply connected and has more than one pixel, it has at least two simple pixels. It can be verified that C can be modified in the neighborhood of a simple p to create a C' such that (a) C' passes through p; (b) C'_I is contained in C_I; and (c) C and C' differ by a strong local deformation. Several examples of this are shown in Figure 16.14. From (a) and (b), it follows that C'_I has strictly fewer pixels than C_I; hence, repeating this process eventually results in a C^* such that C^*_I consists of only one pixel. Such a C^* (see Figure 16.11) is the smallest possible nontrivial curve. Evidently such a C^* differs from the trivial curve (which coincides with C^*_I) by a strong local deformation. Thus we can construct a sequence of strong local deformations that shrinks C (or D) into a trivial curve, and we can reverse the latter sequence to expand a trivial curve into D. ∎

From Corollary 16.1 and the proof of Proposition 16.6, we have the following:

Theorem 16.5 C differs from a trivial curve by a strong deformation in T iff it does not surround a hole in T.

Our main goal in the remainder of this section is to prove that, if C and D both surround a single hole H in T, they differ by a strong deformation in T. To do this, we first need the following:

Lemma 16.1 Let W be any finite 4-connected set of pixels, and let W^* be the dilation of W (see Section 15.1) using structuring element N_8; then any border of W^* is a 4-curve.

16.4 Deformations of Curves

FIGURE 16.15 Neighborhoods of a border pixel.

Proof Let E be a border of T^* (e.g., its Z-border), and let q be any pixel of E. We will show that q can have only two 4-neighbors that are on E; hence, E is a 4-curve (see Exercise 5 in Section 7.6).

Let p and r be the predecessor and successor of q when we traverse the border E (see Algorithm 4.3). Then p and r are 4-adjacent to q, so the neighborhood of q looks like one of the patterns in Figure 16.15 or their rotations by multiples of $90°$, where the 0s are in Z. We must show that no pixel of E other than p or r can be 4-adjacent to q.

The case on the left is immediate, because the other two 4-neighbors of q are 0s. In the case in the middle, suppose the south neighbor q' of q is a border pixel of W^*; then q' is not in W (because the pixels of W are interior pixels of W^*), and some neighbor z of q' other than p, q, or r is 0. p and r are in W^* but not in W; hence, they have 8-neighbors p' and r' that are in W. Let ρ be a shortest 4-path in W from any such p' to any such r'. Because ρ is shortest, nonconsecutive pixels of ρ cannot be the same and cannot be 4-adjacent, and ρ is not 4-adjacent to p or r, except possibly at its endpoints; also, ρ cannot be 4-adjacent to q, because the 4-neighbors of q are not in W. Hence we can complete ρ into a 4-curve κ in W^* by adjoining to it p, q, r, and, if necessary, common 4-neighbors of p and p' and r and r' in W^*. The column of pixels that contains q crosses κ (at q); hence, (as in Section 7.3.2), κ separates q' from q's north neighbor and (because κ is a 4-curve) z and q' are in the same 8-component of the complement of κ. Hence z and q's north neighbor are in different 8-components of the complement of W^*, so q' is on a different border of W^* than q is.

The proof in the case on the right in Figure 16.5 is similar; there are several subcases, depending on whether the east or north neighbor q' of q is a border pixel of W^*. Here the row or column of pixels that contains q may not cross κ, but a nearly horizontal or nearly vertical digital straight line (see Chapter 9) must cross κ. For example, if q' is the east neighbor of q, we can use the nearly horizontal line that consists of the part of q's row to the right of q (including q itself) and the part of r's row to the left of r (including r itself). Evidently, p and r' are on opposite sides of this line, so the line crosses κ. Further details of the proof in this case will not be given here. ∎

Proposition 16.7 Let D and C be curves such that C surrounds D and there are no 0s "between" D and C (i.e., no holes in $D_O \cap C_I$); then D and C differ by a strong deformation in T.

Proof If every pixel of C is on D, we evidently have $D = C$, and we are done. If some pixel p of C is not on D, it is in D_O. Because D separates D_O from D_I, the neighbors of p cannot be in D_I. Thus C can be locally diverted around p through neighbors of p that are in C_I but not in D_I (so that these neighbors cannot be 0s). If this diversion C' is still a curve (see below), it is evidently a strong local deformation of C. Moreover, C' still surrounds D, and the intersection of $C' \cup C'_I$ with D_O has a smaller area than the intersection of $C \cup C_I$ with D_O. Thus, repeating this process eventually gives us a $C^{(n)}$ such that the intersection of $C^{(n)} \cup C_I^{(n)}$ with D_O is empty, which implies that $C^{(n)} = D$.

Let p be a pixel of $C \cap D_O$ such that C cannot be diverted around p into C_I while remaining a curve. Then p must have an 8-neighbor r that is on C (or 4-adjacent to a pixel q of C) such that r or q (respectively) is not close to p in the traversal of C (i.e., is at least three steps away from p). Suppose first that r is on C; then it cannot be a 4-neighbor of p, and, because r is not close to p along C, the common 4-neighbors of p and r cannot be on C. Let r' be one of these common 4-neighbors. r' must be inside C, and it must be on or outside of D, because D cannot pass between p and r; hence, r' cannot be 0. It is not hard to see that at most one of the arcs pr and rp of C (e.g., pr) can be 4-adjacent to r' at a pixel of C that is not close to p or r along C, and, if there are two such pixels u and v, they must be two steps apart along pr. In any case, the arc pr (or the arcs pu and ur or pv and vr) together with r' and the arc rp together with r' must be curves. Evidently, D cannot pass between p and r; hence, one of these curves must still surround D, and there are neither 0s nor pixels of D in the interior(s) of the other(s). By induction on the length of C, the curve that surrounds D can be diverted as in the previous paragraph, and the other curve(s) can be eliminated entirely by a sequence of strong local deformations using the methods described earlier in this section for curves that do not surround holes.

The argument is similar if r is not on C but is 4-adjacent to a pixel q of C. Here r may be a 4-neighbor of p; in this case, each of the arcs pq and qp of C can be 4-adjacent to r at pixels other than their endpoints p and q. However, this can happen at most at two pixels (e.g., u and v). The subarcs of C between p and q and these pixels (e.g., pu, uv, vq), together with r, must be curves, one of which surrounds D. Finally, if r is an 8-neighbor of p, one or both of their common 4-neighbors (r_1 and r_2) may be consecutive to p on C; in any case, because r has only one other 4-neighbor (in addition to r_1, r_2, and q), only one of the arcs pq (or r_1q) or qp (or qr_2) can be 4-adjacent to r at a nonendpoint (e.g., u), and the subarcs between the pixels at which they are adjacent (together with r) must be curves, one of which surrounds D. ∎

The proof of Proposition 16.7 says nothing about possible 0s in D_I; it is valid even if D (and therefore C) surrounds many holes.

We can now prove the following:

16.4 Deformations of Curves

Theorem 16.6 If C and D both surround a single hole H in T, they differ by a strong deformation in T.

Proof We will show in the next paragraph that, for any hole H in T, there is a "smallest" curve D_H that surrounds H (i.e., if C is any curve that surrounds H, then D_H is contained in $C \cup C_I$). (It should be pointed out that there need not exist a curve that surrounds *only* a given hole H. For example, if H has a narrow-necked concavity, there can be another hole K surrounded by 1s that are inside the concavity, and any curve that surrounds H must also surround K.) According to Proposition 16.7, C and D_H differ by a strong deformation, and so do D and D_H, which implies this Theorem.

To see that there is a "smallest" curve surrounding any H, note first that the set of 1s adjacent to the outer border of H (the outer "coborder" of H) is not necessarily a curve, because if H has a narrow concavity, this set of 1s may intersect or touch itself. However, according to Lemma 16.1, if we dilate H using N_8, the outer border of the resulting set H^* is a curve. This outer border consists entirely of 1s of the original picture prior to the dilation. It surrounds H^* and hence, surrounds H. Moreover, if C is any curve that surrounds H, H^* is contained in $C \cup C_I$. Hence the outer border of H^* is the smallest curve that surrounds H. ∎

Proposition 16.8 If C and D surround the same set of holes in T and there are no holes outside of either of them, they differ by a strong deformation in T.

Proof If the outer border of T is a curve (call it E), it surrounds both C and D, and, because there are no holes outside of C and D, there are no 0s between either of them and E. Thus, by the proof of Proposition 16.7, C and D differ by strong deformations from E and hence, from each other. If the outer border of T is not a curve, we can proceed as follows. Temporarily change the 0s in the holes of T to 1s; call the result $T^{(1)}$. Erode $T^{(1)}$ using N_8; call the result $T^{(2)}$. Dilate $T^{(2)}$ using N_8; call the result $T^{(3)}$. Because $T^{(3)}$ is a closing of $T^{(1)}$, it is contained in $T^{(1)}$ and contains all of the original hole pixels, but it has none of them on its border, so we can change the hole pixels back to 0s; call this final result T^*. Any curve C in S must still be contained in T^*, because the interior of C must have survived the erosion, so C must be contained in the redilation. Note that T^* is not necessarily connected, but (except for the holes that were in T) its components must be simply connected. Because C is connected, it must be contained in some component W of T^*, and any holes surrounded by C must also be surrounded by W. Thus, if C and D both surround the same set of holes, they must be contained in the same W. However, because W is the

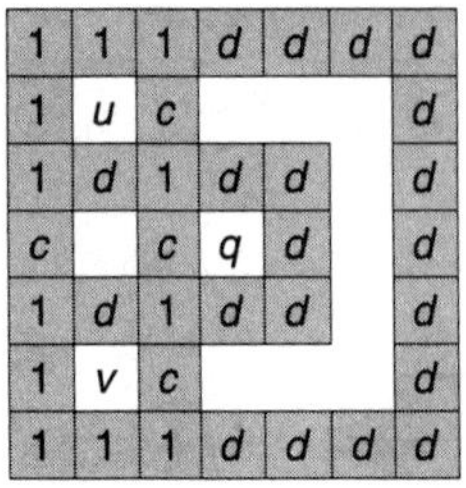

FIGURE 16.16 C and D both surround holes u and v, and hole q is outside of both of them, but they do not differ by a strong deformation. (The cs and ds are as in Figure 16.12.)

> result of a dilation, its outer border is a curve (see Lemma 16.1), and this curve E surrounds both C and D. ∎

According to Theorems 16.5 and 16.6, if C and D surround no holes or surround the same (one) hole, they differ by a strong deformation in T, even if there are other holes outside of them. On the other hand, if C and D both surround the same two holes and there is another hole outside of them, they may not differ by a strong deformation, as we see from the example in Figure 16.16. (Note that, in this example, any curve E that surrounds both C and D also surrounds q; thus, E does not surround the same set of holes that C and D do.) To see that C and D in Figure 16.16 cannot differ by a strong deformation, let $C^* = C \cup C_I$ and $D^* = D \cup D_I$, and note that $C^* \cup D^*$ has a hole that contains a 0 (the pixel q). If C and D differed by a strong deformation, there would exist a sequence of strong local deformations that would make C and D coincide, thus, eliminating the hole. However, 0s cannot cross the hole's coborder, because it consists of 1s that belong to C and D; hence, no sequence of strong local deformations can eliminate the hole.

16.5 Interchangeable Pairs of Pixels

In this section, we discuss local deformations that preserve the topology of a binary picture P and that also preserve the number of 1s in the picture. A pair of adjacent pixels p and q of P is called *interchangeable* if p and q have opposite values (i.e., $p = 1$ and $q = 0$, or vice versa) and interchanging their values preserves the topology of P. Evidently, repeatedly interchanging pairs of interchangeable pixels deforms P while preserving both its topology and its area (its number of 1s). For concreteness, we will use 4-adjacency for 1s and 8-adjacency for 0s, but, because interchangeability is a symmetric concept, our results also hold under the opposite assumptions.

16.5 Interchangeable Pairs of Pixels

Proposition 16.9 Two adjacent pixels $p = 1$ and $q = 0$ are interchangeable if the following conditions are true, where $M_8(p) = N_8(p) \cup \{q\}$ (i.e., with q changed to 1) and $M_8(q) = N_8(q) \cup \{p\}$ (i.e., with p changed to 0):

a) Each 4-component of 1s in $N_8(p)$ that is 4-adjacent to p is 4-adjacent to exactly one 4-component of 1s in $M_8(q)$ that is 4-adjacent to q (and vice versa).

b) Each 8-component of 0s in $N_8(q)$ is 8-adjacent to exactly one 8-component of 0s in $M_8(p)$ (and vice versa).

Proof If p is 4-adjacent to no 4-component of 1s in $N_8(p)$, q can be adjacent to no 4-component of 1s in $M_8(q)$ (and vice versa); hence p before the interchange and q after the interchange are 4-isolated 1s, and interchanging them shifts a singleton 4-component of 1s, but neither destroys nor creates one.

Otherwise, p is 4-adjacent in $N_8(p)$ to the same set of 4-components of 1s that q is 4-adjacent to in $M_8(q)$; hence, the interchange deletes p from each of these components and adds q to it. Thus the interchange does not split a 4-component of 1s, does not merge two such 4-components, and does not change the number of pixels in any such component.

If there are no 0s in $M_8(p)$, there are none in $N_8(q)$, and vice versa; thus, q before the interchange and p after the interchange are 8-isolated 0s, and interchanging them shifts a singleton 8-component of 0s, but neither destroys nor creates one.

Otherwise, q is 8-adjacent to the same set of 8-components of 0s in $N_8(q)$ that p is 8-adjacent to in $M_8(p)$; hence, the interchange deletes q from each of these components and adds p to it. Thus the interchange does not split an 8-component of 0s, does not merge two such 8-components, and does not change the number of pixels in any such component. ∎

The local conditions in Proposition 16.9 are sufficient but not necessary for interchangeability. Indeed, it is not possible to formulate necessary conditions for the interchangeability of p and q using any neighborhood of p and q of bounded size. (By the neighborhood of p of size n, we mean the set of pixels with chessboard distances from p that are at most n.) We show this by constructing two pictures P_1 and P_2 in which p and q have the same neighborhood of any given size n, but p and q are interchangeable in P_1 but not in P_2. For example, consider the following sets of pixels, where $N > n$:

$$S_1 = \{(0, \pm j) : 0 \leq j \leq N\}$$
$$S_2 = \{(i, \pm N) : -N \leq i \leq 0\}$$
$$S_3 = \{(-N, \pm j) : 0 \leq j \leq N\}$$
$$S_4 = \{(i, -1) : 2 \leq i \leq N\}$$
$$S_5 = \{(i, -N) : 0 \leq i \leq N\}$$
$$S_6 = \{(N, j) : -N \leq j \leq 0\}$$

Let P_1 be the picture in which all of these pixels are 1s, and let P_2 be the picture in which all of them are 1s except for those in S_2. Interchanging $(0,0)$ with $(-1,1)$ in P_1 creates a curve (consisting of S_4, S_5, S_6, and the lower half of S_1) and breaks open another curve (consisting of S_1, S_2, and S_3), so that both before and after the interchange, P_1 has one component of 1s with one hole. On the other hand, in P_2, this interchange breaks the 1s into two components and creates a hole that did not previously exist. We have thus, proved the following:

Proposition 16.10 For any n, there exist two pictures containing a pair of adjacent opposite-valued pixels p and q having neighborhoods of size n that are the same in both pictures but that are interchangcable in one picture and not in the other.

A picture P may have no interchangeable pairs of pixels. For example, if $\langle P \rangle$ is a 3×3 hollow square, it is easily seen that none of the 1s of $\langle P \rangle$ is interchangeable with any of its adjacent 0s. Even a simply connected component of $\langle P \rangle$ may not have any 1s that are interchangeable with any of their adjacent 0s. For example, let $\langle P \rangle$ be a singleton 1 surrounded by a 5×5 hollow square of 1s:

1	1	1	1	1
1	0	0	0	1
1	0	1	0	1
1	0	0	0	1
1	1	1	1	1

The central 1 is not interchangeable with any of its adjacent 0s, because interchanging them reduces the number of components of 1s from two to one. However, we will now show that, if $\langle P \rangle$ itself is simply connected, it has at least one interchangeable 1.

Theorem 16.7 If $\langle P \rangle$ is simply connected, at least one pixel of $\langle P \rangle$ is interchangeable with one of its adjacent 0s.

Proof Let (x,y) be the coordinates of the uppermost of the rightmost pixels of $\langle P \rangle$. If $(x, y-1) = 1$ and $(x-1, y) = 0$ or $(x-1, y) = (x, y-1) = (x-1, y-1) = 1$, (x,y) can be interchanged with $(x+1, y-1)$. If $(x, y-1) = (x-1, y) = 1$ and $(x-1, y-1) = 0$, (x,y) can be interchanged with $(x-1, y-1)$, unless doing so would create a hole. Now, if $(x-2, y) = 0$, $(x-1, y)$ can be interchanged with $(x, y+1)$ and if $(x, y-2) = 0$, then $(x, y-1)$ can be interchanged with $(x+1, y)$. Hence, we can assume that $(x-2, y) = (x, y-2) = 1$. Suppose there exists a 0 (call it z) that is not 8-connected to B (the background component of 0s) after the interchange of (x,y) with $(x-1, y-1)$; then, before the interchange, any 8-path from z to B must have passed through $(x-1, y-1)$. An 8-path ρ of 0s through $(x-1, y-1)$ cannot pass through $(x-2, y), (x-1, y), (x, y), (x, y-1)$, or $(x, y-2)$, because they are all 1s; hence, the pixels of ρ preceding and following $(x-1, y-1)$ must be two of $(x-2, y-1), (x-2, y-2)$, and $(x-1, y-2)$ (not necessarily distinct). Because these three pixels are all 8-neighbors of each other, π can be shortened to go directly from the predecessor of $(x-1, y-1)$ to its successor; hence, even after the interchange, z is still 8-connected to B, which is a contradiction.

Thus there exists a pixel of $\langle P \rangle$ that is interchangeable with one of its adjacent 0s unless $(x-1, y) = 1$ and $(x, y-1) = 0$. If $(x-1, y+1) = 1, (x, y)$ would be interchangeable with $(x, y+1)$; hence, $(x-1, y+1) = 0$. We can then interchange (x, y) with $(x-1, y+1)$, unless doing so would create a hole. One way this could happen is if $(x-2, y+1) = 1$ and $(x-2, y) = 0$; we must then have $(x-1, y-1) = 1$ (otherwise, (x, y) and $(x-1, y)$ would not be connected to the other 1s). Then (x, y) is interchangeable with $(x, y-1)$, unless this would create a hole, and this can happen only when $(x, y-2) = (x, y-3) = 1$ and $(x-1, y-2) = 0$, so $(x, y-2)$ can be interchanged with $(x+1, y-3)$. The other way a hole can be created is if $(x-1, y+2) = 1$, so the uppermost pixel in column $(x-1)$ is at height $z > y+1$.

We can repeat the argument in these two paragraphs with $(x-1, z)$ replacing (x, y) to show that some pixel of $\langle P \rangle$ is interchangeable with one of its adjacent 0s, unless the uppermost pixel in column $(x-2)$ is at height $w > z+1$, and so on. However, because $\langle P \rangle$ is finite, this argument cannot be repeated indefinitely; hence, the theorem must be true. ∎

Corollary 16.2 Let S be a simply connected component of 1s in P, let $N^*(S)$ be the set of 0s that are adjacent to pixels of S, and suppose S is "isolated" in the sense that any pixel of $\langle P \rangle$ that is adjacent to a 0 of $N^*(S)$ must be in S. Then at least one pixel of S is interchangeable with one of its adjacent 0s.

Two pictures P and Q will be called *directly IP-equivalent* if P has an interchangeable pair of pixels p and q such that Q is obtained from P by interchanging p and q. P and Q will be called *IP-equivalent* if $P = Q$ or if there exist pictures $P_0, P_1, \ldots, P_n$ such that $P_0 = P$, $P_n = Q$, and P_k is directly IP-equivalent to P_{k-1} ($1 \leq k \leq n$).

If P and Q are IP-equivalent, they must have the same topology and the same number of 1s. The converse is not true in general; there exist pictures that have the same topology and the same number of 1s but that are not IP-equivalent. For example, if the pixels of $\langle P \rangle$ belong to k nonadjacent 3×3 hollow squares, P has no interchangeable pixels; hence, no two such Ps are IP-equivalent, even though they have the same topology (k components of 1s and k holes) and the same number $8k$ of 1s. In the remainder of this section, we will show that the converse is true for pictures with sets of 1s that are simply connected.

Let S be a set of 1s that are on rows $1, \ldots, n$ of a binary picture (numbered from top to bottom), and let S_k be the subset of S in rows $1, \ldots, k$. We call S *k-singular* if the horizontal runs (maximal sequences) of 1s in S_k are all singletons. Evidently, if S is k-singular for some $k > 0$, the 4-components of S_k are all vertical line segments ("towers"). If S is 4-connected and k-singular for some $0 < k < n$, the southernmost pixel of each of these towers must be on row k, and its south neighbor must be 1; if S is 4-connected and n-singular, S itself must be a tower. If S is 4-connected and $(n-1)$-singular, the nth row of S must have a single run r of 1s and the southernmost pixel of each 4-component of S_{n-1} must be 4-adjacent to r.

Lemma 16.2 Let S be 4-connected and k-singular, and let r be a run of 1s on row $k+1$. Then S is IP-equivalent to a k-singular S' such that at most one tower T in S'_k is 4-adjacent to r, and we can make T adjacent to any desired pixel of r. Moreover, S' is the same as S on rows $k+1,\ldots,n$.

Proof If no tower in S_k is 4-adjacent to r, there is nothing to do. Suppose two or more towers in S_k are 4-adjacent to r. Let U and V be two consecutive such towers; suppose without loss of generality that U is west of V and that V is at least as tall as U. Because S is k-singular, U and V cannot be 4-adjacent.

If U and V are more than two pixels apart, they can be merged into a single tower as follows. Interchange the top pixel of U with its southeast neighbor, "slide" it down U (by repeated interchanges with its south neighbor) until it reaches r, slide it along r (by repeated interchanges with its east neighbor) until it reaches V, slide it up V (by repeated interchanges with its north neighbor) until it reaches the top of V, and, finally, make it the top pixel of V by interchanging it with its northeast neighbor. Note that, because S is 4-connected and k-singular, there can be no 1s in S_k between or above U and V so there is nothing to interfere with these interchanges. Repeat this process for the new top pixel of U; keep repeating it until all of the pixels of U have been transferred to the top of V.

If U and V are two pixels apart, we begin by interchanging the top pixel of U with its east neighbor, which makes it 4-adjacent to V. We then slide it up V and make it the top pixel of V, and we repeat the process. In either case, the process eliminates one of the towers in S_k that was 4-adjacent to r. We can do this repeatedly until only one such tower T is left.

Finally, T can be moved so that it is 4-adjacent to any desired pixel of r. For example, to move T eastward, we interchange its top pixel with its southeast neighbor, slide it down T until it reaches r, and slide it eastward along r until it reaches the desired position, where it is now a tower T' of height 1. We can then repeatedly transfer all of the pixels of T to the top of T' by proceeding as above. ∎

Lemma 16.3 Let S be 4-connected and k-singular where $0 \leq k < n-1$ is as large as possible. Then S is IP-equivalent to a k-singular S' that has horizontal runs

16.5 Interchangeable Pairs of Pixels

on rows $k+1$ and $k+2$ that satisfy the following:

1) A run r on row $k+1$ cannot have length 2. If r is a singleton, it must be 4-adjacent to a run on row $k+2$. If r has length ≥ 3, its endpoints must be 4-adjacent to endpoints of runs on row $k+2$, and its interior points can be adjacent only to singletons on row $k+2$.

2) A run s on row $k+2$ can have any length and need not be 4-adjacent to any run on row $k+1$. It can be 4-adjacent to any number of singletons on row $k+1$, and its endpoints can also be 4-adjacent to the endpoints of non-singleton runs on row $k+1$.

Proof Let r be a run of 1s on row $k+1$. According to Lemma 16.2, we can assume that at most one tower T in S_k is 4-adjacent to r.

a) If r is 4-adjacent to no runs of 1s on row $k+2$, we must have $k+1=n$. As pointed out earlier, in this case, r must be the only run of 1s on row n.

b) If r is 4-adjacent to one run s of 1s on row $k+2$, according to Lemma 16.2, we can assume that T is 4-adjacent to a pixel p of r that is 4-adjacent to s. If r is not a singleton, we can move all of its pixels (other than p) to the top of T. (If p is not the leftmost pixel q of r, we interchange q with its northeast neighbor, then slide it along r by repeated interchanges with its east neighbor until it reaches T, then slide it up T by repeated interchanges with its north neighbor until it reaches the top of T and becomes the new top pixel of T. We repeat this process for all of the pixels of r to the left of p and analogously for all of the pixels of r to the right of p.) This converts S into a k-singular S' that is the same as S below row $k+1$ and in which r and T have been converted into a single tower in S_{k+1} 4-adjacent to s. Thus we can assume from now on that r is 4-adjacent to $m>1$ runs $s_1,\ldots,s_m$ of 1s on row $k+2$. If there is a tower T in S_k that is 4-adjacent to r, we can make T 4-adjacent to a pixel of r that is strictly between two of the s_is, as in Lemma 16.2.

c) If two or more pixels of r are 4-adjacent to s_1, the pixels of r to the left of the rightmost pixel q of s_1 can similarly be moved along r until they reach q and can then be used to build a tower in S_{k+1} that is 4-adjacent to q; this is done similarly for s_m, with "rightmost" replaced by "leftmost." These towers can then be moved to the top of T, as in Lemma 16.2.

d) If $s=s_i$ is three or more pixels long, any pixel p of r that is 4-adjacent to a nonendpoint of s can be interchanged with its northeast or northwest neighbor (e.g., the former if p is west of T and the latter if p is east of T). This creates a height-2 tower W that extends upward from s and breaks r into two runs r' and r'', each of which overlaps s; in fact, s is s_m relative to r' and s_1 relative to r''. Note that both W and T are 4-adjacent to either r' or r'', so W can be moved to the top of T. According to condition (c), the pixels of r' that are 4-adjacent to s can then be converted into a tower U that extends upward from the leftmost pixel of s; similarly, the pixels of r'' that are 4-adjacent to s can be converted into a tower V that extends upward from the rightmost pixel of s. Finally, U or V (depending on whether T is 4-adjacent to r' or r'') can be moved to the top of T.

e) Let $s = s_i$ be two pixels long, and suppose there is only one 0 (call it z) between s_i and s_{i-1} on row $k+2$. According to condition (c), we can also assume that $1 < i < m$, so r also overlaps s_{i-1}.

Suppose interchanging z with one of the 1s of r created a hole H. Because z was not originally in a hole, there must have been an 8-path through z from H to the background. Because z's east and west neighbors are 1s, this 8-path must have come to z from its southwest neighbor and gone from z to its southeast neighbor (or vice versa), and the south neighbor t of z must have been 1. However, subsequently interchanging t with z cannot create a hole and cannot disconnect S, because t must have been 4-connected to S through its south neighbor. This interchange merges s_i with s_{i-1} and creates a new run s^* of 1s on row $k+2$ that overlaps r in at least four pixels; we can now proceed as in (d).

Thus, we can assume that interchanging z with one of the 1s of r (e.g., the northeast neighbor of z) does not create a hole. This interchange merges s_i with s_{i-1} to create a new run s^* of 1s on row $k+2$ and splits r into two runs r' and r''. Note that r' now overlaps s^* in two or more pixels; however, s^* is s_m relative to r', so we can proceed as described in (c). Similarly, if r'' overlaps s^* in two or more pixels, we can proceed as in (c), because s^* is s_1 relative to r''.

f) The argument is similar if there is only one 0 between s_i and s_{i+1} on row $k+2$. Hence, there remains only the case where $s = s_i$ is two pixels long and there are two or more 0s between s_i and s_{i-1} and between s_i and s_{i+1}. As in (e), we can assume that r overlaps s_{i-1} and s_{i+1}; thus, the neighborhood of s_i looks like the following, where $a \ldots h$ are 1s of r; q and u are the pixels of s; p is the rightmost pixel of s_{i-1}; v is the leftmost pixel of s_{i+1}; and $w \ldots x$, $y \ldots z$ are 0s:

$$\ldots a\, b \ldots c\, d\, e\, f \ldots g\, h \ldots$$
$$\ldots p\, w \ldots x\, q\, u\, y \ldots z\, v \ldots$$

Suppose interchanging d with x does not create a hole. This interchange breaks r into two runs r' and r'', where s is s_m relative to r' and s_1 relative to r'', so we can proceed as in (c); this can also be done in a similar manner if interchanging e with y does not create a hole.

Suppose interchanging d with x creates a hole. It is not hard to see that the south neighbor of x must be 1 and the south neighbor of q must be 0. If the west neighbor of d (formerly x) is in the hole created when d is interchanged with x, we can instead interchange c with x; it is easily seen that this does not create a hole, so we can again proceed as in (c).

16.5 Interchangeable Pairs of Pixels

Similarly, suppose interchanging e with y creates a hole; then the south neighbor of y must be 1 and the south neighbor of u must be 0. If the east neighbor of e (formerly y) is in the hole that is created when e is interchanged with y, we can instead interchange f with y; it is easily seen that this does not create a hole, so we can again proceed as in (c).

The only remaining possibility is that the south neighbor of q is in the hole H created when d is interchanged with x and the west neighbor of x is not in a hole; in addition, the south neighbor of u is in the hole K created when e is interchanged with y and the east neighbor of y is not in a hole. Evidently, however, the east neighbor of y has to be in H and the west neighbor of x has to be in K, so this situation is impossible.

If r has length 2, it can be 4-adjacent to at most one run s of 1s on row $k+2$. Evidently, one of the pixels of r can be moved to the top of the other one to create a height-2 tower extending upward from s; alternatively, if there is a tower T extending upward from one of the pixels of r, the other pixel of r can be moved to the top of T so that T now extends upward from s. Similarly, according to the condition (b), any r that overlaps only one s can be converted into a tower that extends upward from s; according to the condition (c), the overlaps of r with s_1 and s_m can be reduced to one pixel; and according to the conditions (d), (e), and (f), if any s_i $(1 < i < m)$ is not a singleton, r can be split into pieces, each of which overlaps s_i only at an endpoint. As we have seen, if this process creates more than one tower that extends upward from a given r or s, they can be merged into a single tower. ∎

Lemma 16.4 *If S' is as in Lemma 16.3, it is IP-equivalent to a $(k+1)$-singular S''.*

Proof Let r be a run of 1s on row $k+1$ of S'. If r is not a singleton, the first and last pixels of r have south neighbors that are last (first) pixels of runs of 1s on row $k+2$, and some of the interior pixels of r may also have south neighbors that are singleton 1s on row $k+2$. Let the pixels of r with south neighbors that are 1s be $r_1,\ldots,r_m$; note that r_1 and r_m are the first and last pixels of r and that r_i and r_{i+1} are nonconsecutive pixels of r $(1 \leq i < m)$. We will show that S' is IP-equivalent to an S'' in which all of the pixels of r (except r_m) have been changed to 0s and the runs of 1s on the rows above row $k+1$ are still all singletons. Doing this for every non-singleton r on row $k+1$ converts S' into an S'' in which all of the runs of 1s on row $k+1$ are also singletons, so that S'' is $(k+1)$-singular.

If there is a tower T that extends upward from one of the pixels of r, we first move it (if necessary) so that it extends upward from r_m. Let $s_1,\ldots,s_m$ be the south neighbors of $r_1,\ldots,r_m$; let p and q be any two successive pixels of r, and let s and t be the south neighbors of p and q. We will now show that the successive pixels of r (from left to right) (except for r_m) can be moved either to row $k+2$ or to the top of T and that, when pixels $r_1,\ldots,p$ of r have been moved, pixels $s_1,\ldots,t$ on row $k+2$ are all 1s.

For any i ($1 \leq i < m$), consider the subrun $r_i, \ldots, u, v$ of r (where v is the west neighbor of r_{i+1}), and let p be successively $r_i, \ldots, u$. After all of these ps have been moved as described below, $r_i, \ldots, u$ have become 0s and $s_i, \ldots, s_{i+1}$ are all 1s. We can then interchange v with its northeast neighbor (the north neighbor of r_{i+1}) and move it to the top of T; this completes the moving process for the subrun (call it r').

Suppose the pixels of r' to the west of p have already been moved; thus, the neighborhood of p looks like the following, where $z = 0$:

Suppose interchanging p with z does not create a hole. Because this interchange evidently does not disconnect S, we can move p by simply interchanging it with z.

Suppose interchanging p with z does create a hole; then the neighborhood of p must look like one of the following:

p	1	1		p	1	1
1	z	x	or	1	z	0
0	1	0		0	1	1

In the first case, there is a 4-path of 1s joining z's west neighbor to z's south neighbor such that, if z is changed to 1, the resulting 4-curve separates z's southwest and southeast neighbors. In the second case, there is a 4-path of 1s joining z's west neighbor to z's south and southeast neighbors such that, if z is changed to 1, the resulting 4-curve separates z's east and southwest neighbors.

In the first case, we interchange z with its south neighbor. This interchange does not disconnect S and does not create a hole. We can then move p to the top of T.

In the second case, let the neighborhood of p be as shown in Figure 16.17. If $y = 1$, we can interchange z with its south neighbor and proceed as we did in the first case. If $y = 0$, there are two possibilities:

> a) The 4-path from z's south neighbor to its west neighbor passes through w. In this subcase, we can interchange z with its southeast neighbor and proceed as we did in the first case. Evidently, when we interchange z with its southeast neighbor, the new z and its north and south neighbors cannot be in a hole; if z's southwest neighbor is in a hole, it is not hard to see that, before the interchange, either y must have been in a hole or z

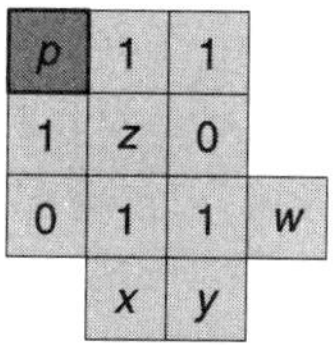

16.5 Interchangeable Pairs of Pixels

FIGURE 16.17 Neighborhood of p in the second case.

(and its east and southwest neighbors) must have been in a hole, which contradicts the fact that S is simply connected.

b) The 4-path from z's south neighbor to its west neighbor passes through x. In this subcase, we can interchange z's south neighbor with its east neighbor and then interchange p with z.

In summary, we can interchange z with p if this does not create a hole; if it does create a hole, we can interchange z with one of its neighbors on row $k+2$ and then move p to the top of T. In either case, we can do this successively for $p = r_i, \ldots, u$ and then move v to the top of T; this changes all of the 1s of r' to 0s, and all of their south neighbors become 1s (note that s_i was already 1). We can repeat this process for the subruns $r_1, \ldots; r_2, \ldots; \ldots; r_{m-1}, \ldots$; this changes all of the 1s of r (except for r_m) to 0s, and all of their south neighbors become 1s. Thus the non-singleton run r of S' has been replaced by the singleton r_m; when this is done for every r, the resulting S'' is $(k+1)$-singular. ∎

Using Lemmas 16.3 and 16.4, we can prove the following:

Lemma 16.5 Let S be simply 4-connected; then S is IP-equivalent to a vertical line segment.

Proof S is IP-equivalent to an S' with it first two rows satisfying Lemma 16.3. Hence, S is IP-equivalent to a 1-singular S'', as in Lemma 16.4. We can then apply Lemma 16.3 to rows 2 and 3 and then Lemma 16.4 to obtain a 2-singular S''. By repeating this process, we eventually obtain an $(n-1)$-singular S'''; this S''' has a single run on row n with a tower extending upward from it, so it is evidently IP-equivalent to a vertical line segment. ∎

It is easy to see that a horizontal line segment is IP-equivalent to a vertical line segment and that interchanges can be used to translate a horizontal line segment vertically or a vertical line segment horizontally. Because interchanges are reversible, we thus, finally have the following:

Theorem 16.8 Let $\langle P \rangle$ and $\langle Q \rangle$ be simply 4-connected and have the same number of pixels; then P and Q are IP-equivalent.

16.6 Deformations of 3D Pictures

In this section, P is a 3D binary picture. We will (as is usually done) use 26-adjacency for $\langle P \rangle$ and 6-adjacency for $\langle \overline{P} \rangle$ and assume that $\langle P \rangle$ is finite. (Less common assumptions are to use 6- for $\langle P \rangle$ and 26- for $\langle \overline{P} \rangle$ or to use 18- instead of 26-.) In what follows, $A(p) = A_{26}(p)$ is the set of voxels that are 26-adjacent to p, and $N(p) = N_{26}(p) = A_{26}(p) \cup \{p\}$.

Definition 16.5 A voxel $p \in \langle P \rangle$ ($p \in \langle \overline{P} \rangle$) is *simple* iff it can be removed (in the cell model) from $\langle P \rangle$ ($\langle \overline{P} \rangle$) by a continuous deformation.

Changing the value of a simple voxel p from 1 to 0 (or vice versa) results in a picture that is topologically equivalent to P. However, changing the value of a nonsimple voxel may also make no changes in the topology of P. As explained in [747], it can remove one tunnel and introduce another one. For example, let P be a "mug with a handle." If we remove one end of the handle where it joins the mug, we lose the tunnel defined by the handle, but at the same time we may create a new tunnel by drilling a hole into the mug; however, topology is preserved by this operation. Note that the voxel that is removed cannot be a simple voxel.

In 2D, p is (8,4)-simple iff p is 8-adjacent to exactly one 8-component of $\langle P \rangle \cap A_8(p)$ and 4-adjacent to exactly one 4-component of $\langle \overline{P} \rangle \cap A_8(p)$; indeed, we used this as the definition of simplicity in Section 16.1. We might think that, in 3D (analogously), p is (26,6)-simple iff it is 26-adjacent to exactly one 26-component of $\langle P \rangle \cap A_{26}(p)$ and 6-adjacent to exactly one 6-component of $\langle \overline{P} \rangle \cap A_{26}(p)$; however, this is not true, because changing the value of such a p from 1 to 0 or vice versa might create or destroy a tunnel, as illustrated in Figure 16.18.

If we use the grid point model, the following are two criteria for 3D simplicity [747]; see [77, 78] for other characterizations.

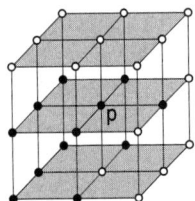

FIGURE 16.18 A nonsimple voxel p in a (26,6) picture; changing p from 1 to 0 creates a tunnel.

16.6 Deformations of 3D Pictures

a) p is 26-adjacent to exactly one 26-component of $\langle P \rangle \cap A_{26}(p)$ and 6-adjacent to exactly one 6-component of $\langle \overline{P} \rangle \cap A_{26}(p)$, and changing the value of p from 1 to 0 (or vice versa) does not change the Euler characteristic of $N(p)$.

b) p is 26-adjacent to exactly one 26-component of $\langle P \rangle \cap A_{26}(p)$ and is 6-adjacent to $\langle \overline{P} \rangle \cap A_{26}(p)$, and any two voxels of $\langle \overline{P} \rangle$ that are 6-adjacent to p are joined by a 6-path that consists of voxels of $\langle \overline{P} \rangle$ that are 18-adjacent to p.

If we use the grid cell model as described in Section 16.1, criteria for simplicity can be formulated in terms of the P-attachment set of p. As in 2D, we define the P-attachment set p_P as follows:

$$\bigcup_{q \in \langle P \rangle, q \neq p} (\vartheta_p \cap \vartheta_q)$$

It can be shown that p is simple iff both p_P and $\vartheta p - p_P$ are nonempty and connected.

Topology is preserved by any sequence of changes of the values of simple voxels from 1 to 0 or vice versa, provided the voxels are simple just before their values are changed; simultaneously changing the values of a set of voxels from 1 to 0 (or vice versa) preserves topology iff the voxels in the set can be arranged in such a sequence. Let $\mathcal{A}$ be an algorithm that consists of a sequence of iterations $\mathcal{A}_j$ at each of which the voxels that satisfy certain conditions have their values simultaneously changed from 1 to 0; the voxels that satisfy these conditions will be called $\mathcal{A}_j$-*simple*. The following [568] is a 3D generalization of Proposition 16.3:

Proposition 16.11 $\mathcal{A}$ preserves topology if, for every j, the following conditions are true:

a) Every $\mathcal{A}_j$-simple 1 is simple.
b) For every set of two or more $\mathcal{A}_j$-simple 1s that is contained in a $1 \times 2 \times 2$ block of voxels, each 1 in the set remains $\mathcal{A}_j$-simple when the values of the other 1s in the set are changed to 0.
c) In any set of mutually 26-adjacent 1s at least one of them is not $\mathcal{A}_j$-simple.

In a 3D picture, an elongated part E of a component of $\langle P \rangle$ can be *plate-like* (i.e., a thick surface with frontiers) or *stick-like* (i.e., a thick arc or curve). The goal of *3D thinning* is to shrink each such E into a union of arcs, curves, or surfaces (with frontiers) that are centrally located in E. A 3D thinning algorithm must preserve topology and must not remove voxels that could be either endpoints of arcs or "rim points" of surfaces with frontiers. We can define an arc endpoint as a voxel of $\langle P \rangle$ that is adjacent to exactly one other voxel of $\langle P \rangle$ (compare with Section 16.3) and a rim point as a voxel of $\langle P \rangle$ that has at least one pair of 6-neighbors in $\langle \overline{P} \rangle$ on opposite sides of it. Many 3D thinning algorithms have been defined; we will not discuss them here except to say that a few are "fully parallel" and a few are of the subfield type, but most are of the subiteration type.

16.7 Deformations of Multivalued Pictures

In this section, we return to the 2D case, but we drop the assumption that P is binary. As we saw in Section 1.2.10, if we divide the values of the pixels in P by $G_{\max}$, the quotients are in the range $[0,1]$, so P defines a fuzzy subset μ of the grid. A transformation that takes such a P into another such picture can be regarded as a deformation of μ. In this section, we study deformations of μ that preserve its connectedness properties (see [218]).

We saw in Section 13.3.3 that a picture P is μ-connected iff its level sets are all connected. This suggests calling a transformation of P topology-preserving if it preserves the topology of every level set of P. In what follows, we will denote the λ-level set of P by P_λ, and we will use 4-connectedness for every $\langle P_\lambda \rangle$ and 8-connectedness for every $\langle \overline{P}_\lambda \rangle$. Let $\overline{P}$ be the picture with the membership function $1 - \mu$. Evidently, the level sets of $\overline{P}$ are the complements of the level sets of P. For any pixel p, we abbreviate $P_{\mu(p)}$ by P_p.

A pixel p is called P-destructible if it is simple in P_p and P-constructible if it is simple in $\overline{P}_p$. We state the following proposition for destructible pixels; the analogous proposition about constructible pixels is obtained by replacing P with $\overline{P}$ and reversing the inequality relations. Let $A_8(p)$ be the 8-adjacency set of p; let $N_8(p) = A_8(p) \cup \{p\}$; let $L(p) = \{q : q \in A_8(p) \land P(q) < P(p)\}$; let $m = \min\{P(q) : q \in L(p)\}$; and let $M = \max\{P(q) : q \in L(p)\}$.

Proposition 16.12 If p is destructible and $P(p)$ is reduced to less than m, the topology of some level set of P is not preserved; if it is reduced to M, the topology of every level set of P is preserved.

Proof Let $P(p)$ be reduced to a value $< m$, and call the resulting picture P'; then the topology of P'_m is different from that of P_m, because P'_m has a new 1-pixel hole at p. If $P(p)$ is reduced to M, there is no change in any P_λ for $\lambda \leq M$ and no change in any P_λ for $\lambda > P(p)$. P_λ does change when $\lambda = P(p)$, but because p is destructible, the topology of P_λ does not change. ∎

Corollary 16.3 Let $l = \max\{P(q) : q \in P \land P(q) < P(p)\}$ be the highest pixel value less than $P(p)$ in P; then, if $P(p)$ is reduced to l, the topology of every level set of P is preserved.

Pixels can be classified in many ways on the basis of their neighbors' relative values; the following is one such classification. Let $c_+(p)$ be the number of 4-components of pixels in $N_8(p)$ with values that are $> P(p)$, and let $c'_+(p)$ be the number of such 4-components with values that are $\geq P(p)$. Similarly, let $c_-(p)$ be the number of 8-components of pixels in $N_8(p)$ with values that are $< P(p)$, and let $c'_-(p)$ be the number of such 8-components with values that are $\leq P(p)$. Thus, P is

16.7 Deformations of Multivalued Pictures

destructible iff $c'_+(p) = c'_-(p) = 1$ and constructible iff $c_+(p) = c_-(p) = 1$. In terms of these numbers, we can also define the following classes of pixels:

1) p is a *peak* if $c'_+(p) = 0$.
2) p is *minimal* if $c_-(p) = 0$; otherwise, p is an *upper pixel*.
3) p is k-*divergent* if $c_-(p) = k > 1$.
4) p is a *pit* if $c'_-(p) = 0$.
5) p is *maximal* if $c_+(p) = 0$; otherwise, p is a *lower pixel*.
6) p is k-*convergent* if $c_+(p) = k > 1$.
7) p is an *interior pixel* if it is both maximal and minimal.
8) p is a *side pixel* if it is both constructible and destructible.
9) p is a *saddle pixel* if it is both convergent and divergent.

In terms of these definitions, it can be shown [218] that any p is of exactly one of the following types:

a) a peak;
b) a pit;
c) an interior pixel;
d) a minimal constructible pixel;
e) a maximal destructible pixel;
f) a minimal convergent pixel;
g) a maximal divergent pixel;
h) a side pixel;
i) a destructible convergent pixel;
j) a constructible divergent pixel; or
k) a saddle pixel.

These concepts can be used to define algorithms for thinning multivalued pictures. Such an algorithm should preserve the topology of the level sets of the picture and should also preserve "ridge endpoints" (the fuzzy generalization of arc endpoints). We call p a ridge endpoint if one pixel or two 4-adjacent pixels in $N_8(p)$ have value $P(p)$ and of all the other pixels in $N_8(p)$ have lower values. If the values of p's 8-neighbors are as follows (so that p has value e),

a	b	c
d	e	f
g	h	i

we call p a diagonal ridge point if a and e are both greater than d and b, c and e are both greater than b and f, i and e are both greater than f and h, or g and e are both greater than h and d. We call p a vertical ridge point if b, e, and h are all greater than d and f, and we call it a horizontal ridge point if d, e, and f are all greater than b and h.

It can be shown that, if p has greater value than at least one of its 8-neighbors but it is not a peak or a ridge point, topology is preserved if $P(p)$ is reduced to the lowest of the values of p's 4-neighbors. This allows for the faster reduction of pixel values than does the reduction method in Proposition 16.12 or Corollary 16.3. Subiteration algorithms can be defined for thinning nonbinary pictures based on these reduction methods.

16.8 Exercises

1. Let the neighborhood of p be as follows:

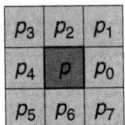

 The cyclic sequence $p_0, \ldots, p_7$ (modulo 8) is a succession of runs of 1s and runs of 0s. "Reduce" this succession by omitting any of p_1, p_3, p_5, or p_7 that is a singleton 0. Prove that p is simple iff the reduced sequence consists of one run of 1s and one run of 0s. Let [1150] the following be true, where v_i is the value of p_i (0 or 1) and the subscripts are modulo 8:

$$n_c(p) = \sum_{i=0,2,4,6} (v_i - v_i v_{i+1} v_{i+2})$$

 Prove that p is simple iff $n_c(p) = 1$. (If we use 8-adjacency for 1s, we must exchange the roles of 1s and 0s in this definition; the adjusted formula for $n_c(p)$ is obtained by replacing each v_i with its Boolean complement [i.e., with $\text{not}(v_i)$].)

2. Let C be a finite simply connected component of $\langle P \rangle$ that contains at least two pixels. Prove that C contains at least two simple pixels.

3. Let C be a component of $\langle P \rangle$ and C' a subset of C. We say that C is simply connected relative to C' if, whenever p and q are separated by C, they are also separated by C'. Prove that, if C' is a nonempty connected subset of C that does not contain the border of C and C is simply connected relative to C', then there is a simple pixel in the border of C that is not in C'.

4. A binary picture P is called *well composed* if any component of either the 1s or the 0s of P is a 4-connected 8-component. Prove that P is well composed iff it

contains no 2×2 block of pixels in which two diagonally adjacent pixels are 1s and the other two are 0s. Prove that any binary picture can be made well composed by a simple deformation. (Hint: Magnify the picture [see Theorem 16.2]; if it contains blocks of 1s that touch at their corners, change these corners to 0s.)

5. For any set of pixels S of P, let S_1 be the set of 1s that are either in S or immediately above or to the left of pixels of S after the shrinking process described at the end of Section 16.2 is applied to P, and let S_0 be the set of such 0s. Prove that, if C is a non-singleton 4-component of 1s, then C_1 is a 4-component of 1s, and, if D is a non-singleton 8-component of 0s, then D_0 is an 8-component of 0s; moreover, if C is 4-adjacent to D, then C_1 is 4-adjacent to D_0. Thus, this shrinking process is topology-preserving, except for singleton components.

6. Prove that, when the shrinking process described at the end of Section 16.2 is applied repeatedly to P, any component C of $\langle P \rangle$ shrinks to a single 1 located at (x_c, y_c), where x_c and y_c are the highest x-coordinate and lowest y-coordinate of any pixel of C. Prove also that the number of steps required for this to happen is the largest city block distance of any pixel of C from (x_c, y_c).

16.9 Commented Bibliography

Simple pixels were studied in [881] under the name "deletable elements." For Exercise 1, see [437, 531, 1150]. Topologic equivalence is discussed in many papers about shrinking and thinning (see below); see also [401, 488, 492, 493, 869].

Two-way simple deformations are treated in [916], which gives two proofs that topologically equivalent pictures differ by a simple deformation. The shrinking method described at the end of Section 16.2 was introduced in [645] (the version of it that uses 8-connectedness for $\langle P \rangle$ and 4-connectedness for $\overline{\langle P \rangle}$ was known earlier; see [733]); see also [29, 489, 492, 493].

Detailed discussions of shrinking and thinning algorithms can be found in [43, 403, 405]; for shrinking, see also [366, 489]. More than 100 2D thinning algorithms have been published; a partial list is [19, 26, 31, 36, 37, 44, 113, 164, 183, 187, 189, 191, 192, 195, 201, 246, 250, 270, 296, 298, 374, 387, 388, 399, 400, 402, 437, 442, 470, 471, 472, 609, 612, 615, 636, 638, 671, 753, 755, 758, 759, 760, 783, 791, 803, 804, 867, 887, 936, 985, 986, 994, 1017, 1018, 1027, 1035, 1036, 1067, 1117, 1135, 1158, 1159, 1160, 1161, 1162, 1163]. For other references on thinning, see [64, 118, 163, 188, 193, 313, 315, 438, 444, 487, 633, 649, 690, 699, 705, 729, 730, 742, 799, 817, 822, 908, 995, 1013, 1074, 1143].

[665] contains a comparative analysis of subiteration and fully parallel thinning algorithms, as well as a summary of topologic criteria for defining thinning algorithms, such as the P-simple points introduced in [75]. Fully parallel thinning algorithms are described in [696, 697, 698].

For simple voxels in 3D binary pictures, see [75, 77, 78, 337, 403, 577, 626, 937, 939, 940]. For the attachment set characterization of simple pixels or voxels and for changing their values "in parallel," see [568]. Algorithms for 3D thinning are described in [76, 79, 110, 369, 395, 482, 662, 676, 677, 678, 679, 680, 681, 682, 750, 793, 794, 795, 827, 941, 1039, 1066].

Deformations of digital curves are discussed in [919]; see also [691]. Interchangeable pairs of opposite-valued pixels are discussed in [920, 923]. For destructible and constructible pixels in multivalued pictures, see [27, 218]; for their application to "thinning," see [45]. See [1094] for the skeletonization of binary and multivalued pictures. For other work on multivalued thinning, see [1, 33, 289, 610, 790, 836, 853, 948, 1046].

CHAPTER **17**

Picture Properties and Spatial Relations

A function that takes pictures into numbers is called a *picture property*; a function that takes k-tuples (e.g., pairs) of pictures into numbers is called a *relation* among (or between) pictures. This chapter defines classes of picture properties, such as predicates, local properties,[1] linear properties, and invariant properties. Particular attention is given to the study of *moments*, which are an important class of linear properties.

A property of a binary picture P can be regarded as a property of the set of 1s $\langle P \rangle$ of P (e.g., as a property of an "object"). Many topologic and metric properties of objects were studied in earlier chapters. In this chapter, we briefly discuss some important topologic and metric relations between pairs of objects.

17.1 Properties

17.1.1 Predicates

A function that takes pictures into $\{0,1\}$ can be regarded as a proposition that is either true or false (a *predicate*) for a given picture, where the value of the function is 1 iff the proposition is true. An example of a predicate that is defined for an arbitrary picture P is "has constant value"; this predicate is true iff all of the pixels of P have the same value. Many useful predicates can be defined for binary pictures P that are true iff $\langle P \rangle$ has some geometric property. Examples are "is connected", "is convex", "is circular", and so on. Predicates can also be defined for special types of binary pictures; examples are "is straight" (if $\langle P \rangle$ is an arc) and "is knotted" (if $\langle P \rangle$ is a curve).

---

1. See also Definition 8.7.

FIGURE 17.1 Are the two exits of this maze connected by a 4-path of white pixels?

Predicates can have very complex definitions. Even if P is binary, there are 2^{mn} possible pictures on an $m \times n$ grid, and $\mathcal{F}(P)$ can be true for any subset of these pictures, so it may be very complicated to determine whether $\mathcal{F}(P)$ is true or false. For more about the complexity of predicates, see [733] and Exercise 2 in Section 17.6. Figure 17.1 illustrates complexity by a popular example; in this maze, the set of black pixels is 4-connected iff there is no 4-path of white pixels between the two "exits."

17.1.2 Local properties

From now on, we will assume that a picture property is a function $\mathcal{F}$ that takes pictures into real numbers. It is convenient to consider properties $\mathcal{F}$ of pictures that are defined on a finite grid $\mathbb{G}_{m,n} \subseteq \mathbb{Z}^2$ and that have pixel values that belong to a fixed finite set of real numbers (not necessarily nonnegative). We will assume that $\mathcal{F}$ is translation-invariant (i.e., for any picture P, $\mathcal{F}(P)$ depends only on the (m,n)-tuple of pixel values of P; it does not depend on the position of $\mathbb{G}_{m,n}$ in $\mathbb{Z}^2$). We will usually assume that P is extended from $\mathbb{G}_{m,n}$ to all of $\mathbb{Z}^2$ by assigning a special value (often 0) to every pixel in $\mathbb{Z}^2 \setminus \mathbb{G}_{m,n}$.

The value of $\mathcal{F}$ depends only on the pixel values in the nonempty finite subset $\mathbb{G}_{m,n}$ of $\mathbb{Z}^2$. The smallest such subset σ (more fully, $\sigma_{\mathcal{F}}$) is called the *set of support* of $\mathcal{F}$; thus the value of $\mathcal{F}$ for any picture P depends only on a "σ-tuple" of values of pixels of P. $\mathcal{F}$ is called a *local property* if the diameter of σ is constant, so it does not depend on m and n and therefore remains small even if m and n are very large. A property that is not local is called *global*. $\sigma_{\mathcal{F}}$ is a uniquely defined subset of $\mathbb{G}_{m,n}$. For example, the property "is there a 3×3 square on which the picture has a constant value" is not local.

Let o be a pixel in a known position relative to σ (compare this with Section 15.1); we will usually assume that o is one of the pixels of σ. For any pixel p, let $\sigma(p)$ be the

17.1 Properties

result of translating σ so o coincides with p; thus $\sigma(o) = \sigma$. Each p thus defines a "translate" $\mathcal{F}_p$ of $\mathcal{F}$ that has a set of support that is $\sigma(p)$, so its value depends on the pixel values in $\sigma(p)$ in the same way that the value of $\mathcal{F}$ depends on the pixel values in σ.

We usually speak about the family of properties $\mathcal{F}_p$ as though it were a single property $\mathcal{F}$ that has a value "at" each pixel p. For example, let the set of support $\mathcal{F}$ be the single pixel $\sigma = \{o\}$; then, for any picture P and any pixel p, the value of $\mathcal{F}_p$ depends only on $P(p)$ (i.e., the value of $\mathcal{F}$ "at p" is some function of $P(p)$). Such an $\mathcal{F}$ is sometimes called a "point property." Similarly, let the set of support of $\mathcal{F}$ be $N_8(o)$. Then, for any picture P and any pixel p, the value of $\mathcal{F}$ "at p" depends only on the pixel values in $N_8(p)$, so $\mathcal{F}$ is a "neighborhood property." A family of properties $\mathcal{F}_p$ (for $p \in \mathbb{G}_{m,n}$) is called *local* iff it is defined by a local property $\mathcal{F}$. Evidently, point properties and neighborhood properties are local properties. For example, the family of properties "is there a 3×3 square centered at pixel p on which P has a constant value" is local.

We call a picture property *semilocal* if its value depends only on the values of some local property at every pixel p. For example, the area $\mathcal{A}(P)$ is computed by counting the number of ps that have value 1. As a less trivial example (see Exercise 6 in Section 6.5), the Euler characteristic of P can be computed by counting the numbers of certain types of 2×2 pixel patterns in P. (Figure 17.2 shows the locations of some of these patterns in a simple image.) Relatively simple binary pictures (see the picture on the cover of [733]) allow us to show that neither the number of connected components of a binary picture nor its number of holes are semilocal properties.

$\mathcal{F}(P)$ may depend only on the set of values of the pixels of P but not on which pixels have these values. Examples of such properties are statistics of the values of the pixels of P: for example, their mean, standard deviation, median, range, min, or max. Because a local property has a value at each pixel of P, we can also compute statistics of the values of local properties of P; such properties are sometimes called

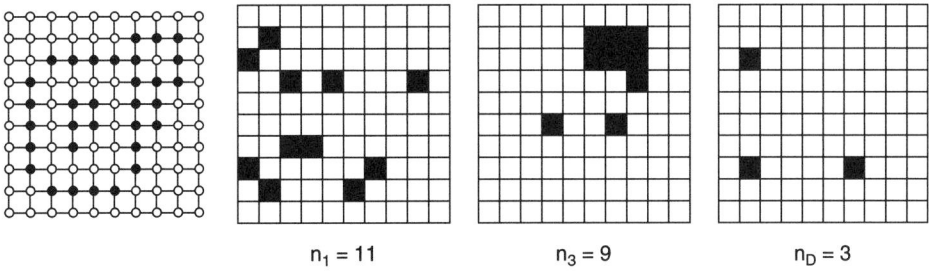

FIGURE 17.2 The black squares in the three pictures on the right are the locations of the upper left corners of the 2×2 blocks of pixels in the picture on the left that contain exactly one black pixel, three black pixels, and two diagonally adjacent black pixels. The number of black squares is shown below each of the three pictures on the right; for the notation, see Exercise 6 in Section 6.5.

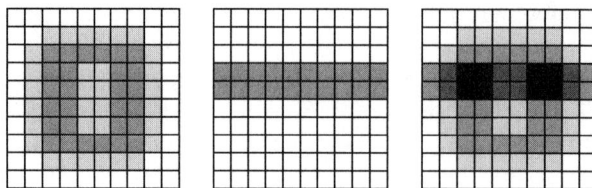

FIGURE 17.3 The addition of two pictures (left and middle), resulting in the picture on the right.

textural properties of P. Local properties and textural properties are studied in books about digital picture analysis (e.g., [911]).

17.1.3 Linear properties

A picture property $\mathcal{F}$ is called *linear* if, for all pictures P and Q (defined on a given finite grid) and all real constants a and b, we have $\mathcal{F}(aP+bQ) = a\mathcal{F}(P)+b\mathcal{F}(Q)$, where addition and scalar multiplication are performed pixel-wise (i.e., for all p in the grid, we have $\mathcal{F}(aP)(p) = a\mathcal{F}(P)(p)$ and $\mathcal{F}(P+Q)(p) = \mathcal{F}(P)(p)+\mathcal{F}(Q)(p)$ [see Figure 17.3]). In what follows, we assume that pixel values can be either positive or negative. We will now prove that, for any linear property $\mathcal{F}$ (defined for pictures on a given finite grid $\mathbb{G}$), there exists a "template" $H_\mathcal{F}$ such that, for any such picture P, $\mathcal{F}(P)$ is the sum of the pixel-wise product of $H_\mathcal{F}$ and P.

Theorem 17.1 For any linear picture property $\mathcal{F}$, there exists a picture $H_\mathcal{F}$ such that $\mathcal{F}(P) = \sum_p H_\mathcal{F}(p)P(p)$ for all pictures P, where the sum is taken over the grid.

Proof For every pixel $p \in \mathbb{G}_{m,n}$, let δ_p be the "unit picture" that has value 1 at p and value 0 elsewhere. Evidently, any picture P is the sum over all $p \in \mathbb{G}_{m,n}$ of $P(p) \cdot \delta_p$. Define $H_\mathcal{F}$ by $H_\mathcal{F}(p) = \mathcal{F}(\delta_p)$ for $p \in \mathbb{G}_{m,n}$; then the linearity of $\mathcal{F}$ implies that $\mathcal{F}(P)$ is the sum over all $p \in \mathbb{G}_{(m,n)}$ of $P(p) \cdot H_\mathcal{F}(p)$. ∎

This theorem is a discrete version of the representation theorem in functional analysis,[2] which states that any linear functional Ag in the space $L^2(\Omega)$ of (complex-valued) measurable functions can be represented in the form $Ag = \langle g, f \rangle$, where the generating function f is uniquely defined by the functional A. (The L^2 scalar product

---

2. This was independently discovered by both M. Fréchet and F. Riesz; see [855].

17.2 Moments

$\langle f,g \rangle$ is defined by the integration of $f(x)g^\star(x)$ on the measurable set Ω.) Note that the proof of Theorem 17.1 is much simpler than the proof in the L^2 space.

17.2 Moments

A variety of useful linear properties are obtained by using $H_\mathcal{F}$s that are digital versions of standard mathematic functions. In this section, we discuss *moment properties* in which $H_\mathcal{F}$ is a monomial $x^i y^j$. We treat the 2D case, but our definitions and results extend straightforwardly to higher dimensions.

17.2.1 Moments of pictures

Let $P(x,y)$ be the value of the pixel of P at location (x,y). The (digital) (i,j) *moment* of P is as follows:

$$m_{ij}(P) = \sum_{x,y} x^i y^j P(x,y)$$

The *order* of $m_{ij}(P)$ is $i+j$. For example, consider the 3×3 picture P with the following pixel values and where the origin of the (x,y) coordinate system is at the (center of the) pixel in the lower left-hand corner.

2	1	1
3	1	0
3	2	1

The first few moments of P are as listed in Table 17.1.

Moments can be given a physical interpretation by regarding the value of a pixel as its "mass" (i.e., regarding P as being composed of a set of point masses located at the grid points). Under this interpretation, $m_{00}(P)$ is the total mass of P (the sum of all of its pixel values) and $m_{02}(P)$ and $m_{20}(P)$ are the moments of inertia of P around the x- and y-axes. The moment of inertia of P around the origin is as follows:

$$m_0(P) = \sum_{x=0}^{m-1}\sum_{y=0}^{n-1}(x^2+y^2)P(x,y) = m_{02}(P) + m_{20}(P)$$

If we substitute $-x$ for x in the definition, of m_{ij}, we obtain the following:

$$\sum_{x=0}^{m-1}\sum_{y=0}^{n-1}(-x)^i y^j P(-x,y) = (-1)^i \sum_{x,y} x^i y^j P(-x,y)$$

TABLE 17.1 The first few moments of the 3 × 3 picture shown in the text.

i	j	m_{ij}
0	0	14
1	0	8
0	1	12
2	0	12
1	1	7
0	2	20

Hence, if P is symmetric around the y-axis (i.e., $P(-x,y) = P(x,y)$ for all x,y), we have $m_{ij}(P) = (-1)^i m_{ij}(P)$, so that, if i is odd, $m_{ij}(P)$ must be 0. Similarly, $m_{ij}(P) = 0$ if P is symmetric around the x-axis and j is odd, and $m_{ij}(P) = 0$ if P is symmetric around the origin (i.e., $P(-x,-y) = P(x,y)$ for all x,y) and $i+j$ is odd. Moments for which i, j, or $i+j$ is odd can thus be regarded as measures of asymmetry about the y-axis, the x-axis, or the origin.

The *centroid* or *center of gravity* of P is the point with coordinates $(\bar{x}, \bar{y})$ where $\bar{x} = m_{10}/m_{00}$ and $\bar{y} = m_{01}/m_{00}$. Thus, the centroid of the 3 × 3 picture shown on p. 541 has coordinates $(\frac{4}{7}, \frac{6}{7})$. If we use a coordinate system that has its origin o' at the centroid, moments of P computed with respect to this system are called *central moments* and are denoted by $\bar{m}_{ij}(P)$. Evidently, $\bar{m}_{00} = m_{00}$, and it is easily verified that $\bar{m}_{10} = \bar{m}_{01} = 0$ for any P.

The line through (u,v) that has slope $\tan\theta$ is $(x-u)\sin\theta = (y-v)\cos\theta$. The moment of inertia of P around this line is as follows:

$$\sum_{x=0}^{m-1}\sum_{y=0}^{n-1}[(x-u)\sin\theta + (y-v)\cos\theta]^2 P(x,y)$$

If we differentiate this expression with respect to u or v and set the result equal to 0, we get the following,

$$\sum_{x=0}^{m-1}\sum_{y=0}^{n-1}[(x-u)\sin\theta + (y-v)\cos\theta]P(x,y) = 0$$

so that the following is also true:

$$m_{10}\sin\theta - m_{01}\cos\theta + m_{00}(v\cos\theta - u\sin\theta) = 0$$

Dividing by m_{00} gives $(\bar{x}-u)\sin\theta - (\bar{y}-v)\cos\theta = 0$; hence the minimum value of the moment of inertia must be as follows:

$$\sum_{x=0}^{m-1}\sum_{y=0}^{n-1}[(x-\bar{x})\sin\theta - (y-\bar{y})\cos\theta]^2 P(x,y)$$

17.2 Moments

This is the moment of inertia around the line of slope θ through $(\overline{x}, \overline{y})$. Thus the line around which the moment of inertia is smallest passes through the centroid; this line is called the *principal axis* of P. To find the slope of the principal axis, take the origin at the centroid; then the moment of inertia of P around the line $y = x \tan \theta$ is as follows:

$$\sum_{x=0}^{m-1}\sum_{y=0}^{n-1}(x \sin \theta - y \cos \theta)^2 P(x,y) = \overline{m}_{20} \sin^2 \theta - \overline{m}_{11} \sin \theta \cos \theta + \overline{m}_{02} \cos^2 \theta$$

Differentiating this with respect to θ and equating to zero gives the following:

$$2\overline{m}_{20} \sin \theta \cos \theta - 2\overline{m}_{11}(\cos^2 \theta - \sin^2 \theta) - 2\overline{m}_{02} \cos \theta \sin \theta = 0$$

This can also be stated as follows:

$$\overline{m}_{20} \sin 2\theta - 2\overline{m}_{11} \cos 2\theta - \overline{m}_{02} \sin 2\theta = 0$$

This results in the following:

$$\tan 2\theta = 2\overline{m}_{11}/(\overline{m}_{20} - \overline{m}_{02})$$

17.2.2 Moments of sets

The (real) (i,j) *moment* of a bounded measurable subset S of the plane is defined by the following, where $i,j \geq 0$ are integers and the *order* of $\mathcal{M}_{ij}(S)$ is $i+j$:

$$\mathcal{M}_{ij}(S) = \iint_S x^i y^j \, dx \, dy$$

Let $G_h(S)$ be the Gauss digitization of S in $\mathbb{Z}_h^2$. We are interested in estimating $\mathcal{M}_{ij}(S)$ by the following (digital) moment, where $h > 0$ is the grid resolution:

$$m_{ij}(S) = \frac{1}{h^{i+j+2}} \sum_{(u,v) \in G_h(S)} u^i v^j$$

For $h = 1$, this is the same as the definition given in Section 17.2.1, because $P = 1$ if $(u,v) \in G(S)$ and $P = 0$ otherwise. In what follows, we assume that S has been magnified to $h \cdot S$, and we use Gauss digitization in the grid with $h = 1$.

In particular, the *area* $\mathcal{A}(S)$ (i.e., its moment $\mathcal{M}_{00}(S)$ of order 0) is estimated by the number of grid points in $G(S)$ (i.e., by $m_{00}(S)$). Similarly, the following centroid

$$\left(\frac{\mathcal{M}_{10}(S)}{\mathcal{M}_{00}(S)}, \frac{\mathcal{M}_{01}(S)}{\mathcal{M}_{00}(S)} \right)$$

is estimated by the following:

$$\left(\frac{m_{10}(S)}{m_{00}(S)}, \frac{m_{01}(S)}{m_{00}(S)} \right)$$

The *orientation* of S is the slope of its axis of least second moment. This is the line from which the integral of the squares of the distances to the points of S is a minimum. This integral is as follows,

$$F(S,\varphi,\rho) = \iint_S d_p^2(x,y,\varphi,\rho) \mathrm{d}x\,\mathrm{d}y$$

where $d_p(x,y,\varphi,\rho)$ is the perpendicular distance from (x,y) to the following line:

$$x\cos\varphi - y\sin\varphi = \rho$$

The orientation of S is the value of φ for which $F(S,\varphi,\rho)$ takes its minimum; we estimate it by replacing integration and the set S by summation and $G(S)$. For a unique minimum to exist, S should have a "main orientation" (i.e., $\mathcal{M}_{20}(S) \neq \mathcal{M}_{02}(S)$). Finally, the *elongatedness* $\Theta(S)$ of S in direction φ is defined as the ratio of the maximum and minimum values of $F(S,\varphi,\rho)$.

17.2.3 Estimation of moments of sets

In this section, we give worst-case error bounds when digital moments are used as estimates of real moments (and related properties) of subsets S of the plane. From now on, we assume that S is convex and that its frontier consists of a finite number of $C^{(3)}$ ("3-smooth") arcs (i.e., arcs that have continuous derivatives up to order 3).

Theorem 17.2 If S is a convex planar set with a frontier that consists of a finite number of $C^{(3)}$ arcs, $\mathcal{M}_{ij}(S)$ can be estimated by $h^{-(i+j+2)} \cdot m_{ij}(h \cdot S)$ for all $i,j \geq 0$ within an error of $\mathcal{O}(h^{-1})$, where h is the grid resolution; this error term is the best possible.

For the proof of this theorem, see [560]. We now give an example showing that this upper bound is the best possible.

Let S be the unit square with vertices $(0,0), (1,0), (1,1), (0,1)$; then we have the following:

$$\mathcal{M}_{ij}(S) = \frac{1}{(i+1)(j+1)}$$

For a given grid resolution h (so that there are h grid points per unit length), the square S_h that has the following vertices

$$(0,0),\ (1+\frac{1}{2h},0),\ (1+\frac{1}{2h},1+\frac{1}{2h}),\ (0,1+\frac{1}{2h})$$

has the same Gauss digitization as S (i.e., $G(S) = G(S_h)$); however, the difference $\mathcal{M}_{ij}(S) - \mathcal{M}_{ij}(S_h)$ is as follows:

17.2 Moments

$$\int_0^1 x^i \, dx \int_1^{1+\frac{1}{2h}} y^j \, dy + \int_1^{1+\frac{1}{2h}} x^i \, dx \int_0^{1+\frac{1}{2h}} y^j \, dy = \frac{i+j+2}{2(i+1)(j+1)} \cdot \frac{1}{h} + \mathcal{O}\left(h^{-2}\right)$$

Thus, for any grid resolution h, there exists a real square S_h such that the Gauss digitization of $h \cdot S_h$ is the same as the Gauss digitization of $h \cdot S$, but the real moments $\mathcal{M}_{ij}(S_h)$ differ from digital moments $\mathcal{M}_{ij}(S)$ by the following, so error $\mathcal{O}(h^{-1})$ is the best possible:

$$\frac{i+j+2}{2(i+1)(j+1)} \cdot \frac{1}{h} + \mathcal{O}\left(h^{-2}\right)$$

If none of the arcs in the frontier of S is straight, application of Theorem 2.4 leads to a lower upper bound on the error:

Theorem 17.3 If S is a bounded planar 3-smooth convex set, $\mathcal{M}_{ij}(S)$ can be estimated by $h^{-(i+j+2)} \cdot m_{ij}(h \cdot S)$ for all $i, j \geq 0$ within an error of the following:

$$\mathcal{O}\left((\log h)^{\frac{47}{22}} \cdot h^{-\frac{15}{11}}\right) \approx \mathcal{O}\left(h^{-1.3636\ldots}\right)$$

In this theorem, the integers i and j are fixed before changing the resolution h. For example, it is not possible to have $i = j = h^2$, because this would imply $i,j \to \infty$ if $h \to \infty$.

The next theorem shows how accurate estimation of the following "basic difference"

$$|\mathcal{M}_{00}(h \cdot S) - h^2 \cdot \mathcal{A}(S)|$$

within error $\mathcal{O}(\kappa(h))$ can be used to improve error bounds for the following higher-order estimates for an n-smooth S:

$$|\mathcal{M}_{ij}(S) - h^{-(i+j+2)} \cdot m_{ij}(h \cdot S)|$$

Let $C_{\kappa(h)}$ be a nonempty family of planar sets such that the following are true:

(i) $|m_{00}(h \cdot S) - h^2 \cdot \mathcal{A}(S)| = \mathcal{O}(\kappa(h))$ for all $S \in C_{\kappa(h)}$.

(ii) If $S \in C_{\kappa(h)}$, any isometric transformation of S also $\in C_{\kappa(h)}$. (A geometric transformation f is called *isometric* if, for any points p and q, we have $d_e(f(p), f(q)) = d_e(p,q)$.)

(iii) Any set that can be constructed by taking finite numbers of unions, intersections, and set differences of sets from $C_{\kappa(h)}$ also belongs to $C_{\kappa(h)}$.

Let C_0 be the smallest such family of planar sets that contains all n-smooth convex bounded sets and that is closed under finite numbers of intersections, unions, and set differences. Let $\kappa(h)$ be such that C_0 is contained in $C_{\kappa(h)}$.

Theorem 17.4 Let S be a planar 3-smooth convex set. Then $\mathcal{M}_{ij}(S)$ can be estimated by $h^{-(i+j+2)} \cdot m_{ij}(h \cdot S)$ within an error of $\mathcal{O}\left(\kappa(h) \cdot h^{-2}\right)$ for all integers $i,j \geq 0$.

The area, centroid coordinates, orientation, and elongatedness of a set S can all be estimated within worst-case errors of the following (where $\varepsilon > 0$),

$$\mathcal{O}\left(h^{-(\frac{15}{11}-\varepsilon)}\right)$$

using the following estimates:

(1) $\frac{1}{h^2} \cdot m_{00}(h \cdot S)$ for $\mathcal{A}(S)$

(2) $\frac{\frac{1}{h} \cdot m_{10}(h \cdot S)}{m_{00}(h \cdot S)}$ and $\frac{\frac{1}{h} \cdot m_{01}(h \cdot S)}{m_{00}(h \cdot S)}$ for $\frac{\mathcal{M}_{10}(S)}{\mathcal{M}_{00}(S)}$ and $\frac{\mathcal{M}_{01}(S)}{\mathcal{M}_{00}(S)}$

(3) $\frac{2 \cdot \overline{m}_{11}(h \cdot S)}{\overline{m}_{20}(h \cdot S) - \overline{m}_{02}(h \cdot S)}$ for $\tan 2\varphi$

(4) $\frac{t_1(h \cdot S) + \sqrt{t_2(h \cdot S)}}{t_1(h \cdot S) - \sqrt{t_2(h \cdot S)}}$ for $\Theta(h \cdot S)$

where the following are given:

$$t_1(h \cdot S) = \overline{m}_{20}(h \cdot S) + \overline{m}_{02}(h \cdot S)$$
$$t_2(h \cdot S) = 4 \cdot (\overline{m}_{11}(h \cdot S))^2 + (\overline{m}_{20}(h \cdot S) - \overline{m}_{02}(h \cdot S))^2$$

17.3 Experimental Evaluation of Moment Estimates

We use four types of sets to experimentally evaluate error bounds on moment estimates. The sets were chosen to allow for the calculation of real moments and to have a variety of shapes.

17.3.1 Square

Our first example is a centered set (i.e., a set with its centroid at the origin). For such a set S, the moments $\mathcal{M}_{ij}(S)$ are zero for odd values of i if S is symmetric with respect to the x-axis and zero for odd values of j if S is symmetric with respect to the y-axis.

Let $S(a)$ be a $2a \times 2a$ isothetic square centered at the origin; see Figure 17.4 (left). Then we have the following for $i,j \geq 0$:

$$\mathcal{M}_{ij}(S(a)) = \int_{-a}^{+a} \int_{-a}^{+a} x^i y^j \, dx \, dy = \frac{(1-(-1)^{i+1})(1-(-1)^{j+1})}{(i+1)(j+1)} \cdot a^{i+j+2}$$

17.3 Experimental Evaluation of Moment Estimates

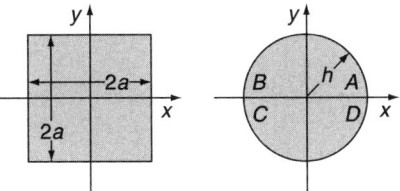

FIGURE 17.4 A centered square and circle.

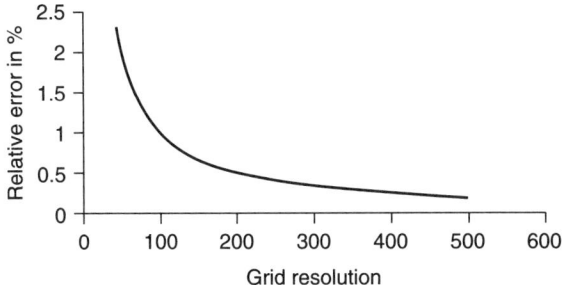

FIGURE 17.5 Relative errors in the estimated zero-order moment of a centered square.

It follows that $\mathcal{M}_{00}(S(a)) = 4a^2$, $\mathcal{M}_{01}(S(a)) = \mathcal{M}_{10}(S(a)) = \mathcal{M}_{11}(S(a)) = 0$, $\mathcal{M}_{02}(S(a)) = \mathcal{M}_{20}(S(a)) = (4/3)a^4$, and so on. As mentioned in Section 17.2.1, $\mathcal{M}_{ij}(S(a)) = 0$ when i or j is odd. The following difference varies between $-(4a+1)$ and $(4a+1)$ when a is in an open interval between two successive (positive) integers:

$$\mathcal{M}_{00}(S(a)) - m_{00}(S(a)) = a^2 \cdot \mathcal{M}_{00}(S(1)) - m_{0,0}(S(a))$$

For arbitrary $i, j \geq 0$, the following is in the interval $[0, g(a)]$ where $g(a) = \mathcal{O}(a^{i+j+1})$; see Theorem 17.4:

$$\kappa(a) = |\mathcal{M}_{ij}(S(a)) - m_{ij}(S(a))| = |a^{i+j+2} \cdot \mathcal{M}_{ij}(S(1)) - m_{ij}(S(a))|$$

Figure 17.5 shows a plot of $\kappa(a)$ for $i = j = 0$.

17.3.2 Disk

We first consider the quarter disk A shown in Figure 17.4 (right) and the moments $\mathcal{M}_{0j}(A)$ where $j = 2t - 1 \leq 0$ is odd. In this case, the integration is simple:

$$\mathcal{M}_{0j}(A) = \int_0^h \left(\int_0^{\sqrt{h^2-x^2}} y^j \, dy \right) dx = \frac{1}{j+1} \int_0^h (h^2 - x^2)^t \, dx$$

Here are two examples:

$$\mathcal{M}_{01}(A) = \frac{1}{2} \int_0^h (h^2 - x^2) \, dx = \frac{1}{3} h^3$$

$$\mathcal{M}_{03}(A) = \frac{1}{2} \int_0^h (h^2 - x^2)^2 \, dx = \frac{4}{15} h^5$$

These values for A can then be used to calculate the values for B, C, and D (see Figure 17.4). For j odd, we have $\mathcal{M}_{0j}(B) = \mathcal{M}_{0j}(A)$ and $\mathcal{M}_{0j}(C) = \mathcal{M}_{0j}(D) = -\mathcal{M}_{0j}(A)$. It follows that, for the half-disk $A \cup B$, we have $\mathcal{M}_{01}(A \cup B) = (2/3)h^3$ and $\mathcal{M}_{03}(A \cup B) = (8/15)h^5$. By symmetry, we have $\mathcal{M}_{0j}(A \cup D) = 0$ for the half-disk $A \cup D$ if j is odd and $\mathcal{M}_{0j}(S) = 0$ for the full disk $S = A \cup B \cup C \cup D$ if j is odd.

The difference between the zero-order discrete moment and the zero-order moment (i.e., the area) of a centered disk (or a disk with its center at a grid point) that has radius h is at most the following; see [460]:

$$\mathcal{O}\left(h^{\frac{131}{208}}\right)$$

It can be shown [562] that the following differences

$$|\mathcal{M}_{i0}(C_1) - m_{i0}(C_1)| \quad \text{and} \quad |\mathcal{M}_{0i}(C_1) - m_{0i}(C_1)|$$

are at most

$$\mathcal{O}(h^{i+\frac{131}{208}})$$

for a disk C_1 that has radius h and center (a, b) where the integers a and b are at least h. An analogous theoretic result for the general difference $|\mathcal{M}_{ij}(C_1) - m_{ij}(C_1)|$ is not yet known. However, experiments in [562] compared the error term with the following:

$$h^{-\frac{285}{208}} = h^{\frac{131}{208}} \cdot h^{-2}$$

The exact values of the real moments of the disk $(x-1)^2 + (y-1)^2 \leq 1$ are as follows (rounded to six digits):

$$\mathcal{M}_{01}(S) = 3.141592 \qquad \mathcal{M}_{02}(S) = 3.926991$$
$$\mathcal{M}_{11}(S) = 3.141592 \qquad \mathcal{M}_{12}(S) = 3.926991$$
$$\mathcal{M}_{23}(S) = 6.675884 \qquad \mathcal{M}_{24}(S) = 9.866564$$

17.3 Experimental Evaluation of Moment Estimates

TABLE 17.2 Errors in approximating $\mathcal{M}_{ij}(S)$ by $h^{-(i+j+2)} \cdot m_{ij}(h \cdot S)$ for different grid resolutions h, where S is the disk $(x-1)^2 + (y-1)^2 \leq 1$.

i	j	h	$\mathcal{M}_{ij}(S) - \dfrac{m_{ij}(h \cdot S)}{h^{i+j+2}}$	$h^{-\frac{285}{208}}$
0	1	10	-0.018407	0.042639
		50	$+0.003992$	0.004699
		100	-0.000007	0.001818
		500	$+0.000200$	0.000200
	2	10	-0.033609	0.042639
		50	$+0.005779$	0.004699
		100	-0.000062	0.001818
		500	$+0.000298$	0.000200
1	1	10	-0.028407	0.042639
		50	$+0.003592$	0.004699
		100	-0.000107	0.001818
		500	$+0.000196$	0.000200
	2	10	-0.043609	0.042639
		50	$+0.005379$	0.004699
		100	-0.000162	0.001818
		500	$+0.000294$	0.000200
2	3	10	-0.129811	0.042639
		50	$+0.009904$	0.004699
		100	-0.001224	0.001818
		500	$+0.000608$	0.000200
	4	10	-0.225556	0.042639
		50	$+0.016346$	0.004699
		100	-0.002193	0.001818
		500	$+0.001014$	0.000200

In Table 17.2, these exact values are compared with the discrete moments calculated for the Gauss-digitized disk.

17.3.3 Isometric quartic frontier segments

We next consider a shape that is more complex than a disk and defined by four equal-length ("isometric") quartic frontier segments; see Figure 17.6. This shape is the compact region with a frontier that consists of four segments of the algebraic curve γ of order 4:

$$\gamma : \left(y - \frac{1}{2}\right)^2 = \left(\frac{1}{2} - \sqrt{1 - |2x - 1|} - \left|x - \frac{1}{2}\right|\right)^2$$

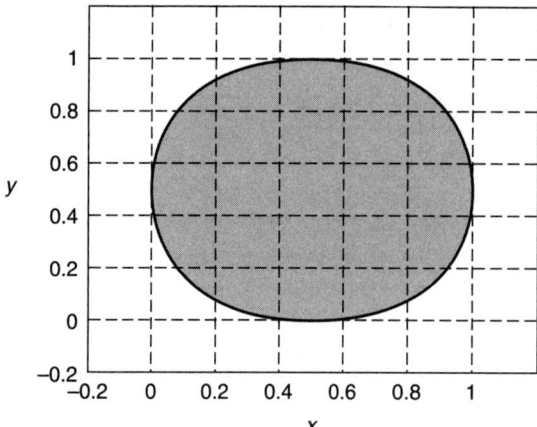

FIGURE 17.6 An isometric quartic frontier in the unit square.

This curve is of number-theoretic interest for at least two reasons:
(1) The function $e(m)$ (see Theorem 13.1) is equal to the following for $m \leq 0$:

$$\frac{12}{\sqrt[3]{4\pi^2}} \cdot m^{\frac{2}{3}} + \mathcal{O}(m^{\frac{1}{3}} \cdot \log m)$$

It was calculated by analyzing maximal "circular" convex grid polygons P_m such as those shown in Figure 13.7 for $m = 7, 8, 9$. It can be shown [1166] that these P_ms converge to γ with respect to the Hausdorff metric d_e:

$$\lim_{m \to \infty} d_e(\frac{1}{m} \cdot P_m, \gamma) = 0$$

(2) The grid polygons P_m are examples of $\mathbf{Z}_m$-*lattice polygons* that have width and height m. It can be shown [57, 1092] that, as $m \to \infty$, almost all convex $\mathbf{Z}_m$-lattice polygons (scaled by factor m^{-1} so that they lie in the unit square $[0,1]^2$) are "very close" (in the Hausdorff metric sense) to γ. The related central limit theorem is proved in [993].

Some numeric estimates of moments of S are given in Table 17.3. The exact values of the real moments of S are as follows (rounded to six digits):

$$\mathcal{M}_{00}(S) = 0.833333 \qquad \mathcal{M}_{01}(S) = 0.416667$$
$$\mathcal{M}_{11}(S) = 0.208333 \qquad \mathcal{M}_{12}(S) = 0.131845$$
$$\mathcal{M}_{23}(S) = 0.057726 \qquad \mathcal{M}_{24}(S) = 0.043308$$

17.3 Experimental Evaluation of Moment Estimates

TABLE 17.3 Errors in approximating $\mathcal{M}_{ij}(S)$ by $h^{-(i+j+2)} \cdot m_{ij}(h \cdot S)$ for different grid resolutions h. The bounded set S is bounded by four equal-length segments of the curve γ.

i	j	h	$\mathcal{M}_{ij}(S) - \dfrac{m_{ij}(h \cdot S)}{h^{i+j+2}}$	$h^{-\frac{15}{11}}$
0	0	1	+0.833333	1
		2	−0.416666	0.388601
		100	+0.001633	0.001873
		100 π	+0.000138	0.000393
	1	1	+0.416666	1
		5	+0.096666	0.111393
		30 π	+0.000132	0.002031
		300	+0.000394	0.000418
1	1	2	−0.104167	0.388601
		e	+0.043449	0.255729
		70	+0.001139	0.003047
		350	+0.000086	0.000339
	2	5	+0.035845	0.111393
		10	+0.004295	0.004328
		121	+0.000266	0.001444
		400	+0.000071	0.000282
2	3	7	+0.008997	0.070403
		14.5	−0.001638	0.026080
		93.3	−0.000170	0.002059
		444	+0.000047	0.000245
	4	12	+0.004634	0.033758
		27	+0.001788	0.011172
		121.22	−0.000148	0.001411
		500	+0.000018	0.000208

17.3.4 Parameterized quartic frontier segments

Finally, consider a more general class of parameterized sets defined by four arcs of the quartic curve $y^2 \leq (mx^2 - m)^2$; see Figure 17.7. This curve too was chosen for number-theoretic reasons.

We recall that the following is at most $\mathcal{O}(h)$, even in cases where straight segments are allowed on the frontier of S:

$$\kappa(h) = |\mathcal{M}_{00}(h \cdot S) - m_{00}(h \cdot S)|$$

$\kappa(h)$ can be smaller than this under additional assumptions about the frontier of S; see, for example, Theorem 2.4. Furthermore [598], if we express $\kappa(h)$ in the form $\mathcal{O}(h^\alpha)$, the statement $|\mathcal{M}_{00}(h \cdot S) - m_{00}(h \cdot S)| = \mathcal{O}(h^\alpha)$ is false for $\alpha < 0.5$. The

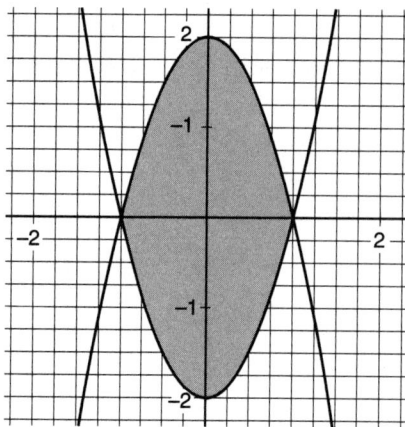

FIGURE 17.7 Four quartic frontier segments for which $m = 2$.

quartic arc $y = \sqrt{x}$ where $x \in [1, h]$ is an example of an arc that has a length of order of magnitude h and that passes through exactly $\lfloor h^{0.5} \rfloor$ grid points.

Tables 17.4 and 17.5 give numeric examples in which $m = 1$ and $m = 4$, where the sets translated by the vectors $\vec{(2,2)}$ and $\vec{(5,2)}$, respectively, contain only points with positive coordinates. The exact values of the real moments in the case $m = 1$ are as follows (rounded to seven significant figures)

$$m_{0,0}(S) = 2.666667 \qquad m_{0,1}(S) = 5.333333$$
$$m_{1,1}(S) = 10.66667 \qquad m_{1,2}(S) = 22.55238$$
$$m_{2,3}(S) = 104.6349 \qquad m_{2,4}(S) = 240.5445$$

The exact values of the real moments in the case $m = 4$ are as follows (rounded to seven significant figures):

$$\mathcal{M}_{00}(S) = 10.66667 \qquad \mathcal{M}_{01}(S) = 53.33333$$
$$\mathcal{M}_{11}(S) = 106.6667 \qquad \mathcal{M}_{12}(S) = 611.3523$$
$$\mathcal{M}_{23}(S) = 8005.587 \qquad \mathcal{M}_{24}(S) = 53289.62$$

The values in both tables are given for the same values of h to illustrate two things:

1) The effect of the sizes of the real moments that are to be estimated: As can be seen from Table 17.5, if these moments have relatively large values, the required precision is not achieved for small values of h as it was in the previous examples; higher resolutions h have to be used. It is more appropriate to use the relative error in such situations. If we use the usual definition of relative

17.3 Experimental Evaluation of Moment Estimates

TABLE 17.4 Errors in approximating $\mathcal{M}_{ij}(S)$ by $h^{-(i+j+2)} \cdot m_{ij}(h \cdot S)$ for different grid resolutions h, where S is the bounded region that has a frontier that consists of four segments of the quartic curve $(y-2)^2 = ((x-2)^2 - 1)^2$.

i	j	h	$\mathcal{M}_{ij}(S) - \dfrac{m_{ij}(h \cdot S)}{h^{i+j+2}}$	$h^{-\frac{15}{11}}$
0	0	10	−0.023333	0.043287
		100	−0.000233	0.001873
		200	+0.000841	0.000728
		800	+0.000190	0.000109
	1	20	+0.002833	0.016821
		80	+0.010520	0.002540
		160	+0.004505	0.000987
		320	+0.000175	0.000383
1	1	30	+0.031111	0.009677
		90	+0.000148	0.002163
		270	+0.001700	0.000483
		540	+0.001467	0.000187
	2	25	−0.150651	0.012408
		75	+0.011268	0.002773
		225	−0.002367	0.000620
		450	−0.000591	0.000240
2	3	40	+0.530125	0.006536
		160	+0.108207	0.000987
		320	+0.042065	0.000383
		640	+0.016145	0.000149
	4	50	+0.517891	0.004821
		200	+0.104890	0.000728
		400	+0.038664	0.000282
		1000	+0.014139	0.000081

error and Theorem 17.4 is applied to regions that have no straight segments on their frontiers, then we have the following:

$$\left| \frac{m_{ij}(h \cdot S)}{\mathcal{M}_{ij}(h \cdot S)} - 1 \right| = \frac{|m_{ij}(h \cdot S) - \mathcal{M}_{ij}(h \cdot S)|}{h^{i+j+2} \cdot \mathcal{M}_{ij}(S)} = \mathcal{O}\left(\frac{1}{h^{\frac{15}{11} - \varepsilon} \cdot \mathcal{M}_{ij}(S)} \right)$$

2) The effect of the elongation of the region: For example, $m = 5$ gives a greater elongation than $m = 2$. No theoretic results on this subject have yet been obtained, but it seems that an increase in elongation leads to an increase in worst-case error.

It might also be of interest to calculate errors in moment estimates for values of m such that both m and h go to infinity (e.g., $m = \log h$, $m = \sqrt{h}$, $m = h^2$).

TABLE 17.5 Errors in approximating $\mathcal{M}_{ij}(S)$ by $h^{-(i+j+2)} \cdot m_{i,j}(h \cdot S)$ for different grid resolutions h, where S is the bounded region whose frontier consists of four segments of the quartic curve $(y-5)^2 = (4(x-2)^2 - 4)^2$.

i	j	h	$\mathcal{M}_{ij}(S) - \dfrac{m_{ij}(h \cdot S)}{h^{i+j+2}}$	$h^{-\frac{15}{11}}$
0	0	10	−0.023333	0.043287
		100	−0.003833	0.001873
		200	−0.000958	0.000728
		800	+0.000215	0.000109
	1	20	−0.079166	0.016821
		80	+0.020052	0.002540
		160	+0.014388	0.000987
		320	+0.006722	0.000383
1	1	30	+0.012222	0.009677
		90	+0.003703	0.002163
		270	+0.005898	0.000483
		540	+0.004355	0.000187
	2	25	−0.940180	0.012408
		75	+0.102672	0.002773
		225	−0.015434	0.000620
		450	−0.020359	0.000240
2	3	40	−4.711709	0.006536
		160	+3.055033	0.000987
		320	+1.433887	0.000383
		640	+0.609743	0.000149
	4	50	−65.552138	0.004821
		200	−8.304786	0.000728
		400	−0.389115	0.000282
		1000	+0.331766	0.000081

17.4 Operations on Pictures and Invariant Properties

Let **P** be the set of pictures P defined on a given grid, and let Φ be a function from **P** into itself. Φ is called a *pixel-wise operation* (or a "point operation") if the value of $(\Phi(P))(p)$ depends only on the value of $P(p)$. It is called a *local operation* if the value of $(\Phi(P))(p)$ depends only on the values of a set of $P(q)$s such that p is one of the qs and $\max_q(d_e(p,q))$ is constant so that it does not depend on the size of the picture. It is called a *geometric operation* if there exists a function g from the grid into itself such that the value of $(\Phi(P))(p)$ depends only on the values of a set of $P(q)$s such that $\max_q((g(p),q))$ is constant and so does not depend on the size of the picture.

Pixelwise and local operations are studied in books about digital picture processing. For more about geometric operations see Chapter 14.

A picture property $\mathcal{F}$ is called *invariant* under Φ if $\mathcal{F}(\Phi(P)) = \mathcal{F}(P)$ for all $P \in \mathbf{P}$. For example, let Φ be the linear pixel-wise operation that takes $P(p)$ into $aP(p)+b$. A ratio of differences of pixel values (e.g., $\frac{P(p)-P(q)}{P(s)-P(t)}$), is invariant under Φ, because the following is true:

$$\frac{[aP(p)+b]-[aP(q)+b]}{[aP(s)+b]-[aP(t)+b]} = \frac{a[P(p)-P(q)]}{a[P(s)-P(t)]} = \frac{P(p)-P(q)}{P(s)-P(t)}$$

The set of values of the pixels of P is invariant under any one-to-one mapping of P into itself; in particular, it is invariant under any one-to-one geometric transformation of P (see Section 14.4). It follows that any $\mathcal{F}(P)$ that depends only on the set of values of P (e.g., any statistic property such as m_{00}) is invariant under any one-to-one Φ. Evidently, $m_{20}(P)$ is invariant under reflection of P in the y-axis; this is similar for $m_{02}(P)$ and the x-axis; and $m_0(P)$ is invariant under reflection of P in the origin or under any one-to-one rotation of P around the origin.

Properties that are invariant under various types of (real) geometric transformations are studied in various branches of geometry (see Section 14.1); however, because most geometric transformations do not take pictures into pictures (see Section 14.5), properties that are invariant under real geometric transformations are at best "approximately invariant" when the transformations are applied to pictures.

17.5 Spatial Relations

Properties of binary pictures can be regarded as properties of the sets of 1s in the pictures (e.g., as properties of [not necessarily connected] "objects"). Relative values of properties of pictures define relations between pictures, and this is similar for relations between objects. In this section, we briefly discuss two types of "spatial" (i.e., geometric) relations between objects: relations of relative position and topologic relations.

17.5.1 Relations of relative position

Quantitative relations of relative position between single pixels can be defined by fuzzy subsets of the grid. For example, if $p = (x,y)$, the degree of membership of $q = (u,v)$ in "above p" can be defined by a fuzzy subset $\mu_{a(p)}$ such that $\mu_{a(p)}(x,v) = 1$ for all $v > y$; $\mu_{a(p)}(u,v)$ decreases monotonically as $|u-x|$ increases; and $\mu_{a(p)}(u,v) = 0$ for $v < y$. Similarly, "near p" can be defined by a fuzzy subset $\mu_{n(p)}$ such that $\mu_{n(p)}(x,y) = 1$ and $\mu_{n(p)}(u,v)$ decreases monotonically as $d((u,v),(x,y))$ increases.

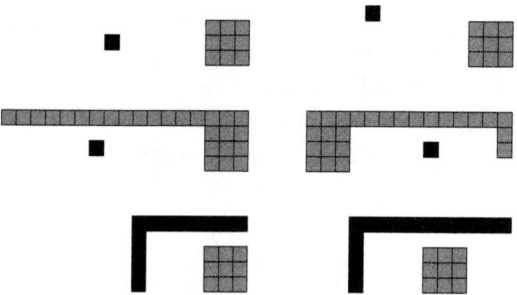

FIGURE 17.8 The difficulty of defining "to the left of": in which of these cases is the black object to the left of the gray object?

It is much harder to define relations of relative position between sets of pixels. To see this, consider Figure 17.8, which illustrates the difficulty of defining "to the left of." Let A be the black object and B the gray object. If we require that every pixel of A be to the left of every pixel of B, we exclude all but the two upper cases; however, excluding the lower left case seems unreasonable. If we require only that every pixel of A be to the left of some pixel of B, we include all but the case on the lower right; however, including the case on the middle right seems unreasonable. If we require that A's centroid be to the left of B's centroid, we include all but the case on the middle right; however, including the case on the lower right seems unreasonable. One proposed definition requires that two conditions be satisfied: A's centroid must be to the left of B's leftmost pixel and A's rightmost pixel must be to the left of B's rightmost pixel. This definition excludes the two middle cases and the lower right case, which is defensible. Another possibility is to require that every pixel of A be to the left of some pixel of B and every pixel of B be to the right of some pixel of A. This excludes all but the two upper cases; it includes the lower left case if the rightmost pixel of A is removed. Note that this definition implies that the "to the left of" relation is transitive, and its inverse is "to the right of." The relation would also be reflexive if we required that every pixel of A be to the left of or in the same column as some pixel of B and that every pixel of B be to the right of or in the same column as some pixel of A. However, a purely coordinate-based definition of this type may not be adequate in all cases; judgments about "to the left of" can be influenced by the interpretation of the picture as a projection of a 3D scene or by the recognition of A and B as familiar objects.

17.5.2 Topologic relations

We recall that any adjacency relation on a grid defines an adjacency relation between disjoint subsets of the grid: S and T are called α-*adjacent* iff some pixel of S is

17.5 Spatial Relations

α-adjacent to some pixel of T. For any such adjacency relation, we can also define quantitative measures of *degree of adjacency* (e.g., in terms of what fraction of the pixels of S are adjacent to pixels of T or vice versa); see Exercise 5.

We have also seen that an adjacency relation defines separation relations between disjoint subsets of the grid. Let U, V, and W be sets of pixels such that U and V are disjoint. We say that W α-*separates* U from V iff any α-path from a pixel of U to a pixel of V must contain a pixel of W. More generally, let μ, ν, and ω be fuzzy subsets of a picture. We say that ω α-separates μ from ν iff, on any α-path $\rho = (p_0, \ldots, p_n)$, there exists a pixel p_i ($1 \leq i < n$) such that $\omega(p_i) \geq \max(\mu(p_0), \nu(p_n))$.

Surroundedness is a special case of separateness; in a binary picture, we say that S surrounds U if it separates U from the infinite background component B of the picture. Quantitative measures of *degree of surroundedness* can also be defined. For example, let $U = \{p\}$, and let W not contain p. The following are two ways of measuring the degree to which W surrounds p in a 2D picture:

a) Evidently, W separates U from B if it separates U from the border ϑB of B, which is finite; let ϑB consist of n pixels. For each $b \in \vartheta B$, let $\overline{pb}$ be a digital straight line segment with endpixels p and b. If m of the n $\overline{pb}$s intersect W, we say that the degree of *v-surroundedness* ("visibility surroundedness") of p by W is m/n.

b) We define the degree of τ-*surroundedness* ("turn surroundedness") of p by W as $\min(\tau(\rho))$, where $\tau(\rho)$ is the total turn of ρ (see Chapter 10, Exercise 6) and the min is taken over all 4-paths ρ from p to ϑB that do not intersect W. If no such ρ exists (i.e., if W 4-surrounds $\{p\}$), we say that the degree of τ-surroundedness of p by W is infinite.

In a 3D picture under (α, α')-adjacency, if the Euler characteristic of a set W is equal to the number of α-components of W minus the number of α'-components of $\overline{W}$, W is called *tunnel-free*. Let W and Z be α-connected sets, each of which is contained in the background component of the other (so they are disjoint). It can be shown that W is tunnel-free iff Z is tunnel-free. Let W^* be a tunnel-free set that contains W. We say that W and Z are *linked* if any such W^* intersects Z. It can be shown that this implies that W and Z are not tunnel-free.

A (simple) 3D curve can be knotted, and a set of pairwise disjoint 3D curves can be linked [768]. A curve can therefore be called a *knot*, and a finite union of pairwise disjoint curves can be called a *link*. A knot or link is called *polygonal* if the curves are (real) polygons, and it is called *digital* if they are isothetic grid polygons. Two knots or links are called (knot-theoretically) *isomorphic* iff there exists an orientation-preserving homeomorphism that maps one of them onto the other. A knot is called *unknotted* if it is isomorphic to a simple planar polygonal curve.

Two simple polygons Q and Q' are said to differ by an *elementary deformation* if there exist three points p, q, and r such that Q intersects the planar triangular region pqr in the line segment pr; Q' intersects pqr in the line segments pq and qr; and Q and Q' are otherwise identical. It can be shown that two polygonal knots or links are isomorphic iff they differ by a finite sequence of elementary deformations. A similar definition can be given for digital knots or links using deformations that take three sides of a rectangle into the fourth side. (Note that this is not the same as the elementary deformations of digital curves in Section 16.4.)

The parallel projection L_θ of a polygonal link L in direction θ (onto a plane perpendicular to θ) is the union of a connected set of straight line segments. We call L_θ *regular* if the following are true:

a) For all but finitely many points p of L_θ, the preimage of p is a single point of L.

b) Each of the exceptional points has a preimage that is an interior point of exactly two polygon sides of L.

It is not hard to see that, for any polygonal link L, there exist directions θ for which L_θ is regular. An exceptional point p in a regular projection of L is called a *crossing point*. The two preimages of p on L have different coordinates with respect to a coordinate axis in direction θ. These two preimages lie on two polygon sides of L; the side with a preimage that has a higher (lower) coordinate is said to cross *over (under)* the other side. It is easy to show that, if some regular projection of a knot K never crosses itself, K must be unknotted.

If, in some regular projection of a link L, some knot K crosses only over or under the other knots (or does not cross them at all), K cannot be linked to the other knots. It can be shown that W is not tunnel-free iff there exist a simple α-curve K contained in W and a simple α'-curve K' contained in $\overline{W}$ such that K and K' are linked.

17.6 Exercises

1. Let $P_1, \ldots, P_n$ be distinct pictures defined on a given grid. Prove that there exist pixels $p_1, \ldots, p_{n-1}$ such that, for any $1 \leq i \neq j \leq n$, we cannot have $P_i(p_1) = P_j(p_1), \ldots, P_i(p_{n-1}) = P_j(p_{n-1})$.

2. Let P be a binary picture defined on an k-pixel grid. Let $\mathcal{F}$ be a predicate on this class of pictures, and let Φ be a Boolean function of k variables such that $\mathcal{F}(P)$ is true iff $\Phi(P) = 1$; such a Φ is called a Boolean expression for $\mathcal{F}$. The length of the shortest Boolean expression for $\mathcal{F}$ is called the length of $\mathcal{F}$. Let $\mathcal{F}_1(\mathcal{F}_1^+)$ be the predicate that is true iff exactly (at least) one pixel of P has value 1. Prove that the length of $\mathcal{F}_1^+$ is k and that the length of $\mathcal{F}_1$ cannot be a linear function of k but is at most k^2.

3. Prove that, if P is symmetric around the line $y = x$ (i.e., $f(x,y) = f(y,x)$ for all x,y), then $m_{ij}(P) = m_{ji}(P)$ for all i,j.

4. Prove that the principal axis is in the direction of the eigenvector of the following matrix, which corresponds to its larger eigenvalue:

$$\begin{pmatrix} \overline{m}_{20} & \overline{m}_{11} \\ \overline{m}_{11} & \overline{m}_{02} \end{pmatrix}$$

5. Let S be a nonempty finite 4-connected subset of $\mathbb{Z}^2$, and let T be a subset of $\mathbb{Z}^2$ that is disjoint from S. Let $\partial_T S$ be the set of pixels of S that are 4-adjacent to pixels of T; note that, if $T = \bar{S} = \mathbb{Z}^2 \setminus S$, $\partial_T S = \partial S$ is the border of S. Define the *degree of adjacency* of S to T as follows:

$$a_T(S) = \frac{\operatorname{card}(\partial_T S)}{\operatorname{card}(\partial S)}$$

Evidently, $0 \leq a_T(S) \leq 1$. For what proper subsets T of $\mathbb{Z}^2$ do we have $a_T(S) = 1$? If T is nonempty, finite, and 4-connected, how is $a_S(T)$ related to $a_T(S)$?

17.7 Commented Bibliography

Properties for which the cardinality or the diameter of σ does not depend on the picture size are studied in [733] and [533, 534].

Exercise 1 is based on [360] and Exercise 2 on [439]. Theorem 17.1 was first applied to digital pictures in [880].

Moments were introduced into picture analysis in [446]; see also [362] and [15]. For elongatedness $\Theta(S)$, see [468, 1113]. (For an "intrinsic" definition of elongatedness, see Section 15.6.2.) For the approximation of real moments by digital moments, see [558, 559, 560, 1113]. [561] gives the upper bound example that follows Theorem 17.2. Section 17.3 reviews [562]. For other types of moments, see [980].

For invariants in geometry, see [559, 560]. For the difficulty of defining "to the left of," see [1127]. For measures of degree of adjacency or surroundedness, see [913]. For digital knots and links, see [338, 768].

List of Algorithms

2D borders, boundaries, or frontiers

- border tracing, 142
- boundary approximation ("marching squares"), 146
- frontier tracing (frontier grid, incidence grid), 187

3D boundaries or frontiers

- frontier tracing (FILL algorithm), 191
- frontier tracing (Artzy-Herman algorithm), 300–301
- boundary approximation (marching cubes), 301–302

adjacency and connectedness

- FILL for component labeling (adjacency grid), 53
- Rosenfeld-Pfaltz component labeling, 54
- local adjacency (incidence grid), 182
- ordered adjacency procedure (multilevel pictures), 183
- switch adjacency (multilevel pictures), 39, 249

2D convex hulls

- Graham scan (set of points), 25
- Sklansky algorithm ("visible" simple polygon), 430
- simple polygon, 429
- set of grid points, 428

minimum-length polygons

- 2D MLP calculation, 344–345
- approximating-sausage algorithm, 351
- 3D MLP approximation (rubber-band algorithm), 385–390

2D curve representation and analysis

- directional encoding, 71
- vertex chain code, 72–75
- local and global length estimation, 302–304
- tangent-based length estimation, 346, 352–353
- corner detection, 363–364
- curvature estimation, 364–366

3D curve analysis

- local and global length estimation, 409–413
- estimation of curvature and torsion, 413

diagrams

- 2D Delaunay (Voronoi) diagrams, 440–443
- watersheds, 449–450

distance

- Hausdorff metric, 86
- two-pass distance transform (chamfer metric), 94–95
- Danielsson distance transform (Euclidean metric), 109

DSS generation and recognition

- Bresenham's algorithm (DSS generation), 63
- 8-DSS based on syntactic characterization (**CHW1982a**), 329–331
- 4-DSS for frontier in incidence grid (**K1990**), 334–336
- 3D DSS recognition, 379
- 8-DSS based on arithmetic geometry (**DR1995**), 378

DPS recognition

- incremental algorithm (**KS2001**), 402–404
- convex-hull based algorithm, 407

graphs

- FILL (connected components), 120
- Dijkstra algorithm (shortest path), 128
- border cycle in an oriented adjacency graph, 142

region analysis (2D)

- medial axis transform, 111–112, 495–496
- discrete integration (area of isothetic polygon), 279–280
- circumscribing rectangle of smallest area, 428–429

region analysis (3D)

- local surface area estimation, 414
- surface area estimation based on normal integration, 418–419
- surface area estimation based on DPS recognition, 421
- mean curvature estimation (**HK2003a**), 423–424

geometric transformations

- value interpolation, 466–477
- magnification, 58, 468, 470–471
- demagnification, 471, 473–474

morphologic operations

- dilation, 481–483
- erosion, 483–485
- opening and closing, 486
- hit-and-miss transforms, 485–486
- fusing of clusters, 492–493

- detection of (intrinsically) elongated object parts, 493–494
- distance transforms and medial axes, 495–496

deformations
- shrinking, 506–509
- thinning (skeletonization), 509–513
- deformation of curves, 532–534
- thinning of multivalued pictures, 533

List of Symbols

Pictures

c, b	cells in a grid
$\mathcal{F}(P)$	feature of a picture P
$G_{\max}$	maximal picture value (maximal gray level)
$\mathbb{G}, \mathbb{G}_{m,n}, \mathbb{G}_{l,m,n}$	grid (the medium on which digital pictures reside)
$P, P(p), P(i,j)$	picture, picture value at pixel $p = (i,j)$
$\langle P \rangle$	set of object pixels (value 1) in binary picture P
p, q, r, s, t	points, locations (grid points or cells in the grid)
Π	plateau
ρ	path or cycle in a subset of a grid
σ	structuring element
u, v, w	picture values, words (e.g., chain codes)
x, y, z	picture coordinates

Digitizations

$G_h(S)$	Gauss digitization of set S
$J_h^-(S), J_h^+(S)$	inner and outer Jordan digitizations of set S
$\kappa(h)$	speed of convergence
h	grid resolution
θ	grid constant, $\theta = 1/h$
$\rho_h(\gamma)$	digitized grid-intersection sequence for arc or curve γ
$R_h(\gamma)$	grid-intersection digitization of arc or curve γ

Digital spaces

A, A_α	adjacency relation
α	metavariable for type of adjacency
$\delta M, \delta_\alpha M$	border of set M for adjacency relation A_α
d_α	metric in grid point model
d_4	Manhattan or city block distance
d_8	chessboard distance
∂_α	metric in grid cell model
e	radius of a neighborhood
Γ, Γ_α	connectedness relation for adjacency relation A or A_α
$M^\nabla, M_\alpha^\nabla$	inner set of set M for adjacency relation A or A_α
N, N_α	smallest nontrivial neighborhood in grid point model
η, η_α	smallest nontrivial neighborhood in grid cell model

General mathematic symbols

a, b, c	real numbers
a, b, ...	vectors
A, B, C	sets, polygons
A, B, ...	matrices
$\mathcal{A}(S)$	area of set S
$\alpha, \beta, \eta, \psi, \phi$	angles, fractional parts of real numbers
$C^{(n)}(S)$	set of functions that are n-times continuously differentiable on S
$\mathbb{C}_n$	set of all m-cells $(m \leq n)$
d	distance
d_e	Euclidean distance
$\mathbb{E}^2, \mathbb{E}^3$	Euclidean plane $[\mathbb{R}^2, d_e]$, Euclidean space $[\mathbb{R}^3, d_e]$
f, g	functions
γ	curve or arc in Euclidean space
h, i, j, k, l, m, n	integers, picture coordinates
$\kappa(p)$	curvature at point p
$\mathcal{L}(\gamma) = \|\gamma\|$	length of curve γ
λ	real number, eigenvalue of matrix
L_m	Minkowski metric, $m = 1, 2, ..., \infty$
L, M	subsets of set S or of grid $\mathbb{G}$
$\mathbb{N}$	the set of all natural numbers $\{0, 1, 2, ...\}$
π	polygonal curve, perimeter of unit circle
Π	polygonal region
$\mathcal{P}(S)$	perimeter of set S
R	relation
$\mathbb{R}$	the set of all real numbers

List of Symbols

$\sigma, \theta, \vartheta$	permutations
S, T	sets
$\mathcal{V}(S)$	volume of set S
x, y, z	rectangular Cartesian coordinates
$\mathbb{Z}$	the set of all integers

Set theory

$\emptyset$	empty set
$\wp(S)$	power set of S
$\wp_{\text{fin}}(S)$	class of all finite subsets of S
$\text{card}(M)$	cardinality of set M
$\aleph_0$	cardinality of $\mathbb{N}$
$\overline{M}$	complementary set $S \setminus M$ of $M \subseteq S$
$M_1 \Delta M_2$	symmetric difference $(M_1 \setminus M_2) \cup (M_2 \setminus M_1)$

Integer arithmetic and asymptotics

$\lfloor a \rfloor$	largest integer smaller than or equal to a
$[a]$	nearest integer to a if uniquely defined, $\lfloor a \rfloor$ otherwise
$\lceil a \rceil$	smallest integer larger than or equal to a
$f_1(n) \approx f_2(n)$	approximate equality between $f_1(n)$ and $f_2(n)$
$a \approx b$	approximate equality $|a-b| < 1$ between $a, b \in \mathbb{R}$
$g(n) = \mathcal{O}(f(n))$	$|g(n)| \leq c|f(n)|$ for all $n \geq m$, some $c > 0$, and some $m \geq 0$
$g(n) = \Omega(f(n))$	$|g(n)| \geq c|f(n)|$ for all $n \geq m$, some $c > 0$, and some $m \geq 0$

Topology in metric spaces

ε	radius of a neighborhood
ϑM	frontier of M defined by ε-neighborhoods
M°	interior $M \setminus \vartheta M$ of M
$M^\bullet$	closure $M \cup \vartheta M$ of M
$U_\varepsilon(p)$	ε-neighborhood of p

Combinatorial topology

$\alpha_0, \alpha_1, \alpha_2$	numbers of vertices, edges, and faces
β_0, β_i	number of components, Betti numbers ($i \geq 0$)
χ	Euler characteristic
$\xi, \xi(p)$	orientation, local circular order

List of Axioms and Properties

A1-A3	adjacency graph,	120
A4	adjacency graph (orientation),	136
C1-C2	abstract complex,	217
D1-D3	digital topology (grid cell model),	192
D1-D3	digital topology (grid point model),	193
E1-E2	Euclidean complex,	214
G1-G8	axiomatic digital geometry,	450–452
H1-H4	hull and near-hull,	419–420
I1-I8	incidence pseudograph,	158–160
M1-M5	metric,	12–13
M4-M5	metric,	79
N1-N3	norm,	78
N2*	norm,	79
T1-T3	topology,	188
V0-V8	vector space.	18

Bibliography

[1] K. Abe, F. Mizutani, and C. Wang. Thinning of gray-scale images with combined sequential and parallel conditions for pixel removal. *IEEE Trans. Systems, Man, Cybernetics*, **24**:294–299, 1994.

[2] S.V. Ablameyko, C. Arcelli, and G. Sanniti di Baja. Hierarchical decomposition of distance labeled skeletons. *Int. J. Pattern Recognition Artificial Intelligence*, **10**:957–970, 1996.

[3] M. Abramowitz and I. Stegun, editors. *Handbook of Mathematical Functions with Formulas, Graphs and Mathematical Tables* (10th printing with corrections). Wiley, New York, 1972.

[4] D.M. Acketa and J.D. Žunić. The number of linear partitions of the (m,n)-grid. *Information Processing Letters*, **38**:163–168, 1991.

[5] D.M. Acketa and J.D. Žunić. On the maximal number of edges of convex digital polygons included into an $m \times m$-grid. *J. Combinatorial Theory*, **A69**:358–368, 1995.

[6] R. Aharoni, G.T. Herman, and M. Loebl. Jordan graphs. *Graphical Models Image Processing*, **58**:345–359, 1996.

[7] N. Ahuja, L.S. Davis, D.L. Milgram, and A. Rosenfeld. Piecewise approximation of pictures using maximal neighborhoods. *IEEE Trans. Computers*, **27**:375–379, 1978.

[8] N. Ahuja and B.J. Schachter. *Pattern Models*. Wiley, New York, 1983.

[9] P.S. Aleksandrov. *Combinatorial Topology*, Volume 1. Graylock Press, Rochester, 1956.

[10] P.S. Aleksandrov. *Combinatorial Topology*, Volume 2. Graylock Press, Rochester, 1957.

[11] J.C. Alexander and A.I. Thaler. The boundary count of digital pictures. *J. ACM*, **18**:105–112, 1971.

[12] J.W. Alexander. A proof of the invariance of certain constants of analysis situs. *Trans. American Mathematical Society*, **16**:148–154, 1915.

[13] P. Alexandroff and H. Hopf. *Topologie* — Erster Band. Julius Springer, Berlin, Germany, 1935.

[14] A. Alpers. Digital topology: Regular sets and root images of the cross-median filter. *J. Mathematical Imaging Vision*, **17**:7–14, 2002.

[15] F.L. Alt. Digital pattern recognition by moments. *J. ACM* **9**:240–258, 1962.

[16] T.M. Amarunnishad and P.P. Das. Estimation of length for digitized straight lines in three dimensions. *Pattern Recognition Letters*, **11**:207–213, 1990.

[17] R.V. Ambartzumjan. *Combinatorial Integral Geometry*. Wiley, Chichester, 1982.

[18] A. Amir and G. Benson. Two-dimensional periodicity and its applications. *Proc. ACM-SIAM Symp. Discrete Algorithms*, 440–452, 1992.

[19] C.J. Ammann and A.G. Sartori-Angus. Fast thinning algorithm for binary images. *Image Vision Computing*, **3**:71–79, 1985.

[20] T.A. Anderson and C.E. Kim. Representation of digital line segments and their preimages. *Computer Vision, Graphics, Image Processing*, **30**:279–288, 1985.

[21] E. Andres. Cercles discrets et rotations discrètes. Ph.D. Thesis, Université Louis Pasteur, Strasbourg, 1992.

[22] E. Andres. Discrete circles, rings and spheres. *Computers & Graphics*, **18**:695–706, 1994.

[23] E. Andres, R. Acharya, and C. Sibata. Discrete analytical hyperplanes. *Graphical Models Image Processing*, **59**:302–309, 1997.

[24] Apollonius of Perga. *Conics*, books I-III. Translation by R.C. Taliaferro. Green Lion Press, Annapolis, Maryland, 1998.

[25] K. Appel and W. Haken. Every planar map is four colorable. Part I. Discharging. *Illinois J. Math.*, **21**:429–490, 1977.

[26] C. Arcelli. Pattern thinning by contour tracing. *Computer Graphics Image Processing*, **17**:130–144, 1981.

[27] C. Arcelli. Topological changes in grey-tone digital pictures. *Pattern Recognition*, **32**:1019–1023, 1999.

[28] C. Arcelli, L.P. Cordella, and S. Levialdi. From local maxima to connected skeletons. *IEEE Trans. Pattern Analysis Machine Intelligence*, **3**:134–143, 1981.

[29] C. Arcelli and S. Levialdi. Parallel shrinking in three dimensions. *Computer Graphics Image Processing*, **1**:21–30, 1972.

[30] C. Arcelli and A. Massarotti. Regular arcs in digital contours. *Computer Graphics Image Processing*, **4**:339–360, 1975.

[31] C. Arcelli and A. Massarotti. On the parallel generation of straight digital lines. *Computer Graphics Image Processing*, **7**:67–83, 1978.

[32] C. Arcelli and G. Ramella. Finding contour-based abstractions of planar patterns. *Pattern Recognition*, **26**:1563–1577, 1993.

[33] C. Arcelli and G. Ramella. Finding grey-skeletons by iterated pixel removal. *Image Vision Computing*, **13**:159–167, 1995.

[34] C. Arcelli and G. Ramella. Sketching a grey-tone pattern from its distance transform. *Pattern Recognition*, **29**:2033–2045, 1996.

[35] C. Arcelli and G. Sanniti di Baja. On the sequential approach to medial line transformation. *IEEE Trans. Systems, Man, Cybernetics*, **8**:139–144, 1978.

[36] C. Arcelli and G. Sanniti di Baja. A thinning algorithm based on prominence detection. *Pattern Recognition*, **13**:225–235, 1981.

[37] C. Arcelli and G. Sanniti di Baja. A width-independent fast thinning algorithm. *IEEE Trans. Pattern Analysis Machine Intelligence*, **7**:463–474, 1985.

[38] C. Arcelli and G. Sanniti di Baja. Computing Voronoi diagrams in digital pictures. *Pattern Recognition Letters*, **4**:383–389, 1986.

[39] C. Arcelli and G. Sanniti di Baja. Finding local maxima in a pseudo-Euclidean distance transform. *Computer Vision, Graphics, Image Processing*, **43**:361–367, 1988.

[40] C. Arcelli and G. Sanniti di Baja. A one-pass two-operation process to detect the skeletal pixels on the 4-distance transform. *IEEE Trans. Pattern Analysis Machine Intelligence*, **11**:411–414, 1989.

[41] C. Arcelli and G. Sanniti di Baja. Ridge points in Euclidean distance maps. *Pattern Recognition Letters*, **13**:237–243, 1992.

[42] C. Arcelli and G. Sanniti di Baja. Euclidean skeleton via centre-of-maximal-disc extraction. *Image Vision Computing*, **11**:163–173, 1993.

[43] C. Arcelli and G. Sanniti di Baja. Skeletons of planar patterns. In T.Y. Kong and A. Rosenfeld, editors. *Topological Algorithms for Digital Image Processing*, 99–143. Elsevier, Amsterdam, The Netherlands, 1996.

[44] C. Arcelli, G. Sanniti di Baja, and P.C.K. Kwok. Parallel pattern compression by octagonal propagation. *Int. J. Pattern Recognition Artificial Intelligence*, **7**:1077–1102, 1993.

[45] C. Arcelli and L. Serino. Parallel reduction operators for gray-tone pictures. *Int. J. Pattern Recognition Artificial Intelligence*, **14**:281–295, 2000.

[46] F. Arrebola, A. Bandera, P. Camacho, and F. Sandoval. Corner detection by local histograms of contour chain code. *Electronics Letters*, **23**:1769–1771, 1997.

[47] F. Arrebola, P. Camacho, A. Bandera, and F. Sandoval. Corner detection and curve representation by circular histograms of contour chain code. *Electronics Letters*, **35**:1065–1067, 1999.

[48] E. Artzy, G. Frieder, and G.T. Herman. The theory, design, implementation and evaluation of a three-dimensional surface detection algorithm. *Computer Vision, Graphics, Image Processing*, **15**:1–24, 1981.

[49] T. Asano, Y. Kawamura, R. Klette, and K. Obokkata. A new approximation scheme for digital objects and curve length estimations. *Proc. Image Vision Computing New Zealand*, pages 26–31, 2000.

[50] T. Asano, Y. Kawamura, R. Klette, and K. Obokkata. Minimum-length polygons in approximation sausages. In: C. Arcelli, L. P. Cordella, and G. Sanniti di Baja, editors. *Visual Form 2001*, 103–112. LNCS 2059, Springer, Berlin, 2001.

[51] T. Asano, Y. Kawamura, R. Klette, and K. Obokkata. Digital curve approximation with length evaluation. *IEICE Trans. Fundamentals Electronics Communication Computer Sciences*, **86**:987–994, 2003.

[52] M. Atallah. A linear-time algorithm for Hausdorff distance between convex polygons. *Information Processing Letters*, **17**:207–209, 1983.

[53] L. Auslander and S.V. Parter. On imbedding graphs in the plane. *J. Mathematics Mechanics*, **10**:517–523, 1961.

[54] R. Ayala, E. Domínguez, A.R. Francés, and A. Quintero. Homotopy in digital spaces. *Discrete Applied Mathematics*, **125**:3–24, 2003.

[55] N. Aziz and S. Bhat. On the computation of integral properties of objects. *Advances Engineering Software*, **12**:174–180, 1990.

[56] P. Bakker, L.J. van Vliet, and P.W. Verbeek. Confidence and curvature estimation of curvilinear structures in 3-D. *Proc. Int. Conf. Computer Vision*, 139–144, 2001.

[57] I. Bárány. The limit shape of convex lattice polygons. *Discrete Computational Geometry*, **13**:279–295, 1995.

[58] E. Barcucci, S. Brunetti, A. Del Lungo, and M. Nivat. Reconstruction of discrete sets from three or more X-rays. *Proc. Italian Conf. Algorithms Complexity*, 199–200. LNCS 1767, Springer, Berlin, Germany, 2000.

[59] E. Barcucci, A. Del Lungo, M. Nivat, and R. Pinzani. Reconstructing convex polyominoes from their vertical and horizontal projections. *Theoretical Computer Science*, **155**:321–347, 1996.

[60] E. Barcucci, A. Del Lungo, M. Nivat, R. Pinzani, and A. Zurli. Reconstructing digital pictures from X-rays. *Proc. Int. Conf. Image Analysis Processing*, 166–173. LNCS 1310, Springer, Berlin, Germany, 1997.

[61] E. Barcucci, A. Del Lungo, and R. Pinzani. Deco polyominoes permutations and random generation. *Theoretical Computer Science*, **159**:29–42, 1996.

[62] R.P. Barneva, V.E. Brimkov, and P. Nehlig. Thin discrete triangular meshes. *Theoretical Computer Science*, **246**:73–105, 2000.

[63] M.F. Barnsley. *Fractals Everywhere*, 2nd edition., Academic Press, Boston, 1993.

[64] O. Baruch. Line thinning by line following. *Pattern Recognition Letters*, **8**:271–276, 1988.

[65] H. Bässmann and P.W. Besslich. Konturorientierte Verfahren in der digitalen Bildverarbeitung. Springer, Berlin, Germany, 1989.

[66] A.L.D. Beckers and A.W.M. Smeulders. A comment on "A note on 'Distance transformations in digital images'." *Computer Vision, Graphics, Image Processing*, **47**:89–91, 1989.

[67] A.L.D. Beckers and A.W.M. Smeulders. Optimization of length measurements for isotropic distance transformations in three dimensions. *CVGIP: Image Understanding*, **55**:296–306, 1992.

[68] S.B.M. Bell, F.C. Holroyd, and D.C. Mason. A digital geometry for hexagonal pixels. *Image Vision Computing*, **7**:194–204, 1989.

[69] E.A. Bender. Convex n-ominoes. *Discrete Mathematics*, **8**:219–226, 1974.

[70] A. Bengtsson and J.-O. Eklundh. Shape representation by multiscale contour approximation. *IEEE Trans. Pattern Analysis Machine Intelligence*, **13**:85–93, 1991.

[71] J.R. Bennett and J.S. MacDonald. On the measurement of curvature in a quantized environment. *IEEE Trans. Computers*, **24**:803–820, 1975.

[72] D.P. Bentz, P.J.P. Pimienta, E.J. Garboczi, and W.C. Carter. Cellular automaton simulations of surface mass transport due to curvature gradients: Simulations of sintering in 3-D. *Synthesis and Proceedings of Ceramics: Scientific Issues*, **249**:413–418, 1992.

[73] C. Berenstein and D. Lavine. On the number of digital straight line segments. *IEEE Trans. Pattern Analysis Machine Intelligence*, **10**:880–887, 1988.

[74] J. Berstel and M. Pocchiola. Random generation of finite Sturmian words. *Discrete Mathematics*, **153**:29–39, 1996.

[75] G. Bertrand. Simple points, topological numbers and geodesic neighborhoods in cubic grids. *Pattern Recognition Letters*, **15**:1003–1011, 1994.

[76] G. Bertrand. A parallel thinning algorithm for medial surfaces. *Pattern Recognition Letters*, **16**:979–986, 1995.

[77] G. Bertrand. A Boolean characterization of three-dimensional simple points. *Pattern Recognition Letters*, **17**:115–124, 1996.

[78] G. Bertrand and G. Malandain. A new characterization of three-dimensional simple points. *Pattern Recognition Letters*, **15**:169–175, 1994.

[79] G. Bertrand and G. Malandain. A note on "Building skeleton models via 3-D medial surface/axis thinning algorithms." *Graphical Models Image Processing*, **57**:537–538, 1995.

[80] G. Bertrand and R. Malgouyres. Some topological properties of surfaces in $\mathbb{Z}^3$. *J. Mathematical Imaging Vision*, **11**:207–221, 1999.

[81] G. Bertrand and R. Malgouyres. Topological properties of discrete surfaces. In: S. Miguet, A. Montanvert and S. Ubéda, editors. *Discrete Geometry for Computer Imagery*, pages 325–336. LNCS 1176, Springer, Berlin, Germany, 1996.

[82] Y. Bertrand, G. Damiand, and C. Fiorio. Topological encoding of 3-D segmented images. In: G. Borgefors, I. Nyström, and G. Sanniti di Baja, editors. *Discrete Geometry for Computer Imagery*, 311–324. LNCS 1953, Springer, Berlin, Germany, 2000.

[83] S. Beucher and F. Meyer. The morphological approach to segmentation: The watershed transformation. In: E.R. Dougherty, editor. *Mathematical Morphology in Image Processing*, 433–481. Marcel Dekker, New York, 1993.

[84] H.L. Beus and S.S.H. Tiu. An improved corner detection algorithm based on chain-coded plane curves. *Pattern Recognition*, **20**:291–296, 1987.

[85] P. Bhattacharya and X. Lu. A width-independent sequential thinning algorithm. *Int. J. Pattern Recognition Artificial Intelligence*, **11**:393–403, 1997.

[86] P. Bhattacharya and A. Rosenfeld. a-convexity. *Pattern Recognition Letters*, **21**:955–957, 2000.

[87] A. Bieniek and A. Moga. An efficient watershed algorithm based on connected components. *Pattern Recognition*, **33**:907–916, 2000.

[88] H. Bieri and W. Nef. Algorithms for the Euler characteristic and related additive functionals of digital objects. *Computer Vision, Graphics, Image Processing*, **28**:166–175, 1984.

[89] N.L. Biggs, E.K. Lloyd, and R.J. Wilson. *Graph Theory 1736–1936*. Clarendon Press, Oxford, 1976.

[90] S. Biswas and S.K. Pal. Approximate coding of digital contours. *IEEE Trans. Pattern Analysis Machine Intelligence*, **18**:1056–1066, 1988.

[91] I. Bloch. Fuzzy geodesic distance in images. In: A. Ralescu and T. Martin, editors. *Fuzzy Logic in Artificial Intelligence, Towards Intelligent Systems*, pages 153–166. LNCS 1188, Springer, Berlin, Germany, 1996.

[92] I. Bloch. On fuzzy distances and their use in image processing under imprecision. *Pattern Recognition*, **32**, 1873–1895, 1999.

[93] H. Blum. A transformation for extracting new descriptors of shape. In W. Wathen-Dunn, editor, *Models for the Perception of Speech and Visual Form*, 362–380. MIT Press, Cambridge, Massachusetts, 1967.

[94] L.M. Blumenthal. *Theory and Applications of Distance Geometry*. Clarendon Press, Oxford, England, 1953.

[95] A. Bogomolny. Digital geometry may not be discrete. *Computer Vision, Graphics, Image Processing*, **43**:205–220, 1988.

[96] J. Böhm and E. Hertel. Jenaer Beiträge zur diskreten Geometrie. *Wiss. Zeitschrift der Friedrich-Schiller-Universität Jena*, **1**:10–21, 1987.

[97] J.-D. Boissonnat and M. Yvinec. *Algorithmic Geometry*. Cambridge University Press, Cambridge, UK, 1998.

[98] J. Bokowski, H. Hadwiger, and J.M. Wills. Eine Ungleichung zwischen Volumen, Oberfläche, und Gitterpunktzahl konvexer Körper im n-dimensionalen euklidischen Raum. *Mathematische Zeitschrift*, **127**:363–364, 1972.

[99] D. Boltcheva. Segmentation Anatomique du Foie à Partir des Repères Internes et Externes. MSc Thesis, Université Louis Pasteur, Strasbourg, Germany, 2003.

[100] V.G. Boltjanskij and V.A. Efremovič. *Anschauliche kombinatorische Topologie*. VEB deutscher Verlag der Wissenschaften, Berlin, Germany, 1986.

[101] G. Bongiovanni, F. Luccio, and A. Zorat. The discrete equation of the straight line. *IEEE Trans. Computers*, **24**:310–313, 1975.

[102] T. Bonnesen. Über das isoperimetrische Defizit ebener Figuren. *Mathematische Annalen*, **91**:252–268, 1922.

[103] G. Borgefors. Distance transformations in arbitrary dimensions. *Computer Vision, Graphics, Image Processing*, **27**:321–345, 1984.

[104] G. Borgefors. Distance transformations in digital images. *Computer Vision, Graphics, Image Processing*, **34**:344–371, 1986.

[105] G. Borgefors. Distance transformations on hexagonal grids. *Pattern Recognition Letters*, **9**:97–105, 1989.

[106] G. Borgefors. A semiregular image grid. *J. Visual Communication Image Representation*, **1**:127–136, 1990.

[107] G. Borgefors. Another comment on "A note on distance transformations in digital images." *CVGIP: Image Understanding*, **54**:301–306, 1991.

[108] G. Borgefors. On digital distance transforms in three dimensions. *Computer Vision Image Understanding*, **64**:368–376, 1996.

[109] G. Borgefors and I. Nyström. Efficient shape representation by minimizing the set of centres of maximal discs/spheres. *Pattern Recognition Letters*, **18**:465–471, 1997.

[110] G. Borgefors, I. Nyström, and G. Sanniti di Baja. Computing skeletons in three dimensions. *Pattern Recognition*, **32**:1225–1236, 1999.

[111] G. Borgefors, I. Nyström, and G. Sanniti di Baja, editors. *Discrete Geometry for Computer Imagery*. LNCS 1953, Springer, Berlin, Germany, 2000.

[112] N. Bourbaki. *Topologie Générale*, 3rd edition. Hermann, Paris, 1961.

[113] N.G. Bourbakis. A parallel-symmetric thinning algorithm. *Pattern Recognition*, **22**:387–396, 1989.

[114] L. Boxer. Digitally continuous functions. *Pattern Recognition Letters*, **15**:833–839, 1994.

[115] L. Boxer and R. Miller. Efficient computation of the Euclidean distance transform. *Computer Vision Image Understanding*, **80**:379–383, 2000; corrigendum, **86**:137–140, 2002.

[116] S.M. Boyles and G.X. Ritter. The encoding of arbitrary two-dimensional geometric configurations. *Int. J. Computer Information Sciences*, **10**:1–24, 1981.

[117] J. Bracquelaire and L. Brun. Image segmentation with topological maps and interpixel representation. *J. Visual Communication Image Representation*, **9**:62–79, 1998.

[118] J.W. Brandt. Convergence and continuity criteria for discrete approximations of the continuous planar skeleton. *CVGIP: Image Understanding*, **59**:116–124, 1994.

[119] J.W. Brandt and V.R. Algazi. Continuous skeleton computation by Voronoi diagram. *CVGIP: Image Understanding*, **55**:329–338, 1992.

[120] E. Breitenberger. Johann Benedict Listing. *Neue Deutsche Biographie*, **14**:700–701, 1952.

[121] J. Bresenham. An incremental algorithm for digital plotting. *Proc. ACM Nat. Conf.*, 1963.

[122] J. Bresenham. Algorithm for computer control of a digital plotter. *IBM Systems J.*, **4**:25–30, 1965.

[123] J. Bresenham. A linear algorithm for incremental digital display of circular arcs. *Comm. ACM*, **20**:100–106, 1977.

[124] A. Bretto. Comparability graphs and digital topology. *Computer Vision Image Understanding*, **82**:33–41, 2001.

[125] H. Breu, J. Gil, D. Kirkpatrick, and M. Werman. Linear time Euclidean distance transform algorithms. *IEEE Trans. Pattern Analysis Machine Intelligence*, **17**:529–533, 1995.

[126] E. Bribiesca. A new chain code. *Pattern Recognition*, **32**:235–251, 1999.

[127] E. Bribiesca. A chain code for representing 3D curves. *Pattern Recognition*, **33**:755–765, 2000.

[128] R. Brice and C.L. Fennema. Scene analysis using regions. *Artificial Intelligence*, **1**:205–226, 1970.

[129] V.E. Brimkov. Digital flatness and related combinatorial problems. CITR, TR-120, Computer Science Department University of Auckland, New Zealand, 2002.

[130] V.E. Brimkov, E. Andres, and R.P. Barneva. Object discretizations in higher dimensions. In G. Borgefors, I. Nyström, and G. Sanniti di Baja, editors., *Discrete Geometry for Computer Imagery*, 210–221. LNCS 1953, Springer, Berlin, 2000.

[131] V.E. Brimkov, E. Andres, and R.P. Barneva. Object discretizations in higher dimensions. *Pattern Recognition Letters*, **23**:623–636, 2002.

[132] V.E. Brimkov and R. Barneva. Honeycomb vs. square and cubic models. *Electronic Notes in Theoretical Computer Science*, **46**, 2001.

[133] V.E. Brimkov and R.P. Barneva. Graceful planes and lines. *Theoretical Computer Science*, **283**:151–170, 2002.

[134] V.E. Brimkov and R.P. Barneva. Connectivity of discrete planes. *Theoretical Computer Science*, to appear.

[135] V.E. Brimkov, R. Barneva, and P. Nehlig. Minimally thin discrete triangulations. In: M. Chen, A. Kaufman, and R. Yagel, editors, *Volume Graphics*, 51–70. Springer, New York, 2000.

[136] E. Brisson. Representing geometric structures in d dimensions: Topology and order. *Discrete Computational Geometry*, **9**:387–426, 1993.

[137] R. Brons. Linguistic methods for description of a straight line on a grid. *Computer Graphics Image Processing*, **2**:48–62, 1974.

[138] L.E.J. Brouwer. Beweis des Jordanschen Kurvensatzes. *Mathematische Annalen*, **69**:169–175, 1910. Also in: *Collected Works*, **2**:377–383. North-Holland, Amsterdam, and American Elsevier, New York, 1976.

[139] L.E.J. Brouwer. Beweis des Jordanschen Satzes für den n-dimensionalen Raum. *Mathematische Annalen*, **71**:314–319, 1911. Also in: *Collected Works*, **2**:489–494. North-Holland, Amsterdam, The Netherlands, and American Elsevier, New York, 1976.

[140] A.M. Bruckstein. Self-similarity properties of digitized straight lines. *Contemporary Mathematics*, **119**:1–20, 1991.

[141] S. Brunetti, A. Daurat, and A. Del Lungo. Approximate X-rays reconstruction of special lattice sets. *Pure Mathematics Applications*, **11**:409–425, 2000.

[142] S. Brunetti and A. Daurat. An algorithm reconstructing lattice convex sets. *Theoretical Computer Science*, **304**:35–57, 2003.

[143] A. Bryant and J. Bryant. Following boundaries of discrete binary objects in space. *Pattern Recognition*, **23**:547–552, 1990.

[144] V. Bryant. *Metric spaces: Iteration and Application*. Cambridge University Press, Cambridge, United Kingdom, 1985.

[145] J.-B. Brzoska, B. Lesaffre, C. Coléou, K. Xu, and R.A. Pieritz. Computation of 3D curvatures on a wet snow sample. *Eur. Phys. J. Ap.*, **7**:45–57, 1999.

[146] G.L.L. Buffon. Essai d'arithmétique morale. Supplément a *L'Histoire Naturelle*, **4**, 1777.

[147] J.W. Bullard, E.J. Garboczi, W.C. Carter, and E.R. Fuller, Jr. Numerical methods for computing interfacial mean curvature. *Computational Materials Science*, **4**:103–116, 1995

[148] T. Bülow and R. Klette. Rubber band algorithm for estimating the length of digitized space curves. *Proc. Int. Conf. Pattern Recognition*, **III**:551–555, IEEE, 2000.

[149] T. Bülow and R. Klette. Digital curves in 3D space and a linear-time length estimation algorithm. *IEEE Trans. Pattern Analysis Machine Intelligence*, **24**:962–970, 2002.

[150] B.H. Bunch. *Mathematical Fallacies and Paradoxes*. Van Nostrand, New York, 1982.

[151] O.P. Buneman. A grammar for the topological analysis of plane figures. In: B. Meltzer and D. Michie, editors. *Machine Intelligence 5*, 383–393. Edinburgh University Press, Edinburgh, Scotland, 1969.

[152] Y.D. Burago and V.A. Zalgaller. *Geometric Inequalities*. Springer, Berlin, Germany, 1988.

[153] J.J. Burckhardt. Zur Geschichte der Entdeckung der 230 Raumgruppen. *Archive History Exact Sciences*, **4**:235–246, 1967/1968.

[154] P. Burt. Fast filter transforms for image processing. *Computer Graphics Image Processing*, **16**:20–51, 1981.

[155] R. Busemann and W. Feller. Krümmungseigenschaften konvexer Flächen. *Acta Mathematica*, **66**:27–45, 1935.

[156] M.A. Butt and P. Maragos. Optimum design of chamfer distance transforms. *IEEE Trans. Image Processing*, **7**:1477–1484, 1988.

[157] L. Buzer. An incremental linear time algorithm for digital line and plane recognition using a linear incremental feasibility problem. In: A. Braquelaire, J.-O. Lachaud, and A. Vialard, editors. *Discrete Geometry for Computer Imagery*, 372–381. Lecture Notes in Computer Science 2301, Springer, Berlin, Germany, 2002.

[158] A.I. Bykov and L.G. Zerkalov. Algorithms for homotopy classification of binary images. *Pattern Recognition*, **29**:565–574, 1996.

[159] V. Caglioti. On the uncertainty of straight lines in digital images. *CVGIP: Graphical Models Image Processing*, **55**:255–270, 1993.

[160] L. Calabi and W.E. Hartnett. Shape recognition, prairie fires, convex deficiencies and skeletons. *American Mathematical Monthly*, **75**:335–342, 1968.

[161] D. Campbell and J. Higgins. Matrix method for finding sets of contiguous non-zero elements in a 2-dimensional array—II. *Pattern Recognition*, **21**:451–453, 1988.

[162] J. Canny and J.H. Reif. New lower bound techniques for robot motion planning problems. *Proc. IEEE Conf. Foundations Computer Science*, 49–60, 1987.

[163] R. Cardoner and F. Thomas. Residuals + directional gaps = skeletons. *Pattern Recognition Letters*, **18**:343–353, 1997.

[164] R.C. Carrasco and M.L. Forcada. A note on the Nagendraprasad–Wang–Gupta thinning algorithm. *Pattern Recognition Letters*, **16**:539–541, 1995.

[165] A. Cauchy. Recherches sur les polyèdres, 2de partie. *Journal de l'Ecole Polytechnique Paris*, **16**:76–86, 1813. Also in: *Œuvres Complètes d'Augustin Cauchy*, ser.2, t.1, 13–25. Gauthier-Villars, Paris, France, 1905.

[166] B. Cavalieri. *Geometria Indivisibilibus Continuorum Nova Quadam Ratione Promota*. Bologna, 1635. (2nd edition, 1653.)

[167] A. Cayley. On contour and slope lines. *London, Edinburgh, and Dublin Philosophical Magazine and Journal of Science*, Ser. 4, **18**:264–268, 1859.

[168] A. Cayley. *An Elementary Treatise on Elliptic Functions*. Deighton, Bell, and Co., Cambridge, United Kingdom, 1876.

[169] R. Cederberg. On the coding, processing, and display of binary images. Linköping University, PhD No. 57, Linköping, Sweden, 1980.

[170] Y. Censor and S. Matej. Binary steering of nonbinary iterative algorithms. In: G.T. Herman and A. Kuba, editors. *Discrete Tomography — Foundations, Algorithms, and Applications*, 285–296. Birkhäuser, Boston, Massachusetts, 1999.

[171] Y.-L. Chang and J.K. Aggarwal. Representing and estimating 3-D lines. *Pattern Recognition*, **28**:1181–1190, 1995.

[172] J.M. Chassery. Connectivity and consecutivity in digital pictures. *Computer Graphics Image Processing*, **9**:294–300, 1979.

[173] J.M. Chassery. Discrete convexity: Definition, parametrization and compatibility with continuous convexity. *Computer Vision, Graphics, Image Processing*, **21**:326–344, 1983.

[174] J.M. Chassery and A. Montanvert. *Géométrie Discrète en Imagerie*. Hermès, Paris, 1991.

[175] S. Chattopadhyay and P.P. Das. A new method of analysis for discrete straight lines. *Pattern Recognition Letters*, **12**:747–755, 1991.

[176] S. Chattopadhyay and P.P. Das. Parameter estimation and reconstruction of digital conics in normal positions. *CVGIP: Graphical Models Image Processing*, **54**:385–395, 1992.

[177] S. Chattopadhyay and P.P. Das. Estimation of the original length of a straight line segment from its digitization in three dimensions. *Pattern Recognition*, **25**:787–798, 1992.

[178] S. Chattopadhyay, P.P. Das, and D. Ghosh Dastidar. Reconstruction of a digital circle. *Pattern Recognition*, **27**:1663–1676, 1994.

[179] B.B. Chaudhuri and A. Rosenfeld. On a metric distance between fuzzy sets. *Pattern Recognition Letters*, **17**:1157–1160, 1996.

[180] B.B. Chaudhuri and A. Rosenfeld. On the computation of the digital convex hull and circular hull of a digital region. *Pattern Recognition*, **31**:2007–2016, 1998.

[181] B.B. Chaudhuri and A. Rosenfeld. A modified Hausdorff distance between fuzzy sets. *Information Sciences*, **118**:159–171, 1999.

[182] D. Chaudhuri, C.A. Murthy, and B.B. Chaudhuri. A modified metric to compute distance. *Pattern Recognition*, **25**:667–677, 1992.

[183] C.S. Chen and W.H. Tsai. A new fast one-pass thinning algorithm and its parallel hardware implementation. *Pattern Recognition Letters*, **11**:471–477, 1990.

[184] L. Chen, D.H. Cooley, and J. Zhang. The equivalence between two definitions of digital surfaces. *Information Sciences*, **115**:201–220, 1999.

[185] L. Chen, G.T. Herman, R.A. Reynolds, and J.K. Udupa. Surface shading in the cuberille environment. *IEEE Computer Graphics Applications*, **5**:33–43, 1985.

[186] M.H. Chen and P.F. Yan. A fast algorithm to calculate the Euler number for binary images. *Pattern Recognition Letters*, **8**:295–297, 1988.

[187] Y.S. Chen. Comments on "A systematic approach for designing 2-subcycle and pseudo 1-subcycle parallel thinning algorithms." *Pattern Recognition*, **25**:1545–1546, 1992.

[188] Y.S. Chen. Hidden deletable pixel detection using vector analysis in parallel thinning to obtain bias-reduced skeletons. *Computer Vision Image Understanding*, **71**:294–311, 1998.

[189] Y.S. Chen and W.H. Hsu. A modified fast parallel algorithm for thinning digital patterns. *Pattern Recognition Letters*, **7**:99–106, 1988.

[190] Y.S. Chen and W.H. Hsu. Parallel algorithm for corner finding on digital curves. *Pattern Recognition Letters*, **8**:47–53, 1988.

[191] Y.S. Chen and W.H. Hsu. A systematic approach for designing 2-subcycle and pseudo-subcycle parallel thinning algorithms. *Pattern Recognition*, **22**:267–282, 1989.

[192] Y.S. Chen and W.H. Hsu. A comparison of some one-pass parallel thinnings. *Pattern Recognition Letters*, **11**:35–41, 1990.

[193] Y.S. Chen and Y.T. Yu. Thinning approach for noisy digital patterns. *Pattern Recognition*, **29**:1847–1802, 1996.

[194] D. Chetverikov and Z. Szabo. A simple and efficient algorithm for detection of high curvature points in planar curves. *Proc. Workshop Austrian Pattern Recognition Group*, Schriftenreihe der Â…CG, 128:175–184, 1999.

[195] R.T. Chin, H.K. Wan, D.L. Stover, and R.D. Iverson. A one-pass thinning algorithm and its parallel implementation. *Computer Vision, Graphics, Image Processing*, **40**:30–40, 1987.

[196] M. Chleík and F. Sloboda. Approximation of surfaces by minimal surfaces with obstacles. Technical report, Institute of Control Theory and Robotics, Slovak Academy of Sciences, Bratislava, Slovakia, 2000.

[197] H.I. Choi, S.W. Choi, H.P. Moon, and N.S. Wee. New algorithm for medial axis transform of plane domain. *Graphical Models Image Processing*, **59**:463–483, 1997.

[198] J. Choi, J. Sellen, and C.-K. Yap. Approximate Euclidean shortest path in 3-space. *Proc. ACM Conf. Computational Geometry*, ACM Press, 41–48, 1994.

[199] S.W. Choi and H.-P. Seidel. Hyperbolic Hausdorff distance for medial axis transform. *Graphical Models*, **63**:369–384, 2001.

[200] S.W. Choi and H.-P. Seidel. Linear one-sided stability of MAT for weakly injective domain. *J. Mathematical Imaging Vision*, **17**:237–247, 2002.

[201] S.S.O. Choy, C.S.T. Choy, and W.C. Siu. New single-pass algorithm for parallel thinning. *Computer Vision Image Understanding*, **62**:69–77, 1995.

[202] W.K. Clifford. The philosophy of the pure sciences. II. The postulates of the science of space. *Contemporary Review*, **25**:360–376, 1874. Also in: Gateway to the Great Books, Vol. 9: Mathematics, Encyclopaedia Britannica, 1963.

[203] D. Coeurjolly, I. Debled-Rennesson, and O. Teytaud. Segmentation and length estimation of 3D discrete curves. In: G. Bertrand, A. Imiya, and R. Klette, editors. *Digital and Image Geometry: Advanced Lectures*, 299–317. Lecture Notes in Computer Science 2243, Springer, Berlin, Germany, 2001.

[204] D. Coeurjolly, F. Flin, O. Teytaud, and L. Tougne. Multigrid convergence and surface area estimation. In: T. Asano, R. Klette, and C. Ronse, editors. *Geometry, Morphology, and Computational Imaging*, 101–119. Lecture Notes in Computer Science 2616, Springer, Berlin, Germany, 2003.

[205] D. Coeurjolly and R. Klette. A comparative evaluation of length estimators. *Proc. Int. Conf. Pattern Recognition*, **IV**:330–334, IEEE, 2002.

[206] D. Coeurjolly and R. Klette. A comparative evaluation of length estimators of digital curves. *IEEE Trans. Pattern Analysis Machine Intelligence*, **26**:252–258, 2004.

[207] D. Coeurjolly, S. Miguet and L. Tougne. Discrete curvature based on osculating circle estimation. In: C. Arcelli, L. P. Cordella, and G. Sanniti di Baja, editors. *Visual Form 2001*, 303–312. Lecture Notes in Computer Science 2059, Springer, Berlin, Germany, 2001.

[208] D. Coeurjolly and O. Teytaud. Multigrid convergence of discrete differential estimators: Discrete tangent and length estimation. Research Report RR-0101, Laboratoire ERIC, Université Lumière, Lyon, France, 2001.

[209] D. Cohen-Or and A. Kaufman. Fundamentals of surface voxelization. *Graphical Models Image Processing*, **87**:453–461, 1995.

[210] D. Cohen-Or and A. Kaufman. 3D line voxelization and connectivity control. *IEEE Computer Graphics Applications*, **17**:80–87, 1997.

[211] D. Cohen-Or, A. Kaufman, and T.Y. Kong. On the soundness of surface voxelizations. In: T.Y. Kong and A. Rosenfeld, editors. *Topological Algorithms for Digital Image Processing*, 181–204. Elsevier, Amsterdam, The Netherlands, 1996.

[212] D. Coquin and P. Bolon. Discrete distance operator on rectangular grids. *Pattern Recognition Letters*, **16**:911–923, 1995.

[213] L.P. Cordella and G. Sanniti di Baja. Geometric properties of the union of maximal neighborhoods. *IEEE Trans. Pattern Analysis Machine Intelligence*, **11**:214–217, 1989.

[214] R. Cori. *Graphes Planaires et Systèmes de Parenthèses.* Centre National de la Recherche Scientifique, Institut Blaise Pascal, Paris, France, 1969.

[215] A.M. Cormack. Early tomography and related topics. In: G.T. Herman and F. Natterer, editors. *Mathematical Aspects of Computerized Tomography*, 1–6. LNM 1497, Springer, Berlin, Germany, 1980.

[216] R. Cotes. Logometria. *Phil. Trans. Roy Soc. Lond.*, **29**, No. 338, 1714. English translation from Latin in R. Gowing, editor. *Roger Cotes — Natural Philosopher*, Appendix 1, 169–171. Cambridge University Press, Cambridge, United Kingdom, 1983.

[217] M. Couprie and G. Bertrand. Tessellations by connection. *Pattern Recognition Letters*, **23**:637–647, 2002.

[218] M. Couprie, F.N. Bezerra, and G. Bertrand. Grayscale image processing using topological operators. SPIE *Vision Geometry VIII*, **3811**:261–272, 1999.

[219] E.M. Coven and G.A. Hedlund. Sequences with minimal block growth. *Mathematical Systems Theory*, **7**:138–153, 1973.

[220] H.S.M. Coxeter. *Regular Polytopes*, 3rd edition, 122–123. Dover, New York, 1973.

[221] H.S.M. Coxeter. *Introduction to Geometry*. Wiley, New York, 1989.

[222] E. Creutzburg, A. Hübler, and O. Sýkora. Geometric methods for on-line recognition of digital straight segments. *Computers Artificial Intelligence*, **7**:253–276, 1988.

[223] E. Creutzburg, A. Hübler, and V. Wedler. On-line Erkennung digitaler Geradensegmente in linearer Zeit. In: R. Klette and J. Mecke, editors. *Proc. Geometric Problems of Image Processing*, Jena University, Germany, 48–65, 1982. (See also: On-line recognition of digital straight line segments. *Proc. Int. Conf. AI Inf. Control Systems Robots*, Smolenice, 42–46, 1982.)

[224] O. Cuisenaire. *Distance transformations: Fast algorithms and applications to medical image processing.* PhD Thesis, Université Catholique de Louvain, Louvain-la-Neuve, France, 1999.

[225] O. Cuisenaire and B. Macq. Fast Euclidean distance transformation by propagation using multiple neighborhoods. *Computer Vision Image Understanding*, **76**:163–172, 1999.

[226] J. Czyzowicz, E. Rivera-Campo, and J. Urrutia. Illuminating rectangles and triangles in the plane. *J. Combinatorial Theory*, **B57**:1–17, 1993.

[227] F. d'Amore and P.G. Franciosa. On the optimal binary plane partition for sets of isothetic rectangles. *Information Processing Letters*, **44**:255–259, 1992.

[228] N.A. Danielsson. Axiomatic discrete geometry. Internal report, Department of Computing, Imperial College, London, United Kingdom, 2002.

[229] P.E. Danielsson. Euclidean distance mapping. *Computer Graphics Image Processing*, **14**:227–248, 1980.

[230] P.E. Danielsson. An improved segmentation and coding algorithm for binary and nonbinary images. *IBM J. Research Development*, **26**:698–707, 1982.

[231] P.P. Das. An algorithm for computing the number of the minimal paths in digital images. *Pattern Recognition Letters*, **9**:107–116, 1989.

[232] P.P. Das. More on path generated digital metrics. *Pattern Recognition Letters*, **10**:25–31, 1989.

[233] P.P. Das. Metricity preserving transforms. *Pattern Recognition Letters*, **10**:73–76, 1989.

[234] P.P. Das. Lattice of octagonal distances in digital geometry. *Pattern Recognition Letters*, **11**:663–667, 1990.

[235] P.P. Das. Counting minimal paths in digital geometry. *Pattern Recognition Letters*, **12**:595–603, 1991.

[236] P.P. Das. A note on "Distance functions in digital geometry." *Information Sciences*, **58**:181–190, 1992.

[237] P.P. Das and P.P. Chakrabarti. Distance functions in digital geometry. *Information Sciences*, **42**:113–136, 1987.

[238] P.P. Das, P.P. Chakrabarti, and B.M. Chatterji. Generalized distances in digital geometry. *Information Sciences*, **42**:51–67, 1987.

[239] P.P. Das and B.N. Chatterji. A note on "Distance transformations in arbitrary dimensions." *Computer Vision, Graphics, Image Processing*, **43**:368–385, 1988.

[240] P.P. Das and B.N. Chatterji. Knight's distance in digital geometry. *Pattern Recognition Letters*, **7**:215–226, 1988.

[241] P.P. Das and B.N. Chatterji. Estimation of errors between Euclidean and m-neighbor distance. *Information Sciences*, **48**:1–26, 1989.

[242] P.P. Das and B.N. Chatterji. Hyperspheres in digital geometry. *Information Sciences*, **50**:73–91, 1990.

[243] P.P. Das and B.N. Chatterji. Octagonal distances for digital pictures. *Information Sciences*, **50**:123–150, 1990.

[244] P.P. Das and J. Mukherjee. Metricity of super-knight's distance in digital geometry. *Pattern Recognition Letters*, **11**:601–604, 1990.

[245] P.P. Das, J. Mukherjee, and B.N. Chatterji. The t-cost distance in digital geometry. *Information Sciences*, **59**:1–20, 1992.

[246] A. Datta and S.K. Parui. A robust parallel thinning algorithm for binary images. *Pattern Recognition*, **27**:1181–1192, 1994.

[247] A. Daurat, A. Del Lungo, and M. Nivat. The medians of discrete sets according to a linear distance. *Discrete Computational Geometry*, **23**:465–483, 2000.

[248] H. Davenport. On a principle of Lipschitz. *J. London Mathematical Society*, **26**:179–183, 1951.

[249] A. Davies and P. Samuels. *An Introduction to Computational Geometry for Curves and Surfaces.* Clarendon Press, Oxford, United Kingdom, 1996.

[250] E.R. Davies and A.P.N. Plummer. Thinning algorithms: A critique and a new methodology. *Pattern Recognition*, **14**:53–63, 1982.

[251] L.S. Davis. Understanding shape: Angles and sides. *IEEE Trans. Computers*, **26**:236–242, 1977.

[252] T.J. Davis. Fast decomposition of digital curves into polygons using the Haar transform. *IEEE Trans. Pattern Analysis Machine Intelligence*, **21**:786–790, 1999.

[253] I. Debled-Rennesson. *Etude et reconnaissance des droites et plans discrets.*

PhD Thesis, Université Louis Pasteur, Strasbourg, Germany, 1995.

[254] I. Debled-Rennesson, J.-L. Remy, and J. Royer-Degli. Segmentation of discrete curves into fuzzy segments. In: A. Del Lungo, V. Di Gesù, and A. Kuba, editors. *Electronic Notes in Discrete Mathematics* (Elsevier), **12**, 2003.

[255] I. Debled-Rennesson and J.-P. Reveillès. A new approach to digital planes. SPIE *Vision Geometry III*, **2356**:12–21, 1994.

[256] I. Debled-Rennesson and J.-P. Reveillès. A linear algorithm for segmentation of digital curves. *Int. J. Pattern Recognition Artificial Intelligence*, **9**:635–662, 1995.

[257] K. Deguchi. Multi-scale curvatures for contour feature extraction. *Proc. Int. Conf. Pattern Recognition*, 1113-1115, IEEE, 1988.

[258] M. Dehn and P. Heegard. Analysis situs. In: *Enzyklopädie der Math. Wiss.* III.1.1:153–220, Teubner, Leipzig, Germany, 1907.

[259] B.N. Delaunay. Sur la sphère vide. *Bull. Academy Science USSR (VII), Classe Sci. Mat. Nat.*, **7**:793–800, 1934.

[260] S. Dellepiane and F. Fontana. Extraction of intensity connectedness for image processing. *Pattern Recognition Letters*, **16**:313–324, 1995.

[261] A. Del Lungo and N. Nivat. Reconstruction of connected sets from two projections. In: G.T. Herman and A. Kuba, editors. *Discrete Tomography — Foundations, Algorithms, and Applications*, 163–188. Birkhäuser, Boston, Massachusetts, 1999.

[262] A. Del Lungo, N. Nivat, and R. Pinzani. The number of convex polyominoes reconstructible from their orthogonal projections. *Discrete Mathematics*, **157**:65–78, 1996.

[263] L. Deniau. Proposition d'un operateur géométrique pour l'analyse et l'identification de signaux et images. PhD Thesis, Université Paris-Sud (Orsay), France, 1997.

[264] R. Descartes. *Discours de la Méthode pour Bien Conduire sa Raison et Chercher la Vérité dans les Sciences. Plus La Dioptrique, et La Géométrie, qui Sont des Essais de cette Méthode.* Leyden, 1637. English translation of the appendix *La Géométrie* by D.E. Smith and M.L. Latham: *The Geometry*. Dover, New York, 1954.

[265] L. Devroye and G.T. Toussaint. A note on linear expected time algorithms for finding convex hulls. *Computing*, **26**:361–366, 1981.

[266] J.L. Diaz-de-Leon S. and J.H. Sossa-Azuela. On the computation of the Euler number of a binary object. *Pattern Recognition*, **29**:471–476, 1996.

[267] V. Di Gesu and C. Valenti. Detection of regions of interest via the pyramid discrete symmetry transform. In: F. Solina, W.G. Kropatsch, R. Klette, and R. Bajcsy, editors. *Advances in Computer Vision*, 129–136. Springer, Vienna, Austria, 1997.

[268] E.W. Dijkstra. A note on two problems in connection with graphs. *Numerische Mathematik*, **1**:269–271, 1959.

[269] M.B. Dillencourt, H. Samet, and M. Tamminen. A general approach to connected-component labeling for arbitrary image representations. *J. ACM*, **39**:253–280, 1992.

[270] G. Dimauro, S. Impedovo, and G. Pirlo. A new thinning algorithm based on controlled deletion of edge regions. *Int. J. Pattern Recognition Artificial Intelligence*, **7**:969–986, 1993.

[271] G.P. Dinneen. Programming pattern recognition. *Proc. Western Joint Computer Conf.*, 94–100, Inst. Electrical Engineers, New York, 1955.

[272] G.A. Dirac. Some Theorems on Abstract Graphs. *Proc. London Mathematical Society*, **2**:69–81, 1952.

[273] G.L. Dirichlet. Über die Reduktion der positiven quadratischen Formen mit drei unbestimmten ganzen Zahlen. *J. Reine Angewandte Mathematik*, **40**:209–227, 1850.

[274] H.-U. Döhler and P. Zamperoni. Compact contour codes for convex binary patterns. *Signal Processing*, **8**:23–39, 1985.

[275] P. Dokládal. *Grey-scale image segmentation: A topological approach.* PhD Thesis, Université de Marne-La-Vallée, France, 2000.

[276] M. Doros. Algorithms for generation of discrete circles, rings, and disks. *Computer Graphics Image Processing*, **10**:366–371, 1979.

[277] M. Doros. On some properties of the generation of discrete circular arcs on a square grid. *Computer Vision Graphics Image Processing*, **28**:377–383, 1984.

[278] L. Dorst and R.P.W. Duin. Spirograph theory: A framework for calculations on digitized straight lines. *IEEE Trans. Pattern Analysis Machine Intelligence*, **6**:632–639, 1984.

[279] L. Dorst and A.W.M. Smeulders. Discrete representations of straight lines. *IEEE Trans. Pattern Analysis Machine Intelligence*, **6**:450–462, 1984.

[280] L. Dorst and A.W.M. Smeulders. Best linear unbiased estimators for properties of digitized straight lines. *IEEE Trans. Pattern Analysis Machine Intelligence*, **8**:276–282, 1986.

[281] L. Dorst and A.W.M. Smeulders. Length estimators for digitized contours. *Computer Vision, Graphics, Image Processing*, **40**:311–333, 1987.

[282] L. Dorst and A.W.M. Smeulders. Discrete straight line segments: Parameters, primitives and properties. *Contemporary Mathematics*, **119**:45–62, 1991.

[283] L. Dorst and A.W.M. Smeulders. Decomposition of discrete curves into piecewise segments in linear time. *Contemporary Mathematics*, **119**:169–195, 1991.

[284] D.Z. Du and D.J. Kleitman. Diameter and radius in the Manhattan metric. *Discrete Computational Geometry*, **5**:351–356, 1990.

[285] D. Dubois and H. Prade. *Fuzzy Sets and Systems: Theory and Applications*. Academic Press, New York, 1980.

[286] R.O. Duda, P.E. Hart, and J.H. Munson. Graphical-data-processing research study and experimental investigation. TR ECOM-01901-26, Stanford Research Institute, Menlo Park, California, March 1967.

[287] J.F. Dufourd and F. Puitg. Functional specification and prototyping with combinatorial maps. *Computational Geometry — Theory and Applications*, **16**:129–156, 2000.

[288] S. Dulucq and D. Gouyou-Beauchamps. Sur les facteurs des suites de Sturm. *Theoretical Computer Science*, **71**:381–400, 1991.

[289] C.R. Dyer and A. Rosenfeld. Thinning algorithms for grayscale pictures. *IEEE Trans. Pattern Analysis Machine Intelligence*, **1**:88–89, 1979.

[290] C.R. Dyer, A. Rosenfeld, and H. Samet. Region representation: Boundary codes from quadtrees. *Comm. ACM*, **23**:171–179, 1980.

[291] D. Eberly. Level set extraction from gridded 2D and 3D data. www.magic-software.com/Documentation, 2002.

[292] D. Eberly. Triangulation by ear clipping. www.magic-software.com/Documentation, 2002.

[293] D. Eberly and J. Lancaster. On gray scale image measurements. I. Arc length and area. *CVGIP: Graphical Models Image Processing*, **53**:538–549, 1991.

[294] D. Eberly and J. Lancaster. On gray scale image measurements. II. Surface area and volume. *CVGIP: Graphical Models Image Processing*, **53**:550–562, 1991.

[295] M.J. Eccles, M.P.C. McQueen, and D. Rosen. Analysis of digitized boundaries of planar objects. *Pattern Recognition*, **9**:31–41, 1977.

[296] U. Eckhardt. A note on Rutovitz' method for parallel thinning. *Pattern Recognition Letters*, **8**:35–38, 1988.

[297] U. Eckhardt and L. Latecki. Topologies for the digital spaces $\mathbb{Z}^2$ and $\mathbb{Z}^3$. *Computer Vision Image Understanding*, **90**:295–312, 2003.

[298] U. Eckhardt and G. Maderlechner. Invariant thinning. *Int. J. Pattern Recognition Artificial Intelligence*, **7**:1115–1144, 1993.

[299] H. Edelsbrunner. *Algorithms in Combinatorial Geometry*. Springer, Heidelberg, Germany, 1987.

[300] H. Edelsbrunner. *Geometry and Topology for Mesh Generation*. Cambridge University Press, Cambridge, United Kingdom, 2001.

[301] J. Edmonds. A combinatorial representation for polyhedral surfaces (abstract). *Notices American Mathematical Society*, **7**:646, 1960.

[302] B. Efron. The convex hull of a random set of points. *Biometrika*, **52**:331–343, 1965.

[303] H. Eggers. Parallel Euclidean distance transformations in $\mathbb{Z}_g^n$. *Pattern Recognition Letters*, **17**:751–757, 1996.

[304] H. Eggers. Two fast Euclidean distance transformations in $\mathbb{Z}^2$ based on sufficient propagation. *Computer Vision Image Understanding*, **69**:106–116, 1998.

[305] T.J. Ellis, D. Proffitt, D. Rosen, and W. Rutkowski. Measurement of the lengths of digitized curved lines. *Computer Graphics Image Processing*, **10**:333–347, 1979.

[306] H. Embrechts and D. Roose. A parallel Euclidean distance transformation algorithm. *Computer Vision Image Understanding*, **63**:15–26, 1996.

[307] P. Erdös. Integral distances. *Bull. American Mathematical Society*, **51**:996, 1945.

[308] P. Erdös, T. Grünwald, and E. Vàzsonyi. Über Euler-Linien unendlicher Graphen. *J. Math. Phys. Mass. Inst. of Tech.*, **17**:59–75, 1938.

[309] L. Euler. Elementa doctrinae solidorum. *Novi Commentarii Acad. Sc. Petrop.*, **4**:104–160, 1758 (for 1752/3). Also in: *Opera Omnia* (Series I), **26**:72–108, 1975.

[310] R. Fabbri, L.F. Estrozi, and L. da Fontoura Costa. On Voronoi diagrams and medial axes. *J. Mathematical Imaging Vision*, **17**:27–40, 2002.

[311] V.N. Faddeeva. *Computational Methods of Linear Algebra*, 154–155. Translated from Russian by C.D. Benster. Dover, New York, 1959.

[312] A. Fam and J. Sklansky. Cellularly straight images and the Hausdorff metric. *Proc. IEEE Conf. Pattern Recognition Image Processing*, 242–247, 1977.

[313] K.C. Fan, D.F. Chen, and M.G. Wen. Skeletonization of binary images with nonuniform width via block decomposition and contour vector matching. *Pattern Recognition*, **31**:823–838, 1998.

[314] O.D. Faugeras, M. Hebert, P. Mussi, and J.D. Boissonnat. Polyhedral approximation of 3-D objects without holes. *Computer Vision, Graphics, Image Processing*, **25**:169–183, 1984.

[315] A. Favre and H. Keller. Parallel syntactic thinning by recoding of binary pictures. *Computer Vision, Graphics, Image Processing*, **23**:99–112, 1983.

[316] J. Feder. Languages of encoded line patterns. *Information Control*, **13**:230–244, 1968.

[317] E.S. Fedorov. The symmetry of regular systems of figures (in Russian). *Proc. Imperial St. Petersburg Mineralogical Society*, Series 2, **28**:1–146, 1891.

[318] E.S. Fedorov. Symmetry in the plane (in Russian). *Proc. Imperial St. Petersburg Mineralogical Society*, Series 2, **28**:345–390, 1891.

[319] L. Fejes Tóth. *Lagerungen in der Ebene, auf der Kugel und im Raum*. Springer, Berlin, Germany, 1953.

[320] H. Fell. Detectable properties of planar figures. *Information Control*, **31**:107–128, 1976.

[321] M.M. Ferguson, Jr. Matrix method for finding sets of contiguous non-zero elements in a 2-dimensional array. *Pattern Recognition*, **19**:73, 1986.

[322] A. Ferreira and S. Ubeda. Computing the medial axis transform in parallel with eight scan operations. *IEEE Trans. Pattern Analysis Machine Intelligence*, **21**:277–282, 1999.

[323] F. Feschet and L. Tougne. Optimal time computation of the tangent of a discrete curve: Application to the curvature. In: G. Bertrand, M. Couprie, and L. Perroton, editors. *Discrete Geometry for Computer Imagery*, 31–40. Lecture Notes in Computer Science 1568, Springer, Berlin, Germany, 1999.

[324] F. Feschet and L. Tougne. On the min DSS problem of closed discrete curves. In: A. Del Lungo, V. Di Gesù, and A. Kuba, editors. *Electronic Notes in Discrete Mathematics* (Elsevier), **12**, 2003.

[325] O. Figueiredo and J.P. Reveillès. A contribution to 3D digital lines. *Proc. Discrete Geometry for Computer Imagery*, Universit d'Auvergne (Laboratoire de Logique, Algorithmique et Informatique de Clermont 1), Clermond-Ferrand, France, 187–198, 1995.

[326] C. Fiorio. *Approche interpixel en analyse d'images, une topologie et des algorithmes de segmentation*. PhD Thesis, Université des Sciences et Techniques du Languedoc, Montpellier, France, 1995.

[327] M. Fischler and R. Bolles. Perceptual organization and curve partitioning. *IEEE Trans. Pattern Analysis Machine Intelligence*, **8**:100–105, 1986.

[328] S. Fisk. Separating point sets by circles, and the recognition of digital disks. *IEEE Trans. Pattern Analysis Machine Intelligence*, **8**:554–556, 1986.

[329] A. Fitzgibbon, M. Pilu, and R.B. Fisher. Direct least square fitting of ellipses. *IEEE Trans. Pattern Analysis Machine Intelligence*, **21**:476–480, 1999.

[330] F. Flin, J.-B. Brzoska, D. Coeurjolly, R.A. Pieritz, B. Lesaffre, C. Coléou, P. Lamboley, O. Teytaud, G. Vignoles, and J.-F. Delesse. Adaptive estimation of normals and surface area for discrete 3D objects: Application to snow binary data from X-ray tomography. *IEEE Trans. Image Processing*, to appear.

[331] F. Flin, J.-B. Brzoska, B. Lesaffre, C. Coléou, and P. Lamboley. Computation of normal vectors of discrete 3D objects: Application to natural snow images from X-ray tomography. *Image Analysis Stereology*, **20**:187–191, 2001.

[332] S. Forchhammer. Digital plane and grid point segments. *Computer Vision, Graphics, Image Processing*, **47**:373–384, 1989.

[333] S. Forchhammer and C.E. Kim. Digital squares. Technical report CS-88-182, Washington State University, Pullman, Washington, 1988.

[334] H.G. Forder. *The Foundations of Euclidean Geometry*. Cambridge University Press, Cambridge, United Kingdom, 1927.

[335] H.G. Forder. Coordinates in geometry. Mathematical Series No. 1, Bulletin No. 41, Auckland University College, New Zealand, 1953.

[336] S. Fourey, T.Y. Kong, and G.T. Herman. Generic axiomatized digital surface-structures. *Electronic Notes in Theoretical Computer Science* (Elsevier), **46**, 2001.

[337] S. Fourey and R. Malgouyres. A concise characterization of 3D simple points. In: G. Borgefors, I. Nyström, and G. Sanniti di Baja, editors. *Discrete Geometry for Computer Imagery*, 27–36. Lecture Notes in Computer Science 1953, Springer, Berlin, Germany, 2000.

[338] S. Fourey and R. Malgouyres. A digital linking number for discrete curves. *Int. J. Pattern Recognition Artificial Intelligence*, **15**:1053–1074, 2001.

[339] J. Françon. Discrete combinatorial surfaces. *Graphical Models Image Processing*, **57**:20–26, 1995.

[340] J. Françon and L. Papier. Polyhedrization of the boundary of a voxel object. In: G. Bertrand, M. Couprie, and L. Perroton, editors. *Discrete Geometry for Computer Imagery*, 425–434. Lecture Notes in Computer Science 1568, Springer, Berlin, Germany, 1999.

[341] J. Françon, J.-M. Schramm, and M. Tajine. Recognizing arithmetic straight lines and planes. In: S. Miguet, A. Montanvert and S. Ubéda, editors. *Discrete Geometry for Computer Imagery*, 141–150. Lecture Notes in Computer Science 1176, Springer, Berlin, Germany, 1996.

[342] H. Freeman. Techniques for the digital computer analysis of chain-encoded arbitrary plane curves. *Proc. Nat. Electronics Conf.*, **17**:421–432, 1961.

[343] H. Freeman. A review of relevant problems in the processing of line-drawing data. In: A. Grasselli, editor. *Automatic Interpretation and Classification of Images*, 155–174. Academic Press, New York, 1969.

[344] H. Freeman. Boundary encoding and processing. In: B. S. Lipkin and A. Rosenfeld, editors. *Picture Processing and Psychopictorics*, 241–263. Academic Press, New York, 1970.

[345] H. Freeman. Computer processing of line-drawing images. *ACM Computing Surveys*, **6**:57–97, 1974.

[346] H. Freeman and L.S. Davis. A corner finding algorithm for chain-coded curves. *IEEE Trans. Computers*, **26**:297–303, 1977.

[347] H. Freeman and R. Shapira. Determining the minimum-area enclosing rectangle of an arbitrary closed curve. *Comm. ACM*, **18**:409–413, 1975.

[348] M. Gaafar. Convexity verification, block-chords and digital straight lines. *Computer Graphics Image Processing*, **6**:361–370, 1977.

[349] N. Gagvani and D. Silver. Parameter-controlled volume thinning. *Graphical Models Image Processing*, **61**:149–164, 1999.

[350] D. Gale. A theorem on flows in networks. *Pacific J. Mathematics*, **7**:1073–1082, 1957.

[351] Z. Galil and K. Park. Truly alphabet-independent two-dimensional pattern matching. *Proc. IEEE Symp. Foundations Computer Science*, 247–256, 1992.

[352] F. Galvin and S.D. Shore. Distance functions and topologies. *American Mathematical Monthly*, **98**:620–623, 1991.

[353] R.J. Gardner and P. Gritzmann. Successive determination and verification of polytopes by their X-rays. *J. London Mathematical Society*, Second Series, **50**:375–391, 1994.

[354] R.J. Gardner and P. Gritzmann. Discrete tomography: Determination of finite sets by x-rays. *Trans. American Mathematical Society*, **349**:2271–2295, 1997.

[355] R.J. Gardner and P. McMullen. On Hammer's X-ray problem. *J. London Mathematical Society*, **21**:171–175, 1980.

[356] M.R. Garey and D.S. Johnson. *Computers and Intractability: A Guide to the Theory of NP-Completeness*. W. H. Freeman, New York, 1983.

[357] J.C.F. Gauss. Disquisitiones generales circa superficies curva. *Commentationes Gottingensis*, **6**, 1823–1827. English translation: *General Investigations of Curved Surfaces*. Introduction by R. Courant, translated from the Latin and German by A. Hiltebeitel and J. Morehead. Raven Press, Hewlett, New York, 1965.

[358] I. Gawehn. Über unberandete zweidimensionale Mannigfaltigkeiten. *Mathematische Annalen*, **98**:321–354, 1927.

[359] Y. Ge and J.M. Fitzpatrick. On the generation of skeletons from discrete Euclidean distance maps. *IEEE Trans. Pattern Analysis Machine Intelligence*, **18**:1055–1066, 1996.

[360] A. Gill. Minimum-scan pattern recognition. *IRE Trans. Information Theory*, **5**:52–57, 1959.

[361] A. Giraldo, A. Gross, and L.J. Latecki. Digitizations preserving shape. *Pattern Recognition*, **32**:365–370, 1999.

[362] V.E. Giuliano, P.E. Jones, G.E. Kimball, R.F. Meyer, and B.A. Stein. Automatic pattern recognition by a Gestalt method. *Information and Control*, **4**:332–345, 1961.

[363] K. Gödel. Über formal unentscheidbare Sätze der Principia Mathematica und verwandter Systeme, I. *Monatshefte für Mathematik und Physik*, **38**:173–198, 1931.

[364] S.C. Goh and C.N. Lee. Counting minimal paths in 3D digital geometry. *Pattern Recognition Letters*, **13**:765–771, 1992.

[365] S.-C. Goh and C.-N. Lee. Counting minimal 18-paths in 3D digital space. *Pattern Recognition Letters*, **14**:39–52, 1993.

[366] M. Gokmen and R.W. Hall. Parallel shrinking algorithms using 2-subfields approaches. *Computer Vision, Graphics, Image Processing*, **52**:191–209, 1990.

[367] A.J. Goldstein. An efficient and constructive algorithm for testing whether a graph can be embedded in a plane. *Proc. Graph Combinatorics Conf.*, Princeton University, Princeton, New Jersey, 1963.

[368] S.W. Golomb. *Polyominoes*. Scribner's, New York, 1965. 2nd edition, Princeton University Press, Princeton, New Jersey, 1996.

[369] W. Gong and G. Bertrand. A note on "Thinning of 3-D images using the Safe Point Thinning Algorithm (SPTA)." *Pattern Recognition Letters*, **11**:499–500, 1990.

[370] R.C. Gonzalez and R.E. Woods. *Digital Image Processing*. 2nd edition, Prentice Hall, Upper Saddle River, New Jersey, 2002.

[371] J.E. Goodman and J. O'Rourke, editors. *Handbook of Discrete and Computational Geometry*. CRC Press, Boca Raton, Florida, 1997.

[372] R. Gordon, R. Bender, and G.T. Herman. Algebraic reconstruction techniques (ART) for three-dimensional electron microscopy and X-ray photography. *J. Theoretical Biology*, **29**:471–481, 1970.

[373] C. Gotsman and M. Lindenbaum. Euclidean Voronoi labeling on the multidimensional grid. *Pattern Recognition Letters*, **16**:409–415, 1995.

[374] V.K. Govindan and A.P. Shivaprasad. A pattern adaptive thinning algorithm. *Pattern Recognition*, **20**:623–637, 1987.

[375] R.L. Graham. An efficient algorithm for determining the convex hull of a finite planar set. *Information Processing Letters*, **7**:175–180, 1972.

[376] F. Gray. Pulse code communication. United States Patent 2,632,058, March 17, 1953.

[377] S.B. Gray. Local properties of binary images in two dimensions. *IEEE Trans. Computers*, **20**:551–561, 1971.

[378] T.N.E. Greville. *Theory and Applications of Spline Functions*. Academic Press, New York, 1969.

[379] D. Gries and I. Stojmenović. A note on Graham's convex hull algorithm. *Information Processing Letters*, **25**:323–327, 1987.

[380] P. Gritzmann and M. Nivat, editors. Discrete tomography: Algorithms and complexity. Dagstuhl Seminar Report 165, Informatik-Zentrum Dagstuhl, Wadern, Germany, 1997.

[381] P. Gritzmann, D. Prangenberg, S. de Vries, and M. Wiegelmann. Success and failure of certain reconstruction and uniqueness algorithms in discrete tomography. *Int. J. Imaging Systems Technology*, **9**:101–109, 1998.

[382] A. Gross and L.J. Latecki. A realistic digitization model of straight lines. *Computer Vision Image Understanding*, **67**:131–142, 1997.

[383] J.L. Gross and S.R. Alpert. The topological theory of current graphs. *J. Combinatorial Theory*, **B17**:218–233, 1974.

[384] A. Grzegorczyk. *An Outline of Mathematical Logic*. Reidel, Dordrecht, The Netherlands, 1974.

[385] W. Guan and S. Ma. A list-processing approach to compute Voronoi diagrams and the Euclidean distance transform. *IEEE Trans. Pattern Analysis Machine Intelligence*, **20**:757–761, 1998.

[386] A. Gueziec and R. Hummel. Exploiting triangulated surface extraction using tetrahedral decomposition. *IEEE Trans. Visualization Computer Graphics*, **1**:328–342, 1995.

[387] Z. Guo and R.W. Hall. Parallel thinning with two-subiteration algorithms. *Comm. ACM*, **32**:359–373, 1989.

[388] Z. Guo and R.W. Hall. Fast fully parallel thinning algorithms. *CVGIP: Image Understanding*, **55**:317–328, 1992.

[389] H. Hadwiger. Über die rationalen Hauptwinkel der Goniometrie. *Elemente der Mathematik*, **1**:98–100, 1946.

[390] H. Hadwiger. Minkowskische Addition und Subtraktion beliebiger Punktmengen und die Theoreme von Erhard Schmidt. *Mathematische Zeitschrift*, **53**:210–218, 1950.

[391] H. Hadwiger. *Altes und Neues über konvexe Körper*. Birkhäuser, Basel, Switzerland, 1955.

[392] H. Hadwiger. *Vorlesungen über Inhalt, Oberfläche, und Isoperimetrie.* Springer, Berlin, Germany, 1957.

[393] H. Hadwiger. Gitterperiodische Punktmengen und Isoperimetrie. *Monatsh. Math.*, **76**:410–418, 1972.

[394] H. Hadwiger and H. Debrunner. *Combinatorial Geometry in the Plane.* Holt, Rinehart and Winston, New York, 1964.

[395] K.J. Hafford and K. Preston, Jr. Three-dimensional skeletonization of elongated solids. *Computer Vision, Graphics, Image Processing*, **27**:78–91, 1984.

[396] H. Hagen, S. Heinz, M. Thesing, and T. Schreiber. Simulation-based modeling. *J. Shape Modeling*, **4**:143–164, 1998.

[397] H. Hakalahti, D. Harwood, and L. Davis. Two-dimensional object recognition by matching local properties of contour points. *Pattern Recognition Letters*, **2**:227–234, 1984.

[398] P. Hall and I.S. Molchanov. Corrections for systematic boundary effects in pixel-based area counts. *Pattern Recognition*, **32**:1519-1528, 1999.

[399] R.W. Hall. Fast parallel thinning algorithms: Parallel speed and connectivity preservation. *Comm. ACM*, **32**:124–131, 1989.

[400] R.W. Hall. Comments on "A parallel-symmetric thinning algorithm" by Bourbakis. *Pattern Recognition*, **25**:435–441, 1992.

[401] R.W. Hall. Tests for connectivity preservation for parallel reduction operators. *Topology and Its Applications*, **46**:199–217, 1992.

[402] R.W. Hall. Optimally small operator supports for fully parallel thinning algorithms. *IEEE Trans. Pattern Analysis Machine Intelligence*, **15**:828–833, 1993.

[403] R.W. Hall. Parallel connectivity-preserving thinning algorithms. In: T.Y. Kong and A. Rosenfeld, editors. *Topological Algorithms for Digital Image Processing*, 145–179. Elsevier, Amsterdam, The Netherlands, 1996.

[404] R.W. Hall and C.Y. Hu. Time-efficient computation of 3D topological functions. *Pattern Recognition Letters*, **17**:1017–1033, 1996.

[405] R.W. Hall, T.Y. Kong, and A. Rosenfeld. Shrinking binary images. In: T.Y. Kong and A. Rosenfeld, editors. *Topological Algorithms for Digital Image Processing*, 31–98. Elsevier, Amsterdam, The Netherlands, 1996.

[406] P.C. Hammer. Problem 2. *Proc. Symp. Pure Mathematics, Vol. VII: Convexity*, 498–499. American Mathematical Society, Providence, Rhode Island, 1963.

[407] J.H. Han and T. Poston. Chord-to-point distance accumulation and planar curvature: A new approach to discrete curvature. *Pattern Recognition Letters*, **22**:1133–1144, 2001.

[408] T.L. Hankins. *Sir William Rowan Hamilton*, 341–343. Johns Hopkins University Press, Baltimore, Maryland, and London, United Kingdom, 1980.

[409] R.M. Haralick and L.G. Shapiro. *Computer and Robot Vision*, Volume II. Addison-Wesley, Reading, Massachusetts, 1993.

[410] F. Harary. The four color conjecture and other graphical diseases. *Proc. Ann Arbor Graph Theory Conf.*, 1–9, Academic Press, New York, 1969.

[411] F. Harary, R.A. Melter, and I. Tomescu. Digital metrics: A graph-theoretical approach. *Pattern Recognition Letters*, **2**:159–163, 1984.

[412] G.H. Hardy. On the expressions of a number as the sum of two squares. *Quarterly J. Mathematics*, **46**:263–283, 1915. Also in: *Collected Papers of G.H. Hardy*, **2**:243–263. Clarendon Press, Oxford, United Kingdom, 1967.

[413] L.H. Harper. Optimal numberings and isoperimetric problems on graphs. *J. Combinatorial Theory*, **1**:385–393, 1966.

[414] F. Hausdorff. *Mengenlehre*. Gruyter, Berlin, Germany, 1927.

[415] L. Heffter. Über das Problem der Nachbargebiete. *Mathematische Annalen*, **8**:17–20, 1891.

[416] H.J.A.M. Heijmans. Morphological discretization. In: U. Eckhardt, A. Hübler, W. Nagel, and G. Werner, editors. *Geometrical Problems in Image Processing*, 99–106. Akademie Verlag, Berlin, Germany, 1991.

[417] H.J.A.M. Heijmans. *Morphological Image Operators*. Academic Press, Boston, Massachusetts, 1994.

[418] H.J.A.M. Heijmans and C. Ronse. The algebraic basis of mathematical morphology I: Dilations and erosions. *Computer Vision Graphics Image Processing*, **50**:245–295, 1990.

[419] H.J.A.M. Heijmans and A. Toet. Morphological sampling. *CVGIP: Image Understanding*, **54**:384–400, 1991.

[420] A. Held and K. Abe. On approximate convexity. *Pattern Recognition Letters*, **15**:611–618, 1994.

[421] A. Held, K. Abe, and C. Arcelli. Towards a hierarchical contour description via dominant point detection. *IEEE Trans. Systems, Man, Cybernetics*, **24**:942–949, 1994.

[422] E. Helly. Über Mengen konvexer Körper mit gemeinschaftlichen Punkten. *Jahresbericht Deutscher Mathematiker Vereinigung*, **32**:175–176, 1923.

[423] G.T. Herman. Reconstruction of binary patterns from a few projections. In: A. Günther, B. Levrat, and H. Lipps, editors. *Int. Computing Symposium*, 371–378. North-Holland, Amsterdam, The Netherlands, 1973.

[424] G.T. Herman. On topology as applied to image analysis. *Computer Vision, Graphics, Image Processing*, **52**:409–415, 1990.

[425] G.T. Herman. Discrete multidimensional Jordan surfaces. *Contemporary Mathematics*, **119**:85–94, 1991.

[426] G.T. Herman. Discrete multidimensional Jordan surfaces. *Graphical Models Image Processing*, **54**:507–515, 1992.

[427] G.T. Herman. Oriented surfaces in digital spaces. *Graphical Models Image Processing*, **55**:381–396, 1993.

[428] G.T. Herman. Boundaries in digital spaces: Basic theory. In: T.Y. Kong and A. Rosenfeld, editors. *Topological Algorithms for Digital Image Processing*, 233–261. Elsevier, Amsterdam, The Netherlands, 1996.

[429] G.T. Herman. Geometry of digital spaces. SPIE *Vision Geometry VII*, **3454**:2–13, 1998.

[430] G.T. Herman. *Geometry of Digital Spaces*. Birkhäuser, Boston, Massachusetts, 1998.

[431] G.T. Herman and A. Kuba, editors. *Discrete Tomography — Foundations, Algorithms, and Applications*. Birkhäuser, Boston, Massachusetts, 1999.

[432] G.T. Herman and D. Webster. Surfaces of organs in discrete three-dimensional space. In: G. Herman and F. Natterer, editors. *Mathematical Aspects of Computerized Tomography*, **8**:204–224, 1980.

[433] G.T. Herman and D. Webster. A topological proof of a surface tracking algorithm. *Computer Vision, Graphics, Image Processing*, **23**:162–177, 1983.

[434] S. Hermann and R. Klette. Multigrid analysis of curvature estimators. *Proc. Image Vision Computing New Zealand*, 108–112, 2003.

[435] D. Hilbert. Über die stetige Abbildung einer Linie auf ein Flächenstück. *Mathematische Annalen*, **38**:459–460, 1891.

[436] D. Hilbert. *Grundlagen der Geometrie* (mit Suppl. von Paul Bernays), 13th edition (1st edition, 1899, Leipzig, Germany). B.G. Teubner, Stuttgart, Germany, 1987.

[437] C.J. Hilditch. Linear skeletons from square cupboards. In: B. Meltzer and D. Michie, editors. *Machine Intelligence 4*, 403–420. Edinburgh University Press, Edinburgh, Scotland, 1969.

[438] C.J. Hilditch. Comparison of thinning algorithms on a parallel processor. *Image Vision Computing*, **1**:115–132, 1983.

[439] L. Hodes. The logical complexity of geometric properties in the plane. *J. ACM*, **17**:339–347, 1970.

[440] L. Hodes. Discrete approximation of continuous convex blobs. *SIAM J. Applied Mathematics*, **19**:477–485, 1970.

[441] O. Hölder. *Beiträge zur Potentialtheorie*. Diss. Tübingen, Stuttgart, Germany, 1882.

[442] C.M. Holt, A. Stewart, M. Clint, and R.H. Perrott. An improved parallel thinning algorithm. *Comm. ACM*, **30**:156–160, 1987.

[443] J. Hopcroft and R. Tarjan. Efficient planarity testing. *J. ACM*, **21**:549–568, 1974.

[444] G. Hu and Z.N. Li. An X-crossing preserving skeletonization algorithm. *Int. J. Pattern Recognition Artificial Intelligence*, **7**:1031–1053, 1993.

[445] J. Hu and H. Yan. Polygonal approximation of digital curves based on the principles of perceptual organization. *Pattern Recognition*, **30**:701–718, 1997.

[446] M.K. Hu. Visual pattern recognition by moment invariants. *IRE Trans. Information Theory*, **8**:179–187, 1962.

[447] F. Huang, S.K. Wei, and R. Klette. Geometrical fundamentals of polycentric panoramas. *Proc. Int. Conf. Computer Vision*, 560–565, IEEE, 2001.

[448] S.-C. Huang and Y.-N. Sun. Polygonal approximation using genetic algorithms. *Pattern Recognition*, **32**:1409–1420, 1999.

[449] A. Hübler. Diskrete Geometrie für die digitale Bildverarbeitung. Habilitationschrift, Friedrich-Schiller-Universität, Jena, Germany, 1989.

[450] A. Hübler. Motions in the discrete plane. In: A. Hübler, W. Nagel, B.D. Ripley, and G. Werner, editors. *Geometrical Problems in Image Processing*, 29–36. Akademie-Verlag, Berlin, Germany, 1989.

[451] A. Hübler, R. Klette, and K. Voss. Determination of the convex hull of a finite set of planar points within linear time. *Elektronische Informationsverarbeitung Kybernetik*, **17**:121–140, 1981.

[452] A. Hübler, R. Klette, and G. Werner. Shortest path algorithms for graphs of restricted in-degree and out-degree. *Elektronische Informationsverarbeitung Kybernetik*, **18**:141–151, 1982.

[453] P. Hufnagl, A. Schlosser, and K. Voss. Ein Algorithmus zur Konstruktion von Voronoidiagramm und Delaunaygraph. *Bild und Ton*, **38**:241–245, 1985.

[454] S.H.Y. Hung. On the straightness of digital arcs. *IEEE Trans. Pattern Analysis Machine Intelligence*, **7**:1264–1269, 1985.

[455] S.H.Y. Hung and T. Kasvand. Critical points on a perfectly 8- or 6-connected thin binary line. *Pattern Recognition*, **16**:297–306, 1983.

[456] S.H.Y. Hung and T. Kasvand. On the chord property and its equivalences. *Proc. Int. Conf. Pattern Recognition*, 116–119, IEEE, 1984.

[457] D.P. Huttenlocher and L. Kedem. Computing the minimum Hausdorff distance for point sets under translation. *Proc. ACM Symp. Computational Geometry*, 340–349, ACM Press, 1990.

[458] D.P. Huttenlocher and W.J. Rucklidge. A multi-resolution technique for comparing images using the Hausdorff distance. *Proc. IEEE Conf. Computer Vision Pattern Recognition*, 705–706, 1993.

[459] M.N. Huxley. Exponential sums and lattice points. II. *Proc. London Mathematical Society*, **66**:279–301, 1993.

[460] M.N. Huxley. The integer points close to a curve III. In: H. Iwaniec, J. Urbanowicz, and A. Schinzel, editors. *Number Theory in Progress*, 911–940. Walter de Gruyter, Berlin, Germany, 1999.

[461] A. Imiya and U. Eckhardt. The Euler characteristics of discrete objects and discrete quasi-objects. *Computer Vision Image Understanding*, **75**:307–318, 1999.

[462] A. Imiya, M. Saito, K. Tatara, and K. Nakamura. Digital curvature flow and its application for skeletonization. *J. Mathematical Imaging Vision*, **18**:55–68, 2003.

[463] A. Imiya, T. Ueno, and I. Fermin. Discovery of symmetry by voting method. *Engineering Applications Artificial Intelligence*, **15**:161–168, 2002.

[464] M.C. Irwin. Geometry of continued fractions. *American Mathematical Monthly*, **96**:696–703, 1989.

[465] A. Ivić, J. Koplowitz, and J.D. Žunić. On the number of digital convex polygons inscribed into an (m,m)-grid. *IEEE Trans. Information Theory*, **40**:1681–1686, 1994.

[466] A. Jacques. Sur le genre d'une paire de substitutions. *C. R. Acad. Sci. Paris*, **7**:625–627, 1968.

[467] A. Jacques. Constellations et graphes topologiques. *Proc. Colloque Math. Sot. János Bolyai*, 657–672, North-Holland, 1970.

[468] R. Jain, R. Kasturi, and B.G. Schunck. *Machine Vision*. McGraw-Hill, New York, 1995.

[469] M. Y. Jaisimha, R.M. Haralick, and D. Dori. Quantitative performance evaluation of thinning algorithms in the presence of noise. In: C. Arcelli, L.P. Cordella, and G. Sanniti di Baja, editors. *Aspects of Visual Form Processing*, 261–286. World Scientific, Singapore, 1994.

[470] B.K. Jang and R.T. Chin. Analysis of thinning algorithms using mathematical morphology. *IEEE Trans. Pattern Analysis Machine Intelligence*, **12**:541–551, 1990.

[471] B.K. Jang and R.T. Chin. One-pass parallel thinning: Analysis, properties, and quantitative evaluation. *IEEE Trans. Pattern Analysis Machine Intelligence*, **14**:1128–1140, 1992.

[472] B.K. Jang and R.T. Chin. Reconstructable parallel thinning. *Int. J. Pattern Recognition Artificial Intelligence*, **7**:1145–1181, 1993.

[473] L. Janos and A. Rosenfeld. Some results on fuzzy (digital) convexity. *Pattern Recognition*, **15**:379–382, 1982.

[474] L. Janos and A. Rosenfeld. Digital connectedness: An algebraic approach. *Pattern Recognition Letters*, **1**:135–139, 1983.

[475] I.V. Jarnik. Über die Gitterpunkte auf konvexen Kurven. *Mathematische Zeitschrift*, **24**:500–518, 1925.

[476] L. Ji and J. Piper. Fast homotopy-preserving skeletons using mathematical morphology. *IEEE Trans. Pattern Analysis Machine Intelligence*, **14**:653–664, 1992.

[477] E.G. Johnston and A. Rosenfeld. Geometrical operations on digitized pictures. In: B.S. Lipkin and A. Rosenfeld, editors. *Picture Processing and Psychopictorics*, 217–240. Academic Press, New York, 1970.

[478] J.M. Jolion. Stochastic pyramids. *Pattern Recognition Letters*, **24**:1035–1042, 2003.

[479] J.M. Jolion and A. Rosenfeld. *A Pyramid Framework for Early Vision*. Kluwer, Dordrecht, The Netherlands, 1994.

[480] A. Jonas and N. Kiryati. Digital representation schemes for 3-D curves. *Pattern Recognition*, **30**:1803–1816, 1997.

[481] A. Jonas and N. Kiryati. Length estimation in 3-D using cube quantization. *J. Mathematical Imaging Vision*, **8**:215–238, 1998.

[482] P.P. Jonker. Skeletons in N dimensions using shape primitives. *Pattern Recognition Letters*, **23**:677–686, 2002.

[483] C. Jordan. *Cours d'Analyse de l'École Polytechnique*, Paris, France, 1887. Also in: *Œuvres de Camille Jordan*, Vol. 4. Gauthier-Villars, Paris, France, 1961.

[484] C. Jordan. Remarques sur les intégrales définies. *Journal de Mathématiques* (4e Série), **8**:69–99, 1892.

[485] H.W.E. Jung. Über die kleinste Kugel, die eine räumliche Figur einschliesst. *J. Reine Angewandte Mathematik*, **123**:241–257, 1901.

[486] R. Kakarala. On symmetry in digital geometry. In: R. Klette, A. Rosenfeld, and F. Sloboda, editors. *Advances in Digital and Computational Geometry*, 317–339. Springer, Singapore, 1998.

[487] D. Kalles and D.T. Morris. A novel fast and reliable thinning algorithm. *Image Vision Computing*, **11**:588–603, 1993.

[488] C.V. Kameswara Rao, P.E. Danielsson, and B. Kruse. Checking connectivity preservation properties of some types of picture processing operations. *Computer Graphics Image Processing*, **8**:299–309, 1978.

[489] C.V. Kameswara Rao, B. Prasada, and K.R. Sarma. A parallel shrinking algorithm for binary patterns. *Computer Graphics Image Processing*, **5**:265–270, 1976.

[490] L.W. Kantorowitsch and G.P. Akilow. *Funktionalanalysis in normierten Räumen*. Akademie-Verlag, Berlin, Germany, 1964.

[491] A. Kaufmann. *Introduction to the Theory of Fuzzy Subsets*. Academic Press, New York, 1975.

[492] S. Kawai. On the topology preserving property of local parallel operations. *Computer Graphics Image Processing*, **19**:265–280, 1982.

[493] S. Kawai. Topology quasi-preservation by local parallel operations. *Computer Vision, Graphics, Image Processing*, **23**:353–365, 1983.

[494] S.R. Keller. On the surface area of the ellipsoid. *Mathematics of Computation*, **33**:310–314, 1979.

[495] Y. Kenmochi and A. Imiya. Discretization of three-dimensional objects: Approximation and convergence. SPIE *Vision Geometry VII*, **3454**:64–74, 1998.

[496] Y. Kenmochi, A. Imiya, and A. Ichikawa. Discrete combinatorial geometry. *Pattern Recognition*, **30**:1719–1728, 1997.

[497] Y. Kenmochi, A. Imiya, and A. Ichikawa. Boundary extraction of discrete objects. *Computer Vision Image Understanding*, **71**:281–293, 1998.

[498] Y. Kenmochi and R. Klette. Surface area estimation for digitized regular solids. SPIE *Vision Geometry IX*, **4117**:100–111, 2000.

[499] Y. Kenmochi, K. Kotani, and A. Imiya. Marching cubes method with connectivity. *Proc. IEEE Conf. Image Processing*, **4**:361–365, 1999.

[500] E. Keppel. Approximation of complex surfaces by triangulation of contour lines. *IBM J. Research Development*, **19**:2–11, 1975.

[501] N. Keskes and O. Faugeras. Surfaces simples dans $\mathbb{Z}^3$. *Proc. Congrès Reconaissance des Formes et Intelligence Artificielle*, 718–729, AFRIF-AFIA, 1981.

[502] M. Khachan, P. Chenin, and H. Deddi. Polyhedral representation and adjacency graph in n-dimensional digital images. *Computer Vision Image Understanding*, **79**:428–441, 2000.

[503] E. Khalimsky. Pattern analysis of N-dimensional digital images. *Proc. Int. Conf. Systems, Man, Cybernetics*, 1559–1562, IEEE, 1986.

[504] E. Khalimsky. Motion, deformation, and homotopy in finite spaces. *Proc. Int. Conf. Systems, Man, Cybernetics*, 227–234, IEEE, 1987.

[505] E. Khalimsky, R. Kopperman, and P.R. Meyer. Computer graphics and connected topologies on finite ordered sets. *Topology and its Applications*, **36**:1–17, 1990.

[506] V. Khanna, P. Gupta, and C.J. Hwang. Finding connected components in digital images by aggressive reuse of labels. *Image Vision Computing*, **20**:557–568, 2002.

[507] S. Khuller, A. Rosenfeld, and A. Wu. Centers of sets of pixels. *Discrete Applied Mathematics*, **103**:297–306, 2000.

[508] C.E. Kim. On the cellular convexity of complexes. *IEEE Trans. Pattern Analysis Machine Intelligence*, **3**:617–625, 1981.

[509] C.E. Kim. On cellular straight line segments. *Computer Graphics Image Processing*, **18**:369–381, 1982.

[510] C.E. Kim. Digital convexity, straightness, and convex polygons. *IEEE Trans. Pattern Analysis Machine Intelligence*, **4**:618–626, 1982.

[511] C.E. Kim. Three-dimensional digital line segments. *IEEE Trans. Pattern Analysis Machine Intelligence*, **5**:231–234, 1983.

[512] C.E. Kim. Digital disks. *IEEE Trans. Pattern Analysis Machine Intelligence*, **6**:372–374, 1984.

[513] C.E. Kim. Three-dimensional digital planes. *IEEE Trans. Pattern Analysis Machine Intelligence*, **6**:639–645, 1984.

[514] C.E. Kim and A. Rosenfeld. On the convexity of digital regions. *Proc. Int. Conf. Pattern Recognition*, 1010–1015, IEEE, 1980.

[515] C.E. Kim and A. Rosenfeld. Digital straightness and convexity. *Proc. ACM Symp. Theory of Computing*, 80–89, ACM Press, 1981.

[516] C.E. Kim and A. Rosenfeld. Digital straight lines and convexity of digital regions. *IEEE Trans. Pattern Analysis Machine Intelligence*, **4**:149–153, 1982.

[517] C.E. Kim and A. Rosenfeld. Convex digital solids. *IEEE Trans. Pattern Analysis Machine Intelligence*, **4**:612–618, 1982.

[518] C.E. Kim and J. Sklansky. Digital and cellular convexity. *Pattern Recognition*, **15**:359–367, 1982.

[519] C.E. Kim and I. Stojmenovic. On the recognition of digital planes in three-dimensional space. *Pattern Recognition Letters*, **12**:665–669, 1991.

[520] R. Kimmel, N. Kiryati, and A.M. Bruckstein. Sub-pixel distance maps and weighted distance transforms. *J. Mathematical Imaging Vision*, **6**:223–233, 1996.

[521] R. Kimmel, D. Shaked, N. Kiryati, and A.M. Bruckstein. Skeletonization via distance maps and level sets. *Computer Vision Image Understanding*, **62**:382–391, 1995.

[522] N. Kiryati and A. Jonas. Length estimation in 3D using cube quantization. *J. Mathematical Imaging Vision*, **8**:215–238, 1998.

[523] N. Kiryati and O. Kübler. Chain code probabilities and optimal length estimators for digitized three-dimensional curves. *Pattern Recognition*, **28**:361–372, 1995.

[524] N. Kiryati, M. Lindenbaum, and A.M. Bruckstein. Digital or analog Hough transform? *Pattern Recognition Letters*, **12**:291–297, 1991.

[525] N. Kiryati and G. Székely. Estimating shortest paths and minimal distances on digitized three-dimensional surfaces. *Pattern Recognition*, **26**:1623–1637, 1993.

[526] C.O. Kiselman. Regularity properties of distance transformations in image analysis. *Computer Vision Image Understanding*, **64**:390–398, 1996.

[527] C.O. Kiselman. Digital Jordan curve theorem. In: G. Borgefors, I. Nyström, and G. Sanniti di Baja, editors. *Discrete Geometry for Computer Imagery*. Lecture Notes in Computer Science 1953, Springer, Berlin, Germany, 2000.

[528] K. Kishimoto. Characterizing digital convexity and straightness in terms of "length" and "total absolute curvature." *Computer Vision Image Understanding*, **63**:326–333, 1996.

[529] H. Klaasman. Some aspects of the accuracy of the approximated position of a straight line on a square grid. *Computer Graphics Image Processing*, **4**:225–235, 1975.

[530] D.A. Klarner. Cell growth problems. *Canadian J. Mathematics*, **19**:851–863, 1967.

[531] G. Klette. A comparative discussion of distance transforms and simple deformations in digital image processing. *Machine Graphics Vision*, **12**:235–256, 2003.

[532] R. Klette. Wir konstruieren einen vierdimensionalen Würfel. *Wurzel*, Mathematics Department, University of Jena, Germany, **6**:88–91, 1972.

[533] R. Klette. A parallel computer for digital image processing. *Elektronische Informationsverarbeitung Kybernetik*, **15**:237–263, 1979.

[534] R. Klette. Parallel operations on binary pictures. *Computer Graphics Image Processing*, **14**:145–158, 1980.

[535] R. Klette. Hüllen für endliche Punktmengen. In: R. Klette and J. Mecke, editors. *Proc. Geometric Problems of Image Processing*, 74–83, Jena University, Germany, 1982.

[536] R. Klette. *M*-dimensional cellular spaces. Computer Science Dep., TR-1266, University of Maryland, College Park, Maryland, 1983.

[537] R. Klette. On the approximation of convex hulls of finite grid point sets. *Pattern Recognition Letters*, **2**:19–22, 1983.

[538] R. Klette. Grundbegriffe der digitalen Geometrie. *Proc. Automatische Bildverarbeitung*, 94–126, Jena University, Germany, 1983.

[539] R. Klette. Mathematische Probleme der digitalen Bildverarbeitung. *Bild und Ton*, **36**:107–113, 1983.

[540] R. Klette. The m-dimensional grid point space. *Computer Vision, Graphics, Image Processing*, **30**:1–12, 1985.

[541] R. Klette. Measurement of object surface area. *Computer Assisted Radiology*, 147–152, 1998.

[542] R. Klette. Cell complexes through time. SPIE *Vision Geometry IX*, **4117**:134–145, 2000.

[543] R. Klette. Multigrid convergence of geometric features. In: G. Bertrand, A. Imiya, and R. Klette, editors. *Digital and Image Geometry: Advanced Lectures*, 314–333. Lecture Notes in Computer Science 2243, Springer, Berlin, Germany, 2001.

[544] R. Klette. Digital geometry — The birth of a new discipline. In: L.S. Davis, editor. *Foundations of Image Understanding*, 33–71. Kluwer, Boston, Massachusetts, 2001.

[545] R. Klette. Switches may solve adjacency problems. *Proc. Int. Conf. Pattern Recognition*, **III**:907–910, IEEE, 2002.

[546] R. Klette and T. Bülow. Minimum-length polygons in simple cube-curves. In: G. Borgefors, I. Nyström, and G. Sannati di Baja, editors. *Discrete Geometry for Computer Imagery*, 467–478. Lecture Notes in Computer Science 1953, Springer, Berlin, Germany, 2000.

[547] R. Klette and E.V. Krishnamurthy. Algorithms for testing convexity of digital polygons. *Computer Graphics Image Processing*, **16**:177–184, 1981.

[548] R. Klette, K. Schlüns, and A. Koschan. *Computer Vision — Three-Dimensional Data from Images*. Springer, Singapore, 1998.

[549] R. Klette, I. Stojmenović, and J. Žunić. A parametrization of digital planes by least square fits and generalizations. *Graphical Models Image Processing*, **58**:295–300, 1996.

[550] R. Klette and H.-J. Sun. Digital planar segment based polyhedrization for surface area estimation. In: C. Arcelli, L.P. Cordella, and G. Sanniti di Baja, editors. *Visual Form 2001*, 356–366. Lecture Notes in Computer Science 2059, Springer, Berlin, Germany, 2001.

[551] R. Klette and K. Voss. Theoretische Grundlagen der digitalen Bildverarbeitung, Part III: Gebietsnachbarschaftsgraphen. *Bild und Ton*, **39**:45–50, 55, 1986.

[552] R. Klette and K. Voss. Theoretische Grundlagen der digitalen Bildverarbeitung, Part VII: Planare reguläre Gitter und Polygone im Gitter. *Bild und Ton*, **40**:112–118, 1987.

[553] R. Klette and K. Voss. The three basic formulas of oriented graphs. *Pattern Recognition Image Analysis* **1**:385–405, 1991. Also in: Center for Automation Research, TR-305, University of Maryland, College Park, Maryland, 1987.

[554] R. Klette, K. Voss, and P. Hufnagl. Theoretische Grundlagen der digitalen Bildverarbeitung, Part II: Nachbarschaftsstrukturen. *Bild und Ton*, **38**:325–331, 1985.

[555] R. Klette and B. Yip. The length of digital curves. *Machine Graphics Vision*, **9**:673–703, 2000. (Extended verison of: R. Klette, V.V. Kovalevsky, and B. Yip. Length estimation of digital curves. SPIE *Vision Geometry VIII*, **3811**:117–129, 1999.)

[556] R. Klette and P. Zamperoni. Measures of correspondence between binary patterns. *Image Vision Computing*, **5**:287–295, 1987.

[557] R. Klette and P. Zamperoni. *Handbook of Image Processing Operators*. Wiley, Chichester, United Kingdom, 1996.

[558] R. Klette and J. Žunić. Digital approximation of moments of convex regions. *Graphical Models Image Processing*, **61**:274–298, 1999.

[559] R. Klette and J. Žunić. Interactions between number theory and image analysis. SPIE *Vision Geometry IX*, **4117**:210–221, 2000.

[560] R. Klette and J. Žunić. Multigrid convergence of calculated features in image analysis. *J. Mathematical Imaging Vision*, **13**:173–191, 2000.

[561] R. Klette and J. Žunić. Multigrid error bounds for moments of arbitrary order. *Proc. Int. Conf. Pattern Recognition*, **III**:790–793, IEEE, 2000.

[562] R. Klette and J. Žunić. Towards experimental studies of digital moment convergence. SPIE *Vision Geometry IX*, **4117**:12–23, 2000.

[563] B. Klotzek. *Kombinieren, Parkettieren, Färben*. Aulis, Köln, Germany, 1985.

[564] W.R. Knorr. *Evolution of the Elements: A Study of the Theory of Incommensurable Magnitudes and its Significance for Early Greek Geometry*. Reidel, Dordrecht, The Netherlands, 1975.

[565] W.R. Knorr. *The Ancient Tradition of Geometric Problems*. Dover, New York, 1993.

[566] D.E. Knuth. *The Art of Computer Programming*, Vol. 2, 309, 316–332. Addison-Wesley, Reading, Massachusetts, 1969.

[567] T.Y. Kong. A digital fundamental group. *Computers & Graphics*, **13**:159–166, 1989.

[568] T.Y. Kong. On topology preservation in 2-D and 3-D thinning. *Int. J. Pattern Recognition Artificial Intelligence*, **9**:813–844, 1995.

[569] T.Y. Kong. Digital topology. In: L.S. Davis, editor. *Foundations of Image Understanding*, 33–71. Kluwer, Boston, Massachusetts, 2001.

[570] T.Y. Kong. Topological adjacency relations on $\mathbb{Z}^n$. *Theoretical Computer Science*, **283**:3–28, 2002.

[571] T.Y. Kong, R. Kopperman, and P.R. Meyer. A topological approach to digital topology. *American Mathematical Monthly*, **92**:901–917, 1991.

[572] T.Y. Kong, R. Litherland, and A. Rosenfeld. Problems in the topology of binary digital images. In: J. van Mill and G.M. Reed, editors. *Open Problems in Topology*, 376–385. North-Holland, Amsterdam, The Netherlands, 1990.

[573] T.Y. Kong and A.W. Roscoe. Continuous analogs of axiomatized digital surfaces. *Computer Vision, Graphics, Image Processing*, **29**:60–86, 1985.

[574] T.Y. Kong and A.W. Roscoe. A theory of binary digital pictures. *Computer Vision, Graphics, Image Processing*, **32**:221–243, 1985.

[575] T.Y. Kong and A.W. Roscoe. Characterizations of simply-connected finite polyhedra in 3-space. *Bull. London Mathematical Society*, **17**:575–578, 1985.

[576] T.Y. Kong, A.W. Roscoe and A. Rosenfeld. Concepts of digital topology. *Topology and its Applications*, **46**:219–262, 1992.

[577] T.Y. Kong and A. Rosenfeld. Digital topology: Introduction and survey. *Computer Vision, Graphics, Image Processing*, **48**:357–393, 1989.

[578] T.Y. Kong and A. Rosenfeld. If we use 4- or 8-connectedness for both the objects and the background, the Euler characteristic is not locally computable. *Pattern Recognition Letters*, **11**:231-232, 1990.

[579] T.Y. Kong and A. Rosenfeld. Digital topology: A comparison of the graph-based and topological approaches. In: G.M. Reed, A.W. Roscoe, and R.F. Wachter, editors. *Topology and Category Theory in Computer Science*, 273–289. Oxford University Press, Oxford, United Kingdom, 1991.

[580] T.Y. Kong and A. Rosenfeld, editors. *Topological Algorithms for Digital Image Processing*. North-Holland, Amsterdam, The Netherlands, 1996.

[581] T.Y. Kong and A. Rosenfeld. Digital topology: A brief introduction and bibliography. In: T.Y. Kong and A. Rosenfeld, editors., *Topological Algorithms for Digital Image Processing*, 263–292. North-Holland, Amsterdam, The Netherlands, 1996.

[582] T.Y. Kong and J.K. Udupa. A justification of a fast surface tracking algorithm. *Graphical Models Image Processing*, **54**:162–170, 1992.

[583] J. Koplowitz. On the performance of chain codes for quantization of line drawings. *IEEE Trans. Pattern Analysis Machine Intelligence*, **3**:180–185, 1981.

[584] J. Koplowitz and A.M. Bruckstein. Design of perimeter estimators for digitized planar shapes. *IEEE Trans. Pattern Analysis Machine Intelligence*, **11**:611–622, 1989.

[585] J. Koplowitz, M. Lindenbaum, and A.M. Bruckstein. The number of digital straight lines on an $N \times N$ grid. *IEEE Trans. Information Theory*, **36**:192–197, 1990.

[586] J. Koplowitz and S. Plante. Corner detection for chain coded curves. *Pattern Recognition*, **28**:843–852, 1995.

[587] R. Kopperman, P.R. Meyer, and R. Wilson. A Jordan surface theorem for three-dimensional digital spaces. *Discrete Computational Geometry*, **6**:155–161, 1991.

[588] V. Kovalevsky. Bilderkennung mit Hilfe einer Kreisbogenapproximation mit Minmax-Eigenschaften. *Tagungsbericht Akad. Landwirtsch.-Wiss. DDR*, **204**:5–17, 1982.

[589] V. Kovalevsky. Discrete topology and contour definition. *Pattern Recognition Letters*, **2**:281–288, 1984.

[590] V. Kovalevsky. Strukturen der Bildträger und Bilder. *Proc. Automatische Bildverarbeitung*, 122–149, Jena University, Germany, 1986.

[591] V. Kovalevsky. Finite topology as applied to image analysis. *Computer Vision, Graphics, Image Processing*, **46**:141–161, 1989.

[592] V. Kovalevsky. New definition and fast recognition of digital straight segments and arcs. *Proc. Int. Conf. Pattern Recognition*, 31–34, IEEE, 1990.

[593] V. Kovalevsky. Applications of digital straight segments to economical image encoding. In: E. Ahronovitz and C. Fiorio, editors. *Discrete Geometry for Computer Imagery*, 51–62. Lecture Notes in Computer Science 1347, Springer, Berlin, Germany, 1997.

[594] V. Kovalevsky. Algorithms and data structures for computer topology. In: G. Bertrand, A. Imiya, and R. Klette, editors. *Digital and Image Geometry: Advanced Lectures*, 38–58. Lecture Notes in Computer Science 2243, Springer, Berlin, Germany, 2001.

[595] V. Kovalevsky. Curvature in digital 2D images. *Int. J. Pattern Recognition Artificial Intelligence*, **15**:1183–1200, 2001.

[596] V. Kovalevsky. Multidimensional cell lists for investigating 3-manifolds. *Discrete Applied Mathematics*, **125**:25–44, 2003.

[597] V. Kovalevsky and S. Fuchs. Theoretical and experimental analysis of the accuracy of perimeter estimates. In: W. Förstner and S. Ruwiedel, editors. *Robust Computer Vision*, 218–242. Wichmann, Karlsruhe, Germany, 1992.

[598] E. Krätzel. *Zahlentheorie*. VEB Deutscher Verlag der Wissenschaften, Berlin, Germany, 1981.

[599] R. Krishnaswamy and C.E. Kim. Digital parallelism, perpendicularity, and rectangles. *IEEE Trans. Pattern Analysis Machine Intelligence*, **9**:316–321, 1987.

[600] N. Kritikos. Über konvexe Flächen und einschließende Kugeln. *Mathematische Annalen*, **95**:583–587, 1926.

[601] W.G. Kropatsch and H. Tockner. Detecting the straightness of digital curves in $O(N)$ steps. *Computer Vision, Graphics, Image Processing*, **45**:1–21, 1989.

[602] B. Kruse. A fast algorithm for segmentation of connected components in binary images. *Proc. Scandinavian Conf. Image Analysis*, 57–63, 1980.

[603] B. Kruse and C.V.K. Rao. A matched filtering technique for corner detection. *Proc. Int. Conf. Pattern Recognition*, 642–644, IEEE, 1978.

[604] S.R. Kulkarni, S.K. Mitter, T.J. Richardson, and J.N. Tsitsiklis. Local versus nonlocal computation of length of digitized curves. *IEEE Trans. Pattern Analysis Machine Intelligence*, **16**:711–718, 1994.

[605] Z. Kulpa. Area and perimeter measurements of blobs in discrete binary pictures. *Computer Graphics Image Processing*, **6**:434–454, 1977.

[606] Z. Kulpa. On the properties of discrete circles, rings, and disks. *Computer Graphics Image Processing*, **10**:348–365, 1979.

[607] Z. Kulpa. More about areas and perimeters of quantized objects. *Computer Vision, Graphics, Image Processing*, **22**:268–276, 1983.

[608] Z. Kulpa and B. Kruse. Algorithm for circular propagation in discrete images. *Computer Vision, Graphics, Image Processing*, **24**:305–328, 1983.

[609] P. Kumar, D. Bhatnagar, and P.S. Unapathi Rao. Pseudo one pass thinning algorithm. *Pattern Recognition Letters*, **12**:543–555, 1991.

[610] M.K. Kundu, B.B. Chaudhuri, and D. Dutta Majumder. A parallel greytone thinning algorithm. *Pattern Recognition Letters*, **12**:491–496, 1991.

[611] C. Kuratowski. Sur le probleme des courbes gauches en topologie. *Fundamenta Mathematica*, **15**:271–283, 1930. Also in: K. Borsuk, editor. *Selected Papers*, 345–357. PWN — Polish Scientific Publishers, Warszawa, Poland, 1988.

[612] P.C.K. Kwok. A thinning algorithm by contour generation. *Comm. ACM*, **31**:1314–1324, 1988.

[613] J.-O. Lachaud and A. Montanvert. Continuous analogs of digital boundaries: A topological approach to isosurfaces. *Graphical Models*, **62**:129–164, 2000.

[614] I. Lakatos. *Proofs and Refutations: The Logic of Mathematical Discovery*. Cambridge University Press, Cambridge, United Kingdom, 1976.

[615] L. Lam, S.W. Lee, and C.Y. Suen. Thinning methodologies—A comprehensive survey. *IEEE Trans. Pattern Analysis Machine Intelligence*, **14**:869–885, 1992.

[616] E. Landau. *Vorlesungen über Zahlentheorie*. Hirzel, Leipzig, Germany, 1927. Reprint: Chelsea, New York, 1955.

[617] E. Landau. *Ausgewählte Abhandlungen zur Gitterpunktlehre*. Deutscher Verlag der Wissenschaften, Berlin, Germany, 1962.

[618] D.J. Langridge. On the computation of shape. In: S. Watanabe, editor. *Frontiers of Pattern Recognition*, 347–366. Academic Press, London, United Kingdom, 1972.

[619] C. Lantuejoul and F. Maisonneuve. Geodesic methods in quantitative image analysis. *Pattern Recognition*, **17**:177–187, 1984.

[620] L. Latecki. Topological connectedness and 8-connectedness in digital pictures. *CVGIP: Image Understanding*, **57**:261–262, 1993.

[621] L. Latecki. Multicolor well-composed pictures. *Pattern Recognition Letters*, **16**:425–431, 1995.

[622] L.J. Latecki. 3D well-composed pictures. *Graphical Models Image Processing*, **59**:164–172, 1997.

[623] L.J. Latecki. *Discrete Representation of Spatial Objects in Computer Vision*. Kluwer, Dordrecht, The Netherlands, 1998.

[624] L. Latecki, C. Conrad, and A. Gross. Preserving topology by a digitization process. *J. Mathematical Imaging Vision*, **8**:131–159, 1998.

[625] L. Latecki, U. Eckhardt, and A. Rosenfeld. Well-composed sets. *Computer Vision Image Understanding*, **61**:70–83, 1995.

[626] L. Latecki and C.M. Ma. An algorithm for a 3D simplicity test. *Computer Vision Image Understanding*, **63**:388–393, 1996.

[627] L.J. Latecki and A. Rosenfeld. Supportedness and tameness: Differentialless geometry of plane curves. *Pattern Recognition*, **31**:607–622, 1998.

[628] D.F. Lawden. *Elliptic Functions and Applications*, 100–102. Springer, New York, 1989.

[629] C.-N. Lee, T. Poston, and A. Rosenfeld. Representation of orthogonal regions by vertices. *CVGIP: Graphical Models Image Processing*, **53**:149–156, 1991.

[630] C.-N. Lee, T. Poston, and A. Rosenfeld. Winding and Euler numbers in 2D and 3D digital images. *CVGIP: Graphical Models Image Processing*, **53**:522–537, 1991.

[631] C.-N. Lee, T. Poston, and A. Rosenfeld. Holes and genus of 2D and 3D digital images. *CVGIP: Graphical Models Image Processing*, **55**:20–47, 1993.

[632] C.-N. Lee and A. Rosenfeld. Simple connectivity is not locally computable for connected 3D images. *Computer Vision, Graphics, Image Processing*, **51**:87–95, 1990.

[633] C.Y. Lee and P.S.P. Wang. A simple and robust thinning algorithm. *Int. J. Pattern Recognition Artificial Intelligence*, **13**:357–366, 1999.

[634] D.T. Lee. Medial axis transformation of a planar shape. *IEEE Trans. Pattern Analysis Machine Intelligence*, **4**:363–369, 1982.

[635] H.C. Lee and K.S. Fu. Using the FFT to determine digital straight line chain codes. *Computer Graphics Image Processing*, **18**:359–368, 1982.

[636] S.W. Lee, L. Lam, and C.Y. Suen. A systematic evaluation of skeletonization algorithms. *Int. J. Pattern Recognition Artificial Intelligence*, **7**:1203–1225, 1993.

[637] T.C. Lee, R.L. Kashyap, and C.N. Chu. Building skeleton models via 3-D medial surface/axis thinning algorithms. *Graphical Models Image Processing*, **56**:462–478, 1994.

[638] Y.H. Lee and S.J. Horng. Optimal[ly] computing the chessboard distance transform on parallel processing systems. *Computer Vision Image Understanding*, **73**:374–390, 1999.

[639] S. Lefschetz. *Introduction to Topology*. Princeton University Press, Princeton, New Jersey, 1949.

[640] A.-M. Legendre. *Traite de Fonctions Élliptiques*, tome 1. Huzard-Courchier, Paris, 1825.

[641] R.B. Leighton et al. Mariner 6 television pictures: First report. *Science*, **165**:684–690, 1969.

[642] A. Lenoir. *Des outils pour les surfaces discrètes*. PhD Thesis, Université de Caen, France, 1999.

[643] A. Lenoir. Fast estimation of mean curvature on the surface of a 3D discrete object. *ISMRA*, **6**:303–312, 2001.

[644] A. Lenoir, R. Malgouyres, and R. Revenu. Fast computation of the normal vector field of the surface of a 3-D discrete object. In: S. Miguet, A. Montanvert, and S. Ubéda, editors. *Discrete Geometry for Computer Imagery*, 101–112. Lecture Notes in Computer Science 1176, Springer, Berlin, Germany, 1996.

[645] G. Levi and U. Montanari. A gray-weighted skeleton. *Information Control*, **17**:62–91, 1970.

[646] S. Levialdi. On shrinking binary picture patters. *Comm. ACM*, **15**:7–10, 1972.

[647] N. Levitt. The Euler characteristic is the unique locally determined numerical homotopy invariant of finite complexes. *Discrete Computational Geometry*, **7**:59–67, 1992.

[648] F. Leymarie and M.D. Levine. Fast raster scan distance propagation on the discrete rectangular lattice. *CVGIP: Image Understanding*, **55**:84–94, 1992.

[649] A.-J. Lhuilier. Mémoire sur la polyédrométrie, contenant une démonstration directe du théoréme d'Euler sur les polyédres, et un examen de diverses exceptions auxquelles ce théoréme est assujetti (extrait par M. Gergonne). *Annales de Mathématiques Pures et Appliquées par Gergonne* **III**:169ff., 1812.

[650] B. Li and C.Y. Suen. A knowledge-based thinning algorithm. *Pattern Recognition*, **24**:1211–1221, 1991.

[651] S.X. Li and M.H. Loew. Analysis and modeling of digitized straight-line segments. *Proc. Int. Conf. Pattern Recognition*, 294–296, IEEE, 1988.

[652] P. Lienhardt. Topological models for boundary representation: A comparison with n-dimensional generalized maps. *Computer Aided Design*, **23**:59–81, 1991.

[653] T. Lindeberg. *Scale-Space Theory in Computer Vision*. Kluwer, Dordrecht, The Netherlands, 1994.

[654] M. Lindenbaum. Compression of chain codes using digital straight line sequences. *Pattern Recognition Letters*, **7**:167–171, 1988.

[655] M. Lindenbaum and A.M. Bruckstein. On recursive, $O(N)$ partitioning of a digitized curve into digital straight segments. *IEEE Trans. Pattern Analysis Machine Intelligence*, **15**:949–953, 1993.

[656] M. Lindenbaum and J. Koplowitz. A new parametrization of digital straight lines. *IEEE Trans. Pattern Analysis Machine Intelligence*, **13**:847–852, 1991.

[657] M. Lindenbaum, J. Koplowitz, and A.M. Bruckstein. On the number of digital straight lines on an $N \times N$ grid. In: *Proc. IEEE Conf. Computer Vision Pattern Recognition*, 610–615, 1988.

[658] T.K. Linh and A. Imiya. Nonlinear optimization for polygonalization. *IEICE Technical Report*, **102**:43–48, 2003.

[659] R. Lipschitz. Asymptotische Gesetze gewisser zahlentheoretischer Functionen. *Monatsbericht Königl. Akademie der Wissenschaften zu Berlin*, 174–184, 1865.

[660] J.B. Listing. Vorstudien zur Topologie. Göttinger Studien, 1. Abteilung math. und naturw. Abh., 811–875, 1847. Several missing proofs were later published by P.G. Tait, On knots. *Proc. Roy. Soc. Edinburgh*, **9**:306–317, 1875–1878. A more recent review: A. Tripodi, L'introduzione alla topologia di Johann Benedict Listing. *Mem. Acad. Naz. Sci. Lett. Arti Modena*, **13**:3–14, 1971.

[661] J.B. Listing. Der Census räumlicher Complexe oder Verallgemeinerungen des Euler'schen Satzes von den Polyëdern. *Abhandlungen der Mathematischen Classe der Königlichen Gesellschaft der Wissenschaften zu Göttingen*, **10**:97–182, 1861 and 1862.

[662] H.-C. Liu and M.D. Srinath. Corner detection from chain code. *Pattern Recognition*, **23**:51–68, 1990.

[663] S. Lobregt, P.W. Verbeek, and F.C.A. Groen. Three-dimensional skeletonization: Principle and algorithm. *IEEE Trans. Pattern Analysis Machine Intelligence*, **2**:75–77, 1980.

[664] A.F. Lochovsky. Algorithms for realtime component labelling of images. *Image Vision Computing*, **6**:21–27, 1988.

[665] J. Loeb. Communication theory of transmission of simple drawings. In: W. Jackson, editor. *Communication Theory*, 317–327. Butterworths, London, United Kingdom, 1953.

[666] C. Lohou. Contribution à l'analyse topologique des images: Etude d'algorithmes de squelettisation pour images 2D et 3D, selon une approche topologie digitale ou topologie discrète. PhD Thesis, Université de Marne-La-Vallée, France, 2001.

[667] W.E. Lorensen and H.E. Cline. Marching cubes - A high-resolution 3D surface construction algorithm. *Computer Graphics*, **21**:163–169, 1987.

[668] M. Lothaire, editor. *Combinatorics on Words*, 2nd edition. Cambridge University Press, Cambridge, United Kingdom, 1987.

[669] M. Lothaire, editor. *Algebraic Combinatorics on Words*. Cambridge University Press, Cambridge, United Kingdom, 2002.

[670] Augusta Ada, Countess of Lovelace. Note F to her translation (1843) of L.F. Menabrea, *Sketch of the Analytical Engine Invented by Charles Babbage*. In: P. and E. Morrison, editors. *Charles Babbage and his Calculating Engines*, p. 281. Dover, New York, 1961.

[671] R. Lowen. Convex fuzzy sets. *Fuzzy Sets Systems*, **3**:291–310, 1980.

[672] H.E. Lu and P.S.P. Wang. A comment on "A fast parallel algorithm for thinning digital patterns." *Comm. ACM*, **29**:239–242, 1986.

[673] E. Luczak and A. Rosenfeld. Distance on a hexagonal grid. *IEEE Trans. Computers*, **25**:532–533, 1976.

[674] R. Lumia. A new three-dimensional connected components algorithm. *Computer Vision, Graphics, Image Processing*, **23**:207–217, 1983.

[675] R. Lumia, L. Shapiro, and O. Zuniga. A new connected components algorithm for virtual memory computers. *Computer Vision, Graphics, Image Processing*, **22**:287–300, 1983.

[676] W.F. Lunnon and P.A.B. Pleasants. Characterization of two-distance sequences. *J. Australian Mathematical Society*, **A53**:198–218, 1992.

[677] C.M. Ma. On topology preservation in 3D thinning. *CVGIP: Image Understanding*, **59**:328–339, 1994.

[678] C.M. Ma. A 3D fully parallel thinning algorithm for generating medial faces. *Pattern Recognition Letters*, **16**:83–87, 1995.

[679] C.M. Ma. Connectivity preservation of 3D 6-subiteration thinning algorithms. *Graphical Models Image Processing*, **58**:382–386, 1996.

[680] C.M. Ma and M. Sonka. A fully parallel 3D thinning algorithm and its applications. *Computer Vision Image Understanding*, **64**:420–433, 1996.

[681] C.M. Ma and S.Y. Wan. Parallel thinning algorithms on 3D (18,6) binary images. *Computer Vision Image Understanding*, **80**:364–378, 2000.

[682] C.M. Ma, S.Y. Wan, and H.K. Chang. Extracting medial curves on 3-D images. *Pattern Recognition Letters*, **23**:895–904, 2002.

[683] C.M. Ma, S.Y. Wan, and J.D. Lee. Three-dimensional topology preserving reduction on the 4-subfields. *IEEE Trans. Pattern Analysis Machine Intelligence*, **24**:1594–1605, 2002.

[684] M. Maes. Digitization of straight line segments: Closeness and convexity. *Computer Vision, Graphics, Image Processing*, **52**:297–305, 1990.

[685] P. Magillo. *Spatial operations on multiresolution cell complexes*. PhD Thesis, Università di Genova, Italy, 1999.

[686] G. Malandain, G. Bertrand, and N. Ayache. Topological segmentation of discrete surfaces. *Int. J. Computer Vision*, **10**:183–197, 1993.

[687] G. Malandain and S. Fernández-Vidal. Euclidean skeletons. *Image Vision Computing*, **16**:317–327, 1998.

[688] R. Malgouyres. Graphs generalizing closed curves with linear construction of the Hamiltonian cycle. *Theoretical Computer Science*, **143**:189–249, 1995.

[689] R. Malgouyres. There is no local characterization of separating and thin objects in $\mathbb{Z}^2$. *Theoretical Computer Science*, **116**:303-308, 1996.

[690] R. Malgouyres. A definition of surfaces of $\mathbb{Z}^3$: A new 3D discrete Jordan theorem. *Theoretical Computer Science*, **186**:1–41, 1997.

[691] R. Malgouyres. Local characterization of strong surfaces within strongly separating objects. *Pattern Recognition Letters*, **19**:341–349, 1998.

[692] R. Malgouyres. Homotopy in two-dimensional digital images. *Theoretical Computer Science*, **230**:221-233, 2000.

[693] R. Malgouyres. Computing the fundamental group in digital spaces. *Int. J. Pattern Recognition Artificial Intelligence*, **15**:1075–1088, 2001.

[694] R. Malgouyres and G. Bertrand. Complete local characterization of strong 26-surfaces: Continuous analogs for strong 26-surfaces. *Int. J. Pattern Recognition Artificial Intelligence*, **13**:465–484, 1999.

[695] R. Malgouyres and A. Lenoir. Topology preservation within digital surfaces. *Graphical Models*, **62**:71–84, 2002.

[696] H.v. Mangoldt and K. Knopp. *Einführung in die höhere Mathematik* (III. Band, 12. Auflage). Hirzel, Leipzig, Germany, 1965.

[697] A. Manzanera, T.M. Bernard, F. Prêteux, and B. Longuet. Ultra-fast skeleton based on an isotropic fully parallel algorithm. In: G. Bertrand, M. Couprie, and L. Perroton, editors. *Discrete Geometry for Computer Imagery*, 313-324. Lecture Notes in Computer Science 1568, Springer, Berlin, Germany, 1999.

[698] A. Manzanera, T.M. Bernard, F. Prêteux, and B. Longuet. A unified mathematical framework for a compact and fully parallel n-D skeletonisation procedure. SPIE *Vision Geometry VIII*, **3811**:57–68, 1999.

[699] A. Manzanera and T.M. Bernard. A coherent collection of 2D parallel thinning algorithms. Ecole Nationale Supérieure de Techniques Avancées, Laboratoire d'Electronique de Informatique,TR 02-002, Paris, France, 2002.

[700] S. Marchand-Maillet and Y.M. Sharaiha. Discrete convexity, straightness, and the 16-neighborhood. *Computer Vision Image Understanding*, **66**:316–329, 1997.

[701] S. Marchand-Maillet and Y.M. Sharaiha. Skeleton location and evaluation based on local digital width in ribbon-like images. *Pattern Recognition*, **30**:1855–1865, 1997.

[702] S. Marchand-Maillet and Y.M. Sharaiha. Euclidean ordering via chamfer distance calculations. *Computer Vision Image Understanding*, **73**:404–413, 1999.

[703] S. Marchand-Maillet and Y.M. Sharaiha. *Binary Digital Image Processing — A Discrete Approach*. Academic Press, San Diego, California, 2000.

[704] M. Marji. *On the detection of dominant points on digital planar curves*. PhD Thesis, Wayne State University, Detroit, Michigan, 2003.

[705] G.E. Martin. *Transformation Geometry: An Introduction to Symmetry*. Springer, New York, 1992.

[706] M.P. Martinez-Perez, J. Jimenez, and J.L. Navalon. A thinning algorithm based on contours. *Computer Vision, Graphics, Image Processing*, **39**:186–201, 1987.

[707] G. Matheron. *Random Sets and Integral Geometry*. Wiley, New York, 1975.

[708] G. Matheron. Examples of topological properties of skeletons. In: J. Serra, editor. *Mathematical Morphology and Image Analysis II: Theoretical Advances*, 217–237. Academic Press, London, United Kingdom, 1988.

[709] G. Matheron and J. Serra. The birth of mathematical morphology. In: H. Talbot and R. Beare, editors. *Proc. Int. Society for Math. Morphology*, 1–16, 2002.

[710] C.R.F. Maunder. *Algebraic Topology*, 2nd edition. Cambridge University Press, Cambridge, United Kingdom, 1980. Reprint: Dover, New York, 1996.

[711] J.C. Maxwell. On hills and dales. *London, Edinburgh, and Dublin Philosophical Magazine and Journal of Science*, Series 4, **40**:421–427, 1870.

[712] S. Mazurkiewicz. Sur les lignes de Jordan. *Fundamenta Mathematica*, **1**:166–209, 1920. Also in: *Travaux de Topologie et ses Applications*, 77-113. PWN — Polish Scientific Publishers, Warszawa, Poland, 1969.

[713] M.D. McIlroy. Best approximate circles on integer grids. *ACM Trans. Graphics*, **2**:237–263, 1983.

[714] A.M. McIvor. A cubic facet based method for estimating the principal quadric. *Proc. Image Vision Computing New Zealand*, 136–141, 1998.

[715] A.M. McIvor and R.J. Valkenburg. A comparison of local surface geometry estimation methods. *Machine Vision Applications*, **10**:17–26, 1997.

[716] G. Medioni and Y. Yasumoto. Corner detection and curve representation using cubic B-splines. *Computer Vision, Graphics, Image Processing*, **39**:267–278, 1987.

[717] N. Megiddo. Linear programming in linear time when the dimension is fixed. *J. ACM*, **31**:114–127, 1984.

[718] A.J.H. Mehnert and P.T. Jackway. On computing the exact Euclidean distance transform on rectangular and hexagonal grids. *J. Mathematical Imaging Vision*, **11**:223–230, 1999.

[719] R.A. Melter. Some characterizations of city block distance. *Pattern Recognition Letters*, **6**:235–240, 1987.

[720] R.A. Melter. Convexity is necessary—A correction. *Pattern Recognition Letters*, **8**:59, 1988.

[721] R.A. Melter. A survey of digital metrics. *Contemporary Mathematics*, **119**:95–106, 1991.

[722] R.A. Melter and A. Rosenfeld. New views of linearity and connectedness in digital geometry. *Pattern Recognition Letters*, **10**:9–16, 1989.

[723] R.A. Melter, I. Stojmenovic, and J. Zunic. A new characterization of digital lines by least square fits. *Pattern Recognition Letters*, **14**:83–88, 1993.

[724] R.A. Melter and I. Tomescu. Path generated digital metrics. *Pattern Recognition Letters*, **1**:151–154, 1983.

[725] R.A. Melter and I. Tomescu. Metric bases in digital geometry. *Computer Vision, Graphics, Image Processing*, **25**:113–121, 1984.

[726] K. Menger. *Kurventheorie*. Teubner, Leipzig, Germany, 1932.

[727] J.B. Meusnier. Mémoire sur la courbure des surfaces. *Mém. des savans étrangers*, **10**:477–510, 1776.

[728] F. Meyer. Topographic distance and watershed lines. *Signal Processing*, **38**:113–125, 1994.

[729] F. Mignosi. On the number of factors of Sturmian words. *Theoretical Computer Science*, **82**:71–84, 1991.

[730] E.H. Milun, D.K.W. Walters, and Y. Li. General ribbon-based thinning algorithms for stylus-generated images. *Computer Vision Image Understanding*, **76**:267–277, 1999.

[731] E.H. Milun, D.K.W. Walters, Y. Li, and B. Atanacio. General ribbons: A model for stylus-generated images. *Computer Vision Image Understanding*, **76**:259–266, 1999.

[732] H. Minkowski. Volumen und Oberfläche. *Mathematische Annalen*, **57**:447–495, 1903.

[733] H. Minkowski. *Geometrie der Zahlen*. Teubner, Leipzig, Germany, 1910.

[734] M. Minsky and S. Papert. *Perceptrons—An Introduction to Computational Geometry*. MIT Press, Cambridge, Massachusetts, 1969.

[735] A.F. Möbius. Über die Bestimmung des Inhaltes eines Polyëders. *Leipziger Sitzungsberichte math.phys. Classe* **17**, 1867. Also in: *Gesammelte Werke*, **2**:473–512. Hirzel, Stuttgart, Germany, 1886. (Facsimile edition: Dr. Martin Saendig oHG, Wiesbaden, Germany, 1967.)

[736] A.N. Moga. Parallel image component labelling with watershed transformations. *IEEE Trans. Pattern Analysis Machine Intelligence*, **19**:441–450, 1997.

[737] F. Mokhtarian and A. Mackworth. Scale-based description and recognition of planar curves and two-dimensional shapes. *IEEE Trans. Pattern Analysis Machine Intelligence*, **8**:34–43, 1986.

[738] U. Montanari. A method for obtaining skeletons using a quasi-Euclidean distance. *J. ACM*, **15**:600-624, 1968.

[739] U. Montanari. A note on minimal length polygonal approximation to a digitized contour. *Comm. ACM*, **13**:41–47, 1970.

[740] U. Montanari. On limit properties of digitization schemes. *J.ACM*, **17**:348–360, 1970.

[741] C. Montani, R. Scanteni, and R. Scopigno. Discretized marching cubes. *Proc. IEEE Visualization*, 281–286, 1994.

[742] B.S. Moon. A representation of digitized patterns and an edge tracking thinning method. *Pattern Recognition*, **34**:2155–2161, 2001.

[743] E.F. Moore. The shortest path through a maze. *Proc. Int. Symp. Switching Theory*, II:285–292. Harvard University Press, Cambridge, Massachusetts, 1959.

[744] G.A. Moore. Automatic scanning and computer processes for the quantitative analysis of micrographs and equivalent subjects. In: G.C. Cheng, R.S. Ledley, D.K. Pollock, and A. Rosenfeld, editors. *Pictorial Pattern Recognition*, 275–326. Thompson, Washington, DC, 1968.

[745] P. Moreau. Transformations et modélisation de dégradés dans le plan discret. PhD Thesis, Université Bordeaux - 1, France, 1995.

[746] F. Morgan. *Geometric Measure Theory*, 3rd edition. Academic Press, San Diego, California, 2000.

[747] D.G. Morgenthaler and A. Rosenfeld. Surfaces in three-dimensional digital images. *Information Control*, **51**:227–247, 1981.

[748] M. Morse and G.A. Hedlund. Symbolic dynamics II: Sturmian sequences. *American J. Mathematics*, **61**:1–42, 1940.

[749] J.C. Mott-Smith. Medial axis transformations. In: B.S. Lipkin and A. Rosenfeld, editors. *Picture Processing and Psychopictorics*, 266–278. Academic Press, New York, 1970.

[750] J. Mukherjee and B.N. Chatterji. Thinning of 3-D images using the Safe Point Thinning Algorithm (SPTA). *Pattern Recognition Letters*, **10**:167–173, 1989.

[751] J. Mukherjee, M. Aswatha Kumar, P.P. Das, and B.N. Chatterji. Use of medial axis transforms for computing normals at boundary points. *Pattern Recognition Letters*, **23**:1649–1656, 2002.

[752] J. Mukherjee, P.P. Das, M. Aswatha Kumar, and B.N. Chatterji. On approximating Euclidean metrics by digital distances in 2D and 3D. *Pattern Recognition Letters*, **21**:573–582, 2000.

[753] J. Mukherjee, P.P. Das, and B.N. Chatterji. On connectivity issues of ESPTA. *Pattern Recognition Letters*, **11**:643–648, 1990.

[754] J.C. Mullikin. The vector distance transform in two and three dimensions. *Graphical Models Image Processing*, **54**:526–535, 1992.

[755] I.S.N. Murthy and K.J. Udupa. A search algorithm for skeletonization of thick patterns. *Computer Graphics Image Processing*, **3**:247–259, 1974.

[756] J.P. Mylopoulos and T. Pavlidis. On the topological properties of quantized spaces. I. The notion of dimension. *J. ACM*, **18**:239–246, 1971.

[757] J.P. Mylopoulos and T. Pavlidis. On the topological properties of quantized spaces. II. Connectivity and order of connectivity. *J. ACM*, **18**:247–254, 1971.

[758] N.J. Naccache and R. Shinghal. SPTA: A proposed algorithm for thinning binary patterns. *IEEE Trans. Systems, Man, Cybernetics*, **14**:409–418, 1984.

[759] N.J. Naccache and R. Shinghal. An investigation into the skeletonization approach of Hilditch. *Pattern Recognition*, **17**:279–284, 1984.

[760] N.J. Naccache and R. Shinghal. In: response to "A comment on an investigation into the skeletonization approach of Hilditch." *Pattern Recognition*, **19**:111, 1986.

[761] P.F.M. Nacken. Chamfer metrics in mathematical morphology. *J. Mathematical Imaging Vision*, **4**:233–253, 1994.

[762] Y. Nakagawa and A. Rosenfeld. A note on the use of local min and max operations in digital picture processing. *IEEE Trans. Systems, Man, Cybernetics*, **8**:899–901, 1978.

[763] A. Nakamura and K. Aizawa. Digital circles. *Computer Vision, Graphics, Image Processing*, **26**:242–255, 1984.

[764] A. Nakamura and K. Aizawa. Digital images of geometric pictures. *Computer Vision, Graphics, Image Processing*, **30**:107–120, 1985.

[765] A. Nakamura and K. Aizawa. Digital squares. *Computer Vision, Graphics, Image Processing*, **49**:357–368, 1990.

[766] A. Nakamura and K. Aizawa. Some results concerning connected fuzzy digital pictures. *Pattern Recognition Letters*, **12**:335–341, 1991.

[767] A. Nakamura and A. Rosenfeld. Digital calculus. *Information Sciences*, **98**:83–98, 1997.

[768] A. Nakamura and A. Rosenfeld. Digital knots. *Pattern Recognition*, **33**:1541–1553, 2000.

[769] D. Nassimi and S. Sahni. Finding connected components and connected ones on a mesh-connected parallel computer. *SIAM J. Computing*, **9**:744–757, 1980.

[770] R. Neumann and G. Teisseron. Extraction of dominant points by estimation of the contour fluctuations. *Pattern Recognition*, **35**:1447–1462, 2002.

[771] M.H.A. Newman. *Elements of the Topology of Plane Sets of Points*. Cambridge University Press, London, United Kingdom, 1939. (2nd edition, 1954.)

[772] J.R. Newman. William Kingdon Clifford. *Scientific American*, **188**(2):78–84, 1953.

[773] W.M. Newman and R.F. Sproull. *Principles of Interactive Computer Graphics*, 2nd edition. McGraw Hill, New York, 1979.

[774] C.W. Niblack, P.B. Gibbons, and D.W. Capson. Generating skeletons and centerlines from the distance transform. *Graphical Models Image Processing*, **54**:420–437, 1992.

[775] G.M. Nielson and B. Hamann. The asymptotic decider: Resolving the ambiguity in marching cubes. *Proc. IEEE Visualization*, 83–91, 1991.

[776] G.M. Nielson, A. Huang, and S. Sylvester. Approximating normals for marching cubes applied to locally supported isosurfaces. *Proc. IEEE Visualization*, 459–466, 2002.

[777] F. Nilsson and P.E. Danielsson. Finding the minimal set of maximum disks for binary objects. *Graphical Models Image Processing*, **59**:55–60, 1997.

[778] L. Noakes and R. Kozera. More-or-less uniform sampling and lengths of curves. *Quarterly J. Applied Mathematics*, to appear.

[779] L. Noakes, R. Kozera, and R. Klette. Length estimation for curves with different samplings. In: G. Bertrand, A. Imiya, and R. Klette, editors. *Digital and Image Geometry: Advanced Lectures*, 334–346. Lecture Notes in Computer Science 2243, Springer, Berlin, Germany, 2001.

[780] D. Nogly and M. Schladt. Digital topology on graphs. *CVGIP: Image Understanding*, **63**:394–396, 1996.

[781] S.P. Novikov and A.T. Fomenko. *Basic Elements of Differential Geometry and Topology* (in Russian). Nauka, Moscow, Russia, 1987.

[782] H. Ogawa. Corner detection on digital curves based on local symmetry of the shape. *Pattern Recognition*, **22**:351–357, 1989.

[783] L. O'Gorman. $k \times k$ thinning. *Computer Vision, Graphics, Image Processing*, **51**:195–215, 1990.

[784] D.A. O'Handley and W.D. Green. Recent developments in digital image processing at the Image Processing Laboratory at the Jet Propulsion Laboratory. *Proc. IEEE*, **60**:821–828, 1972.

[785] N. Okabe, J. Toriwaki, and T. Fukumura. Paths and distance functions on three-dimensional digitized pictures. *Pattern Recognition Letters*, **1**:205–212, 1983.

[786] O. Ore. *Theory of Graphs*. American Mathematical Society, Providence, Rhode Island, 1962.

[787] W. F. Osgood. On the existence of the Green's function for the most general simply connected plane region. *Trans. American Mathematical Society*, **1**:310–314, 1900; **2**:484–485, 1901.

[788] R. Osserman. The isoperimetric inequality. *Bulletin American Mathematical Society*, **84**:1182–1238, 1978.

[789] J. Pach and P.K. Agarwal. *Combinatorial Geometry*. Wiley, New York, 1995.

[790] S.K. Pal. Fuzzy skeletonization of an image. *Pattern Recognition Letters*, **10**:17–23, 1989.

[791] S. Pal and P. Bhattacharya. Analysis of template matching thinning algorithms. *Pattern Recognition*, **25**:497–505, 1992.

[792] S.K. Pal and A. Rosenfeld. A fuzzy medial axis transformation based on fuzzy disks. *Pattern Recognition Letters*, **12**:585–590, 1991.

[793] K. Palagyi. A 3-subiteration thinning algorithm for extracting medial surfaces. *Pattern Recognition Letters*, **23**:663–675, 2002.

[794] K. Palagyi and A. Kuba. A 3D 6-subiteration thinning algorithm for extracting medial lines. *Pattern Recognition Letters*, **19**:613–627, 1998.

[795] K. Palagyi and A. Kuba. A parallel 3D 12-subiteration thinning algorithm. *Graphical Models Image Processing*, **61**:199–221, 1999.

[796] L. Papier. Polyédrisation et visualisation d'objets discrets tridimensionnels. PhD Thesis, Université Louis Pasteur, Strasbourg, Germany, 2000.

[797] L. Papier and J. Francon. Evaluation de la normale au bord d'un object discret 3D. *Revue de CFAO et d'Informatique Graphique*, **13**:205–226, 1998.

[798] A.S. Parchomenko. *Was ist eine Kurve?* VEB Deutscher Verlag der Wissenschaften, Berlin, Germany, 1957.

[799] S.K. Parui. A parallel algorithm for decomposition of binary objects through skeletonization. *Pattern Recognition Letters*, **12**:235–240, 1991.

[800] N. Passat. Caractérisation des Empreintes Anatomiques sur le Foie. MSc Thesis, Université Louis Pasteur, Strasbourg, Germany, 2002.

[801] M. Paterson and F. Yao. Optimal binary space partitions for orthogonal objects. *J. Algorithms*, **13**:99–113, 1992.

[802] T. Pavlidis. *Structural Pattern Recognition*. Springer, New York, 1977.

[803] T. Pavlidis. A thinning algorithm for discrete binary images. *Computer Graphics Image Processing*, **12**:142–157, 1980.

[804] T. Pavlidis. An asynchronous thinning algorithm. *Computer Graphics Image Processing*, **20**:133-157, 1982.

[805] T. Pavlidis. *Algorithms for Graphics and Image Processing*. Computer Science Press, Rockville, Maryland, 1982.

[806] G. Peano. Sur une courbe que remplit toute une aire plane. *Mathematische Annalen*, **36**:157–160, 1890. English translation by H.C. Kennedy. *Selected Works of Giuseppe Peano*, 143–148. George Allen & Unwin, London, United Kingdom, 1973.

[807] D. Pedoe. Notes on the history of geometrical ideas: Homogeneous coordinates. *Mathematics Magazine*, **48**:215–217, 1975.

[808] S. Peleg and A. Rosenfeld. A min-max medial axis transformation. *IEEE Trans. Pattern Analysis Machine Intelligence*, **3**:208–210, 1981.

[809] L. Perroton. A new 26-connected objects surface tracking algorithm and its related PRAM version. *Int. J. Pattern Recognition Artificial Intelligence*, **9**:719–734, 1995.

[810] J.L. Pfaltz and A. Rosenfeld. Computer representation of planar regions by their skeletons. *Comm. ACM*, **10**:119–122, 125, 1967.

[811] S. Pham. Digital straight segments. *Computer Vision, Graphics, Image Processing*, **36**:10–30, 1986.

[812] S. Pham. Parallel, overlapped, and intersected digital straight lines. *The Visual Computer*, **4**:247–258, 1988.

[813] S. Pham. Digital circles with non-lattice point centers. *The Visual Computer*, **9**:1–24, 1992.

[814] O. Philbrick. A study of shape recognition using the medial axis transform. Air Force Cambridge Research Laboratory, Cambridge, Massachusetts, 1966.

[815] T.Y. Phillips and A. Rosenfeld. A method for curve partitioning using arc-chord distance. *Pattern Recognition Letters*, **5**:285–288, 1987.

[816] G. Pick. Geometrisches zur Zahlenlehre. *Sitzungsberichte des deutschen naturw.-med. Vereins für Böhmen "Lotos" in Prag*, **47**:315–323, 1899.

[817] J. Piper. Efficient implementation of skeletonisation using interval coding. *Pattern Recognition Letters*, **3**:389–397, 1985.

[818] J. Piper and E. Granum. Computing distance transformations in convex and non-convex domains. *Pattern Recognition*, **20**:599–615, 1987.

[819] I. Pitas. Performance analysis of morphological Voronoi tessellation algorithms. In: R. Klette, A. Rosenfeld, and F. Sloboda, editors. *Advances in Digital and Computational Geometry*, 227–254. Springer, Singapore, 1998.

[820] H. Poincaré. Analysis situs. *J. Ecole Polytechnique*, **1**:1–121, 1895. Also in: Œuvres, **6**:193–288. Gauthier-Villars, Paris, 1953.

[821] T. Poston, T.-T. Wong, and P.-A. Heng. Multiresolution isosurface extraction with adaptive skeleton climbing. *Computer Graphics Forum*, **17**:137–148, 1998.

[822] V. Poty and S. Ubeda. A parallel thinning algorithm using $K \times K$ masks. *Int. J. Pattern Recognition Artificial Intelligence*, **7**:1183–1202, 1993.

[823] F.P. Preparata and M.I. Shamos. *Computational Geometry: An Introduction*. Springer, New York, 1985. (3rd edition, 1990.)

[824] F. Preteux. Watershed and skeleton by influence zones: A distance-based approach. *J. Mathematical Imaging Vision*, **1**:239–255, 1992.

[825] J.M.S. Prewitt. Object enhancement and extraction. In: B.S. Lipkin and A. Rosenfeld, editors. *Picture Processing and Psychopictorics*, 75–14. Academic Press, New York, 1970.

[826] D. Proffitt and D. Rosen. Metrication errors and coding efficiency of chain-encoding schemes for the representation of lines and edges. *Computer Graphics Image Processing*, **10**:318–332, 1979.

[827] C. Pudney. Distance-ordered homotopic thinning: A skeletonization algorithm for 3D digital images. *Computer Vision Image Understanding*, **72**:404–413, 1998.

[828] X. Qi and X. Li. A 3D surface tracking algorithm. *Computer Vision Image Understanding*, **64**:147–156, 1996.

[829] K. Qian, S. Cao, and P. Bhattacharya. Gray image skeletonization with hollow preprocessing using distance transformation. *Int. J. Pattern Recognition Artificial Intelligence*, **13**:881–892, 1999.

[830] H. Rademacher and O. Toeplitz. *Proben Mathematischen Denkens für Liebhaber der Mathematik*. 2nd edition, Julius Springer, Berlin, Germany, 1933. English translation by H. Zuckerman. *The Enjoyment of Mathematics*. Princeton University Press, Princeton, New Jersey, 1957.

[831] I. Ragnemalm. Generation of Euclidean distance maps. PhD Thesis, Linköping University, Sweden, 1990.

[832] I. Ragnemalm. Contour processing distance transforms. In: V. Cantoni, L. Cordella, S. Levialdi, and G. Sanniti di Baja, editors. *Progress in Image Analysis and Processing*, 204–212. World Scientific, Singapore, 1990.

[833] I. Ragnemalm. Neighborhoods for distance transformations using ordered propagation. *CVGIP: Image Understanding*, **56**:399–409, 1992.

[834] I. Ragnemalm. The Euclidean distance transformation in arbitrary dimensions. *Pattern Recognition Letters*, **14**:883–888, 1993.

[835] B. Randell. *The Origins of Digital Computers—Selected Papers*, 3rd edition, 5-6. Springer, Berlin, Germany, 1982.

[836] V. Ranwez and P. Soille. Order independent homotopic thinning for binary and grey tone anchored skeletons. *Pattern Recognition Letters*, **23**:687–702, 2002.

[837] B.K. Ray and K.S. Ray. An algorithm for detecting dominant points and polygonal approximation of digitized curves. *Pattern Recognition Letters*, **13**:849–856, 1992.

[838] P. Reche, C. Urdiales, A. Bandera, C. Trazegnies, and F. Sandoval. Corner detection by means of contour local vectors. *Electronics Letters*, **38**:699–701, 2002.

[839] D.H. Redelmeier. Counting polyominoes: Yet another attack. *Discrete Mathematics*, **36**:191–203, 1981.

[840] G.M. Reed. On the characterization of simple closed surfaces in three-dimensional digital images. *Computer Vision, Graphics, Image Processing*, **25**:226–235, 1984.

[841] G.M. Reed and A. Rosenfeld. Recognition of surfaces in three-dimensional images. *Information Control*, **53**:108–120, 1982.

[842] J.E. Reeve. On the volume of lattice polyhedra. *Proc. London Mathematical Society*, **7**:378–395, 1957.

[843] G.B. Reggiori. Digital computer transformations for irregular line drawings. Technical Report 403-22, New York University, New York, 1972.

[844] K. Reidemeister. *Topologie der Polyeder und kombinatorische Topologie der Komplexe*. Akad. Verlagsgesellschaft Geest & Portig, Leipzig, Germany, 1938. (2nd edition, 1953.)

[845] E. Remy and E. Thiel. Computing 3D medial axis for chamfer distances. In: G. Borgefors, I. Nyström, and G. Sanniti di Baja, editors. *Discrete Geometry for Computer Imagery*, 418–430. Lecture Notes in Computer Science 1953, Springer, Berlin, Germany, 2000.

[846] E. Remy and E. Thiel. Medial axis for chamfer distances: Computing look-up tables and neighbourhoods in 2D or 3D. *Pattern Recognition Letters*, **23**:649–661, 2002.

[847] A. Rényi and R. Sulanke. Über die konvexe Hülle von n zufällig gewählten Punkten, Parts I and II. *Zeitschrift für Wahrscheinlichkeitstheorie*, **2**:75–84, 1963, and **3**:138–147, 1964.

[848] J.-P. Reveillès. Géométrie discrète, calcul en nombres entiers et algorithmique. Thèse d'état, Université Louis Pasteur, Strasbourg, Germany, 1991.

[849] J.-P. Reveillès. Combinatorial pieces in digital lines and planes. SPIE *Vision Geometry IV*, **2573**:23–34, 1995.

[850] F. Rhodes. Some characterizations of the chessboard metric and the city block metric. *Pattern Recognition Letters*, **11**:669–675, 1990.

[851] F. Rhodes. Discrete Euclidean metrics. *Pattern Recognition Letters*, **13**:623–628, 1992.

[852] F. Rhodes. On the metrics of Chaudhuri, Murthy and Chaudhuri. *Pattern Recognition*, **28**:745–752, 1995.

[853] S. Riazanoff, B. Cervelle, and J. Chorowicz. Parametrisable skeletonization of binary and multi-level images. *Pattern Recognition Letters*, **11**:25–33, 1990.

[854] M. Richter. Some concepts of convexity in gridpoint spaces. In: W. Nagel, editor. *Proc. Geometric Problems of Image Processing*, 87–91, Jena University, Germany, 1985.

[855] F. Riesz and B. Sz.-Nagy. *Leçons d'Analyse Fonctionelle,* 3rd edition. Gauthier-Villars, Paris, France, and Akadémiai Kiado, Budapest, Hungary, 1955.

[856] G. Ringel and J.W.T. Youngs. Solution of the Heawood map-coloring problem. *Proc. Nat. Acad. Sci. U.S.A.*, **60**:438–445, 1968.

[857] W. Rinow. *Lehrbuch der Topologie*. Deutscher Verlag der Wissenschaften, Berlin, Germany, 1975.

[858] B.D. Ripley. *Spatial Statistics*. Wiley, New York, 1981.

[859] S. Rital, A. Bretto, D. Aboutajdine, and H. Cherifi. Application of adaptive hypergraph model to impulsive noise detection. *Proc. Computer Analysis of Images and Patterns*, 555-562. Lecture Notes in Computer Science 2124, Springer, Berlin, Germany, 2001.

[860] G.X. Ritter and J.N. Wilson. *Handbook of Computer Vision Algorithms in Image Algebra*, 2nd edition. CRC Press, Boca Raton, Florida, 2001.

[861] J.C. Roberts. An overview of rendering from volume data including surface and volume rendering. Technical Report, University of Kent at Canterbury, United Kingdom, 1993.

[862] N. Robertson, D.P. Sanders, P.D. Seymour, and R. Thomas. A new proof of the four colour theorem. *Electronic Research Announcements American Mathematical Society*, **2**:17–25, 1996.

[863] J.J. Robinson. Line symmetry of convex digital regions. *Computer Vision Image Understanding*, **64**:263–285, 1996.

[864] C. Ronse. A simple proof of Rosenfeld's characterization of digital straight line segments. *Pattern Recognition Letters*, **3**:323–326, 1985.

[865] C. Ronse. Definitions of convexity and convex hulls in digital images. *Bulletin de la Société Mathématique de Belgique*, **37**:71–85, 1985.

[866] C. Ronse. Criteria for approximation of linear and affine functions. *Archiv der Mathematik*, **46**:371–384, 1986.

[867] C. Ronse. A strong chord property for 4-connected convex digital sets. *Computer Vision, Graphics, Image Processing*, **35**:259–269, 1986.

[868] C. Ronse. A topological characterization of thinning. *Theoretical Computer Science*, **43**:31–41, 1986.

[869] C. Ronse. Minimal test patterns for connectivity preservation in parallel thinning algorithms for binary digital images. *Discrete Applied Mathematics*, **21**:67–79, 1988.

[870] C. Ronse. A bibliography on digital and computational convexity (1961–1988). *IEEE Trans. Pattern Analysis Machine Intelligence*, **11**:181–190, 1989.

[871] C. Ronse. A note on the approximation of linear and affine functions: The case of bounded slope. *Archiv der Mathematik*, **54**:601–609, 1990.

[872] C. Ronse. Set-theoretical algebraic approaches to connectivity in continuous or digital spaces. *J. Mathematical Imaging Vision*, **8**:41–58, 1998.

[873] C. Ronse and P.A. Devijver. *Connected Components in Binary Images: The Detection Problem*. Wiley, New York, 1984.

[874] C. Ronse, H.J.A.M. Heijmans. The algebraic basis of mathematical morphology II: Openings and closings. *CVGIP: Image Understanding*, **54**:74–97, 1991.

[875] C. Ronse and M. Tajine. Discretization in Hausdorff space. *J. Mathematical Imaging Vision*, **12**:219–242, 2000.

[876] C. Ronse and M. Tajine. Hausdorff discretization for cellular distances, and its relation to cover and supercover discretizations. *J. Visual Communication Image Representation*, **12**:169–200, 2001.

[877] D. Rosen. On the area and boundaries of quantized objects. *Computer Graphics Image Processing*, **13**:94–98, 1980.

[878] D. Rosen. Errors in digitized area measurements: Circles and rectangles. *Pattern Recognition Letters*, **13**:613–621, 1992.

[879] B. Rosenberg. Computing dominant points on simple shapes. *Int. J. Man-Machine Studies*, **6**:1–12, 1974.

[880] A. Rosenfeld. *Picture Processing by Computer*. Academic Press, New York, 1969.

[881] A. Rosenfeld. Connectivity in digital pictures. *J. ACM*, **17**:146–160, 1970.

[882] A. Rosenfeld. Arcs and curves in digital pictures. *J. ACM*, **20**:81–87, 1973.

[883] A. Rosenfeld. Digital straight line segments. *IEEE Trans. Computers*, **23**:1264–1269, 1974.

[884] A. Rosenfeld. Compact figures in digital pictures. *IEEE Trans. Systems, Man, Cybernetics*, **4**:221–223, 1974.

[885] A. Rosenfeld. A note on perimeter and diameter in digital pictures. *Information Control*, **24**:384–388, 1974.

[886] A. Rosenfeld. Adjacency in digital pictures. *Information Control*, **26**:24–33, 1974.

[887] A. Rosenfeld. A characterization of parallel thinning algorithms. *Information Control*, **29**:286–291, 1975.

[888] A. Rosenfeld. A converse to the Jordan curve theorem for digital curves. *Information Control*, **29**:292–293, 1975.

[889] A. Rosenfeld. Geodesics in digital pictures. *Information Control*, **36**:74–84, 1978.

[890] A. Rosenfeld. Clusters in digital pictures. *Information Control*, **39**:19–33, 1978.

[891] A. Rosenfeld. *Picture Languages*, Chapter 2 on digital geometry. Academic Press, New York, 1979.

[892] A. Rosenfeld. Fuzzy digital topology. *Information Control*, **40**:76–87, 1979.

[893] A. Rosenfeld. Digital topology. *American Mathematical Monthly*, **86**:621–630, 1979.

[894] A. Rosenfeld. Three-dimensional digital topology. *Information Control*, **50**:119–127, 1981.

[895] A. Rosenfeld. On connectivity properties of grayscale pictures. *Pattern Recognition*, **16**:47–50, 1983.

[896] A. Rosenfeld. Some notes on digital triangles. *Pattern Recognition Letters*, **1**:147–150, 1983.

[897] A. Rosenfeld. A note on "geometric transforms" of digital sets. *Pattern Recognition Letters*, **1**:223–225, 1983.

[898] A. Rosenfeld. Parallel image processing using cellular arrays. *IEEE Computer*, **16**(1):62–67, 1983.

[899] A. Rosenfeld, editor. *Multiresolution Image Processing and Analysis*. Springer, Berlin, Germany, 1984.

[900] A. Rosenfeld. The fuzzy geometry of image subsets. *Pattern Recognition Letters*, **4**:311–317, 1984.

[901] A. Rosenfeld. Distances between fuzzy sets. *Pattern Recognition Letters*, **3**:229–233, 1985.

[902] A. Rosenfeld. Measuring the sizes of concavities. *Pattern Recognition Letters*, **3**:71–75, 1985.

[903] A. Rosenfeld. "Continuous" functions on digital pictures. *Pattern Recognition Letters*, **4**:177–184, 1986.

[904] A. Rosenfeld. A note on average distances in digital sets. *Pattern Recognition Letters*, **5**:281–283, 1987.

[905] A. Rosenfeld. Fuzzy rectangles. *Pattern Recognition Letters*, **11**:677-679, 1990.

[906] A. Rosenfeld. Fuzzy plane geometry: Triangles. *Pattern Recognition Letters*, **15**:1261–1264, 1994.

[907] A. Rosenfeld. Digital geometry—Introduction and bibliography. In: R. Klette, A. Rosenfeld, and F. Sloboda, editors. *Advances in Digital and Computational Geometry*, 1–85. Springer, Singapore, 1998.

[908] A. Rosenfeld and L.S. Davis. A note on thinning. *IEEE Trans. Systems, Man, Cybernetics*, **6**:226–228, 1976.

[909] A. Rosenfeld and S. Haber. The perimeter of a fuzzy set. *Pattern Recognition*, **18**:125–130, 1985.

[910] A. Rosenfeld and E. Johnston. Angle detection on digital curves. *IEEE Trans. Computers*, **22**:875–878, 1973.

[911] A. Rosenfeld and A.C. Kak. *Digital Picture Processing*. Academic Press, New York, 1976. (2nd edition, two volumes, 1982.)

[912] A. Rosenfeld and C.E. Kim. How a digital computer can tell whether a line is straight. *American Mathematical Monthly*, **89**:230–235, 1982.

[913] A. Rosenfeld and R. Klette. Degree of adjacency or surroundedness. *Pattern Recognition*, **18**:169–177, 1985.

[914] A. Rosenfeld and R. Klette. Digital straightness. *Electronic Notes in Theoretical Computer Science* (Elsevier), **46**, 2001.

[915] A. Rosenfeld and T.Y. Kong. Connectedness of a set, its complement, and their common boundary. *Contemporary Mathematics*, **119**:125–128, 1991.

[916] A. Rosenfeld, T.Y. Kong, and A. Nakamura. Topology-preserving deformations of two-valued digital pictures. *Graphical Models Image Processing*, **60**:24–34, 1998.

[917] A. Rosenfeld, T.Y. Kong, and A.Y. Wu. Digital surfaces. *CVGIP: Graphical Models Image Processing*, **53**:305–312, 1991.

[918] A. Rosenfeld and R.A. Melter. Digital geometry. Center for Automation Research, TR-323, University of Maryland, College Park, Maryland, 1987.

[919] A. Rosenfeld and A. Nakamura. Local deformations of digital curves. *Pattern Recognition Letters*, **18**:613–620, 1997.

[920] A. Rosenfeld and A. Nakamura. Two simply connected sets that have the same area are IP-equivalent. *Pattern Recognition*, **35**:537–541, 2002.

[921] A. Rosenfeld and J.L. Pfaltz. Sequential operations in digital picture processing. *J. ACM*, **13**:471–494, 1966.

[922] A. Rosenfeld and J.L. Pfaltz. Distance functions on digital pictures. *Pattern Recognition*, **1**:33–61, 1968.

[923] A. Rosenfeld, P.K. Saha, and A. Nakamura. Interchangeable pairs of pixels in two-valued digital images. *Pattern Recognition*, **34**:1833–1865, 2001.

[924] A. Rosenfeld and J.S. Weszka. An improved method of angle detection on digital curves. *IEEE Trans. Computers*, **24**:940–941, 1975.

[925] A. Rosenfeld and A.Y. Wu. "Digital geometry" on graphs. *Contemporary Mathematics*, **119**:129–136, 1991.

[926] A. Rosenfeld and A.Y. Wu. Geodesic convexity in discrete spaces. *Information Sciences*, **80**:127–132, 1994.

[927] B.A. Rosenfeld and I.M. Jaglom. Mehrdimensionale Räume. In: P.S. Alexandroff, A.I. Markuschewitsch, and A.J. Chintschin, editors. *Enzyklopädie der Elementarmathematik*, Vol. 5, 369–371. Deutscher Verlag der Wissenschaften, Berlin, Germany, 1971.

[928] P.L. Rosin. Techniques for assessing polygonal approximations of curves. *IEEE Trans. Pattern Analysis Machine Intelligence*, **19**:659–666, 1997.

[929] P.L. Rosin and G.A.W. West. Salience distance transforms. *Graphical Models Image Processing*, **57**:483–521, 1995.

[930] U. Rösler and G. Schwarze. Maskenverfahren zur Manipulation von binären Rasterbildern linienförmiger Objekte. *Elektronische Informationsverarbeitung Kybernetik*, **16**:171–184, 1980.

[931] J. Rothstein and C. Weiman. Parallel and sequential specification of a context sensitive language for straight line grids. *Computer Graphics Image Processing*, **5**:106–124, 1976.

[932] W.S. Rutkowski and A. Rosenfeld. A comparison of corner-detection techniques for chain-coded curves. Computer Science Dep., TR-623, University of Maryland, College Park, Maryland, 1978.

[933] D. Rutovitz. Data structures for operations on digital images. In: G.C. Cheng, R.S. Ledley, D.K. Pollock, and A. Rosenfeld, editors. *Pictorial Pattern Recognition*, 105–133. Thompson, Washington, DC, 1968.

[934] H.J. Ryser. Combinatorial properties of matrices of zeros and ones. *Canadian J. Mathematics*, **9**:371–377, 1957.

[935] H. Sagan. *Space-Filling Curves*. Springer, Berlin, Germany, 1994.

[936] P.K. Saha, B. Chanda, and D. Dutta Majumder. A single scan boundary removal thinning algorithm for 2-D binary object[s]. *Pattern Recognition Letters*, **14**:173–179, 1993.

[937] P.K. Saha and B.B. Chaudhuri. Detection of 3-D simple points for topology preserving transformations with application to thinning. *IEEE Trans. Systems, Man, Cybernetics*, **16**:1028–1032, 1994.

[938] P.K. Saha and B.B. Chaudhuri. A new approach to computing the Euler characteristic. *Pattern Recognition*, **28**:1955–1963, 1995.

[939] P.K. Saha and B.B. Chaudhuri. 3D digital topology under binary transformation with applications. *Computer Vision Image Understanding*, **63**:418–429, 1996.

[940] P.K. Saha, B.B. Chaudhuri, B. Chanda, and D. Dutta Majumder. Topology preservation in 3D digital space. *Pattern Recognition*, **27**:295–300, 1994.

[941] P.K. Saha, B.B. Chaudhuri, and D. Dutta Majumder. A new shape preserving parallel thinning algorithm for 3D digital images. *Pattern Recognition*, **30**:1939–1955, 1997.

[942] P.K. Saha and A. Rosenfeld. The digital topology of sets of convex voxels. *Graphical Models*, **62**:343–352, 2000.

[943] P.K. Saha and A. Rosenfeld. Determining simplicity and computing topological change in strongly normal partial tilings of $\mathbb{R}^2$ or $\mathbb{R}^3$. *Pattern Recognition*, **33**:105–118, 2000.

[944] P.K. Saha and A. Rosenfeld. Local and global topology preservation in locally finite sets of tiles. *Information Sciences*, **137**:303–311, 2001.

[945] P.K. Saha, F.W. Wehrli, and B.R. Gomberg. Fuzzy distance transform: Theory, algorithms, and applications. *Computer Vision Image Understanding*, **86**:171–190, 2002.

[946] T. Saito and J. Toriwaki. New algorithms for Euclidean distance transformation of an n-dimensional digitized picture with applications. *Pattern Recognition*, **27**:1551–1565, 1994.

[947] T. Saito and J. Toriwaki. A sequential thinning algorithm for three dimensional digital pictures using the Euclidean distance transformation. *Proc. Scandinavian Conf. Image Analysis*, 507–516, 1995.

[948] E. Salari and P. Siy. The ridge-seeking method for obtaining the skeleton[s] of digital images. *IEEE Trans. Systems, Man, Cybernetics*, **14**:524–528, 1984.

[949] H. Samet. *Applications of Spatial Data Structures*. Addison-Wesley, Reading, Massachusetts, 1993.

[950] F.J. Sanchez-Marin. The curvature function evolved in scale space as a representation of biological shapes. *Computers Biology Medicine*, **27**:77–85, 1997.

[951] P. V. Sankar. Grid intersect quantization schemes for solid object digitization. *Computer Graphics Image Processing*, **8**:25–42, 1978.

[952] P.V. Sankar and E.V. Krishnamurthy. On the compactness of subsets of digital pictures. *Computer Graphics Image Processing*, **8**:136–143, 1978.

[953] G. Sanniti di Baja. Well-shaped, stable, and reversible skeletons from the (3,4)-distance transform. *J. Visual Communication Image Representation*, **5**:107–115, 1994.

[954] G. Sanniti di Baja and S. Svensson. A new shape descriptor for surfaces in 3D images. *Pattern Recognition Letters*, **23**:703–711, 2002.

[955] G. Sanniti di Baja and E. Thiel. (3,4)-weighted skeleton decomposition for pattern representation and description. *Pattern Recognition*, **27**:1039–1049, 1994.

[956] G. Sanniti di Baja and E. Thiel. Skeletonization algorithm running on path-based distance maps. *Image Vision Computing*, **14**:47–57, 1996.

[957] L.A. Santaló. Complemento a la nota: Un teorema sôbre conjuntos de paralelipipedos de aristas paralelas. *Publ. Inst. Mat. Univ. Nac. Litoral*, **2**:49–60, 1940.

[958] L.A. Santaló. *Integral Geometry and Geometric Probability*. Addison-Wesley, Reading, Massachusetts, 1976.

[959] G. Schaeffer. *Conjugaison d'arbres et cartes combinatoires aleatoires*. PhD Thesis, L'Université Bordeaux - 1, France 1998.

[960] W. Scherrer. Ein Satz über Gitter und Volumen. *Mathematische Annalen*, **86**:99–107, 1922.

[961] W. Scherrer. Die Einlagerung eines regulären Vielecks in ein Gitter. *Elemente der Mathematik*, **1**:97–98, 1946.

[962] L. Schläfli. Theorie der Vielfachen Kontinuität. *Neue Denkschriften der allgemeinen schweizerischen Gesellschaft für die gesamten Naturwissenschaften*, **38**:1–237, 1901.

[963] W.M. Schmidt. Volume, surface area, and the number of integer points covered by a convex set. *Archiv der Mathematik*, **23**:537-543, 1972.

[964] R. Schneider and W. Weil. *Integralgeometrie*, 226. Teubner, Stuttgart, Germany, 1992.

[965] U. Schnell. Lattice inequalities for convex bodies and arbitrary lattices. *Monatshefte der Mathematik*, **116**:331–337, 1993.

[966] A. Schönflies. *Kristallsysteme und Kristallstruktur*. Teubner, Leipzig, Germany, 1891.

[967] W.J. Schroeder, J.A. Zarge, and W.E. Lorensen. Decimation of triangle meshes. *Computer Graphics*, **26**:65–70, 1992.

[968] H.A. Schwarz. Sur une définition erronée de l'aire d'une surface courbe. *Gesammelte mathematische Abhandlungen*, **2**:309–311, 1890.

[969] J. Serra. *Image Analysis and Mathematical Morphology*. Academic Press, London, United Kingdom, 1982.

[970] J. Serra. Mathematical morphology for Boolean lattices. In: J. Serra, editor. *Image Analysis and Mathematical Morphology, II: Theoretical Advances*, 37–58. Academic Press, London, United Kingdom, 1988.

[971] S.A. Shafer. *Shadows and Silhouettes in Computer Vision*. Kluwer, Boston, Massachusetts, 1985.

[972] D. Shaked and A.M. Bruckstein. Pruning medial axes. *Computer Vision Image Understanding*, **69**:156–169, 1998.

[973] D. Shaked, J. Koplowitz, and A.M. Bruckstein. Star-shapedness of digitized planar shapes. *Contemporary Mathematics*, **119**:137–158, 1991.

[974] M.I. Shamos. *Computational geometry*. PhD Thesis, Yale University, New Haven, Connecticut, 1978.

[975] B. Shapiro, J. Pisa, and J. Sklansky. Skeleton generation from x,y boundary sequences. *Computer Graphics Image Processing*, **15**:136–153, 1981.

[976] L.G. Shapiro. Connected component labeling and adjacency graph construction. In: T.Y. Kong and A. Rosenfeld, editors. *Topological Algorithms for Digital Image Processing*, 1–30. Elsevier, Amsterdam, The Netherlands, 1996.

[977] Y.M. Sharaiha and N. Christofides. A graph-theoretic approach to distance transformations. *Pattern Recognition Letters*, **15**:1035–1041, 1994.

[978] Y.M. Sharaiha and P. Garat. A compact chord property for digital arcs. *Pattern Recognition*, **26**:799–803, 1993.

[979] Y.M. Sharaiha and P. Garat. Digital straightness and the skeleton property. *Pattern Recognition Letters*, **16**:417–423, 1995.

[980] J. Shen, W. Shen, and D. Shen. On geometric and orthogonal moments. *Int. J. Pattern Recognition Artificial Intelligence*, **14**:875–894, 2000.

[981] H. Shi. Image algebra techniques for binary image component labeling with local operators. *J. Mathematical Imaging Vision*, **5**:159–170, 1995.

[982] F.Y. Shih and J.J. Liu. Size-invariant four-scan Euclidean distance transformation. *Pattern Recognition*, **31**:1761–1766, 1998.

[983] F.Y.C. Shih and O.R. Mitchell. A mathematical morphology approach to Euclidean distance transformation. *IEEE Trans. Image Processing*, **1**:197–204, 1992.

[984] F.Y. Shih and C.C. Pu. A skeletonization algorithm by maxima tracking on [the] Euclidean distance transform. *Pattern Recognition*, **28**:331–341, 1995.

[985] F.Y. Shih and W.T. Wong. Fully parallel thinning with tolerance to boundary noise. *Pattern Recognition*, **27**:1677–1695, 1994.

[986] F.Y. Shih and W.T. Wong. A new safe-point thinning algorithm based on (the) mid-crack code tracing. *IEEE Trans. Systems, Man, Cybernetics*, **25**:370–378, 1995.

[987] F.Y. Shih and H. Wu. Optimization on Euclidean distance transformation using grayscale morphology. *J. Visual Communication Image Representation*, **3**:104–114, 1992.

[988] S. Shlien. Segmentation of digital curves using linguistic techniques. *Computer Vision, Graphics, Image Processing*, **22**:277–286, 1983.

[989] R. Shonkwiler. Computing the Hausdorff set distance in linear time for any L_p point distance. *Information Processing Letters*, **38**:201–207, 1991.

[990] R. Shoucri, R. Benesch, and S. Thomas. Note on the determination of a digital straight line from chain codes. *Computer Vision, Graphics, Image Processing*, **29**:133–139, 1985.

[991] W. Sierpinski. *Elementary Theory of Numbers*. Państwowe Wydawnictwo Naukowe, Warszawa, Poland, 1964.

[992] H. Simon, K. Kunze, K. Voss, and W.R. Herrmann. *Automatische Bildverarbeitung in Medizin und Biologie*. Steinkopff, Dresden, Germany, 1975.

[993] Y.G. Sinai. Probabilistic approach to analyze the statistics of convex polygonal curves (in Russian). *Funksional Anal. Appl.*, **28**:41–48, 1994.

[994] R.M.K. Sinha. Comments on "Fast thinning algorithm for binary images." *Image Vision Computing*, **4**:57–58, 1986.

[995] A. Sirjani and G.R. Cross. On representation of a shape's skeleton. *Pattern Recognition Letters*, **12**:149–154, 1991.

[996] W. Skarbek. Generalized Hilbert scan in image printing. In: R. Klette and W.G. Kropatsch, editors. *Theoretical Foundations of Computer Vision*, 47–58. Akademie Verlag, Berlin, Germany, 1992.

[997] S. Skiena. *Implementing Discrete Mathematics*. Addison-Wesley, Redwood City, California, 1990.

[998] J. Sklansky. Recognition of convex blobs. *Pattern Recognition*, **2**:3–10, 1970.

[999] J. Sklansky. Measuring concavity on a rectangular mosaic. *IEEE Trans. Computers*, **21**:1355–1364, 1972.

[1000] J. Sklansky, R.L. Chazin, and B.J. Hansen. Minimum-perimeter polygons of digitized silhouettes. *IEEE Trans. Computers*, **21**:260–268, 1972.

[1001] J. Sklansky and D.F. Kibler. A theory of nonuniformly digitized binary pictures. *IEEE Trans. Systems, Man, Cybernetics*, **6**:637–647, 1976.

[1002] F. Sloboda and B. Zaťko. On one-dimensional grid continua in $\mathbb{R}^2$. Technical report, Institute of Control Theory and Robotics, Slovak Academy of Sciences, Bratislava, Slovakia, 1996.

[1003] F. Sloboda and B. Zaťko. On polyhedral form for surface representation. Technical report, Institute of Control Theory and Robotics, Slovak Academy of Sciences, Bratislava, Slovakia, 2000.

[1004] F. Sloboda and B. Zaťko. On approximation and representation of surfaces in implicit form. Unpublished manuscript, Slovak Academy of Sciences, Bratislava, Slovakia, 2002.

[1005] F. Sloboda, B. Zaťko, and R. Klette. On the topology of grid continua. SPIE *Vision Geometry VII*, **3454**:52–63, 1998.

[1006] F. Sloboda, B. Zaťko, and J. Stoer. On approximation of planar one-dimensional continua. In: R. Klette, A. Rosenfeld, and F. Sloboda, editors. *Advances in Digital and Computational Geometry*, 113–160. Springer, Singapore, 1998.

[1007] A.W.M. Smeulders and L. Dorst. Decomposition of discrete curves into piecewise segments in linear time. *Contemporary Mathematics*, **119**:169–195, 1991.

[1008] J.T. Smith. *Methods of Geometry*. Wiley, New York, 2000.

[1009] M.B. Smyth. Region-based discrete geometry. *J. Universal Computer Science*, **6**:447–459, 2000.

[1010] P. Soille. Spatial distribution from contour lines: An efficient methodology based on distance transformations. *J. Visual Communication Image Representation*, **2**:138–150, 1991.

[1011] P. Soille. Generalized geodesy via geodesic time. *Pattern Recognition Letters*, **15**:1235–1240, 1994.

[1012] P. Soille. *Morphological Image Analysis: Principles and Applications*, 2nd edition. Springer, Heidelberg, Germany, 2003.

[1013] J.H. Sossa. An improved parallel algorithm for thinning digital patterns. *Pattern Recognition Letters*, **10**:77–80, 1989.

[1014] M. Spivak. *A Comprehensive Introduction to Differential Geometry*. Publish or Perish, Boston, Massachusetts, 1975 (three volumes).

[1015] R.C. Staunton. An analysis of hexagonal thinning algorithms and skeletal shape representation. *Pattern Recognition*, **29**:1131–1146, 1996.

[1016] L.A. Steen and J.A. Seebach. *Counterexamples in Topology*, 2nd edition. Springer, New York, 1978. Reprint: Dover, New York, 1995.

[1017] R. Stefanelli. A comment on an investigation into the skeletonization approach of Hilditch. *Pattern Recognition*, **19**:13–14, 1986.

[1018] R. Stefanelli and A. Rosenfeld. Some parallel thinning algorithms for digital pictures. *J. ACM*, **18**:255–264, 1971.

[1019] J. Steiner. Einfache Beweise der isoperimetrischen Hauptsätze. *J. Reine Angewandte Mathematik*, **18**:289–296, 1838.

[1020] J. Steiner. Über parallele Flächen. *Monatsbericht preussischen Akademie Wissenschaften*, 114–118, 1840. Also in: *Gesammelte Werke*, Vol. 2, 173–176, 1882.

[1021] J. Steiner. Sur le maximum et le minimum des figures dans le plan, sur la sphére et dans l'espace en géneral. *J. Reine Angewandte Mathematik*, **24**:93–152, 1842.

[1022] H. Steinhaus. Praxis der Rektifikation und zum Längenbegriff. *Berichte Sächsischen Akad. Wiss. Leipzig*, **82**:120–130, 1930.

[1023] H. Steinhaus. Sur un théoréme de M.V. Jarník. *Colloq. Math.*, **1**:1–5, 1947.

[1024] H. Steinhaus. *Mathematical Snapshots*. Oxford University Press, New York, 1960.

[1025] E. Steinitz. Beiträge zur Analysis. *Sitzungsberichte der Berliner Mathematischen Gesellschaft*, **7**:29–49, 1908.

[1026] E. Steinitz. *Vorlesungen über die Theorie der Polyeder*. Julius Springer, Berlin, Germany, 1934.

[1027] A. Stewart. A one-pass thinning algorithm with interference guards. *Pattern Recognition Letters*, **15**:825–832, 1994.

[1028] I. Stojmenovic and R. Tosic. Digitization schemes and the recognition of digital straight lines, hyperplanes, and flats in arbitrary dimensions. *Contemporary Mathematics*, **119**:197–212, 1991.

[1029] L.N. Stout. Two discrete forms of the Jordan Curve Theorem. *American Mathematical Monthly*, **95**:332–336, 1988.

[1030] S. Straszewicz. Über eine Verallgemeinerung des Jordanschen Kurvensatzes. *Fundamenta Mathematica*, **4**:128–135, 1923.

[1031] D.J. Struik. *A Source Book in Mathematics, 1200-1800*, 209–219. Harvard University Press, Cambridge, Massachusetts, 1969.

[1032] Y. Suenaga, T. Kamae, and T. Kobayashi. A high-speed algorithm for the generation of straight lines and circular arcs. *IEEE Trans. Computers*, **28**:729–736, 1979.

[1033] I.E. Sutherland. Sketchpad: A man-machine graphical communication system. *Proc. AFIP Spring Joint Computer Conf.*, 329–346, 1963.

[1034] S. Suzuki and K. Abe. Topological structural analysis of digitized binary images by border following. *Computer Vision, Graphics, Image Processing*, **30**:32–46, 1985.

[1035] S. Suzuki and K. Abe. Binary picture thinning by an iterative parallel two-subcycle operation. *Pattern Recognition*, **20**:297–307, 1987.

[1036] S. Suzuki, N. Veda, and J. Sklansky. Graph-based thinning for binary images. *Int. J. Pattern Recognition Artificial Intelligence*, **7**:1009–1030, 1993.

[1037] S. Svensson and G. Borgefors. Digital distance transforms in 3D images using information from neighborhoods up to $5 \times 5 \times 5$. *Computer Vision Image Understanding*, **88**:24–53, 2002.

[1038] S. Svensson, G. Borgefors, and I. Nyström. On reversible skeletonization using anchor-points from distance transforms. *J. Visual Communication Image Representation*, **10**:379–397, 1999.

[1039] S. Svensson, I. Nyström, and G. Sanniti di Baja. Curve skeletonization of surface-like objects in 3D images guided by voxel classification. *Pattern Recognition Letters*, **23**:1419-1426, 2002.

[1040] S. Svensson and G. Sanniti di Baja. Using distance transforms to decompose 3D discrete objects. *Image Vision Computing*, **20**:529-540, 2002.

[1041] P.G. Tait. Johann Benedict Listing. *Nature*, **27**:316–317, 1883.

[1042] M. Tajine and A. Daurat. On local definitions of length of digital curves. In: G. Sanniti di Baja, S. Svensson, and I. Nyström, editors. *Discrete Geometry for Computer Imagery*, 114–123, Lecture Notes in Computer Science 2886, Springer, Berlin, Germany.

[1043] M. Tajine and C. Ronse. Topological properties of Hausdorff discretization, and comparison to other discretization schemes. *Theoretical Computer Science*, **283**:243–268, 2002.

[1044] M. Tajine, D. Wagner, and C. Ronse. Hausdorff discretization and its comparison to other discretization schemes. In: G. Bertrand, M. Couprie, and L. Perroton, editors. *Discrete Geometry for Computer Imagery*, 399–410. Lecture Notes in Computer Science 1568, Springer, Berlin, Germany, 1999.

[1045] S. Tanimoto and T. Pavlidis. A hierarchical data structure for picture processing. *Computer Graphics Image Processing*, **4**:104–119, 1975.

[1046] Z.S.G. Tari, J. Shah, and H. Pien. Extraction of shape skeletons from grayscale images. *Computer Vision Image Understanding*, **66**:133–146, 1997.

[1047] G.J. Tee. Up with determinants! *IMAGE*, **30**:5–9, 2003.

[1048] P. Tellier and I. Debled-Rennesson. 3D discrete normal vectors. In: G. Bertrand, M. Couprie, and L. Perroton, editors. *Discrete Geometry for Computer Imagery*, 447–458. Lecture Notes in Computer Science 1568, Springer, Berlin, Germany, 1999.

[1049] P. Thanisch, B.V. McNally, and A. Robin. Linear time algorithm for finding a picture's connected components. *Image Vision Computing*, **2**:191–197, 1984.

[1050] A.H. Thiessen and J.C. Alter. Climatological data for July, 1911: District No. 10, Great Basin. *Monthly Weather Review*, 1082–1089, 1911.

[1051] S. Thompson and A. Rosenfeld. Discrete, nonlinear curvature-dependent contour evolution. *Pattern Recognition*, **31**:1949–1959, 1998.

[1052] T. Thong. A symmetric linear algorithm for line segment generation. *Computers & Graphics*, **6**:15–17, 1982.

[1053] L. Thurfjell, E. Bengtsson, and B. Nordin. A new three-dimensional connected components labeling algorithm with simultaneous object feature extraction capability. *CVGIP: Graphical Models Image Processing*, **54**:357–364, 1992.

[1054] P.J. Toivanen. New geodesic distance transforms for gray-scale images. *Pattern Recognition Letters*, **17**:437–450, 1996.

[1055] J. Toriwaki, N. Kato, and T. Fukumura. Parallel local operations for a new distance transformation of a line pattern and their applications. *IEEE Trans. Systems, Man, Cybernetics*, **9**:628–643, 1979.

[1056] J. Toriwaki and K. Mori. Distance transformations and skeletonizations of 3D pictures and their applications to medical images. In: G. Bertrand, A. Imiya, and R. Klette, editors. *Digital and Image Geometry: Advanced Lectures*, 412–428. Lecture Notes in Computer Science 2243, Springer, Berlin, Germany, 2001.

[1057] J.I. Toriwaki, M. Tanaka, and T. Fukumura. A generalized distance transformation of a line pattern with gray values and its applications. *Computer Graphics Image Processing*, **20**:319–346, 1982.

[1058] J. Toriwaki, S. Yokoi, T. Yonekura, and T. Fukumura. Topological properties and topology-preserving transformation of a three-dimensional binary picture. *Proc. Int. Conf. Pattern Recognition*, 414–419, IEEE, 1982.

[1059] J. Toriwaki and T. Yonekura. Local patterns and connectivity indexes in a three dimensional digital picture. *Forma*, **17**:275–291, 2002.

[1060] G. Tourlakis. Homological methods for the classification of discrete Euclidean structures. *SIAM J. Applied Mathematics*, **33**:51–54, 1977.

[1061] G. Tourlakis and J. Mylopoulos. Some results in computational topology. *J. ACM*, **20**:430–455, 1973.

[1062] G.T. Toussaint. Pattern recognition and geometrical complexity. *Proc. Int. Conf. Pattern Recognition*, 1324–1346, IEEE, 1980.

[1063] G.T. Toussaint and J.A. McAlear. A simple $O(n \log n)$ algorithm for finding the maximum distance between two finite planar sets. *Pattern Recognition Letters*, **1**:21–24, 1982.

[1064] A. Tripodi. Sviluppi della topologia secondo Johann Benedict Listing. *Accad. Naz. Sci. Lett. Arti Modena Atti Mem.*, **13**:5–14, 1971.

[1065] D.M. Tsai, H.T. Hou, and H.J. Sou. Boundary-based corner detection using eigenvalues of covariance matrices. *Pattern Recognition Letters*, **20**:31–40, 1999.

[1066] Y.F. Tsao and K.S. Fu. A parallel thinning algorithm for 3-D pictures. *Computer Graphics Image Processing*, **17**:315–331, 1981.

[1067] Y.F. Tsao and K.S. Fu. A general scheme for constructing skeleton models. *Information Sciences*, **27**:53–87, 1982.

[1068] Y.F. Tsao and K.S. Fu. Stochastic skeleton modeling of objects. *Computer Vision, Graphics, Image Processing*, **25**:348–370, 1984.

[1069] A.W. Tucker. An abstract approach to manifolds. *Annals Mathematics*, **34**:191–243, 1933.

[1070] J. Turner. A graph-theoretical model for periodic discrete structures. *Proc. Ann Arbor Graph Theory Conf.*, 155–160, 1969.

[1071] W. Tutschke. *Grundlagen der reellen Analysis*. VEB Deutscher Verlag der Wissenschaften, Berlin, Germany, 1971.

[1072] W.T. Tutte. "What is a map?" In: *New Directions in the Theory of Graphs*, 309–325. Academic Press, New York, 1973.

[1073] W.T. Tutte. *Graph Theory*. Addison-Wesley, Reading, 1984.

[1074] S. Ubeda. A parallel thinning algorithm using (the) bounding boxes techniques. *Int. J. Pattern Recognition Artificial Intelligence*, **7**:1103–1114, 1993.

[1075] J.K. Udupa. Multidimensional digital boundaries. *Graphical Models Image Processing*, **56**:311–323, 1994.

[1076] J.K. Udupa. Connected, oriented, closed boundaries in digital spaces: Theory and algorithms. In: T.Y. Kong and A. Rosenfeld, editors. *Topological Algorithms for Digital Image Processing*, 205–231. Elsevier, Amsterdam, The Netherlands, 1996.

[1077] J.K. Udupa and V.G. Ajjanagadde. Boundary and object labelling in three-dimensional images. *Computer Vision, Graphics, Image Processing*, **51**:355–369, 1990.

[1078] J.K. Udupa, S.N. Srihari, and G.T. Herman. Boundary detection in multidimensions. *IEEE Trans. Pattern Analysis Machine Intelligence*, **4**:41–50, 1982.

[1079] E.E. Underwood. *Quantitative Stereology*. Addison-Wesley, Reading, Massachusetts, 1970.

[1080] S.H. Unger. A computer oriented toward spatial problems. *Proc. IRE*, **46**:1744–1750, 1958.

[1081] P. Urysohn. Über die allgemeinen Cantorischen Kurven. Annual meeting, Deutsche Mathematiker Vereinigung, Marbourg, Germany, 1923.

[1082] P. Urysohn. Mémoire sur les multiplicités Cantoriennes. *Fundamenta Mathematica*, **6**:30–130, 1925.

[1083] B.L. van der Waerden. *Geometry and Algebra in Ancient Civilizations*. Springer, Berlin, Germany, 1983.

[1084] M.L.P. van Lierop, C.W.A.M. van Overveld, and H.M.M. van de Wetering. Line rasterization algorithms that satisfy the subset line property. *Computer Vision, Graphics, Image Processing*, **41**:210–228, 1988.

[1085] O. Veblen. Theory of plane curves in non-metrical analysis situs. *Trans. American Mathematical Society*, **6**:83–98, 1905.

[1086] O. Veblen. Decomposition of an n-space by a polyhedron. *Trans. American Mathematical Society*, **13**:65–72, 1912.

[1087] O. Veblen. *The Cambridge Colloquium 1916 — Part II: Analysis Situs*. American Mathematical Society, New York, 1922.

[1088] P. Veelaert. On the flatness of digital hyperplanes. *J. Mathematical Imaging Vision*, **3**:205–221, 1993.

[1089] P. Veelaert. Digital planarity of rectangular surface segments. *IEEE Trans. Pattern Analysis Machine Intelligence*, **16**:647–652, 1994.

[1090] P. Veelaert. Geometric constructions in the digital plane. *J. Mathematical Imaging Vision*, **11**:99–118, 1999.

[1091] P. Verbeek, L. van Vliet, and J. van de Weijer. Improved curvature and anisotropy estimation for curved line bundles. *Proc. Int. Conf. Pattern Recognition*, 528–533, IEEE, 1998.

[1092] A.M. Vershik. Limit shape of convex lattice polygons and related topics (in Russian). *Funksional Anal. Appl.*, **28**:16–25, 1994.

[1093] B.J.H. Verwer. Local distances for distance transformations in two and three dimensions. *Pattern Recognition Letters*, **12**:671–682, 1991.

[1094] B.J.H. Verwer, L.J. Van Vliet, and P.W. Verbeek. Binary and grey-value skeletons: Metrics and algorithms. *Int. J. Pattern Recognition Artificial Intelligence*, **7**:1287–1308, 1993.

[1095] B.J.H. Verwer, P.W. Verbeek, and S.T. Dekker. An efficient uniform cost algorithm applied to distance transforms. *IEEE Trans. Pattern Analysis Machine Intelligence*, **11**:425–429, 1989.

[1096] A. Vialard. Geometrical parameter(s) extraction from discrete paths. In: S. Miguet, A. Montanvert and S. Ubéda, editors. *Discrete Geometry for Computer Imagery*, 24–35. Lecture Notes in Computer Science 1176, Springer, Berlin, Germany, 1996.

[1097] L. Vincent. Exact Euclidean distance function by chain propagations. *IEEE Conf. Computer Vision Pattern Recognition*, 520–525, 1991.

[1098] L. Vincent and P. Soille. Watersheds in digital spaces: An efficient algorithm based on immersion simulations. *IEEE Trans. Pattern Analysis Machine Intelligence*, **13**:583–598, 1991.

[1099] J. Vittone. Caractérisation et reconnaissance de droites et plans en géométrié discréte. PhD Thesis, Université Joseph Fourier, Grenoble, France, 1999.

[1100] J. Vittone and J.M. Chassery. Recognition of digital naive planes and polyhedrization. In: G. Borgefors, I. Nyström, and G. Sanniti di Baja, editors. *Discrete Geometry for Computer Imagery*, 296–307. Lecture Notes in Computer Science 1953, Springer, Berlin, Germany, 2000.

[1101] G.F. Voronoi. Nouvelles applications des paramètres continus à la théorie des formes quadratiques. *J. Reine Angewandte Mathematik*, **133**:97–178, 1907, **134**:198–287, 1908, and **136**:67–181, 1909.

[1102] K. Voss. Digitalisierungseffekte in der automatischen Bildverarbeitung. *Elektronische Informationsverarbeitung Kybernetik*, **11**:469–477, 1975.

[1103] K. Voss. Theoretische Grundlagen der digitalen Bildverarbeitung, Part V: Planare Strukturen und homogene Netze. *Bild und Ton*, **39**:303–307, 1986.

[1104] K. Voss. *Theoretische Grundlagen der digitalen Bildverarbeitung*. Akademie-Verlag, Berlin, Germany, 1988.

[1105] K. Voss. Coding of digital straight lines by continued fractions. *Computers Artificial Intelligence*, **10**:75–80, 1991.

[1106] K. Voss. Images, objects, and surfaces in $\mathbf{Z}^n$. *Intl. J. Pattern Recognition Artificial Intelligence*, **5**:797–808, 1991.

[1107] K. Voss. *Discrete Images, Objects, and Functions in Z^n*. Springer, Berlin, Germany, 1993.

[1108] K. Voss. Discrete integral geometry and stochastic images. In: R. Klette, A. Rosenfeld, and F. Sloboda, editors. *Advances in Digital and Computational Geometry*, 87–111. Springer, Singapore, 1998.

[1109] K. Voss, P. Hufnagl, and R. Klette. Theoretische Grundlagen der digitalen Bildverarbeitung, Part I: Einleitung. *Bild und Ton*, **38**:299–303, 1985.

[1110] K. Voss and R. Klette. On the maximal number of edges of convex digital polygons included into a square. *Computers Artificial Intelligence*, **1**:549–558, 1982.

[1111] K. Voss and R. Klette. Theoretische Grundlagen der digitalen Bildverarbeitung, Part IV: Orientierte Nachbarschaftsstrukturen. *Bild und Ton*, **39**:213–219, 1986.

[1112] K. Voss and K. Roth. Discontinuous point sets and digital curves in the two-dimensional orthogonal lattice. *Computers Artificial Intelligence*, **3**:539–549, 1984.

[1113] K. Voss and H. Süsse. *Adaptive Modelle und Invarianten für zweidimensionale Bilder*. Shaker, Aachen, Germany, 1995.

[1114] A.M. Vossepoel. A note on "Distance transformations in digital images." *Computer Vision, Graphics, Image Processing*, **43**:88–97, 1988.

[1115] A.M. Vossepoel and A.W.M. Smeulders. Vector code probability and metrication error in the representation of straight lines of finite length. *Computer Graphics Image Processing*, **20**:347–364, 1982.

[1116] D. Wagner, M. Tajine, and C. Ronse. An approach to discretization based on Hausdorff metric. In: H.J.A.M. Heijmans and J.B.T.M. Roerdink, editors. *Mathematical Morphology and its Applications to Image and Signal Processing IV*, 67-74. Kluwer, Amsterdam, The Netherlands, 1998.

[1117] P.S.P. Wang and Y.Y. Zhang. A fast and flexible thinning algorithm. *IEEE Trans. Computers*, **38**:741–745, 1989.

[1118] S. Wang, A. Rosenfeld, and A.Y. Wu. A medial axis transformation for gray scale pictures. *IEEE Trans. Pattern Analysis Machine Intelligence*, **4**:419–421, 1982.

[1119] X. Wang and G. Bertrand. Some sequential algorithms for a generalized distance transformation based on Minkowski operations. *IEEE Trans. Pattern Analysis Machine Intelligence*, **14**:1114–1121, 1992.

[1120] E. Welzl. Smallest enclosing disks (balls and ellipsoids). In: H. Maurer, editor. *New Results and New Trends in Computer Science*, 359–370. Lecture Notes in Computer Science 555, Springer, Berlin, Germany, 1991.

[1121] M. Werman, A.Y. Wu, and R.A. Melter. Recognition and characterization of digitized curves. *Pattern Recognition Letters*, **5**:207–213, 1987.

[1122] H. Weyl. *Symmetry*. Princeton University Press, Princeton, New Jersey, 1952.

[1123] J.H.C. Whitehead. On the realizability of homotopy groups. *Annals Mathematics*, **50**:261–263, 1949. Also in: *The Mathematical Works of J.H.C. Whitehead*, Vol. 3, 221–223. Pergamon Press, Oxford, United Kingdom, 1962.

[1124] P.D. Whiting and J.A. Hillier. A method for finding the shortest route through a road network. *Operational Research Quarterly*, **11**:37–40, 1960.

[1125] J. Wilhelms and A.V. Gelder. Octrees for faster isosurface generation. *ACM Trans. Graphics*, **11**:201–227, 1992.

[1126] J. Wilhelms and A.V. Gelder. Topological considerations in isosurface generation. Technical Report UCSC-CRL-94-31, University of California, Santa Cruz, California, 1994.

[1127] P.H. Winston. Learning structural descriptions from examples. In: P. H. Winston, editor. *The Psychology of Computer Vision*, 157–209. McGraw-Hill, New York, 1975.

[1128] D. Wood. An isothetic view of computational geometry. In: G.T. Toussaint, editor. *Computational Geometry*, 429–459. North Holland, Amsterdam, The Netherlands, 1985.

[1129] D.W. Woodard. On two-dimensional analysis situs with special reference to the Jordan curve theorem. *Fundamenta Mathematicae*, **13**:121–145, 1929.

[1130] S. Wolfram. *The Mathematica Book*, 4th edition. Cambridge University Press, Cambridge, United Kingdom, 1999.

[1131] M. Worring and A.W.M. Smeulders. Digital curvature estimation. *Computer Vision, Graphics, Image Processing*, **58**:366–382, 1993.

[1132] M. Worring and A.W.M. Smeulders. Digitized circular arcs: Characterization and parameter estimation. *IEEE Trans. Pattern Analysis Machine Intelligence*, **17**:587–598, 1995.

[1133] M.W. Wright, R. Cipolla, and P.J. Giblin. Skeletonization using an extended Euclidean distance transform. *Image Vision Computing*, **13**:367–375, 1995.

[1134] A.Y. Wu and A. Rosenfeld. Geodesic visibility in graphs. *Information Sciences*, **108**:5–12, 1998.

[1135] K.Y. Wu and W.H. Tsai. A new one-pass parallel thinning algorithm for binary images. *Pattern Recognition Letters*, **13**:715–723, 1992.

[1136] L. Wu. On the Freeman's conjecture about the chain code of a line. *Proc. Int. Conf. Pattern Recognition*, 32–34, IEEE, 1980.

[1137] L.D. Wu. On the chain code of a line. *IEEE Trans. Pattern Analysis Machine Intelligence*, **4**:347–353, 1982.

[1138] L.D. Wu and F.L. Weng. Chain code for a line segment and formal language. *Proc. Int. Conf. Pattern Recognition*, 1124–1126, IEEE, 1986.

[1139] X. Wu and J. Rokne. On properties of discretized convex curves. *IEEE Trans. Pattern Analysis Machine Intelligence*, **2**:217–223, 1989.

[1140] F. Wyse. A special topology for the integers. *American Mathematical Monthly*, **77**:1119, 1970.

[1141] G. Wyvill, C. McPheeters, and B. Wyvill. Data structures for soft objects. *The Visual Computer*, **2**:227–234, 1986.

[1142] Y. Xia. Skeletonization via the realization of the fire front's propagation and extinction in digital binary shapes. *IEEE Trans. Pattern Analysis Machine Intelligence*, **11**:1076–1086, 1989.

[1143] W. Xu and C. Wang. CGT: A fast thinning algorithm implemented on a sequential computer. *IEEE Trans. Systems, Man, Cybernetics*, **17**:847–851, 1987.

[1144] R. Yagel, D. Cohen, and A. Kaufman. Normal estimation in 3D discrete space. *The Visual Computer*, **8**:278–291, 1992.

[1145] I.M. Yaglom and V.G. Boltjanskij. *Convex Figures* (in Russian). Gosudarstv. Izdat. Tehn.-Teor. Lit., Moscow, Russia, 1951. English translation by P.J. Kelly and L.F. Walton, Holt, Rinehart and Winston, New York, 1961.

[1146] M. Yamashita and N. Honda. Distance functions defined by variable neighborhood sequences. *Pattern Recognition*, **17**:509–513, 1984.

[1147] M. Yamashita and T. Ibaraki. Distances defined by neighborhood sequences. *Pattern Recognition*, **19**:237–246, 1986.

[1148] L. Yan and D. Shiran. *Chinese Mathematics: A Concise History*, translated by N. Crossley and W.C. Lun, 66–68. Clarendon Press, Oxford, United Kingdom, 1987.

[1149] B. Yip and R. Klette. Angle counts for isothetic polygons and polyhedra. *Pattern Recognition Letters*, **24**:1275–1278, 2003.

[1150] S. Yokoi, J. Toriwaki, and T. Fukumura. An analysis of topological properties of digitized binary pictures using local features. *Computer Graphics Image Processing*, **4**:63–73, 1975.

[1151] S. Yokoi, J. Toriwaki, and T. Fukumura. On generalized distance transformation of digitized pictures. *IEEE Trans. Pattern Analysis Machine Intelligence*, **3**:424–443, 1981.

[1152] D.P. Young, R.G. Melvin, M.B. Bieterman, F.T. Johnson, and S.S. Samant. A locally refined rectangular grid finite element method: Application to computational fluid dynamics and computational physics. *J. Computational Physics*, **82**:1–66, 1991.

[1153] J.W.T. Youngs. Minimal imbeddings and the genus of a graph. *J. Mathematics Mechanics*, **12**:303–315, 1963.

[1154] L. Yu and R. Klette. An approximative calculation of relative convex hulls for surface area estimation. *Proc. Image Vision Computing New Zealand*, 69–74, 2001.

[1155] J. Yuan and C.Y. Suen. An optimal $O(n)$ algorithm for identifying line segments from a sequence of chain codes. *Pattern Recognition*, **28**:635–646, 1995.

[1156] P.C. Yuen and G.C. Feng. A novel method for parameter estimation of [a] digital arc. *Pattern Recognition Letters*, **17**:929–938, 1996.

[1157] L. Zadeh. Fuzzy sets. *Information Control*, **8**:338–353, 1965.

[1158] T.Y. Zhang and C.Y. Suen. A fast parallel algorithm for thinning digital patterns. *Comm. ACM*, **27**:236–239, 1984.

[1159] Y.Y. Zhang. Redundancy of parallel thinning. *Pattern Recognition Letters*, **18**:27–35, 1997.

[1160] Y.Y. Zhang and P.S.P. Wang. Analytical comparison of thinning algorithms. *Int. J. Pattern Recognition Artificial Intelligence*, **7**:1227–1246, 1993.

[1161] Y.Y. Zhang and P.S.P. Wang. A new parallel thinning methodology. *Int. J. Pattern Recognition Artificial Intelligence*, **8**:999–1011, 1994.

[1162] Y.Y. Zhang and P.S.P. Wang. Analysis and design of parallel thinning algorithms—A generic approach. *Int. J. Pattern Recognition Artificial Intelligence*, **9**:735–752, 1995.

[1163] R.W. Zhou, C. Quek, and G.S. Ng. A novel single-pass thinning algorithm and an effective set of performance criteria. *Pattern Recognition Letters*, **16**:1267–1275, 1995.

[1164] P. Zhu and P.M. Chirlain. On critical point detection of digital shapes. *IEEE Trans. Pattern Analysis Machine Intelligence*, **17**:737–748, 1995.

[1165] J. Žunić. On digital convex polygons. In: R. Klette, A. Rosenfeld, and F. Sloboda, editors. *Advances in Digital and Computational Geometry*, 255–283. Springer, Singapore, 1998.

[1166] J. Žunić. Limit shape of convex lattice polygons having the minimal L_∞ diameter w.r.t. the number of their vertices. *Discrete Mathematics*, **187**:245–254, 1998.

[1167] J. Žunić. On the number of digital discs. *J. Mathematical Imaging Vision*, to appear.

Index

4-DSS, 314
4-ray
  digital, 314
  lower digital, 314
  upper digital, 314

A

accumulation
  point, 232
adjacency
  1-, 39
  4-, 39
  in a graph, 118
  invalid, 122
  topological, 206
  valid, 122
adjacency graph, 122
  oriented, 137
  oriented planar, 142
  regular oriented, 146
adjacency model
  grid cell, 43
  grid point, 43
adjacency pair, 118
adjacency relation, 118
adjacency set, 43, 118, 125
adjacency structure, 118
adjacent, 161, 448
  0-, 39
  1-, 38
  4-, 9, 38
  6-, 42
  8-, 9, 39
  18-, 42
  26-, 42
  α-, 42
  edge-, 261
  i-, 160
  k-, 48
  P-, 40
  (σ, α)-, 40
  strongly edge-, 265
  switch, 40
  vertex, 265
  Voronoi, 122
algorithm
  linear time off-line, 353
  linear time on-line, 353
  off-line DSS, 328
  on-line DSS, 328
angle
  slope, 273
angular value, 80
approximation
  8-, 451
  theory, 21
arc, 127
  irreducible 8-, 311
  polygonal, 13
  simple, 234
  simple, in the grid cell topology, 240
arc length, 271
area, 67, 269, 276
array
  1D periodic, 396
  2D periodic, 396
  eventually periodic, 397
  Sturmian, 397
  tiled, 396

645

axiom system, 456
axioms
   of incidence, 457
axis
   principal, 543

B

background
   closed, 169
   open, 169
balanced, 397
   infinite word, 324-325
   set of words, 323
   set of arrays, 397
basic segment
   of an infinite word, 312
basis, 223
   countable, 195
   of a set of translations, 459
   of a topological space, 195
   of a topology, 85
   of an array, 396
Betti number, 224, 226
border, 123, 165
   α-, 123
border cycle
   inner, 148
   outer, 143
border point, 259
boundary, 144, 168
branch point, 234
branching index, 234
Bresenham's algorithm, 63
bridge
   in a graph, 131
Buffon's needle problem, 28

C

catchment basin, 449
Cavalieri's principle, 296
cell
   0-, 36, 175, 216
   1-, 37, 216
   2-, 36
   3-, 36
   convex, 219
   Dirichlet, 439
   i-, 175
   n-, 101
   Voronoi, 122, 439
cells
   adjacent, 48
center
   of a graph, 131
chain
   code, 61, 309, 471
   frontier, 168
   i-, 168
chord property, 316
   compact, 318
chordal triangle property, 393
chromatic number, 134
circle
   digital, 72
   unit, 87
circuit, 13, 127
class cardinality, 173
Clifford algebra, 18
closed set, 188
closing, 30, 124, 486
closure, 85, 167, 194
coborder, 125
code
   syntactic, of a DSS, 329
cogeodetic
   α-, 104
completion
   digitally-convex, 437
   in an incidence pseudograph, 162
complex
   abstract, 222
   Euclidean, 224
   geometric, 216
   one-dimensional geometric, 216
   surface, 217
   two-dimensional geometric, 217
complexity
   of a word, 322
component, 9, 47, 120, 163, 194
   α-, 120
   background, 47, 245
   border, 123
   complementary, 120
   complementary α-, 120
   inner, 123
   of a graph, 19, 127

regular, of a curve, 235
   tunnel-free, 557
computer graphics, 1
concavity, 450
congruence class, 50
connected
   α-, 120
connectedness
   degree of, 21
connectivity, 217, 257
   of a triangulation, 256
continuity, 209
continuum, 232
   one-dimensional, 232
   one-dimensional at a point, 232
contraction, 134
   of a path into a single point, 214
convergence, 113
   multigrid, 35, 70
   of words, 323
   speed of, 23, 71
coordinate system, 459
   Cartesian, 11
core, 161
coset, 225
count
   boundary, 176
   incidence, 172
cover, 144
cross product, 282
crossing number
   of a graph, 133
cube
   unit, 87
curvature, 281
   Gaussian, 291, 293
   mean, 293
   multiscale, 363
curvature chain, 363
curve, 233
   0-, 513
   1-, 513
   3-smooth, 69
   $C^{(2)}$-regular, 271
   elementary, 216, 234
   elementary, in the grid cell topology, 240
   Hilbert, 8
   in the grid cell topology, 237
   Jordan, 231

   Peano, 8, 232
   rectifiable, 65
   simple, 234
   simple α-, 243
   smooth Jordan, 273
   smooth space, 281
   Urysohn-Menger, 233
cut node, 131
cycle, 134, 137
   atomic, 140
   border, 140
   i-, 169
   of a translation, 458

D

d-continuous at p, 462
dart, 151
decomposition, 492
decomposition vertex, 236
deficit
   isoperimetric, 28
degree
   of a node, 126
   of adjacency, 557, 559
   of closeness, 92
   of connectedness, 21
   of surroundedness, 557
Delaunay triangulation, 440
deMorgan's rules, 196
Descartes, 34
Descartes-Euler polyhedron theorem, 133, 218
determinant, 277
deviation
   relative, 58
diagonal
   main, 315, 399
diagram
   d_α-Voronoi, 443-444
   Delaunay, 440
   dual Voronoi, 440
   in a metric space, 439
   nearest neighbor, 441
   Voronoi, 440
diameter
   intrinsic, 344
   of a graph, 129

difference
    Minkowski, 29
    symmetric, 88
    digital plane, 391
    irrational, 391
    quadrant, 391
    rational, 391
    segment, 391
digital ray
    irrational, 311
    rational, 311
digital topology, 21
digitization, 1, 56
    by dilation, 436
    Gauss, 56
    grid-intersection, 60
    inner Jordan, 59
    outer Jordan, 59
digraph, 118
Dijkstra's algorithm, 128
dilation, 30, 124
dimension, 18, 459
    in an adjacency structure, 206
    index, 160, 223
    of a triangulation, 222
direction, 457
    of a simple arc, 255
disk
    centered, 69
    digital, 64
    unit, 89
distance
    between two points, 77
    extrinsic α-, 98
    hexagonal, 93
    in a graph, 127
    intrinsic, 344
    intrinsic α-, 98
    main diagonal, 400
    value-weighted, 105
    weighted, 94
distance field, 99
distance function, 12
divide, 449
divide line
    of a surface, 449
domain of influence, 65
downstream, 448
DPS, 392, 399

DSL property, 319
DSS, 315
DSS property, 320
DSS-based length estimator, 343

E

eccentricity, 129
edge, 13, 19, 118
    assigned to a cycle, 141
    directed, 118
    directed invalid, 141
    in a picture, 487
    undirected invalid, 141
eigenvalue, 293
element
    structuring, 482
ellipsoid
    surface area, 288
embedding, 150
    combinatorial, 135
encoding
    directional, 61
end node, 130
endpixel, 46
endpoint, 95, 234
    of a line segment, 459
endvoxel, 46
equivalence
    topological, 211, 215
Erlangener Programm, 455
erosion, 30, 124
error
    estimation, 22
Euler characteristic, 137, 173, 216
Euler's formula, 133
Euler's totient function, 433
even set of 3D grid points, 393
exterior
    unbounded, 217

F

face
    external, 134
    internal, 134
    of a planar graph, 133
factor
    of a word, 311

Index

right special, of an infinite word, 323
family of sets
   discrete, 456
Farey series, 326
feature
   local, 491
Fibonacci word, 323
filling, 51
first fundamental form, 286
forest, 131
Frenet formulae, 275
Frenet frame, 273
   3D, 282
frontier, 85, 194
   in a triangulation, 255
   L-, 232
   point, 85
   smooth, 64
   tracing algorithm, 189
function
   characteristic, 27
   signum, 93
fundamental group, 214

G

gap, 264
gapfree, 264
   β-, 264
Gauss map, 290
Gaussian image, 291
genus, 21, 256
geodesic, 95, 127
geometry, 13, 455
   affine, 14
   analytical, 455
   combinatorial, 455
   computational, 25
   descriptive, 455
   digital, 33, 35
   digital grid, 459
   discrete, 456
   elliptic, 455
   Euclidean, 14, 455
   hyperbolic, 455
   integral, 29
   non-Euclidean, 455
   perspective, 455
   projective, 16, 455
   similarity, 14
gluing
   of surfaces, 256
good pair, 246
gradient, 285
Graham's Scan, 25
graph, 19, 118
   bipartite, 131
   complete, 131
   complete bipartite, 132
   directed, 118
   Eulerian, 129
   geometric dual, 135
   Hamiltonian, 130
   isomorphic, 132
   k-strong, 131
   planar, 133
   Platonic, 266
   self-dual, 135
   underlying, 118
   weighted, 127
graph metric, 127
graph theory, 19
grid, ii, 1, 2
   2D, 459
   hexagonal, 30, 140
   orthogonal, 140
   regular, 30
   triangular, 30, 140
grid cell model, 37
   topology, 201
grid constant, 12
grid cube, 36
grid edge, 2, 36
grid line, 37, 38
grid model, 37
grid plane, 38
grid point, 2, 30, 35
grid point model, 37
grid resolution, 38
grid square, 2, 36
grid vertex, 2, 35
group, 477
   factor, 226
   free cyclic, 215
   frieze, 461
   fundamental, 214
   homology, 226
   wallpaper, 461

H

handle, 252
height
　of a word, 323
　of an array, 397
Hessian matrix, 294
hole, 143
　8-, 246
　closed, 169
　improper, 143
　open, 169
　proper, 143
homeomorphism, 209
homomorphism, 225
homotopy, 214, 215
hull, 427
　8-, 452
　convex, 427, 459
　digital 8-convex, 437
　Gabriel pseudo-, 451
　near-, 428
　pseudo-, 428

I

image analysis, 1
in-degree
　of a node, 127
in-faces, 301
incidence, 159
　grid cell, 45
　in a graph, 118
　model
　　n-, 160
　of a cell in an abstract complex, 222
　relation, 159
　structure, 159
inequality
　isoperimetric, 28
　Minkowski, 77
　Schwarz, 78
inner point, 259
integrability condition, 292
integrable, 296
integral point
　of a ray, 312
integration
　discrete column-wise, 279

intercept
　approximate, 394
　of digital ray, 310
interior, 85, 194
interpolation
　bicubic spline, 468
　bilinear, 467
　nearest-neighbor, 466
　zero-order, 466
interval
　closed, 194
　open, 194
invariance
　topological, 211
involution, 152
isometry
　local, 292
isomorphism, 39
isomorphy, 39
isoperimetric deficit, 28
isosurface, 301

J

Jacobian matrix, 277
Jordan-Veblen curve theorem, 236

K

Klein bottle, 222
knot
　digital, 557

L

labelling, 53
Lagrange estimate, 272
layer, 301
　in a tree, 131
leaf
　of a tree, 131
length, 346
　of a path, 125
　of a word, 311
　value-weighted, 105
length estimate
　tangent-based, 346
length estimator

approximating sausage, 349
BLUE, 347
corner count, 348
letter
    in a word, 311
    nonsingular, 318
    singular, 318
level
    in a tree, 131
limit point, 113
line, 457
    arithmetic, 313
    naive, 313
    oriented, 458
    standard, 313
line segment, 459
lines
    of support, 436
    supporting, 313
linguistic technique
    for DSS recognition, 328
Lipschitz condition, 463
Listing band, 255
loop, 118

M

magnification, 58
manifold
    hole-free n-, 251
    Jordan, 258
    n-, 251
map
    combinatorial, 151
    n-dimensional combinatorial, 152
mapping
    continuous, 209
marching cubes algorithm, 301
matching theorem, 172
mathematical morphology, 30
metric, 12
    binary, 84
    chessboard, 90
    city block, 90
    Euclidean, 13, 77
    forest, 113
    Hausdorff, 86
    Manhattan, 90
    Minkowski, 79
    regular integer-valued, 82
metrics
    topologically equivalent, 85
Minkowski difference, 29
    sum, 29
MLP, 343
    approximating sausage, 351
MLP-based length estimates, 345
module
    unitary, 82
moment, 537
    (i, j), 543
morphism, 314
    nonerasing, 314
morphology
    mathematical, 29, 481
motion
    rigid, 478
multigraph, 118

N

needle problem
    Buffon's, 28
n-gon, 253
neighbor
    proper, 456
    Voronoi, 440
neighborhood, 9, 44, 118
    algebraic, 456
    cyclic, in a triangulation, 254
    e-, 91
    ε-, 85, 86
    smallest, 91
    topological, 194
node, 19, 118
    border, 123, 165
    central, 129
    connectivity, 131
    i-, 160
    initial, 118
    inner, 122, 165
    invalid, 168
    isolated, 130
    marginal, 160
    principal, 160
    terminal, 118

nodes
   connected, 19, 119, 127
   merging of two, 134
noise, 487
   cleaning, 487
   removal, 487
   salt-and-pepper, 487
n-omino, 50
norm, 78
   Minkowski, 79
   semi-, 79
normal, 286, 394

O

omino
   n-, 50
open set, 188
opening, 30, 124, 486
operation
   local, 106
orbit, 153
order
   local circular, 137
   of a group, 461
   partial, 195
   reduced local circular, 140
ordered adjacency procedure, 183
orientation, 137, 269, 544
   coherent, 255
   of a triangle, 255
origin
   of a coordinate system, 11
out-degree
   of a node, 125
out-faces, 301

P

P-equivalence class, 40
P-equivalent, 40
parabola
   digital, 64
parameterization
   regular, 270, 282
patch
   isometric, 291
   Monge, 283
   planar surface, 283

path, 9, 19, 46, 119, 125
   α-, 95, 103, 120
   Eulerian, 129
   Hamiltonian, 129
   homotopic, 214
   initiation of a, 136
   monotonic, 446
   of sets, 222
   parameterized, 214
   polygonal, 222
   shortest, 127
   total weight of a, 125
   zero-homotopic, 214
pendant edge, 130
perimeter, 67
period, 396
   of a finite word, 312
   of an infinite word, 312
periodic, 311
periodicity, 312
   approximate, 339
Pick's formula, 279
picture, 1
   1D, 2
   2D, 2
   3D, 2
   binary, 4
   digital, 1
   multivalued, 4
   resolution, 5
   well-composed binary, 534
picture analysis, 1
picture size, 6
pictures
   congruent, 478
   similar, 478
pixel, 1, 2
   simple, 499
pixels
   connected, 47
plane
   arithmetic, 394
   cellular digital, 399
   digital, 13, 399
   Euclidean, 13
   graceful, 398
   naive, 394
   projective, 222
   separating, 394

standard digital, 394
supporting, 393, 402
tangent, 290
plane segment
digital, 399
plateau, 446
point
bifurcation, 251
branch, 234
concave, 274
convex, 274
elliptic, 292
frontier, of a surface, 252
hyperbolic, 292
interior, of a surface, 252
of inflection, 274
parabolic, 292
planar, 292
regular, 228
singular, 234, 270, 281
points
connected, 19
polygon
digitization, of a DSS, 331
grid, 23
isothetic, 23
minimum-length, 343
regular, 23
simple, 13
simple grid, 148
weak digitization, 333
polyhedron, 20, 216, 221
grid, 23
isothetic, 23
polyomino, 50
chiral, 50
fixed, 50
free, 50
poset, 195
position
in an array, 396
prefix
in a word, 312
principal normal, 282
product
cross, 282
of two paths, 214
scalar, 79
vector, 282

weak scalar, 80
projection, 475
property
global, 538
local, 538
semilocal, 539
topological invariant, 211
pseudograph, 118
incidence, 161
monotonic incidence, 161
pyramid, 471

R

radius
α-, 108
of a graph, 129
of torsion, 282
ravine
of a surface, 449
ray, 459
digital, 310
lower digital, 312
upper digital, 312
rectangle
circumscribing, 436
reduction operation, 319
region, 20, 121, 169
α-, 123
planar, 276
smooth, 487
region adjacency graph, 125
region detection, 51
regions
homeomorphic, 211
relation
adjacency, 118
bounded-by, 222
connectedness, 9
incidence, 14, 159
irreflexive, 117
parallelism, 457
reflexive, 117
symmetric, 117
representation
geometric, 2D incidence grid, 45
geometric, 3D incidence grid, 46
resolution
geometric, 15

ridge
  of a surface, 449
root
  of a tree, 131
run
  in a binary picture, 9
  in a word, 318

S

scalar, 18
scan
  Hilbert, 8
  of a grid, 6
  Peano, 8
Schönflies-Brouwer theorem, 236
segment
  digital 4-straight line, 314
  digital straight line, 314
segmentation, 490
separation, 142
  α-, 264
set
  adjacent, 125
  α-inner, 123
  bounded, 12, 85
  closed, 85, 166, 194
  column-convex, 438
  compact, 85
  complete, 163
  complete, in the grid point topology, 200
  connected, 9
  continuously connected, 31
  convex, 23, 427, 458
  digitally convex, 64, 436
  discrete, 456
  fuzzy, 27
  inner, 123, 165
  λ-level, 27
  open, 85, 166, 194
  polygonally connected, 31
  row-convex, 438
  simply-connected, 21, 214
  topologically connected, 194
  totally disconnected, 206
sets
  incident, 14, 44
shape, 269
shape factor, 28

shortest path problem, 128
side
  improper, 219
  k-, 219
signature
  of a spirograph, 326
significance measure, 362
similarity relation, 40
simple
  pixel, 508
  voxel, 530, 531
simplex, 216
skeleton, 509
  linear, 215
skeleton climbing
  adaptive, 306
skeletonization, 509
Sklansky's algorithm, 429
slope
  atomic, 321
  of a digital ray, 309
  of a word, 323
solid
  Platonic, 130
space
  Aleksandroy, 194
  digital, 13, 33
  Euclidean, 13
  Hausdorff, 195
  homeomorphic, 209
  isotopic, 212
  Kolmogorov, 195
  locally compact, 251
  metric, 12
  n-dimensional Euclidean, 77
  T_0-, 195
  T_1-, 195
  T_2-, 195
  topological, 194
  unbounded metric, 83
  vector, 18
speed, 269
sphere
  Gaussian, 290
  open n-, 244
  unit, 87
splitting formula, 321
square
  digital, 64

unit, 87
star, 201
step codes, 391
stereology, 30, 287
straight line
   cellular, 310
   digital, 64, 310
straight line segment
   cellular, 310
   digital, 62, 64
strip
   Listing, 255
   Möbius, 255
structure
   adjacency, 118
   incidence, 159
Sturmian
   array, 397
   word, 322, 323
subgraph, 119
subgraphs
   disjoint, 119
subspace
   topological, 194
substitution, 313
substructure, 140
subword, 311
suffix
   in a word, 312
   of an array, 397
sum
   Minkowski, 29
supporting lines, 315
surface, 252
   α-, 258
   digital, 258
   hole-free, 251
   hole-free simple 26, 258
   Jordan, 252
   nonorientable, 255
   orientable, 255
   simple 26, 258
   simple hole-free, 252
   simple, with r contours, 252
   strong, 267
   with frontiers, 252
surface normal, 290
surface patch
   digital, 258

surface pixel, 265
switch adjacency, 40

T

thickness
   arithmetic, 394
Thiessen polygons, 439
thin
   set of pixels, 488
thinning, 509
   3D, 531
threshold, 490
thresholding, 490
Thue-Morse word, 323
tiling, 253
   regular, 253
time complexity, 25
topology, 20, 85, 193
   alternating, 198, 202
   degenerate, 194
   digital, 22, 198, 199
   Euclidean, 85, 194
   grid cell, 202
   grid point, 198
   induced, 194
   inherited, 194
   point-set, 193
   poset, 195
torsion, 282
total border strength, 372
total turn, 373
transform
   hit-and-miss, 485
transformation
   affine, 478
   identity, 477
   isometric, 545
translation, 457
   atomic, 459
   cyclic, 458
tree, 131
   minimum spanning, 446
   rooted, 131
   simple 2-, 187
triangle, 132
   oriented, 255

triangulation, 221
　by ear clipping, 306
　of a 2-manifold, 252
　pure 2D, 222
　strongly connected 2D, 222
tripod
　of a 3-cell, 399
tube, 381

U

upstream, 448

V

vector, 18
　binormal, 282
　periodicity, 396
　symmetry, 396
　tangent, 290
　unit normal, 273
　unit tangent, 273
vector product, 282
vector space, 17
　finite-dimensional, 18
vectors
　orthogonal, 80
vertex, 13
　concave, 280
　convex, 280
　decomposition, 236
volume, 295
Voronoi adjacent, 440
Voronoi cell, 122
voxel, 1, 2
　orientable surface, 258
　simple, 530, 531
　surface, 258
voxels
　connected, 47

W

watershed, 449
watershed line, 449
weight
　of an edge, 127
width, 315
　arithmetical, 313
word, 311
　aperiodic, 312
　empty, 311
　eventually periodic, 312
　infinite, 312
　mechanical, 324
　reducible, 318